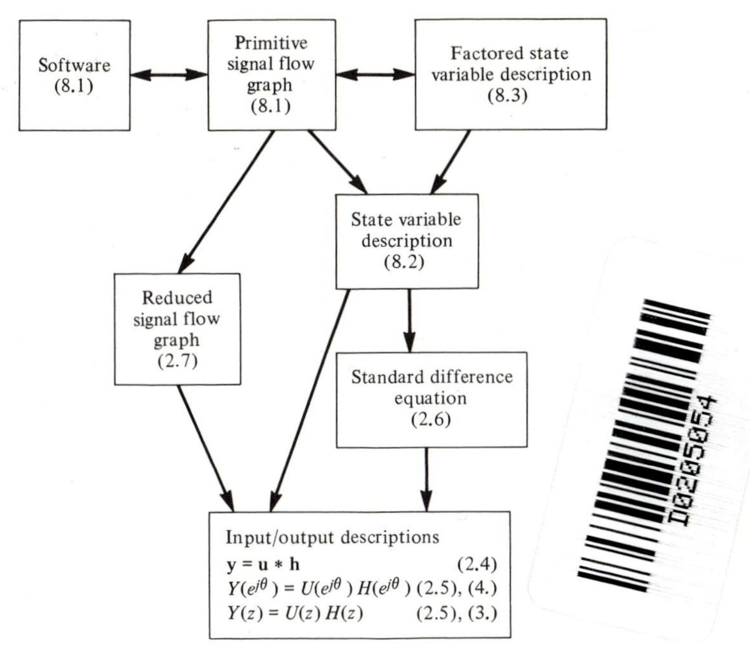

A hierarchy of filter descriptions

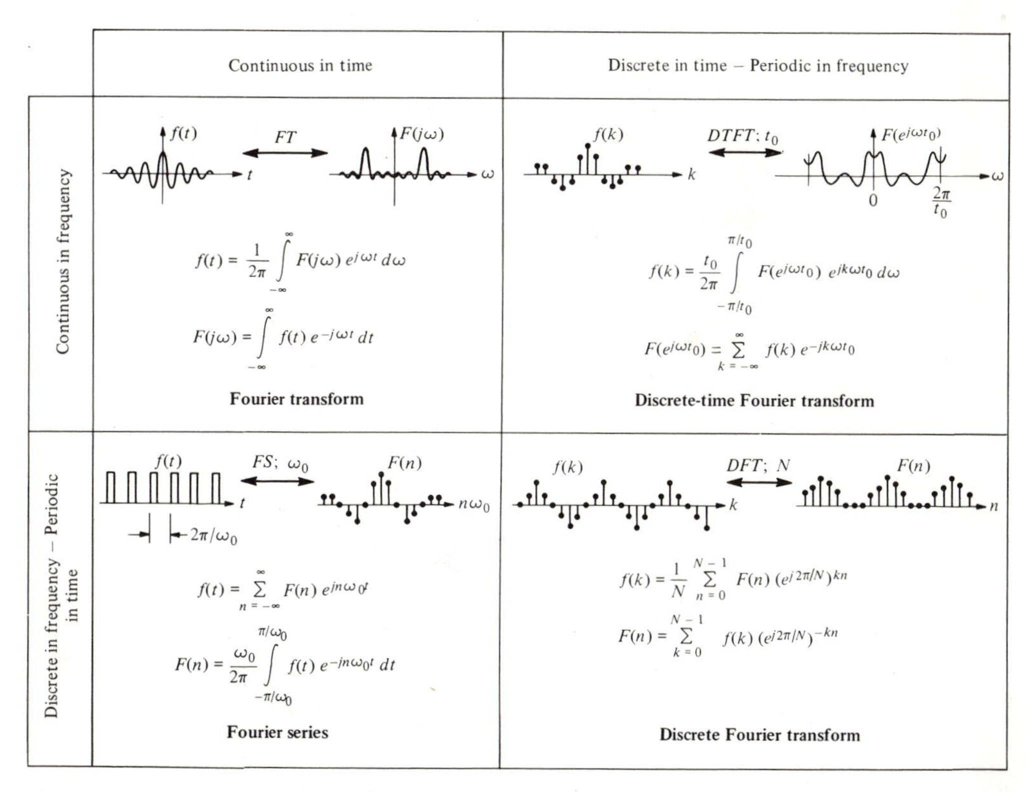

Four classes of Fourier transform

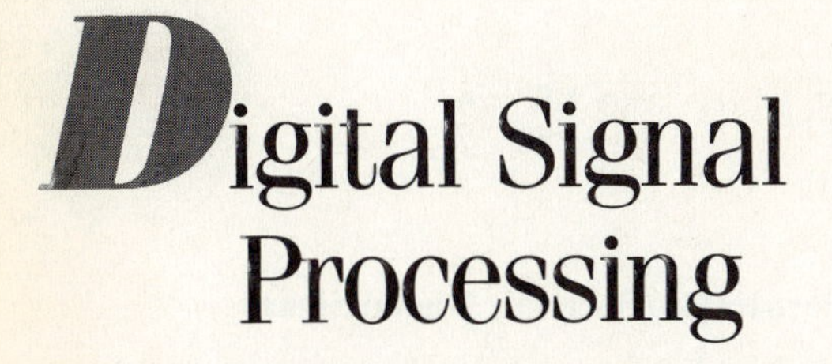

Digital Signal Processing

RICHARD A. ROBERTS CLIFFORD T. MULLIS

University of Colorado at Boulder

ADDISON-WESLEY PUBLISHING COMPANY

Reading, Massachusetts • Menlo Park, California • Don Mills, Ontario • Wokingham, England • Amsterdam • Sydney • Singapore • Tokyo • Madrid • Bogotá • Santiago • San Juan

To

Sandy,
Judy, Natalie, Jennifer

Judy,
Mark, John, Michele, Molly, Emily, Mary, and Megan

This book is in the **Addison-Wesley Series in Electrical Engineering: Digital Signal Processing**

Sponsoring Editor	Tom Robbins
Production Supervisor	Mary Coffey
Production Supervision	Cobb/Dunlop Publisher Services
Text Design	Patti Williams
Cover Design	Dick Hannus
Manufacturing Supervisor	Hugh Crawford

Library of Congress Cataloging in Publication Data

Roberts, Richard A., 1935-
Digital signal processing.
Bibliography: p.
Includes index.
1. Signal processing—Digital techniques. I. Mullis, Clifford T., 1943- . II. Title.
TK5102.5.R525 1987 621.38'043 85-30638
ISBN 0-201-16350-0

Reprinted with corrections May, 1987

ABCDEFGHIJ-MA-8987

Preface

Digital signal processing is as old as the numerical procedures invented by Newton and Gauss in the 17th and 19th centuries. New technologies in the past 20 years have decreased the cost of digital hardware, and its speed has increased to such an extent that digital signal processing has replaced a great deal of analog signal processing.

This book presents an introduction to digital signal processing, a field of study created by the great interest in the design and application of numerical algorithms resulting from the change in signal processing that the new technologies have brought about.

ORGANIZATION

This book is used in a two-semester sequence at the University of Colorado. The first seven chapters present the basic introductory material on linear, discrete-time systems and discuss methods for solving the approximation problem for digital filters. Also included in the first seven chapters is a discussion of fast algorithms for computing the discrete-Fourier transform. The prerequisite for this material is a course in linear systems theory including exposure to both discrete-time and continuous-time systems. Chapters 1-3 serve mainly as a review. Fourier analysis has always played a primary role in the analysis of linear systems and signals, and this is true in the case of linear, discrete-time systems. Chapter 4 discusses the Fourier analysis of both continuous and discrete-time systems. This serves to unify the underlying concepts in Fourier analysis. Because the discrete-Fourier transform (DFT) can be finitely parametrized and developed using finite-dimensional matrices, the discrete-Fourier transform can provide intuition for any form of Fourier analysis. Chapter 5 considers the design of fast algorithms for the discrete-Fourier transform. This material is concerned with the inner workings of the input-output function defined by DFT and as such it could fit into the second part of the book which considers the design of algorithms or structures for DSP. However, many instructors prefer to present the fast algorithms for the DFT immediately after presenting the theory. In Chapters 6 and 7 we discuss the approximation problem for digital filters. Chapter 6 is a discussion of the design of FIR and IIR filters. In Chapter 7 we consider the design of least-squares digital filters. One might view Chapter 6 as a circuit theorist's design of digital filters and Chapter 7 as a communication theorist's design of filters. Although the material in Chapter 7 is self-contained, some background with probability theory and random processes is helpful.

As technology continues to decrease the cost and increase the speed of digital hardware, digital filters will be implemented more and more in special-purpose computing resources. Thus it is important to develop theories that allow the designer to tailor structures and algorithms for arbitrary computational resources. Furthermore, we can increase the effectiveness of algorithms by correctly matching the algorithm to the available computational resources. Chapters 5, 8, 9, and 10 discuss this aspect of design in some depth. Chapter 8 introduces descriptions that explicitly define the internal computation of a given input-output function. Chapter 9 discusses the practical implications of using finite length registers in digital filters and presents a theory of how to design structures which minimize these adverse effects. Chapter 10 looks at other structures for digital filtering that optimize other criteria associated with the actual computation of the filter. Included in this discussion is a class of structures known as orthogonal filters, which are useful building blocks for VLSI implementations.

Chapter 11 presents a discussion of spectral analysis of finite data, a topic very important in applications. This chapter requires some background in random processes and probability theory. Although it represents a slight departure from the rest of the book, many of the techniques used in earlier sections are again employed.

The last four chapters constitute a second-semester course in digital signal processing. The first seven chapters (with the exception of Chapter 5) deal with input-output characteristics. The emphasis of the last portion of the book is on algorithms and structures. It is possible to include material from Chapter 8 in a first course. In this way the reader would be exposed to both input-output and internal descriptions of digital processing tasks.

USE OF THIS BOOK

Although we have ordered the material according to our personal preference, many alternative presentations might be used. The figure below depicts the partial ordering of the chapters. We use this material for two courses in DSP. The first covers the material in Chapters 2, 3, 4, 5, 6, and 7. The second course includes the material in Chapters 8, 9, 10, and 11. One could use this material in several ways. For example, Chapters 4, 5, 7, and 11 could be used as the basis of a course in power spectrum estimation. If

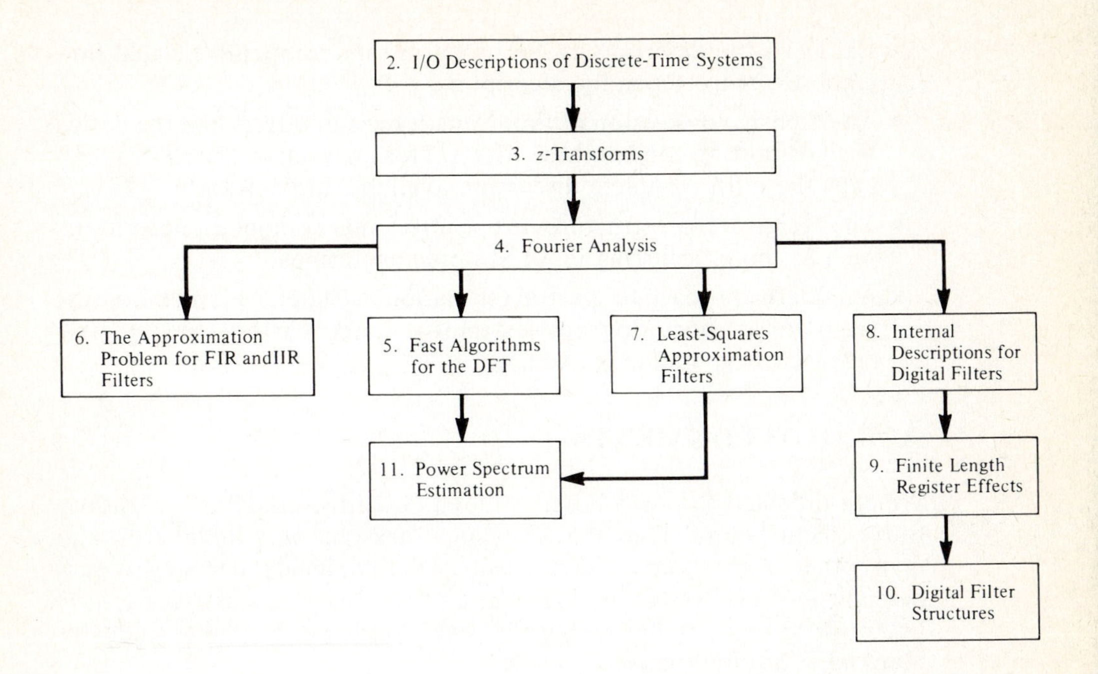

your curriculum includes only one course in DSP, then Chapters 2, 3, 4, 5, 6, and 8 might be appropriate. A second course in DSP could be based on Chapters 7, 8, 9, 10, and 11. A course in algorithmic design could be based on Chapters 5, 8, 9, and 10.

FEATURES

This book possesses certain unique aspects which we believe are useful in obtaining a fundamental understanding of digital signal processing:

- An integrated discussion of Fourier analysis in Chapter 4. This chapter discusses the DFT and its inverse as matrix transformations emphasizing the orthogonality and shift-invariant properties of the transform.
- The design of arbitrary digital-to-digital frequency transformations in Chapter 6. Usually these transformations are tabulated for a few cases. The material in Chapter 6 allows the designer to easily fashion arbitrary transformations.
- A concise coverage of deterministic and statistical least-squares digital filter designs in Chapter 7.
- A general discussion of signal flow graphs, state variable representation, and software descriptions for digital filters. This material includes a generalization of state variable representations, called the factored

state variable descriptions, for any arbitrary computable signal flow graph or, more generally, algorithm.

- An extensive discussion of finite length register effects and the design of algorithms to combat these effects. This material in Chapter 9 is perhaps the most complete discussion available in book form.
- The design of filter structures to optimize other computational criteria, such as those occurring in VLSI implementations.
- A modern approach to spectral estimation in Chapter 11, including the traditional periodogram estimator and various parametric autoregressive methods (AR).

ACKNOWLEDGMENTS

We are indebted to our colleagues Lloyd Griffiths, Louis Scharf, Sidney Burrus, Shu Leung, Tom Parks, Leland Jackson, Alan Steinhardt, and many others with whom we have discussed digital signal processing. We appreciate the reviews of this material and the lively discussion we have obtained from our students over the years. We hope the reader finds this material as stimulating as we have.

Clifford T. Mullis
Richard A. Roberts

Boulder

Contents

8 INTERNAL DESCRIPTIONS FOR DIGITAL FILTERS 287

APPENDIX 8A SPECTRAL REPRESENTATION AND FUNCTIONS OF A MATRIX 331

9 FINITE LENGTH REGISTER EFFECTS IN FIXED POINT DIGITAL FILTERS 343

CHAPTER 1

Digital Signal Processing

1.1 Introduction

Digital signal processing is a field of study concerned with the processing of information represented in digital form. Certain techniques in the field can be traced back to numerical algorithms performed in the seventeenth and eighteenth century. However, the advent of modern high-speed digital computing devices has caused a revolution in applications of the theory to a variety of problems. Signal processing is used in such areas as biomedical data processing, digital audio, sonar and radar processing, speech processing, data communication, reliable data storage of computerized information, seismic signal processing, and a host of other applications. One of the most interesting aspects of digital signal processing is the wide variety of applications. This has served to create a vitality in the field that is often missing in other scientific fields of study.

Signal processing is not restricted to one-dimensional data. There are many multidimensional processing applications. Image processing, processing of signals from antenna arrays, and similar applications are as important as one-dimensional problems. However, because of space limitations we shall confine our study to the one-dimensional theory.

Digital signal processing has become an increasingly significant field because of the technology associated with digital computers and digital devices. A digital computer used to process signals offers a tremendous advantage in flexibility. Furthermore, the use of numerical or digital algorithms makes for easy accessibility of the processing. It's not like having to design and troubleshoot a specialized piece of analog hardware to do a particular processing problem. Real-time digital processing is now possible except for those applications involving very large signal bandwidths.

Analog filtering and processing influenced early developments in digital signal processing. It was natural at first to use digital signal processing to approximate a given analog processing scheme. As time went on scientists quickly realized that the digital schemes were more flexible and did not possess certain constraints imposed by analog techniques. New and more sophisticated algorithms could be implemented digitally. Even traditional processing techniques like spectral analysis could be implemented with more accuracy and with better resolution. As technology continues to increase the speed and decrease the cost of digital devices, more and more digital schemes become practical. Thus the theory and practice of digital signal processing becomes more significant as time goes on. It seems that the importance of digital signal processing is accelerating. And the fields of application are still expanding.

This book takes the point of view that digital processing schemes are a field of study unto themselves. They may have begun as approximations to analog schemes but they have long since evolved into an autonomous discipline. We also take the point of view that more and more implementations of digital algorithms will be in terms of special purpose hardware rather than as software for a general purpose computer. This may not seem to be a fundamental difference from other books in the field and, in a certain sense, it is not. However, this philosophical point of view does steer our treatment of digital signal processing toward an interest in the structure of digital algorithms. For this reason we use matrix formulations when we can, and we've chosen topics after the fundamental material that emphasize the structure of the processing. Digital signal processing is a large field. We have attempted to cover what we consider to be essential in one-dimensional processing. We also have attempted to include material that is technology independent although we know that new technology can create new important areas of study in digital signal processing, as evidenced by the current interest in new digital processing algorithms for VLSI.

Digital signal processing is an exciting field of study because it is highly applicable and yet is rich in its dependence on certain rather profound mathematics. It brings together aspects from many fields of study. Pedagogically, it is very useful in developing an understanding in many fields of study, such as advanced linear systems, Fourier analysis, state variable theory, and approximation and optimization theory. We cannot, of course, develop all these aspects as fully as they deserve. However, we hope the reader is able to appreciate the many facets of digital signal processing from this presentation and is encouraged to further extend the frontiers of this exciting field.

1.2 AN OUTLINE OF THE BOOK

Our viewpoint of digital signal processing, as mentioned previously, is bent toward structures and algorithms. This has influenced the kinds of descriptions we have used to describe digital processors. Chapters 2, 3, 4, 6, and 7 consider digital processors from an input/output description only. In Chapter 2, we introduce several descriptions for digital filters including the unit-pulse response, the frequency response or transfer function, the standard difference equation, and signal flow graphs. We then develop the properties of these descriptions and point out how they are interrelated. In Chapter 3, we introduce the transform theory associated with discrete-time systems, namely the z-transform. The z-transform plays the same role for discrete-time systems as the Laplace transform does for continuous-time systems. Perhaps the most important concept used in z-transforms is the idea of poles and zeros. The intuition provided by using poles and zeros in the z-plane is often invaluable. In Chapter 4, we present Fourier analysis for discrete-time systems. Fourier analysis for discrete-time systems includes two types of transforms. They are known as the discrete-time Fourier transform (DTFT) and the discrete-Fourier transform (DFT). The DFT is the Fourier transform that is always digitally computed. For this reason, it is a very important topic. Our presentation of the DFT is based on a matrix formulation and is intended to unite many concepts common among Fourier type transforms.

Chapters 6 and 7 present the so-called approximation problem for digital filters. In Chapter 6, we assume that specifications are given in terms that do not explicitly depend on the signals to be filtered. Generally, the specifications in Chapter 6 are expressed in frequency domain terms, e.g., a bandpass filter with certain tolerances on the transition width of the band edges and a minimum stopband attenuation. The problem in Chapter 6 is to design a digital filter transfer function to approximate the given specifications. In Chapter 7, the same kind of objective is desired, namely, the transfer function or unit-pulse response function for a digital filter. In this chapter the specifications are presented in terms of the properties of the signals we wish to filter. The resulting approximation problems are formulated as least square optimization problems.

Chapters 5, 8, 9, and 10 discuss and study concepts and problems that are associated with the structure or algorithm of a digital processing task. These four chapters study a different aspect of digital processing: How do we efficiently compute the output sample for a given input/output characteristic? The answer to this question is becoming increasingly significant as the cost of digital components continues to decrease. The answer depends on what we mean by "efficiently compute." In Chapter 5, the problem is to efficiently compute the DFT by the use of so-called fast algorithms. Here the algorithm is chosen to minimize the number of computations. In Chapters 9 and 10, we use the descriptions introduced in Chapter 8 to design algorithms for digital filters that minimize roundoff noise, eliminate limit cycles, reduce coefficient sensitivity, efficiently raise or lower the sampling rate, and are well suited for VLSI implementation. These chapters all depend heavily on matrix descriptions of the process, because matrix descriptions can be factored to completely describe any algorithm. This kind of design is not possible with input/output descriptions.

Chapter 11 presents the important subject of spectral estimation. This chapter begins with a discussion of traditional nonparametric methods based on Fourier analysis of the spectrum. We then introduce so-called all-pole models for the spectrum and briefly introduce pole-zero models.

As mentioned previously a good understanding of digital signal processing requires mathematics from a variety of fields. We rely on concepts in linear systems theory, Fourier analysis, state variable theory, and optimization theory. We have added appendices on subjects that the reader may find useful.

CHAPTER 2

Discrete-Time Signals and Systems

2.1 INTRODUCTION

Digital signal processing is the theory associated with processing digital signals. A signal is something that conveys information such as the magnetic domains on a floppy diskette, the pressure variations of an acoustic transducer, or the electrical signals generated by the heart muscles in an electrocardiogram (EKG). Signals can be *continuous-time* or *discrete-time* signals depending on whether the independent variable, time, takes on a continuum or a set of discrete values, respectively. Discrete-time signals are thus sequences of real or complex numbers. Often these numbers are obtained by sampling a continuous-time signal although they may arise naturally in other contexts. In practical applications of digital signal processing, we use *digital signals*, which are quantized both in amplitude and in time (or the independent variable). Amplitude quantization is necessary if we are to use digital devices to store and process the information.

Digital signal processing is the study of discrete-time systems that process digital signals in order to extract or modify the original information as embodied in a given sequence. The processing can take on a variety of forms. We may wish to process digital signals in order to reduce noise that has been added to some desired signal in transmission through a noisy channel such as occurs in long-distance telephone communication. Or we may wish to process information in such a way that we can reduce the amount of information we have to store to perform a given task, e.g., data compression of speech.

In this chapter we consider the basic concepts of discrete-time signals and systems. We are primarily concerned with discrete-time systems that are *linear* and *shift-invariant* (LSI). This special class of systems is very useful and important in applications of digital signal processing. There is a large amount of mathematical theory that we can bring to bear on the analysis and synthesis of these systems. This chapter is concerned with the analysis of LSI discrete-time systems. The goal of this text is to present theories for the design or synthesis of digital processing structures. However, it is only after we gain an understanding of the analysis of these systems that we can study the synthesis problem.

A linear, shift-invariant system has an efficient characterization involving either a *unit-pulse response sequence* (in the time domain) or *frequency response function* (in the frequency domain). These are introduced in Sections 2.4 and 2.5, respectively. This duality of description between time and frequency domains should be familiar to anyone who has seen "transforms" applied to differential or difference equations. This dual description is a very powerful tool that provides two separate bases for intuition, and allows us to formulate questions in two different ways before attempting an answer. For example, the question "What does the filter do to high frequency signals?" is best answered from a consideration of the frequency response function. The question "How can we simulate this filter on a microcomputer?" is best answered given a suitably detailed time-domain description. The question "What is the average delay between the input signal and the output signal?" could, very likely, involve both domains.

There are many ways to describe a LSI discrete-time system. Some methods are more detailed than others. Descriptions such as the unit-pulse response and the frequency response function specify only the input-output relationship of the system. Other descriptions specify the input-output relation and some information concerning the internal structure of the filter. In the design of digital filters some questions require more information about that system than is contained in the input-output relation. For example, if we wish to understand the effect of a coefficient variation on the system's frequency response function, we must have information on how the coefficient is related to that function. The various descriptions form a hierarchy involving different levels of detail. It is a major goal of this book to describe this hierarchy. There is no "best" system description. We should always select the lowest level (least detailed) description which is sufficient to answer the question at hand.

In the hierarchy of descriptions, there are clear, absolute, lowest level descriptions. These are the input-output descriptions such as the unit-pulse response sequence and frequency response function. However, there is no highest level description. It is always possible to add another layer of detail which becomes more and more dependent on the technology being used. We have tried to stop just short of technological detail.

In the first seven chapters (with the exception of Chapter 5) we shall be concerned with input-output characterizations of discrete-time systems. These descriptions are sufficient until we begin to ask questions concerning the detailed computations in a particular signal processing task. For example, when we ask for the best algorithm for a digital filter that will minimize the effects of finite-length registers in the filter, we must look at more than the input-output relationships. We shall postpone a detailed discussion of these internal descriptions until they are needed in Chapter 8.

2.2 DISCRETE-TIME SIGNALS

We need some definitions to begin. A *signal* is a sequence of real or complex numbers. We shall denote a complete sequence by either $\mathbf{y}$ or $\{y(k)\}$. Thus

$$\mathbf{y} = \{y(k):k = \cdots -2, -1, 0, 1, 2, \ldots\}.$$

Often, the sequence $\mathbf{y}$ is obtained from a continuous signal $y(t)$ by sampling uniformly in time with a time interval between samples of, say, t_0. In this case, the signal $\mathbf{y} = \{y(kt_0)\}$. Usually we drop the explicit dependence on the sampling interval t_0. Three examples of discrete-time signals are shown in Fig. 2.2.1. In particular, the *unit-pulse sequence* is the signal $\boldsymbol{\delta}$, where

$$\delta(k) = \begin{cases} 1, & k = 0 \\ 0, & \text{otherwise} \end{cases}. \tag{2.2.1}$$

We can manipulate signals (sequences) in various ways. The sum of two signals is

$$\mathbf{w} = \mathbf{x} + \mathbf{y} \quad \Rightarrow \quad w(k) = x(k) + y(k).$$

Multiplication of a signal by a scalar multiplies each value by the same scalar, i.e.,

$$\mathbf{w} = c\mathbf{y} \quad \Rightarrow \quad w(k) = cy(k).$$

If a signal is delayed by m time units, then $y(k)$ becomes $y(k - m)$. We shall use the notation $\mathbf{S}^m$ to denote a delay shift operator of m, i.e.,

$$(\mathbf{S}^m\mathbf{y})(k) = y(k - m). \tag{2.2.2}$$

[We can also advance the signal by changing the sign of m in Eq. (2.2.2).]

One useful representation of an arbitrary sequence $\mathbf{w}$ is a decomposition into a sum of displaced unit-pulse sequences, multiplied by the scalar values of the sequence $\mathbf{w}$. Using Eq. (2.2.1), we have the identity

$$w(k) = \sum_{m=-\infty}^{\infty} w(m)\delta(k - m).$$

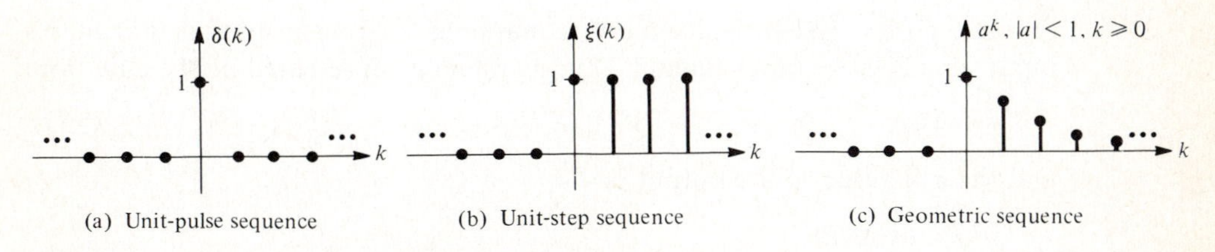

FIGURE 2.2.1 Examples of discrete-time signals.

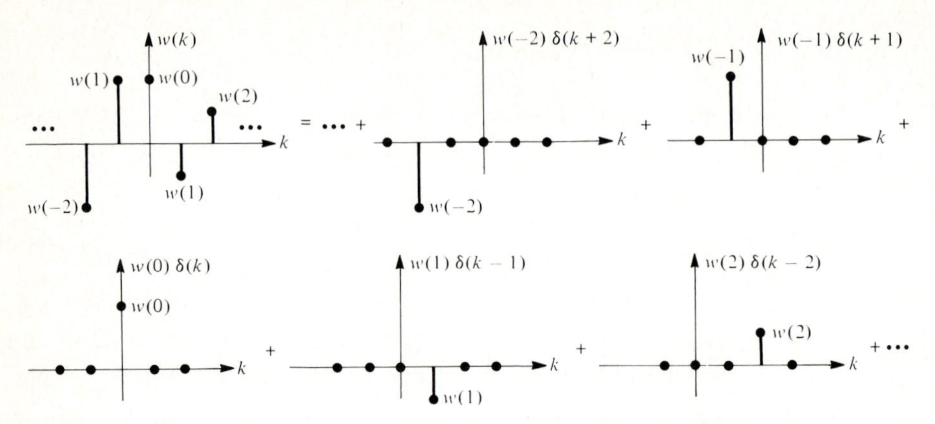

FIGURE 2.2.2 A decomposition of an arbitrary sequence *u* using unit-pulse sequences.

Using Eq. (2.2.2), this generates the signal (sequence) representation

$$\mathbf{w} = \sum_{m=-\infty}^{\infty} w(m)\mathbf{S}^m \boldsymbol{\delta}. \tag{2.2.3}$$

This decomposition is shown in Fig. 2.2.2.

We can specify a sequence in two ways. We can specify a rule for calculating the kth value of the sequence. For example, we might define a sequence $\mathbf{y}$ with values

$$y(k) = \begin{cases} (\frac{1}{3})^k, & k \geqslant 0 \\ 0, & k < 0 \end{cases}.$$

We can also explicitly list the values of the sequence. For example,

$$\mathbf{y} = \{\ldots, 0, -1, 3.2, 2, \underset{\uparrow}{4}, 5, -2.7, 0, \ldots\}.$$

We use the arrow to denote the $k = 0$ term. We shall use the convention that if the arrow is omitted, the first term is the $k = 0$ term and all values of the sequence before the $k = 0$ term are zero.

2.3 DISCRETE-TIME SYSTEMS

A discrete-time system is, formally, a mapping or transformation that maps input signals into output signals. Thus a system is represented by the equation

$$\mathbf{y} = \mathbf{T}(\mathbf{u}), \tag{2.3.1}$$

and the kth value of the output is

$$y(k) = (\mathbf{T}(\mathbf{u}))(k).$$

If $\mathbf{x}$ and $\mathbf{y}$ are signals and a and b are scalars, then $a\mathbf{x} + b\mathbf{y}$ is the signal having values

$$(a\mathbf{x} + b\mathbf{y})(k) = ax(k) + by(k). \tag{2.3.2}$$

The system $\mathbf{T}$ is *linear* if for all signals $\mathbf{x}$, $\mathbf{y}$ and scalars a, b, we have

$$\mathbf{T}(a\mathbf{x} + b\mathbf{y}) = a\mathbf{T}(\mathbf{x}) + b\mathbf{T}(\mathbf{y}). \tag{2.3.3}$$

If a system is linear, then linear combinations of inputs map into linear combinations of outputs. In particular, sums of inputs map into sums of outputs (superposition) and a scalar multiple of an input maps into a scalar multiple of its output.

A simple example of a linear system is a system that is a linear combination of the present and two past inputs. This system is described by the equation

$$y(k) = u(k) + b_1 u(k-1) + b_2 u(k-2).$$

A system is *shift-invariant* if for all input signals $\mathbf{u}$ the delay operator $\mathbf{S}$ and the system transformation $\mathbf{T}$ commute. That is,

$$\mathbf{T}(\mathbf{S}(\mathbf{u})) = \mathbf{S}(\mathbf{T}(\mathbf{u})). \tag{2.3.4}$$

Equation (2.3.4) implies also that $\mathbf{T}$ and $\mathbf{S}^m$ commute, i.e.,

$$\mathbf{T}(\mathbf{S}^m(\mathbf{u})) = \mathbf{S}^m(\mathbf{T}(\mathbf{u})). \tag{2.3.5}$$

Equation (2.3.4) is depicted schematically in Fig. 2.3.1. Equations (2.3.4) and (2.3.5) imply that if the response to $\{u(k)\}$ is $\{y(k)\}$, then the response to $\{u(k \pm m)\}$ is $\{y(k \pm m)\}$.

A *causal system* is a system whose present output depends only on past and present inputs, i.e.,

$$u_1(m) = u_2(m) \text{ for } m \leqslant k \quad \Rightarrow \quad (\mathbf{T}\mathbf{u}_1)(m) = (\mathbf{T}\mathbf{u}_2)(m) \text{ for } m \leqslant k.$$

A linear, shift-invariant system is causal if and only if the response to the unit-pulse sequence $\{\delta(k)\}$ is zero for $k < 0$. For this reason, we call a sequence *causal* if its values are zero for $k < 0$.

A system is *bounded-input, bounded-output stable* (BIBO) if the output $\mathbf{y}$ is bounded for a bounded input $\mathbf{u}$. In symbols, if the input satisfies $|u(k)| \leqslant R_1$ for all k, then there exists a real number R_2 for which the output satisfies $|y(k)| \leqslant R_2$ for all k. We shall characterize BIBO stability in other ways in Section 2.4.

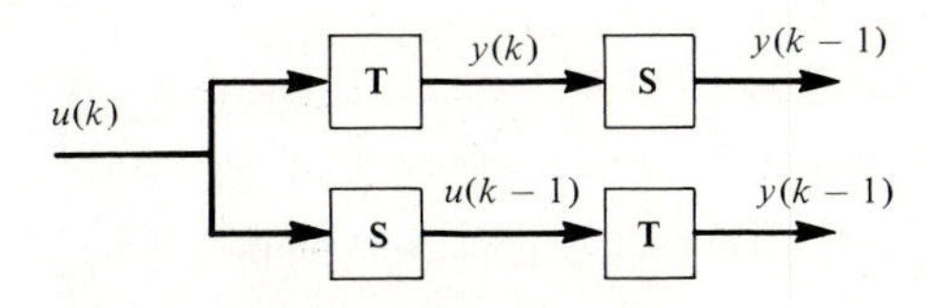

FIGURE 2.3.1 A characterization of shift invariance.

The primitive elements of linear, shift-invariant discrete-time systems are devices that add, multiply, and store numbers (unit-delays). A system is *finite order* if it is composed of a finite number of these primitive building blocks. All practical systems are finite order. However, we will have occasion to discuss systems that are not of finite order, such as an ideal low-pass filter.

It helps to visualize a digital system in various ways. The next example suggests we can think of a digital system as a hardware or block diagram, a software program, or as a set of equations. The descriptions are all equivalent, and each has its own applications.

EXAMPLE 2.3.1

Three descriptions of a digital filter.

In Figure 2.3.2 are three equivalent descriptions of a digital filter. The block diagram is meant to suggest a hardware description, although it is, of course, an abstraction. The unit-delay is a memory element, such as storage register. There is an underlying clock (not shown) that synchronizes the movement of the data through the filter. The software description is a program for a general purpose computer. Given the input sequence **u**, the output sequence **y** is calculated step by step, one output for each pass of the loop. The third description is a mathematical description of the filter. This description really consists of an infinite number of equations, one for each value of the index k.

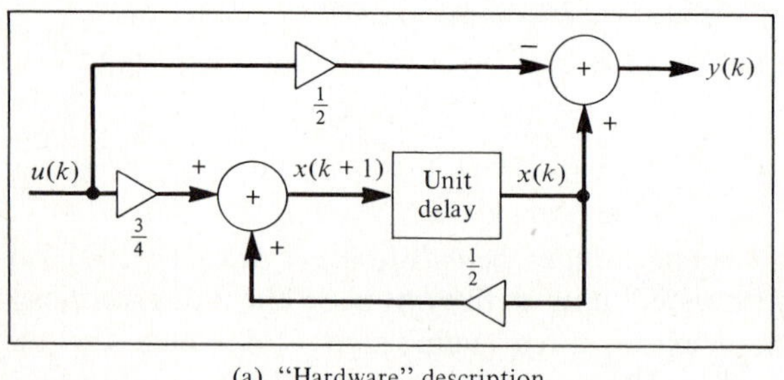

(a) "Hardware" description

```
1 READ u
  y ← x − .5 * u
  WRITE y
  x ← .5 * x + .75 * u
  Go To 1
```

(b) Software description

$$x(k+1) = \tfrac{1}{2}\, x(k) + \tfrac{3}{4}\, u(k)$$

$$y(k) = x(k) - \tfrac{1}{2}\, u(k)$$

(c) Mathematical description

FIGURE 2.3.2 **Three descriptions of a digital filter.**

2.4 THE UNIT-PULSE RESPONSE CHARACTERIZATION OF LSI SYSTEMS

If a discrete-time linear system is LSI, then the I/O function $\mathbf{y} = \mathbf{T}(\mathbf{u})$ has other equivalent characterizations that are often more useful.

The Unit-Pulse Response Sequence

Certain useful input sequences are used in characterizing an LSI system. One of these is the unit-pulse sequence. The unit-pulse response sequence $\mathbf{h}$ is, by definition, the output of a filter when the input is $\mathbf{u} = \boldsymbol{\delta}$. Thus

$$\mathbf{h} = \mathbf{T}(\boldsymbol{\delta}). \tag{2.4.1}$$

EXAMPLE 2.4.1

Direct evaluation or simulation of the unit-pulse response.

Suppose we have a digital filter as shown in Fig. 2.4.1. The output satisfies the recursion

$$y(k) = u(k) + \tfrac{1}{2}y(k-1), \qquad k \geqslant 0.$$

If we set $\mathbf{u} = \boldsymbol{\delta}$, then $\mathbf{y} = \mathbf{h}$ (by definition). Thus we have

$$h(k) = \delta(k) + \tfrac{1}{2}h(k-1), \qquad k \geqslant 0$$

We can find $h(k)$ in the preceding equation by direct evaluation. Thus

$$h(0) = \delta(0) + \tfrac{1}{2}h(-1) = 1.$$

[We assume that $h(-1)$, the initial loading in the unit delay, is zero.] Continuing, we find

$$h(1) = \delta(1) + \tfrac{1}{2}h(0) = \tfrac{1}{2},$$
$$h(2) = \delta(2) + \tfrac{1}{2}h(1) = (\tfrac{1}{2})^2,$$
$$\vdots$$
$$h(k) = (\tfrac{1}{2})^k.$$

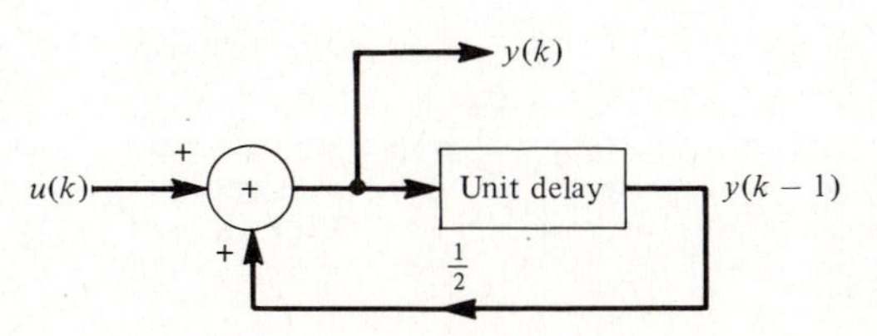

FIGURE 2.4.1.

For this simple example, direct evaluation leads to the closed form expression for the unit-pulse response

$$h(k) = \begin{cases} (\frac{1}{2})^k, & k \geqslant 0 \\ 0, & k < 0 \end{cases}.$$

Convolution and the Unit-Pulse Response Sequence

We can represent any arbitrary sequence **u** as a sum of shifted unit-pulse sequences each multiplied by the appropriate sequence values of **u** as was shown in Eq. (2.2.3). Thus **u** can be written in the form

$$\mathbf{u} = \sum_{m=-\infty}^{\infty} u(m)\mathbf{S}^m \boldsymbol{\delta},$$

or

$$u(k) = \sum_{m=-\infty}^{\infty} u(m)\delta(k-m). \tag{2.4.2}$$

This representation is a key idea. We have, by definition, that the response to $\boldsymbol{\delta}$ is $\mathbf{h}$. By linearity, the response to $c\boldsymbol{\delta}$ is $c\mathbf{h}$. By shift-invariance, the response to $c\mathbf{S}^m\boldsymbol{\delta}$ is $c\mathbf{S}^m\mathbf{h}$. This is shown schematically in Fig. 2.4.2.

Using linearity, the response to a sum of terms of the form in Eq. (2.4.2) is

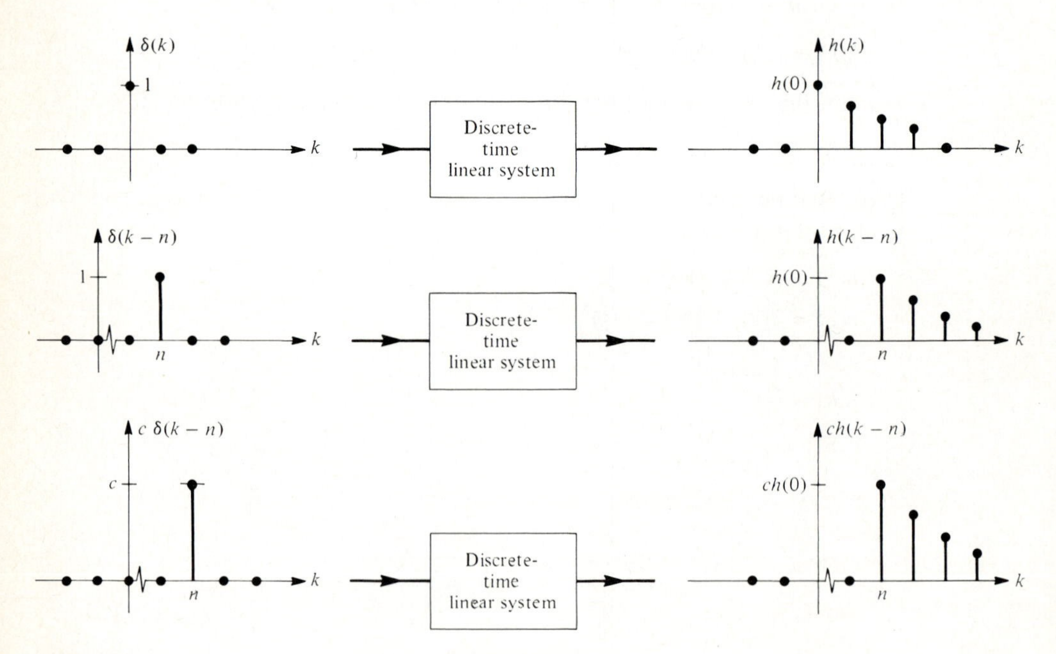

FIGURE 2.4.2.

the sum of responses to each term. Thus

$$
\begin{aligned}
\mathbf{y} = \mathbf{Tu} = \mathbf{T}\left(\sum_{m=-\infty}^{\infty} u(m)\mathbf{S}^m\boldsymbol{\delta}\right) & \qquad \begin{pmatrix}\text{representation}\\ \text{Eq. (2.4.2)}\end{pmatrix} \\
= \sum_{m=-\infty}^{\infty} u(m)\mathbf{TS}^m\boldsymbol{\delta} & \qquad \begin{pmatrix}\text{linearity}\\ \text{Eq. (2.3.3)}\end{pmatrix} \\
= \sum_{m=-\infty}^{\infty} u(m)\mathbf{S}^m\mathbf{T}\boldsymbol{\delta} & \qquad \begin{pmatrix}\text{shift-invariance}\\ \text{Eq. (2.3.5)}\end{pmatrix} \\
= \sum_{m=-\infty}^{\infty} u(m)\mathbf{S}^m\mathbf{h} & \qquad \begin{pmatrix}\text{definition of } \mathbf{h}\\ \text{Eq. (2.4.1)}\end{pmatrix}
\end{aligned} \tag{2.4.3}
$$

This is a sequence equation. If we evaluate the sequences in Eq. (2.4.3) at time k, using Eq. (2.2.2), we have

$$y(k) = \sum_{m=-\infty}^{\infty} u(m)h(k-m). \tag{2.4.4}$$

The combination of the sequences $\mathbf{u}$ and $\mathbf{h}$ in Eq. (2.4.4) is called *convolution.* It is a sequence product and the standard notation is

$$\mathbf{y} = \mathbf{u} * \mathbf{h},$$

which means

$$y(k) = \sum_{m=-\infty}^{\infty} u(m)h(k-m). \tag{2.4.5}$$

Notice that the convolution of two sequences is another sequence, and that the value of $\mathbf{u} * \mathbf{h}$ at time k is the sum of all products of one value of $\mathbf{h}$ and one value of $\mathbf{u}$ for which the respective indices add to k. One can show that convolution is commutative and associative, i.e.,

$$\mathbf{u} * \mathbf{h} = \mathbf{h} * \mathbf{u}, \tag{2.4.6}$$

$$\mathbf{g} * (\mathbf{u} * \mathbf{h}) = (\mathbf{g} * \mathbf{u}) * \mathbf{h}. \tag{2.4.7}$$

The output of an initially quiescent LSI system is the input convolved with the unit-pulse response sequence. Thus $\mathbf{h}$ completely characterizes the I/O properties of the filter.

Properties of the Unit-Pulse Response

In general the unit-pulse response of a linear system (not necessarily shift-invariant) is a function of the response at the kth instant to a unit-pulse applied at the mth instant. That is,

$$h(k,m) = (\mathbf{TS}^m\boldsymbol{\delta})(k) \qquad \begin{pmatrix}\text{response at } k\text{th instant to a unit pulse}\\ \text{applied at the } m\text{th instant}\end{pmatrix} \tag{2.4.8}$$

In general, we can express the output of a linear discrete-time system in the form

$$y(k) = \sum_{n=-\infty}^{\infty} h(k, m)u(m). \tag{2.4.9}$$

In this case, the system is characterized by a matrix $\mathbf{H}$ with elements $h(k, m)$:

$$\mathbf{H} = [h(k, m)]. \tag{2.4.10}$$

If the system is shift-invariant, then $h(k, m)$ is a function only of $(k - m)$, i.e.,

$$h(k, m) = h(k - m). \tag{2.4.11}$$

The $\mathbf{H}$ matrix in this case is *Toeplitz*, i.e., the entries of $\mathbf{H}$ along diagonals are equal:

$$\mathbf{H} = \begin{bmatrix} h(0) & h(-1) & h(-2) & \cdots \\ h(1) & h(0) & h(-1) & \cdots \\ h(2) & h(1) & h(0) & \cdots \\ & \vdots & & \end{bmatrix}. \tag{2.4.12}$$

If the system is also causal, then the system cannot respond before an input is applied. Thus

$$h(k, m) = 0, \qquad m > k. \tag{2.4.13}$$

This implies that $\mathbf{H}$ is lower triangular. The combination of Eqs. (2.4.11) and (2.4.13) implies a matrix $\mathbf{H}$ of the form

$$\mathbf{H} = \begin{bmatrix} h(0) & 0 & 0 & \cdots \\ h(1) & h(0) & 0 & \cdots \\ h(2) & h(1) & h(0) & \cdots \\ & \vdots & & \end{bmatrix}. \tag{2.4.14}$$

If both $\mathbf{u}$ and $\mathbf{h}$ are causal sequences, then each sum in Eq. (2.4.5) is finite.

EXAMPLE 2.4.2

Convolution with a FIR filter response.

A filter whose unit-pulse response sequence $\mathbf{h}$ has finitely many nonzero values is called a *finite-impulse-response* (FIR) filter. Consider a filter for which $h(k) = 0$ for $k < 0$ and for $k > 3$. Then

$$y(k) = (\mathbf{u} * \mathbf{h})(k) = \sum_{l=k-3}^{k} u(l)h(k - l). \tag{2.4.15}$$

A block diagram of such a filter is shown in Fig. 2.4.3. Schematically, we may visualize this finite convolution as a "sliding sum of products" or "moving average" as follows.

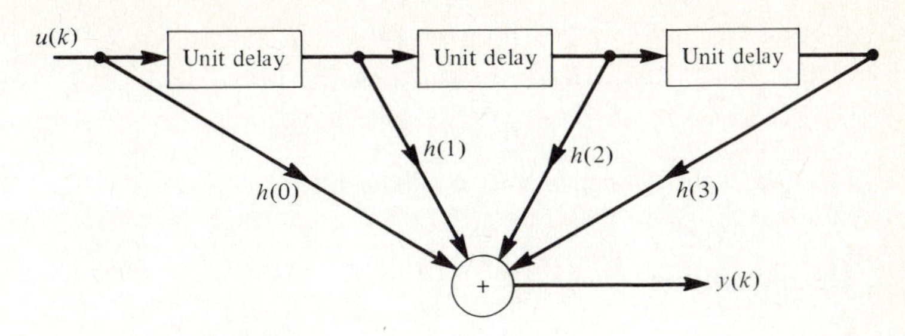

FIGURE 2.4.3 An FIR filter.

Data: ... $u(-3)$ $u(-2)$ $u(-1)$ $u(0)$ $u(1)$ $u(2)$ ...

Unit-pulse response: $h(3)$ $h(2)$ $h(1)$ $h(0)$ →
(reversed and shifted)

Output at $k = 1$: $y(1) = h(0)\,u(1) + h(1)\,u(0) + h(2)\,u(-1) + h(3)\,u(-2)$

In order to calculate the output $y(k)$, one slides the weights so that $h(0)$ lines up with $u(k)$ and then takes the sum of products between the two sequences **u** and **h**, which is time reversed.

EXAMPLE 2.4.3

Convolution of two geometric sequences.

For any finite $n \geqslant 0$, the partial sum of a geometric series is

$$\sum_{l=0}^{n} a^l = \frac{1 - a^{n+1}}{1 - a}, \qquad a \neq 1. \tag{2.4.16}$$

The infinite sum converges provided $|a| < 1$. Suppose the unit-pulse response of a system and the input are given by

$$h(k) = \begin{cases} a^k, & k \geqslant 0 \\ 0, & k < 0 \end{cases}, \qquad u(k) = \begin{cases} b^k, & k \geqslant 0 \\ 0, & k < 0 \end{cases}. \tag{2.4.17}$$

Then, with the help of Eq. (2.4.16), we may compute the output as

$$y(k) = (\mathbf{u} * \mathbf{h})(k) = \begin{cases} 0, & k < 0 \\ \dfrac{a^{k+1} - b^{k+1}}{a - b}, & k \geqslant 0, \quad a \neq b, \\ (k+1)a^k, & k \geqslant 0, \quad a = b. \end{cases} \tag{2.4.18}$$

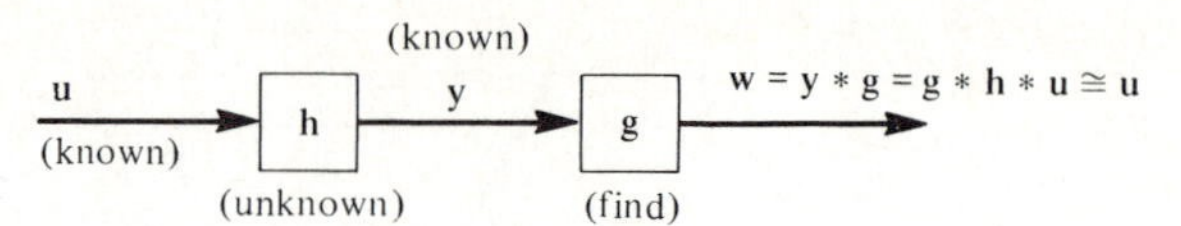

FIGURE 2.4.4 **Design of a deconvolution filter.**

Deconvolution

In the previous characterization, we are given the input **u** and system pulse response **h** and asked to find the output **y**. There are applications in which we are given the output **y** and either **u** or **h** and asked to find the unknown sequence. **y**, **u**, and **h** are related by the equation, $\mathbf{y} = \mathbf{u} * \mathbf{h}$. In deconvolution problems, we seek to recover either **u** or **h** from the given information.

One context in which this problem arises is a situation in which one is attempting to characterize a "channel" or transmission path. Suppose we wish to determine the transmission properties of a certain medium. We could envision exciting the medium with a known input **u**, measuring the output **y**, and then obtaining **h** from knowledge of **u** and **y** and the equation $\mathbf{y} = \mathbf{u} * \mathbf{h}$.

Another variation of the deconvolution problem is to find an LSI system with unit-pulse response **g** so that when **y** is an input to the system **g**, the output approximates **u**. This is shown in Fig. 2.4.4. The idea here is that if we can find **g** so that $\mathbf{h} * \mathbf{u} * \mathbf{g} = \mathbf{u}$, then clearly $\mathbf{h} * \mathbf{g} = \boldsymbol{\delta}$. In this application we are primarily interested in "undoing" the effects of **h** on the input **u**. This problem is sometimes called the *channel equalization problem*. The sequence **g** is said to be the *inverse sequence* for **h** (Robinson 1976).

EXAMPLE 2.4.4

Finding an inverse sequence.

Suppose **h** is the sequence shown in Fig. 2.4.5. Find the inverse sequence **g** so that $\mathbf{h} * \mathbf{g} = \boldsymbol{\delta}$. Assume **g** is causal. We have

$$g(k)h(0) + g(k-2)h(2) = \delta(k),$$

and so

$$g(k) - \tfrac{1}{2}g(k-2) = \delta(k). \tag{2.4.19}$$

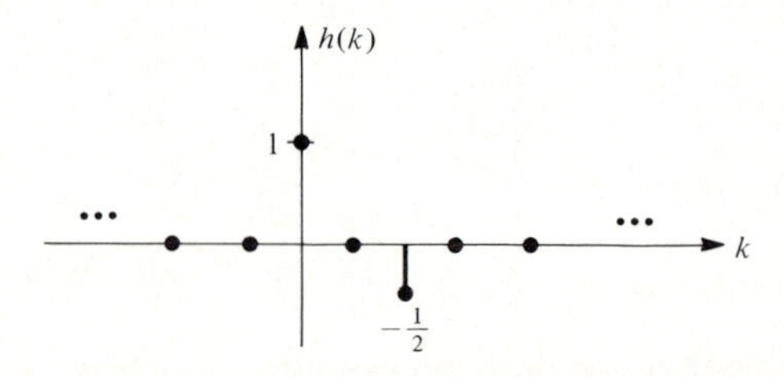

FIGURE 2.4.5.

By direct evaluation, beginning at $k = 0$, we find

$$\begin{aligned}
&g(0) - \tfrac{1}{2}g(-2) = 1 \Rightarrow g(0) = 1 \\
&g(1) - \tfrac{1}{2}g(-1) = 0 \Rightarrow g(1) = 0 \\
&g(2) - \tfrac{1}{2}g(0) = 0 \quad \Rightarrow g(2) = \tfrac{1}{2} \\
&g(3) = 0 \\
&g(4) = (\tfrac{1}{2})^2 \\
&\vdots
\end{aligned}$$

Thus we can express $g(k)$ as

$$g(k) = \begin{cases} \tfrac{1}{2}[(1/\sqrt{2})^k + (-1/\sqrt{2})^k], & k \geqslant 0 \\ 0, & k < 0 \end{cases}.$$

We shall explain shortly how to solve equations of the form (2.4.19) without resorting to a direct evaluation.

Characterization of BIBO Stability

A filter is BIBO stable if a bounded input sequence implies the output sequence is also bounded. Since a linear, shift-invariant filter is characterized by its unit-pulse response sequence, the property of BIBO stability must depend only on the sequence **h**. We shall use the following standard notations:

$$\|\mathbf{h}\|_\infty = \max\{|h(k)| : -\infty < k < \infty\}, \tag{2.4.20}$$

$$\|\mathbf{h}\|_1 = \sum_{k=-\infty}^{\infty} |h(k)|, \tag{2.4.21}$$

$$\|\mathbf{h}\|_2 = \left(\sum_{k=-\infty}^{\infty} |h(k)|^2\right)^{1/2}. \tag{2.4.22}$$

These are called "norms." Roughly speaking, a norm measures the "magnitude" or "size" of something. To each norm there is an associated "linear normed space" of sequences:

$$l_\infty = \{\mathbf{h} : \|\mathbf{h}\|_\infty < \infty\}, \tag{2.4.23}$$

$$l_1 = \{\mathbf{h} : \|\mathbf{h}\|_1 < \infty\}, \tag{2.4.24}$$

$$l_2 = \{\mathbf{h} : \|\mathbf{h}\|_2 < \infty\}. \tag{2.4.25}$$

Using this terminology, a filter is BIBO stable if for any input $\mathbf{u} \in l_\infty$ (this means **u** is bounded), then

$$\mathbf{y} = \mathbf{u} * \mathbf{h} \in l_\infty$$

(which means **y** is bounded).

Consider a filter with unit-pulse response **h**, driven by a bounded input. Then

$$\begin{aligned}|y(k)| = |(\mathbf{u} * \mathbf{h})(k)| &= \left|\sum_{l=-\infty}^{\infty} h(l)u(k-l)\right| \\ &\leqslant \sum_{l=-\infty}^{\infty} |h(l)||u(k-l)| \\ &\leqslant \sum_{l=-\infty}^{\infty} |h(l)|\,\|\mathbf{u}\|_{\infty} = \|\mathbf{h}\|_1\|\mathbf{u}\|_{\infty}\end{aligned} \tag{2.4.26}$$

This last expression is independent of the time variable k. Therefore, the filter will be BIBO stable provided $\mathbf{h} \in l_1$. To prove the converse, suppose the filter is BIBO stable. We wish to show the unit-pulse response sequence **h** is an element of l_1, i.e. $\|\mathbf{h}\|_1$ is finite. To do this we choose an input $u(k)$, as follows

$$u(k) = \begin{cases} \dfrac{h(-k)}{|h(-k)|}, & h(k) \neq 0 \\ 0, & \text{otherwise} \end{cases}. \tag{2.4.27}$$

Clearly $\|\mathbf{u}\|_{\infty} = 1$. Now if we compute the output at time $k = 0$, we obtain

$$y(0) = \sum_{l=-\infty}^{\infty} h(l)u(-l) = \sum_{l} \frac{h^2(l)}{|h(l)|} = \sum_{l} |h(l)| = \|\mathbf{h}\|_1. \tag{2.4.28}$$

Equation (2.4.28) states that $\|\mathbf{h}\|_1$ is equal to $y(0)$, which must be finite. Thus BIBO stability implies $\mathbf{h} \in l_1$.

Our principal results can be summarized as follows. If a filter is linear, shift-invariant, and BIBO stable, then

$$\begin{aligned} &(1)\quad \mathbf{h} = \mathbf{T}(\boldsymbol{\delta}) \in l_1, \\ &(2)\quad \mathbf{y} = \mathbf{T}(\mathbf{u}) = \mathbf{u} * \mathbf{h}, \\ &(3)\quad \|\mathbf{y}\|_{\infty} \leqslant \|\mathbf{h}\|_1\|\mathbf{u}\|_{\infty}. \end{aligned} \tag{2.4.29}$$

2.5 THE FREQUENCY RESPONSE FUNCTION

Just as the unit-pulse sequence served as the basic input in the characterization of the convolutional sum as an I/O description for a linear, shift-invariant system, the complex exponential sequence $\{e^{jk\theta}\}$ plays the same role for another equally important I/O description. Consider the steady-state output of a linear, shift-invariant system with input $u(k) = e^{jk\theta}$. From Eq. (2.4.5) we have

$$y(k) = \sum_{l-=\infty}^{\infty} h(l)u(k-l).$$

Substituting for $u(k-l)$, we obtain

$$y(k) = \sum_{l=-\infty}^{\infty} h(l)e^{j(k-l)\theta}$$
$$= e^{jk\theta}\left[\sum_{l=-\infty}^{\infty} h(l)e^{-jl\theta}\right]$$
$$= e^{jk\theta}[H(e^{j\theta})]. \tag{2.5.1}$$

Equation (2.5.1) states that the output is an exponential of the same frequency θ as the input, but modified by the complex function $H(e^{j\theta})$, given by

$$H(e^{j\theta}) = \sum_{k=-\infty}^{\infty} h(k)e^{-jk\theta}. \tag{2.5.2}$$

This complex function of the frequency θ is called the *frequency response function* of the filter. $H(e^{j\theta})$ is called the *discrete-time Fourier transform* (DTFT) of the sequence **h**. Notice that for each input frequency θ, we obtain the same frequency at the output. Only the phase and amplitude of the complex sinusoid are changed. This is only true in the case of linear, shift-invariant systems.

The frequency response function defines how every sinusoid (real or complex) is modified in phase and amplitude by the filter. Thus the input $e^{jk\theta}$ produces an output $H(e^{j\theta})e^{jk\theta}$. This output can be written

$$y(k) = |H(e^{j\theta})|\, e^{j(k\theta + \arg H(e^{j\theta}))}. \tag{2.5.3}$$

If the input is $\cos(k\theta) = \text{Re}\{e^{jk\theta}\}$, the output is

$$y(k) = \text{Re}\{e^{jk\theta}H(e^{j\theta})\}$$
$$= |H(e^{j\theta})|\cos(k\theta + \arg H(e^{j\theta})). \tag{2.5.4}$$

These equations allow one to interpret the magnitude response, $|H(e^{j\theta})|$, and the phase response, $\arg H(e^{j\theta})$, functions. The magnitude response measures the coefficient of increase or decrease in the amplitude of the input sinusoid. The phase measures the shift in phase of the input sinusoid between input and output. (See Fig. 2.5.1.)

The frequency response function $H(e^{j\theta})$ is regarded as a complex function of the real variable θ. (Our notation is not, strictly speaking, correct. We should write $G(\theta) = H(e^{j\theta})$ if it were necessary to emphasize the true state of affairs.) Now the complex exponential is periodic of period 2π (it moves continuously around the unit circle in a counterclockwise direction, coming full circle every 2π radians). It follows that the frequency response function is also 2π-periodic as a function of θ. The same must therefore be true of the magnitude and phase response functions. If the coefficients $\{h(k)\}$ are all real, then the complex conjugate of $H(e^{j\theta})$ is

$$H^*(e^{j\theta}) = H(e^{-j\theta}) \tag{2.5.5}$$

and therefore,

$$|H(e^{j\theta})|^2 = H(e^{j\theta})H(e^{-j\theta}). \tag{2.5.6}$$

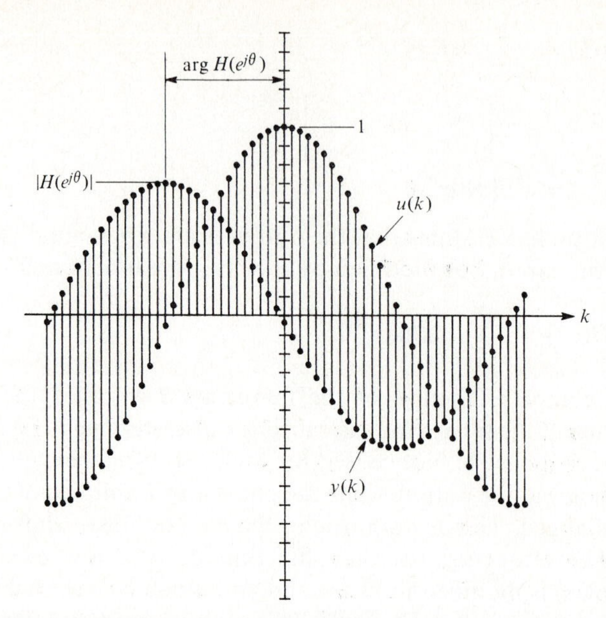

FIGURE 2.5.1 An interpretation of magnitude and phase function at a given frequency.

It follows that $|H(e^{j\theta})|$ is an even function of θ and arg $H(e^{j\theta})$ is an odd function of θ (up to 2π increments). Thus a plot of the magnitude and phase of $H(e^{j\theta})$ on the interval $[0, \pi]$ is sufficient to determine $H(e^{j\theta})$ for all θ. This is the standard way to display the frequency response function, and one can find such plots in manufacturer's data books as well as in textbooks.

The frequency response above $\theta = \pi$ must be the mirror image of the response below π, and thus this frequency is called the *foldover frequency*.

EXAMPLE 2.5.1

Evaluation of the frequency response function for an FIR filter.

Consider an FIR filter as shown in Fig. 2.5.2. The input-output equation for this filter is of the form

$$y(k) = h(0)u(k) + h(1)u(k-1) + h(2)u(k-2) + h(3)u(k-3).$$

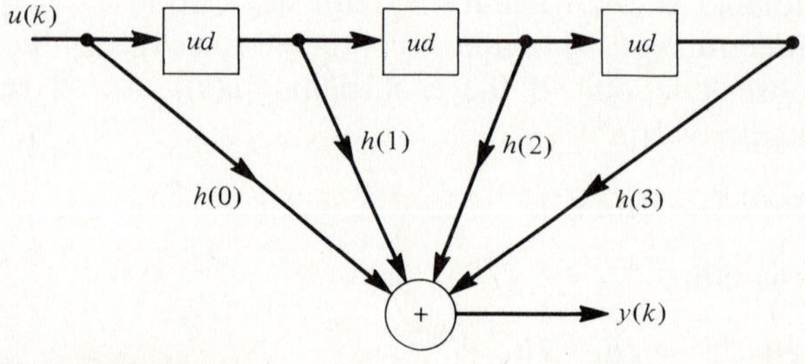

$u(k)$ ud ud ud $h(1)$ $h(2)$ $h(0)$ $h(3)$ + $y(k)$

FIGURE 2.5.2 An FIR filter.

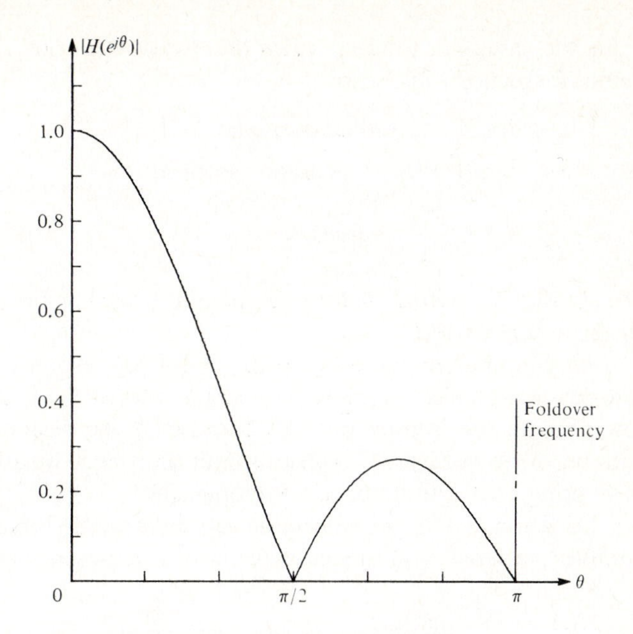

FIGURE 2.5.3 The magnitude response of a simple FIR filter.

The frequency response function is obtained from Eq. (2.5.2) as

$$H(e^{j\theta}) = \sum_{k=0}^{3} h(k)e^{-jk\theta}.$$

Suppose that the "tap weights" $h(k)$, $k = 0, 1, 2, 3$ are all equal to $\frac{1}{4}$. The resulting output is merely the average of the present and past values. Thus the filter acts to "smooth" the input. In other words, the frequency response of the filter attenuates high frequencies. This is shown by plotting the magnitude response $|H(e^{j\theta})|$ vs θ in Fig. 2.5.3 where

$$H(e^{j\theta}) = \frac{1}{4}\sum_{k=0}^{3} e^{-jk\theta} = \frac{e^{-j3\theta/2}}{2}\,[\cos(\theta/2) + \cos(\tfrac{3}{2}\theta)].$$

Often samples of an input sequence are derived from a continuous time signal $\tilde{u}(t)$. The time between samples (known as the sampling period) is t_0 seconds. Suppose the continuous time input is $\tilde{u}(t) = \cos\omega t = \cos 2\pi f t$. Then the corresponding sequence values are

$$\tilde{u}(kt_0) = u(k) = \cos(\omega k t_0) = \cos(k\theta).$$

In our previous discussion, we used the "normalized" frequency variable θ, which is related to the unnormalized frequency (ω or f) by

$$\theta = \omega t_0 = 2\pi f t_0. \tag{2.5.7}$$

Thus the sampling time t_0 scales the actual frequency. The normalized and actual frequency ranges are

$$-\pi \leqslant \theta \leqslant \pi \qquad \text{(radians/sample)} \tag{2.5.8}$$

$$-\pi/t_0 \leqslant \omega \leqslant \pi/t_0 \qquad \text{(radians/second)} \tag{2.5.9}$$

$$-\frac{1}{2t_0} \leqslant f \leqslant \frac{1}{2t_0} \qquad \text{(hertz).} \tag{2.5.10}$$

We see that the actual foldover frequency, $0.5t_0^{-1}$, is inversely proportional to the sampling period.

All digital filters have a periodic frequency response. When one designs a low-pass digital filter, it is low-pass on the interval $[-1/2t_0, 1/2t_0]$ Hz. It is not low-pass on the infinite interval because of the periodicity of H. In applications, this is not usually a serious limitation since we are generally interested in filtering over a finite band of frequencies.

Equation (2.5.2) can be generalized by formally replacing $e^{j\theta}$ by a general complex variable z. In this case, we have a representation

$$H(z) = \sum_{k=-\infty}^{\infty} h(k)z^{-k}. \tag{2.5.11}$$

$H(z)$ is called the *transfer function* of the system. This representation is valid whenever the sum in Eq. (2.5.11) converges absolutely. $H(z)$ is said to be the z-transform of the sequence $\mathbf{h}$. Chapter 3 presents a more complete discussion of this I/O characterization of a filter. The frequency response function $H(e^{j\theta})$ is merely the transfer function $H(z)$ evaluated on the unit circle $|z| = 1$ in the complex z-plane. (This is analogous to the situation for continuous-time systems. In that case the frequency response function $H(j\omega)$ is the system transfer function $H(s)$, the Laplace transform of the impulse response $\mathbf{h}$, evaluated on the $j\omega$ axis in the complex s-plane.)

One of the important properties of the transform domain is the transformation of the convolution sum. If we define the transform of the output as $Y(z)$, then we can write

$$Y(z) = \sum_{k=-\infty}^{\infty} y(k)z^{-k}. \tag{2.5.12}$$

But $y(k)$ is given by

$$y(k) = \sum_{m=-\infty}^{\infty} h(m)u(k-m). \tag{2.5.13}$$

Substituting Eq. (2.5.13) into Eq. (2.5.12), we obtain

$$\begin{aligned} Y(z) &= \sum_{k=-\infty}^{\infty} z^{-k} \sum_{m=-\infty}^{\infty} h(m)u(k-m)(z^m z^{-m}) \\ &= \sum_{m=-\infty}^{\infty} h(m)z^{-m} \sum_{k=-\infty}^{\infty} u(k-m)z^{-(k-m)} \\ &= H(z)U(z). \end{aligned} \tag{2.5.14}$$

Thus the convolution of the sequences **h** and **u** is transformed into the product of the transforms $H(z)$ and $U(z)$. And from Eq. (2.5.14) we obtain the output DTFT as

$$Y(e^{j\theta}) = Y(z)|_{z=e^{j\theta}} = H(e^{j\theta})U(e^{j\theta}). \tag{2.5.15}$$

Equations (2.5.15) and (2.5.14) represent a second I/O description of a linear, shift-invariant system. In Chapters 3 and 4, we discuss these two descriptions in greater detail.

EXAMPLE 2.5.2

Finding the transfer function of digital filter.

Normally the calculation of $H(z)$ or $H(e^{j\theta})$ using the unit-pulse response sequence and Eq. (2.5.2) is difficult. A simpler method is to use the property that if the input $u(k) = z^k$, then the output of a linear, shift-invariant system is $z^k H(z)$. [This is merely Eq. (2.5.1) with z substituted for $e^{j\theta}$.] For example, consider Example 2.3.1. The time domain equations are

$$x(k+1) = \tfrac{1}{2}x(k) + \tfrac{3}{4}u(k),$$
$$y(k) = x(k) - \tfrac{1}{2}u(k).$$

Using Eq. (2.5.1), we make the substitutions $u(k) = z^k$, $x(k) = z^k X(z)$ and $y(k) = z^k H(z)$. Substituting into the time domain equations, we obtain

$$X(z)z^{k+1} = \tfrac{1}{2}X(z)z^k + \tfrac{3}{4}z^k,$$
$$H(z)z^k = X(z)z^k - \tfrac{1}{2}z^k.$$

Hence multiplying through by z^{-k}, we obtain

$$X(z) = \frac{\frac{3}{4}}{z - \frac{1}{2}}$$

and so

$$H(z) = \frac{1 - z/2}{z - \frac{1}{2}}.$$

Computation of the Frequency Response Function

If one has a finite-order filter, the frequency response function must have the form

$$H(e^{j\theta}) = \frac{\sum_{k=0}^{n} b_k e^{-jk\theta}}{\sum_{k=0}^{n} a_k e^{-jk\theta}}, \qquad a_0 = 1 \tag{2.5.16}$$

The coefficients a_k and b_k are real. Computing the magnitude and phase of such a function can be a formidable endeavor. We shall present three approaches to this computation. Each has applications for which it is the method of choice.

A GRAPHICAL APPROACH, USING POLES AND ZEROS. We shall assume that the transfer function $H(z)$ is rational, i.e., the ratio of the polynomials $\hat{a}(z)$ and $\hat{b}(z)$:

$$H(z) = \hat{b}(z)/\hat{a}(z). \tag{2.5.17}$$

The roots of the polynomial $\hat{b}(z)$ are called the zeros of H since $\hat{b}(\lambda) = 0$ implies $H(\lambda) = 0$ [assuming $\hat{a}(\lambda) \neq 0$]. The roots of $\hat{a}(z)$ are called the poles of $H(z)$, and if $\hat{a}(\lambda) = 0$ [with $\hat{b}(\lambda) \neq 0$], then $H(\lambda)$ is infinite. The poles, zeros, and a gain factor g parameterize $H(z)$ since we can write

$$H(z) = g \prod_{i=1}^{n} (z - z_i) \Big/ \prod_{i=1}^{n} (z - p_i) \tag{2.5.18}$$

with z_i the zeros of $H(z)$ and p_i the poles of $H(z)$. To obtain the poles and zeros, we must factor the polynomials $\hat{a}(z)$ and $\hat{b}(z)$. Once they are known, we can use Eq. (2.5.18) to evaluate the magnitude and phase of $H(z)$ at $z = e^{j\theta}$:

$$|H(e^{j\theta})| = g \prod_{i=1}^{n} |(e^{j\theta} - z_i)| \Big/ \prod_{i=1}^{n} |(e^{j\theta} - p_i)|, \tag{2.5.19}$$

$$\arg H(e^{j\theta}) = \sum_{i=1}^{n} [\arg(e^{j\theta} - z_i) - \arg(e^{j\theta} - p_i)]. \tag{2.5.20}$$

Each pole and zero contributes an independent term: additive for phase and multiplicative for magnitude. These individual contributions may be obtained graphically using a *pole-zero diagram*, as shown in Fig. 2.5.4.

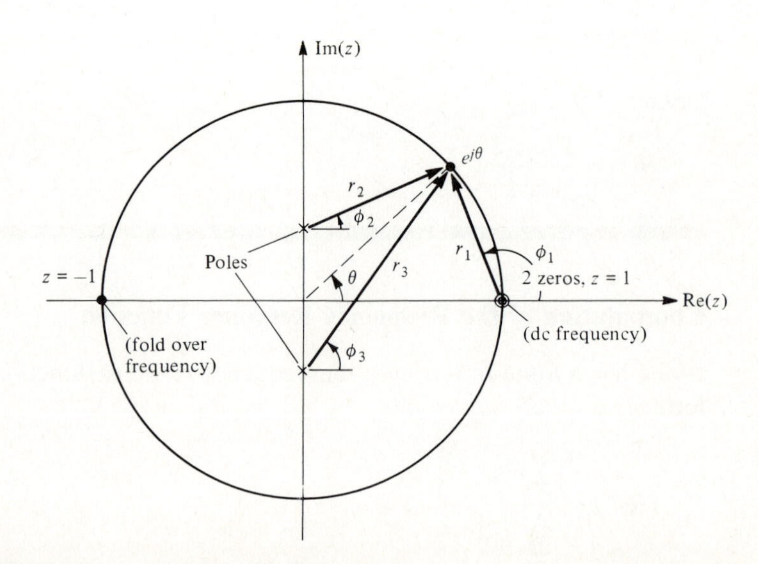

FIGURE 2.5.4 A pole-zero diagram.

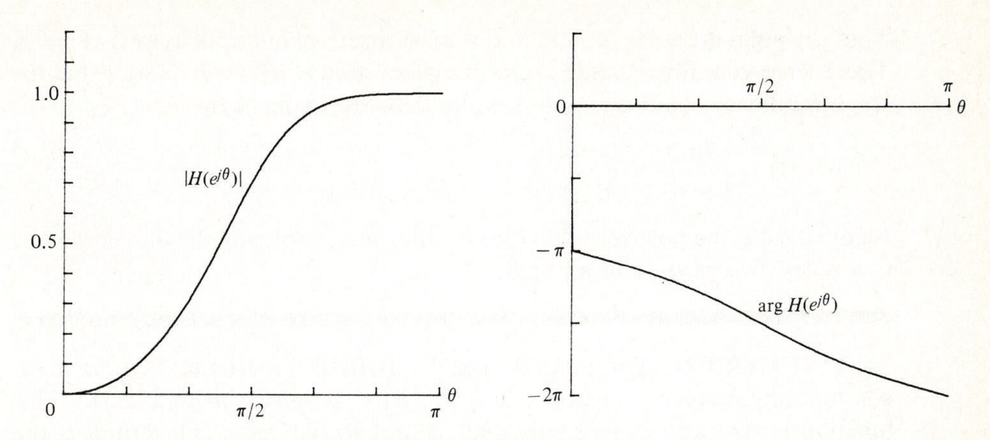

FIGURE 2.5.5 A plot of the amplitude and phase response for the pole-zero diagram of Figure 2.5.4.

Consider a term of the form $\mathbf{e}^{j\theta} - \mathbf{z}_i$. Interpreting each term $\mathbf{e}^{j\theta}$ and $-\mathbf{z}_i$ as vectors, we see that the vector represented by $(\mathbf{e}^{j\theta} - \mathbf{z}_i)$ is a vector from $\mathbf{z}_i$ to the point $\mathbf{e}^{j\theta}$ (which is on the circle $|z| = 1$ with angle θ). Thus the magnitude expression is a ratio of the distances from the zero locations to $\mathbf{e}^{j\theta}$ over the distances from the pole locations to $\mathbf{e}^{j\theta}$. The phase of $H(z)|_{z=e^{j\theta}}$ is the sum of the angles from the zeros to $\mathbf{e}^{j\theta}$ minus the sum of the angles from the poles to $\mathbf{e}^{j\theta}$. Thus, for example, in Fig. 2.5.4 the magnitude and phase of $H(z)$ are given by

$$|H(e^{j\theta})| = gr_1^2/(r_2 r_3),$$
$$\arg H(e^{j\theta}) = 2\phi_1 - (\phi_2 + \phi_3).$$

A sketch of the resulting frequency response is shown in Fig. 2.5.5.

One can also gain insight into the filter synthesis problem from the pole-zero diagram. If a frequency θ_0 is to be rejected, simply place a zero at $\mathbf{e}^{j\theta_0}$. If the frequency θ_0 is to be emphasized, place a pole inside the circle, near $\mathbf{e}^{j\theta_0}$. Zeros pull the response down; poles pull it up.

EXAMPLE 2.5.3

Finding the frequency response of a Butterworth digital filter.

Consider the following second-order filter (a Butterworth high-pass filter):

$$y(k) + a_2 y(k-2) = b_0\{u(k) - 2u(k-1) + u(k-2)\}, \tag{2.5.21}$$

where

$$a_2 = 3 - 2\sqrt{2}, \qquad b_0 = 1 - \sqrt{2}\big/2.$$

Thus

$$\hat{a}(z) = z^2 + a_2 = [z + j(\sqrt{2} - 1)][z - j(\sqrt{2} - 1)],$$
$$\hat{b}(z) = b_0(z^2 - 2z + 1) = b_0(z - 1)^2.$$

Thus the poles are $\pm j(\sqrt{2} - 1)$, and there is a zero of multiplicity two at $z = 1$. These zeros guarantee that the zero frequency gain is $H(1) = 0$. Notice that the "high-frequency" gain is ($z = -1$ is the foldover frequency)

$$H(-1) = \frac{4b_0}{1 + a_2} = 1.$$

Figure 2.5.4 is the pole-zero diagram for this filter, depicting the unit circle, the two poles, two zeros, and an angle θ.

A NUMERICAL APPROACH, USING HORNER'S RULE. We wish to compute the frequency response function $H(e^{j\theta})$ as given in Eq. (2.5.16). This function is the ratio of two polynomials and so our basic calculation is the evaluation of a polynomial on $|z| = 1$. Horner's rule is a very old algorithm for the efficient evaluation of a polynomial (Conte 1972). Consider the polynomial

$$q(z) = q_0 z^n + q_1 z^{n-1} + \cdots + q_n. \tag{2.5.22}$$

Suppose we wish to evaluate $q(z)$ at some given value of z. The following algorithm does it with exactly n multiplies and n additions.

Horner's algorithm

Initialization: $x(0) = 0$

Loop: $x(k + 1) = zx(k) + q_k$

End: $x(n + 1) = q(z)$ **(2.5.23)**

EXAMPLE 2.5.4

Evaluation of a polynomial using Horner's algorithm.

As an example of this calculation, consider the evaluation of the polynomial

$$q(z) = 3z^3 + 4z^2 - 2z + 5$$

for $z = 2$. Beginning with $x(0) = 0$, we have

$$x(1) = zx(0) + q_0 = 2(0) + 3 = 3,$$
$$x(2) = zx(1) + q_1 = 2(3) + 4 = 10,$$
$$x(3) = zx(2) + q_2 = 2(10) - 2 = 18,$$

and so

$$q(z)|_{z=2} = x(4) = zx(3) + q_3 = 2(18) + 5 = 41.$$

There are several variations of this algorithm. The following generalization is useful for the numerical computation of $H(e^{j\theta})$. It allows us to evaluate any polynomial q at values z and z^{-1} simultaneously.

Let $\mathbf{x}(0)$ be the zero vector. Consider the matrix recursion equivalent of Horner's algorithm (2.5.23):

$$\mathbf{x}(k+1) = \mathbf{A}\mathbf{x}(k) + \mathbf{B}q_k. \tag{2.5.24}$$

Carrying through this recursion until $k + 1 = n + 1$, one can show in a straightforward manner that

$$\mathbf{x}(n+1) = q(\mathbf{A})\mathbf{B},$$

where $q(\mathbf{A})$ is the matrix polynomial

$$q(\mathbf{A}) = q_0\mathbf{A}^n + q_1\mathbf{A}^{n-1} + \cdots + q_n\mathbf{I}.$$

The matrix $q(\mathbf{A})$ can be simplified and expressed in terms of its eigenvalues. By clever choice of $\mathbf{A}$, we can obtain both $q(z)$ and $q(z^{-1})$ from the expression $\mathbf{x}(n+1) = q(\mathbf{A})\mathbf{B}$. We leave the proof of this result for a problem in Chapter 8. The variation of Horner's rule that will accomplish the double evaluation is the following:

Initialization: $\mathbf{x}(0) = \begin{bmatrix} 0 \\ 0 \end{bmatrix}, \quad v = \dfrac{z + z^{-1}}{2}, \quad \mathbf{A} = \begin{bmatrix} 0 & 1 \\ -1 & 2v \end{bmatrix}, \quad \mathbf{B} = \begin{bmatrix} 0 \\ 1 \end{bmatrix}$

Loop: $\mathbf{x}(k+1) = \mathbf{A}\mathbf{x}(k) + \mathbf{B}q_k$

End: $\begin{bmatrix} q(z) \\ q(z^{-1}) \end{bmatrix} = \begin{bmatrix} -z^{-1} & 1 \\ -z & 1 \end{bmatrix} \mathbf{x}(n+1). \quad$ **(2.5.25)**

Since the scalar components of the loop equation are

$$x_1(k+1) = x_2(k),$$
$$x_2(k+1) = (2v)x_2(k) - x_1(k) + q_k,$$

this algorithm evaluates the polynomial q at both z and z^{-1} using approximately n multiples and $2n$ additions.

EXAMPLE 2.5.5

Evaluation of a polynomial using a matrix variant of Horner's algorithm.

Consider the evaluation of the polynomial of Example 2.5.4 at $z = 2$ and $z^{-1} = \frac{1}{2}$. Thus

$$q(z) = 3z^3 + 4z^2 - 2z + 5,$$

with $z = 2$ so that $v = (2 + \frac{1}{2})/2 = \frac{5}{4}$ and $\mathbf{A} = \begin{bmatrix} 0 & 1 \\ -1 & \frac{5}{2} \end{bmatrix}$.

Then

$$\mathbf{x}(1) = \mathbf{A}\begin{bmatrix}0\\0\end{bmatrix} + \begin{bmatrix}0\\1\end{bmatrix}3 = \begin{bmatrix}0\\3\end{bmatrix},$$

$$\mathbf{x}(2) = \mathbf{A}\begin{bmatrix}0\\3\end{bmatrix} + \begin{bmatrix}0\\1\end{bmatrix}4 = \begin{bmatrix}3\\\frac{23}{2}\end{bmatrix},$$

$$\mathbf{x}(3) = \mathbf{A}\begin{bmatrix}3\\\frac{23}{2}\end{bmatrix} + \begin{bmatrix}0\\1\end{bmatrix}(-2) = \begin{bmatrix}\frac{23}{2}\\\frac{95}{4}\end{bmatrix},$$

$$\mathbf{x}(4) = \mathbf{A}\begin{bmatrix}\frac{23}{2}\\\frac{95}{4}\end{bmatrix} + \begin{bmatrix}0\\1\end{bmatrix}5 = \begin{bmatrix}\frac{95}{4}\\\frac{423}{8}\end{bmatrix}.$$

Thus to find $q(z)$ and $q(z^{-1})$, we compute

$$\begin{bmatrix}q(z)\\q(z^{-1})\end{bmatrix}\Bigg|_{z=2} = \begin{bmatrix}-z^{-1} & 1\\-z & 1\end{bmatrix}\mathbf{x}(4)\Bigg|_{z=2} = \begin{bmatrix}41\\\frac{43}{8}\end{bmatrix}$$

and so

$$q(z)|_{z=2} = 41, \qquad q(z^{-1})|_{z=2} = \tfrac{43}{8}.$$

We can use this result to obtain an efficient algorithm for the evaluation of the frequency response function. For frequency response, we are interested in the particular complex number $z = e^{j\theta}$. In this case v is given by

$$v = \frac{z + z^{-1}}{2}\Bigg|_{z=e^{j\theta}} = \cos\theta. \tag{2.5.26}$$

Assuming the coefficients of the polynomial $q(z)$ are real, we have

$$q^*(z) = q(z^{-1}).$$

Suppose that we represent $q(e^{j\theta})$ in terms of its real and imaginary parts as

$$q(e^{j\theta}) = \alpha(e^{j\theta}) + j\beta(e^{j\theta}). \tag{2.5.27}$$

We can use the algorithm (2.5.25) to evaluate Eq. (2.5.27) as follows

Begin: $\quad \mathbf{x}(0) = \begin{bmatrix}0\\0\end{bmatrix}$

Loop: $\quad \mathbf{x}(k+1) = \begin{bmatrix}0 & 1\\-1 & 2\cos\theta\end{bmatrix}\mathbf{x}(k) + \begin{bmatrix}0\\1\end{bmatrix}q_k.$

End:

$$\begin{bmatrix}-z^{-1} & 1\\-z & 1\end{bmatrix}\mathbf{x}(n+1)\Bigg|_{z=e^{j\theta}} = \begin{bmatrix}-e^{-j\theta} & 1\\-e^{j\theta} & 1\end{bmatrix}\mathbf{x}(n+1) = \begin{bmatrix}\alpha + j\beta\\\alpha - j\beta\end{bmatrix}.$$

Therefore

$$\alpha(e^{j\theta}) = x_2(n+1) - x_1(n+1)\cos\theta,$$
$$\beta(e^{j\theta}) = x_1(n+1)\sin\theta. \tag{2.5.28}$$

All the operations in algorithm (2.5.28) are *real*, even though the argument of the polynomial is the complex number $e^{j\theta}$. Thus the formidable expression (2.5.16) can be reduced to two applications of this algorithm to obtain

$$\hat{b}(e^{j\theta}) = \alpha_b(e^{j\theta}) + j\beta_b(e^{j\theta}), \tag{2.5.29}$$

$$\hat{a}(e^{j\theta}) = \alpha_a(e^{j\theta}) + j\beta_a(e^{j\theta}). \tag{2.5.30}$$

The desired result is then

$$H(e^{j\theta}) = \frac{1}{\alpha_a^2 + \beta_a^2}(\alpha_a - j\beta_a)(\alpha_b + j\beta_b). \tag{2.5.31}$$

A description of the algorithm is given in Fig. 2.5.6.

AN ANALYTICAL APPROACH. Suppose $H(z)$ is a rational transfer function having real coefficients. Then the square of the magnitude response is given by

$$|H(e^{j\theta})|^2 = H(e^{j\theta})H(e^{-j\theta}) = H(z)H(z^{-1})|_{z=e^{j\theta}}. \tag{2.5.32}$$

For a variety of problems, including the design of filters having a specified magnitude response, and some optimal filtering problems, it is useful to factor the square of the magnitude response $|H(e^{j\theta})|^2$ in the form

$$H(z)H(z^{-1}) = G\left(\frac{z+z^{-1}}{2}\right) = G(v). \tag{2.5.33}$$

Evaluation of polynomials with real coefficients at $e^{j\theta}$

Given: θ, n, $q[0, n]$

To compute: α, β **satisfying** $\alpha + j\beta = \sum_{k=0}^{n} q_k e^{j(n-k)\theta}$

Initialization:
$c = \cos(\theta)$
$s = \sin(\theta)$
$a = 2c$
$x_1 = 0$
$x_2 = 0$

Body: For $k = 0$ to n, Do
$x_3 \leftarrow q_k - x_1 + ax_2$
$x_1 \leftarrow x_2$
$x_2 \leftarrow x_3$
(end loop on k)
$\alpha = x_2 - cx_1$
$\beta = sx_1$

FIGURE 2.5.6.

It is not *a priori* clear that this is possible. But it is, and the function G is also rational. It follows that the square of the magnitude response is a rational function of $\cos\theta$:

$$|H(e^{j\theta})|^2 = G(\cos\theta). \tag{2.5.34}$$

On the interval $[0, \pi]$ the function $\cos\theta$ is monotone, decreasing from 1 to -1. This allows one to attempt a design of $H(z)$ to achieve a given magnitude response by first designing $G(v)$, positive on the interval $[-1, 1]$. Finding $H(z)$ given $G(v)$ is not easy, and involves the factorization of the numerator and denominator. However finding $G(v)$ given $H(z)$ is not difficult. We may use algorithm (2.5.36) to accomplish it, operating on the numerator and denominator polynomials independently.

Suppose we are given the coefficients of the polynomial $q(z)$. We seek a polynomial $p(v)$ for which

$$p(v) = q(z)q(z^{-1}), \qquad v = \frac{z + z^{-1}}{2}. \tag{2.5.35}$$

Using algorithm (2.5.28), together with the identity

$$q(z)q(z^{-1}) = \tfrac{1}{2}[q(z),\ q(z^{-1})]\begin{bmatrix} 0 & 1 \\ 1 & 0 \end{bmatrix}\begin{bmatrix} q(z) \\ q(z^{-1}) \end{bmatrix},$$

we develop the following procedure for finding $p(v)$.

$$\begin{aligned}
&\textbf{Initialization:} \quad \mathbf{x}(0, v) = \begin{bmatrix} 0 \\ 0 \end{bmatrix} \\
&\textbf{Loop:} \quad \mathbf{x}(k+1, v) = \begin{bmatrix} 0 & 1 \\ -1 & 2v \end{bmatrix}\mathbf{x}(k, v) + \begin{bmatrix} 0 \\ 1 \end{bmatrix} q_k \\
&\textbf{End:} \quad p(v) = \mathbf{x}(n+1, v)^T \begin{bmatrix} 1 & -v \\ -v & 1 \end{bmatrix} \mathbf{x}(n+1, v)
\end{aligned} \tag{2.5.36}$$

EXAMPLE 2.5.6
Computing $|H(e^{j\theta})|^2$.

Consider the filter of Example (2.5.2) for which

$$\hat{a}(z) = z^2 + 3 - 2\sqrt{2}, \qquad \hat{b}(z) = (1 - \sqrt{2}/2)(z^2 - 2z + 1).$$

We seek the polynomial $p_a(v) = \hat{a}(z)\hat{a}(z^{-1})$. Employing algorithm (2.5.36), we compute

$$\mathbf{x}(1) = \begin{bmatrix} 0 \\ 1 \end{bmatrix}, \quad \mathbf{x}(2) = \begin{bmatrix} 1 \\ 2v \end{bmatrix}, \quad \mathbf{x}(3) = \begin{bmatrix} 2v \\ 4v^2 + 2 - 2\sqrt{2} \end{bmatrix},$$

and so

$$p_a(v) = \mathbf{x}^T(3)\begin{bmatrix} 1 & -v \\ -v & 1 \end{bmatrix}\mathbf{x}(3) = 4(3 - 2\sqrt{2})(v^2 + 1).$$

Similarly

$$p_b(v) = 2(3 - 2\sqrt{2})(v - 1)^2$$

Therefore, with $v = (z + z^{-1})/2$, we obtain

$$H(z)H(z^{-1}) = p_b(v)/p_a(v) = \frac{(v - 1)^2}{2(v^2 + 1)}$$

and so

$$|H(e^{j\theta})|^2 = \frac{(\cos\theta - 1)^2}{2(\cos^2\theta + 1)}$$

Summary • We now have two basic I/O descriptions for linear, shift-invariant filters. In the time domain, $\mathbf{y} = \mathbf{h} * \mathbf{u}$. Thus $\mathbf{h}$ completely characterizes the filter I/O map and is the response to the standard test input $\boldsymbol{\delta}$. In the frequency domain, $Y(e^{j\theta}) = H(e^{j\theta})U(e^{j\theta})$. Here $H(e^{j\theta})$ characterizes the filter I/O map and can be measured by using the standard test input $u(k) = e^{j\theta k}$, $-\infty < k < \infty$.

In the laboratory, measuring the unit-pulse response sequence can characterize the filter, but a single experiment can be easily corrupted by noise. On the other hand, one can measure the magnitude and phase of $H(e^{j\theta})$ very precisely, even in the presence of noise (by averaging). However, this latter experiment must be repeated for several frequencies.

2.6 THE STANDARD DIFFERENCE EQUATION

In Section 2.5, we introduced the transfer function $H(z)$ defined as

$$H(z) = \sum_{k=-\infty}^{\infty} h(k)z^{-k} \tag{2.6.1}$$

In most applications $H(z)$ is a rational function of z and can be expressed as the ratio of two polynomials in z as

$$H(z) = \frac{\hat{b}(z)}{\hat{a}(z)} = \frac{b_0 + b_1 z^{-1} + \cdots + b_n z^{-n}}{1 + a_1 z^{-1} + \cdots + a_n z^{-n}}. \tag{2.6.2}$$

The integer n is called the *order* of the filter. It is not necessary that the numerator and denominator order be the same. Depending on the design of a filter, $\hat{b}(z)$ or $\hat{a}(z)$ may, in fact, be of zero order. Notice that $H(z)$ is parameterized by $2n + 1$ real coefficients, namely $\{a_1, a_2, \ldots, a_n, b_0, b_1, \ldots, b_n\}$.

Define two sequences $\hat{\mathbf{a}}$ and $\hat{\mathbf{b}}$ as follows:

$$\begin{aligned} \hat{\mathbf{a}} &= \{1, a_1, a_2, \ldots, a_n\}, \\ \hat{\mathbf{b}} &= \{b_0, b_1, b_2, \ldots, b_n\}. \end{aligned} \tag{2.6.3}$$

From Eq. (2.5.14), we have

$$H(z) = \frac{Y(z)}{U(z)} = \frac{\hat{b}(z)}{\hat{a}(z)}. \tag{2.6.4}$$

Cross multiplying in Eq. (2.6.4), we obtain

$$Y(z)\hat{a}(z) = U(z)\hat{b}(z) \tag{2.6.5}$$

Recall that multiplication of transforms is equivalent to convolution of the corresponding sequences. Thus Eq. (2.6.5) in the time-domain has the representation

$$\mathbf{y} * \hat{\mathbf{a}} = \mathbf{u} * \hat{\mathbf{b}}. \tag{2.6.6}$$

If we evaluate Eq. (2.6.6) at k, we obtain the so-called *standard-difference equation* (SDE)

$$\begin{aligned} y(k) + a_1 y(k-1) + \cdots + a_n y(k-n) = b_0 u(k) \\ + b_1 u(k-1) + \cdots + b_n u(k-n). \end{aligned} \tag{2.6.7}$$

Equation (2.6.7) is a recursion relating the input sequence $\mathbf{u}$ and the output $\mathbf{y}$. Notice that the SDE is really an infinite number of equations, one for each value of the index k. One method of solving these equations is to use direct evaluation or *simulation* as the next example demonstrates.

EXAMPLE 2.6.1

Simulation of a simple SDE.

Consider the first-order SDE

$$y(k) + \beta y(k-1) = u(k), \qquad k \geqslant 0$$

where β is a constant with $|\beta| < 1$. Solving for $y(k)$, we have

$$y(k) = u(k) - \beta y(k-1).$$

Assume the input is $u(k) = 1$, $k \geqslant 0$ and the initial value of $y(-1) = \alpha$, where α is a constant. Then we have

$$y(0) = u(0) - \beta y(-1) = 1 - \beta\alpha.$$

Knowing $y(0)$ and $u(1)$, we can compute $y(1)$ as

$$y(1) = u(1) - \beta y(0) = 1 - \beta + \beta^2\alpha.$$

Continuing in a similar manner, we find that

$$y(2) = 1 - \beta + \beta^2 - \beta^3\alpha,$$

and, in general, $y(k)$ is

$$y(k) = 1 - \beta + \beta^2 - \beta^3 + \cdots + \alpha(-\beta)^{k+1}.$$

This is a solution of the original SDE for $u(k) = \xi(k)$ and initial condition $y(-1) = \alpha$. This method of solution is generally not satisfactory because it is not in closed form. Notice that in order to obtain a solution we must have knowledge of:

1. The input $u(k)$, $k \geqslant 0$
2. The initial conditions [in this case $y(-1)$].

Solving the SDE

We are interested in solutions to the nth-order linear difference equation with constant coefficients of the form

$$\sum_{i=0}^{n} a_i y(k-i) = \sum_{i=0}^{n} b_i u(k-i), \qquad a_0 = 1, \qquad k \geqslant 0. \tag{2.6.8}$$

In general, Eq. (2.6.8) by itself does not uniquely specify the I/O description of a linear, shift-invariant system. There is a family of solutions. To any solution of Eq. (2.6.8) one can add a solution to the homogeneous equation (HE), obtained by setting the right-hand side of Eq. (2.6.8) equal to zero, and the sum will satisfy Eq. (2.6.8). In symbols, suppose $\mathbf{y}^0$ is any solution of the HE. Then $\mathbf{y}^0$ satisfies the equation

$$HE: \quad \hat{\mathbf{a}} * \mathbf{y}^0 = \mathbf{0}, \tag{2.6.9}$$

or

$$HE: \quad \sum_{i=0}^{n} a_i y^0(k-i) = 0, \qquad a_0 = 1. \tag{2.6.10}$$

And so it follows that

$$\hat{\mathbf{a}} * (\mathbf{y} + \mathbf{y}^0) = \hat{\mathbf{b}} * \mathbf{u}. \tag{2.6.11}$$

The way in which we choose the solution $\mathbf{y}^0$ will therefore affect the output obtained from the SDE representation of our system. As we shall show, $\mathbf{y}^0$ contains arbitrary constants. The way in which we evaluate these constants determines the complete solution to the SDE.

Characterization of All Solutions to the HE

The solutions to the HE are needed to construct a complete solution to Eq. (2.6.8). Suppose $\mathbf{y}^1$ is a solution to the SDE in Eq. (2.6.8). Consider any other solution $\mathbf{y}^2$ to the SDE. Then $\mathbf{y}^2 - \mathbf{y}^1$ must satisfy the HE. Therefore, if we can identify *all* solutions to the HE and *one* solution to the SDE, we can characterize the set of all solutions to the SDE.

There is an n-dimensional linear space of solutions to the HE. We can show this by first finding n linearly independent "basic" solutions. Then we show that *every* solution to the HE is a linear combination of these basic solutions.

From the coefficients of the HE in Eq. (2.6.10), we extract the polynomial $\hat{a}(z)$ known as the *characteristic polynomial*:

$$\hat{a}(z) = 1 + a_1 z^{-1} + \cdots + a_n z^{-n}. \tag{2.6.12}$$

The n roots $\lambda_1, \ldots, \lambda_n$ of $\hat{a}(z)$ define the basic solutions to the HE. To see this, consider the identity

$$\sum_{i=0}^{n} a_i z^{k-i} = z^k \hat{a}(z) \tag{2.6.13}$$

which holds for all integers k and all complex numbers z. Suppose λ is a root of $\hat{a}(z)$, so that $\hat{a}(\lambda) = 0$. Substituting λ for z in Eq. (2.6.13) gives

$$\sum_{i=0}^{n} a_i \lambda^{k-i} = \lambda^k \hat{a}(\lambda) = 0. \tag{2.6.14}$$

Comparing Eqs. (2.6.14) and (2.6.10), it is clear that $y^0(k) = \lambda^k$ is a solution to the HE. Furthermore, any of the n roots of $\hat{a}(z)$ provide solutions to the HE. If the roots are all distinct, then there are n solutions of the form $y_i(k) = \lambda_i^k$, $1 \leqslant i \leqslant n$. Every solution to the HE in this case is of the form

$$y^0(k) = \sum_{i=1}^{n} c_i \lambda_i^k. \tag{2.6.15}$$

The case of repeated roots is treated in a similar manner. The n basic solutions used to construct the general homogeneous solution depend on the multiplicity of the roots. For example, if λ is a repeated root of multiplicity two, then $\hat{a}(\lambda) = \hat{a}'(\lambda) = 0$. If we differentiate both sides of Eq. (2.6.14), we find

$$k\lambda^{k-1}\hat{a}(\lambda) + \lambda^k \hat{a}'(\lambda) = \sum_{i=0}^{n} a_i \lambda^{k-i-1}(k-i) = 0. \tag{2.6.16}$$

Comparing Eqs. (2.6.16) and (2.6.10), we see that $k\lambda^{k-1}$ is also a solution to the HE. In general, for a root of multiplicity m, this reasoning will show that there are m solutions (associated with this root) of the form $y_i(k) = k^{i-1}\lambda^{k-i+1}$, $1 \leqslant i \leqslant m$.

We have identified n basic solutions to the homogeneous equation (2.6.10), assuming that n is not zero. Thus we have an n parameter family of solutions to the HE of the form of Eq. (2.6.17) below, where $\mathbf{y}_i$ is the ith basic solution. Must every solution to the HE have this form? Suppose that one specifies n consecutive values of a solution; $y^0(m+1), \cdots, y^0(m+n)$, for some time m. Such a specification is called *initial data.* Starting from the initial data, the sequence can be extended forward (and backward) using the simulation procedure of Example 2.6.1. Therefore the set of all solutions to the HE is also an n parameter family since any solution can be parameterized by its initial data. This is the same family as the one generated by Eq. (2.6.17) if and only if the coefficients $c_1, \cdots, c_n$ are in one-to-one correspondence with the initial data. Using Eq. (2.6.17) evaluated at the times for which the initial data is specified, one gets a linear equation relating these two n dimensional vectors. The coefficient matrix $\mathbf{V}$ has elements

$$V_{ij} = y_i(m+j), \qquad 1 \leqslant i, j \leqslant n.$$

Because of the form of the basic solutions, this matrix is a generalized Vandermonde matrix and is known to have a nonzero determinant. Consequently there is precisely one set of coefficients for each set of initial data.

In summary, the basic solutions to the HE are of the form:

(1) Distinct roots generate basic solutions λ^k.

(2) Roots of multiplicity m generate m basic solutions λ^k, $k\lambda^{k-1}$, . . . , $k^{m-1}\lambda^{k-m+1}$.

Every solution to the HE is a linear combination of the n basic solutions, i.e.,

$$y^0(k) = \sum_{i=1}^{n} c_i y_i(k). \tag{2.6.17}$$

The question remains as to how we should choose the constants $c_1, \ldots, c_n$ in Eq. (2.6.17) so as to obtain a unique homogeneous solution and thereby obtain the complete solution to the SDE in Eq. (2.6.8).

Choosing the Constants of the Homogeneous Solution

To calculate the constants in Eq. (2.6.17), we need additional information. The SDE is not sufficient to determine the constants. This is why the SDE by itself does not uniquely specify the I/O description of a linear, shift-invariant system. There are two ways to ensure that the SDE specifies a unique I/O description. We can impose the condition that the system be *causal* and specify initial conditions to find the constants associated with the solution to the HE. This is called an *initial value problem* and is generally the model one assumes. However, an alternative is to require *system stability* and choose the constants in the solution to the HE to ensure stability. The system may no longer be causal in this case. (In most filter design problems we shall require both system stability and causality.) The following example illustrates these two methods of choosing the solution to the HE.

EXAMPLE 2.6.2

Choosing the HE solution.

Suppose we have the SDE

$$y(k) - a_1 y(k-1) = \delta(k), \qquad |a_1| < 1. \tag{2.6.18}$$

By direct evaluation, assuming zero initial conditions and $k \geqslant 0$, we have

$$y(0) = \delta(0) + a_1 y(-1) = 1,$$
$$y(1) = \delta(1) + a_1 y(0) = a_1,$$
$$y(2) = \delta(2) + a_1 y(1) = a_1^2,$$
$$\vdots$$

and so a solution to the original SDE is

$$y(k) = \begin{cases} 0, & k < 0 \\ a_1^k, & k \geqslant 0 \end{cases}.$$

This solution is shown in Fig. 2.6.1. What are the solutions to the HE? Trying the form $y^0(k) = cr^k$ in the HE, we obtain

$$cr^k - a_1 cr^{k-1} = 0,$$

which implies that

$$cr^k(1 - a_1 r^{-1}) = 0.$$

Assuming $cr^k \neq 0$ means that for cr^k to be a solution to the HE, the term $1 - a_1 r^{-1}$ must be zero. Thus,

$$1 - a_1 r^{-1} = 0 \qquad \Rightarrow \qquad r = a_1.$$

Thus a solution to the HE is

$$y^0(k) = ca_1^k \quad \text{for all } k,$$

and the value of c is arbitrary. Thus solutions to the original SDE are of the form

$$y(k) = y^0(k) + a_1^k \xi(k), \qquad -\infty < k < \infty$$

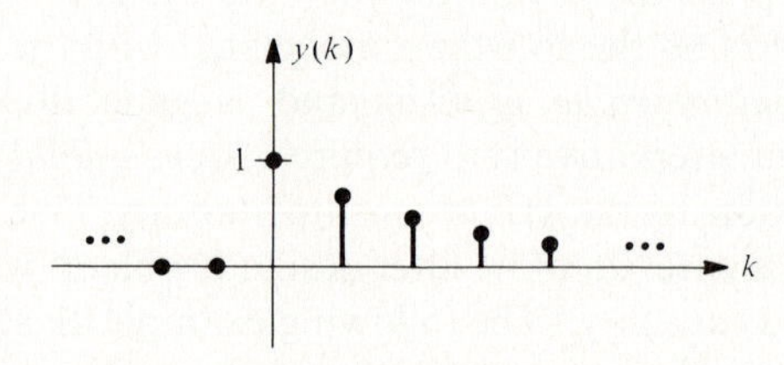

FIGURE 2.6.1.

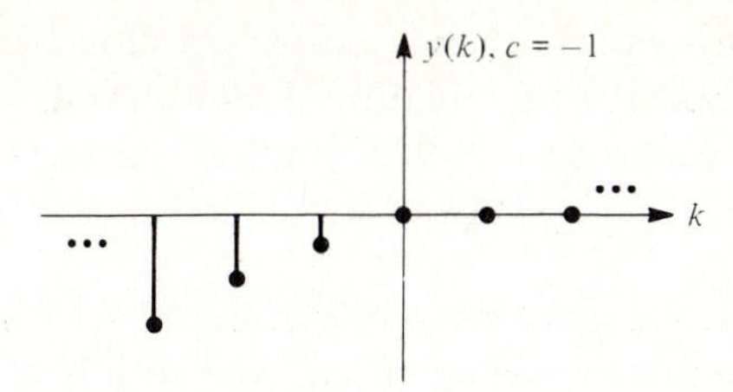

FIGURE 2.6.2.

where the sequence $\{\xi(k)\}$ is used to initiate the nonzero portion of a_1^k for $k \geqslant 0$. Now we can choose the value of c in the solution to the HE as we wish. Suppose, for example, we choose $c = -1$. Then the solution to the original SDE is

$$y(k) = -a_1^k + a_1^k \xi(k) = \begin{cases} -a_1^k, & k < 0 \\ 0, & k \geqslant 0 \end{cases}.$$

For $|a_1| < 1$, this solution is depicted in Figure 2.6.2. Notice that if $|a_1| > 1$, then this solution would have decayed to zero as k approaches $-\infty$. That is, for $|a_1| > 1$ the choice of $c = -1$ yields a complete solution that decays to zero but is noncausal.

Clearly the ambiguity in the solution $y(k)$ can be resolved by choosing the constant c. Generally c is determined by the initial values associated with the SDE. Suppose that we assume $y(-1) = 0$ and the system output is causal. Then we can solve for the constant c using

$$y(-1) = 0 = ca_1^{-1} + a_1^{-1}\xi(-1) = ca_1^{-1}, \qquad ca_1^{-1} = 0 \qquad \Rightarrow \qquad c = 0.$$

Thus $c = 0$, and the complete solution is

$$y(k) = a_1^k \xi(k) = \begin{cases} 0, & k < 0 \\ a_1^k, & k \geqslant 0 \end{cases}.$$

Thus we obtain different solutions for the output of the system described by the same SDE. The true solution requires additional information to evaluate the constants of the solution to the HE. This additional information comes in one of two forms:

(1) The solution is causal and initial values of the solution are specified;

(2) The solution is stable, i.e., $y(k)$ approaches 0 as $|k|$ approaches ∞, but the solution may be noncausal.

The Unit-Pulse Response from the SDE

The basic solutions of the HE can be used to characterize the unit-pulse response sequence **h**. If we assume that the SDE has an input $\mathbf{u} = \boldsymbol{\delta}$ and the solution $\mathbf{y} = \mathbf{h}$ is causal, then a unique solution $\mathbf{h}$ can be found using n initial conditions obtained by simulation. For $\mathbf{h}$ causal, we have

$$\sum_{i=0}^{n} a_i h(k-i) = \sum_{i=0}^{n} b_i \delta(k-i), \quad a_0 = 1, \quad k \geqslant 0. \tag{2.6.19}$$

Now for $k > n$, the right-hand side of Eq. (2.6.19) is zero. Thus for $k > n$, Eq. (2.6.19) is a homogeneous equation, and so $\mathbf{h}$ is of the form

$$h(k) = \begin{cases} 0, & k < 0 \\ b_0, & k = 0 \\ \sum_{i=1}^{n} c_i y_i(k), & k > 0 \end{cases} \tag{2.6.20}$$

where the $y_i(k)$ are the basic solutions to the HE. The initial conditions $h(1), h(2), \ldots, h(n)$ needed to determine c_i, $i = 1, 2, \ldots, n$ are obtained from Eq. (2.6.19) by direct evaluation. Note that $h(0)$ cannot, in general, be used as an initial condition because the right-hand side includes a unit-pulse at $k = n$ and we must include all of the delayed unit-pulses $\delta(k-i)$, $i = 0, 1, \ldots, n$ acting on the filter.

EXAMPLE 2.6.3

Finding **h** *from the SDE.*

Consider the following digital filter. This filter is described by the SDE

$$y(k) + \tfrac{1}{2}y(k-2) = 3u(k) + u(k-2).$$

The characteristic equation is

$$\hat{a}(z) = z^2 + \tfrac{1}{2}$$

with roots equal to $\pm j\sqrt{\tfrac{1}{2}}$. The unit-pulse response is of the form

$$h(k) = c_{-1}(j\sqrt{\tfrac{1}{2}})^k + c_2(-j\sqrt{\tfrac{1}{2}})^k.$$

The complex roots are more conveniently expressed in polar form as $\rho e^{\pm j\phi}$. If $z_1, z_2 = a \pm jb$, then

$$\rho = (a^2 + b^2)^{1/2}, \qquad \phi = \tan^{-1}(b/a).$$

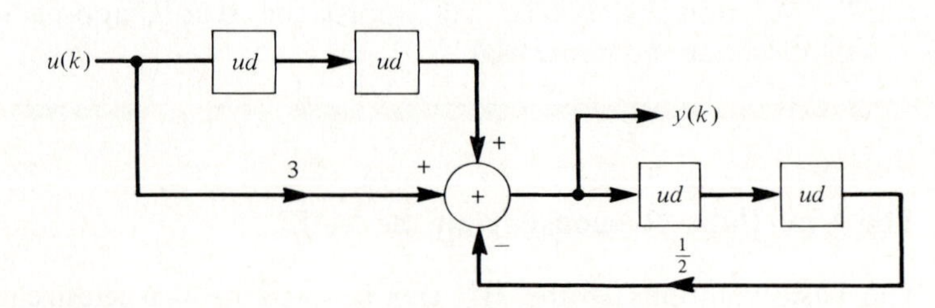

FIGURE 2.6.3.

Thus

$$h(k) = c_1 \rho^k e^{jk\varphi} + c_2 \rho^k e^{-jk\phi}.$$

In this case $\rho = 1/\sqrt{2}$ and $\phi = \pi/2$. The unit-pulse response can thus be written

$$h(k) = \left(\frac{1}{\sqrt{2}}\right)^k \left\{c_1' \cos\left(\frac{k\pi}{2}\right) + c_2' \sin\left(\frac{k\pi}{2}\right)\right\}.$$

The constants c_1' and c_2' are obtained from the SDE with $\mathbf{u} = \boldsymbol{\delta}$ and $\mathbf{y} = \mathbf{h}$. Thus

$$h(k) = 3\delta(k) + \delta(k-2) - \tfrac{1}{2}h(k-2).$$

If $\mathbf{h}$ is causal, we have

$$h(0) = 3\delta(0) + \delta(-2) - \tfrac{1}{2}h(-2) = 3,$$
$$h(1) = 3\delta(1) + \delta(-1) - \tfrac{1}{2}h(-1) = 0,$$
$$h(2) = 3\delta(2) + \delta(0) - \tfrac{1}{2}h(0) = -\tfrac{1}{2}.$$

Thus the constants c_1' and c_2' are found using the equations

$$h(1) = 0 = c_1'\left(\frac{1}{\sqrt{2}}\right)\cos\frac{\pi}{2} + c_2'\left(\frac{1}{\sqrt{2}}\right)\sin\frac{\pi}{2},$$

$$h(2) = -\tfrac{1}{2} = c_1'(\tfrac{1}{2})\cos\pi + c_2'(\tfrac{1}{2})\sin\pi.$$

From these equations, we find $c_1' = 1$, $c_2' = 0$. Thus the causal unit-pulse response is

$$h(k) = \begin{cases} 0, & k < 0 \\ 3, & k = 0. \\ \left(\dfrac{1}{\sqrt{2}}\right)^k \cos\dfrac{k\pi}{2}, & k > 0 \end{cases}$$

In this example, the $k = 0$ term cannot be included in the general expression for $k > 0$.

Generally we assume the unit-pulse response is causal, resulting in of solution demonstrated in the previous example. However, one can also require stability of $\mathbf{h}$ to obtain a unique unit-pulse response. Suppose in the previous example the SDE was of the form

$$y(k) + 2y(k-2) = 3u(k) + u(k-1).$$

Now the causal solution is unstable because of the multiplier 2 (instead of $\tfrac{1}{2}$). We can obtain a stable solution if we allow the index k to run from 0 to $-\infty$, i.e., we consider noncausal solutions. The unit-pulse response is of the form

$$h(k) = c_1(\sqrt{2})^k \cos\left(\frac{k\pi}{2}\right) + c_2(\sqrt{2})^k \sin\left(\frac{k\pi}{2}\right)$$

To evaluate the constants c_1 and c_2, we again use direct evaluation of the SDE with a slight modification. We write

$$h(k+2) + 2h(k) = 3\delta(k+2) + \delta(k+1)$$

or

$$h(k) = \tfrac{1}{2}\{3\delta(k+2) + \delta(k+1) - h(k+2)\}. \tag{2.6.21}$$

The use of Eq. (2.6.21) is necessitated by the fact that we are considering solutions for $k < 0$.

Although we do not often encounter the use of noncausal unit-pulse responses, we should keep in mind they can be of practical value. Whenever data is processed in nonreal time (such as occurs in data stored on a magnetic tape or diskette), the possibility of using noncausal filters exists. The idea of future and past time in these applications is really of no consequence.

The Frequency Response Function from the SDE

The complete solution to the SDE consists of two parts: the complete solution to the HE (made up of a linear combination of the basic solutions) and a so-called particular solution to the SDE. The particular solution can be used to determine the frequency response function $H(e^{j\theta})$. A particular solution is any solution of the SDE with the right-hand side not equal to zero.

EXAMPLE 2.6.4

Finding the frequency response function from the SDE.

Consider a digital filter described by the SDE

$$y(k) + \tfrac{1}{2}y(k-2) = u(k).$$

What is the frequency response function for this filter? Every homogeneous solution is of the form

$$y^0(k) = \left(\frac{1}{\sqrt{2}}\right)^k \left\{c_1 \cos\left(\frac{k\pi}{2}\right) + c_2 \sin\left(\frac{k\pi}{2}\right)\right\}.$$

In order to determine the frequency response function, we set the input $u(k) = e^{jk\theta}$ and seek a steady-state solution for this input. We assume a solution of the form $y^p(k) = c_3 e^{jk\theta}$, where c_3 is an arbitrary multiplier not a function of k. Substituting into the original SDE, we have

$$c_3 e^{jk\theta} + \tfrac{1}{2}c_3 e^{j(k-2)\theta} = e^{jk\theta}.$$

Simplifying, we obtain

$$e^{jk\theta}\{c_3 + \tfrac{1}{2}c_3 e^{-2j\theta}\} = e^{jk\theta}.$$

Thus for $y^p(k) = c_3 e^{k\theta}$ to be a solution to the SDE, the term in brackets must be unity. This implies c_3 is equal to

$$c_3 = \frac{1}{1 + \frac{1}{2}e^{-2j\theta}}.$$

A particular solution for the input $e^{jk\theta}$ is thus

$$y^p(k) = \left[\frac{1}{1 + \frac{1}{2}e^{-2j\theta}}\right] e^{jk\theta}.$$

That is, the steady-state response to an input $e^{jk\theta}$ is the same complex sinusoid modified in phase and amplitude by the frequency response function $H(e^{j\theta})$, where

$$H(e^{j\theta}) = \frac{1}{1 + \frac{1}{2}e^{-2j\theta}}.$$

Stability

BIBO stability requires that the unit-pulse response sequence **h** be an l_1 sequence, i.e.,

$$\sum_{k=-\infty}^{\infty} |h(k)| < \infty.$$

For $k > 0$, the unit-pulse response sequence satisfies the homogeneous equation. Thus for the case of distinct roots, **h** has the form

$$h(k) = \sum_{i=1}^{n} c_i \lambda_i^k, \qquad \text{for } k > 0. \tag{2.6.22}$$

If the unit-pulse response is known to be causal, then Eq. (2.6.22) requires that $|\lambda_i| < 1, i = 1, 2, \ldots, n$ for **h** to be an l_1 sequence. Recall from Section 2.5 the roots of $\hat{a}(z)$ are precisely the poles of the system. Thus in the case of causal **h**, all poles must lie inside $|z| = 1$ for BIBO stability.

If the causality of **h** is not required and there are no roots satisfying $|\lambda_i| = 1$, then we can obtain a unique l_1 unit-pulse response sequence in the following manner: Choose the n basic solutions that make up **h** as follows:

1. If $|\lambda_i| < 1$, use the causal term $c_i \lambda_i^k \xi(k)$
2. If $|\lambda_i| > 1$, use the anti-causal term $c_i \lambda_i^k \xi(-k)$

In this way, **h** is made up of a sum of causal sequences which decay to zero at $k = \infty$ and noncausal sequences which decay to zero at $k = -\infty$. The resultant sum is l_1. This is discussed in more detail in Section 3.5.

Notice that the usual definition of stability, i.e., "all roots of $\hat{a}(z)$ satisfy $|\lambda_i| < 1, i = 1, 2, \ldots, n$" is true *only* if we require causality of the system.

2.7 SIGNAL FLOW GRAPHS

In our discussions of discrete-time systems, we shall often have occasion to combine systems to form a larger system or to break apart a large system into smaller pieces. In some applications of digital signal processing it is necessary to know how the actual computation of the output is accomplished. This includes knowing what internal variables are used, the order of their use, what variables must be stored for future use, and other detailed knowledge of the computation. One convenient tool for combining systems and for specifying details of the computation is the signal flow graph.

A signal flow graph (SFG) is a representation for a connection of systems specified by their transfer function or, sometimes called, their *transmittance* function. The SFG consists of branches and nodes as shown in Fig. 2.7.1. The rules are:

1. Nodes represent signals.
2. Branches represent systems. Branches are directed with an input side and an output side.
3. For each node with incoming branches, there is an equation. The signal at the node is equal to the sum of products of the form $T_i e_i$, where T_i is the transmittance function of the ith branch and e_i is the signal at the input to the branch. In our application T_i is the transfer function of a system $H_i(z)$ and e_i is the z-transform of the input signal, $U_i(z)$. Often the branches are labeled with the most elemental of discrete-time systems, namely, unit delays with a transfer function z^{-1} and constant gains with a transfer function equal to the gain g.

Every SFG is completely equivalent to a set of equations having the form shown in Fig. 2.7.1. It's clear that we can manipulate the SFG in any way consistent with algebraic manipulation of the equations. Some elementary rules for SFG manipulations are given in Fig. 2.7.2. To demonstrate how one can use the SFG reduction rules consider the following example.

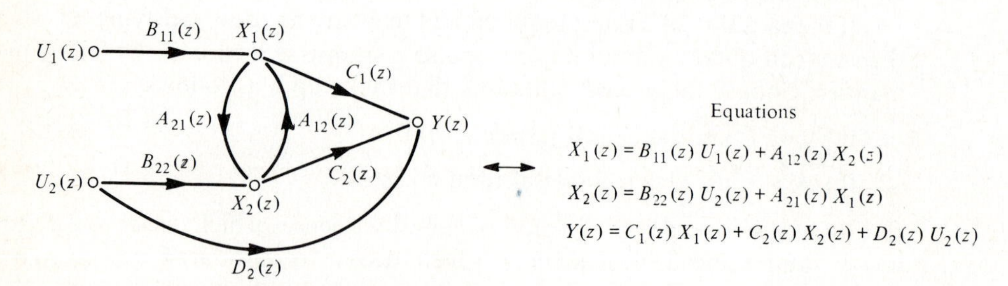

FIGURE 2.7.1 A signal flow graph with two input nodes, two internal nodes, and one output node. The equivalent set of equations is also given.

Rule	Original SFG	Equivalent SFG
(1) Cascade transformation	$U \xrightarrow{A} X \xrightarrow{B} Y$	$U \xrightarrow{AB} Y$
(2) Parallel transformation	$U \to Y$ via branches A and B	$U \xrightarrow{A+B} Y$
(3) Elimination of a node	$U_1 \xrightarrow{A} X$, $U_2 \xrightarrow{B} X$, $X \xrightarrow{C} Y_1$, $X \xrightarrow{D} Y_2$	$U_1 \xrightarrow{AC} Y_1$, $U_2 \xrightarrow{BC} Y_1$, $U_1 \xrightarrow{AD} Y_2$, $U_2 \xrightarrow{BD} Y_2$
(4) Elimination of a branch	$U_1 \xrightarrow{A} X$, $U_2 \xrightarrow{B} X$, $X \xrightarrow{C} Y_1$, $X \xrightarrow{D} Y_2$	$U_1 \xrightarrow{AC} Y_1$, $U_2 \xrightarrow{BC} Y_1$, $U_1 \xrightarrow{A} X$, $U_2 \xrightarrow{B} X$, $X \xrightarrow{D} Y_2$
(5) Loop elimination	$U \xrightarrow{A} \circ \xrightarrow{B} Y$, $Y \xrightarrow{C} \circ$	$U \to Y$: $A(\frac{B}{1 - BC})$
(6) Self loop elimination	$U \xrightarrow{A} X \xrightarrow{C} Y$, self loop B at X	$U \to Y$: $\frac{AC}{1 - B}$
(7) Elimination of cascade loops	$U \xrightarrow{1} X_1 \xrightarrow{A} X_2 \xrightarrow{C} Y$, $X_2 \xrightarrow{B} X_1$, $Y \xrightarrow{D} X_2$	$U \xrightarrow{A} X_2 \xrightarrow{C} Y$, self loop $AB + CD$ at X_2

FIGURE 2.7.2 Some rules for SFG reduction.

EXAMPLE 2.7.1

Reduction of an SFG to find overall system transfer function.

Consider the discrete-time system shown in Fig. 2.7.3. What is the overall transfer function of this system? The equations for this SFG are given by

(1) $X_1(z) = U(z) + H_3(z)X_2(z) + H_4(z)Y(z)$

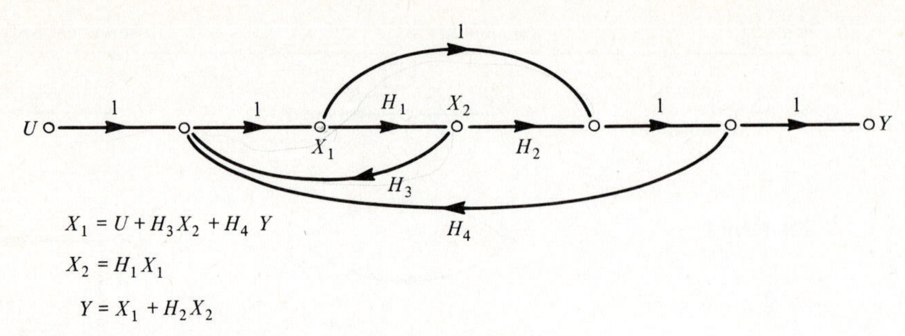

FIGURE 2.7.3 An SFG for a discrete-time system.

(2) $X_2(z) = H_1(z)X_1(z)$

(3) $Y(z) = X_1(z) + H_2(z)X_2(z)$

Notice that the node representing $X_2(z)$ is the main stumbling block to reducing the SFG. Suppose we eliminate $X_2(z)$ by substituting $X_2(z)$ from Eq. (2) into (1) and (3). The resulting equations are

(1′) $X_1(z) = U(z) + H_3(z)H_1(z)X_1(z) + H_4(z)Y(z)$

(3′) $Y(z) = X_1(z) + H_1(z)H_2(z)X_1(z).$

In terms of the SFG, we have moved the input of branch $H_3(z)$ "through" the system $H_1(z)$ creating a self-loop at node $X_1(z)$ as shown in Fig. 2.7.4. Using rule (6) from Fig. 2.7.2, we can reduce this SFG to the SFG of Fig. 2.7.5. Using rule (5) from Fig. 2.7.2, we finally obtain an SFG of the form shown in Fig. 2.7.6.

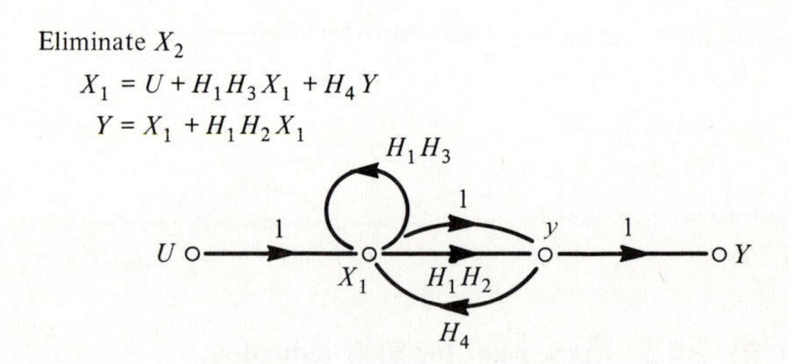

FIGURE 2.7.4 Step 1.

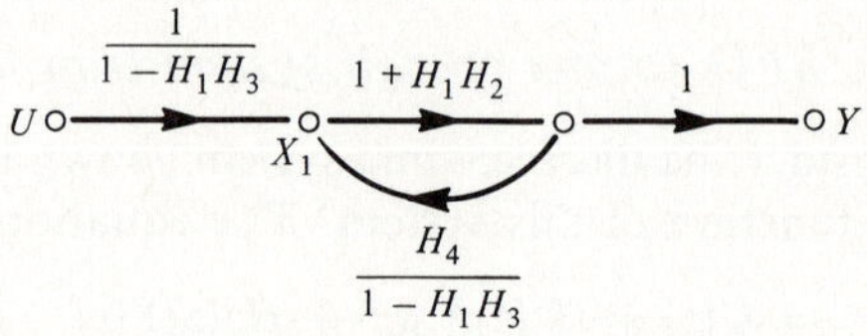

$$U \quad \frac{1}{1 - H_1H_3} \quad X_1 \quad 1 + H_1H_2 \quad 1 \quad Y \qquad \frac{H_4}{1 - H_1H_3}$$

FIGURE 2.7.5 Step 2.

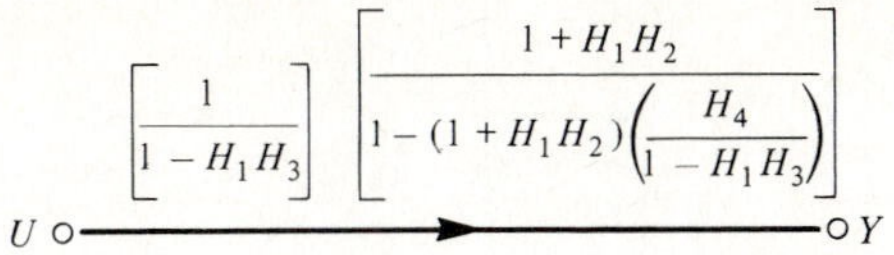

$$U \circ \longrightarrow \left[\frac{1}{1 - H_1H_3}\right] \left[\frac{1 + H_1H_2}{1 - (1 + H_1H_2)\left(\frac{H_4}{1 - H_1H_3}\right)}\right] \longrightarrow \circ\, Y$$

FIGURE 2.7.6 Step 3.

Simplifying the cascade connection, we find that the transfer function of the entire system is

$$H(z) = \frac{Y(z)}{U(z)} = \frac{1 + H_1(z)H_2(z)}{1 - H_1(z)H_3(z) - H_4(z) - H_1(z)H_2(z)H_4(z)}.$$

This same result is, of course, directly obtainable from Eqs. (1′) and (3′) by eliminating the internal variable $X_1(z)$.

We shall have occasion to use SFGs in the remainder of the book to specify filters and systems for various examples.

2.8 A HIERARCHY OF DISCRETE-TIME SYSTEM DESCRIPTIONS

We have introduced four system descriptions in this introductory chapter. Three are input-output descriptions: the unit-pulse response sequence $\mathbf{h}$, the transfer function or frequency response function $H(z)$ and $H(e^{j\theta})$, respectively, and the SDE. The fourth description is the SFG, which contains more detail than the I/O descriptions. In our study of discrete-time systems, we shall use several descriptions depending on the problem we wish to solve. These descriptions form a hierarchy. At the lowest level of this hierarchy are the I/O descriptions such as $\mathbf{y} = \mathbf{h} * \mathbf{u}$ or $Y(z) = H(z)U(z)$. Higher-level descriptions contain more detail about how the system is actually implemented. We order the descriptions in an obvious way: description 1 is "higher" than description 2 if description 1 contains more detail than description 2. One tries to employ the lowest-level description sufficient for the problem at hand. In Chapter 6, we study the problem of obtaining a transfer function that meets certain design criteria. For this problem the lowest-level descriptions suffice. However, when we implement a digital filter having a given transfer function in either software or hardware, we shall need more detail.

Figure 2.8.1 is a representation of the hierarchy of filter descriptions used in this book. It assumes finite order, single-input, single-output filters even though most of the descriptions are easily extended to multiple-input, multiple-output filters. An arrow from one block to another means that the first description determines the second. The numbers indicate the section where the relevant discussion may be found.

Notice that the I/O descriptions $\mathbf{y} = \mathbf{h} * \mathbf{u}$ and $Y(e^{j\theta}) = H(e^{j\theta})U(e^{j\theta})$ are the lowest descriptions in this hierarchy and contain the least detail. They are a

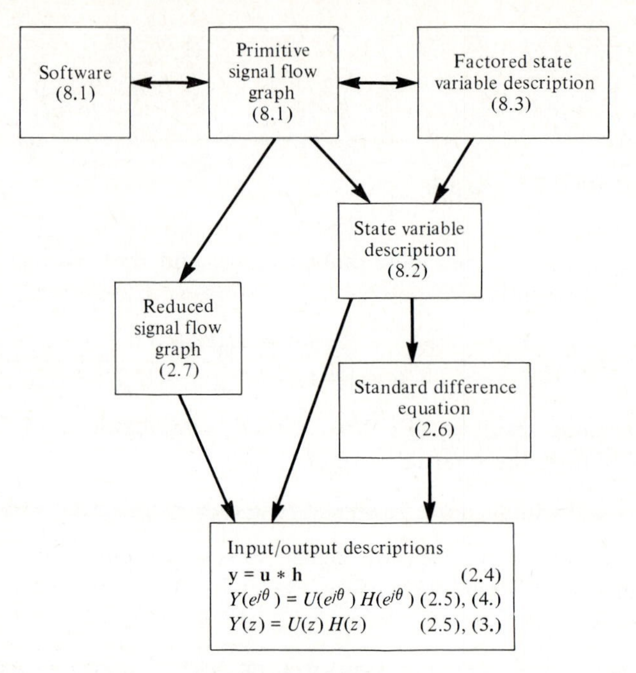

FIGURE 2.8.1 A hierarchy of filter descriptions.

"black box" characterization of a digital filter and are concerned only with calculation of the output given the input. At the top of this hierarchy are the descriptions labeled primitive signal flow graphs, software realization, and factored state variables. These descriptions are all equivalent and specify not only how one calculates the output for a given input but also a great deal of information on precisely how the calculation is performed. This information includes the intermediate variables used in the calculation and the order in which these variables are calculated.

Coming down in the hierarchy from descriptions with more detail to those with less detail is essentially unique. In other words, going from high-level descriptions to lower-level descriptions results in a unique lower-level description. The reverse direction is not unique. There are many possible (final) descriptions when one begins with a less detailed description and moves to a more detailed description. For example, for every I/O description $\mathbf{y} = \mathbf{h} * \mathbf{u}$, there are an infinite number of signal flow graphs with the same I/O characteristic. We can use this freedom in the design of digital filters to optimize parameters concerned with the actual calculations performed. For example, we can choose a state variable description that specifies certain internal connections to minimize effects due to finite length registers. Or perhaps in the implementation of digital filters in very large system integration (VLSI) we might choose a signal flow graph for the filter that reduces communication between logic elements to a minimum. The point is that we can specify an I/O characteristic like $H(z)$ or the unit-pulse response $\mathbf{h}$ and manipulate the internal structure of the filter to optimize other parameters or criteria connected with the actual computation. By using only I/O descriptions of the filter, we are able to study these computational effects and implementation criteria only indirectly. This is the reason we need more than just I/O descriptions in the design of digital filters.

PROBLEMS

2.1. The following equations describe the I/O characteristic of a discrete time system. Which of these systems are:
(i) linear, (ii) shift (or time) invariant, (iii) causal, (iv) BIBO stable?

a) $y(k) = 2^k u(k)$

b) $y(k) = u(k) + 3.2u(k-1) + 10u(k-2)$

c) $y(k) = u(k) \cdot u(k-2)$

d) $y(k) = \cos[u(k)], \qquad |u(k)| < 1$

2.2. For each sequence $\mathbf{h}$ shown, find and sketch (i) $\mathbf{h} * \mathbf{h}$ and (ii) $\mathbf{h} * \mathbf{h} * \mathbf{h}$.

a) $\mathbf{h} = \{0, 1, 0, 0, \ldots\}$

b) $\mathbf{h} = \{1, \underset{\uparrow}{0}, 1\}$

c) $\mathbf{h} = \{-1, \underset{\uparrow}{0}, 1\}$

d) $\mathbf{h} = \{1, 1, 1, \ldots\}$

2.3. Find the causal unit-pulse response sequence for LSI systems described by the following SDEs.

a) $y(k) - y(k-1) + \frac{1}{4}y(k-2) = u(k)$

b) $y(k) - y(k-1) + \frac{1}{4}y(k-2) = u(k) + u(k-2)$

c) $y(k) + \frac{1}{16}y(k-2) = u(k)$

d) $y(k) + \frac{1}{16}y(k-2) = u(k) + u(k-1) - u(k-2)$

2.4. Find the frequency response functions for the LSI systems described by the SDEs of Problem 2.3.

2.5. Consider the SDE given below. Find the output for the given inputs.

SDE: $y(k) + \frac{1}{4}y(k-2) = u(k) - u(k-1)$

a) $u(k) = \xi(k)$

b) $u(k) = (k+1)\xi(k)$

c) $u(k) = [2^k + 2^{-k}]\xi(k)$

d) $u(k) = k2^k\xi(k)$

e) $u(k) = \cos(3k)\xi(k)$

f) $u(k) = k\cos(3k)\xi(k)$

2.6. A causal, linear, shift-invariant filter is described by the SDE

$$y(k) = u(k) - au(k-1) + by(k-1).$$

Find the value of a (with $a \neq b$) such that the amplitude response of the filter is unity for all frequencies. This is called an all-pass filter.

2.7. The input to a stable filter is $u(k) = \cos(k\theta)$. Suppose one plots the pair of points $u(k)$, $y(k)$ in the plane (in analogy to the classical analog technique of driving the horizontal and vertical inputs of an oscilloscope with the input and output of some linear system, e.g., an amplifier). (1) Show that these pairs of points lie on an ellipse as shown below. (2) Show how one can measure the magnitude and phase of the system's frequency response $H(e^{j\omega})$ using the figure. (3) Give necessary and sufficient conditions that the set of all pairs $u(k)$, $y(k)$ will occupy a finite number of points on this ellipse.

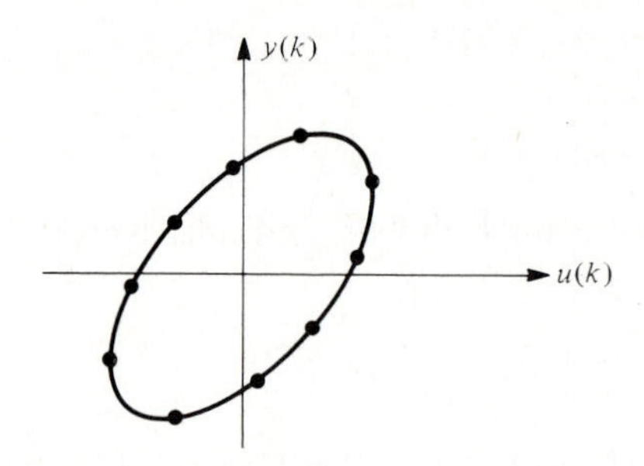

FIGURE 2P.7.

2.8. Sketch an SFG of a filter whose unit-pulse response is given by

a) $\{1, \frac{1}{3}, \frac{1}{9}, \ldots, (\frac{1}{3})^k, \ldots\}$

b) $\{1, 1, \frac{1}{3}, \frac{1}{3}, \frac{1}{9}, \frac{1}{9}, \ldots\}$

2.9. In the following FIR filter, we wish to pass a constant input without attenuation and to completely eliminate a sinusoidal input at 60 Hz. Assume the foldover frequency is 500 Hz. Find the coefficients of the filter shown below. Also find and sketch the resulting magnitude and phase response of the filter.

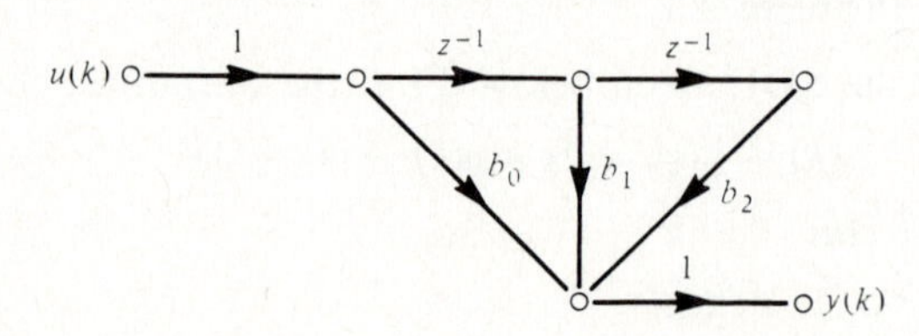

FIGURE 2P.9.

2.10. Consider an FIR filter with an I/O characteristic of the form

$$y(k) = \frac{1}{n}\left[\sum_{l=k-(n-1)}^{k} u(l)\right]$$

2.24. Consider a digital filter with transfer function

$$H(z) = \frac{z^{-1} - a}{1 - az^{-1}}, \qquad |a| < 1.$$

This is called an all-pass filter.

a) Sketch an SFG for this filter.

b) Obtain an SDE for this filter.

c) Find and sketch the unit-pulse response **h**.

d) Find and sketch the frequency response function.

2.25. Find the unit-pulse response **h** for the following filter.

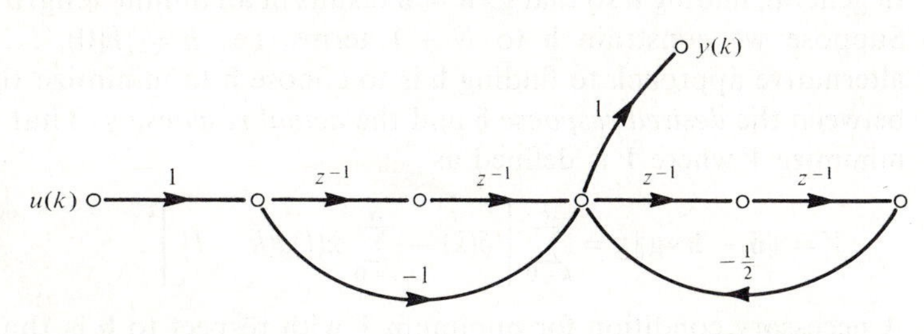

FIGURE 2P.25.

2.26. Consider the following SFG of a filter.

a) Choose c_1, c_2, and d so that the unit-pulse response begins with $h(0) = 1$, $h(1) = 1$, $h(2) = 1$.

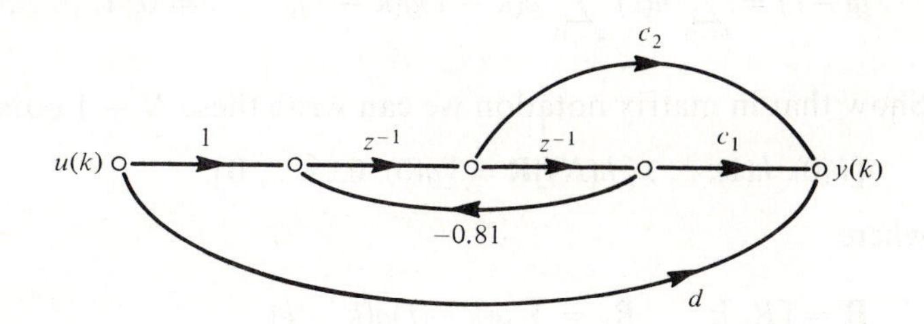

FIGURE 2P.26.

Choose c_1, c_2, and d so that the steady-state response to $\xi(k)$ is $y(\infty) = 1$, but the steady-state response to the input $\cos(k\pi/2)$ is zero.

c) Choose c_1, c_2, and d so that the numerator polynomial in the $H(z)$ is $\hat{b}(z) = z^2 + z + 1$.

d) Choose c_1, c_2, and d so that the frequency response function has magnitude identically one, i.e.,

$$|H(e^{j\theta})| = 1 \quad \text{for all } \theta.$$

Intuitively, this kind of filter acts to "smooth" the input. Thus it must be some form of low-pass filter. Plot the magnitude and phase of the frequency response for the cases $n = 2$, 3, and 4. For arbitrary n, identify the poles and zeros of the transfer function $H(z)$ of this filter.

2.11. Let $\{u(k)\}$ be a data sequence. A least-squares straight-line approximation to the n data points $u(k)$, $u(k-1), \ldots, u(k-n+1)$ is

$$\hat{u}(l) = al + b,$$

where the parameters a and b are chosen to minimize

$$V(a, b) = \sum_{l=k-n+1}^{k} [al + b - u(k)]^2$$

The minimizing a and b depend on k. Let $y(k) = a(k)k + b(k)$. Show that the system whose input is $\{u(k)\}$ and whose output is $\{y(k)\}$ is linear and shift-invariant, and give a closed form expression for its unit-pulse response.

2.12. Show that the unit-pulse response for the system with SFG shown below is the unit-step sequence

$$\xi(k) = \begin{cases} 1, & k \geq 0 \\ 0, & k < 0 \end{cases}.$$

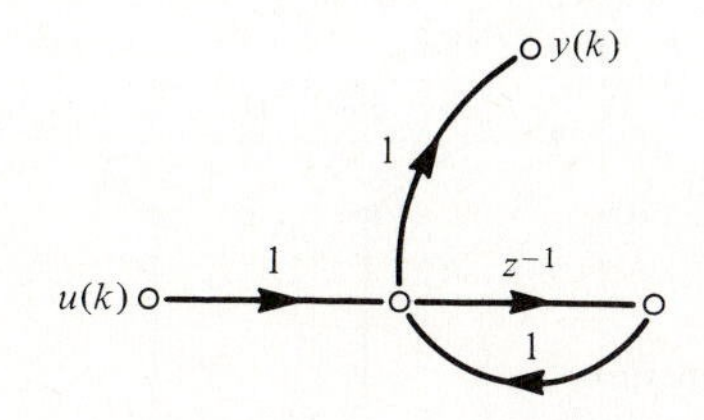

FIGURE 2P.12.

Use this fact to show that if $\{h(k)\}$ is the unit-pulse response sequence for a given filter, then the step response of the filter can be obtained by passing $h(k)$ through the above filter. What is the result?

2.13. The step response of a filter is $g(k)$, [the response to the input $u(k) = \xi(k)$]. If the step response of a given filter is

$$g(k) = (\tfrac{1}{2})^k \xi(k) + (\tfrac{1}{8})^k \xi(k-1),$$

then find the unit-pulse response $h(k)$ of this filter.

2.14. Let $\hat{a}(z) = z^2 + a_1 z + a_2$ be the denominator polynomial of a transfer function $H(z)$. On the parameter plane with axes (a_1, a_2), find the following sets:

a) The set for which $\hat{a}(z)$ has complex roots.

b) The set for which $\hat{a}(z)$ has complex roots with $|\lambda_i| < 1$ (λ_i are the roots).

c) The set for which $\hat{a}(z)$ has real roots with $|\lambda_i| < 1$.

d) The stability triangle for which the poles of the filter result in a stable filter—the union of sets in (b) and (c).

2.15. An ideal low-pass filter has the unit-pulse response sequence

$$h(k) = \frac{\omega_0}{\pi}\,\mathrm{sinc}(k\omega_0); \quad \mathrm{sinc}\,(x) \overset{\Delta}{=} \frac{\sin x}{x}.$$

Is this filter

(a) causal, (b) BIBO stable, (c) of finite-order?

2.16. Using poles and zeros and a graphical interpretation of frequency response, design by trial and error a 60-Hz notch filter to meet the following amplitude response specification. Use two complex poles and two complex zeros in your design. Sketch the amplitude response of your design. Give an SFG and a software realization of the final design.

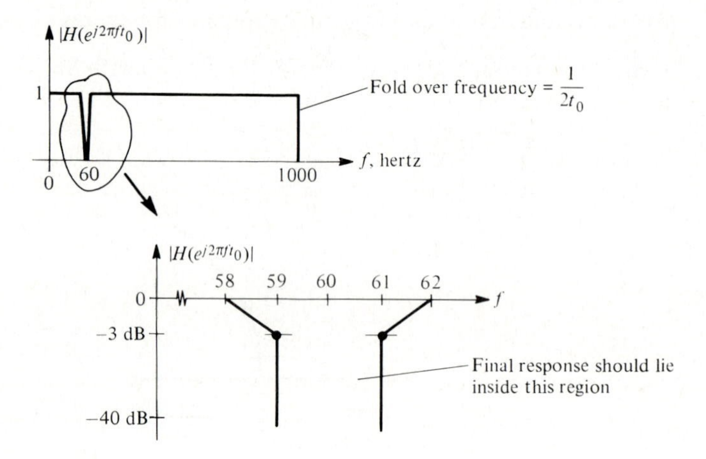

FIGURE 2P.16.

2.17. Linear phase FIR filters are easy to design by imposing symmetry in the unit-pulse response sequence. Suppose N is odd. Show that if

$$h(k) = h(N-k), \qquad k = 0, 1, 2, \ldots, \frac{N-1}{2},$$

then $H(e^{j\theta})$ has a linear phase response.

2.18. Linear phase IIR filters are difficult to design. Suppose $\mathbf{h}$ is the unit-pulse response of an arbitrary IIR filter. Show that the following scheme always produces a resultant filter with linear phase. Assume $\mathbf{h}$ is real and causal.

Step 1 Calculate $w(k) = (\mathbf{h} * \mathbf{u})(k)$

Step 2 Calculate $v(k) = (\mathbf{h} * \mathbf{w}^-)(k); \qquad w^-(k) \overset{\Delta}{=} w(-k)$

Step 3 Calculate $y(k) = v(-k)$

(*Hint:* Show that the filter with $\mathbf{u}$ as an input and $\mathbf{y}$ as an output has a zero phase characteristic.)

2.19. An input $(\frac{1}{2})^k \xi(k)$ produces the steady-state output $(k/2)(\frac{1}{2})^k \xi(k)$. Find th pulse response sequence for this filter. Sketch an SFG of this filter.

2.20. Suppose a certain digital filter has a unit-pulse response $\mathbf{g}$. We wish to c another filter with unit-pulse response $\mathbf{h}$ so that $\mathbf{g} * \mathbf{h} = \boldsymbol{\delta}$. Find $\mathbf{h}$ if $\mathbf{g}$ has response

$$g(k) = (\tfrac{1}{2})^k \xi(k).$$

Comment on the structure of $\mathbf{h}$. Sketch an SFG for this filter.

2.21. Repeat the previous problem if $\mathbf{g}$ has the pulse response $\{g(k)\} = \{1, 0.9\}$, i is an FIR filter. Comment on the structure of $\mathbf{h}$ and sketch an SFG for

2.22. In general, finding $\mathbf{h}$ so that $\mathbf{g} * \mathbf{h} = \boldsymbol{\delta}$ results in an infinite length sequence fo Suppose we constrain $\mathbf{h}$ to $N+1$ terms, i.e., $h = \{h(0), \ldots, h(N)\}$. alternative approach to finding $\mathbf{h}$ is to choose $\mathbf{h}$ to minimize the error ene between the *desired response* $\boldsymbol{\delta}$ and the *actual response* $\mathbf{y}$. That is, we find l minimize V where V is defined as

$$V = \|\boldsymbol{\delta} - \mathbf{h} * \mathbf{g}\|_2^2 = \sum_{k=0}^{\infty}\left[\delta(k) - \sum_{l=0}^{N} h(l)g(k-l)\right]^2.$$

A necessary condition for minimum V with respect to $\mathbf{h}$ is that

$$\frac{\partial V}{\partial h(i)} = 0, \qquad i = 0, 1, \ldots, N.$$

Show that this condition can be written in the form

$$g(-i) = \sum_{l=0}^{N} h(l) \sum_{k=0}^{\infty} g(k-i)g(k-l), \qquad i = 0, 1, \ldots, N.$$

Show that in matrix notation we can write these $N+1$ equations in the form

$$[h(0), h(1), \ldots, h(N)]\mathbf{R} = [g(0), 0, \ldots, 0],$$

where

$$\mathbf{R} = [R_{li}], \qquad R_{li} = \sum_k g(k-i)g(k-l).$$

Find the equations for $\mathbf{h}$ if $\mathbf{g} = \{1, a\}$. Choose $a = \frac{1}{2}$ and plot the frequenc responses for $\mathbf{g}$ and $\mathbf{h}$ for $N = 4$. [See Orfanidis (1985) and Chapter 7.]

2.23. An input sequence

$$u(k) = \begin{cases} (\frac{1}{2})^{-k}, & k \leq 0 \\ 0, & k > 0 \end{cases}$$

produces a frequency domain (steady-state) output of

$$Y(e^{j\theta}) = \frac{2\cos\theta}{2 - \cos\theta - j\sin\theta}$$

for an LSI system. Find the frequency response function $H(e^{j\theta})$ of this syst

CHAPTER 3

The z-Transform

The z-transform serves the discrete-time or sequence domain in a manner analogous to the Laplace transform for the continuous-time domain. In the same way the Laplace transform is a generalization of the Fourier integral, the z-transform is a generalization of the discrete-time Fourier transform (DTFT). Fourier analysis of sequences using the DTFT restricts the domain of analysis to the unit circle, $|z| = 1$. The z-transform expands the domain of analysis to an annulus in the complex z-plane which generally includes $|z| = 1$. We shall discuss the z-transform for two classes of sequences: (1) sequences possibly nonzero for all indices in which case the transform is called the *bilateral z-transform* and; (2) sequences that are one-sided (causal or anti-causal) resulting in the *unilateral z-transform.*

3.1 THE BILATERAL z-TRANSFORM

The (bilateral) z-transform of a sequence $\mathbf{f}$ is defined as

$$F(z) = Z\{\mathbf{f}\} = \sum_{k=-\infty}^{\infty} f(k)z^{-k}, \tag{3.1.1}$$

where z is a complex variable. We shall adopt the convention of omitting the reference to bilateral in our discussions and indicate only the unilateral z-transform explicitly. One important property of the power series $F(z)$ is the set of values z for which the sum (3.1.1) *converges absolutely*, i.e., the region of the z-plane for which $\sum_k |f(k)z^{-k}| < \infty$.

This set of values in the z-plane is called the *region of convergence*, ROC.

EXAMPLE 3.1.1

Calculation of a z-transform.

Consider the sequence $u(k) = (1/2)^k \xi(k)$. Its z-transform is

$$U(z) = \sum_{k=-\infty}^{\infty} (\tfrac{1}{2})^k \xi(k) z^{-k} = \sum_{k=0}^{\infty} (\tfrac{1}{2})^k z^{-k} = \frac{1}{1 - \dfrac{z^{-1}}{2}}\,.$$

The z-transform of $\mathbf{u}$ is $U(z)$ with an ROC $|\frac{1}{2}z^{-1}| < 1$ or $|z| > 1/2$ since this is the condition for convergence of the geometric series.

Why is knowledge of the ROC for a z-transform important? The answer to this question has to do with the uniqueness of the inverse transform. If we transform a two-sided sequence $\mathbf{f}$ and obtain $F(z)$, the inverse transform of $F(z)$ is *not unique unless we specify the ROC*. The following example illustrates the problem.

EXAMPLE 3.1.2

Nonuniqueness of the z-transform.

Consider two sequences $\mathbf{x}$ and $\mathbf{y}$ defined as

$$x(k) = \begin{cases} 0, & k < 0 \\ \alpha^k, & k \geqslant 0 \end{cases}, \qquad y(k) = \begin{cases} -\alpha^k, & k < 0 \\ 0, & k \geqslant 0 \end{cases},$$

calculating the z-transforms, we obtain

$$X(z) = \sum_{k=0}^{\infty} \alpha^k z^{-k} = \frac{1}{1 - \alpha z^{-1}}, \qquad |z| > |\alpha|$$

$$Y(z) = \sum_{k=-\infty}^{-1} [-\alpha^k z^{-k}] = \frac{1}{1 - \alpha z^{-1}}, \qquad |z| < |\alpha|.$$

Thus distinct sequences give rise to the same z-transforms. The only way we can resolve the ambiguity is to specify the ROC of the z-transform. In this example, $Y(z)$ converges absolutely for $|z| < |\alpha|$ and $X(z)$ converges absolutely for $|z| > |\alpha|$. Knowing the ROC we are then able to represent $1/(1 - \alpha z^{-1})$ in the correct power series to obtain the corresponding sequence.

Another interpretation of this ambiguity can be explained by the fact that a rational function of z can be expanded in positive and/or negative powers of z, thus leading to sequences with negative and/or positive indices. The ROC specifies which expansion in powers of z is valid.

For sequences whose values are zero for negative indices, the ROC is of the form $|z| > R$ since

$$\sum_{k=0}^{\infty} |f(k)z^{-k}| = \sum_{k=0}^{\infty} |f(k)(re^{j\theta})^{-k}| = \sum_{k=0}^{\infty} |f(k)|r^{-k} < \infty. \tag{3.1.2}$$

The infinite sum in Eq. (3.1.2) converges provided $|f(k)| \leqslant MR^k$, $k > 0$ where M and R are positive numbers; and r is chosen so that $(R/r) < 1$, which implies $|z| > R$. The smallest number R for which the sequence $\mathbf{f}$ can be so bounded is denoted R_-. Thus in Example 3.1.2, $R_- = |\alpha|$ for the sequence $\mathbf{x}$.

Notice that for positive indices k, the sequence z^{-k} is large near $z = 0$ and so to obtain convergence we must bound z values away from the origin, which implies an ROC of the form $|z| > R_-$.

Similarly, for sequences whose values are zero for positive indices, the ROC is of the form $|z| < R$. We have

$$\sum_{k=-\infty}^{0} |f(k)z^{-k}| = \sum_{k=0}^{\infty} |f(-k)(re^{j\theta})^k| = \sum_{k=0}^{\infty} |f(-k)|r^k < \infty. \tag{3.1.3}$$

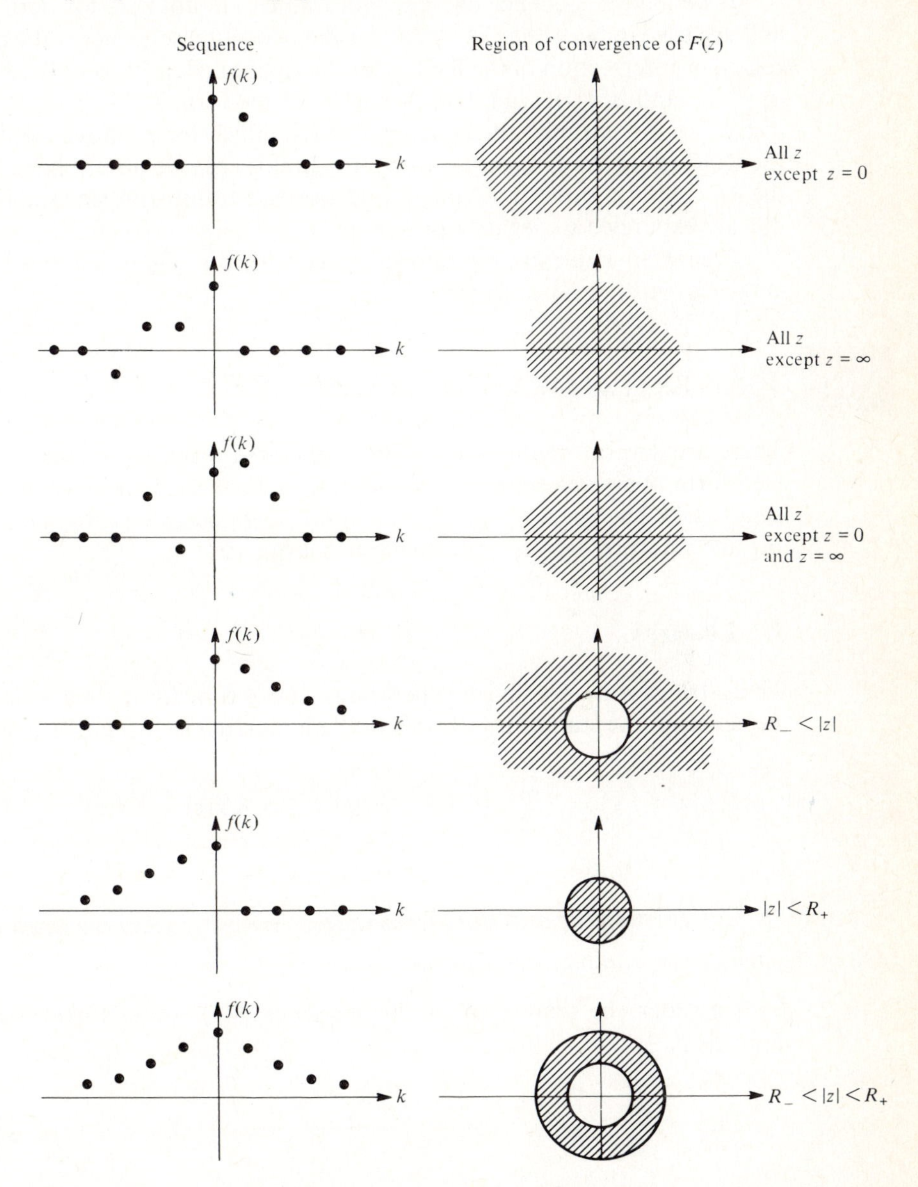

FIGURE 3.1.1 Typical sequences and the corresponding ROC for their z-transforms.

The infinite sum in Eq. (3.1.3) converges provided $|f(k)| \leqslant MR^k$, $k < 0$ where M and R are positive numbers and r is chosen so that $(r/R) < 1$, which implies $|z| < R$. The largest number R for which the sequence **f** can be so bounded is denoted R_+. In Example 3.1.2, $R_+ = |\alpha|$ for the sequence **y**. For negative k, z^{-k} blows up for z values away from the origin. Thus we must bound the sum (3.1.3) by restricting the z values close to the origin implying an ROC of the form $|z| < R_+$.

If we have a sequence having values which are not zero for both positive and negative indices, then the ROC for the two-sided sequence is the common region or intersection of the ROCs, for the right and left parts of the sequence, $|z| < R_-$ and $|z| > R_+$ implying an ROC of the form $R_- < |z| < R_+$. If our discussion is restricted to sequences which vanish for negative indices, then the ROC is always of the form $|z| > R_-$. In this case we do not have to retain the ROC information to obtain unique inverse z-transforms since all functions of z are expanded in negative powers of z.

Figure 3.1.1 depicts typical sequences with the regions of absolute convergence for their z-transforms.

3.2 PROPERTIES OF THE z-TRANSFORM

There are several methods for obtaining closed-form expressions for the z-transform of a given sequence in addition to direct evaluation of the defining sum (3.1.1). The properties discussed in this section can be valuable in obtaining and interpreting the z-transform of a sequence.

1. Linearity

The z-transform is a linear operation. The z-transform of a sum of two sequences is the sum of transforms of the individual sequences. Thus we have

$$Z\{a\mathbf{x} + b\mathbf{y}\} = \sum_{k=-\infty}^{\infty} (ax(k) + by(k))z^{-k} = aX(z) + bY(z). \tag{3.2.1}$$

EXAMPLE 3.2.1

Use of linearity in obtaining z-transforms.

Consider the transform of the sequence $f(k) = \cos(k\theta)\xi(k)$. From the definition,

$$F(z) = \sum_{k=0}^{\infty} \cos(k\theta)z^{-k} = \sum_{k=0}^{\infty} \left(\frac{e^{j\theta k}z^{-k} + e^{-j\theta k}z^{-k}}{2}\right)$$

$$= \frac{1}{2}\sum_{k=0}^{\infty}\left[\left(\frac{e^{j\theta}}{z}\right)^k + \left(\frac{e^{-j\theta}}{z}\right)^k\right].$$

Since this is a geometric series, we have

$$F(z) = \frac{\frac{1}{2}}{1 - e^{j\theta}z^{-1}} + \frac{\frac{1}{2}}{1 - e^{-j\theta}z^{-1}} = \frac{1 - z^{-1}\cos\theta}{1 - 2z^{-1}\cos\theta + z^{-2}}$$

2. Shifting

Assume the sequence **f** has transform $F(z)$. What is the transform of the sequence $\{f(k \pm n)\}$, corresponding to the original sequence delayed by n (using the minus sign) or advanced by n (using the plus sign)? By definition

$$Z\{f(k \pm n)\} = \sum_{k=-\infty}^{\infty} f(k \pm n)z^{-k} = \sum_{m=-\infty}^{\infty} f(m)z^{-m \pm n} = z^{\pm n}F(z).$$

Thus we have the result

$$\{f(k \pm n)\} \leftrightarrow z^{\pm n}F(z). \tag{3.2.2}$$

In the case of one-sided sequences, we shall employ the *unilateral* z-transform defined by

$$F_u(z) = Z_u\{\mathbf{f}\} = \sum_{k=0}^{\infty} f(k)z^{-k}. \tag{3.2.3}$$

If we now shift the sequence, we must add or subtract terms because of the one-sided sum in Eq. (3.2.3). Thus for delayed sequences we have

$$\begin{aligned} Z_u\{f(k - n)\} &= \sum_{k=0}^{\infty} f(k - n)z^{-k} = \sum_{m=-n}^{\infty} f(m)z^{-m}z^{-n} \\ &= f(-n) + f(-n + 1)z^{-1} + \cdots + f(-1)z^{-n+1} \\ &\quad + z^{-n}\sum_{m=0}^{\infty} f(m)z^{-m}. \end{aligned}$$

Thus we have

$$\{f(k - n)\} \leftrightarrow f(-n) + f(-n + 1)z^{-1} + \cdots + f(-1)z^{-n+1} + z^{-n}F_u(z). \tag{3.2.4}$$

In a similar manner we can show

$$\{f(k + n)\} \leftrightarrow -f(0)z^n - f(1)z^{n-1} - \cdots - f(n - 1)z + z^nF_u(z). \tag{3.2.5}$$

EXAMPLE 3.2.2

Solving a difference equation with z-transforms.

Find the output of the discrete-time system with the difference equation $y(k) + \frac{1}{2}y(k - 1) = u(k) + u(k - 1)$ with initial conditions $y(-1)$ and $u(-1) = 0$. Assume the input sequence is $u(k) = (\frac{1}{2})^k\xi(k)$. Taking z-transforms of both sides of the difference equation, we obtain

$$Y_u(z) + \tfrac{1}{2}[z^{-1}Y_u(z) + y(-1)] = U_u(z) + z^{-1}U_u(z) + u(-1). \tag{3.2.6}$$

where $U_u(z) = Z_u\{(1/2)^k\xi(k)\} = 1/(1 - z^{-1}/2)$, $u(-1) = 0$. Substituting into Eq. (3.2.6) yields

$$Y_u(z) + \tfrac{1}{2}[z^{-1}Y_u(z) + y(-1)] = \frac{1}{1 - \dfrac{z^{-1}}{2}}(1 + z^{-1}).$$

Solving for $Y_u(z)$, we obtain

$$Y_u(z) = \frac{1 + z^{-1}}{\left(1 - \dfrac{z^{-1}}{2}\right)\left(1 + \dfrac{z^{-1}}{2}\right)} - \frac{1}{2}\frac{y(-1)}{(1 + \frac{1}{2}z^{-1})}$$

The output transform $Y_u(z)$ consists of a sum of two terms. The first term on the right is the response due to the input sequence **u** acting on the system with zero initial conditions $[y(-1) = 0]$. The second term on the right is the response of the system due to the initial condition.

3. Convolution

One of the primary reasons transforms are used in the analysis of linear systems is to transform a convolutional algebra into a multiplicative algebra. Suppose $\mathbf{u} \leftrightarrow U_u(z)$ and $\mathbf{h} \leftrightarrow H_u(z)$. Let $\mathbf{y} = \mathbf{u} * \mathbf{h}$. Then

$$Y_u(z) = U_u(z)H_u(z) \tag{3.2.7}$$

and

$$Y(z) = U(z)H(z). \tag{3.2.8}$$

To prove Eq. (3.2.8), we write it in the form

$$\begin{aligned} Y(z) &= \sum_{k=-\infty}^{\infty} y(k)z^{-k} = \sum_{k=-\infty}^{\infty}\left(\sum_{m=-\infty}^{\infty} u(m)h(k - m)\right)z^{-k}z^{m}z^{-m} \\ &= \sum_{m=-\infty}^{\infty} u(m)z^{-m} \sum_{k=-\infty}^{\infty} h(k - m)z^{-(k-m)} = U(z)H(z) \end{aligned}$$

with an ROC the intersection of the ROCs for $U(z)$ and $H(z)$. Equations (3.2.7) and (3.2.8) form the basis of input-output calculation using transforms. To find the output **y** of a system **h** for an input **u**, we find the transforms $H(z)$ and $U(z)$ [or $H_u(z)$ and $U_u(z)$], multiply them, and then perform an inverse transform on the product to obtain the output sequence **y**.

The system transfer function $H(z)$ is defined as

$$H(z) = \frac{Y(z)}{U(z)}. \tag{3.2.9}$$

If the input $\mathbf{u}$ is the unit pulse sequence $\boldsymbol{\delta}$, then the output is

$$Y(z) = H(z). \tag{3.2.10}$$

In other words, the transfer function $H(z)$ is the z-transform of the unit-pulse response sequence $\mathbf{h}$ (as we discussed in Chapter 2).

For a cascade of two systems with unit-pulse response sequences $\mathbf{h}_1$ and $\mathbf{h}_2$, the output is

$$\mathbf{y} = \mathbf{u} * \mathbf{h}_1 * \mathbf{h}_2.$$

Let $\mathbf{w}$ be the output of the first system, $\mathbf{u} * \mathbf{h}_1$. In the transform domain

$$W(z) = U(z)H_1(z) \tag{3.2.11}$$

and so, since $\mathbf{y} = \mathbf{w} * \mathbf{h}_2$, we obtain

$$Y(z) = U(z)H_1(z)H_2(z). \tag{3.2.12}$$

Equation (3.2.12) generalizes immediately to m cascaded systems.

4. Scaling

Let $\mathbf{f} \leftrightarrow F(z)$, with a ROC $R_- < |z| < R_+$. If a is any nonzero number (real or complex), then the z-transform of $a^k f(k)$ is $F(z/a)$ with a ROC, $|a|R_- < |z| < |a|R_+$. For example, in Example 3.2.1 we found the transform pair

$$\cos(k\theta)\xi(k) \leftrightarrow \frac{1 - z^{-1}\cos\theta}{1 - 2z^{-1}\cos\theta + z^{-2}}, \qquad |z| > 1.$$

The transform of a damped cosine $0.5^k \cos(k\theta)\xi(k)$ is thus given by

$$0.5^k \cos k\theta\xi(k) \leftrightarrow \frac{1 - 0.5z^{-1}\cos\theta}{1 - z^{-1}\cos\theta + 0.25z^{-2}}, \qquad |z| > 0.5$$

5. Multiplication by k

Let $\mathbf{f} \leftrightarrow F(z)$ and suppose we wish to find the transform of $\{kf(k)\}$. From the definition we have

$$Z\{kf(k)\} = \sum_{k=-\infty}^{\infty} kf(k)z^{-k} = z\sum_{k=-\infty}^{\infty} kf(k)z^{-k-1} = -z\frac{d}{dz}F(z). \tag{3.2.13}$$

The interchange of summation and differentiation in Eq. (3.2.13) is valid because one can differentiate or integrate a power series term by term within its region of convergence. We can generalize Eq. (3.2.13) to include multiplication by k^n. Thus we have

$$\{k^n f(k)\} \leftrightarrow -z\frac{d}{dz}\left(\cdots -z\frac{d}{dz}(F(z))\cdots\right). \tag{3.2.14}$$

EXAMPLE 3.2.3

Using property (3.2.14).

We have

$$\alpha^k \xi(k) \leftrightarrow \frac{1}{1 - \alpha z^{-1}}.$$

From Eq. (3.2.13), we find that

$$k\alpha^k \xi(k) \leftrightarrow -z \frac{d}{dz}\left(\frac{1}{1 - \alpha z^{-1}}\right) = \frac{\alpha z^{-1}}{(1 - \alpha z^{-1})^2}.$$

Thus dividing by α in the sequence and the transform, we have

$$k\alpha^{k-1}\xi(k) \leftrightarrow \frac{z^{-1}}{(1 - \alpha z^{-1})^2}.$$

Using the shifting property, we obtain (multiply by z)

$$(k + 1)\alpha^k \xi(k + 1) \leftrightarrow \frac{1}{(1 - \alpha z^{-1})^2}. \tag{3.2.15}$$

The sequence $(k + 1)\alpha^k \xi(k + 1)$ is zero for $k < 0$ since $k + 1$ is zero when $k = -1$ even though $\xi(k + 1)$ is not zero. Continuing, we have

$$\frac{1}{2} k(k + 1)\alpha^{k-1}\xi(k) \leftrightarrow \frac{z^{-1}}{(1 - \alpha z^{-1})^3},$$

from which it follows that

$$\frac{1}{2}(k + 1)(k + 2)\alpha^k \xi(k + 1) \leftrightarrow \frac{1}{(1 - \alpha z^{-1})^3}. \tag{3.2.16}$$

Similarly, we can show

$$\begin{aligned} -\alpha^k \xi(-k - 1) &\leftrightarrow \frac{1}{1 - \alpha z^{-1}} \\ &\vdots \\ -\frac{1}{2}(k + 1)(k + 2)\alpha^k \xi(-k - 1) &\leftrightarrow \frac{1}{(1 - \alpha z^{-1})^3}. \end{aligned} \tag{3.2.17}$$

Transforms of the form $1/(1 - \alpha z^{-1})^n$ are useful in obtaining the inverse transform using the partial fraction expansion technique. Table 3.2.1 summarizes many of the important properties of the z-transform. Property 4 is established using the inversion integral for obtaining a sequence from its transform. We shall discuss these two properties after we have discussed the techniques for performing the inverse transform. Table 3.2.2 summarizes some of the important z-transform pairs.

TABLE 3.2.1 Properties of the z-transform.

1. Linearity	$\alpha\mathbf{f} + \beta\mathbf{g}$	$\longleftrightarrow \alpha F(z) + \beta G(z)$
2. Shifting	$f(k \pm n)$	$\longleftrightarrow z^{\pm n}F(z)$
	$f(k + n)$	$\overset{Z_u}{\longleftrightarrow} z^n F_u(z) - f(0)z^n - f(1)z^{n-1} - \cdots - f(n-1)z$
	$f(k - n)$	$\overset{Z_u}{\longleftrightarrow} z^{-n}F_u(z) + f(-1)z^{-n+1} + \cdots + f(-n)$
3. Convolution	$\mathbf{f} * \mathbf{g}$	$\longleftrightarrow F(z)G(z)$
4. Multiplication	$\mathbf{f} \cdot \mathbf{g}$	$\longleftrightarrow \dfrac{1}{2\pi j}\oint_{\Gamma} \dfrac{F(\lambda)G(z/\lambda)}{\lambda}\, d\lambda$
5. Scaling	$a^k f(k)$	$\longleftrightarrow F(z/a)$
6. Multiplication by k^n	$k^n f(k)$	$\longleftrightarrow \left(-z\dfrac{d}{dz}\right)^n (F(z))$

TABLE 3.2.2 z-Transform pairs.

$f(k)$	$F(z)$	ROC
1. $\xi(k)$	$\dfrac{1}{1 - z^{-1}}$	$\lvert z\rvert > 1$
2. $k\xi(k)$	$\dfrac{z^{-1}}{(1 - z^{-1})^2}$	$\lvert z\rvert > 1$
3. $\alpha^k \xi(k)$	$\dfrac{1}{1 - \alpha z^{-1}}$	$\lvert z\rvert > \alpha$
4. $\alpha^{\lvert k\rvert}$	$\dfrac{1 - \alpha^2}{(1 - \alpha z)(1 - \alpha z^{-1})}$	$\lvert\alpha\rvert < \lvert z\rvert < \dfrac{1}{\lvert\alpha\rvert}$
5. $\dfrac{1}{k}$, $k > 0$ and 0, $k \leqslant 0$	$-\ln(1 - z^{-1})$	$\lvert z\rvert > 1$
6. $\cos(\alpha k)\xi(k)$	$\dfrac{1 - z^{-1}\cos(\alpha)}{1 - 2z^{-1}\cos(\alpha) + z^{-2}}$	$\lvert z\rvert > 1$
7. $\sin(\alpha k)\xi k)$	$\dfrac{z^{-1}\sin(\alpha)}{1 - 2z^{-1}\cos(\alpha) + z^{-2}}$	$\lvert z\rvert > 1$
8. $\left[c\cos(\alpha k) + \dfrac{d + c\cos(\alpha)}{\sin(\alpha)}\sin(\alpha k)\right]\xi(k)$	$\dfrac{c + dz^{-1}}{1 - 2z^{-1}\cos(\alpha) + z^{-2}}$	$\lvert z\rvert > 1$

Vector Sequences

Thus far we have applied the z-transform calculus to scalar sequences. We can, equally well, apply this calculus to vector valued sequences. Consider, for example, the state equation of a shift-invariant system.

$$\mathbf{x}(k+1) = \mathbf{A}\mathbf{x}(k) + \mathbf{B}u(k) \tag{3.2.18}$$

where $\mathbf{x}$ is $n \times 1$, $\mathbf{A}$ is $n \times n$, u is scalar, and $\mathbf{B}$ is $n \times 1$. Let $\mathbf{X}_u(z)$ and $\mathbf{U}_u(z)$ be the z-transforms of $\mathbf{x}$ and $\mathbf{u}$, respectively. The z-transform of a vector valued sequence is obtained by transforming each vector component. Taking transforms of both sides of Eq. (3.2.18) yields

$$z\mathbf{X}_u(z) - z\,\mathbf{x}(0) = \mathbf{A}\mathbf{X}_u(z) + \mathbf{B}U_u(z)$$

or

$$[z\mathbf{I} - \mathbf{A}]\mathbf{X}_u(z) = z\,\mathbf{x}(0) + \mathbf{B}U_u(z).$$

Solving for the vector $\mathbf{X}_u(z)$,

$$\mathbf{X}_u(z) = [z\mathbf{I} - \mathbf{A}]^{-1} z\,\mathbf{x}(0) + [z\mathbf{I} - \mathbf{A}]^{-1}\mathbf{B}U_u(z). \tag{3.2.19}$$

If the input sequence $\mathbf{u}$ is zero, then Eq. (3.2.19) reduces to

$$\mathbf{X}_u(z) = [\mathbf{I} - \mathbf{A}z^{-1}]^{-1}\mathbf{x}(0). \tag{3.2.20}$$

For a zero input, we can show, using Eq. (3.2.18), that the state $\mathbf{x}(k)$ is given by

$$\mathbf{x}(k) = \mathbf{A}^k\mathbf{x}(0), \text{ for } k > 0. \tag{3.2.21}$$

Comparing Eqs. (3.2.20) and (3.2.21), we conclude that

$$\{\mathbf{A}^k \xi(k)\} = Z^{-1}\{(\mathbf{I} - \mathbf{A}z^{-1})^{-1}\}. \tag{3.2.22}$$

This is a method of calculating the important function of a matrix, $\mathbf{A}^k$. (See Appendix 8A.)

3.3 INVERSION OF THE z-TRANSFORM

There are several methods for finding the sequence $\mathbf{f}$ corresponding to a given z-transform $F(z)$ with a given ROC. Essentially we must expand the transform in a power series in z or z^{-1} or both and obtain the sequence values $f(k)$ from the coefficients of z^{-k}. Most often in digital filtering $F(z)$ is a rational function in z (or z^{-1}). In this case, partial fraction expansions can be used to find the corresponding sequences.

The ROC defines whether the z-transform should be expressed in terms of positive or negative powers of z or a combination of the two. If $|z| > R$, then $F(z)$ must be expressed in powers of z^{-1}. If $|z| < R$, use positive powers of z, and if $R_2 < |z| < R_1$, use both positive and negative powers of z. In this latter case, decompose $F(z)$ into $F_1(z) + F_2(z)$, where $F_1(z)$ converges for $|z| > R_2$ and $F_2(z)$ converges for $|z| < R_1$.

Direct Division

If $F(z)$ is given in the form of a rational function of z, we can simply divide the denominator into the numerator to obtain a power series in z or z^{-1}. And while this method is simple in concept, unless $F(z)$ is relatively simple, it is usually not possible to obtain a closed form expression for the sequence values.

EXAMPLE 3.3.1

Finding an inverse z-transform by long division.

Find the pulse response sequence **h** of the second-order digital filter shown in Fig. 3.3.1 by obtaining the transfer function $H(z)$ and then inverting $H(z)$ using direct division. The difference equation is of the form

$$y(k) - \frac{3}{4}\,y(k-1) + \frac{1}{8}\,y(k-2) = u(k) + 2u(k-1), \qquad k \geqslant 0.$$

Taking z-transforms of both sides we obtain

$$Y_u(z) - \frac{3}{4}z^{-1}Y_u(z) + \frac{1}{8}z^{-2}Y_u(z) = U_u(z) + 2z^{-1}U_u(z).$$

The transfer function is thus

$$H_u(z) = \frac{Y_u(z)}{U_u(z)} = \frac{1 + 2z^{-1}}{1 - \frac{3}{4}z^{-1} + \frac{1}{8}z^{-2}} = \frac{z^2 + 2z}{z^2 - \frac{3}{4}z + \frac{1}{8}}.$$

By long division (in negative powers of z), we obtain

$$H_u(z) = 1 + \frac{11}{4}z^{-1} + \frac{31}{16}z^{-2} + \frac{71}{64}z^{-3} + \cdots.$$

Thus

$$\mathbf{h} = \left\{1, \frac{11}{4}, \frac{31}{16}, \frac{71}{64}, \ldots\right\}.$$

This method does not result in closed form expressions, but can be used for numerical work.

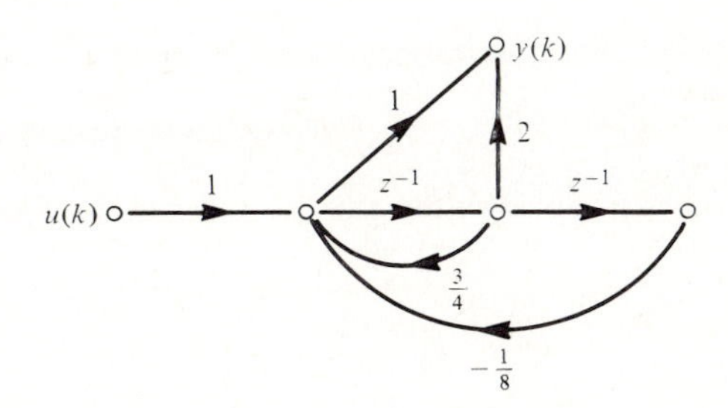

FIGURE 3.3.1.

Partial Fraction Expansions

A more useful method for finding the sequence **f** corresponding to a rational $F(z)$ is to perform a partial-fraction expansion of $F(z)$. We then identify each of the terms of the expansion with a sequence. There are many forms of partial fraction expansions and there are several techniques one can use in obtaining the parameters of the expansion. If $F(z)$ converges for $|z| > R$ and contains simple roots in the denominator polynomial (simple poles), inversion by partial fraction is straightforward. If $F(z)$ converges in an annulus $R_1 < |z| < R_2$ and contains multiple poles, inversion by partial fractions is more difficult.

Assume $F(z)$ is the ratio of two polynomials in z.

$$F(z) = \frac{N(z)}{D(z)} = \frac{b_0 + b_1 z + \cdots + b_m z^m}{1 + a_1 z + \cdots + a_n z^n} \tag{3.3.1}$$

If $m \geqslant n$, then simply divide the numerator by the denominator and obtain a form

$$F(z) = \sum_{k=0}^{m-n} c_k z^k + \frac{N_1(z)}{D(z)}. \tag{3.3.2}$$

$N_1(z)$ is guaranteed to have an order less than the order of the denominator.

Next find the poles of $F(z)$. [The *poles* of $F(z)$ are roots of $D(z) = 0$. The *zeros* of $F(z)$ are roots of $N(z) = 0$.] In the expansion each term has the form $cz/(z - p_i)$. For a pole of multiplicity r, we use r terms of the form $c_i z^i/(z - p)^i$, $i = 1, 2, \ldots, r$. Table 3.3.1 summarizes the allowed terms and the corresponding sequences. Once the form of the terms in the partial fraction is defined, we solve for the unknown coefficients c_i, $i = 1, 2, \ldots, n$. If for a pole p_i, $F_i(z)$ converges for some $|z| > p_i$, then the corresponding sequence is found in column (1) of Table 3.3.1; if $F_i(z)$ converges for some $|z| < p_i$, then the corresponding sequence is obtained from column (2).

TABLE 3.3.1 Inverse transforms for partial fraction terms.

Partial Fraction Term, $F_i(z)$	Corresponding Sequence, $\mathbf{f}_i$	
	(1) $F_i(z)$ converges for $\lvert z\rvert > R_-$	(2) $F_i(z)$ converges for $\lvert z\rvert < R_+$
$\dfrac{z}{z-p}$	$p^k\xi(k)$	$-p^k\xi(-k-1)$
$\dfrac{z^2}{(z-p)^2}$	$(k+1)p^k\xi(k)$	$-(k+1)p^k\xi(-k-1)$
$\vdots$	$\vdots$	$\vdots$
$\dfrac{z^n}{(z-p)^n}$	$\dfrac{1}{(n-1)!}(k+1)\cdots(k+n-1)p^k\xi(k)$	$\dfrac{-1}{(n-1)!}(k+1)\cdots(k+n-1)p^k\xi(-k-1)$

EXAMPLE 3.3.2

Inverting a z-transform by partial fractions.

Consider the z-transform

$$F(z) = \frac{2z^4 + 4z^3 - 14.5z^2 - 44.5z - 33.5}{z^3 + 1.5z^2 - 8.5z - 15}, \qquad 2.6 < |z| < 2.9.$$

The numerator is of higher degree than the denominator. Thus we divide the denominator into the numerator, giving

$$F(z) = 2z + 1 + F_1(z) = 2z + 1 + \frac{z^2 - 6z - 18.5}{z^3 + 1.5z^2 - 8.5z - 15}.$$

Factoring the denominator (using synthetic division, for example) we find poles at $p_1 = -2$, $p_2 = -2.5$, and $p_3 = 3$. Thus we have for $F_1(z)$.

$$F_1(z) = \frac{z^2 - 6z - 18.5}{(z + 2)(z + 2.5)(z - 3)} = c_0 + \frac{c_1 z}{z + 2} + \frac{c_2 z}{z + 2.5} + \frac{c_3 z}{z - 3}. \tag{3.3.3}$$

Combining terms on the right-hand side of Eq. (3.3.3) and equating coefficients of like powers of z, we find the constants $c_0 = 1.233$, $c_1 = -0.5$, $c_2 = -0.4$, and $c_3 = -0.333$. Thus

$$F(z) = 2z + 2.233 - \frac{0.5z}{z + 2} - \frac{0.4z}{z + 2.5} - \frac{0.333z}{z - 3}.$$

The ROC is given as $2.6 < |z| < 2.9$. Thus the poles at -2 and -2.5 correspond to causal sequences. The term with the pole at 3 corresponds to an anticausal sequence. From Table 3.3.1, we obtain

$$f(k) = 2\delta(k - 1) + 2.2333\,\delta(k) - \left[\frac{1}{2}(-2)^k + \frac{2}{5}(-2.5)^k\right]\xi(k) + \frac{1}{3}(3)^k\xi(-k - 1) \tag{3.3.4}$$

The Inversion Integral

A third method of inverting a z-transform $F(z)$ is based on the *Cauchy integral formula* (Churchill 1948). Consider the z-transform

$$F(z) = \sum_{k=-\infty}^{\infty} f(k)z^{-k} \tag{3.3.5}$$

in the region of the z-plane for which the series converges. Select a circular contour C, centered at the origin and lying in the ROC. Multiply both sides of Eq. (3.3.5) by z^{i-1} and integrate on C:

$$\oint_C z^{i-1}F(z)\,dz = \sum_{k=-\infty}^{\infty} \oint_C f(k)z^{i-1-k}\,dz. \tag{3.3.6}$$

The interchange of integration and summation is allowed because the sum in Eq. (3.3.5) converges uniformly. The *Cauchy integral lemma* states

$$\oint_C z^{\nu}\,dz = \begin{cases} 2\pi j, & \nu = -1 \\ 0, & \text{otherwise} \end{cases}. \tag{3.3.7}$$

Thus all the terms on the right in Eq. (3.3.6) vanish except for $k = i$. Thus Eq. (3.3.6) reduces to

$$f(k) = \frac{1}{2\pi j}\oint_C z^{k-1}F(z)\,dz. \tag{3.3.8}$$

For rational z-transforms contour integrals such as Eq. (3.3.8) are often evaluated by finding the residues of the integrand, i.e.,

$$f(k) = \frac{1}{2\pi j}\oint_C z^{k-1}F(z)\,dz = \sum (\text{residues of } z^{k-1}F(z) \text{ at poles inside } C). \tag{3.3.9}$$

If $z^{k-1}F(z)$ is rational and has a pole p_i of multiplicity r, the residue at this pole is

$$\text{Residue}\,\{z^{k-1}F(z) \;@\; p_i\} = \frac{1}{(r-1)!}\frac{d^{r-1}}{dz^{r-1}}\phi(z)\bigg|_{z=p_i}, \tag{3.3.10}$$

where $\phi(z)$ is defined by

$$\phi(z) = (z - p_i)^r \cdot [z^{k-1}F(z)]. \tag{3.3.11}$$

Equation (3.3.8) is valid for all values of k. However, for $k < 0$, there is a multiple-order pole at $z = 0$ whose order depends on k. In this case the right-hand side of Eq. (3.3.9) is difficult to evaluate. It is more convenient to evaluate Eq. (3.3.8) in two parts, one for $k \geqslant 0$ and one for $k < 0$. Thus

$$f(k) = \frac{1}{2\pi j}\oint_C z^{k-1}F(z)\,dz\bigg|_{k\geqslant 0} + \frac{1}{2\pi j}\oint_C z^{k-1}F(z)\,dz\bigg|_{k<0}. \tag{3.3.12}$$

Use Eqs. (3.3.9) and (3.3.10) to find $f(k)$ for $k \geqslant 0$. For $k < 0$, choose a closed contour C' as shown in Fig. 3.3.2. The contour C' encloses all poles of $z^{k-1}F(z)$ which lie outside C. The integrals over C_2 and C_4 cancel each other. And so we have

$$\frac{1}{2\pi j}\oint_{C'} z^{k-1}F(z)\,dz = \frac{1}{2\pi j}\oint_{C_1} z^{k-1}F(z)\,dz - \frac{1}{2\pi j}\oint_{C_3} z^{k-1}F(z)\,dz. \tag{3.3.13}$$

The integral over C_1 in Eq. (3.3.13) is zero provided $F(z)$ is rational and the numerator degree is less than the denominator degree. In this case for all $k \leqslant 0$

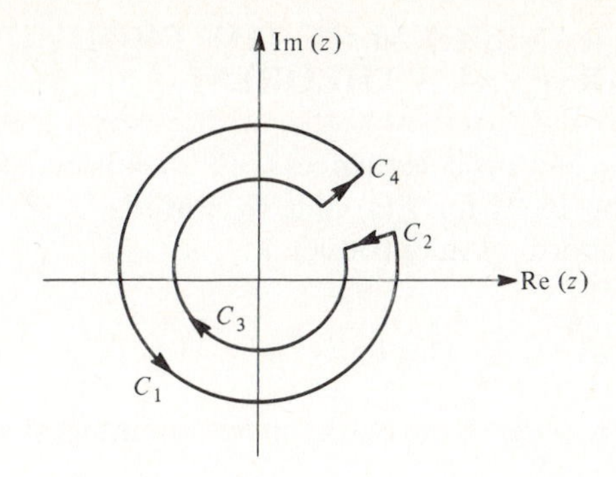

FIGURE 3.3.3.

$$\frac{1}{2\pi j}\oint_C z^{k-1}F(z)\,dz = -\frac{1}{2\pi j}\oint_{C'} z^{k-1}F(z)\,dz$$
$$= -\sum(\text{Residues of } z^{k-1}F(z) \text{ at poles outside } C.) \quad \textbf{(3.3.14)}$$

EXAMPLE 3.3.3

Inverting a z-transform using residues.

Consider the z-transform of Example 3.3.2.

$$F_1(z) = \frac{z^2 - 6z - 18.5}{(z+2)(z+2.5)(z-3)}, \qquad 2.5 < |z| < 3.$$

The contour C is chosen in the annulus $2.5 < |z| < 3$. For $k \geqslant 0$ the residue of $z^{k-1}F_1(z)$ must be evaluated at the simple poles $z = -2$ and $z = -2.5$. In addition,

$$z^{k-1}F_1(z) = \frac{(z^2 - 6z - 18.5)z^k}{z(z+2)(z+2.5)(z-3)}.$$

For $k = 0$, the residue must be found at $z = 0$.

$$\begin{aligned} f(k)|_{k\geqslant 0} &= \text{Res}\{z^{k-1}F(z)\}|_{z=-2} + \text{Res}\{z^{k-1}F(z)\}|_{z=-2.5} \\ &\quad + \text{Res}\{z^{k-1}F(z)\}|_{z=0} \\ &= \left[-\frac{1}{2}(-z)^k - \frac{2}{5}\left(\frac{5}{2}\right)^k\right]\xi(k) + \frac{37}{30}\delta(k). \end{aligned}$$

For $k < 0$, $f(k)$ is obtained from the negative of the residue at the pole $z = 3$:

$$f(k)|_{k<0} = -\text{Res}\{z^{k-1}F(z)\}|_{z=3} = +\frac{1}{3}3^k\xi(-k-1).$$

And so

$$f(k) = \left(-\frac{1}{2}(-2)^k - \frac{2}{5}\left(\frac{5}{2}\right)^k\right)\xi(k) + \frac{37}{30}\delta(k) + \frac{1}{3}3^k\xi(-k-1)$$

3.4 THE z-TRANSFORM OF THE PRODUCT fg AND PARSEVAL'S THEOREM

Let **f** and **g** be causal sequences with z-transforms $F(z)$ and $G(z)$, respectively. Assume the ROC for $F(z)$ and $G(z)$ are $|z| > R_f$ and $|z| > R_g$, respectively. The z-transform of the product is

$$Z_u\{\mathbf{f}\cdot\mathbf{g}\} = \sum_{k=0}^{\infty} f(k)g(k)z^{-k} \tag{3.4.1}$$

for ROC $|z| > R_c$. Now by the inversion integral we can express the sequence **f** as

$$f(k) = \frac{1}{2\pi j}\oint_\Gamma s^{k-1}F(s)\,ds. \tag{3.4.2}$$

Substituting Eq. (3.4.2) into Eq. (3.4.1) yields

$$Z_u\{\mathbf{f}\cdot\mathbf{g}\} = \frac{1}{2\pi j}\sum_{k=0}^{\infty}\oint_\Gamma g(k)s^{k-1}F(s)z^{-k}\,ds \tag{3.4.3}$$

Now Eq. (3.4.1) converges uniformly for $|z| > R_c$ and so we can exchange the summation and integration in Eq. (3.4.3) to obtain

$$Z_u\{\mathbf{f}\cdot\mathbf{g}\} = \frac{1}{2\pi j}\oint_\Gamma s^{-1}F(s)\sum_{k=0}^{\infty} g(k)(zs^{-1})^{-k}\,ds. \tag{3.4.4}$$

Since $G(z)$ is the z-transform of **g**, Eq. (3.4.4) can be expressed as

$$Z_u\{\mathbf{f}\cdot\mathbf{g}\} = \frac{1}{2\pi j}\oint_\Gamma s^{-1}F(s)G\left(\frac{z}{s}\right)ds. \tag{3.4.5}$$

The magnitude of z must be chosen so that there is an annular region between the singularities of $s^{-1}F(s)$ and $G(z/s)$. The contour Γ is taken in this region, $[R_f, |z|/R_g]$. If $|z| = 1$ is contained in the ROC, then Eq. (3.4.5) can be expressed as

$$Z_u\{\mathbf{f}\cdot\mathbf{g}\} = \frac{1}{2\pi j}\oint_\Gamma s^{-1}F(s)G\left(\frac{1}{s}\right)ds. \tag{3.4.6}$$

If we let $\mathbf{f} = \mathbf{g}$, then Eq. (3.4.6) reduces to the discrete form of Parseval's relation:

$$Z_u\{\mathbf{f}\cdot\mathbf{f}\} = \sum_{k=0}^{\infty} f^2(k)z^{-k} = \frac{1}{2\pi j}\oint_\Gamma z^{-1}F(z)F(z^{-1})\,dz. \tag{3.4.7}$$

If we let Γ be the contour $|z| = 1$, then $z = e^{j\theta}$ and Eq. (3.4.7) becomes the familiar form of Parseval's relation used in Fourier analysis:

$$\sum_{k=0}^{\infty} f^2(k) = \frac{1}{2\pi}\int_{-\pi}^{\pi}|F(e^{j\theta})|^2\,d\theta. \tag{3.4.8}$$

More generally, from Eq. (3.4.6) we can write

$$\sum_{k=0}^{\infty} f(k)g^*(k) = \frac{1}{2\pi}\int_{-\pi}^{\pi} F(e^{j\theta})G^*(e^{j\theta})\,d\theta. \tag{3.4.9}$$

EXAMPLE 3.4.1

Using Parseval's relation to calculate $\|\mathbf{f}\|_2^2$.

As a simple application of Eq. (3.4.7), consider the evaluation of $\sum_{k=0}^{\infty} f^2(k)$ using the right-hand side of Eq. (3.4.7) for a sequence **f** given by

$$f(k) = a^k \xi(k), \qquad |a| < 1.$$

Then

$$F(z) = \frac{1}{1 - az^{-1}}, \qquad |z| > |a|$$

and so

$$F(z)F(z^{-1}) = \frac{1}{(1 - az^{-1})(1 - az)}, \qquad |a| < |z| < \frac{1}{|a|}.$$

Using residues to evaluate

$$\frac{1}{2\pi j}\oint_{\Gamma} \frac{z^{-1}}{(1 - az^{-1})(1 - az)}\,dz$$

around $|z| = 1$, we find one enclosed pole at $z = a$ with residue $1/(1 - a^2)$. And so

$$\sum_{k=0}^{\infty} f^2(k) = \sum_{k=0}^{\infty} (a^k)^2 = \frac{1}{1 - a^2}.$$

(In this example direct evaluation of $\sum_{k=0}^{\infty} f^2(k)$ is probably easier.)

3.5 INVERSION OF z-TRANSFORMS, CAUSALITY, AND BIBO STABILITY

In Section 2.6 we discussed the solution to the SDE. We obtained a unique solution by assuming the solution was causal and formed an initial value problem with initial conditions determined by the initial state of the internal registers (unit delay elements). There are other ways to ensure unique solutions to an SDE, as this section discusses. If we take an SDE of the form (2.6.8) and take z-transforms of both sides, we obtain an expression of the form

$$Y(z) + \sum_{i=1}^{u} a_i Y(z)z^{-i} = \sum_{i=0}^{n} b_i U(z)z^{-i}. \tag{3.5.1}$$

Suppose that the input sequence $\mathbf{u}$ is the unit-pulse sequence δ so that $U(z) = 1$. In this case the output $Y(z)$ is the transfer function $H(z)$ so that from Eq. (3.5.1) we can write

$$H(z) = \sum_{i=0}^{n} b_i z^{-i} \bigg/ \left(1 + \sum_{i=1}^{u} a_i z^{-i}\right). \tag{3.5.2}$$

To find the unit-pulse response sequence $\mathbf{h}$, we must invert the z-transform $H(z)$. Factoring the numerator and denominator polynomials of Eq. (3.5.2), we obtain

$$H(z) = g \frac{\prod_{i=1}^{n} (z - z_i)}{\prod_{i=1}^{n} (z - p_i)}, \tag{3.5.3}$$

where the p_i are roots of the polynomial $1 + \sum_{i=1}^{n} a_i z^{-1} = \hat{a}(z)$. We now perform a partial fraction expansion of Eq. (3.5.3). Once we have obtained a partial fraction expansion, we can easily invert terms of the form $c_i z/(z - p_i)$. Notice, however, without more information we have two choices for the sequence that corresponds to the z transform $c_i z/(z - p_i)$. Thus we are immediately confronted with a nonunique solution to Eq. (3.5.1). We can remove the ambiguity in several ways.

If we require a *causal output sequence*, then all the terms in the partial fraction expansion of $H(z)$ must be expanded in negative powers of z. The z-transform $c_i z/(z - p_i)$ is thus inverted according to Eq. (3.5.4).

$$\frac{c_i z}{z - p_i} \leftrightarrow c_i p_i^k \xi(k). \tag{3.5.4}$$

Since causality is required of the output sequence, stability of the system is ensured only if $|p_i| < 1$, $i = 1, 2, \ldots, n$.

We can also remove the ambiguity in the inversion of $c_i/(z - p_i)$, $i = 1, 2, \ldots, n$ by requiring *stability* of the system. In this case, each of the sequences resulting from the inversion of the term $c_i z/(z - p_i)$ must result in a sequence that decays to zero as $k \to \pm\infty$. There are two cases:

1. If $|p_i| < 1$, then invert $c_i z/(z - p_i)$ as in Eq. (3.5.4). As $k \to \infty$, the sequence $c_i p_i^k \xi(k) \to 0$.
2. If $|p_i| > 1$, then invert $c_i z/(z - p_i)$ according to Eq. (3.5.5) (see Table 3.3.1):

$$\frac{c_i z}{z - p_i} \leftrightarrow -p_i^k c_i \xi(-k - 1) \tag{3.5.5}$$

The sequence $-p_i^k c_i \xi(-k - 1)$ is zero for $k \geqslant 0$. As $k \to -\infty$, the sequence $-p_i^k c_i \xi(-k - 1) \to 0$.

Thus we can remove the ambiguity in the SDE by requiring BIBO system stability or causality of the output sequence. Of course, if the ROC of the z-transform is specified, this also removes any ambiguity in the solution to Eq. (3.5.1). Generally, we restrict our solutions to causal outputs from BIBO stable systems. Notice a system pole outside $|z| = 1$ does *not necessarily* imply an unstable system. The system is unstable only if we also require causality. Also we do not allow poles on $|z| = 1$ in order to ensure that the unit-pulse response sequence is absolutely summable, i.e., $\mathbf{h} \in l_1$.

3.6 FREQUENCY RESPONSE AND $H(z)$

In Chapter 2 we found the frequency response of a digital filter in terms of the unit-pulse response sequence **h** or from the difference equation as

$$H(e^{j\theta}) = \sum_k h(k)e^{-jk\theta} = \sum_{i=0}^{n} b_i e^{-ji\theta} \Big/ \left(1 + \sum_{i=1}^{n} a_i e^{-ji\theta}\right), \tag{3.6.1}$$

where

$$y(k) = \sum_{i=0}^{n} b_i u(k-i) - \sum_{i=1}^{n} a_i y(k-i).$$

Comparing Eq. (3.6.1) with the z-transform of the unit-pulse response sequence, we see that if the z-transform is evaluated on $z = 1$, we have

$$H(e^{j\theta}) = H(z)|_{z=e^{j\theta}} = \sum_k h(k)e^{-jk\theta}. \tag{3.6.2}$$

The frequency response function $H(e^{j\theta})$ is the transfer function evaluated around the unit circle in the z-plane.

One of the useful interpretations used in conjunction with transfer functions is the concept of poles and zeros of a system. An intuitive way of displaying a system's transfer function is to plot the system's poles and zeros. With the additional assumption that the system is causal, the poles and zeros of the system uniquely define (to within a gain constant) the operation of the system. An excellent example of this simple, intuitive insight into the character of a discrete-time system is the graphical determination of the frequency response function as discussed in Section 2.5.

EXAMPLE 3.6.1

Calculating the frequency response from $H(z)$.

In Chapter 6 several design methods for obtaining a transfer function $H(z)$ from a given set of specifications are presented. In this example we have designed a band reject filter with center frequency (normalized) of $\pi/2$ radians. The resulting transfer function is

$$H(z) = \frac{0.6298(1 + 2z^{-2} + z^{-4})}{1 + 1.1224z^{-2} + 0.3966z^{-4}}.$$

The corresponding frequency response function is

$$H(e^{j\theta}) = \frac{0.6298(1 + 2e^{-2j\theta} + e^{-4j\theta})}{1 + 1.1224e^{-2j\theta} + 0.3966e^{-4j\theta}}$$

This function can be evaluated using the algorithms given in Section 2.5. The resulting response, magnitude, and phase is shown in Fig. 3.6.1.

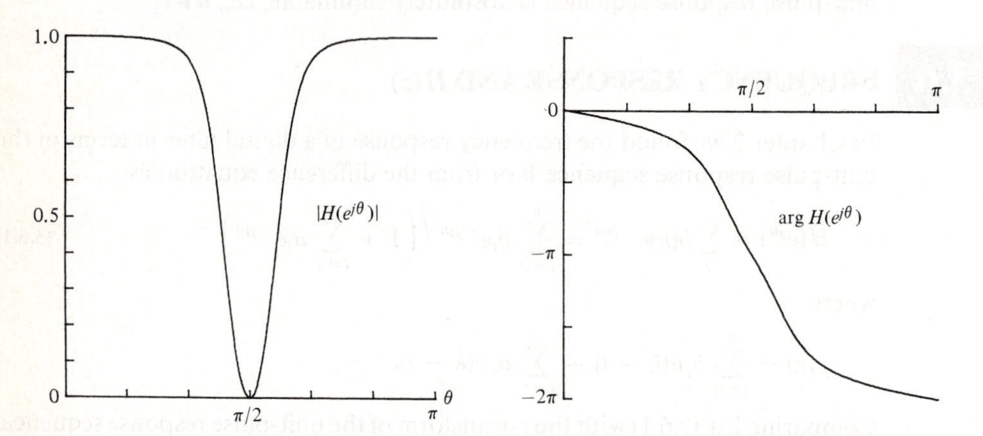

FIGURE 3.6.1 The magnitude and phase response of a band reject filter.

3.7 STRUCTURAL FORMS OF $H(z)$

Digital filters can be realized in an infinite number of ways. The freedom one has in choosing a particular realization can be used to optimize some performance criterion that is generally related to minimizing finite register effects (see Chapter 9) or some measure of hardware or computational complexity. Generally, the discussion of structures is best analyzed in terms of models that explicitly specify the structure of a filter. Our most detailed description for this purpose is the factored state variable description discussed in Chapter 8. However, many structures can be obtained directly from the transfer function $H(z)$.

Suppose the transfer function $H(z)$ is given by the ratio of two polynomials in z^{-1} in the form

$$H(z) = \frac{Y(z)}{U(z)} = \frac{\sum_{i=0}^{n} b_i z^{-i}}{1 + \sum_{i=1}^{n} a_i z^{-i}}. \tag{3.7.1}$$

The corresponding difference equation is found by cross multiplying the terms in Eq. (3.7.1). We obtain

$$Y(z)\left[1 + \sum_{i=1}^{n} a_i z^{-i}\right] = U(z) \sum_{i=0}^{n} b_i z^{-i}. \tag{3.7.2}$$

Taking the inverse-z transform on both sides yields

$$y(k) = \sum_{i=0}^{n} b_i u(k-i) - \sum_{i=1}^{n} a_i y(k-i). \tag{3.7.3}$$

[This is merely the inverse process given by Eqs. (2.6.5) and (2.6.7).] An SFG of Eq. (3.7.3) is easily obtained, as shown in Fig. 3.7.1. This SFG or structure is known as a *direct form 1*. It uses separate delays for the two sums in Eq. (3.7.3). This structure is simple to realize because the parameters of the structure are exactly the same as the parameters in the transfer function $H(z)$ or the difference equation (3.7.3).

We can reduce the number of delays in Fig. 3.7.1 to n by considering $H(z)$ a cascade of two structures. Write $H(z)$ in the form

$$H(z) = H_1(z)H_2(z) = \left[\frac{1}{1 + \sum_{i=1}^{n} a_i z^{-i}}\right]\left[\sum_{i=0}^{n} b_i z^{-i}\right]. \tag{3.7.4}$$

Denote the transform of the output $H_1(z)$ as $W(z)$, which is thus the input to $H_2(z)$. Thus

$$H_1(z) = \frac{W(z)}{U(z)} \quad \text{and} \quad H_2(z) = \frac{Y(z)}{W(z)}. \tag{3.7.5}$$

Or, in terms of difference equations, we write

$$\begin{aligned} w(k) &= u(k) - \sum_{i=1}^{n} a_i w(k-i), \\ y(k) &= \sum_{i=0}^{n} b_i w(k-i). \end{aligned} \tag{3.7.6}$$

Figure 3.7.2 depicts an SFG that has $\mathbf{u}$ as an input and $\mathbf{w}$ as an output. In Fig. 3.7.2 notice stored in the unit delays we have the values $w(k), w(k-1), \ldots, w(k-n)$. Thus to form the output $y(k)$, using Eq. (3.7.6), we merely have to form the sum $\sum_{i=0}^{n} b_i w(k-i)$ as shown in Fig. 3.7.3. This is called a direct form 2 structure. It has the minimum number of multipliers and delay elements for a transfer function given by Eq. (3.7.1). (This structure is also known as the

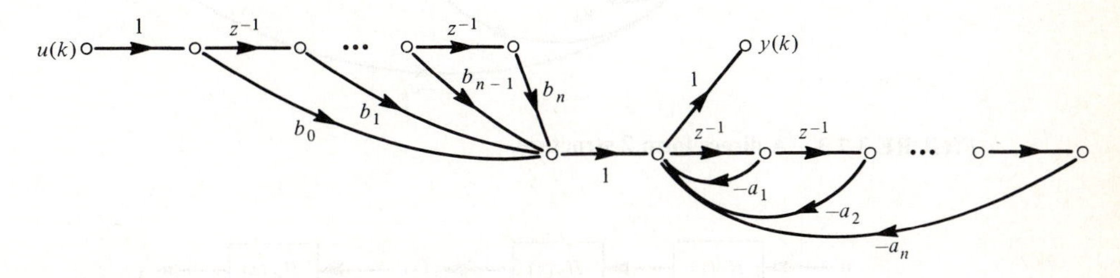

FIGURE 3.7.1 A direct form 1 SFG of Eq. (3.7.1).

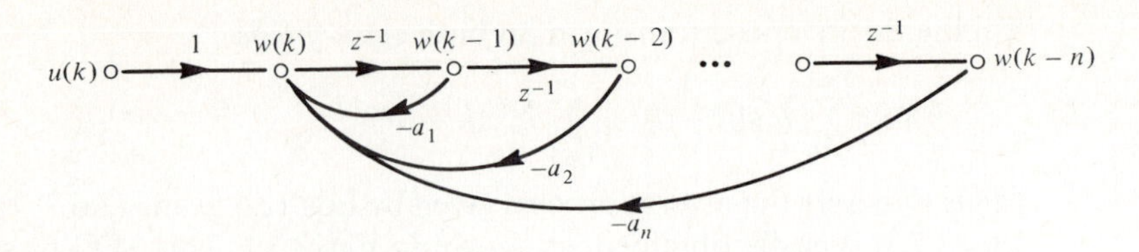

FIGURE 3.7.2 Generation of w.

canonic form digital filter. More on these structures can be found in Chapters 9 and 10.) Direct form structures are not generally realized as in Fig. 3.7.3 because of finite register effects. Instead $H(z)$ in Eq. (3.7.1) is broken into first and second-order sections that are then placed in cascade or parallel connection to realize $H(z)$.

For example, suppose we write $H(z)$ in the following form.

$$H(z) = g \prod_{i=1}^{n} (z - z_i) \Big/ \prod_{i=1}^{n} (z - p_i) = g \prod_{i=1}^{K} H_i(z), \tag{3.7.7}$$

where $H_i(z)$ is either a second- or first-order filter section. That is,

$$H_i(z) = \frac{1 + b_{1i}z^{-1} + b_{2i}z^{-2}}{1 + a_{1i}z^{-1} + a_{2i}z^{-2}} \tag{3.7.8}$$

or

$$H_i(z) = \frac{1 + b_{1i}z^{-1}}{1 + a_{1i}z^{-1}}. \tag{3.7.9}$$

Then Eq. (3.7.7) describes a cascade of first- and second-order sections as depicted in Fig. 3.7.4.

Notice that Eq. (3.7.7) specifies only the gross structure of the filter. There

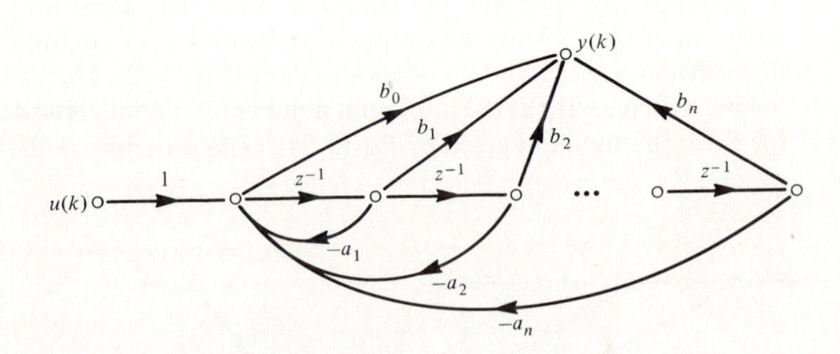

FIGURE 3.7.3 A direct form 2 structure.

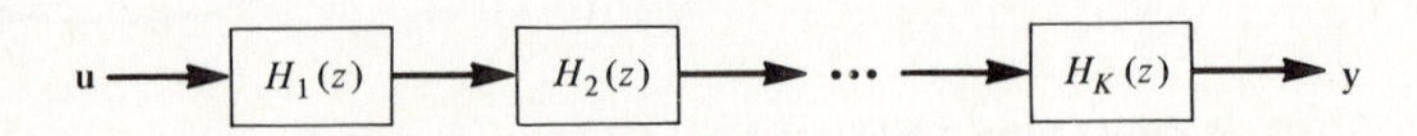

FIGURE 3.7.4 A cascade structure.

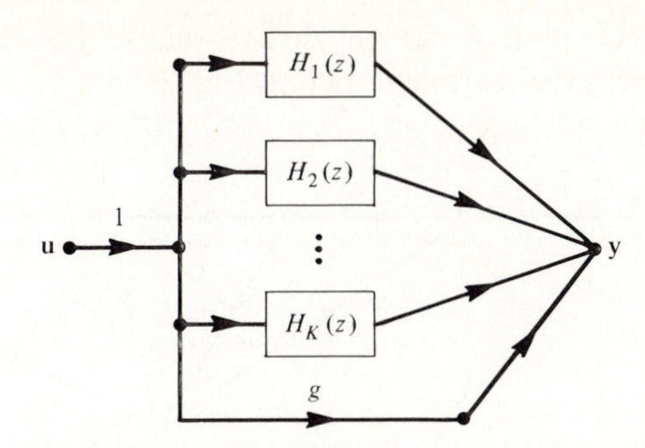

FIGURE 3.7.5 **A parallel structure of first and second order sections.**

are still many variations within the cascade structure constraint. For example, the ordering of the individual sections will affect the finite register performance of the filter. Also the pairing of poles and zeros within the individual sections is important in practical implementations. These ideas are discussed in more detail in Chapter 9.

Another filter structure can be obtained by expressing $H(z)$ in terms of a partial fraction expansion. Suppose we write $H(z)$ in the form

$$H(z) = g + \sum_{i=1}^{K} H_i(z), \tag{3.7.10}$$

where $H_i(z)$ are of the form

$$H_i(z) = \frac{b_{1i}z^{-1} + b_{2i}z^{-2}}{1 + a_{1i}z^{-1} + a_{2i}z^{-2}} \tag{3.7.11}$$

or

$$H_i(z) = \frac{b_{1i}z^{-1}}{1 + a_{1i}z^{-1}}. \tag{3.7.12}$$

This realization is called a parallel form and is sketched in Fig. 3.7.5. In both cascade and parallel forms the realization of the individual sections can be accomplished in a variety of ways. (See Chapter 9.)

These two filter structures are examples of realizations obtainable by factoring the transfer function $H(z)$. This method of obtaining new structures is limited in its applications. We shall discuss more general methods of obtaining new structures in Chapters 9 and 10.

EXAMPLE 3.7.1

Cascade realization of an FIR filter.

Consider the class of FIR filters that have transfer functions of the form given in Eq. (3.7.13):

$$H(z) = \sum_{k=0}^{N-1} h(k)z^{-k}. \tag{3.7.13}$$

The most common structure used to realize Eq. (3.7.13) is the tapped-delay or direct form realization shown in Fig. 3.7.6.

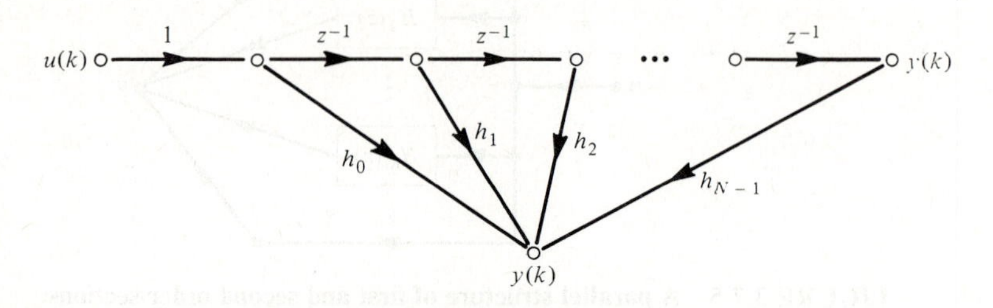

FIGURE 3.7.6 Direct form structure for a FIR filter.

We could also realize Eq. (3.7.13) in terms of a cascade of first- and second-order sections by expressing $H(z)$ in the form

$$H(z) = \prod_{i=1}^{K} H_i(z), \tag{3.7.14}$$

where

$$H_i(z) = b_{0i} + b_{1i}z^{-1} + b_{2i}z^{-2} \tag{3.7.15}$$

or

$$H_i(z) = b_{0i} + b_{1i}z^{-1}. \tag{3.7.16}$$

A network corresponding to Eq. (3.7.14) is shown in Fig. 3.7.7.

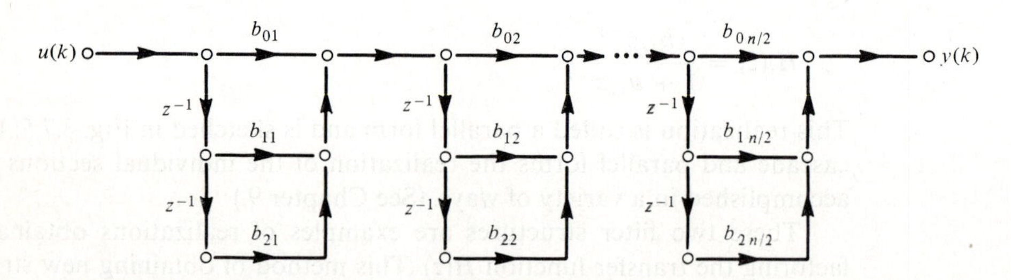

FIGURE 3.7.7 A cascade-form structure for an FIR filter.

FIR Lattice Structures

Another very useful structure for realizing FIR filters is the so-called ladder or lattice structure (Gray 1973). This structure has found wide application in speech synthesis and spectral estimation. One method of obtaining an FIR lattice for a given transfer function $H(z)$ is to use the following recursive method. We shall use a simple example to demonstrate the method. Define the nth-order transfer function for which we wish to obtain a lattice realization as $H_n(z)$. Suppose we have the second-order filter $H_2(z) = 1 - \frac{3}{4}z^{-1} + \frac{1}{8}z^{-2}$ as shown in direct form in Fig. 3.7.8. Define two polynomials in z as

$$p_k(z) = z^k H_k(z), \tag{3.7.17}$$

$$\tilde{p}_k(z) = z^k p_k(z^{-1}). \tag{3.7.18}$$

In this example we have

$$p_2(z) = z^2 - \tfrac{3}{4}z + \tfrac{1}{8},$$

$$\tilde{p}_2(z) = 1 - \tfrac{3}{4}z + \tfrac{1}{8}z^2.$$

The kth-order FIR lattice structure is shown in Fig. 3.7.9. This structure is defined by the parameters $\{\eta_0 = 1, \eta_1, \ldots, \eta_k\}$, which are obtained from the polynomial $p_k(z)$ evaluated at $z = 0$.

By recursively finding the transfer function $H_k(z)$ as a new parameter η_k is added, we can express the new transfer function $H_k(z)$ in terms of the previous transfer function $H_{k-1}(z)$. This procedure is most easily carried out in terms of the polynomials $p_k(z)$ and $\tilde{p}_k(z)$. The recursion in terms of these polynomials is given by

$$\begin{bmatrix} p_{k-1}(z) \\ \tilde{p}_{k-1}(z) \end{bmatrix} = \frac{1}{1 - \eta_k^2} \begin{bmatrix} z^{-1} & -\eta_k z^{-1} \\ -\eta_k & 1 \end{bmatrix} \begin{bmatrix} p_k(z) \\ \tilde{p}_k(z) \end{bmatrix}. \tag{3.7.19}$$

Thus in our numerical example we have

$$\eta_2 = p_2(0) = \tfrac{1}{8}.$$

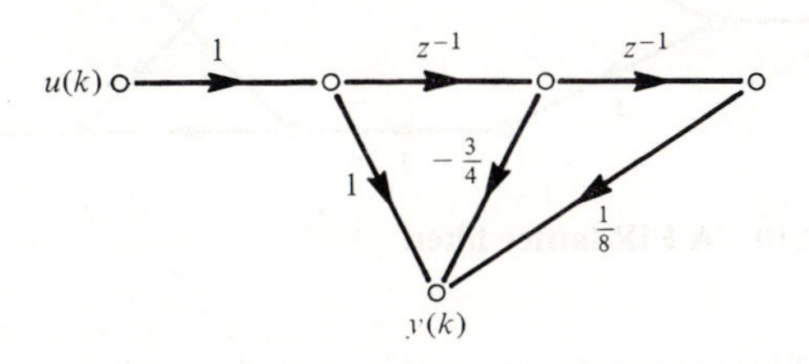

FIGURE 3.7.8 A second-order FIR filter.

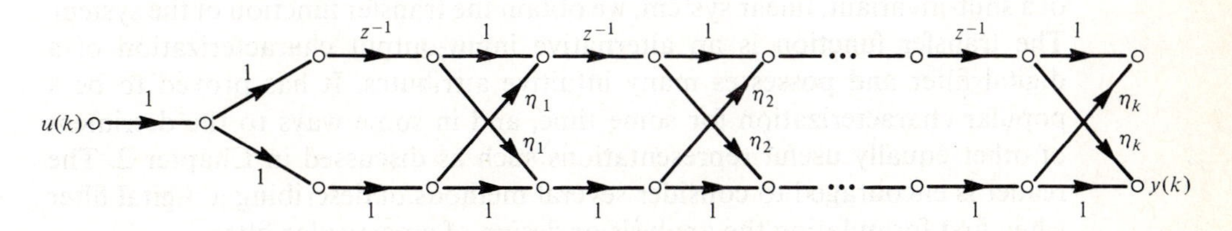

FIGURE 3.7.9 A FIR n^{th}-order lattice structure.

Using Eq. (3.7.19), we obtain $p_1(z)$ and $\tilde{p}_1(z)$ as

$$\begin{bmatrix} p_1(z) \\ \tilde{p}_1(z) \end{bmatrix} = \frac{1}{1-(\frac{1}{8})^2} \begin{bmatrix} z^{-1} & -\frac{1}{8}z^{-1} \\ -\frac{1}{8} & 1 \end{bmatrix} \begin{bmatrix} z^2 - \frac{3}{4}z + \frac{1}{8} \\ 1 - \frac{3}{4}z + \frac{1}{8}z^2 \end{bmatrix}.$$

After some algebra, we obtain

$$p_1(z) = z - \tfrac{2}{3},$$
$$\tilde{p}_1(z) = 1 - \tfrac{2}{3}z.$$

Then we can obtain η_1 from $p_1(z)$ as

$$\eta_1 = p_1(0) = -\tfrac{2}{3}.$$

And as a check we find

$$\begin{bmatrix} p_0(z) \\ \tilde{p}_0(z) \end{bmatrix} = \frac{1}{1-(\frac{2}{3})^2} \begin{bmatrix} z^{-1} & \frac{2}{3}z^{-1} \\ \frac{2}{3} & 1 \end{bmatrix} \begin{bmatrix} z - \frac{2}{3} \\ 1 - \frac{2}{3}z \end{bmatrix} = \begin{bmatrix} 1 \\ 1 \end{bmatrix}.$$

The lattice corresponding to the direct form of Fig. 3.7.8 is shown below in Fig. 3.7.10.

As a verification that this lattice corresponds to the filter with transfer function $H_2(z) = 1 - \frac{3}{4}z^{-1} + \frac{1}{8}z^{-2}$, we can compute the unit-pulse response sequence of the lattice structure. The result is $\{1, \eta_1 + \eta_1\eta_2, \eta_2\} = \{1, -\frac{3}{4}, \frac{1}{8}\}$, which is, of course, identical to unit-pulse response sequence obtained from the structure in Fig. 3.7.8.

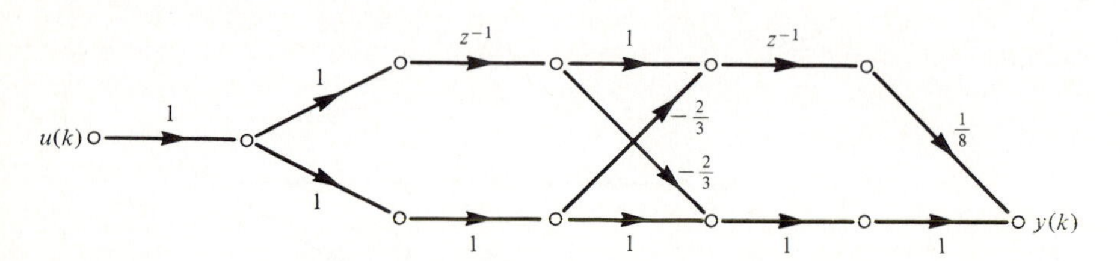

FIGURE 3.7.10 A FIR lattice filter.

Summary • The z-transform is the transform domain representation appropriate for linear, shift-invariant, discrete-time systems. One of the advantages of transform domain analysis is that the convolution process is transformed into a multiplication process. By transforming the unit-pulse response sequence of a shift-invariant, linear system, we obtain the transfer function of the system. The transfer function is an alternative input-output characterization of a digital filter and possesses many intuitive attributes. It has proved to be a popular characterization for some time, and in some ways to the detriment of other equally useful representations such as discussed in Chapter 2. The reader is encouraged to consider several methods of describing a digital filter when first formulating the analysis or design of a particular filter.

PROBLEMS

3.1. Find the transfer function $H(z)$ for each of the following unit-pulse reponse sequences **h**.

a) $h(k) = (\frac{1}{2})^k \xi(k)$
b) $h(k) = (\frac{1}{2})^k \xi(-k-1)$
c) $h(k) = [(\frac{1}{2})^k + 2(\frac{1}{3})^k]\xi(k)$
d) $h(k) = \cos\left(\frac{k\pi}{8}\right)\xi(k)$

3.2. Find the causal, unit-pulse response sequence **h** for each of the following transfer functions $H(z)$.

a) $H(z) = 1/(z - \frac{1}{4})$
b) $H(z) = 1/(z - \frac{1}{2})(z - \frac{1}{4})$
c) $H(z) = z^{-2}/(z + \frac{1}{2})$
d) $H(z) = z^{-1}/(z^2 + \frac{1}{2})$

3.3. Repeat Problem 3.2 but find the anticausal unit-pulse response sequences **h**.

3.4. Find the BIBO stable, unit-pulse responses for each of the following transfer functions $H(z)$.

a) $H(z) = 1/(z - \frac{1}{2})(z + \frac{1}{4})$
b) $H(z) = 1/(z - \frac{1}{2})(z^2 + \frac{3}{8}z + \frac{1}{32})$

3.5. Solve the following difference equations using z-transforms. Assume zero initial conditions unless otherwise specified.

a) $y(k) - 2y(k-1) = (\frac{1}{2})^k \xi(k), \qquad y(-1) = 1$
b) $y(k) - 2y(k-1) + y(k-2) = \xi(k)$
c) $y(k) - \frac{1}{4}y(k-2) = (\frac{1}{2})^k \xi(k); \qquad y(-1) = y(-2) = 1$
d) $y(k) - \frac{1}{4}y(k-2) = \cos\left(\frac{k\pi}{2}\right)\xi(k)$

3.6. Sketch an SFG for each of the systems in the previous problem.

3.7. Find the z-transform of each of the following sequences by using properties of the transform and a table of z-transforms.

a) $\alpha^k \xi(k) * \alpha^k \xi(k)$
b) $\alpha^k \xi(k) * \alpha^k \xi(k) * \cdots * \alpha^k \xi(k)$, m terms
c) $\alpha^{-k} \xi(-k)$
d) $\alpha^k \cos(k\theta)\xi(k)$
e) $k \cos(k\theta)\xi(k)$
f) $k\alpha^k \cos(k\theta)\xi(k)$

3.8. Find the inverse z-transform for the following:

a) $F(z) = 1/(z - \frac{1}{2})(z - \frac{1}{4}), \qquad |z| < \frac{1}{4}$
b) $F(z) = 1/(z - \frac{1}{2})(z - \frac{1}{4}), \qquad \frac{1}{4} < |z| < \frac{1}{2}$

c) $F(z) = 1/(z - \frac{1}{2})(z - \frac{1}{4})$, $|z| > \frac{1}{2}$
d) $F(z) = 1/(z - \frac{1}{2})(z - 4)$, $\mathbf{f} \in l_2$
e) $F(z) = -z/(z - \frac{1}{2})^2(z - 4)$, $\mathbf{f} \in l_2$

3.9. The *deterministic autocorrelation sequence* corresponding to a sequence $\mathbf{x}$ is defined as

$$r_x(l) = \sum_{m=-\infty}^{\infty} x(m)x(m+l)$$

a) Express $r_x(l)$ as the convolution of two sequences.
b) Find the z-transform of $r_x(l)$ in terms of the z-transform of $\mathbf{x}$.
c) Suppose $x(k) = \alpha^k \xi(k)$, $|\alpha| < 1$. Find $r_x(l)$ and $Z\{r_x(l)\}$.
d) If $R_x(z) = Z\{r_x(l)\} = \dfrac{1}{(l - \beta z^{-1})(1 - \beta z)}$, then find $r_x(l)$. Assume $|\beta| < 1$.

3.10. If three filters with the unit-pulse response $h(k) = \alpha^k \xi(k)$ are cascaded, find the unit-pulse response for the cascaded system.

3.11. Consider the second-order all-pole filter shown in the SFG with $|\alpha| < 1$. Find the step-response of the filter, i.e., the response to $\mathbf{u} = \boldsymbol{\xi}$. What is the steady-state response (the response for $n \to \infty$)?

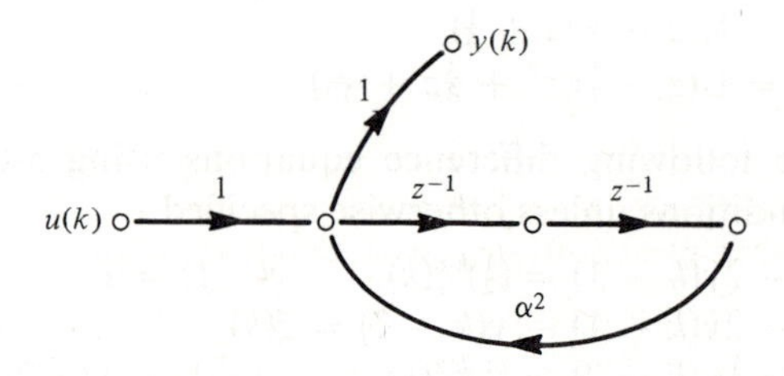

FIGURE 3P.11.

3.12. Using the step response, find the causal unit-pulse response for the previous problem. Check your result by calculating $\mathbf{h}$ from $H(z)$ or the SDE for the filter.

3.13. Design a digital low-pass filter by trial and error using a graphical interpretation of frequency response. Use no more than 4 complex poles and 4 complex zeros. Assume the cutoff frequency of the filter is 100 Hz with a sampling frequency of 2000 Hz. Obtain an attenuation of at least 40 dB in the stop band of the filter. The width between the passband and the stopband is called the transition width of the filter. Try to make this transition width as narrow as you can.

3.14. The transfer function of a digital filter is

$$H(z) = \frac{(z+1)^3}{2z(3z^2+1)}.$$

Find a cascade and parallel decomposition in terms of

a) first-order filter sections
b) second- and first-order filter sections.

Is there any advantage in choosing a) or b)?

3.15. The transfer function of an FIR filter is $H(z) = 1 + (\frac{1}{4})z^{-1} + (\frac{3}{4})z^{-2}$. Find a lattice realization of this filter.

3.16. Repeat the previous problem for the following FIR filter, $H(z) = 4 + (\frac{1}{2})z^{-1} + 2z^{-2} + (\frac{1}{3})z^{-3}$.

3.17. An important calculation in digital filters is the calculation of integrals of the form

$$\frac{1}{2\pi j}\oint_{|z|=1} H(z)H(z^{-1})\frac{dz}{z} \quad \text{or} \quad \frac{1}{2\pi j}\oint_{|z|=1} H(z)G(z^{-1})\frac{dz}{z}$$

By Parseval's theorem, these integrals can also be calculated using $\sum_k h^2(k)$ or $\sum_k g(k)h(k)$. (See Section 4.3.) Suppose

$$H(z) = \frac{\hat{b}(z)}{\hat{a}(z)} = \frac{b_0(z+1)^2}{z^2 + \frac{1}{4}}.$$

Find $\sum_k h^2(k)$ using two methods. First calculate $\sum_k h^2(k)$ by evaluating the integral

$$\frac{1}{2\pi j}\oint H(z)H(z^{-1})\frac{dz}{z}$$

using residues. Second, calculate $\sum_k h^2(k)$ directly by first obtaining $h(k)$ from $H(z)$. Which method of calculation do you prefer?

3.18. A stable, causal filter $H(z)$ is *minimum phase* if $H(\lambda) = 0$ implies $|\lambda| < 1$. A minimum phase filter has the least phase lag among all causal filters with the

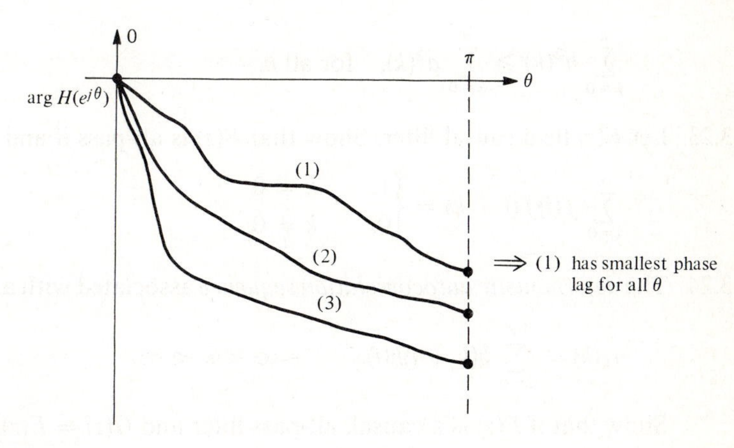

FIGURE 3P.18.

same magnitude response. (See Fig. 3P.18.) Let

$$H(z) = F(z)(1 + \alpha z^{-1}),\ G(z) = F(z)(\alpha + z^{-1}),$$

where $0 < \alpha < 1$.

a) Show that $|H(e^{j\theta})| = |G(e^{j\theta})|$.
b) Which filter, H or G, has the smallest phase lag? [Hint: Sketch the phase responses of $(z^{-1} + \alpha)$ and $(1 + z^{-1}\alpha)$.]

3.19. *All-pass filters* can be interpreted as generalized delay elements. An all-pass filter $F(z)$ is characterized by $|F(e^{j\theta})| = 1$ for all θ. Show that the following is a representation for all-pass filters:

$$F(z) = \pm \prod_{i=1}^{m} \left[\frac{z^{-1} - \lambda_i^*}{1 - \lambda_i z^{-1}}\right], \qquad |\lambda_i| < 1.$$

3.20. A perfect delay element, $F(z) = z^{-m}$, has amplitude response unity and a linear phase response, i.e., $\arg F(e^{j\theta}) = \arg e^{-j\theta m} = -m\theta$. The *group delay* is the negative derivative of the phase response,

$$-\frac{d}{d\theta}(-\theta m) = +m.$$

One definition of *average delay* is $\sum_{k=0}^{\infty} kh^2(k)/\sum_{k=0}^{\infty} h^2(k)$, where $h(k)$ is the unit-pulse response of the filter. Show that for an mth-order all-pass filter (see Problem 3.19) the average delay is m.

3.21. Let $H(z)$ be a causal minimum phase filter. Let $F(z)$ be the order m all-pass filter of problem 3.19, and let $G(z) = F(z)H(z)$. Show that the average delay of $G(z)$ is greater than that of $H(z)$.

3.22. $H(z)$ and $G(z)$ are stable, causal filters and $|H| = |G|$. $H(z)$ is minimum phase. Show that

$$\sum_{k=0}^{n} h^2(k) \geqslant \sum_{k=0}^{n} g^2(k), \quad \text{for all } n.$$

3.23. Let $F(z)$ be a causal filter. Show that $F(z)$ is all-pass if and only if

$$\sum_{i=0}^{\infty} f(i)f(i+k) = \begin{cases} 1, & k = 0 \\ 0, & k \neq 0 \end{cases}.$$

3.24. The *deterministic autocorrelation sequence* associated with a causal filter $H(z)$ is

$$r_h(k) = \sum_{i=0}^{\infty} h(k+i)h(i), \qquad -\infty < k < \infty.$$

Show that if $F(z)$ is a causal, all-pass filter and $G(z) = F(z)H(z)$, then

$$r_g(k) = r_h(k) \quad \text{for all } k.$$

3.25. This problem uses the results of problems 3.23 and 3.24. Let $G(z)$ be a given stable, causal filter. Let $H_n(z)$ be an order n finite impulse response filter

$$H_n(z) = \sum_{k=0}^{n} h_n(k)z^{-k},$$

and let $F_n(z) = H_n(z)G(z)$. The *least-squares inverse* problem is to choose $H_n(z)$ to minimize

$$\alpha_n = \sum_{k=0}^{\infty} [\delta(k) - f_n(k)]^2 = \|\delta - \mathbf{h}_n * \mathbf{g}\|^2.$$

In other words, $H_n(z)$ should approximate $G^{-1}(z)$.

a) Show that $H_n(z)$ satisfies the "normal equations"

$$\sum_{i=0}^{n} h_n(i)r_g(j - i) = \begin{cases} g(0), & j = 0 \\ 0, & 1 \leqslant j \leqslant n \end{cases}.$$

b) What are $F_n(z)$, $H_n(z)$ and α_n when $G(z)$ is all-pass?

c) Show that if $G(z)$ is minimum phase, then α_n approaches zero as n approaches infinity.

d) Show that if $G(z)$ is not minimum phase, then $F_n(z)$ tends to an all-pass filter as n approaches infinity.

3.26. Suppose $H(z)$ and $G(z)$ are both causal, stable, minimum phase filters. Which of the following are also minimum phase?

a) $H^{-1}(z)$
b) $H(z)G(z)$
c) $H(z)/G(z)$
d) $H(z) + G(z)$
e) $H(G(z))$

3.27. Suppose $H(z)$ and $G(z)$ are both causal, stable, all-pass filters. Which of the following are also all-pass filters? Which are causal?

a) $H^{-1}(z)$
b) $H(z)G(z)$
c) $H(z)/G(z)$
d) $H(z) + G(z)$
e) $H(G(z))$

3.28. An nth-order Butterworth filter has the frequency response function

$$|H(e^{j\theta})|^2 = \frac{1}{1 + \left(\dfrac{1 - \cos\theta}{1 + \cos\theta}\right)^n}.$$

a) Sketch this function from 0 to π for large n.

b) Find the poles and zeros of $H(z)$, assuming stability.

c) Compute, in closed form,

$$\sum_{k=0}^{\infty} kh^2(k)$$

as a function of n.

d) Show that the average delay is approximately $n/2$. (See Problem 3.20.)

3.29. The following problem is called the *polynomial spectral factorization problem* and will be used in Chapter 7. Suppose

$$Q(z) = \sum_{k=-n}^{n} q(k)z^{-k}$$

has real coefficients satisfying $q(k) = q(-k)$ for $1 \leqslant k \leqslant n$, and $Q(e^{j\theta})$ is real and positive for all θ. Prove that there exists a polynomial

$$\hat{a}(z) = \sum_{k=0}^{n} a_k z^{-k}$$

satisfying $\hat{a}(z)\hat{a}(z^{-1}) = Q(z)$ for all z, and having roots with $|\lambda| < 1$.

CHAPTER 4

Fourier Analysis of Discrete-Time Signals and Systems

4.1 INTRODUCTION

Almost every branch of engineering and science uses Fourier methods. The words "frequency," "period," "phase," and "spectrum" are important parts of an engineer's vocabulary. The basic idea, the decomposition of signals into orthogonal trigonometric basis functions, is a natural and powerful tool which is used in a vast number of applications. We call such tools *Fourier methods* because of the ideas of Joseph Fourier (1768–1830) and his determination to get them accepted, but the origins of Fourier analysis predate Fourier. The Fourier series inversion formula was known to Leonard Euler in 1777. There is even evidence to the effect that the first "fast Fourier transform" was invented by Carl Friedrich Gauss in 1805, two years before Fourier's first (and unsuccessful) attempt at publication (Heidelman, Johnson, and Burrus 1984).

In Chapter 2, the discrete time Fourier transform (DTFT) was introduced because it arises naturally as the frequency response function of a digital filter. The input-output relation

$$\mathbf{y} = \mathbf{h} * \mathbf{u} \tag{4.1.1}$$

which relates the input, output, and unit pulse response sequences becomes

$$Y(e^{j\theta}) = H(e^{j\theta})U(e^{j\theta}) \tag{4.1.2}$$

after the DTFT is applied. Thus the apparently complicated convolution operation has been transformed into multiplication.

It might seem that we should be content with studying only the DTFT, since digital signal processing must necessarily deal with discrete-time signals. Although it is true that the DTFT is our basic tool, it must be understood in context: many discrete-time signals which exist in a digital processor began life as continuous-time signals. For such signals the appropriate transform is not the DTFT.

Figure 4.1.1 exhibits four separate Fourier transforms, each of which is completely self-contained and appropriate to its own class of signals. There are

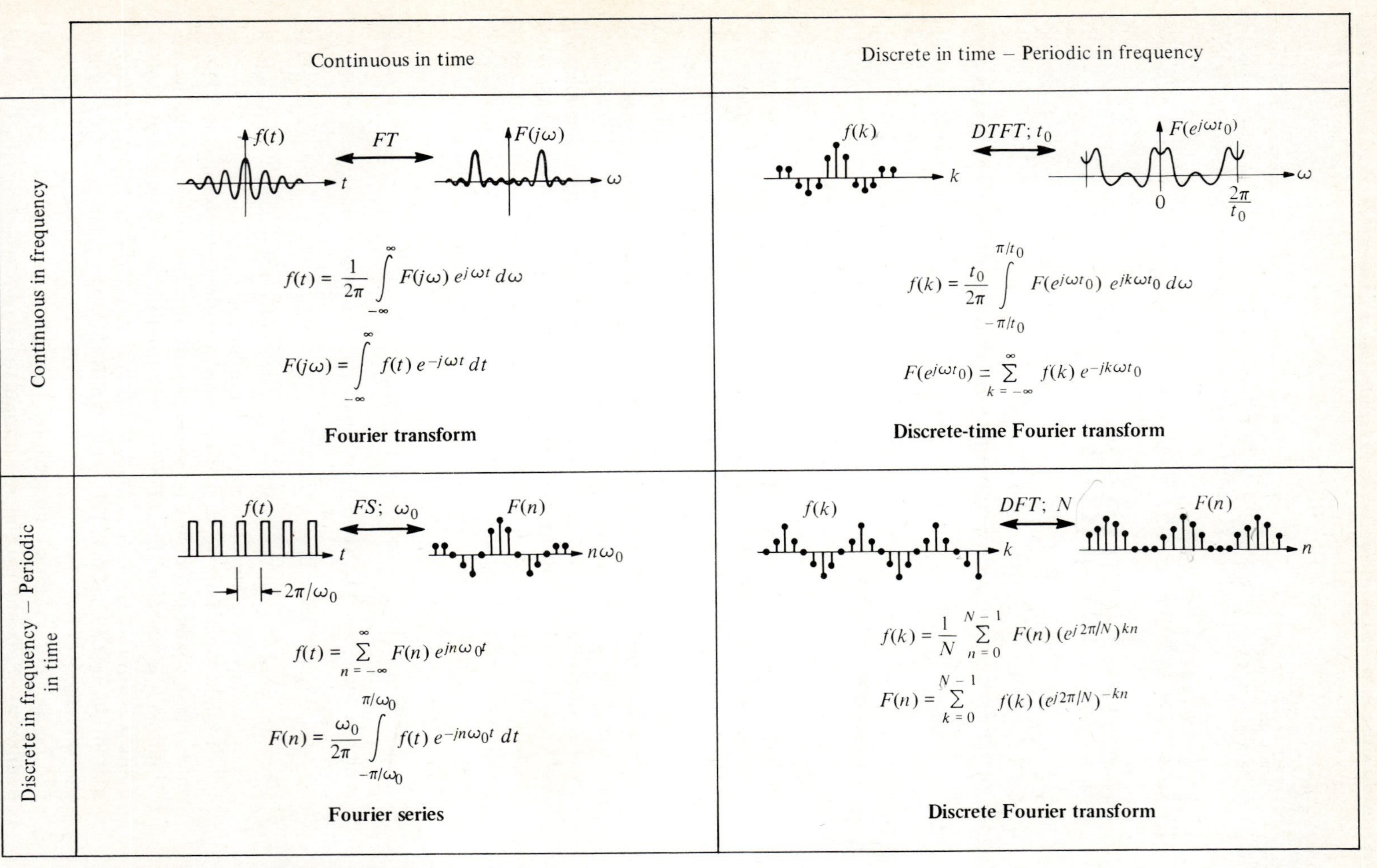

FIGURE 4.1.1 Four classes of Fourier transforms.

only two questions that one need ask in order to decide which transform is appropriate. Is the signal continuous in time, or discrete? And is the signal periodic? The four combinations of yes and no provide a characterization of the appropriate transform.

If one attempts to use the wrong transform for a given signal, he may well succeed in getting an answer, since the four transforms have similar properties. But this usually leads to some rather unnatural signal modeling which can sometimes be deceptive. We shall see that the transforms themselves are not difficult to understand and to use. But problems such as sampling, aliasing, and the relation of the fast Fourier transform to the true Fourier transform generally involve signals from separate classes, and here one must be careful. Sections 4.6, 4.7, and 4.8 deal with these issues.

We shall use the following notation. If $\alpha + j\beta$ is a complex number, then $(\alpha + j\beta)^* = \alpha - j\beta$ is its complex conjugate. Matrices and vectors are given boldface symbols. If $\mathbf{A}$ is a matrix, then $\mathbf{A}^T$ is the "transpose" of $\mathbf{A}$ having elements

$$(\mathbf{A}^T)_{ij} = \mathbf{A}_{ji}. \tag{4.1.3}$$

The complex conjugate transpose of a matrix $\mathbf{A}$ with complex elements is $\mathbf{A}^*$, where

$$(\mathbf{A}^*)_{ij} = (\mathbf{A}_{ji})^*. \tag{4.1.4}$$

The functions *sinc* and *rect* are defined by

$$\operatorname{sinc}(x) = \begin{cases} 1, & \text{if } x = 0 \\ \sin(x)/x, & \text{otherwise} \end{cases}. \tag{4.1.5}$$

$$\operatorname{rect}(x) = \begin{cases} 1, & \text{if } |x| \leqslant 1 \\ 0, & \text{otherwise} \end{cases}. \tag{4.1.6}$$

4.2 THE DISCRETE-TIME FOURIER TRANSFORM

If a signal is discrete in time and has finite energy, then the appropriate Fourier transform is the DTFT. The *energy* of the discrete-time signal $\mathbf{f}$ is

$$\|\mathbf{f}\|_2^2 = \sum_{k=-\infty}^{\infty} |f(k)|^2, \tag{4.2.1}$$

and the set of all finite energy sequences is denoted l_2. Now a finite energy sequence cannot be periodic unless it is identically zero, because the infinite sum in Eq. (4.2.1) would diverge. Thus $\mathbf{f}$ is discrete-time but not periodic, and from Fig. 4.1.1 the proper transform is the DTFT.

In this section we shall use the angle variable $\theta = \omega t_0$ in the DTFT, and will drop explict references to t_0 like the one in Table 4.1.1. In Section 4.6, it will be necessary to use t_0 as a sampling period. With this in mind, the DTFT of a signal $\mathbf{f} \in l_2$ is defined to be

$$F(e^{j\theta}) = \sum_{k=-\infty}^{\infty} f(k)e^{-jk\theta}. \tag{4.2.2}$$

We will use the shorthand

$$f(k) \quad \overset{DTFT}{\longleftrightarrow} \quad F(e^{j\theta}) \tag{4.2.3}$$

for the signal $\mathbf{f}$ and its transform, and call the two a *DTFT-pair*.

The notation $F(e^{j\theta})$ is used in preference to $F(\theta)$ in order to be consistent with the z-transform, which was defined in Chapter 3 and is

$$F(z) = \sum_{k=-\infty}^{\infty} f(k)z^{-k}. \tag{4.2.4}$$

The connection between these is that the DTFT is $F(z)$ restricted to the unit circle, *i.e.*, $z = e^{j\theta}$. This automatically forces the DTFT to be 2π-periodic as a function of θ, and we have indicated in Fig. 4.1.1 that a signal which is discrete in one domain must be periodic in the other.

Now there are some subtle differences between the z-transform and the DTFT which we will point out, even though the differences rarely lead to trouble. A z-transform has a *region of convergence* (ROC). In order for the unit circle to be part of the ROC for $F(z)$, we must have

$$\sum_k |f(k)e^{-jk\theta}| = \sum_k |f(k)| = \|\mathbf{f}\|_1 < \infty. \tag{4.2.5}$$

This means that $\mathbf{f} \in l_1$, which is the set of all sequences satisfying $\|\mathbf{f}\|_1 < \infty$. Now there are sequences which are in l_2 but not in l_1. One example is

$$f(k) = \begin{cases} 0, & k \leqslant 0 \\ 1/k, & k > 0. \end{cases} \tag{4.2.6}$$

For such sequences the Fourier transform has to be given a special interpretation, which is outside the scope of this book. For sequences $\mathbf{f}$ which *are* in l_1, we can regard the DTFT as the restriction of the z-transform to the unit circle, and $F(e^{j\theta})$ is then bounded, using Eq. (4.2.5):

$$|F(e^{j\theta})| \leqslant \sum_k |f(k)e^{-jk\theta}| = \|\mathbf{f}\|_1. \tag{4.2.7}$$

EXAMPLE 4.2.1

Using the z-transform to evaluate the DTFT.

Figure 4.2.1 displays three signals having z-transforms

$$F_1(z) = z + 1 + z^{-1},$$
$$F_2(z) = z^{-2}F_1(z),$$
$$F_3(z) = z^{-1}\left[\frac{0.9 + z^{-1}}{1 + 0.9z^{-1}}\right] F_1(z). \tag{4.2.8}$$

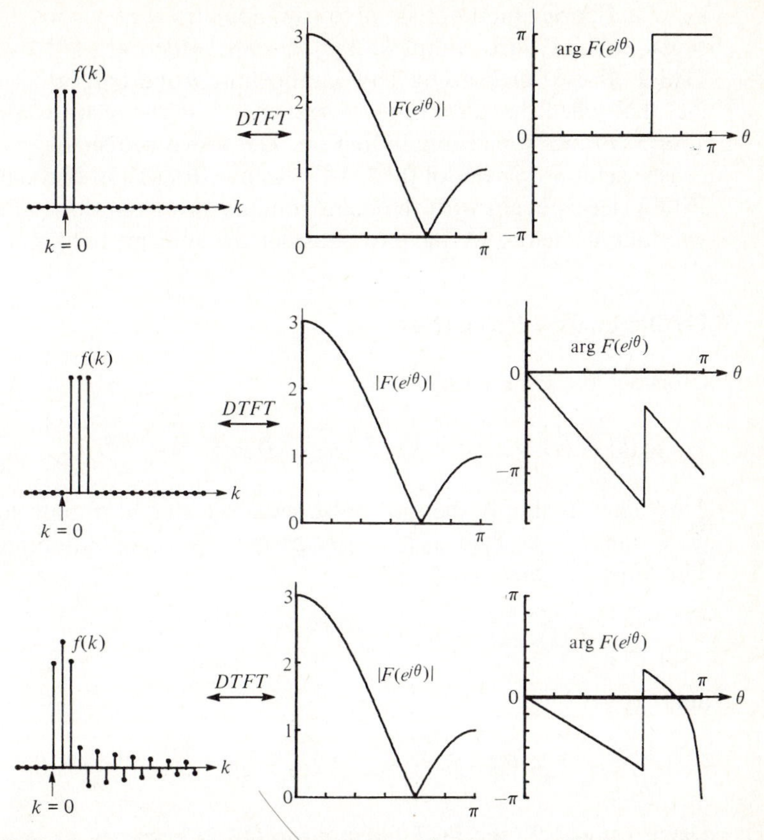

FIGURE 4.2.1 Three sequences having identical amplitude spectra. At top: $F(z) = z + 1 + z^{-1}$, Middle: $F(z) = z^{-1} + z^{-2} + z^{-3}$, Bottom: $F(z) = (0.9 + 1.9z^{-1} + 1.9z^{-2} + z^{-3})/(1. + 0.9z^{-1})$.

The DTFT can be easily numerically evaluated from these functions using the algorithm for evaluating polynomials at $z = e^{j\theta}$ given in Chapter 2. Plots of the DTFT of real sequences are often done in the format shown in the figure. If

$$f(k) \quad \overset{DTFT}{\longleftrightarrow} \quad F(e^{j\theta}),$$

then the function $|F(e^{j\theta})|$ is called the *magnitude spectrum* (or *amplitude spectrum*) of the signal **f**. The function arg $F(e^{j\theta})$ is called the *phase spectrum* of **f**. These functions are plotted against θ, typically when **f** is the unit pulse response of a filter. For real sequences, the magnitude spectrum is an even function and the phase spectrum is an odd function. Consequently the plots are made only for $0 < \theta < \pi$.

The signals in Eqs. (4.2.8) were chosen so that the ratio $F_i(z)/F_j(z)$ is all-pass. Therefore these signals have identical amplitude spectra. The phase spectra differ by the phase of the all-pass filters.

All Fourier methods involve trigonometric expansions. The "basis functions" involved are complex exponentials, which are $e^{jk\theta}$ in the case of the DTFT. These functions have two important properties: orthogonality, and the fact that when they are shifted in either k or θ, the result is merely a complex constant times the original function. These two properties give rise to all the other useful properties of the DTFT. In our discussion of the properties of the DTFT, let us begin with the orthogonality property alone and see how far it will take us before we have to consider the shift properties.

Orthogonality Properties

Consider the DTFT pair

$$d_m(k) = \delta(k - m) \quad \overset{DTFT}{\longleftrightarrow} \quad D_m(e^{j\theta}) = e^{-jm\theta}. \tag{4.2.9}$$

This signal is simply the unit pulse sequence shifted in time so that the pulse position is $k = m$. Two such signals are orthogonal in both time and frequency. The inner products are

$$\sum_{k=-\infty}^{\infty} d_m(k)d_n(k) = \delta(m - n), \tag{4.2.10}$$

and

$$\frac{1}{2\pi}\int_{-\pi}^{\pi} D_m(e^{j\theta})[D_n(e^{j\theta})]^* \, d\theta = \delta(m - n). \tag{4.2.11}$$

Using Eq. (4.2.9), the DTFT definition (4.2.2) becomes

$$F(e^{j\theta}) = \sum_m f(m)D_m(e^{j\theta}),$$

and we can now use the orthogonality property (4.2.11), to establish the inversion formula for the DTFT:

$$\frac{1}{2\pi}\int_{-\pi}^{\pi} F(e^{j\theta})[D_n(e^{j\theta})]^* \, d\theta = \sum_m \left[f(m) \frac{1}{2\pi}\int_{-\pi}^{\pi} D_m(e^{j\theta})[D_n(e^{j\theta})]^* \, d\theta \right]$$

$$= \sum_m f(m)\delta(n - m) = f(n).$$

In the usual form, this is

$$f(k) = \frac{1}{2\pi}\int_{-\pi}^{\pi} F(e^{j\theta})e^{jk\theta} \, d\theta. \tag{4.2.12}$$

Thus the inversion formula is a consequence of the orthogonality of the trigonometric basis functions.

There is another property of the DTFT that plays a central role in least-squares problems, which will be studied in Chapter 7. Consider the inner product of two signals **f** and **g**:

$$\sum_m f(m)[g(m)]^* = \sum_m \left[\frac{1}{2\pi}\int_{-\pi}^{\pi} F(e^{j\theta})e^{jm\theta}\,d\theta[g(m)]^*\right]$$

$$= \frac{1}{2\pi}\int_{-\pi}^{\pi} F(e^{j\theta})\left[\sum_m [g(m)]^* e^{jm\theta}\right] d\theta$$

$$= \frac{1}{2\pi}\int_{-\pi}^{\pi} F(e^{j\theta})[G(e^{j\theta})]^*\,d\theta. \tag{4.2.13}$$

The right-hand side of this equation is, by definition, the inner product of F and G. Therefore the transformation from time domain sequences to frequency domain functions preserves the inner products between signals. The equations (4.2.10) and (4.2.11) are special cases. Linear transformations with this property are called *unitary* transformations.

If we take $\mathbf{f} = \mathbf{g}$ in Eq. (4.2.13) we get the *Parseval relation*:

$$\|\mathbf{f}\|_2^2 = \sum_k |f(k)|^2 = \frac{1}{2\pi}\int_{-\pi}^{\pi} |F(e^{j\theta})|^2\,d\theta. \tag{4.2.14}$$

The energy of the signal is a "sum-squared" formula in both domains.

EXAMPLE 4.2.2

Least-squares FIR filters.

It is usually the case that if a designer has more than one goal, then they will conflict in some way. Consider a filter design problem with two such goals. First, the pulse response must have finite length, i.e.,

$$H(z) = \sum_{k=0}^{n} h(k)z^{-k}. \tag{4.2.15}$$

Second, the frequency response $H(e^{j\theta})$ should approximate a desired response $G(e^{j\theta})$, whose inverse DTFT is not necessarily a finite length sequence. We must then resort to a closest fit. Suppose that we choose H to minimize the integral squared error

$$V(\mathbf{h}) = \frac{1}{2\pi}\int_{-\pi}^{\pi} |H(e^{j\theta}) - G(e^{j\theta})|^2\,d\theta. \tag{4.2.16}$$

This is a natural way to express the quality of approximation for the frequency response, but it is not obvious how to deal with the constraints that $h(k) = 0$ for $k < 0$ and $k > n$. Using the Parseval relation (4.2.14) with $\mathbf{f} = \mathbf{h} - \mathbf{g}$, an alternate definition of V is obtained. This is

$$V(\mathbf{h}) = \|\mathbf{h} - \mathbf{g}\|_2^2 = \sum_k |h(k) - g(k)|^2. \tag{4.2.17}$$

It is now clear that the best fit is obtained when

$$h(k) = \begin{cases} g(k), & 0 \leqslant k \leqslant n, \\ 0, & \text{otherwise.} \end{cases} \tag{4.2.18}$$

Shift Properties

When a signal **u** is passed through m unit delays, then the output signal **y** has values

$$y(k) = u(k - m). \tag{4.2.19}$$

This is called a *time shift*. In order to discover what the frequency domain characterization of a time shift might be, let us apply the DTFT.

$$\begin{aligned} Y(e^{j\theta}) &= \sum_k y(k)e^{-jk\theta} \\ &= \sum_k u(k - m)e^{-j(k-m)\theta}e^{-jm\theta} \\ &= e^{-jm\theta}U(e^{j\theta}). \end{aligned} \tag{4.2.20}$$

In other words, the frequency response function for the filter which delays signals m time steps is $e^{-jm\theta}$. We will use the following shorthand description for the effect that time shifts have on DTFT pairs:

$$f(k - m) \quad \xleftrightarrow{DTFT} \quad e^{-jm\theta}F(e^{j\theta}). \tag{4.2.21}$$

Here, it is understood that k is the time variable.

The DTFT is a *linear* transformation. This means that a linear combination of time signals transforms into the same linear combination of DTFT's. If

$$f(k) \quad \xleftrightarrow{DTFT} \quad F(e^{j\theta})$$

$$g(k) \quad \xleftrightarrow{DTFT} \quad G(e^{j\theta})$$

then

$$\alpha f(k) + \beta g(k) \quad \xleftrightarrow{DTFT} \quad \alpha F(e^{j\theta}) + \beta G(e^{j\theta}), \tag{4.2.22}$$

for arbitrary complex numbers α and β. This generalizes to linear combinations involving many components, and in the limit, to integrals. Let **g** be a finite energy signal. If we apply the linear combination

$$\sum_m g(m)[\cdot]$$

to both sides of the relation (4.2.21), then we get

$$\sum_m g(m)f(k-m) \quad \overset{DTFT}{\longleftrightarrow} \quad \left[\sum_m g(m)e^{-jm\theta}\right] F(e^{j\theta}),$$

or

$$(\mathbf{g} * \mathbf{f})(k) \quad \overset{DTFT}{\longleftrightarrow} \quad G(e^{j\theta})F(e^{j\theta}). \tag{4.2.23}$$

Thus *convolution in time* becomes *multiplication in frequency.*

EXAMPLE 4.2.3

The ideal low-pass filter.

Convolution is the basic time domain operation for filters. Thus the input-output maps are

$$\mathbf{y} = \mathbf{h} * \mathbf{u} \quad \overset{DTFT}{\longleftrightarrow} \quad Y(e^{j\theta}) = H(e^{j\theta})U(e^{j\theta}). \tag{4.2.24}$$

The pulse response sequence **h** and frequency response function H form a DTFT pair. Suppose that we want a filter which passes only low frequency sinusoids. To be precise, it should have the property that

$$u(k) = e^{jk\theta} \quad \Rightarrow \quad y(k) = \begin{cases} u(k), & |\theta| < \theta_1, \\ 0, & \theta_1 < |\theta| < \pi. \end{cases} \tag{4.2.25}$$

Such a filter is called an ideal low-pass filter with cutoff frequency θ_1. Low frequency sinusoids go through the filter unchanged. High frequency sinusoids are completely blocked. However, we know that sinusoidal inputs lead to sinusoidal outputs:

$$u(k) = e^{jk\theta} \quad \Rightarrow \quad y(k) = H(e^{j\theta})e^{jk\theta}.$$

Therefore the ideal low-pass filter has the frequency response function

$$H(e^{j\theta}) = \begin{cases} 1, & |\theta| < \theta_1 \\ 0, & \theta_1 < |\theta| < \pi. \end{cases} \tag{4.2.26}$$

Using the inversion formula (4.2.12), we can compute the pulse response:

$$h(k) = \frac{1}{2\pi}\int_{-\theta_1}^{\theta_1} e^{jk\theta}\, d\theta = \frac{\theta_1}{\pi}\,\mathrm{sinc}(k\theta_1). \tag{4.2.27}$$

The pair **h**, H for the ideal low-pass filter is shown in Fig. 4.2.2. Since **h** is not causal (and cannot be made causal by shifting to the right) the ideal low-pass filter cannot be used for real time applications. Notice that the zero crossings of the function $\mathrm{sinc}(\theta_1 x)$ (which is the envelope for **h**) are spaced at intervals π/θ_1. The limiting case is $\theta_1 = \pi$ for which H is identically one, and **h**

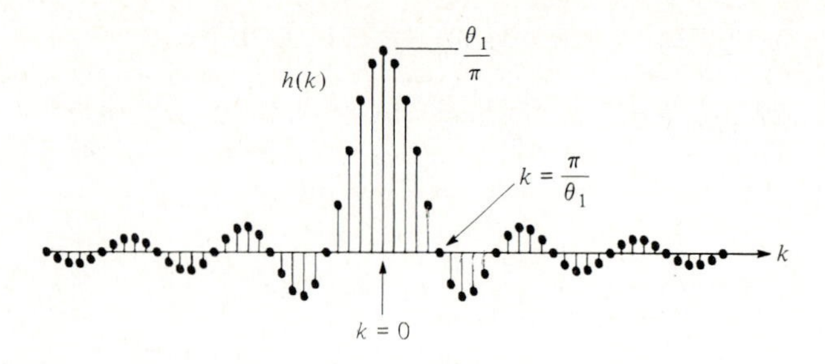

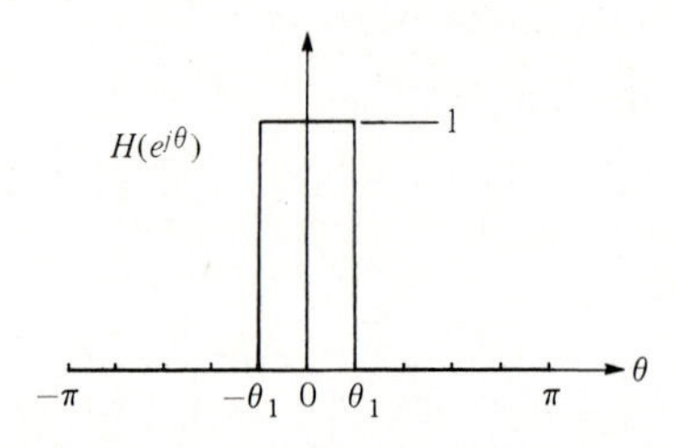

FIGURE 4.2.2 The unit-pulse response h and the frequency response function $H(e^{j\theta})$ for an ideal low-pass filter.

becomes the unit pulse sequence [since it samples sinc($k\pi$) at the zero crossings].

Suppose that the input signal $\mathbf{u}$ has finite energy. Then $\mathbf{y}$ (the output of the low-pass filter) will be that part of $\mathbf{u}$ which lies in the band $[-\theta_1, \theta_1]$. The energy of $\mathbf{y}$ is that part of the energy of $\mathbf{u}$ contained in that band. Using the Parseval relation (4.2.14) this is

$$\begin{aligned}\|\mathbf{y}\|_2^2 &= \frac{1}{2\pi}\int_{-\pi}^{\pi} |Y(e^{j\theta})|^2\, d\theta \\ &= \frac{1}{2\pi}\int_{-\pi}^{\pi} |H(e^{j\theta})U(e^{j\theta})|^2\, d\theta \\ &= \frac{1}{2\pi}\int_{-\theta_1}^{\theta_1} |U(e^{j\theta})|^2\, d\theta. \end{aligned} \tag{4.2.28}$$

In general, that part of the energy of $\mathbf{u}$ which lies in a band B is the integral of $|U(e^{j\theta})|^2$ over B (divided by 2π). For this reason, given any DTFT pair

$$u(k) \quad \overset{DTFT}{\longleftrightarrow} \quad U(e^{j\theta}),$$

the positive real function $|U(e^{j\theta})|^2$ is called the *energy spectral density* for the signal $\mathbf{u}$.

FREQUENCY SHIFTS. Fourier transforms have an amazing degree of symmetry. For every property there is usually a dual property obtained by reversing the roles of time and frequency. A good example of this is the frequency shift dual to the time shift property (4.2.21). This is

$$e^{jk\phi}f(k) \quad \overset{DTFT}{\longleftrightarrow} \quad F(e^{j(\theta-\phi)}), \tag{4.2.29}$$

and may be verified by considering

$$\sum_k [f(k)e^{jk\phi}]e^{-jk\theta} = \sum_k f(k)e^{-jk(\theta-\phi)}.$$

This property is often called *modulation* since it involves multiplication by sinusoids in the time domain which results in a translation of the spectrum. Since $\cos(k\phi) = (e^{jk\phi} + e^{-jk\phi})/2$, we have

$$\cos(k\phi)f(k) \quad \overset{DTFT}{\longleftrightarrow} \quad \frac{1}{2}[F(e^{j(\theta-\phi)}) + F(e^{j(\theta+\phi)})]. \tag{4.2.30}$$

The frequency shift property (4.2.29) leads to a dual to the convolution-multiplication property. Let $G(e^{j\phi})$ be the DTFT of a finite energy signal **g**. Apply the integral operator

$$\frac{1}{2\pi}\int_{-\pi}^{\pi} G(e^{j\phi})[\,\cdot\,]\,d\phi$$

to both sides of the frequency shift relation (4.2.29). We obtain

$$\left[\frac{1}{2\pi}\int_{-\pi}^{\pi} G(e^{j\phi})e^{jk\phi}\,d\phi\right]f(k) \quad \overset{DTFT}{\longleftrightarrow} \quad \frac{1}{2\pi}\int_{-\pi}^{\pi} G(e^{j\phi})F(e^{j(\theta-\phi)})d\phi, \tag{4.2.31}$$

or

$$g(k)f(k) \quad \overset{DTFT}{\longleftrightarrow} \quad (G * F)(e^{j\theta}). \tag{4.2.32}$$

In the time domain we have multiplication of signals, and in the frequency domain we have convolution [defined in Eq. (4.2.31)].

EXAMPLE 4.2.4

Windowing a time signal.

Let **u** be a finite energy signal. We want to compute $U(e^{j\theta})$ but we know the values $u(k)$ only in the range $-L \leq k \leq L$. What happens to the true DTFT if we assume that the unknown values of **u** are all zero? This can be analyzed by multiplying **u** with a rectangular window. Set

$$y(k) = w(k)u(k), \tag{4.2.33}$$

where

$$w(k) = \begin{cases} 1, & |k| \leq L \\ 0, & \text{otherwise.} \end{cases} \tag{4.2.34}$$

The DTFT of **y** should approximate that of **u**. Using the multiplication-convolution property (4.2.32), the approximation becomes

$$Y(e^{j\theta}) = \frac{1}{2\pi} \int_{-\pi}^{\pi} U(e^{j\phi}) W(e^{j(\theta-\phi)}) d\phi. \tag{4.2.35}$$

The ideal would be $W(e^{j\theta}) = 2\pi\delta(\theta)$, since this would give $Y = U$. But what is the actual shape of W? We have

$$W(z) = \sum_{k=-L}^{L} z^{-k} = z^{-L} \sum_{k=0}^{2L} z^k$$
$$= z^{-L} \frac{1 - z^{2L+1}}{1 - z} = \frac{z^{(2L+1)/2} - z^{-(2L+1)/2}}{z^{1/2} - z^{-1/2}}.$$

Therefore

$$W(e^{j\theta}) = \frac{\sin\left[\left(L + \frac{1}{2}\right)\theta\right]}{\sin\left[\frac{1}{2}\theta\right]}. \tag{4.2.36}$$

This DTFT pair is shown in Fig. 4.2.3. The height of the main lobe of W is $2L + 1$, which is the number of known values of **u**. The width of the main lobe is $4\pi/(2L + 1)$. As L increases, the width decreases and the height increases. The effect is to better approximate the delta function. The integration in Eq. (4.2.35) will smooth or "blur" the actual DTFT $U(e^{j\theta})$, if L is not large.

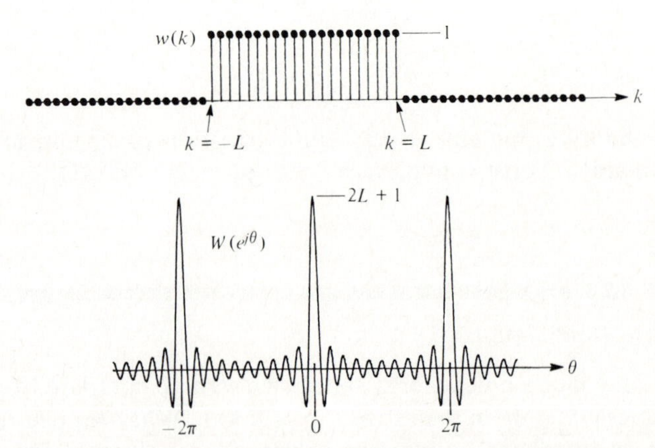

FIGURE 4.2.3 **The window sequence $w(k)$ and its discrete-time Fourier transform $W(e^{j\theta})$.**

Symmetry Properties

Suppose that the signal **f** is real. Then

$$[F(e^{j\theta})]^* = \sum_k f(k)[e^{-jk\theta}]^* = \sum_k f(k)e^{jk\theta} = F(e^{-j\theta}). \tag{4.2.37}$$

Let us call a complex valued function $g(\theta)$ *Hermitian* if $g(-\theta) = [g(\theta)]^*$ for all θ. Then Eq. (4.2.37) says that if **f** is real, then F must be a Hermitian function of θ. In general if a signal is real in one domain, it is Hermitian in the other. If it is pure imaginary in one domain, then it is anti-Hermitian in the other. These properties, together with the others we have discussed are summarized in Table 4.2.1.

Let **h** be a real-valued unit pulse response sequence. Then the frequency response function $H(e^{j\theta})$ satisfies

$$[H(e^{j\theta})]^* = H(e^{-j\theta})$$

which implies that the magnitude response

$$|H(e^{-j\theta})| = |H(e^{j\theta})|$$

is an even function, and the phase response

$$\arg H(e^{-j\theta}) = -\arg H(e^{j\theta})$$

is an odd function. Thus it is never necessary to plot these functions over the full range $-\pi$ to π, but only from 0 to π.

The function $H(e^{-j\theta})$ is the DTFT of the *time reversal* of **h**. In other words

$$f(k) = h(-k) \quad \xleftrightarrow{DTFT} \quad F(e^{j\theta}) = H(e^{-j\theta}). \tag{4.2.38}$$

EXAMPLE 4.2.5

Nontrivial all-pass filters.

Consider a filter whose unit pulse response **f** satisfies the following:

$$\begin{cases} f(k) \text{ is real.} \\ \mathbf{f} \text{ is causal, } i.e., \ f(k) = 0 \text{ for } k < 0. \\ |F(e^{j\theta})| = 1 \text{ for all } \theta. \\ f^2(0) < 1. \end{cases} \tag{4.2.39}$$

We shall call such a filter a nontivial all-pass filter. The Parseval relation forces **f** to have total energy equal to one,

$$\|\mathbf{f}\|^2 = 1. \tag{4.2.40}$$

And since

$$F(e^{j\theta})[F(e^{j\theta})]^* = F(e^{j\theta})F(e^{-j\theta}) = 1,$$

we have

$$F(e^{j\theta}) = 1/F(e^{-j\theta}). \tag{4.2.41}$$

Thus if we set $g(k) = f(-k)$, then $\mathbf{g} * \mathbf{f} = \boldsymbol{\delta}$. The inverse of **f** is the time reversal of **f**, which is anticausal. The condition $f^2(0) < 1$ means that

$$\sum_{k=1}^{\infty} f^2(k) = 1 - f^2(0) > 0, \tag{4.2.42}$$

so that **f** cannot itself be the unit pulse sequence.

A filter **h** is *nonminimum phase* if it can be factored as

$$\mathbf{h} = \mathbf{h}_0 * \mathbf{f}, \tag{4.2.43}$$

TABLE 4.2.1 Properties of the discrete-time Fourier transform.

Linearity

$$\alpha f(k) + \beta g(k) \quad \overset{DTFT}{\longleftrightarrow} \quad \alpha F(e^{j\theta}) + \beta G(e^{j\theta})$$

Orthogonality properties

$$\sum_{k=-\infty}^{\infty} f(k)[g(k)]^* = \frac{1}{2\pi}\int_{-\pi}^{\pi} F(e^{j\theta})[G(e^{j\theta})]^* \, d\theta$$

$$\sum_{k=-\infty}^{\infty} |f(k)|^2 = \frac{1}{2\pi}\int_{-\pi}^{\pi} |F(e^{j\theta})|^2 \, d\theta$$

Shift properties

$$f(k-m) \quad \overset{DTFT}{\longleftrightarrow} \quad e^{-jm\theta}F(e^{j\theta})$$

$$\sum_{m=-\infty}^{\infty} f(k-m)g(m) \quad \overset{DTFT}{\longleftrightarrow} \quad F(e^{j\theta})G(e^{j\theta})$$

$$e^{jk\phi}f(k) \quad \overset{DTFT}{\longleftrightarrow} \quad F(e^{j(\theta-\phi)})$$

$$f(k)g(k) \quad \overset{DTFT}{\longleftrightarrow} \quad \frac{1}{2\pi}\int_{-\pi}^{\pi} F(e^{j(\theta-\phi)})G(e^{j\phi})\, d\phi$$

Symmetry properties

$$\text{Im}[f(k)] = 0 \quad \Leftrightarrow \quad F(e^{-j\theta}) = [F(e^{j\theta})]^*$$

$$f(-k) = [f(k)]^* \quad \Leftrightarrow \quad \text{Im}[F(e^{j\theta})] = 0$$

$$\text{Re}[f(k)] = 0 \quad \Leftrightarrow \quad F(e^{-j\theta}) = -[F(e^{j\theta})]^*$$

$$f(-k) = -[f(k)]^* \quad \Leftrightarrow \quad \text{Re}[F(e^{j\theta})] = 0$$

where $\mathbf{h}_0$ is causal and $\mathbf{f}$ satisfies the four conditions (4.2.39). The magnitude of H agrees with that of H_0 since the magnitude of F is unity. But the phase is additive:

$$\arg H(e^{j\theta}) = \arg H_0(e^{j\theta}) + \arg F(e^{j\theta}). \tag{4.2.44}$$

A nonminimum phase filter cannot have a finite energy causal inverse, as we shall now show. This fact becomes important in the least-squares filter problems in Chapter 7.

Suppose there is a finite energy causal sequence $\mathbf{g}$ for which

$$\boldsymbol{\delta} = \mathbf{h} * \mathbf{g} = \mathbf{h}_0 * \mathbf{f} * \mathbf{g}. \tag{4.2.45}$$

Then

$$H_0(e^{j\theta})G(e^{j\theta}) = 1/F(e^{j\theta}) = F(e^{-j\theta}),$$

which means that $\mathbf{h}_0 * \mathbf{g}$ is the time reversal of $\mathbf{f}$, which cannot be causal. Therefore no such $\mathbf{g}$ exists.

Table 4.2.1 summarizes the properties we have discussed in this section.

4.3 FOURIER SERIES

If a signal is continuous time and periodic, then the appropriate Fourier transform is the Fourier series (FS). If

$$f(t) \equiv f(t + t_0), \tag{4.3.1}$$

then we say that f is "t_0-periodic." (The symbol "$\equiv$" means that the equation holds for all values of t, and the statement (4.3.1) is called an *identity*.) A t_0-periodic signal cannot have finite energy unless it is zero, but it will have finite *power*:

$$\text{Power}(f) = \lim_{T\to\infty} \frac{1}{2T} \int_{-T}^{T} |f(t)|^2 \, dt. \tag{4.3.2}$$

Because of the periodicity, the power is known if one period of f is known, and

$$\text{Power}(f) = \frac{1}{t_0} \int_{0}^{t_0} |f(t)|^2 \, dt. \tag{4.3.3}$$

In 1738, Daniel Bernoulli introduced a trigonometric representation for a function in two variables for the purpose of simplifying the wave equation for a vibrating string. Bernoulli used a sin-cosine series. The *exponential* Fourier series representation is

$$f(t) = \sum_{n=-\infty}^{\infty} F(n)e^{jn\omega_0 t}. \tag{4.3.4}$$

Now this function is periodic with period t_0 where

$$\omega_0 t_0 = 2\pi. \tag{4.3.5}$$

Bernoulli later asserted that "every function" $f(t)$ has such a representation on the interval $0 < t < t_0$ (Kline 1972, Chapters 22 and 28). This belief was shared by Fourier but disputed by the mathematicians of his time, and a famous controversy ensued which revolutionized mathematics. The formula

$$F(n) = \frac{1}{t_0} \int_0^{t_0} f(t) e^{-jn\omega_0 t} \, dt \tag{4.3.6}$$

for the Fourier coefficients was known to Clairaut in 1757, and to Euler in 1777 who treated the subject in the modern way, using the orthogonality of the trigonometric functions. The integral in Eq. (4.3.6) can be done over any period $t_1 \leqslant t \leqslant t_1 + t_0$, since the integrand is t_0-periodic.

We shall call the periodic signal $f(t)$ and the sequence of Fourier

TABLE 4.3.1 Properties of Fourier series, [$\omega_0 t_0 = 2\pi$].

Linearity

$$\alpha f(t) + \beta g(t) \quad \overset{FS:\omega_0}{\longleftrightarrow} \quad \alpha F(n) + \beta G(n)$$

Orthogonality properties

$$\frac{1}{t_0} \int_0^{t_0} f(t)[g(t)]^* dt = \sum_{n=-\infty}^{\infty} F(n)[G(n)]^*$$

$$\frac{1}{t_0} \int_0^{t_0} |f(t)|^2 \, dt = \sum_{n=-\infty}^{\infty} |F(n)|^2$$

Shift properties

$$f(t-\tau) \quad \overset{FS:\omega_0}{\longleftrightarrow} \quad e^{-jn\omega_0 \tau} F(n)$$

$$\frac{1}{t_0} \int_0^{t_0} f(t-\tau) g(\tau) d\tau \quad \overset{FS:\omega_0}{\longleftrightarrow} \quad F(n)G(n)$$

$$e^{jm\omega_0 t} f(t) \quad \overset{FS:\omega_0}{\longleftrightarrow} \quad F(n-m)$$

$$f(t)g(t) \quad \overset{FS:\omega_0}{\longleftrightarrow} \quad \sum_{m=-\infty}^{\infty} F(n-m)G(m)$$

Symmetry properties

$$\text{Im}[f(t)] = 0 \quad \Leftrightarrow \quad F(-n) = [F(n)]^*$$

$$f(-t) = [f(t)]^* \quad \Leftrightarrow \quad \text{Im}[F(n)] = 0$$

$$\text{Re}[f(t)] = 0 \quad \Leftrightarrow \quad F(-n) = -[F(n)]^*$$

$$f(-t) = -[f(t)]^* \quad \Leftrightarrow \quad \text{Re}[F(n)] = 0$$

coefficients $F(n)$ a *Fourier series pair*, and use the notation

$$f(t) \quad \xleftrightarrow{FS;\omega_0} \quad F(n). \tag{4.3.7}$$

Note that one cannot reconstruct the signal from the sequence without knowing ω_0, and that is the reason for explicitly stating it. Again we see that a signal which is periodic in one domain is discrete in the other.

The FS and DTFT transforms are mathematically equivalent, if one exchanges the roles of time and frequency and makes some minor notational adjustments. If

$$g(k) \quad \xleftrightarrow{DTFT;t_1} \quad G(e^{j\omega t_1}), \tag{4.3.8}$$

and we set

$$\omega_0 = t_1, \tag{4.3.9}$$

$$f(t) = G(e^{j\omega t_1})|_{\omega = t}, \tag{4.3.10}$$

$$F(n) = g(k)|_{k = -n}, \tag{4.3.11}$$

then

$$f(t) \quad \xleftrightarrow{FS;\omega_0} \quad F(n). \tag{4.3.12}$$

It follows that there is a dual set of FS properties, which is given in Table 4.3.1. Notice that the Parseval relation deals with the average power of the signal $f(t)$, rather than its energy. The frequency $\omega_0/2\pi = 1/t_0$ is called the *fundamental frequency* of $f(t)$.

4.4 THE FOURIER TRANSFORM

If a signal **f** is defined for all real numbers, then it is called a continuous time (or *analog*) signal. The energy of such a signal is

$$\|\mathbf{f}\|^2 = \int_{-\infty}^{\infty} |f(t)|^2 \, dt. \tag{4.4.1}$$

The appropriate Fourier method for finite energy, continuous time signals is the *Fourier integral transform* (FT). This is defined as

$$F(j\omega) = \int_{-\infty}^{\infty} f(t) e^{-j\omega t} \, dt, \tag{4.4.2}$$

and we use the notation

$$f(t) \quad \xleftrightarrow{FT} \quad F(j\omega) \tag{4.4.3}$$

to indicate that these two form a Fourier transform pair. The spectrum F is also a function of a continuous parameter. We use the notation $F(j\omega)$ rather than $F(\omega)$ in order to be consistent with the *Laplace transform* which has the definition

$$F(s) = \int_{-\infty}^{\infty} f(t)e^{-st}\,dt. \tag{4.4.4}$$

If the signal **f** is both L_1 and L_2 [that is both absolutely integrable and finite energy], then the region in the complex plane for which $F(s)$ is defined contains the imaginary axis. Under these conditions the FT is the restriction of $F(s)$ to $s = j\omega$.

EXAMPLE 4.4.1

Fourier transform from the Laplace transform.

The canonical signal from which most entries in Laplace transform tables can be derived is the causal signal

$$f(t) = \xi(t)e^{-at}, \qquad a > 0 \tag{4.4.5}$$

where ξ is the unit step function. This has Laplace transform

$$F(s) = \int_{0}^{\infty} e^{-(a+s)t}\,dt = \frac{1}{s+a}. \tag{4.4.6}$$

Thus

$$\xi(t)e^{-at} \quad \stackrel{FT}{\longleftrightarrow} \quad \frac{1}{j\omega + a}. \tag{4.4.7}$$

The FT inversion rule is

$$f(t) = \frac{1}{2\pi}\int_{-\infty}^{\infty} F(j\omega)e^{j\omega t}\,d\omega. \tag{4.4.8}$$

This is, for finite energy signals, actually a "limit in the mean" formula (Bachman 1964, p. 24). The convergence need not be pointwise.

The Fourier transform is "self-dual." That is to say that $f(t)$ and $F(j\omega)$ are both nonperiodic functions of continuous parameters. If one exchanges the roles of time and frequency, then the duality property in Table 4.4.1 is the result.

EXAMPLE 4.4.2

Using the duality property.

An analog ideal low-pass filter would have frequency response function

$$H(j\omega) = \mathrm{rect}(\omega/\omega_1). \tag{4.4.9}$$

TABLE 4.4.1 **Properties of the Fourier transform.**

Linearity

$$\alpha f(t) + \beta g(t) \quad \overset{FT}{\longleftrightarrow} \quad \alpha F(j\omega) + \beta G(j\omega)$$

Orthogonality properties

$$\int_{-\infty}^{\infty} f(t)[g(t)]^* dt = \frac{1}{2\pi} \int_{-\infty}^{\infty} F(j\omega)[G(j\omega)]^* d\omega$$

$$\int_{-\infty}^{\infty} |f(t)|^2 dt = \frac{1}{2\pi} \int_{-\infty}^{\infty} |F(j\omega)|^2 d\omega$$

Shift properties

$$f(t - t_1) \quad \overset{FT}{\longleftrightarrow} \quad e^{-j\omega t_1} F(j\omega)$$

$$\int_{-\infty}^{\infty} f(t - t_1) g(t_1) dt_1 \quad \overset{FT}{\longleftrightarrow} \quad F(j\omega) G(j\omega)$$

$$e^{j\omega_1 t} f(t) \quad \overset{FT}{\longleftrightarrow} \quad F(j\omega - j\omega_1)$$

$$f(t) g(t) \quad \overset{FT}{\longleftrightarrow} \quad \frac{1}{2\pi} \int_{-\infty}^{\infty} F(j\omega - j\omega_1) G(j\omega_1) d\omega_1$$

Symmetry properties

$$\mathrm{Im}[f(t)] = 0 \quad \Leftrightarrow \quad F(-j\omega) = [F(j\omega)]^*$$

$$f(-t) = [f(t)]^* \quad \Leftrightarrow \quad \mathrm{Im}[F(j\omega)] = 0$$

$$\mathrm{Re}[f(t)] = 0 \quad \Leftrightarrow \quad F(-j\omega) = -[F(j\omega)]^*$$

$$f(-t) = -[f(t)]^* \quad \Leftrightarrow \quad \mathrm{Re}[F(j\omega)] = 0$$

Duality

$$g(t) = F(jt) \quad \overset{FT}{\longleftrightarrow} \quad 2\pi f(-\omega) = G(j\omega)$$

Time/frequency scaling ($\omega_0 > 0$)

$$g(t) = f(\omega_0 t) \quad \overset{FT}{\longleftrightarrow} \quad G(j\omega) = \frac{1}{\omega_0} F(j\omega/\omega_0)$$

Using the inversion formula, we have

$$h(t) = \frac{1}{2\pi} \int_{-\omega_1}^{\omega_1} e^{j\omega t} d\omega = \frac{\omega_1}{\pi} \operatorname{sinc}(\omega_1 t).$$

These form the FT pair

$$h(t) = \frac{\omega_1}{\pi} \operatorname{sinc}(\omega_1 t) \quad \overset{FT}{\longleftrightarrow} \quad H(j\omega) = \operatorname{rect}(\omega/\omega_1). \tag{4.4.10}$$

Using the duality property, we obtain the dual FT pair

$$g(t) = \mathrm{rect}(t/t_1) \quad \xleftrightarrow{FT} \quad G(j\omega) = 2t_1 \,\mathrm{sinc}(\omega t_1). \tag{4.4.11}$$

(We have also replaced ω_1 with t_1.)

The time-frequency scaling property is important to understand when one considers the problem of selecting an appropriate sampling period for digital processing. Consider the family of FT pairs

$$f(\omega_0 t) \quad \xleftrightarrow{FT} \quad \frac{1}{\omega_0} F(j\omega/\omega_0),$$

indexed by the parameter ω_0. As ω_0 is increased, the time signal is squeezed together while the spectrum is stretched. Thus short duration time signals will have wide bandwidths.

EXAMPLE 4.4.3

The Gaussian pulse.

Let

$$f(t) = \frac{1}{\sqrt{2\pi}} e^{-t^2/2}.$$

This signal has the same form as a Gaussian probability density and is called a *Gaussian pulse.* The Fourier transform can be evaluated using complex integration techniques. The result is the FT pair

$$f(t) = \frac{1}{\sqrt{2\pi}} e^{-t^2/2} \quad \xleftrightarrow{FT} \quad F(j\omega) = e^{-\omega^2/2}.$$

Thus Gaussian pulses have Gaussian spectra. Using the time-frequency scaling property, we get a family of pairs:

$$f(t) = \frac{1}{\sqrt{2\pi}} e^{-t^2/(2\sigma^2)} \quad \xleftrightarrow{FT} \quad F(j\omega) = \sigma e^{-\omega^2\sigma^2/2}. \tag{4.4.12}$$

We shall see these two examples again in Section 4.8, where the DFT approximation to a Fourier transform spectrum is studied. For this application, good approximations require that the duration of the signal be small in both time and frequency. But an FT property which is well known in modern physics places a limit on simultaneous resolution in both domains. This is the *uncertainty principle* (Bracewell 1978).

If we define *duration* by

$$\sigma_t = \left[\int t^2 |f(t)|^2 dt \Big/ \int |f(t)|^2 dt \right]^{1/2},$$

$$\sigma_\omega = \left[\int \omega^2 |F(j\omega)|^2 d\omega \Big/ \int |F(j\omega)|^2 d\omega \right]^{1/2},$$

then the time-bandwidth product is lower bounded:

$$\sigma_t \sigma_\omega \geqslant \tfrac{1}{2}. \tag{4.4.13}$$

Thus these two quantities cannot be simultaneously reduced. Furthermore, the only signals which meet the bound are the Gaussian pulses of Eqs. (4.4.12). It would seem then that the Gaussian pulses are more tolerant of sampling. We shall give some evidence for this in Section 4.8.

4.5 THE DISCRETE FOURIER TRANSFORM

If a discrete-time signal is periodic, then the appropriate transform is the discrete Fourier transform (DFT). Thus one can think of it as a discrete-time version of Fourier series. Although the class of discrete-time periodic signals does not seem to be very useful, the DFT turns out to be extremely important in digital signal processing. The publication (Cooley and Tukey 1965) of a fast algorithm for the DFT created a whole new set of digital signal processing applications. This algorithm came to be known as the *fast Fourier transform* (FFT). (See Heidelman, Johnson, and Burrus 1984.)

There are three good reasons for studying the DFT. First, it can be efficiently computed. Second, it has a large number of applications including the approximation of other transforms, filter design, and fast convolution for FIR filtering. Third, it is the only Fourier transform which can be finitely parameterized. One can study the DFT, and all its properties using matrix algebra.

Let $\mathbf{f}$ be N-periodic, and let W_N be the principal Nth root of unity:

$$W_N = e^{j2\pi/N}. \tag{4.5.1}$$

The DFT is defined by

$$F(n) = \sum_{k=0}^{N-1} f(k) W_N^{-kn}. \tag{4.5.2}$$

Like Fourier series, the sum is over one period of the signal. Since the signal is discrete-time, it is characterized by its values over one period. The values of $F(n)$ will also repeat since

$$W_N^N = 1,$$

and therefore

$$F(n+N) = \sum_{k=0}^{N-1} f(k) W_N^{-kn} [W_N^{-N}]^k$$
$$= \sum_{k=0}^{N-1} f(k) W_N^{-kn} = F(n).$$

Thus both **f** and **F** are characterized by N values each. Because of this, we will introduce the following notation:

$$\mathbf{f} = \begin{bmatrix} f(0) \\ \vdots \\ f(N-1) \end{bmatrix}, \qquad \mathbf{F} = \begin{bmatrix} F(0) \\ \vdots \\ F(N-1) \end{bmatrix}. \tag{4.5.3}$$

The single bold symbol **f** will mean either the N-dimensional vector in Eqs. (4.5.3) or the entire periodic signal, depending on the context.

In matrix form, the DFT equation (4.5.2) becomes

$$\mathbf{F} = \mathbf{V}\mathbf{f} \tag{4.5.4}$$

where **V** is the $N \times N$ matrix with elements

$$V_{nk} = W_N^{-kn}, \qquad 0 \leqslant k,\ n \leqslant N-1. \tag{4.5.5}$$

V is the matrix representation of the DFT.

EXAMPLE 4.5.1

Using the DFT to sample the DTFT of a finite-length sequence.

Although the DFT is appropriate to periodic signals, it can also be applied to finite length sequences. Let

$$g(k) \quad \overset{DTFT}{\longleftrightarrow} \quad G(e^{j\theta}), \tag{4.5.6}$$

where $g(k) = 0$ for $k < 0$ and $k \geqslant N$. Set

$$f(k + mN) = g(k) \tag{4.5.7}$$

for all m and $0 \leqslant k < N$. This is the periodic extension of **g**. Then from the definition of the DFT in Eq. (4.5.2), the DFT of **f** consists of samples of the DTFT of **g**:

$$F(n) = G(e^{j2\pi n/N}) = G(W_N^n). \tag{4.5.8}$$

One can get more samples of $G(e^{j\theta})$ by using a DFT of size $M > N$ and setting $f(k) = 0$ for $N \leqslant k < M$. This is called "zero-padding."

Orthogonality Properties of the DFT

Let $z = W_N^{m-n}$, and consider

$$[\mathbf{V}^*\mathbf{V}]_{mn} = \sum_{k=0}^{N-1} [\mathbf{V}^*]_{mk} V_{kn} = \sum_{k=0}^{N-1} W_N^{mk} W_N^{-nk}$$

$$= \sum_{k=0}^{N-1} z^k = \begin{cases} N, & z = 1 \\ \dfrac{1 - z^N}{1 - z}, & \text{otherwise.} \end{cases}$$

Now $z = 1$ if $m = n$ and $z^N = 1$ if $m \neq n$. Therefore

$$[\mathbf{V}^*\mathbf{V}]_{mn} = N\,\delta(m - n),$$

or

$$\mathbf{V}^*\mathbf{V} = N\,\mathbf{I}. \tag{4.5.9}$$

The columns of $\mathbf{V}$ are seen to be orthogonal, and the matrix $\mathbf{U} = N^{-1/2}\mathbf{V}$ is *unitary* ($\mathbf{U}^*\mathbf{U} = \mathbf{I}$). Equation (4.5.9) generates the inversion rule for the DFT, and the Parseval relation. Since

$$\mathbf{V}^{-1} = \frac{1}{N}\mathbf{V}^*, \tag{4.5.10}$$

the inversion rule is

$$\mathbf{f} = \mathbf{V}^{-1}\mathbf{F} = \frac{1}{N}\mathbf{V}^*\mathbf{F}, \tag{4.5.11}$$

or

$$f(k) = \frac{1}{N}\sum_{n=0}^{N-1} F(n) W_N^{kn}. \tag{4.5.12}$$

Now consider inner products. Let

$$\mathbf{f} \quad \xleftrightarrow{DFT} \quad \mathbf{F} = \mathbf{V}\mathbf{f},$$

$$\mathbf{g} \quad \xleftrightarrow{DFT} \quad \mathbf{G} = \mathbf{V}\mathbf{g}.$$

Then

$$\frac{1}{N}\sum_{n=0}^{N-1} F(n)[G(n)]^* = \frac{1}{N}\mathbf{G}^*\mathbf{F} = \frac{1}{N}\mathbf{g}^*\mathbf{V}^*\mathbf{V}\mathbf{f}$$

$$= \mathbf{g}^*\mathbf{f} = \sum_{k=0}^{N-1} f(k)[g(k)]^*. \tag{4.5.13}$$

Again, Eq. (4.5.9) was used to eliminate **V**. With $\mathbf{g} = \mathbf{f}$, we have the Parseval relation

$$\frac{1}{N}\sum_{n=0}^{N-1} |F(n)|^2 = \sum_{k=0}^{N-1} |f(k)|^2. \tag{4.5.14}$$

Shift Properties of the DFT

Let **f** be an N-periodic sequence. Suppose that we shift **f** one time step to obtain **g** with $g(k) = f(k-1)$. This merely delays the sequence, and **g** will still be N-periodic. But think about what this fact says about the *delay* or *shift* operator. The set of all N-periodic sequences forms a linear subspace of the space of all sequences and has dimension N. Since this subspace is invariant under shifts, there must be a representation for the shift operator restricted to the subspace. This will be no more than an $N \times N$ matrix. Let us develop this idea. Each N-periodic sequence is represented by the vector containing its values over the range $0 \leqslant k < N$. Thus, for the signals **f** and **g**, we have

$$\mathbf{g} = \begin{bmatrix} g(0) \\ g(1) \\ \vdots \\ g(N-1) \end{bmatrix} = \begin{bmatrix} f(-1) \\ f(0) \\ \vdots \\ f(N-2) \end{bmatrix} = \begin{bmatrix} f(N-1) \\ f(0) \\ \vdots \\ f(N-2) \end{bmatrix} = \mathbf{R}^{-1} \begin{bmatrix} f(0) \\ f(1) \\ \vdots \\ f(N-1) \end{bmatrix}. \tag{4.5.15}$$

The matrix $\mathbf{R}^{-1}$ is the characterization of the unit delay for N-periodic sequences. (We have used $\mathbf{R}^{-1}$ rather than $\mathbf{R}$ to suggest a correspondence to z^{-1}, the delay operator for the z-transform.) We can identify the elements of $\mathbf{R}^{-1}$ by setting **f** to unit vectors, which pick out columns of $\mathbf{R}^{-1}$. The result is

$$\mathbf{R}^{-1} = \left[\begin{array}{cccc|c} 0 & \cdots & & 0 & 1 \\ \hline 1 & & & & 0 \\ & \cdot & 0 & & \\ & & \cdot & & \vdots \\ 0 & & & \cdot & \\ & & & 1 & 0 \end{array}\right]. \tag{4.5.16}$$

This is a permutation matrix and operates on column vectors by pushing each element down and rotating the overflow back into the top position. For this reason it is called the *rotation matrix*. Since it is a permutation, it is *unitary*, i.e., $\mathbf{R}^T = \mathbf{R}^{-1}$ and $\mathbf{R}\mathbf{R}^T = \mathbf{I}$. Also, if N rotations are performed then there is no net change. Consequently

$$\mathbf{R}^N = \mathbf{I}, \tag{4.5.17}$$

and the rotation matrix is an Nth root of the identity matrix.

Now consider the nth row of the DFT matrix **V**:

$$\mathbf{V}^{(n)} = [1, W_N^{-n}, W_N^{-2n}, \cdots, W_N^{-(N-1)n}]. \tag{4.5.18}$$

Using Eq. (4.5.16), multiply on the right by $\mathbf{R}^{-1}$ to get

$$\begin{aligned}\mathbf{V}^{(n)}\mathbf{R}^{-1} &= [W_N^{-n}, W_N^{-2n}, \cdots, W_N^{-(N-1)n}, 1] \\ &= W_N^{-n}\mathbf{V}^{(n)}.\end{aligned} \tag{4.5.19}$$

In the last component, we have used the fact that $W_N^{-Nn} = 1$. Equation (4.5.19) is an eigenvector-eigenvalue equation. Each row of $\mathbf{V}$ is an eigenvector of $\mathbf{R}^{-1}$, and the eigenvalues of $\mathbf{R}^{-1}$ are the N distinct Nth roots of unity W_N^{-n}. These numbers are equally spaced around the unit circle and separated by $2\pi/N$ radians.

If we stack the row vector equations in (4.5.19) vertically, we get the following matrix equation.

$$\mathbf{V}\mathbf{R}^{-1} = \mathbf{D}^{-1}\mathbf{V}, \tag{4.5.20}$$

where

$$\mathbf{D}^{-1} = \text{Diag}\{1, W_N^{-1}, \cdots, W_N^{-(N-1)}\}. \tag{4.5.21}$$

Equation (4.5.20) contains the time shift properties of the DFT. For delays of m steps, it generalizes to

$$\mathbf{V}\mathbf{R}^{-m} = \mathbf{D}^{-1}\mathbf{V}\mathbf{R}^{-(m-1)} = \cdots = \mathbf{D}^{-m}\mathbf{V}. \tag{4.5.22}$$

Set $g(k) = f(k - m)$, or $\mathbf{g} = \mathbf{R}^{-m}\mathbf{f}$. Then

$$\mathbf{G} = \mathbf{V}\mathbf{g} = \mathbf{V}\mathbf{R}^{-m}\mathbf{f} = \mathbf{D}^{-m}\mathbf{V}\mathbf{f} = \mathbf{D}^{-m}\mathbf{F}. \tag{4.5.23}$$

The elements of the DFT vector in this equation are $G(n) = W_N^{-mn}F(n)$. Consequently the time shift property is

$$f(k - m) \quad \xleftrightarrow{DFT;\,N} \quad W_N^{-mn}F(n). \tag{4.5.24}$$

This holds for all m, if we keep in mind the fact that $\mathbf{f}$ is N-periodic. In the frequency domain, W_N^{-mn} is also N-periodic as a function of m. Now apply the linear combination

$$\sum_{m=0}^{N-1} g(m)[\cdot] \tag{4.5.25}$$

to both sides of the DFT pair (4.5.24). The result is the convolution-multiplication property

$$\sum_{m=0}^{N-1} g(m)f(k - m) \quad \xleftrightarrow{DFT;\,N} \quad G(n)F(n). \tag{4.5.26}$$

CIRCULAR CONVOLUTION AND CIRCULANTS. The time shift and convolution-multiplication properties assume N-periodic signals. In practice, however, the DFT is applied to vectors and only N values are stored in memory. Since the left-hand sides of the DFT pairs (4.5.24) and (4.5.26) involve time values outside the range $0 \leqslant k < N$, an indexing mechanism must be invented to refer to sequence elements which are in range and have the same values by virtue of periodicity.

Let k be any integer. Then the integer

$$m = k \bmod N \tag{4.5.27}$$

is that integer for which $0 \leqslant m < N$ and $m - k$ is a multiple of N. This means that for N-periodic sequences,

$$f(k) = f(k \bmod N), \quad \text{for all } k. \tag{4.5.28}$$

One can also use this device to describe the elements of $\mathbf{R}^{-1}$ as

$$[\mathbf{R}^{-1}]_{km} = \delta(m + 1 - k \bmod N). \tag{4.5.29}$$

We can now define a convolution for vectors which is consistent with the convolution of periodic sequences in the relation (4.5.26). This is called *circular convolution* and is defined by

$$[\mathbf{f} \circledast \mathbf{g}](k) = \sum_{m=0}^{N-1} g(m) f(k - m \bmod N). \tag{4.5.30}$$

The indices on the right-hand side are now all in range.

EXAMPLE 4.5.2

Circular convolution.

Take $N = 3$ and set $\mathbf{y} = \mathbf{f} \circledast \mathbf{g}$. Then

$$\begin{aligned} y(0) &= g(0)f(0) + g(1)f(2) + g(2)f(1), \\ y(1) &= g(0)f(1) + g(1)f(0) + g(2)f(2), \\ y(2) &= g(0)f(2) + g(1)f(1) + g(2)f(0). \end{aligned} \tag{4.5.31}$$

One can put these equations into vector form as follows:

$$\begin{bmatrix} y(0) \\ y(1) \\ y(2) \end{bmatrix} = \begin{bmatrix} f(0) & f(2) & f(1) \\ f(1) & f(0) & f(2) \\ f(2) & f(1) & f(0) \end{bmatrix} \begin{bmatrix} g(0) \\ g(1) \\ g(2) \end{bmatrix}. \tag{4.5.32}$$

The matrix containing the elements of $\mathbf{f}$ in Eq. (4.5.32) is called a *circulant* matrix. Circular convolution, and multiplication by a circulant matrix are equivalent. Let us develop this connection a bit more for the insight it provides into what all Fourier transforms actually do.

Given the N-dimensional vector $\mathbf{f}$, let us define

$$\hat{f}(z) = \sum_{k=0}^{N-1} f(k) z^{-k}. \tag{4.5.33}$$

Then the elements of the DFT of $\mathbf{f}$ are

$$F(n) = \hat{f}(W_N^n) \tag{4.5.34}$$

[see Eq. (4.5.2)]. With $N = 3$, the matrix in Eq. (4.5.32) is

$$f(0)\mathbf{I} + f(1)\mathbf{R}^{-1} + f(2)\mathbf{R}^{-2} = \hat{f}(\mathbf{R}).$$

In general $\hat{f}(\mathbf{R})$ is a matrix with elements

$$[\hat{f}(\mathbf{R})]_{km} = f(k - m \bmod N). \tag{4.5.35}$$

A matrix whose elements depend only on the difference of the indices mod N is called a *circulant matrix*. Every circulant matrix is polynomial in $\mathbf{R}^{-1}$ and column zero of the matrix contains the polynomial coefficients. One can also write

$$\hat{f}(\mathbf{R}) = [\mathbf{f}, \mathbf{R}^{-1}\mathbf{f}, \cdots, \mathbf{R}^{-(N-1)}\mathbf{f}]. \tag{4.5.36}$$

Each column is the circular shift of the previous column. Multiplication by $\hat{f}(\mathbf{R})$ is equivalent to circular convolution:

$$\hat{f}(\mathbf{R})\mathbf{g} = \mathbf{f} \circledast \mathbf{g}, \tag{4.5.37}$$

since

$$[\hat{f}(\mathbf{R})\mathbf{g}](k) = \sum_{m=0}^{N-1} [\hat{f}(\mathbf{R})]_{km} g(m) = \sum_{m=0}^{N-1} f(k - m \bmod N) g(m).$$

Equation (4.5.37) can be used to show that

$$\hat{f}(\mathbf{R})\hat{g}(\mathbf{R}) = \hat{g}(\mathbf{R})\hat{f}(\mathbf{R}) = \widehat{(\mathbf{f} \circledast \mathbf{g})}(\mathbf{R}). \tag{4.5.38}$$

Circulant matrices form an *algebra*. That means that the sum of two circulants is circulant and the product of two circulants is circulant. In fact the algebra of circulants is a "matrix representation" for the convolution algebra.

What happens when the DFT is used as a coordinate transformation? Apply the linear combination (4.5.25) to Eq. (4.5.22). The result is

$$\mathbf{V}\left[\sum_{m=0}^{N-1} g(m)\mathbf{R}^{-m}\right] = \left[\sum_{m=0}^{N-1} g(m)\mathbf{D}^{-m}\right]\mathbf{V},$$

or

$$\mathbf{V}\,\hat{g}(\mathbf{R}) = \hat{g}(\mathbf{D})\mathbf{V}. \tag{4.5.39}$$

Combining this with Eq. (4.5.38), we have

$$\mathbf{V}[\hat{g}(\mathbf{R})\hat{f}(\mathbf{R})]\mathbf{V}^{-1} = \hat{g}(\mathbf{D})\hat{f}(\mathbf{D}). \tag{4.5.40}$$

Now, using the definitions of $\mathbf{D}$ in Eq. (4.5.21) and of $\hat{f}(z)$ in Eq. (4.5.33), we see that $\hat{f}(\mathbf{D})$ is diagonal, and the diagonal elements are none other than the DFT values:

$$\hat{f}(\mathbf{D}) = \text{Diag}\{F(0), F(1), \cdots, F(N-1)\}. \tag{4.5.41}$$

The DFT matrix $\mathbf{V}$, used as a coordinate transformation in Eq. (4.5.40), converts the algebra of circulants (which represents circular convolution) into the algebra of diagonal matrices (whose multiplication is simply products of

TABLE 4.5.1 Properties of the discrete Fourier transform.

Linearity: $\mathbf{V}[\alpha\mathbf{f} + \beta\mathbf{g}] = \alpha\mathbf{V}\,\mathbf{f} + \beta\mathbf{V}\,\mathbf{g}$

$$\alpha f(k) + \beta g(k) \quad \overset{DFT;\,N}{\longleftrightarrow} \quad \alpha F(n) + \beta G(n)$$

Orthogonality properties: $\mathbf{V}^*\mathbf{V} = N\,\mathbf{I}$

$$\sum_{k=0}^{N-1} f(k)[g(k)]^* = \frac{1}{N}\sum_{n=0}^{N-1} F(n)[G(n)]^*$$

$$\sum_{k=0}^{N-1} |f(k)|^2 = \frac{1}{N}\sum_{n=0}^{N-1} |F(n)|^2$$

Time shift property: $\mathbf{V}\mathbf{R}^{-1} = \mathbf{D}^{-1}\mathbf{V}$

$$f(k - m \bmod N) \quad \overset{DFT;\,N}{\longleftrightarrow} \quad W_N^{-mn}F(n)$$

$$\sum_{m=0}^{N-1} g(m) f(k - m \bmod N) \quad \overset{DFT;\,N}{\longleftrightarrow} \quad F(n)G(n)$$

Frequency shift property: $\mathbf{V}\mathbf{D} = \mathbf{R}^{-1}\mathbf{V}$

$$f(k)W_N^{km} \quad \overset{DFT;\,N}{\longleftrightarrow} \quad F(n - m \bmod N)$$

$$f(k)g(k) \quad \overset{DFT;\,N}{\longleftrightarrow} \quad \frac{1}{N}\sum_{m=0}^{N-1} G(m)F(n - m \bmod N)$$

Symmetry properties

$$\mathrm{Im}[f(k)] = 0 \quad \Leftrightarrow \quad F(-n \bmod N) = [F(n)]^*$$

$$f(-k \bmod N) = [f(k)]^* \quad \Leftrightarrow \quad \mathrm{Im}[F(n)] = 0$$

$$\mathrm{Re}[f(k)] = 0 \quad \Leftrightarrow \quad F(-n \bmod N) = -[F(n)]^*$$

$$f(-k \bmod N) = -[f(k)]^* \quad \Leftrightarrow \quad \mathrm{Re}[F(n)] = 0$$

diagonal elements). This is the matrix version of the convolution-multiplication property.

There is a dual set of properties to the time shift properties. These are the frequency shift properties, and derive from the dual to Eq. (4.5.20). That dual is

$$\mathbf{V}\mathbf{D} = \mathbf{R}^{-1}\mathbf{V} \tag{4.5.42}$$

(see Problem 4.23). The DFT properties are listed in Table 4.5.1.

4.6 SAMPLING AND GENERALIZED CONVOLUTION

In many applications there will be present in the same system signals of more than one type. For each class of signals there is an appropriate Fourier transform. If there is a relation between the signals, there must be a corresponding relation between their transforms. In this section we shall develop these relations.

There are three types of applications that will be considered. The first of these is *sampling.* One can sample in either domain, time or frequency. The sampling operation produces a signal which is more discrete than the original, and typically there will be some loss of information. The other two types of applications generalize the convolution-multiplication and multiplication-convolution properties that all four of the transforms enjoy. The difference is that now there will be two signals involved which are not of the same type. For each of these three types of applications, we will give a table containing the relations between the transform pairs involved.

Sampling and Aliasing

Sampling is a necessary part of any system which has analog input signals, but does digital processing. Let

$$f(t) \quad \xleftrightarrow{FT} \quad F(j\omega) \tag{4.6.1}$$

be an analog signal and let

$$g(k) = t_0 f(kt_0) \quad \xleftrightarrow{DTFT; t_0} \quad G(e^{j\omega t_0}) \tag{4.6.2}$$

be a discrete-time signal derived from $f(t)$ by sampling with period t_0. Now the relation between **f** and **g** in the time domain is quite simple. The interesting question is what relation does G have to F? Our goal is to uncover this relation.

There is a straightforward approach to this. One starts with the DTFT G on the left-hand side and uses first the definition of $g(k)$ and then the inverse FT:

$$\begin{aligned} G(e^{j\omega t_0}) &= \sum_k t_0 f(kt_0) e^{-j\omega k t_0} \\ &= t_0 \sum_k \frac{1}{2\pi} \int_{-\infty}^{\infty} F(j\phi) e^{j(\phi-\omega)kt_0} d\phi \\ &= \frac{t_0}{2\pi} \int_{-\infty}^{\infty} F(j\phi) \left[\sum_k e^{j(\phi-\omega)kt_0} \right] d\phi. \end{aligned}$$

But the sum inside the integral does not converge, and consequently the expression is meaningless. This is an example where exchanging the order of the integral and the sum is not permitted. There is a technique which yields the relation we are seeking, but is indirect. (This method may be used to derive any of the 8 relations in Table 4.6.1.)

Let us start with two observations. First, with

$$\omega_0 t_0 = 2\pi, \tag{4.6.3}$$

then

$$e^{j(\omega + n\omega_0)kt_0} = e^{j\omega kt_0} \tag{4.6.4}$$

for all integers n. Second, if

$$g(k) = \frac{t_0}{2\pi}\int_0^{\omega_0} G(e^{j\omega t_0})e^{j\omega kt_0}d\omega = \frac{t_0}{2\pi}\int_0^{\omega_0} H(e^{j\omega t_0})e^{j\omega kt_0}d\omega \tag{4.6.5}$$

for all integers k, then $G = H$ (at least in the L_2 sense).

Beginning with the sampling relation

$$g(k) = t_0 f(kt_0),$$

express $g(k)$ using the DTFT inversion rule and $f(kt_0)$ using the FT inversion rule. The result is

$$\frac{t_0}{2\pi}\int_0^{\omega_0} G(e^{j\omega t_0})e^{j\omega kt_0}d\omega = \frac{t_0}{2\pi}\int_{-\infty}^{\infty} F(j\omega)e^{j\omega kt_0}d\omega.$$

Recognizing that $e^{j\omega kt_0}$ is periodic with period ω_0, let us break the FT integral into periods:

$$\begin{aligned}\frac{t_0}{2\pi}\int_{-\infty}^{\infty} F(j\omega)e^{j\omega kt_0}d\omega &= \frac{t_0}{2\pi}\sum_n \int_{n\omega_0}^{(n+1)\omega_0} F(j\omega)e^{j\omega kt_0}d\omega \\ &= \frac{t_0}{2\pi}\int_0^{\omega_0} \sum_n [F(j\omega + jn\omega_0)e^{j(\omega + n\omega_0)kt_0}]d\omega.\end{aligned}$$

Now use our first observation, Eq. (4.6.4) to get

$$\frac{t_0}{2\pi}\int_0^{\omega_0} G(e^{j\omega t_0})e^{j\omega kt_0}d\omega = \frac{t_0}{2\pi}\int_0^{\omega_0}\left[\sum_n F(j\omega + jn\omega_0)\right]e^{j\omega kt_0}\,d\omega.$$

From the second observation, it follows that

$$G(e^{j\omega t_0}) = \sum_{n=-\infty}^{\infty} F(j\omega + jn\omega_0). \tag{4.6.6}$$

This equation is the frequency domain consequence of sampling in time.

Sampling destroys information. In the time domain, the samples give no indication, in general, of what the signal $f(t)$ does between samples. In the frequency domain, the infinite ω axis is cut into strips of length ω_0, which are then superimposed and the spectral components summed to form G. This destructive operation is called *aliasing*. The set $\{\omega + n\omega_0$, n an integer$\}$ is called the set of *aliases* of ω. Knowing only the sum of F over all the aliases of ω, one cannot in general reconstruct $F(j\omega)$. The signals and their spectra are depicted in Fig. 4.6.1.

The following example gives an alternate approach to understanding the sampling-aliasing phenomenon.

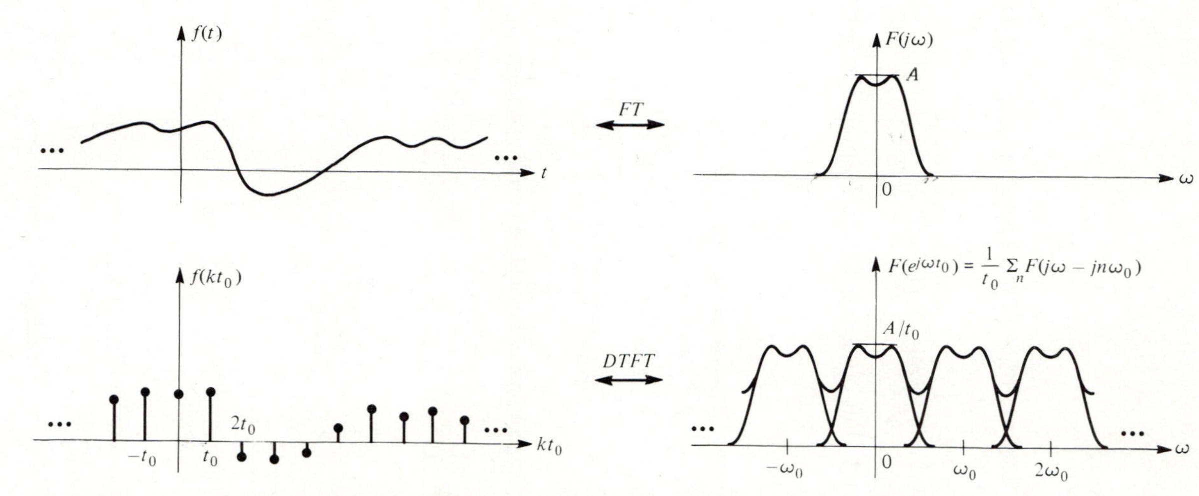

FIGURE 4.6.1 **A sketch of the aliased spectrum corresponding to the spectrum of the discrete-time signal $\{f(kt_0)\}$.**

EXAMPLE 4.6.1

Nonideal sampling.

Let

$$f(t) \xleftrightarrow{FT} F(j\omega), \tag{4.6.7}$$

$$g(t) \xleftrightarrow{FS;\,\omega_0} G(n). \tag{4.6.8}$$

What happens if we multiply a finite energy analog signal (FT) with a periodic analog signal (FS)? Let

$$y(t) = f(t)g(t) = \sum_n G(n)e^{jn\omega_0 t} f(t). \tag{4.6.9}$$

Use the frequency shift property of the FT applied to $[e^{jn\omega_0 t}f(t)]$ and linearity to get

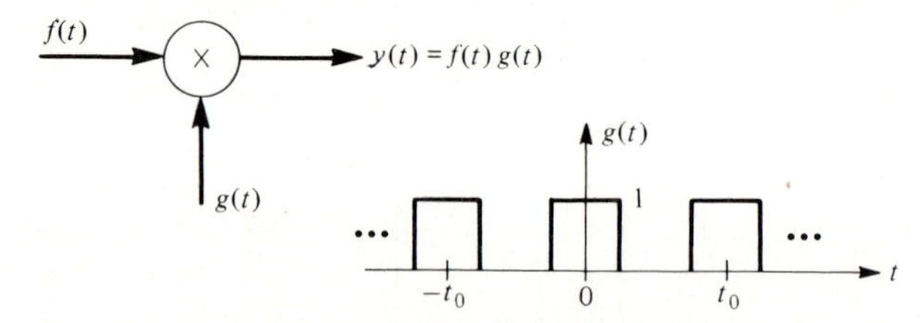

FIGURE 4.6.2 **An example of a nonideal sampling scheme.**

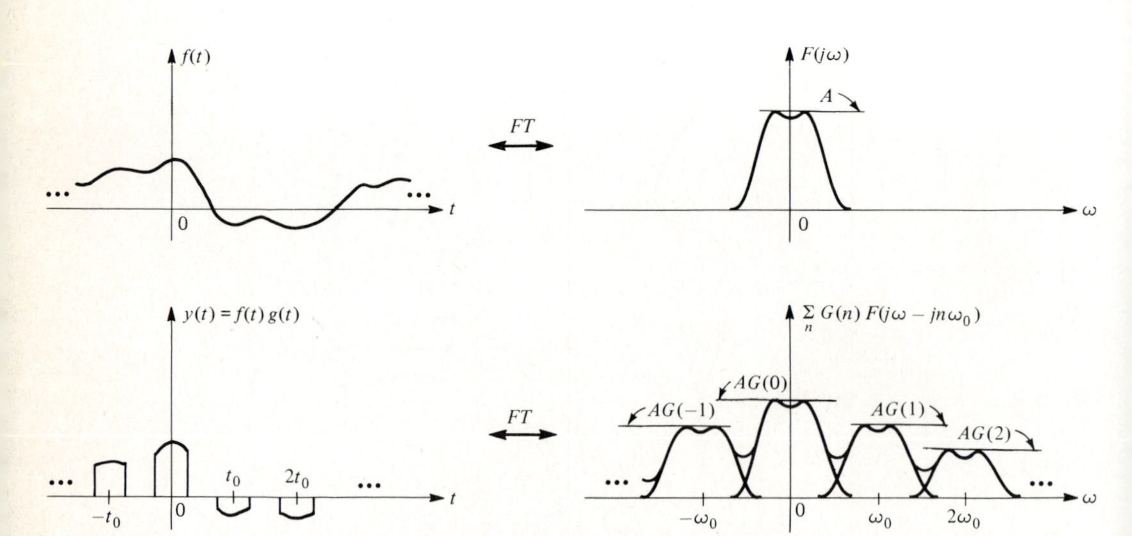

FIGURE 4.6.3 **The spectrum of a sampled waveform using nonideal sampling.**

$$Y(j\omega) = \sum_n G(n)F(j\omega - jn\omega_0). \tag{4.6.10}$$

Now this is not really sampling at all. It is *multiplication in time*, and a kind of *convolution in frequency* (in fact it is one entry in Table 4.6.3). However, if $g(t)$ is a narrow pulse train, as shown in Fig. 4.6.2, then the effect is like that of sampling since that part of the signal $f(t)$ which lies between pulses is destroyed. An actual application of this scheme is time multiplexing of signals (a second signal could be given the time slots for which $y(t) = g(t)f(t)$ is zero). See Problem 4.35.

TABLE 4.6.1 Sampling relations.

Signal class	Sampling in time, aliasing in frequency	Sampling in frequency, aliasing in time
$f(t)$	$g(k) = t_0 f(kt_0)$	$h(t) = \sum_{k=-\infty}^{\infty} f(t + kt_0)$
$FT \updownarrow$	$DTFT \updownarrow\ t_0 = \frac{2\pi}{\omega_0}$	$FS \updownarrow\ \omega_0 = \frac{2\pi}{t_0}$
$F(j\omega)$	$G(e^{j\omega t_0}) = \sum_{n=-\infty}^{\infty} F(j\omega + jn\omega_0)$	$H(n) = \frac{\omega_0}{2\pi} F(jn\omega_0)$
$f(t)$	$g(k) = \frac{1}{N} f\left(k\frac{t_0}{N}\right)$	$h(t) = \sum_{k=0}^{N-1} f\left(t + k\frac{t_0}{N}\right)$
$FS \updownarrow\ \omega_0 = \frac{2\pi}{t_0}$	$DFT \updownarrow\ N$	$FS \updownarrow\ N\omega_0$
$F(n)$	$G(n) = \sum_{m=-\infty}^{\infty} F(n + mN)$	$H(n) = NF(nN)$
$f(k)$	$g(k) = Nf(kN)$	$h(k) = \sum_{m=-\infty}^{\infty} f(k + mN)$
$DTFT \updownarrow\ t_0 = \frac{2\pi}{\omega_0}$	$DTFT \updownarrow\ Nt_0$	$DFT \updownarrow\ N$
$F(e^{j\omega t_0})$	$G(e^{j\omega N t_0}) = \sum_{m=0}^{N-1} F(e^{j\left(\omega + m\frac{\omega_0}{N}\right)t_0})$	$H(n) = F(W_N^n)$
$f(k)$	$g(k) = Mf(kM)$	$h(k) = \sum_{m=0}^{L-1} f(k + mM)$
$DFT \updownarrow\ N = LM$	$DFT \updownarrow\ L$	$DFT \updownarrow\ M$
$F(n)$	$G(n) = \sum_{m=0}^{M-1} F(n + mL)$	$H(n) = F(nL)$

The spectrum $Y(j\omega)$ for the "chopped" output signal $y(t)$ is shown in Fig. 4.6.3. The forms of the right-hand sides of Eqs. (4.6.10) and (4.6.6) are very similar. They become identical when the Fourier coefficients $G(n)$ are all equal to one. This is what happens in the limit as the pulse width approaches zero, and the pulse heights approach infinity so that $g(t)$ becomes an impulse train. (See Example 4.6.2 below.)

The sampling-aliasing relation (4.6.6) is the result of sampling a FT type signal in the time domain. One can also reverse the roles of time and frequency, and sample in the frequency domain. The dual is an aliasing-sampling relation. With 4 types of transforms and 2 domains in which to sample, there are 8 possibilities. These can all be derived using the steps involved in the derivation of Eq. (4.6.6). The resulting transform pairs are given in Table 4.6.1. If one samples a periodic signal, then there should be an integer number of samples per period. This could be called *synchronous sampling*. If one samples a signal which is already discrete, then the operation is called *subsampling*. In order to see what application an aliasing-sampling relation might have, let us consider the following example.

EXAMPLE 4.6.2

A periodic pulse train from one rectangular pulse.

Consider the Fourier transform pair (4.4.11). This signal is a single rectangular pulse of width $2t_1$ in the time domain:

$$q(t) = \frac{1}{2t_1}\,\text{rect}(t/t_1) \quad \overset{FT}{\longleftrightarrow} \quad Q(j\omega) = \text{sinc}(\omega t_1). \tag{4.6.11}$$

Sampling this signal in the frequency domain aliases the signal in the time domain. If the period $t_0 = 2\pi/\omega_0$ is greater than $2t_1$, then this is simply a periodic extension of $q(t)$:

$$p(t) = \sum_{k=-\infty}^{\infty} q(t - kt_0)$$

$$FS \updownarrow \omega_0$$

$$P(n) = \frac{\omega_0}{2\pi} Q(jn\omega_0) = \frac{1}{t_0}\,\text{sinc}(n\omega_0 t_1). \tag{4.6.12}$$

The signal $p(t)$ is a periodic pulse train of period t_0, pulse width $2t_1$, and pulse height $1/(2t_1)$. If we multiply Eq. (4.6.12) by t_0 and let t_1 approach zero, then we get the fictitious Fourier series pair

$$p(t) = t_0 \sum_k \delta(t - kt_0) \quad \overset{FS;\,\omega_0}{\longleftrightarrow} \quad P(n) = 1. \tag{4.6.13}$$

Now if we use this in Example 4.6.1, we get the fictitious Fourier transform pair

$$y(t) = \sum_k t_0 f(kt_0)\delta(t - kt_0) \quad \overset{FT}{\longleftrightarrow} \quad Y(j\omega) = \sum_n F(j\omega - jn\omega_0). \tag{4.6.14}$$

This spectrum is the aliased spectrum in Eq. (4.6.6). Thus one can think of the aliased spectrum $Y(j\omega)$ as the Fourier transform of the impulse train $y(t)$ in Eq. (4.6.14), or as the DTFT of the sequence of samples $g(k) = t_0 f(kt_0)$ in Eq. (4.6.6).

EXAMPLE 4.6.3

Subsampling.

Let **f** be a finite energy discrete-time signal, and consider sampling **f** by taking every Nth sample. This produces another finite energy discrete-time signal **g**.

$$f(k) \quad \overset{DTFT; t_0}{\longleftrightarrow} \quad F(e^{j\omega t_0}),$$

$$g(k) = Nf(kN) \quad \overset{DTFT; Nt_0}{\longleftrightarrow} \quad G(e^{j\omega N t_0}).$$

If the assumed sampling period for **f** is t_0, then the assumed sampling period for **g** must be Nt_0. This is called *subsampling*. From Table 4.6.1, we have

$$G(e^{j\omega N t_0}) = \sum_{m=0}^{N-1} F(e^{j(\omega + m\omega_0/N)t_0}). \tag{4.6.15}$$

This is aliasing in the frequency domain, except there are only N terms in the sum. Since F is ω_0-periodic, N shifts of F through ω_0/N will reproduce it.

Take $N = 2$, and

$$F(z) = \sum_{k=0}^{\infty} f(k)z^{-k} = \frac{1}{1 - 2r\cos(\theta_0)z^{-1} + r^2 z^{-2}}.$$

Then with $g(k) = 2f(2k)$, we have

$$G(z) = \sum_{k=0}^{\infty} 2f(2k)z^{-k} = \frac{2[1 + r^2 z^{-1}]}{1 - 2r^2\cos(2\theta_0)z^{-1} + r^4 z^{-2}}.$$

These sequences and their spectra are shown in Fig. 4.6.4. If the sampling period is t_0 for the sequence **f**, then the sampling period for **g** should be $2t_0$. Thus we set $z = e^{j\omega t_0}$ for F, but we set $z = e^{j\omega 2t_0}$ for G. As a result, F is periodic with period ω_0 (as a function of ω) but G has period $\omega_0/2$.

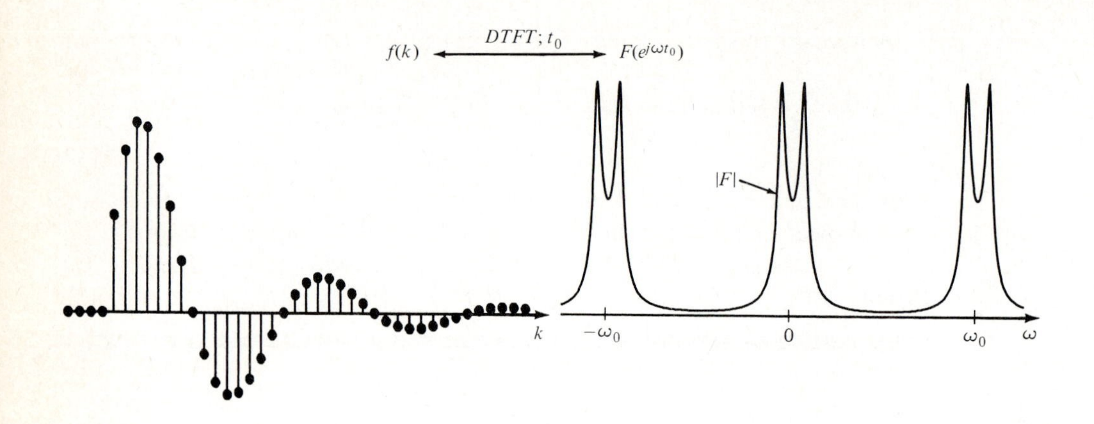

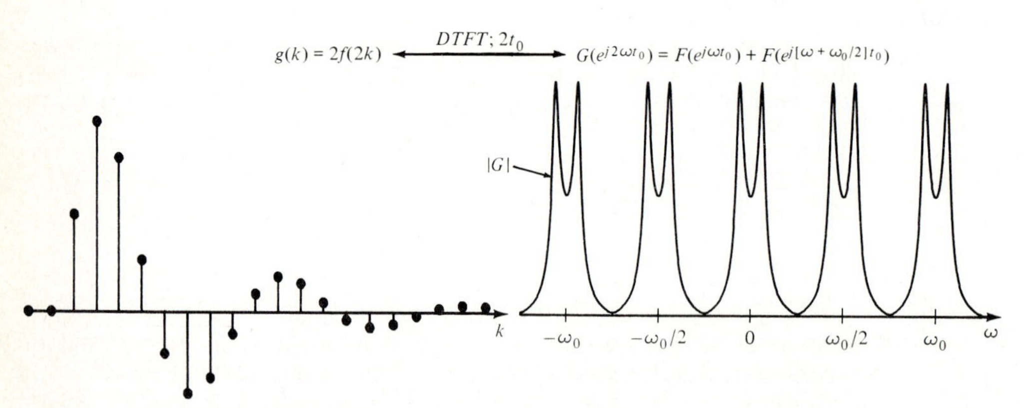

FIGURE 4.6.4 Subsampling with $N=2$. In this example, $F(z)=[1-2r\cos(\theta_0)z^{-1}+r^2z^{-2}]^{-1}$, with $r=0.9$, $\theta_0=\pi/8$.

Generalized Convolution-Multiplication

For each of the four signal classes we have studied, there is a natural convolution operation in the time domain, which leads to multiplication in the frequency domain. There are 6 combinations of 2 signal classes. For each one of these 6 combinations, one can define a convolution operation which involves one signal from each of the two signal classes. The resulting signal will belong to one of the 4 classes and have a spectrum which involves the multiplication of the two original spectra, possibly with sampling. These relations are much easier to derive than the sampling-aliasing relations. Just as we have done for each transform alone, the convolution-multiplication properties involve only a time shift property, and linearity. An important application is signal reconstruction, which is the subject of the following example.

EXAMPLE 4.6.4

Approximate reconstruction of an analog signal from samples.

Suppose that we are given a discrete-time signal

$$g(k) \quad \xleftrightarrow{DTFT;\,t_0} \quad G(e^{j\omega t_0}), \tag{4.6.16}$$

and wish to construct a continuous-time signal

$$y(t) \quad \xleftrightarrow{FT} \quad Y(j\omega), \tag{4.6.17}$$

for which

$$g(k) = t_0 y(kt_0). \tag{4.6.18}$$

This operation is an "inverse" of the sampling operation. Now for filtering applications, this reconstruction process should be linear and have a form of shift-invariance. In Chapter 2, we saw that these properties lead to the convolution operation.

Suppose that we apply the unit pulse sequence $\boldsymbol{\delta}$ to our "reconstruction filter." Call the resulting continuous time signal

$$r(t) \quad \xleftrightarrow{FT} \quad R(j\omega). \tag{4.6.19}$$

By *shift invariance*, we mean that if we apply a shifted unit pulse sequence, for which the pulse position is m, to the reconstruction filter, then the output should be $r(t - mt_0)$. Thus, by linearity, the input sequence $\mathbf{g}$ will produce the output signal

$$y(t) = \sum_k g(k)r(t - kt_0). \tag{4.6.20}$$

This is our reconstruction filter, and it amounts to a kind of convolution between the DTFT signal $\mathbf{g}$ and the FT signal $\mathbf{r}$ to produce an FT signal $\mathbf{y}$.

The goal of the reconstruction filter was to produce a signal $\mathbf{y}$ whose samples satisfy $g(k) = t_0 y(kt_0)$. Moreover, this should be true for every sequence $\mathbf{g}$. Suppose we let $\mathbf{g} = \boldsymbol{\delta}$. Then Eq. (4.6.20) gives $y(t) = r(t)$. Thus $g(k) = t_0 y(kt_0)$ implies

$$\delta(k) = t_0 r(kt_0). \tag{4.6.21}$$

In other words, the samples of the reconstruction filter pulse response should be $1/t_0$ at $k = 0$, and zero for all other values of k. But if this is true, then sampling Eq. (4.6.20) produces the desired property; namely Eq. (4.6.18).

The spectrum $Y(j\omega)$ can be obtained quickly, using linearity and the time shift property of the FT. From Eq. (4.6.20)

$$Y(j\omega) = \sum_{k=-\infty}^{\infty} g(k)[e^{-j\omega k t_0} R(j\omega)] = G(e^{j\omega t_0})R(j\omega). \tag{4.6.22}$$

This is a form of multiplication in the frequency domain.

A commonly used choice for $r(t)$ is the rectangular pulse shown in Fig. 4.6.5. This is called a "zero-order hold." One simply takes the output of a digital-to-analog converter (or DAC), after the settling time has been reached, and holds it constant for one sampling period. The output signal $y(t)$ is then piecewise constant. The spectra for these three signals are shown in Fig. 4.6.6.

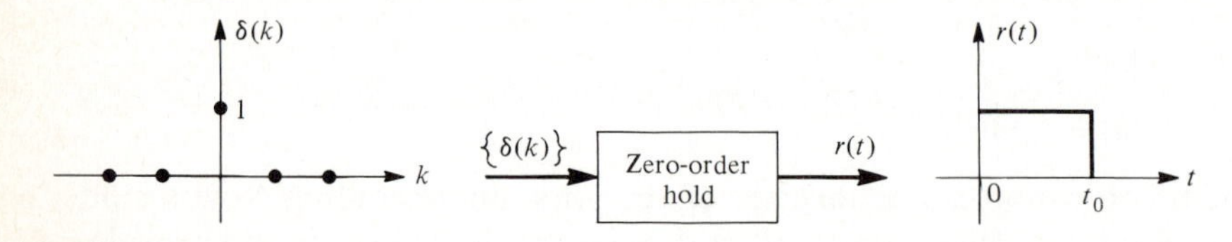

FIGURE 4.6.5 A reconstruction filter for sequences, sometimes called a "zero-order hold" filter.

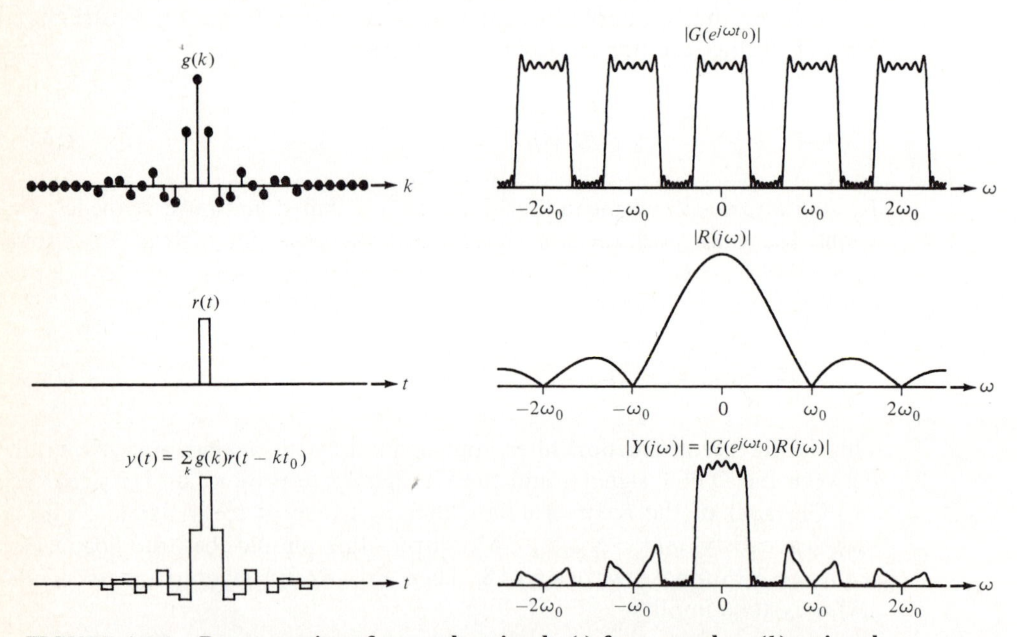

FIGURE 4.6.6 Reconstruction of an analog signal $y(t)$ from samples $g(k)$, using the zero-order hold $r(t)$.

Table 4.6.2 gives the 6 convolution-multiplication relations for the possible combination of signal classes. The reconstruction filter in Example 4.6.4 is the FT/DTFT combination.

TABLE 4.6.2 **Generalized convolution-multiplication.**

Signal types

$$f(t) \overset{FT}{\longleftrightarrow} F(j\omega)$$

$$p(t) \overset{FS;\,\omega_0}{\longleftrightarrow} P(n)$$

$$g(k) \overset{DTFT;\,t_0}{\longleftrightarrow} G(e^{j\omega t_0})$$

$$q(k) \overset{DFT;\,N}{\longleftrightarrow} Q(n)$$

FT/FS combination (Analog filter with periodic input)

$$y(t) = \int_{-\infty}^{\infty} p(\tau) f(t-\tau) d\tau \quad \overset{FS;\,\omega_0}{\longleftrightarrow} \quad Y(n) = P(n) F(jn\omega_0)$$

FT/DTFT combination (Reconstruction filters)

$$y(t) = \sum_{k=-\infty}^{\infty} g(k) f(t - kt_0) \quad \overset{FT}{\longleftrightarrow} \quad Y(j\omega) = G(e^{j\omega t_0}) F(j\omega)$$

FT/DFT combination

$$y(t) = \sum_{k=-\infty}^{\infty} q(k) f\left(t - \frac{2\pi k}{N\omega_0}\right) \quad \overset{FS;\,\omega_0}{\longleftrightarrow} \quad Y(n) = \frac{\omega_0}{2\pi} Q(n) F(jn\omega_0)$$

FS/DTFT combination (No relation between ω_0 and t_0 here)

$$y(t) = \sum_{k=-\infty}^{\infty} g(k) p(t - kt_0) \quad \overset{FS;\,\omega_0}{\longleftrightarrow} \quad Y(n) = G(e^{jn\omega_0 t_0}) P(n)$$

FS/DFT combination (Waveform synthesizers)

$$y(t) = \sum_{k=0}^{N-1} q(k) p\left(t - \frac{2\pi k}{N\omega_0}\right) \quad \overset{FS;\,\omega_0}{\longleftrightarrow} \quad Y(n) = Q(n) P(n)$$

DTFT/DFT combination (Digital filter with periodic input)

$$y(k) = \sum_{m=-\infty}^{\infty} q(m) g(k-m) \quad \overset{DFT;\,N}{\longleftrightarrow} \quad Y(n) = Q(n) G(W_N^n)$$

EXAMPLE 4.6.5

An analog filter with a periodic input.

The FT/FS combination in Table 4.6.2 characterizes the situation of an analog filter

$$h(t) \overset{FT}{\longleftrightarrow} H(j\omega), \tag{4.6.23}$$

[whose impulse response is $h(t)$ and frequency response is $H(j\omega)$] driven by a periodic input signal. If the input is

$$u(t) = e^{jn\omega_0 t}, \tag{4.6.24}$$

then the output is

$$\begin{aligned} y(t) = (\mathbf{h} * \mathbf{u})(t) &= \int_0^\infty h(\tau)e^{jn\omega_0(t-\tau)}d\tau \\ &= H(jn\omega_0)e^{jn\omega_0 t}. \end{aligned} \tag{4.6.25}$$

TABLE 4.6.3 **Generalized multiplication-convolution.**

Signal types

$$f(t) \xleftrightarrow{FT} F(j\omega)$$

$$p(t) \xleftrightarrow{FS;\,\omega_0} P(n)$$

$$g(k) \xleftrightarrow{DTFT;\,t_0} G(e^{j\omega t_0})$$

$$q(k) \xleftrightarrow{DFT;\,N} Q(n)$$

FT/FS combination (Modulation, time multiplexing, non-ideal sampling)

$$y(t) = f(t)p(t) \xleftrightarrow{FT} Y(j\omega) = \sum_{n=-\infty}^{\infty} P(n)F(j\omega - jn\omega_0)$$

FT/DTFT combination (Analysis of DFT approximation to FT in Section 4.8)

$$y(k) = g(k)f(kt_0) \xleftrightarrow{DTFT;\,t_0} Y(e^{j\omega t_0}) = \frac{1}{2\pi}\int_{-\infty}^{\infty} G(e^{j\phi t_0})F(j\omega - j\phi)d\phi$$

FT/DFT combination

$$y(k) = q(k)f(kt_0) \xleftrightarrow{DTFT;\,t_0} Y(e^{j\omega t_0}) = \frac{1}{Nt_0}\sum_{n=-\infty}^{\infty} Q(n)F\left(j\omega - j\frac{2\pi n}{Nt_0}\right)$$

FS/DTFT combination

$$y(k) = g(k)p(kt_0) \xleftrightarrow{DTFT;\,t_0} Y(e^{j\omega t_0}) = \sum_{n=-\infty}^{\infty} P(n)G(e^{j(\omega - n\omega_0)t_0})$$

FS/DFT combination

$$y(k) = q(k)p\left(\frac{2\pi k}{N\omega_0}\right) \xleftrightarrow{DFT;\,N} Y(n) = \sum_{m=-\infty}^{\infty} P(m)Q(n-m)$$

DTFT/DFT combination (Discrete-time modulation)

$$y(k) = q(k)g(k) \xleftrightarrow{DTFT;\,t_0} Y(e^{j\omega t_0}) = \frac{1}{N}\sum_{n=0}^{N-1} Q(n)G\left(e^{j\left(\omega - \frac{2\pi n}{Nt_0}\right)t_0}\right)$$

Thus, with input

$$g(t) = \sum_n G(n)e^{jn\omega_0 t}, \qquad (4.6.26)$$

the output is (by linearity)

$$y(t) = \sum_n [G(n)H(jn\omega_0)]e^{jn\omega_0 t}. \qquad (4.6.27)$$

This signal is periodic, and has Fourier coefficients

$$Y(n) = G(n)H(jn\omega_0). \qquad (4.6.28)$$

Generalized Multiplication-Convolution

The transform pairs in Table 4.6.2 all have duals. These involve multiplication in the time domain and a form of convolution in the frequency domain, and are listed in Table 4.6.3. Each transform pair in the table can be developed using only a frequency shift property and linearity. The transform pair in Example 4.6.1, Eqs. (4.6.9) and (4.6.10) is the FT/FS combination in Table 4.6.3.

4.7 EQUIVALENT ANALOG FILTERS AND THE SAMPLING THEOREM

Many applications of digital signal processing involve analog signals. Suppose that we have an analog input signal. This signal is sampled, and the digital representation of the signal may then be processed. (It could also be stored on tape or optical disk.) The next step is to pass the discrete-time signal through a digital filter. Finally, to get an analog output, a reconstruction filter is used, as in Example 4.6.4. The original signal has gone through the three operations shown along the lower path in Fig. 4.7.1. In this section we shall consider the composition of these three operations and ask two questions. Can the entire process be equivalent to an analog filter? And under what circumstances can the original analog input signal be perfectly reproduced from its samples?

Let the analog input signal in Fig. 4.7.1 have finite energy. This is

$$u(t) \quad \xleftrightarrow{FT} \quad U(j\omega). \qquad (4.7.1)$$

The first operation of the three is sampling. This produces the discrete-time signal

$$\tilde{u}(k) = u(kt_0) \quad \xleftrightarrow{DTFT; t_0} \quad \tilde{U}(e^{j\omega t_0}) = \frac{1}{t_0}\sum_n U(j\omega + jn\omega_0), \qquad (4.7.2)$$

where

$$\omega_0 t_0 = 2\pi. \qquad (4.7.3)$$

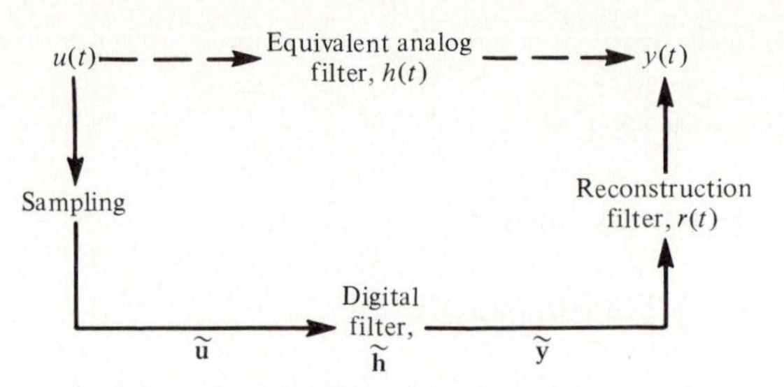

FIGURE 4.7.1 Sampling, filtering, and reconstruction.

The sampling operation produces aliasing in the frequency domain.

The second operation is to pass $\tilde{\mathbf{u}}$ through a discrete-time, linear, shift-invariant filter. The output of this filter is

$$\tilde{y}(k) = [\tilde{\mathbf{h}} * \tilde{\mathbf{u}}](k) \quad \overset{DTFT;\, t_0}{\longleftrightarrow} \quad \tilde{Y}(e^{j\omega t_0}) = \tilde{H}(e^{j\omega t_0})\tilde{U}(e^{j\omega t_0}). \tag{4.7.4}$$

Finally, the analog output $y(t)$ is obtained by passing $\tilde{\mathbf{y}}$ through a reconstruction filter. If the unit pulse response of this device is $r(t)$, then using Eq. (4.6.20), the output is

$$y(t) = \sum_k \tilde{y}(k) r(t - kt_0) \quad \overset{FT}{\longleftrightarrow} \quad Y(j\omega) = \tilde{Y}(e^{j\omega t_0}) R(j\omega). \tag{4.7.5}$$

In the frequency domain, the composition of these three operations is described by combining the right-hand sides of Eqs. (4.7.5), (4.7.4), and (4.7.2). This is

$$Y(j\omega) = \frac{1}{t_0} \tilde{H}(e^{j\omega t_0}) \sum_{n=-\infty}^{\infty} R(j\omega) U(j\omega + jn\omega_0). \tag{4.7.6}$$

Equivalent Analog Filtering

Now let us consider the first question involving these operations. When is the composition of sampling, digital filtering, and reconstruction equivalent to an analog filtering operation? Such a filter would produce the relations

$$y(t) = \int_{-\infty}^{\infty} h(\tau) u(t - \tau) d\tau \quad \overset{FT}{\longleftrightarrow} \quad Y(j\omega) = H(j\omega) U(j\omega), \tag{4.7.7}$$

in time and frequency. Thus the two are equivalent if and only if

$$H(j\omega) U(j\omega) = \frac{1}{t_0} \tilde{H}(e^{j\omega t_0}) \left[\sum_{n=-\infty}^{\infty} R(j\omega) U(j\omega + jn\omega_0) \right]. \tag{4.7.8}$$

This equation must hold as an identity in ω for all input spectra $U(j\omega)$ that we are likely to encounter. The problem is the aliasing on the right-hand side; the terms in the summation with $n \neq 0$ lead to a frequency shift of the input, which is something that a linear, shift-invariant filter cannot do. A sufficient condition for the existence of an equivalent analog filter is that

$$R(j\omega)U(j\omega + jn\omega_0) = 0 \qquad \text{for } n \neq 0. \tag{4.7.9}$$

The sum in Eq. (4.7.8) then reduces to one term and the resulting analog frequency response function is

$$H(j\omega) = \frac{1}{t_0}\tilde{H}(e^{j\omega t_0})R(j\omega). \tag{4.7.10}$$

Notice, however, that the conditions (4.7.9) place constraints on the input signal.

Suppose that the input signal $u(t)$ and reconstruction filter unit pulse response $r(t)$ are both band limited to $\omega_0/2$ radians per second:

$$|\omega| \geqslant \omega_0/2 \qquad \Rightarrow \qquad U(j\omega) = R(j\omega) = 0. \tag{4.7.11}$$

Then Eqs. (4.7.9) will hold. In practice, these are the conditions that one tries for. If the input is not band limited, but one is interested only in the low frequency part of the signal, then the condition (4.7.11) can be imposed by passing $u(t)$ through an analog low-pass filter with cutoff less than $\omega_0/2$. Such a filter is called an *anti-aliasing filter.*

The Shannon Sampling Theorem

Under what circumstances do the samples $u(kt_0)$ of an analog signal completely determine the signal? This question is closely related to the question of the existence of an equivalent analog filter. For if we can achieve

$$Y(j\omega) = H(j\omega)U(j\omega) = U(j\omega), \tag{4.7.12}$$

then $\mathbf{y} = \mathbf{u}$. But $\mathbf{y}$ is constructed from the samples of $\mathbf{u}$, and therefore $\mathbf{u}$ is determined by its samples.

Assume, as we did in condition (4.7.11), that $\mathbf{u}$ is band limited to $\omega_0/2$ radians per second. Take $\tilde{H}(z) = 1$, and

$$r(t) = \text{sinc}(\omega_0 t/2) \qquad \overset{FT}{\longleftrightarrow} \qquad R(j\omega) = t_0 \ \text{rect}(2\omega/\omega_0). \tag{4.7.13}$$

[See Eq. (4.4.10).] Then $\mathbf{r}$ is also band limited, and

$$H(j\omega) = \frac{1}{t_0}\tilde{H}(e^{j\omega t_0})R(j\omega) = \text{rect}(2\omega/\omega_0). \tag{4.7.14}$$

In the low frequency band $|\omega| < \omega_0/2$, $H(j\omega) = 1$. Therefore Eq. (4.7.12) holds in the low frequency band. But outside the band, $U(j\omega) = 0$ (by assumption) and Eq. (4.7.12) holds as well. Therefore $Y(j\omega) = U(j\omega)$ for all ω.

Now $y(t)$ is generated by the reconstruction filter in Eqs. (4.7.5). Since

$$\tilde{H}(z) = 1,$$

$$\tilde{y}(k) = \tilde{u}(k) = u(kt_0).$$

Therefore

$$y(t) = u(t) = \sum_k u(kt_0) r(t - kt_0),$$

or

$$u(t) = \sum_{k=-\infty}^{\infty} u(kt_0) \operatorname{sinc}\left(\frac{\omega_0}{2}[t - kt_0]\right), \tag{4.7.15}$$

where $\omega_0 t_0 = 2\pi$.

This result is known as the *Shannon-Whittaker sampling theorem.* In

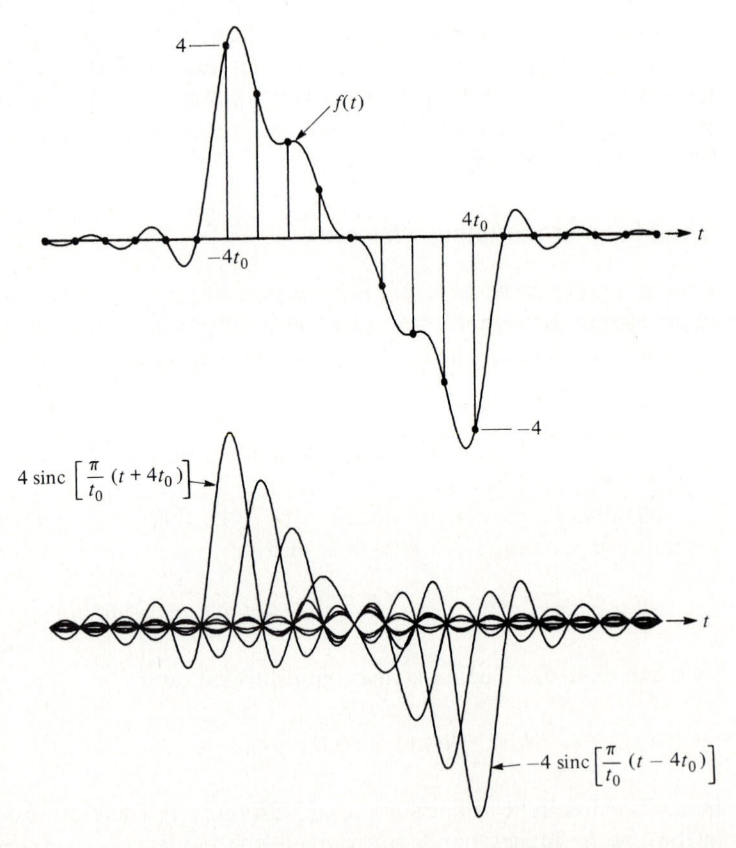

FIGURE 4.7.2 **The signal $f(t)$ is bandlimited to π/t_0 radians per second and has samples $\{f(kt_0)\} = \{\cdots 0, 0, 4, 3, 2, 1, 0, -1, -2, -3, -4, 0, 0, \cdots\}$. Then $f(t)$ has the representation (4.7.15). The sinc functions in the sum are shown in the lower part of the figure.**

summary, if

$$U(j\omega) = 0 \qquad \text{for } |\omega| > \omega_0/2,$$

then the signal $u(t)$ is completely determined by its samples, using the reconstruction formula (4.7.15).

EXAMPLE 4.7.1

Reconstruction of a band-limited signal from its samples.

Suppose a signal $f(t)$ is band limited to π/t_0 radians per second. It is then completely determined by its samples $f(kt_0)$. Suppose that these samples are

$$f(kt_0) = \begin{cases} -k, & \text{if } |k| \leqslant 4 \\ 0, & \text{otherwise.} \end{cases}$$

Then there are only 8 nonzero terms in the reconstruction formula (4.7.15). Figure 4.7.2 exhibits the signal $f(t)$, and the individual components of the sum in Eq. (4.7.15).

Practical Reconstruction

In practice, one cannot employ the reconstruction described in Eq. (4.7.15) because the unit pulse response in Eq. (4.7.13) is not causal. The simplest thing to do is to use the zero-order hold of Example 4.6.4, but this will introduce distortion in the reconstruction. Another source of difficulty is the analog low-pass anti-aliasing filter. An analog filter with sharp cutoff tends to be expensive and at the mercy of parameter variations. Both of these difficulties can be minimized by *oversampling*.

Suppose that the analog signal $u(t)$ is band limited to $\omega_0/2 = \pi/t_0$ radians per second. Then the sampling theorem requires that we sample at the rate of at least $1/t_0$ samples per second. This is called the *Nyquist rate*. Sampling at a rate $1/t_1$ which is greater than the Nyquist rate is called *oversampling*. In practice, this is typical.

Now suppose that $u(t)$ is not band limited but we are interested only in that part of it which lies in the band $|\omega| < \omega_0/2$. Let us use a low-pass filter $G(j\omega)$ which satisfies

$$\begin{aligned} G(j\omega) &\approx 1 \qquad \text{for } |\omega| < \omega_0/2, \\ G(j\omega) &\approx 0 \qquad \text{for } |\omega| > \omega_1/2. \end{aligned} \tag{4.7.16}$$

This is our anti-aliasing filter. If the transition band $\omega_0/2 < |\omega| < \omega_1/2$ is large, then this filter can be inexpensive. Similarly, let us use a reconstruction filter for which

$$\begin{aligned} R(j\omega) &\approx t_1 \qquad \text{for } |\omega| < \omega_0/2, \\ R(j\omega) &\approx 0 \qquad \text{for } |\omega| > \omega_1/2. \end{aligned} \tag{4.7.17}$$

Then the output $y(t)$ [after anti-aliasing, sampling at a rate of $1/t_1$ samples per second, digital filtering using $\tilde{H}(z)$, and reconstruction] is

$$Y(j\omega) = \frac{1}{t_1} \tilde{H}(e^{j\omega t_1})G(j\omega)R(j\omega)U(j\omega). \tag{4.7.18}$$

With oversampling, the filters G and R can be inexpensive. One can then design a *digital* filter $\tilde{H}$ with cutoff at $\omega_0/2$ of high quality. The output $\mathbf{y}$ will then be that part of $\mathbf{u}$ which lies in the band $|\omega| < \omega_0/2$. This oversampling technique is used in compact audio disk playback systems.

4.8 THE DFT APPROXIMATION TO THE FOURIER TRANSFORM

If $f(t)$ is a finite energy signal, then its Fourier transform is

$$F(j\omega) = \int_{-\infty}^{\infty} f(t)e^{-j\omega t}dt. \tag{4.8.1}$$

Suppose that we have taken N samples of the signal, beginning at $t = 0$ and spaced with sampling period t_0. What is a reasonable estimate of $F(j\omega)$ based only on these N samples? The question is equivalent to the numerical integration problem. How do you estimate an integral based on samples of the integrand? For example, the "rectangle-rule" approximation to the integral in Eq. (4.8.1) is

$$\hat{F}(j\omega) = \sum_{k=0}^{N-1} t_0 f(kt_0)e^{-j\omega k t_0}. \tag{4.8.2}$$

If we take N samples of $\hat{F}(j\omega)$ at multiples of the radian frequency

$$\omega_0 = \frac{2\pi}{Nt_0}, \tag{4.8.3}$$

we get

$$\hat{F}(jn\omega_0) = \sum_{k=0}^{N-1} t_0 f(kt_0)W_N^{-nk}. \tag{4.8.4}$$

The frequency spacing ω_0 was deliberately chosen so that

$$e^{j\omega_0 t_0} = e^{j2\pi/N} = W_N. \tag{4.8.5}$$

Thus the right-hand side of Eq. (4.8.4) involves a DFT. That is,

$$d(k) = t_0 f([k \bmod N]t_0) \quad \xleftrightarrow{DFT;N} \quad D(n) = \hat{F}(jn\omega_0). \tag{4.8.6}$$

Because the DFT is invertible, either set of parameters (the N values of $\mathbf{d}$ or the N values of $\mathbf{D}$) characterize the approximation $\hat{F}$.

The DFT can therefore be used to estimate the Fourier transform of a signal. But how good is the approximation? We shall consider this issue in the remainder of this section.

Notice that the relation between t_0 and ω_0 in Eq. (4.8.3) differs from what we have used in previous sections by the factor N. Let us use the following notation for this section only:

$$T_0 = 2\pi/\omega_0 = Nt_0,$$
$$\Omega_0 = 2\pi/t_0 = N\omega_0. \tag{4.8.7}$$

T_0 is the duration of the time interval over which $f(t)$ was sampled. Ω_0 is the period of $\hat{F}$, and the width of the frequency band over which we have N useful samples. Thus Ω_0 becomes the ω_0 of the sampling theorem in Section 4.7.

The approximation $\hat{F}(j\omega)$ is obtained from $F(j\omega)$ via two steps, each of which "destroys" part of the information in the original spectrum. First, the signal $f(t)$ is sampled. Second, only N samples are kept. Let

$$f(t) \quad \xleftrightarrow{FT} \quad F(j\omega). \tag{4.8.8}$$

Sampling this signal leads to an aliased spectrum (see Table 4.6.1):

$$g(k) = t_0 f(kt_0) \quad \xleftrightarrow{DTFT;\, t_0} \quad G(e^{j\omega t_0}) = \sum_{n=-\infty}^{\infty} F(j\omega + jn\Omega_0). \tag{4.8.9}$$

Keeping only samples 0 through $N-1$ may be modeled as a windowing operation. Let

$$w(k) = \begin{cases} 1, & 0 \leqslant k \leqslant N-1 \\ 0, & \text{otherwise} \end{cases}$$
$$\updownarrow \; DTFT \tag{4.8.10}$$
$$W(e^{j\theta}) = e^{-j[(N-1)/2]\theta} \frac{\sin(N\theta/2)}{\sin(\theta/2)}.$$

Now, multiplication in time leads to convolution in frequency, and so

$$h(k) = g(k)w(k)$$
$$\updownarrow \; DTFT;\, t_0$$
$$H(e^{j\omega t_0}) = \frac{t_0}{2\pi} \int_0^{\Omega_0} G(e^{j\phi t_0}) W(e^{j(\omega-\phi)t_0})\, d\phi.$$

The sequence $\mathbf{h}$ is nonzero only for $0 \leqslant k \leqslant N-1$. Its values are the same as one period of the signal $\mathbf{d}$ in the DFT pair (4.8.6).

The DTFT $H(e^{j\omega t_0})$ is equivalent to $\hat{F}(j\omega)$, since

$$H(e^{j\omega t_0}) = \sum_{k=-\infty}^{\infty} t_0 f(kt_0) w(k) e^{-j\omega k t_0}$$
$$= \sum_{k=0}^{N-1} t_0 f(kt_0) e^{-j\omega k t_0} = \hat{F}(j\omega). \tag{4.8.12}$$

To characterize $\hat{F}(j\omega)$ in terms of $F(j\omega)$, let us work backwards from this result.

$$\begin{aligned}\hat{F}(j\omega) = H(e^{j\omega t_0}) &= \frac{t_0}{2\pi}\int_0^{\Omega_0} G(e^{j\phi t_0})W(e^{j(\omega-\phi)t_0})d\phi \\ &= \frac{t_0}{2\pi}\int_0^{\Omega_0}\sum_{n=-\infty}^{\infty} F(j\phi + jn\Omega_0)W(e^{j(\omega-\phi)t_0})d\phi \\ &= \frac{1}{2\pi}\int_{-\infty}^{\infty} F(j\phi)[t_0 W(e^{j(\omega-\phi)t_0})]d\phi. \end{aligned} \quad \textbf{(4.8.13)}$$

The last step "unwraps" the aliasing, and is made possible by the Ω_0-periodicity of W. The final expression may be recognized as convolution in the frequency domain of the FT. Therefore, our characterization of the approximation becomes

$$\hat{F}(j\omega) = (F * Q)(j\omega) = \frac{1}{2\pi}\int_{-\infty}^{\infty} F(j\phi)Q(j\omega - j\phi)d\phi, \quad \textbf{(4.8.14)}$$

where the function Q is defined by

$$Q(j\omega) = t_0 W(e^{j\omega t_0}) = t_0 e^{-j\omega[(N-1)/2]t_0}\left[\frac{\sin(\omega T_0/2)}{\sin(\omega t_0/2)}\right]. \quad \textbf{(4.8.15)}$$

The magnitude of this window is depicted in Fig. 4.8.1. Note that the phase shift in $Q(j\omega)$ corresponds to a time shift of $(N-1)t_0/2$ which is the midpoint of the set of sampling times. In general, if the samples start at t_1 (rather than zero) then

$$\hat{F}(j\omega) = (F * P)(j\omega), \quad \textbf{(4.8.16)}$$

where

$$P(j\omega) = e^{-j\omega t_1}Q(j\omega). \quad \textbf{(4.8.17)}$$

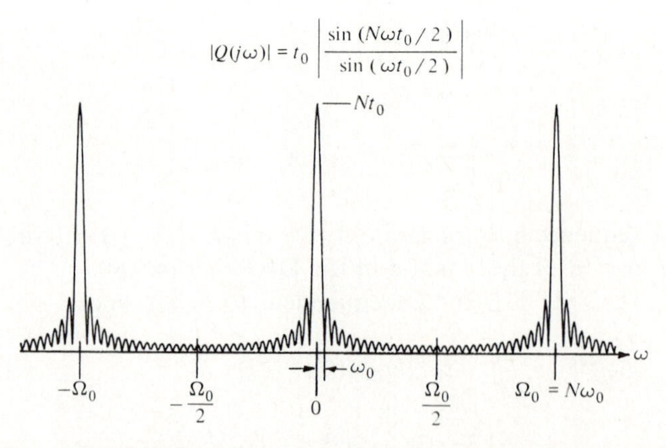

FIGURE 4.8.1 The periodic window function that characterizes the DFT approximation $\hat{F}(j\omega) = (F*Q)(j\omega)$.

Now since $Q(j\omega)$ is Ω_0-periodic, so is $\hat{F}(j\omega)$. Therefore, the approximation is usable at best only in the interval

$$|\omega| < \Omega_0/2 = \pi/t_0. \tag{4.8.18}$$

The frequency $\Omega_0/2$ is called the *foldover frequency*. To eliminate aliasing in this interval, the original spectrum $F(j\omega)$ should be bandlimited to $\Omega_0/2$. However, even if this is true, the approximation $\hat{F}(j\omega)$ will be blurred, due to the width of the main lobe of $Q(j\omega)$, which is $2\omega_0$ wide. Thus resolution is limited to roughly $2\omega_0$ radians per second. Note that to achieve greater bandwidth, we must increase Ω_0 (which is to say that we must decrease t_0). To increase resolution, we must decrease ω_0, which is to say that we must increase N for a fixed t_0. As t_0 approaches zero, the main lobes of $Q(j\omega)$ spread out. As N approaches infinity, $Q(j\omega)$ approaches an impulse train.

Choosing t_0 and N

Suppose we want an approximation $\hat{F}(j\omega)$ with certain resolution and bandwidth constraints. In particular, suppose the true spectrum $F(j\omega)$ is approximately bandlimited to $\omega_{\max}$ and we desire a resolution of Δ or better. This means that the foldover frequency $\Omega_0/2$ must be greater than $\omega_{\max}$ to prevent aliasing. Resolution is limited to roughly $2\omega_0$ radians per second and so

$$\Omega_0/2 > \omega_{\max}, \qquad 2\omega_0 < \Delta. \tag{4.8.19}$$

These inequalities are, in turn, equivalent to the following pair of inequalities on N and t_0:

$$\begin{aligned} \Omega_0/2 = \pi/t_0 > \omega_{\max} \quad &\Rightarrow \quad t_0 < \pi/\omega_{\max}, \\ 2\omega_0 = 2\left[\frac{2\pi}{Nt_0}\right] < \Delta \quad &\Rightarrow \quad N > \frac{4\omega_{\max}}{\Delta}. \end{aligned} \tag{4.8.20}$$

On the other hand, assume that $F(j\omega)$ is bandlimited to $\omega_{\max}$ and $f(t)$ is of duration $t_{\max}$. Then we require

$$\Omega_0/2 > \omega_{\max}, \qquad T_0/2 > t_{\max}. \tag{4.8.21}$$

These inequalities are equivalent to the following inequalities on t_0 and N:

$$\begin{aligned} \Omega_0/2 = \pi/t_0 > \omega_{\max} \quad &\Rightarrow \quad t_0 < \pi/\omega_{\max}, \\ T_0/2 = Nt_0/2 > t_{\max} \quad &\Rightarrow \quad N > \frac{2t_{\max}}{t_0} > \frac{2t_{\max}\omega_{\max}}{\pi}. \end{aligned} \tag{4.8.22}$$

The quantity

$$2t_{\max}f_{\max} = t_{\max}\omega_{\max}/\pi$$

is called the *time-bandwidth product* of the signal.

The inequality $N > (2t_{\max})(2\omega_{\max})/(2\pi)$ is reminiscent of the uncertainty principle. It prompts one to conjecture that perhaps the approximation $\hat{F}(j\omega)$

is better for signals which come close to the bound imposed by the uncertainty principle. Only Gaussian pulses meet the bound. We shall consider two examples to test this conjecture.

EXAMPLE 4.8.1

Approximation of the FT of a square pulse using the DFT.

Let $f(t)$ be the square pulse of Fig. 4.8.2. The pulse is centered at $t = 0$ so that its even symmetry will produce a real Fourier Transform. Let us also sample $f(t)$ symmetrically over an interval that includes the pulse (this signal is strictly time limited). This forces a starting time for sampling to be

$$t_1 = -\frac{(N-1)t_0}{2}, \tag{4.8.23}$$

in Eqs. (4.8.16) and (4.8.17). Consequently

$$P(j\omega) = t_0 \frac{\sin(\omega T_0/2)}{\sin(\omega t_0/2)} \tag{4.8.24}$$

is real. For the square pulse in Fig. 4.8.3,

$$\hat{F}(j\omega) = (F * P)(j\omega) = t_0 \frac{\sin(m\omega t_0/2)}{\sin(\omega t_0/2)}, \tag{4.8.25}$$

where m is the number of nonzero samples of $f(t)$. Roughly,

$$mt_0 \approx 1, \tag{4.8.26}$$

since the width of the pulse is one. Substitution of (4.8.26) into (4.8.25) yields

$$\hat{F}(j\omega) \approx t_0 \frac{\sin(\omega/2)}{\sin(\omega t_0/2)} = \text{sinc}[\omega/2]/\text{sinc}[\omega t_0/2]. \tag{4.8.27}$$

Therefore,

$$F(j\omega) \approx \hat{F}(j\omega)\,\text{sinc}[\omega t_0/2]. \tag{4.8.28}$$

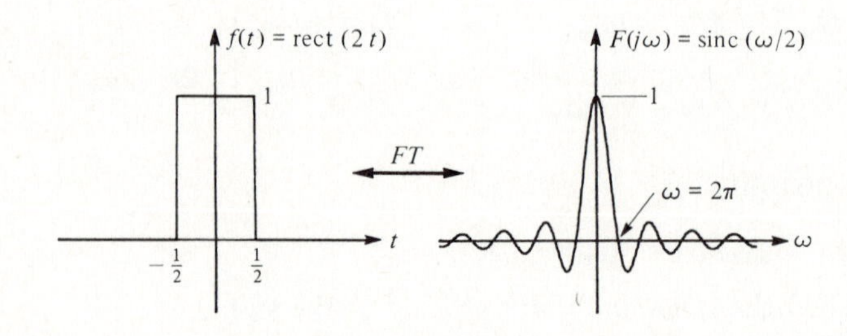

FIGURE 4.8.2.

For this example, the constraint $Nt_0 \geqslant 2t_{\max}$ is easy to meet since $f(t)$ is time limited. In view of (4.8.28), the approximation is very good when ωt_0 is much less than 2π.

Figure 4.8.3 depicts numerical results for two cases: $N = 21$ and $t_0 = 0.1$ and $N = 21$ and $t_0 = 0.2$. In both cases the numerical approximation is valid only over the region

$$|\omega| < \Omega_0/2 = \pi/t_0. \tag{4.8.29}$$

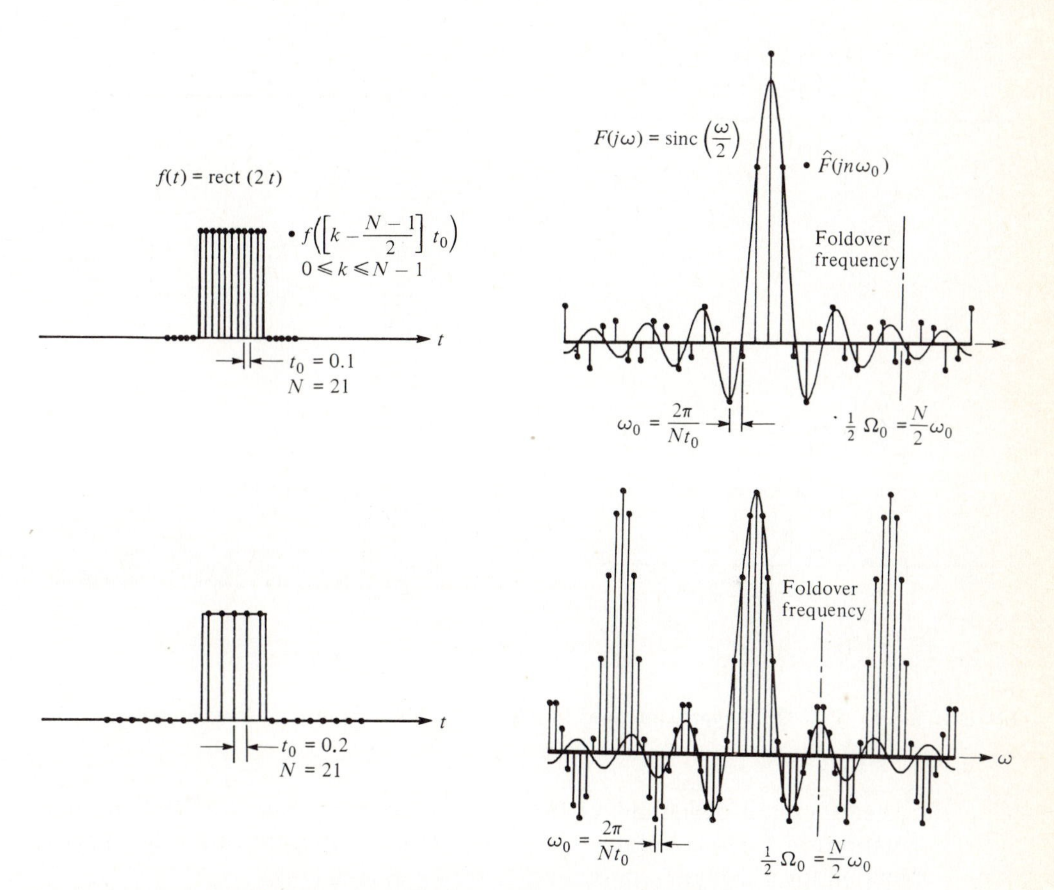

FIGURE 4.8.3 **Two DFT approximations to the Fourier transform of a square pulse.**

EXAMPLE 4.8.2

Approximation of the FT of a Gaussian pulse using the DFT.

Consider next the case of a Gaussian pulse $f(t)$. The FT pair (4.4.12) is

$$f(t) = (2\pi)^{-1/2} e^{-t^2/2} \quad \overset{FT}{\longleftrightarrow} \quad F(j\omega) = e^{-\omega^2/2}. \tag{4.8.30}$$

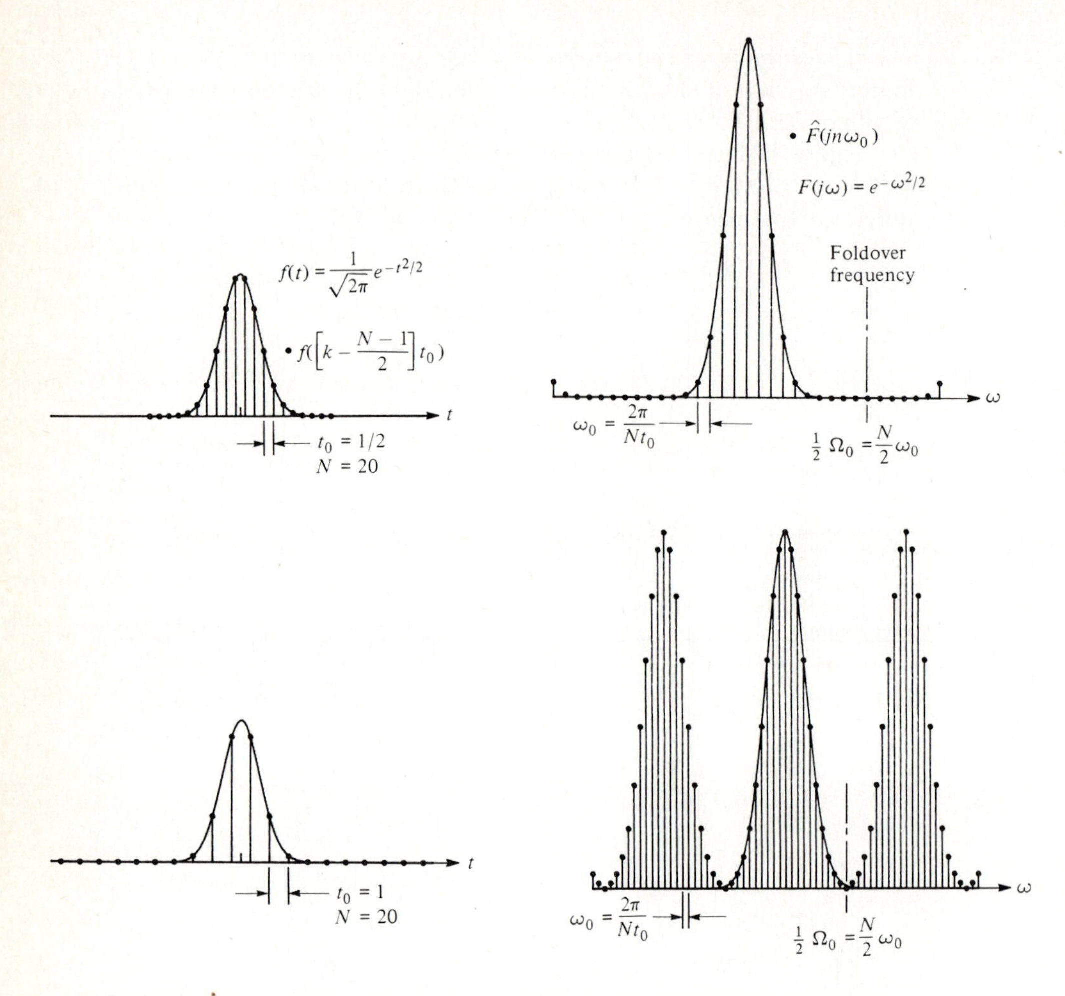

FIGURE 4.8.4 **Two DFT approximations to the Fourier transform of a Gaussian pulse.**

The "standard deviation" of $f(t)$ is $\sigma = 1$. In Fig. 4.8.4, two DFT approximations of $F(j\omega)$ are plotted. Note the marked improvement of the approximation for a Gaussian pulse over that of a square pulse.

4.9 FIR FILTER IMPLEMENTATION BASED ON THE DFT

A discrete-time filter

$$H(z) = \sum_{k=0}^{L-1} h(k) z^{-k} \tag{4.9.1}$$

whose pulse response $\mathbf{h}$ has length $L < \infty$ is called an *FIR filter*. The input/output description for this filter is

$$y(k) = \sum_{m=0}^{L-1} h(m)u(k-m), \tag{4.9.2}$$

which amounts to the convolution of an infinite length sequence $\mathbf{u}$ with a finite length sequence $\mathbf{h}$. As it stands, there are L multiplications per output value implicit in Eq. (4.9.2). This computational cost can be decreased, using efficient algorithms for computing the DFT. In order to do this, however, the aperiodic convolution in Eq. (4.9.2) must be decomposed into a sequence of circular convolution segments. In this section, we will show how this can be done.

Let $\mathbf{v}$ be an N-periodic sequence with $N > L$. Then $\mathbf{h} * \mathbf{v}$ is N-periodic, and

$$\begin{aligned}(\mathbf{h} * \mathbf{v})(k) &= \sum_{m=0}^{L-1} h(m)v(k-m) = \sum_{m=0}^{N-1} h(m)v(k-m)\\ &= [\mathbf{h} \circledast \mathbf{v}](k \bmod N).\end{aligned} \tag{4.9.3}$$

Thus the aperiodic convolution agrees with the circular convolution so long as the length of $\mathbf{h}$ is less than the period. Let $\mathbf{u}$ be the infinite length input sequence for Eq. (4.9.2). Construct the N-periodic sequence

$$v(k) = u(k \bmod N), \tag{4.9.4}$$

which is the periodic extension of the values $u(0)$ through $u(N-1)$. How do the aperiodic sequence $\mathbf{y} = \mathbf{h} * \mathbf{u}$ and the periodic sequence $\mathbf{x} = \mathbf{h} * \mathbf{v}$ compare? Consider the representation for convolution that was introduced in Chapter 2, in which the sequence $\mathbf{h}$ is reversed and then placed above $\mathbf{u}$. In order to compare $\mathbf{x}$ and $\mathbf{y}$, let us exhibit both $\mathbf{u}$ and $\mathbf{v}$ and the reversal of $\mathbf{h}$:

$$\begin{array}{cc|ccccc|cc} u(-2) & u(-1) & u(0) & u(1) & \cdots & & u(N-1) & u(N) & u(N+1) \\ & & & & & \boxed{h(L-1)\cdots h(0)} & & & \\ u(N-2) & u(N-1) & u(0) & u(1) & \cdots & & u(N-1) & u(0) & u(1) \end{array} \tag{4.9.5}$$

The sequence $\mathbf{u}$ is at the top, and the N-periodic sequence $\mathbf{v}$ is at the bottom. The position of $h(0)$ is the value of k at which we are going to evaluate the convolution. The values $x(k)$ and $y(k)$ will agree if all the values of $\mathbf{u}$ and $\mathbf{v}$ agree along the mask containing $\mathbf{h}$. The boundaries for which this is true are shown, and we have

$$x(k) = y(k) \quad \text{if} \quad L-1 \leqslant k \leqslant N-1. \tag{4.9.6}$$

The number of elements involved here is

$$M = N - L + 1. \tag{4.9.7}$$

Thus, we can evaluate M elements of $\mathbf{y} = \mathbf{h} * \mathbf{u}$ using a period N circular convolution, since the elements of $\mathbf{x}$ can be computed using Eq. (4.9.3).

Now any M values of $\mathbf{y}$ can be obtained in this way by replacing Eq. (4.9.4) with

$$v(k) = u[(k-m) \bmod N]. \tag{4.9.8}$$

We need only change the selection of those values of **u** which participate. We will need to take vectors containing N values of **u**, which are spaced M apart, since we may keep only M values. Thus define

$$\mathbf{f}_m = \begin{bmatrix} u(mM) \\ u(mM+1) \\ \vdots \\ u(mM+N-1) \end{bmatrix}. \tag{4.9.9}$$

This vector will be circularly convolved with

$$\mathbf{h} = \begin{bmatrix} h(0) \\ h(1) \\ \vdots \\ h(L-1) \\ 0 \\ \vdots \\ 0 \end{bmatrix}. \tag{4.9.10}$$

Now let

$$\mathbf{g}_m = \mathbf{h} \circledast \mathbf{f}_m \quad \xleftrightarrow{DFT;\,N} \quad G_m(n) = H(W_N^n)F_m(n). \tag{4.9.11}$$

Then the last M values of the vector $\mathbf{g}_m$ are usable. We can compute this vector using a DFT, followed by multiplication (in the frequency domain), followed by an inverse DFT:

$$\mathbf{g}_m = \mathbf{V}^{-1}\mathbf{D}\mathbf{V}\mathbf{f}_m, \tag{4.9.12}$$

where

$$\mathbf{D} = \text{diag}\{H(1), \text{H}(\text{W}_N), \cdots, H(W_N^{N-1})\}. \tag{4.9.13}$$

Since the DFT's ($\mathbf{V}$ and $\mathbf{V}^{-1}$) can be efficiently computed, this may lead to a decrease in computation.

The method we have outlined is called the *overlap-save method.* In summary, one constructs blocks of input data of size N, as in Eq. (4.9.9). Then the operations in Eq. (4.9.12) are performed, yielding the output data $\mathbf{g}_m$. The elements of this vector with indices $L-1 \leqslant k \leqslant N-1$ are then placed in an output buffer, to be read out in the proper order.

Summary • We have introduced four classes of signals and their appropriate Fourier transforms. For discrete-time signals and systems, either the DFT or DTFT is appropriate; but the properties of all four transforms are similar. In Sections 4.2 through 4.5, each transform was presented alone. In Section 4.6 we considered problems that involved signals from separate classes, and the last three sections contain important applications.

The DFT is of special importance in digital signal processing because it is the only transform that can always be computed, and it can be used to approximate any of the other three. Because of Fast Fourier transform algorithms which have been developed to efficiently compute DFT's, many new digital signal processing applications are now feasible. These fast algorithms are the subject of Chapter 5.

PROBLEMS

4.1 Find the DTFT of each of the following sequences.

a) $f(k) = 2^{-k}\xi(k)$

b) $f(k) = \exp(-3k)\cos(4k)\xi(k)$

c) $f(k) = \begin{cases} \alpha^k, & k = 0, 2, 4, 6, \ldots \\ 0, & \text{otherwise} \end{cases}, \qquad 0 < \alpha < 1.$

d) $f(k) = k\cos(k\theta_0)\xi(k)\xi(N-k)$

4.2 Find the inverse DTFT of each of the following spectra.

a) $F(e^{j\theta}) = [e^{-j2\theta} - 1]/[e^{-j2\theta} - 4]$

b) $F(e^{j\theta}) = [1 - e^{-jN\theta}]/[1 - e^{-j\theta}]$

c) $F(e^{j\theta}) = [1 + 0.5\cos(\theta) - 0.5j\sin(\theta)]^{-1}$

d) $F(e^{j\theta}) = [1 - \frac{1}{2}e^{-j\theta} - \frac{3}{4}e^{-j2\theta}]/[1 - \frac{1}{4}e^{-j\theta} - \frac{1}{8}e^{-j2\theta}]$

e) $F(e^{j\theta}) = [1 + 2\cos(2\theta)]e^{-j3\theta}$

4.3 Find the inverse DTFT for

a) $F(e^{j\theta}) = [1 - ae^{-j\theta}]/[a - e^{-j\theta}], \quad a^2 < 1$

b) $F(e^{j\theta}) = [a - e^{-j\theta}]/[1 - ae^{-j\theta}], \quad a^2 < 1$

c) $F(e^{j\theta}) = \dfrac{1 - ae^{-j\theta}}{a - e^{-j\theta}} \dfrac{b - e^{-j\theta}}{1 - be^{-j\theta}}, \quad a^2 < 1, \qquad b^2 < 1.$

4.4 Find the energy in the signals of problem 4.3.

4.5 A digital filter is described by the standard difference equation

$$y(k) - 0.25\,y(k-2) = u(k).$$

Find $G(e^{j\theta})$ so that $\mathbf{g} * \mathbf{y} = \mathbf{u}$, for any input sequence $\mathbf{u}$.

4.6 Let $\mathbf{h}$ be the unit pulse response of a digital filter. Let $\mathbf{y} = \mathbf{h} * \boldsymbol{\xi}$ be the step response. Express $Y(e^{j\theta})$ in terms of $H(e^{j\theta})$.

4.7 Consider the following two all-pass filters:

$$H(e^{j\theta}) = e^{-j2\theta} \qquad \text{(linear phase response)},$$

$$G(e^{j\theta}) = \frac{e^{-j\theta} - 0.8}{1 - 0.8e^{-j\theta}} \qquad \text{(nonlinear phase response)}.$$

a) Plot the phase response of each filter.

b) Plot the pulse response of each filter.

c) Plot two periods of the steady-state "square-wave" responses $\mathbf{h} * \mathbf{u}$ and $\mathbf{g} * \mathbf{u}$, where

$$u(k) = \begin{cases} 1, & \sin(k\pi/5) > 0 \\ 0, & \sin(k\pi/5) = 0\,. \\ -1, & \sin(k\pi/5) < 0 \end{cases}$$

4.8 Start with the ideal low-pass filter described in Eqs. (4.2.26) and (4.2.27), and use the DTFT frequency shift property to obtain the unit pulse response of the ideal band-pass filter

$$G(e^{j\theta}) = \begin{cases} 1, & \theta_1 < |\theta| < \theta_2 \\ 0, & \theta_2 < |\theta| < \pi, \qquad 0 < |\theta| < \theta_1\,. \end{cases}$$

4.9 Start with the rectangular window sequence described by Eqs. (4.2.34) and (4.2.36), and use the time and frequency shift properties to compute the DTFT of the sequence

$$f(k) = \begin{cases} \cos(k\theta_0), & 0 \leqslant k \leqslant 2L \\ 0, & \text{otherwise.} \end{cases}$$

4.10 Use the DTFT pair in Example 4.2.4, and the convolution-multiplication property to find the DTFT of the triangular window sequence

$$f(k) = \begin{cases} 2L + 1 - |k|, & |k| \leqslant 2L \\ 0, & \text{otherwise.} \end{cases}$$

4.11 Let $W_L(e^{j\theta})$ be the frequency domain window function defined by Eq. (4.2.36). Compute

$$G(e^{j\theta}) = \frac{1}{2\pi} \int_{-\pi}^{\pi} W_L(e^{j\phi}) W_M(e^{j(\theta-\phi)}) d\phi.$$

4.12 Let

$$f_\phi(k) = [\phi/\pi]\mathrm{sinc}(k\phi).$$

Let $0 < \alpha < \beta < \pi$. Show that $\mathbf{f}_\alpha * \mathbf{f}_\beta = \mathbf{f}_\alpha$.

4.13 Suppose that $\mathbf{f}$ is a real, causal sequence with $f(0) > 0$. Does there always exist a real, causal sequence $\mathbf{g}$ for which $\mathbf{g} * \mathbf{g} = \mathbf{f}$? Find $g(k)$ for $0 \leqslant k \leqslant 10$ when $\mathbf{f} = \{1, 1, 0, 0, \cdots\}$.

4.14 Let $h(k) = [\theta_0/\pi]\mathrm{sinc}(k\theta_0)$ be the unit pulse response sequence for an ideal low-pass filter with cutoff $\theta_0 = \pi/1000$. Find an input sequence $\mathbf{u}$ such that: (1) $u(k) \leqslant 1$ for all k; (2) $\mathbf{u}$ is causal; (3) $y(k) = (\mathbf{h} * \mathbf{u})(k) > 10^6$ for some k. What do you conclude from this?

4.15 Let $\mathbf{h}$ be a unit pulse response sequence of digital filter. Find an expression for the *average delay*

$$d = \left[\sum_{k=0}^{\infty} kh^2(k)\right] \Big/ \left[\sum_{k=0}^{\infty} h^2(k)\right]$$

in terms of $H(e^{j\theta})$.

4.16 Let **g** be a given finite energy sequence and let θ_0 be a given angle. Find the sequence **f** for which the energy spectral density at θ_0 is greatest [$|F(e^{j\theta_0})|^2$ is maximized] subject to the constraints $|f(k)| \leq |g(k)|$ for all k. (The values of $f(k)$ need not be real.)

4.17 Let $p(t)$ be the periodic signal in Eq. (4.6.12). Choosing t_0 and t_1 appropriately, use the FS Parseval relation to compute

$$\sum_{n=0}^{\infty} \left(\frac{1}{2n+1} \right)^2 .$$

4.18 Let

$$f(t) \quad \xleftrightarrow{FS;\,\omega_0} \quad F(n)$$

$$g(t) \quad \xleftrightarrow{FS;\,\omega_0} \quad G(n)$$

where

$$G(n) = \begin{cases} F(n), & n = m,\ m \pm N,\ m \pm 2N, \ldots \\ 0, & \text{otherwise.} \end{cases}$$

a) Express $g(t)$ in terms of $f(t)$.
b) In general, what is the fundamental frequency of $g(t)$?

4.19 Using definitions of duration in Section 4.4, compute or estimate σ_t and σ_ω for

a) $f(t) = \mathrm{rect}(t/t_0)$,
b) $g(t) = (\mathbf{f} * \mathbf{f})(t)$,
c) $h(t) = e^{-t}\xi(t)$.

4.20 Let $x(k) = \cos(k\theta_0)$ for all k. Let

$$y(k) = x(k \bmod N) \quad \xleftrightarrow{DFT;\,N} \quad Y(n).$$

a) Find $Y(n)$ analytically.
b) Plot all N values of Y for three cases:

$$N = 10, \qquad \theta_0 = 0.6\pi,$$
$$N = 10, \qquad \theta_0 = 0.7\pi,$$
$$N = 10, \qquad \theta_0 = 0.75\pi.$$

4.21 Find the amplitude limited N-periodic discrete-time signal with greatest power at frequency m. In other words, find

$$x(k) \quad \xleftrightarrow{DFT;\,N} \quad X(n)$$

to maximize $|X(m)|$ subject to $|x(k)| \leq 1$ for all k.

4.22 Let $\mathbf{R}^{-1}$ be the matrix defined in Eq. (4.5.16). Suppose that $\mathbf{G}$ is an $N \times N$ matrix satisfying $\mathbf{GR} = \mathbf{RG}$. Show that $\mathbf{G}$ must be circulant.

4.23 Derive the frequency shift Eq. (4.5.42) for the DFT.

4.24 Let $\mathbf{V}$ be the $N \times N$ DFT matrix in Eq. (4.5.5). Show that $\mathbf{V}^4 = N^2\mathbf{I}$. What does this say about the eigenvalues of $\mathbf{V}$?

4.25 Let $f(k) \overset{DTFT}{\longleftrightarrow} F(e^{j\theta})$ and let

$$g(k) = \begin{cases} f(k), & k \text{ even} \\ 0, & k \text{ odd.} \end{cases}$$

Find $G(e^{j\theta})$ using the appropriate sampling relation in Table 4.6.1. Then find $G(e^{j\theta})$ by noting that $g(k) = q(k)f(k)$ and using the appropriate relation in Table 4.6.3.

4.26 Using the technique outlined in Eqs. (4.6.1) through (4.6.6), derive the DTFT aliasing in time, sampling in frequency relation in Table 4.6.1.

4.27 Let

$$f(t) \overset{FT}{\longleftrightarrow} F(j\omega)$$

$$g(t) = f(t)\,\text{rect}(2t/t_0) \overset{FT}{\longleftrightarrow} G(j\omega)$$

$$h(t) = \sum_n g(t - nt_0) \overset{FS;\,\omega_0}{\longleftrightarrow} H(n)$$

where $\omega_0 t_0 = 2\pi$. Compute $H(n)$ in terms of $F(j\omega)$.

4.28 Digital-to-analog converters are sometimes used with microprocessors to synthesize periodic signals that are piecewise constant as shown in Fig. 4P.28.

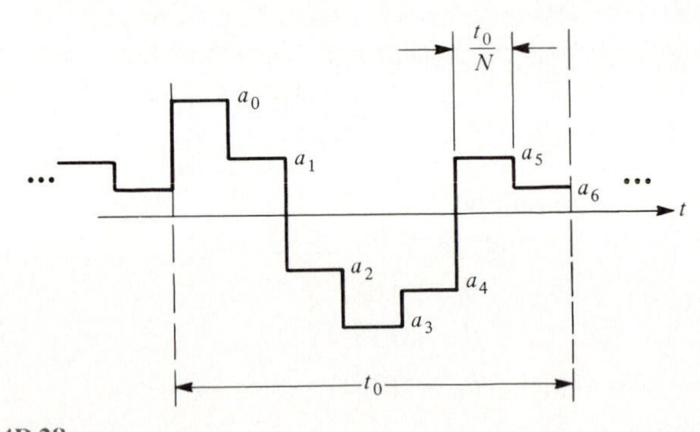

FIGURE 4P.28.

Analyze this waveform generation method using the following signals:

$$p(t) \overset{FS;\,\omega_0}{\longleftrightarrow} P(n)$$

$$y(t) = \sum_{k=0}^{N-1} a_k p(t - kt_0/N) \overset{FS;\,\omega_0}{\longleftrightarrow} Y(n)$$

where $\omega_0 = 2\pi/t_0$.

a) What are $p(t)$, $P(n)$ for piecewise constant $y(t)$?

b) Given a general $p(t)$, find the matrix $\mathbf{G}$ that relates the output Fourier coefficients and the ADC output levels:

$$\begin{bmatrix} Y(0) \\ Y(1) \\ \vdots \\ Y(N-1) \end{bmatrix} = \mathbf{G} \begin{bmatrix} a_0 \\ a_1 \\ \vdots \\ a_{N-1} \end{bmatrix}.$$

c) Express $\det(\mathbf{G})$ in terms of N, $P(0), \ldots, P(N-1)$. Under what conditions does $\det(\mathbf{G}) = 0$?

This problem is the basis for some VLSI modem devices. One can generate a high quality sinusoid by choosing the a_k carefully, and passing $y(t)$ through an inexpensive analog low-pass filter.

4.29 Let $\mathbf{u}$ be an N-periodic sequence and let

$$h(k) \xleftrightarrow{DTFT} H(e^{j\theta})$$

be the unit pulse response of a digital filter.

a) Show that $\mathbf{y} = \mathbf{h} * \mathbf{u}$ is N-periodic.

b) Show that there is a matrix $\mathbf{G}$ for which

$$\begin{bmatrix} y(0) \\ y(1) \\ \vdots \\ y(N-1) \end{bmatrix} = \mathbf{G} \begin{bmatrix} u(0) \\ u(1) \\ \vdots \\ u(N-1) \end{bmatrix}.$$

c) Find the elements of $\mathbf{G}$ and show that $\mathbf{G}$ is circulant.

d) Since $\mathbf{u}$ and $\mathbf{y}$ are N-periodic they have DFT vectors $\mathbf{U}$ and $\mathbf{Y}$. Show that $\mathbf{Y} = \mathbf{F}\mathbf{U}$ where $\mathbf{F}$ is an $N \times N$ diagonal matrix. What are the diagonal elements of $\mathbf{F}$?

e) Show that if $|H(e^{j\theta})| = 1$ for all θ, then $\mathbf{G}\mathbf{G}^T = \mathbf{I}$ and $\mathbf{F}\mathbf{F}^* = \mathbf{I}$.

4.30 Let $f(t) \xleftrightarrow{FT} F(j\omega)$ have energy

$$E_f = \int_{-\infty}^{\infty} |f(t)|^2 dt < \infty.$$

Let $g(k) = t_0 f(kt_0)$ have energy

$$E_g = \sum_k |g(k)|^2.$$

Show that $E_g = t_0 E_f$ if $F(j\omega) = 0$ for $|\omega| > \pi/t_0$.

4.31 (Asynchronous sampling.) Let $f(t)$ be a finite energy signal. Let $t_1 > t_0$, and let $g_i(k) = f(kt_i)$ for $i = 0, 1$.

a) Approximate $\mathbf{g}_1$ given $\mathbf{g}_0$, *i.e.*, find b_{km} so that

$$g_1(k) \approx \sum_m b_{km} g_0(m).$$

b) Under what conditions is the approximation exact?

4.32 In digital audio systems a procedure known as *interpolation* is sometimes used. Let $\mathbf{u}$ be an input sequence. System A inserts $N - 1$ zeros between each input value to produce

$$v(k) = \sum_{m} u(m)\delta(k - mN).$$

System B is an ideal low-pass filter and produces $\mathbf{y} = \mathbf{h} * \mathbf{v}$.

a) Derive expression for $Y(e^{j\theta})$ and $V(e^{j\theta})$ in terms of $U(e^{j\theta})$.

b) For $N = 4$, pick a spectrum $U(e^{j\theta})$ and sketch it and $V(e^{j\theta})$. Choose the cutoff frequency of $H(z)$ to obtain exact bandlimited interpolation, and then sketch $Y(e^{j\theta})$.

c) Why would this procedure be useful in audio systems?

4.33 There are "sampling theorems" with reconstruction formulas analogous to Eq. (4.7.15) for each of the entries in Table 4.6.1. This is a sampling theorem for FS signals. Suppose that $f(t)$ is t_0-periodic and has Fourier coefficients $F(n)$ with $F(n) = 0$ for $|n| > N$. Find a function $p(t)$ for which the following reconstruction formula is valid:

$$f(t) = \sum_{k=0}^{2N} f\left(\frac{kt_0}{2N+1}\right) p\left(t - \frac{kt_0}{2N+1}\right).$$

Show that $p(t)$ is t_0-periodic and find its Fourier coefficients $P(n)$.

4.34 Let $H(z)$ be the FIR filter of Eq. (4.9.1) and let $N \geqslant L$. Show that the filter is completely determined by the N samples $H(W_N^n)$, $0 \leqslant n \leqslant N - 1$ of its frequency response, by exhibiting functions $F_n(z)$ so that

$$H(z) = \sum_{n=0}^{N-1} H(W_N^n) F_n(z).$$

Compute the inverse DFT of the samples.

4.35 Two analog signals $f(t)$ and $g(t)$ are inputs to an analog multiplexer as shown in Fig. 4P.35. Find $Y(j\omega)$ in terms of $F(j\omega)$ and $G(j\omega)$. Under what conditions can $f(t)$ be exactly recovered from $y(t)$? How could this be accomplished?

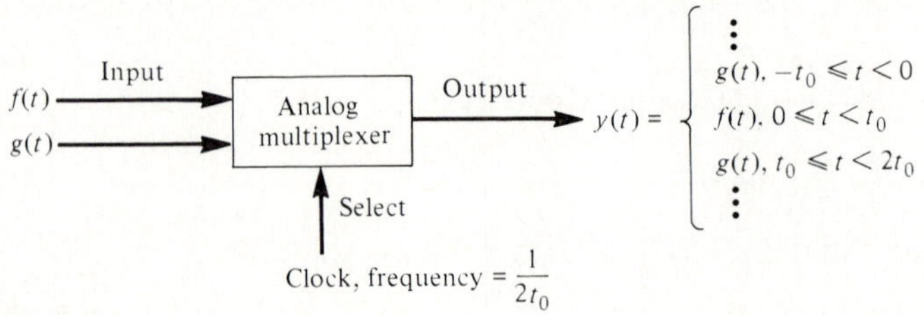

FIGURE 4P.35.

4.36 Let

$$f(t) \overset{FT}{\longleftrightarrow} F(j\omega),$$

$$p(t) \overset{FS:\omega_0}{\longleftrightarrow} P(n).$$

Assume that $f(t)$ is bandlimited, *i.e.*, $F(j\omega) = 0$ for $|\omega| > \omega_0/2$. Assume that $P(n) = 0$ for $|n| < L$ but that $P(L) = 1$. Let $y(t) = f(t)p(t)$.

a) Sketch $Y(j\omega)$.

b) Invent a system which can produce $f(t)$ if the input is $y(t)$. You may assume that $p(t)$ and $f(t)$ are both real valued.

4.37 Let $u(t) \overset{FT}{\longleftrightarrow} U(j\omega)$, and let $f(k) = u(kt_0)$. Consider two reconstruction filter unit pulse response functions

$$r_0(t) = \begin{cases} 1, & |t| < t_0/2 \\ 0, & \text{otherwise} \end{cases},$$

$$r_1(t) = \frac{1}{t_0}\int_{-\infty}^{\infty} r_0(t-\tau)r_0(\tau)d\tau.$$

a) Find $R_0(j\omega)$ and $R_1(j\omega)$.

b) Express $F(e^{j\theta})$ in terms of $U(j\omega)$.

c) For $i = 0, 1$ let

$$y_i(t) = \sum_k f(k)r_i(t - kt_0) \quad \overset{FT}{\longleftrightarrow} \quad Y_i(j\omega).$$

Express $Y_0(j\omega)$ and $Y_1(j\omega)$ in terms of $U(j\omega)$, $R_0(j\omega)$, $R_1(j\omega)$.

d) Evaluate the samples $y_i(kt_0)$ for $i = 0, 1$ in terms of $\mathbf{u}$, $\mathbf{r}_0$, $\mathbf{r}_1$.

e) Let $u(t)$ be a Gaussian pulse and let $t_0 = \sigma/5$. Sketch $r_0(t)$, $r_1(t)$, $u(t)$, $y_0(t)$, $y_1(t)$.

4.38 Suppose that $f(t)$ is real and $F(j\omega) \neq 0$ only if $N\omega_0 < |\omega| < (N+1)\omega_0$. Let $\omega_0 t_0 = 2\pi$, and let

$$g(t) = \int_{-\infty}^{\infty} f(\tau)\cos(N\omega_0\tau)\text{sinc}[\omega_0(t-\tau)/2]d\tau.$$

a) Can $f(t)$ be reconstructed from the samples $f(kt_0)$?

b) Can $f(t)$ be constructed from the samples $f(kt_0/2)$?

c) Can $f(t)$ be reconstructed from the samples $g(kt_0/2)$?

For each affirmative answer, describe the reconstruction operation.

4.39 Let

$$f(t) \quad \overset{FT}{\longleftrightarrow} \quad F(j\omega),$$

$$g(k) = t_0 \sum_{m=-\infty}^{\infty} f(kt_0 - mT_0).$$

Show that $\mathbf{g}$ is N-periodic if $T_0 = Nt_0$. Under this assumption, express the DFT values $G(n)$ in terms of $F(j\omega)$. Use the notation in Eqs. (4.8.7).

4.40 Among all continuous-time signals for which $f(0) = 1$, f is continuous at $t = 0$, and $F(j\omega) = 0$ for $|\omega| > \omega_0/2$, find the signal having least energy.

CHAPTER 5

Fast Algorithms for the Discrete Fourier Transform

5.1 INTRODUCTION

Chapter 4 discussed the Fourier analysis of discrete-time systems and signals by use of the DTFT and the DFT. One of the important aspects of this theory is the actual computation associated with the DFT since it is this transform that is, in practice, computed. This chapter is concerned with what we shall loosely call "fast algorithms" for the DFT and convolution. There are many facets to this theory. We shall present here only a brief look into this large and interesting field of study.

Recall that the DFT of the sequence $\{x(k)\}$, $k = 0, 1, \ldots, N-1$ is given by

$$X(n) = \sum_{k=0}^{N-1} x(k) W_N^{-nk}, \qquad n = 0, 1, \ldots, N-1 \tag{5.1.1}$$

where

$$W_N = \exp\left(j\frac{2\pi}{N}\right). \tag{5.1.2}$$

The inverse DFT (or IDFT) of the spectrum $\{X(n)\}$, $n = 0, 1, \ldots, N-1$ is similarly given by

$$x(k) = \frac{1}{N}\sum_{n=0}^{N-1} X(n) W_N^{nk}, \qquad k = 0, 1, \ldots, N-1. \tag{5.1.3}$$

In matrix notation, the sequence $\{X(n)\}$ is given by

$$\mathbf{X} = \mathbf{V}\mathbf{x}, \tag{5.1.4}$$

where $\mathbf{V}$ is the DFT matrix given by

$$\mathbf{V} = [V_{nk}], \qquad n, k = 0, 1, \ldots, N-1 \tag{5.1.5}$$

having elements

$$V_{nk} = W_N^{-nk} = \exp\left(-j\frac{2\pi nk}{N}\right). \tag{5.1.6}$$

In general, the computation of **X** from **x** in Eq. (5.1.4) requires N^2 multiplications since **V** is an $N \times N$ matrix. However, because of the character of W_N^{-nk}, there are symmetries contained in **V** that dramatically reduce the N^2 multiplies implied by Eq. (5.1.4). The entry in the nth row and kth column of **V** is the Nth root of unity raised to the $-nk$ power. Because $W_N^N = 1$, W_N^{-nk} is identical to W_N^{-l} where

$$l = nk \bmod N. \tag{5.1.7}$$

Using Eq. (5.1.7) in the case of $N = 8$, for example, results in a DFT matrix **V** with powers of W_N^{-1} given in the following matrix **P**.

$$\mathbf{P} = \begin{bmatrix} 0 & 0 & 0 & 0 & 0 & 0 & 0 & 0 \\ 0 & 1 & 2 & 3 & 4 & 5 & 6 & 7 \\ 0 & 2 & 4 & 6 & 0 & 2 & 4 & 6 \\ 0 & 3 & 6 & 1 & 4 & 7 & 2 & 5 \\ 0 & 4 & 0 & 4 & 0 & 4 & 0 & 4 \\ 0 & 5 & 2 & 7 & 4 & 1 & 6 & 3 \\ 0 & 6 & 4 & 2 & 0 & 6 & 4 & 2 \\ 0 & 7 & 6 & 5 & 4 & 3 & 2 & 1 \end{bmatrix}. \tag{5.1.8}$$

The entries in **P** are values of l from Eq. (5.1.7) needed in the DFT matrix **V** to compute an $N = 8$ DFT.

One can view fast algorithms for the DFT as a factorization of the DFT matrix **V**. We shall present a number of factorizations.

5.2 POWER-OF-2 FFT ALGORITHMS

Suppose the number of data points in **x** is a power of 2, i.e., $N = 2^\nu$. The following is a simple derivation of a fast algorithm for the DFT called *decimation-in-time* fast Fourier transform (FFT).

Divide the original data index set into odd and even indices (since N is divisible by 2). Write the DFT [Eq. (5.1.1)] in the form

$$\begin{aligned} X(n) &= \sum_{k=0}^{N/2-1} [x(2k)W_N^{-2kn} + x(2k+1)W_N^{-(2k+1)n}] \\ &= \sum_{k=0}^{N/2-1} x(2k)W_N^{-2kn} + W_N^{-n} \sum_{k=0}^{N/2-1} x(2k+1)W_N^{-2kn}, \end{aligned}$$

$$n = 0, 1, \ldots, N-1. \tag{5.2.1}$$

Now W_N^{-2kn} can be written in the form

$$W_N^{-2kn} = \left[\exp\left(j\frac{2\pi}{N}\right)\right]^{-2kn} = \left[\exp\left(j\frac{2\pi}{N/2}\right)\right]^{-kn} = W_{N/2}^{-kn}. \tag{5.2.2}$$

Equation (5.2.2) is a key step since it permits one to write Eq. (5.2.1) as

$$X(n) = G(n) + W_N^{-n}H(n), \qquad n = 0, 1, \ldots, \frac{N}{2} - 1. \tag{5.2.3}$$

Here $G(n)$ and $H(n)$ are $N/2$-point DFT's. These $N/2$-point DFTs are thus $N/2$ periodic. This allows us to calculate $X(n)$ for $n = N/2, N/2 + 1, \ldots, N - 1$ via

$$X(n) = G\left(n - \frac{N}{2}\right) + W_N^{-n}H\left(n - \frac{N}{2}\right), \qquad n = \frac{N}{2}, \frac{N}{2} + 1, \ldots, N - 1. \tag{5.2.4}$$

One can view the decomposition in Eqs. (5.2.3) and (5.2.4) schematically as shown in Fig. 5.2.1 (assuming $N = 2^3 = 8$). We need only combine $G(n)$ and $H(n)$ using the appropriate weights W_N^{-n} or "twiddle factors."

How many computations are required using one stage of decomposition? Each $N/2$-point DFT requires by brute force $(N/2)^2$ multiplies. Multiplication by W_N^{-n} in Eqs. (5.2.3) and (5.2.4) requires N multiplies. Thus the total is $N + 2(N/2)^2 = N + (N^2/2)$ multiplies versus N^2 for brute force computation of the original DFT.

Since N is a power of 2, we can repeat the decimation process for both sequences **g** and **h**, where

$$g(l) = x(2l), \quad h(l) = x(2l + 1), \qquad l = 0, 1, \ldots, \frac{N}{2} - 1. \tag{5.2.5}$$

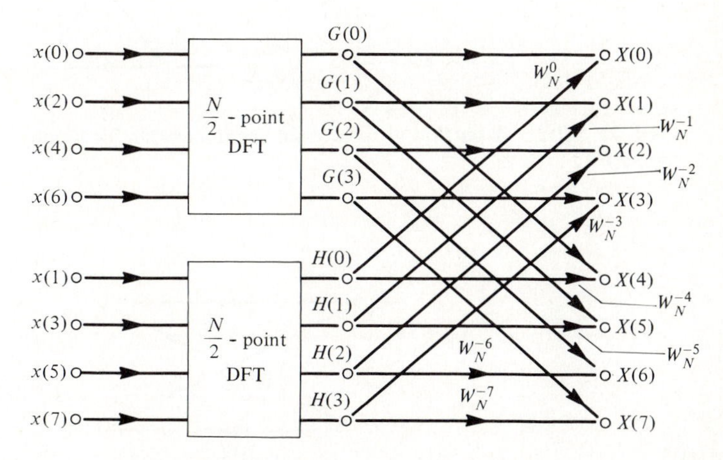

FIGURE 5.2.1 An SFG of the first stage of a decimation-in-time $N = 2^3$-point FFT.

Thus write $G(n)$ in the form

$$
\begin{aligned}
G(n) &= \sum_{k=0}^{N/2-1} g(k)W_{N/2}^{-kn} = \sum_{k=0}^{N/4-1} [g(2k)W_{N/2}^{-2nk} + g(2k+1)W_{N/2}^{-(2k+1)n}] \\
&= \sum_{k=0}^{N/4-1} g(2k)W_{N/4}^{-nk} + W_{N/2}^{-n} \sum_{k=0}^{N/4-1} g(2k+1)W_{N/4}^{-nk} \\
&= \begin{cases} R(n) + W_{N/2}^{-n}S(n), & n = 0, 1, \ldots, \dfrac{N}{4} - 1 \\ R\left(n - \dfrac{N}{4}\right) + W_{N/2}^{-n}S\left(n - \dfrac{N}{4}\right), & n = \dfrac{N}{4}, \dfrac{N}{4} + 1, \ldots, \dfrac{N}{2} - 1 \end{cases}
\end{aligned}
\tag{5.2.6}
$$

[$H(n)$ can be expressed similarly]. $G(n)$ is now expressed as a sum of two $N/4$-point DFTs. The SFG for computing $G(n)$ is shown in Fig. 5.2.2 for the case $N = 8$. Notice again we used the relationship $W_{N/2}^{-2nk} = W_{N/4}^{-nk}$ in this decomposition.

How many computations are involved using two stages of decomposition? Each $N/4$-point DFT requires $(N/4)^2$ multiplies, and there are four of them. The multiplication by $W_{N/2}^{-n}$, $n = 0, 1, \ldots, N/2 - 1$ for both $G(n)$ and $H(n)$

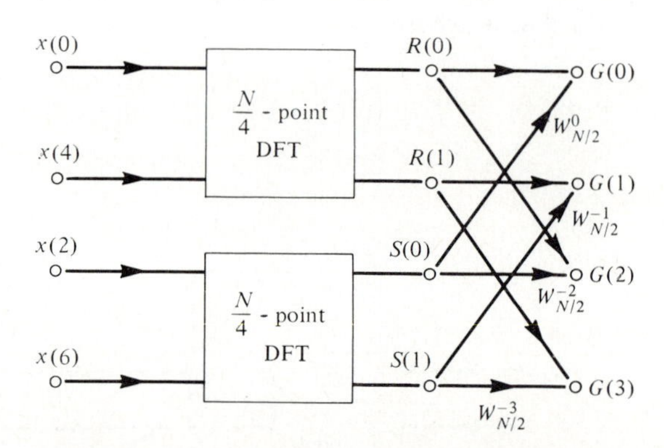

FIGURE 5.2.2 **A portion of the second decomposition for a power-of-2 decimation-in-time FFT.**

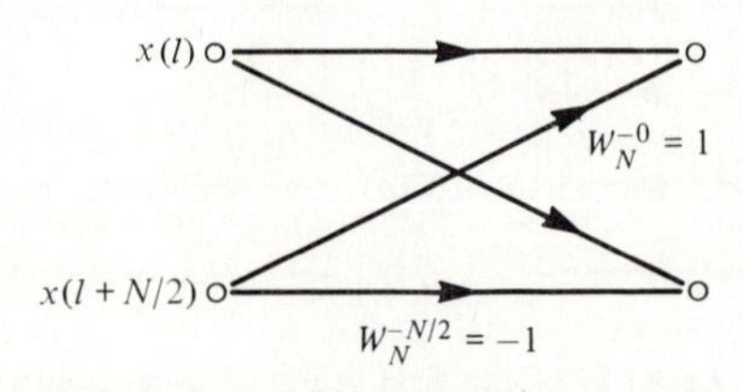

FIGURE 5.2.3 **A signal flow graph (SFG) for a two-point DFT.**

results in N multiplies. Combining $\mathbf{G}$ and $\mathbf{H}$ to form $\mathbf{X}$ also requires N multiplies. The total is thus $2N + 4(N/4)^2$ versus N^2. We can continue this process for $\log_2 N = \log_2 2^\nu = \nu$ stages. The final stage is a two-point DFT of the form shown in Fig. 5.2.3 and requires no multiplies! Thus the only multiplies are those required to combine the smaller-point DFTs. There are $\log_2 N$ stages each requiring N multiplies for a total of $N \log_2 N$ multiplies for the power-of-2 FFT. (These multiplications are complex, and thus correspond to four real multiplications.) Figure 5.2.4 is a complete SFG for the case $N = 2^3$. Based on the SFG, it is not difficult to deduce the matrix factorization of $\mathbf{V}$ that this fast algorithm implies. This factorization is given in Eq. (5.2.7). The product of the four 8×8 matrices in Eq. (5.2.7), and the SFG of Fig. 5.2.4 are equivalent. The matrix $\mathbf{E}$ represents the input data permutations needed for the following matrix multiply which consists of 4 two-point DFTs contained in $\mathbf{V}_{2T}$.

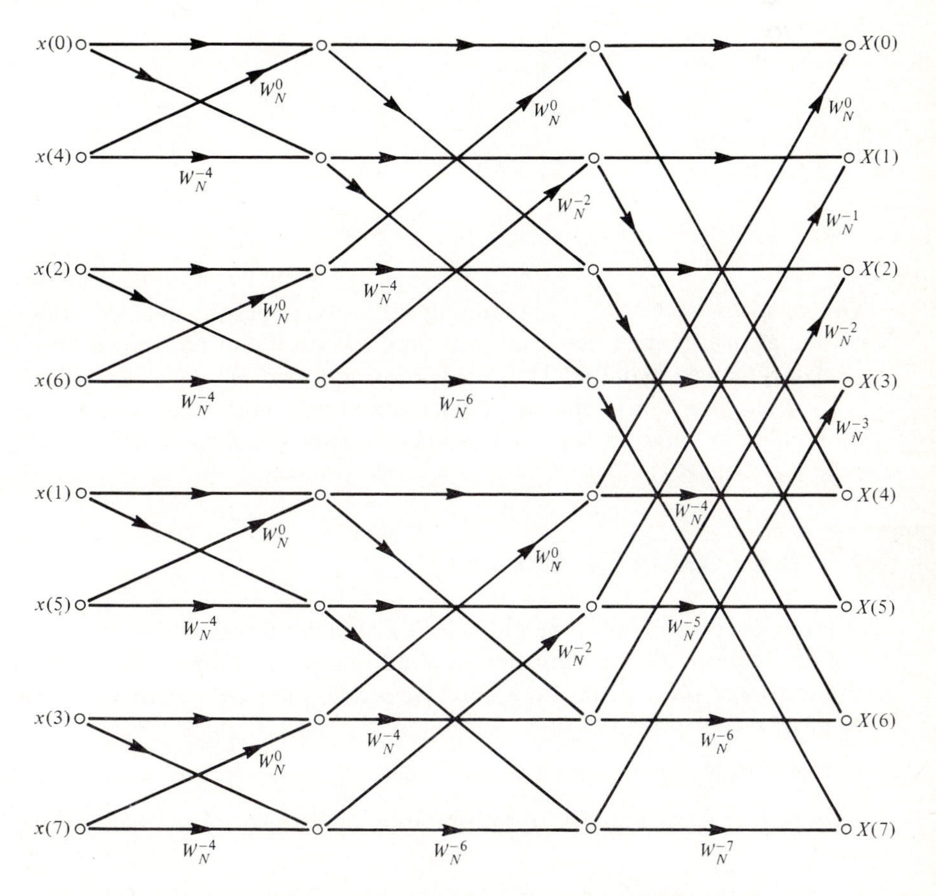

FIGURE 5.2.4 A SFG for the decimation-in-time FFT for $N = 2^3$ stages.

$$\mathbf{V} = \mathbf{V}_{8T}\,\mathbf{V}_{4T}\,\mathbf{V}_{2T}\,\mathbf{E}$$

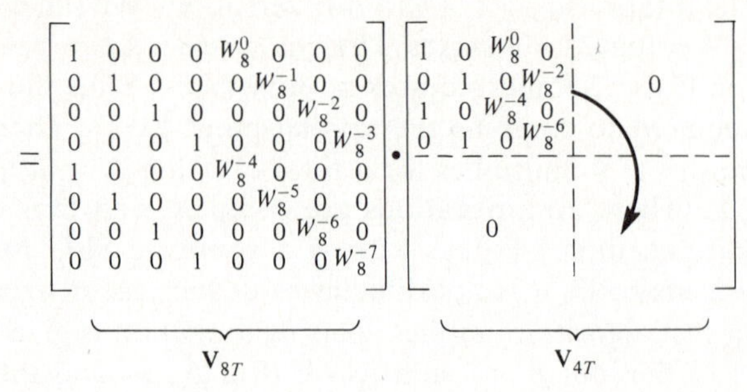

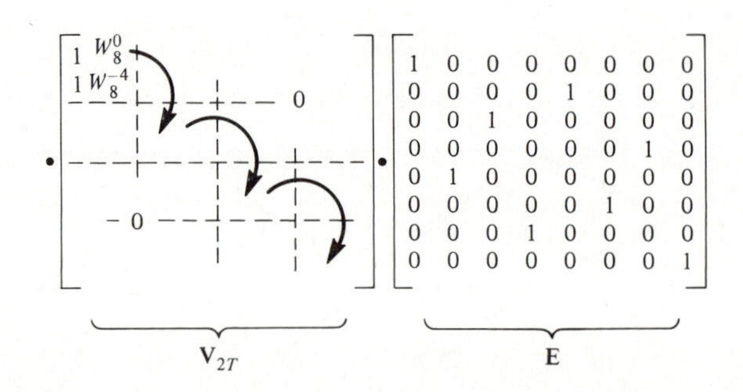

(5.2.7)

The matrices $\mathbf{V}_{8T}$, $\mathbf{V}_{4T}$, $\mathbf{V}_{2T}$ represent the three stages of so-called butterfly computations. Notice that each row contains only two nonzero multiplies, implying that each output is formed from only two input values as in Fig. 5.2.3. There are N W_N^{-n} multiplying factors in each component matrix and a total of $\log_2 N$ stages, resulting in $N \log_2 N$ complex multiplies. ($\mathbf{V}_{2T}$ contains only ± 1's and 0's.)

5.3 OTHER FORMS OF THE FFT

We can use properties of the DFT matrix $\mathbf{V}$ to derive other algorithms using the decimation-in-time algorithm as a starting point. Consider, for example, the transpose of $\mathbf{V}$. From Eq. (5.1.6), we see that the DFT matrix $\mathbf{V}$ is equal to its own transpose, i.e.,

$$\mathbf{V}^T = \mathbf{V}. \tag{5.3.1}$$

Substituting from Eq. (5.2.7), we obtain another form of the factored DFT matrix:

$$\begin{aligned} \mathbf{V} = \mathbf{V}^T &= (\mathbf{V}_{8T}\mathbf{V}_{4T}\mathbf{V}_{2T}\mathbf{E})^T \\ &= \mathbf{E}^T\mathbf{V}_{2T}^T\mathbf{V}_{4T}^T\mathbf{V}_{8T}^T. \end{aligned} \tag{5.3.2}$$

The right-hand side of Eq. (5.3.2) represents another algorithm for computing the DFT. What are the properties of this new algorithm? To answer this question, we need to examine the form of the matrices $\mathbf{V}_{iT}^T$. In each original matrix $\mathbf{V}_{iT}$, each row and column contains two nonzero elements at rows and columns k and $k + i/2$, $k = 1, 2, \ldots, N$, respectively. Now the transpose of such a matrix again places two nonzero elements at exactly the same positions as in the original matrix. Thus the butterfly calculations of the original matrix remain butterfly calculations in the transpose case. Only the multiplier values are changed. What about the matrix $\mathbf{E}^T$? The matrix $\mathbf{E}$ in Eq. (5.2.7) represents a special kind of permutation called bit reversal, which we will discuss in more detail later. The bit reversal permutation is such that $\mathbf{E}^T = \mathbf{E}$. Thus the algorithm of the right-hand side of Eq. (5.3.2) can be expressed as

$$\mathbf{V} = \mathbf{E}^T\mathbf{V}_{2T}^T\mathbf{V}_{4T}^T\mathbf{V}_{8T}^T = \mathbf{E}\mathbf{V}_{2F}\mathbf{V}_{4F}\mathbf{V}_{8F}. \tag{5.3.3}$$

This algorithm is shown in SFG form in Fig. 5.3.1 (see problem 5.17). The presence of $\mathbf{E}$ on the left in Eq. (5.3.3) means the output spectrum $\mathbf{X}$ is permuted (bit-reversed) rather than the input data $\mathbf{x}$, as in Fig. 5.2.4. The algorithm of Eq.

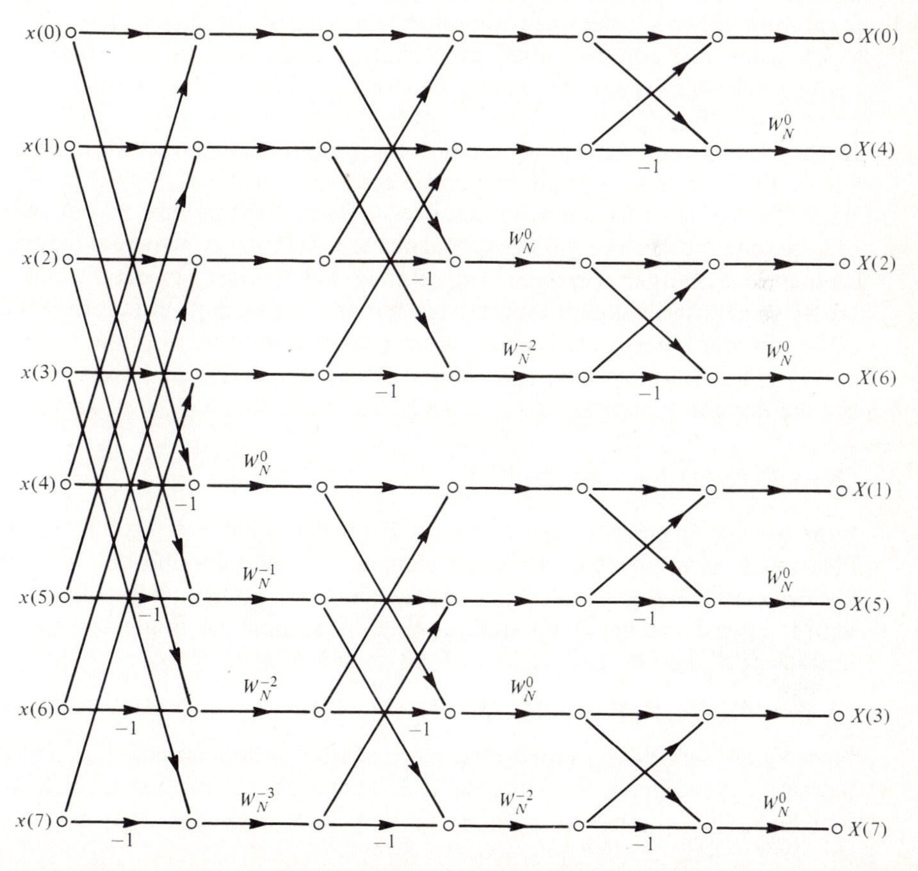

FIGURE 5.3.1 **A SFG for the decimation-in-frequency FFT for $N = 2^3$ stages.**

(5.3.3) is known as *decimation-in-frequency* and contains exactly the same number of calculations as the decimation-in-time algorithm.

We can generate other equivalent algorithms in a similar manner using matrix manipulations of **V**. For example, Eq. (4.5.10) yields an expression for the DFT matrix **V** as

$$\mathbf{V} = N(\mathbf{V}^*)^{-1} = N\mathbf{E}^{-1}(\mathbf{V}_{2T}^*)^{-1}(\mathbf{V}_{4T}^*)^{-1}(\mathbf{V}_{8T}^*)^{-1}. \tag{5.3.4}$$

The right-hand side of Eq. (5.3.4) appears to be a new algorithm for the DFT. Actually it is equivalent to Eq. (5.3.3). (See Problem 5.1.)

Bit Reversal Permutation

The matrix **E** in Eqs. (5.2.7) and (5.3.3) is a special permutation called bit reversal. If the index of the sequence $\{x(k)\}$, k, is expressed in binary, then the bit reversal permutation has an index k' that reverses the order of the binary representation. Thus if $k = 8]_{10}$, then its binary representation is $k = 100]_2$. The bit-reversed permutation is $k' = 001]_2 = 1]_{10}$.

The bit-reversal permutation **E** is factored out of the butterfly matrices in order that the computations of the DFT can be accomplished in place. By permuting either **X** or **x**, one can perform the computations in Eqs. (5.2.7) or (5.3.3) using one complex array of N storage registers (and one additional register). That this is possible can be seen from the SFGs of the two algorithms, or equivalently, from their matrix representations in Eqs. (5.2.7) and (5.3.3). In the matrix representation, in-place computation is characterized by the fact that to compute $x'(i)$ and $x'(j)$ one needs only $x(i)$ and $x(j)$ of the previous stage. To see that this is always possible in Eqs. (5.2.7) and (5.3.3), one needs only to note that for every off-diagonal element at (i, j), say, there is a diagonal element and another corresponding off-diagonal element at (j, i). Thus there is always a butterfly computation that computes the outputs at indexes i and j using the multipliers at (i, i) and (i, j) for output i and the multipliers at (j, i) and (j, j) for output j. One can show that the bit-reversal computation can also be done in place. (See Problem 5.2.)

Other Permutations and the FFT

Suppose one generates a permutation **P** so that $\mathbf{P}^T\mathbf{P} = \mathbf{I}$, with **P** not the identity. **P** must contain one unity element in each row and column. One can insert $\mathbf{P}^T\mathbf{P}$ in Eqs. (5.2.7) or (5.3.3) to generate any number of fast algorithms with the same number of computations. For example, we might choose permutations $\mathbf{P}_1$ and $\mathbf{P}_2$ and write the DFT matrix **V** as

$$\mathbf{V} \doteq \mathbf{V}_{8T}\mathbf{V}_{4T}\mathbf{P}_1^T\mathbf{P}_1\mathbf{V}_{2T}\mathbf{P}_2^T\mathbf{P}_2\mathbf{E}. \tag{5.3.5}$$

Now $\mathbf{V}_{4T}\mathbf{P}_1^T$ and $\mathbf{P}_1\mathbf{V}_{2T}$ represent a reordering of the columns of $\mathbf{V}_{4T}$ and the rows of $\mathbf{V}_{2T}$, respectively. Thus Eq. (5.3.5) is another algorithm for the DFT with the same number of calculations. The only change is that the data is rearranged between stages. One could, for example, choose $\mathbf{P}_2$ so that $\mathbf{P}_2\mathbf{E}$ is $\mathbf{I}$ and then group $\mathbf{P}_1\mathbf{V}_{2T}\mathbf{P}_2^T$ together to obtain an algorithm that has natural ordering in both $\mathbf{X}$ and $\mathbf{x}$. Is there an advantage to these algorithms? Probably not, in most cases. Since the matrices $\mathbf{P}_1$ and $\mathbf{P}_2$ are not unique, there are many different algorithms that differ only in the manner in which the intermediate data vector is addressed. In hardware, this is some wiring pattern between stages. In software, it is not clear, with one possible exception (discussed later), why one would choose a particular permutation other than the in-place computation.

One particularly interesting form is such that the algorithm has exactly the same structure for each matrix factor. Only the branch multipliers change from stage to stage. An SFG with this property is shown in Fig. 5.3.2. This algorithm was originally derived by Singleton (1967).

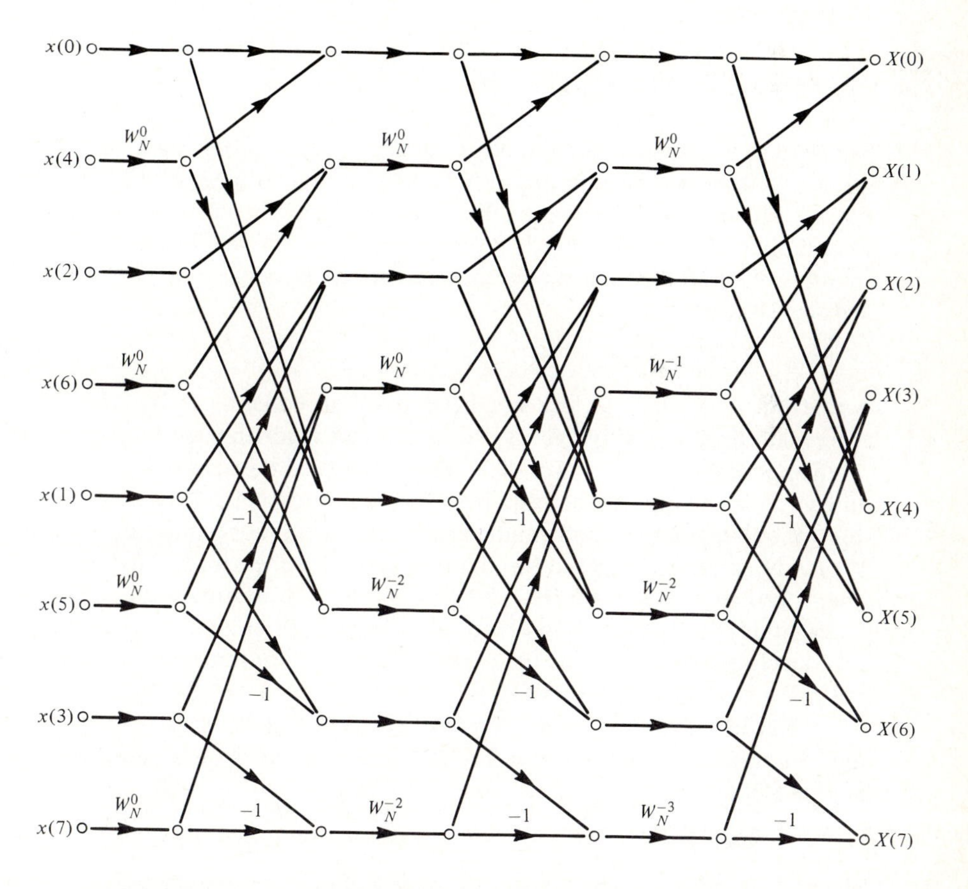

FIGURE 5.3.2 **A SFG of a FFT algorithm with the same structure for each stage. ($N = 2^3$).**

Algorithms for the Inverse DFT

The IDFT is the inverse of Eq. (5.1.4):

$$\mathbf{x} = \mathbf{V}^{-1}\mathbf{X}. \tag{5.3.6}$$

The inverse DFT matrix is given by Eq. (4.5.10), i.e.,

$$\mathbf{V}^{-1} = \frac{1}{N}\mathbf{V}^*. \tag{5.3.7}$$

Thus any algorithm for the DFT can be converted to an algorithm for the IDFT by merely taking the conjugate transpose of the original DFT matrix and then dividing the result by N. Since $\mathbf{V}^*$ differs from $\mathbf{V}$ only in the signs of the exponents of powers of W_N, this modification is easy. The algorithm is essentially the same as the original DFT algorithm. Thus any fast algorithm for the DFT can be used for the IDFT through the use of Eq. (5.3.7). (See Problem 5.5.)

5.4 MORE FAST ALGORITHMS

We have thus far presented the basic concepts for fast DFT algorithms based on the decimation-in-time algorithms and modifications of this algorithm obtained by matrix or equivalent SFG transformations. We have also assumed the number of data points N is a power of 2, i.e., $N = 2^\nu$. The efficient computation of the DFT is a direct consequence of N being a product of many factors. That is,

$$N = N_1 N_2 N_3 \cdots N_\nu, \tag{5.4.1}$$

where N_i, $i = 1, 2, \ldots, \nu$, are not necessarily all equal to 2. It is the highly composite property of N that is crucial for fast algorithms.

Suppose N is not a power of 2. We can proceed in various ways. One can add zeros to the data to increase N to the next power of 2. This zero-padding does not alter the basic information contained in the original data sequence. It merely serves to increase the number of frequency samples by essentially interpolating between the frequency points for the original value of N. Or, alternatively, if we do not wish to zero-pad the original data sequence, we can develop fast algorithms for other values of N that are highly composite but not necessarily a power of 2.

What is the essential component of these fast algorithms? Suppose N is highly composite of the form in Eq. (5.4.1). Represent N as a product of two factors. Thus

$$N = LM, \tag{5.4.2}$$

where $M = N_2 N_3 \cdots N_\nu$. Now, for $N = LM$, we can decompose the input data sequence into L sequences each containing M data samples (or vice versa). Thus we can express the DFT in the form

$$\begin{aligned}
X(n) &= \sum_{n=0}^{N-1} x(k)W_N^{-nk} \\
&= \sum_{m=0}^{M-1} x(Lm)W_N^{-Lmn} + \sum_{m=0}^{M-1} x(Lm+1)W_N^{-n}W_N^{-Lmn} \\
&\quad + \cdots + \sum_{m=0}^{M-1} x(Lm+L-1)W_N^{-(L-1)n}W_N^{-Lmn} \\
&= \sum_{l=0}^{L-1} W_N^{-nl} \sum_{m=0}^{M-1} x(Lm+l)W_N^{-Lmn} \\
&= \sum_{l=0}^{L-1} W_N^{-nl}\left[\sum_{m=0}^{M-1} x(Lm+l)W_M^{-mn}\right].
\end{aligned} \tag{5.4.3}$$

Equation (5.4.3) expresses the DFT in the form of a sum of L different M-point DFTs. If M is also a product of several factors, we can proceed to decompose the M-point DFTs further as in the power-of-2 case.

Another interpretation for Eq. (5.4.3) is obtained by observing that $x(l + mL)$, $l = 0, 1, \ldots, L - 1, m = 0, 1, \ldots, M - 1$ is a transformation of the original data sequence $x(k)$, $k = 0, 1, \ldots, N - 1$ from a one-dimensional sequence into a two-dimensional array $\mathbf{y}$ where

$$y(l, m) = x(l + mL) = x(k) \tag{5.4.4}$$

and

$$k = l + mL \tag{5.4.5}$$

with

$$l = 0, 1, \ldots, L - 1; \qquad m = 0, 1, \ldots, M - 1.$$

For example, suppose $N = LM = 3 \cdot 4 = 12$. The mapping (5.4.4) transforms the vector $\mathbf{x}$ into the array $\mathbf{y}$ where

$$\mathbf{y} = \begin{bmatrix} y(0,0) = x(0) & y(0,1) = x(3) & y(0,2) = x(6) & y(0,3) = x(9) \\ y(1,0) = x(1) & y(1,1) = x(4) & y(1,2) = x(7) & y(1,3) = x(10) \\ y(2,0) = x(2) & y(2,1) = x(5) & y(2,2) = x(8) & y(2,3) = x(11) \end{bmatrix}. \tag{5.4.6}$$

Performing a DFT on the array $\mathbf{y}$ results in another array $\mathbf{Y}$ with coordinates, say, (q, p) for which the output index is related by a mapping

$$n = qM + p, \tag{5.4.7}$$

where $q = 0, 1, \ldots, L - 1$ and $p = 0, 1, \ldots, M - 1$. Thus the output DFT vector $X(n) = Y(q, p)$, which is the DFT of $y(l, m)$. Substituting Eqs. (5.4.5) and (5.4.7) into the definition of the DFT yields

$$X(n) = Y(q, p) = \sum_{l=0}^{L-1} \sum_{m=0}^{M-1} y(l, m)W_N^{-nk}, \tag{5.4.8}$$

where

$$\begin{aligned} W_N^{-nk} &= W_N^{-(l+mL)(qM+p)} \\ &= W_N^{-lqM} W_N^{-mqML} W_N^{-mLp} W_N^{-lp} \\ &= W_L^{-lq} W_M^{-mp} W_N^{-lp}. \end{aligned} \tag{5.4.9}$$

We have used the fact that $ML = N$ and so the second term $W_N^{-mqN} = 1$. Substituting Eq. (5.4.9) in Eq. (5.4.8) gives us

$$Y(q, p) = \sum_{l=0}^{L-1} W_L^{-lq} W_N^{-lp} \sum_{m=0}^{M-1} y(l, m) W_M^{-mp}. \tag{5.4.10}$$

Now the inner sum in Eq. (5.4.10) represents an M-point DFT on the rows of $\mathbf{y}$ resulting in a new array $\mathbf{y}_1$, indexed by l and p only, and so we have

$$Y(q, p) = \sum_{l=0}^{L-1} W_L^{-lq} W_N^{-lp} y_1(l, p). \tag{5.4.11}$$

Suppose we now combine the twiddle factors W_N^{-lp} with the corresponding entries in the array $\mathbf{y}_1$. Call the resultant array $y_2(l, p) = y_1(l, p) W_N^{-lp}$. Then Eq. (5.4.11) becomes

$$Y(q, p) = \sum_{l=0}^{L-1} y_2(l, p) W_L^{-lq}. \tag{5.4.12}$$

Equation (5.4.12) is an L-point DFT on the columns of the array $\mathbf{y}_2$. Thus the original DFT calculation is decomposed into L separate M-point DFTs on the rows of the data array, followed by an element-by-element complex multiplication on each entry of the array, and finally M separate L-point DFTs on the columns of the resultant array.

In this procedure it is important to choose the input and output maps as in Eqs. (5.4.5) and (5.4.7), i.e.,

$$\begin{aligned} k &= l + mL \\ n &= qM + p \end{aligned} \qquad 0 \leqslant l, q < L; \qquad 0 \leqslant m, p < M. \tag{5.4.13}$$

The reason for this particular choice is explained more fully in Section 5.5. However, one reason for the choice is the simplification that occurs in the term W_N^{-nk} as described in Eq. (5.4.9). The particular maps in Eq. (5.4.13) not only serve to simplify the DFT calculation, but they also explain the concept of index reversal (of which bit reversal is a special case). Consider the example of Eq. (5.4.6) for $N = LM = 3 \cdot 4$. The input array corresponds to inserting $\{x(k)\}$ into the array $\mathbf{y}$ by columns. Supposing we now withdraw the output DFT vector $\{X(n)\}$ by columns from the output array $\mathbf{Y}$, we obtain the following correspondence:

$$\mathbf{x}: \begin{bmatrix} 0 \\ 1 \\ 2 \\ 3 \\ 4 \\ 5 \\ 6 \\ 7 \\ 8 \\ 9 \\ 10 \\ 11 \end{bmatrix} \xrightarrow[(5.4.5)]{} \mathbf{y}: \begin{bmatrix} 0 & 3 & 6 & 9 \\ 1 & 4 & 7 & 10 \\ 2 & 5 & 8 & 11 \end{bmatrix} \xrightarrow[(5.4.12)]{\text{DFT}} \mathbf{Y}: \begin{bmatrix} 0' & 1' & 2' \\ 3' & 4' & 5' \\ 6' & 7' & 8' \\ 9' & 10' & 11' \end{bmatrix} \xrightarrow[(5.4.7)]{} \mathbf{X}: \begin{bmatrix} 0' \\ 3' \\ 6' \\ 9' \\ 1' \\ 4' \\ 7' \\ 10' \\ 2' \\ 5' \\ 8' \\ 11' \end{bmatrix}. \tag{5.4.14}$$

If we insert **x** into **y** by columns and withdraw **X** from **Y** by columns, then the output sequence is in index reversed order from **x**. That is, by reversing the indices (i, j) of the output array **Y**, one obtains the correct order for the output DFT vector. This is a generalization of the bit reversal that occurs for $N = 2^{\nu}$.

Let's summarize the steps of this procedure for the case of $N = LM$.

Step 1. Map the data vector into a rectangular array **y**.

Step 2. Compute L separate M-point DFTs on the rows of **y**.

Step 3. Perform an element-by-element multiplication on the array with the twiddle factors.

Step 4. Compute M separate L-point DFTs on the columns of this array to form the array **Y**.

Step 5. Map **Y** into an output vector **X**.

Notice that the calculation in Eq. (5.4.10) can also be performed in reverse order; i.e.,

$$Y(q, p) = \sum_{m=0}^{M-1} W_M^{-mp} \sum_{l=0}^{L-1} y(l, m) W_L^{-lq} W_N^{-lp}. \tag{5.4.15}$$

In this case the computation proceeds by replacing each array element $y(l, m)$ with $y(l, m)W_N^{-lp}$ (p has same range as m). We then compute an M-point DFT on each column, followed by an L-point DFT on each row. The number of calculations is exactly as before.

How many calculations (complex multiplies) are required for this formulation $N = LM$? Referring to Eqs. (5.4.10)–(5.4.12), we require the computation of L DFTs of M-points, plus M DFTs of L-points, plus LM multiplies by the twiddle factors. The total number of complex multiplies is thus

$$LM^2 + ML^2 + LM = LM(M + L + 1). \tag{5.4.16}$$

If $N = N_1 N_2 \cdots N_v$, then Eq. (5.4.16) generalizes to

$$\sum_{i=1}^{v} NN_i + (v - 1)N. \tag{5.4.17}$$

For $N = 2^v$, the summation term in Eq. (5.4.17) is not present since there are no complex multiplies for a two-point DFT.

5.5 PRIME FACTOR AND RELATED ALGORITHMS

The essential idea in Section 5.4 is to transform a vector into a two-or-higher dimensional array to reduce the number of computations in the DFT. We have seen that the mappings in Eq. (5.4.13) can be chosen to reduce calculations. We can generalize these ideas. In particular, it's useful to generalize the mappings that transform $\mathbf{x} \to \mathbf{y}$ and $\mathbf{Y} \to \mathbf{X}$.

The mappings of (5.4.13) are a special case of the mapping

$$k = lK_1 + mK_2 \bmod N. \tag{5.5.1}$$

This mapping is cyclic in k because k is evaluated mod N. This periodicity is important in applications because with the DFT and IDFT all sequences are N periodic. The essential question is what are the conditions involving the integers K_1 and K_2 in Eq. (5.5.1) necessary for the mapping to be one-to-one. We obviously wish to map the vector $\mathbf{x}$ into an array $\mathbf{y}$ in such a way that we obtain exactly the same values in $\mathbf{y}$ as are in $\mathbf{x}$. Now, assuming N is composite, i.e.,

$$N = LM, \tag{5.5.2}$$

there are two cases of interest. Case 1 is when L and M are relatively prime; i.e., they have no common factors. Case 2 is when L and M have a common factor, say, f. We shall use the notation

$$\begin{aligned} &\text{case 1: GCD}(L, M) = 1, \\ &\text{case 2: GCD}(L, M) = f. \end{aligned} \tag{5.5.3}$$

The map (5.5.1) can be cyclic in either index l or m or both. The map is cyclic in l if and only if $K_1 = c_1 M$, c_1 an integer. Similarly, the map is cyclic in m if and only if $K_2 = c_2 L$, c_2 an integer. One can show (Burrus 1977) that for case 1 to have a one-to-one mapping, it must be cyclic in l or m or both. For case 2, the mapping must be cyclic in only one index, l or m. It may appear somewhat mysterious that the mappings transforming $\mathbf{x}$ into an array $\mathbf{y}$ and the array $\mathbf{Y}$ into $\mathbf{X}$ affect the number of computations. But if we recall that it's really the periodicities and symmetries in the DFT matrix $\mathbf{V}$ that we are using to reduce computations, then these mappings can be viewed as a way to re-arrange the computations to take advantage of these symmetries. The key to understanding how calculations are affected by these mappings lies in the calculation of the term W_N^{-nk} in Eq. (5.4.9) when

$$k = lK_1 + mK_2, \qquad n = qK_3 + pK_4, \qquad \bmod N. \tag{5.5.4}$$

Consider the DFT given by

$$X(n) = \sum_{n=0}^{N-1} x(k)W_N^{-nk}. \tag{5.5.5}$$

Let the mappings for the two arrays be defined by Eqs. (5.5.4) where $0 \leqslant l, q < L$ and $0 \leqslant m, p < M$. Substituting Eqs. (5.5.4) into (5.5.5) yields

$$Y(q, p) = X(qK_3 + pK_4) = \sum_{l,m}\sum y(l, m)W_N^{-(lK_1 + mK_2)(qK_3 + pK_4)} \tag{5.5.6}$$

with $y(l, m) = x(lK_1 + mK_2)$. Expanding the exponential W_N^{-nk} term in Eq. (5.5.6), we have

$$W_N^{-nk} = W_N^{-lqK_1K_3 - mqK_2K_3 - lpK_1K_4 - mpK_2K_4}. \tag{5.5.7}$$

For efficiency, we must require the calculation of Eq. (5.5.6) to result in a nested calculation of the form

$$Y(q, p) = \sum_l \left[\sum_m y(l, m)g_1(l, m, p)\right] g_2(l, q, p). \tag{5.5.8}$$

For this nesting to occur the term in Eq. (5.5.7) containing mq must be unity; i.e.,

$$W_N^{-mqK_2K_3} = 1. \tag{5.5.9}$$

This is true provided K_2K_3 is an integer multiple of $N = LM$ and so K_2 and K_3 must be of the form

$$K_2 = c_2L, \qquad K_3 = c_3M. \tag{5.5.10}$$

There are two cases. Let's consider case 2 in which $\mathrm{GCD}(L, M) = f$. Then the maps must be cyclic in exactly one of the variables. In this case, Eq. (5.5.4) reduces to

$$n = K_1l + c_2Lm, \qquad k = c_3Mq + K_4p. \tag{5.5.11}$$

The simplest form of Eq. (5.5.11) is for $K_1 = K_4 = c_2 = c_3 = 1$ resulting in the maps of Eq. (5.4.13). The DFT then reduces to the calculation in Eq. (5.4.15) and is the basis of the decimation-in-time algorithm. If in Eq. (5.5.11) we use $K_1 = c_1M$, $K_4 = c_4L$ with $c_1 = c_4 = K_2 = K_3 = 1$, then we can reverse the nesting in Eq. (5.5.8) and obtain the decimation-in-frequency algorithm.

Let's suppose that L and M are relatively prime, i.e., GCD $(L, M) = 1$. Algorithms in this case are called *prime factor algorithms* (PFA). In this case, it is possible to use Eq. (5.5.9) and also set

$$K_1 = c_1M, \qquad K_4 = c_4L.$$

In this case the term $W_N^{-K_1K_4lp}$ in Eq. (5.5.7) is also unity. This gives for Eq. (5.5.6) the following

$$Y(q, p) = \sum_l \left[\sum_m y(l, m) W_M^{-mpc_2c_4L} \right] W_L^{-lqc_1c_3M}. \tag{5.5.12}$$

Equation (5.5.12) permits one to not only nest the calculations of the DFT, but completely eliminates any twiddle factors. It is also trivial to reverse the nesting. Depending on the choice of the constants $c_i, i = 1, 2, 3, 4$, we can obtain many fast algorithms for the DFT.

A Fast Algorithm by Good

Good (1971) considered using as constants in Eq. (5.5.12) the following

$$c_1 = c_2 = 1, \quad c_3 = \frac{1}{M} \bmod L, \quad c_4 = \frac{1}{L} \bmod M. \tag{5.5.13}$$

This results in a transform of the form (for two factors)

$$Y(q, p) = \sum_l \left[\sum_m y(l, m) W_M^{-mp} \right] W_L^{-lq}. \tag{5.5.14}$$

Equation (5.5.14) is a true two-dimensional DFT with no intermediate twiddle factors. We could represent Eq. (5.5.14) in matrix notation as

$$\mathbf{Y} = \mathbf{V}_L \mathbf{y} \mathbf{V}_M, \tag{5.5.15}$$

where $\mathbf{V}_L$ and $\mathbf{V}_M$ represent DFT matrices with components W_L^{-lq} and W_M^{-mp}, respectively, and with dimensions that are consistent with the array $\mathbf{y}$.

EXAMPLE 5.5.1

Suppose we consider a case in which $N = 2 \cdot 3$. The PFA or Good algorithm for $L = 2$ and $M = 3$ is given by Eq. (5.5.14). Substituting, we have

$$Y(q, p) = \sum_{l=0}^{1} W_2^{-lq} \sum_{m=0}^{2} W_3^{-mp} y(l, m), \qquad \begin{array}{l} 0 \leq q < 2 \\ 0 \leq p < 3 \end{array}$$

(3-point DFT for each l)

$$= \sum_{l=0}^{L} W_2^{-lq} y_1(l, p).$$

(2-point DFT for each p)

This decomposition is depicted in Fig. 5.5.1.

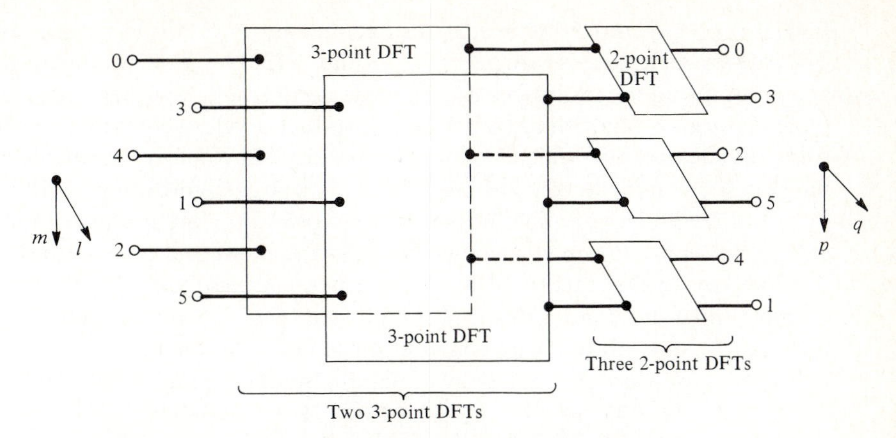

FIGURE 5.5.1 **A PFA schematic for $N = LM = 2 \cdot 3 = 6$.**

The mappings used to obtain Eq. (5.1.12) result in

$$\begin{aligned} k &= lM + mL \bmod N, \\ n &= q + p \bmod N. \end{aligned} \quad \textbf{(5.5.16)}$$

The mapping for k is called the *Chinese remainder theorem* (Ore 1948). One can also obtain Eq. (5.5.12) by using the simplest map for n and the Chinese remainder theorem for n.

There are a multitude of possible fast algorithms based on the approach presented here. The PFAs are simpler in some sense than the common factor algorithms (CFA) because they contain no twiddle factors. However, the restriction that the number of data points N be a highly composite number with relatively prime factors is somewhat restrictive in practice. It does not allow, for example, a case in which N is highly composite but contains two factors that are identical or multiples of each other.

5.6 REDUCED MULTIPLIER FFT ALGORITHMS

Winograd (1976) has developed a class of fast algorithms for the DFT called the *Winograd-Fourier transform* (WFT) that require approximately one-third the multiplies of a power-of-2 FFT for data sequences of over 1000 points. These algorithms use the PFA to convert the DFT into multidimensional prime length DFTs as explained in Section 5.5. These small N DFTs are then converted to cyclic convolutions to obtain small-N DFT algorithms requiring order N multiplies. Small-N algorithms are currently known for $N = \{2, 3, 4, 5, 7, 8, 9, 11, 13, 16, 17, 19\}$ and are given in Blahut (1985).

Small-N algorithms can be cast in a form that decomposes the DFT matrix $\mathbf{V}$ as

$$\mathbf{V} = \mathbf{SGH}. \quad \textbf{(5.6.1)}$$

In Eq. (5.6.1), **S** and **H** represent output and input additions, respectively. They are sparse matrices with entries of 0's and ± 1's. **G** is $M \times M$ and diagonal and contains all the multiplies. Given these good small-N algorithms, we can build up large-N algorithms (when N is a product of relatively prime numbers) using the ideas of Section 5.5. Winograd has also shown how to rearrange the terms in the composite algorithm so that the large-N algorithm is of the same form as Eq. (5.6.1); i.e., all additions are grouped at the input and output and all multiplies are grouped together between the input and output additions.

The key to this theory is the development of good small-N algorithms. As mentioned previously, this is accomplished by converting a small-N DFT into a cyclic convolution (the convolution of periodic sequences). The theory proves that for a length N DFT the minimum number of multiplies is $2N - K$, where K is the number of integral factors of N (including 1 and N). This theory is based on the Chinese remainder theorem for polynomials and will not be developed here. We refer the interested reader to Blahut (1985) for more details.

5.7 FAST ALGORITHMS FOR APERIODIC CONVOLUTION OF TWO FINITE SEQUENCES

Consider the aperiodic convolution of two finite length sequences **f** and **g** of length N. The result is a sequence **y** of length $2N - 1$ where

$$y(k) = \sum_{m=0}^{k} g(m)f(k-m), \qquad k = 0, 1, \ldots, 2N-2. \tag{5.7.1}$$

We assume **f** and **g** have indices $0, 1, \ldots, N - 1$. Another way of displaying the calculation in Eq. (5.7.1) is shown in Fig. 5.7.1. To find the values $y(k)$, $k = 0, 1, \ldots, 2N - 2$, one merely adds the values between the dotted lines. Clearly, a brute force calculation of Eq. (5.7.1) requires N^2 multiplies, the number of entries in the $N \times N$ matrix in Fig. 5.7.1.

Another method of calculating Eq. (5.7.1) is to use the DFT as discussed previously. Since the product of the DFTs of **g** and **f** results in a circular convolution of **g** and **f**, we must add $N - 1$ zeros to each sequence to form two new sequences $\tilde{\mathbf{f}}$ and $\tilde{\mathbf{g}}$ where

$$\tilde{f}(k) = \begin{cases} f(k), & 0 \leqslant k \leqslant N-1 \\ 0, & N \leqslant k \leqslant 2N-1 \end{cases} \qquad \tilde{g}(k) = \begin{cases} g(k), & 0 \leqslant k \leqslant N-1, \\ 0, & N \leqslant k \leqslant 2N-1. \end{cases} \tag{5.7.2}$$

Now the product of the DFTs of $\tilde{\mathbf{f}}$ and $\tilde{\mathbf{g}}$, $\tilde{\mathbf{F}}\tilde{\mathbf{G}}$, results in **Y**. The inverse DFT of **Y** is **y**, the aperiodic convolution of **f** and **g** repeated periodically. The number of multiplies needed to compute **Y** is $2N$, plus the multiplies used in computation of $\tilde{\mathbf{F}}$ and $\tilde{\mathbf{G}}$ which is $2 \times (2N)\log_2(2N)$. To find **y** from **Y** we must perform an inverse DFT resulting in $(2N)\log_2(2N)$ multiplies for a grand total of $[(2N) + 3 \times (2N)\log_2(2N)]$ multiplies.

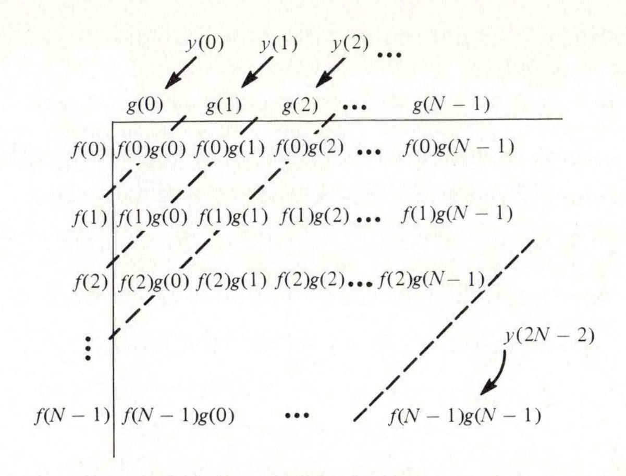

FIGURE 5.7.1 **A method for finding the values in the aperiodic convolution of f and g.**

Another method of calculating an aperiodic convolution is based on using the Lagrange interpolation formula. The sequences **f** and **g** have z-transforms

$$F(z) = f(0) + f(1)z^{-1} + f(2)z^{-2} + \cdots + f(N-1)z^{-(N-1)}$$
$$G(z) = g(0) + g(1)z^{-1} + g(2)z^{-2} + \cdots + g(N-1)z^{-(N-1)}. \tag{5.7.3}$$

Now the z-transform of the aperiodic convolution $\mathbf{f} * \mathbf{g} = \mathbf{y}$ is

$$Y(z) = F(z)G(z), \tag{5.7.4}$$

where

$$Y(z) = y(0) + y(1)z^{-1} + y(2)z^{-2} + \cdots + y(2N-2)z^{-(2N-2)}, \tag{5.7.5}$$

and the coefficients $y(i)$, $i = 0, 1, 2, \ldots, 2N - 2$ are given by

$$y(i) = \sum_{j=0}^{i} f(j)g(i-j), \qquad i = 0, 1, \ldots, 2N-2. \tag{5.7.6}$$

The polynomial $Y(z)$ is order $2N - 2$ and thus is uniquely specified by $2N - 1$ points. If one has $2N - 1$ samples of the polynomial Y, then the polynomial can be expressed using the Lagrange interpolation formula as

$$Y(z) = \sum_{i=0}^{2N-2} \tilde{Y}_i \prod_{\substack{k=1 \\ k \neq i}}^{2N-2} \left[\frac{z^{-1} - p_k^{-1}}{p_i^{-1} - p_k^{-1}}\right], \tag{5.7.7}$$

where

$$\tilde{Y}_i = Y(p_i) = F(p_i)G(p_i), \qquad i = 0, 1, \ldots, 2N-2. \tag{5.7.8}$$

To compute the polynomial $Y(z)$ (and thus the $y(i)$, $i = 0, 1, \ldots, 2N - 2$) requires $2N - 1$ multiplies in Eq. (5.7.8). There are additional calculations in Eq. (5.7.7) and in the calculation of $F(p_i)$ and $G(p_i)$. These calculations can be simplified by the choice of the values p_i. Example 5.7.1 illustrates the calculations involved.

EXAMPLE 5.7.1

Suppose **f** and **g** are $N = 2$ length sequences where

$$F(z) = f(0) + f(1)z^{-1}, \qquad G(z) = g(0) + g(1)z^{-1}.$$

Then their aperiodic linear convolution is given by

$$\begin{aligned} Y(z) = F(z)G(z) &= (f(0) + f(1)z^{-1})(g(0) + g(1)z^{-1}) \\ &= f(0)g(0) + (f(0)g(1) + f(1)g(0))z^{-1} + f(1)g(1)z^{-2} \\ &= y(0) + y(1)z^{-1} + y(2)z^{-2}. \end{aligned}$$

Direct computation of the coefficients of $Y(z)$ requires four multiplies ($N^2 = 4$). Using Eqs. (5.7.7) and (5.7.8) we should be able to do it in [2(2)-1] = 3 multiplies.

Suppose we choose the p_i in this case to be -1, ∞, and 1. Then we have

$$\begin{aligned} \tilde{Y}_0 &= F(-1)G(-1) = (f(0) - f(1))(g(0) - g(1)), \\ \tilde{Y}_1 &= F(\infty)G(\infty) = f(0)g(0), \\ \tilde{Y}_2 &= F(1)G(1) = (f(0) + f(1))(g(0) + g(1)). \end{aligned}$$

And so the polynomial $Y(z)$ is given by

$$\begin{aligned} Y(z) &= \tilde{Y}_0 \frac{z^{-1}(z^{-1} - 1)}{(-1)(-2)} + \tilde{Y}_1 \frac{(z^{-1} + 1)(z^{-1} - 1)}{(1)(-1)} + \tilde{Y}_2 \frac{z^{-1}(z^{-1} + 1)}{(1)(2)} \\ &= \tilde{Y}_1 + \frac{1}{2}(\tilde{Y}_2 - \tilde{Y}_0)z^{-1} + \left(\frac{1}{2}(\tilde{Y}_2 + \tilde{Y}_0) - \tilde{Y}_1\right)z^{-2}. \end{aligned}$$

Thus if we disregard the multiplication by $\frac{1}{2}$ in the calculation of the coefficients $y(i)$, $i = 0, 1, 2$, the number of multiplies is indeed three. (The multiplication by $\frac{1}{2}$ is a right shift of a single bit if one uses digital computation with 2's complement number representations.)

This scheme of calculating aperiodic convolutions in $(2N - 1)$ multiplies (not counting the multiplies needed to simplify the Lagrange interpolation formula) is the best one can do. This algorithm is essentially the so-called Cook-Toom algorithm [Blahut 1985].

5.8 ADDITIONAL COMMENTS ON THE USE OF THE FFT

FFT programs are often written to operate on complex data sequences. This allows one to use the same program for both the DFT and IDFT. If the input data sequence is real, the code is somewhat inefficient because complex arithmetic is always assumed. This disadvantage can be overcome by combining two real sequences into a single complex sequence.

Suppose **f** and **g** are two real N-point sequences. Form a complex sequence **h** as

$$\mathbf{h} = \mathbf{f} + j\mathbf{g}. \tag{5.8.1}$$

By linearity we have

$$\mathbf{h} \quad \overset{DFT;N}{\longleftrightarrow} \quad \mathbf{H}, \tag{5.8.2}$$

where

$$H(n) = F(n) + jG(n), \qquad n = 0, 1, \ldots, N-1. \tag{5.8.3}$$

Now we can imbed both N length sequences, **h** and **H**, into N periodic sequences because of the periodicity induced by the circular shift property of the DFT. Using the N periodicity of **H** and the symmetry properties of the DFT, we have

$$\{h^*(k) = f(k) - jg(k)\} \quad \overset{DFT;N}{\longleftrightarrow} \quad \{H^*(N-n)\}, \tag{5.8.4}$$

where

$$H^*(N-n) = F^*(N-n) - jG^*(N-n) = F(n) - jG(n). \tag{5.8.5}$$

From Eqs. (5.8.3) and (5.8.5), it is clear that

$$F(n) = \frac{H(n) + H^*(N-n)}{2} \tag{5.8.6a}$$

$$G(n) = \frac{H(n) - H^*(N-n)}{2j}. \tag{5.8.6b}$$

Thus we can find the DFTs of the real sequences **f** and **g** using the DFT of the complex sequence **h**.

We can use this same idea to compute the DFT of a real sequence of length $2N$. Suppose **x** is a $2N$-length sequence. Define $f(k) = x(2k)$ and $g(k) = x(2k+1)$ with $h(k) = f(k) + jg(k)$, $k = 0, 1, 2, \ldots, N-1$. Now the N-point DFTs of **f** and **g** are computed as in Eq. (5.8.6). To find the total DFT of the sequence **x** we combined the DFTs of **f** and **g** using

$$\begin{aligned} X(n) &= F(n) + W_{2N}^{-n}G(n), \qquad n = 0, 1, \ldots, N-1, \\ X(n+N) &= F(n) - W_{2N}^{-n}G(n), \qquad n = 0, 1, \ldots, N-1. \end{aligned} \tag{5.8.7}$$

There is a computational savings using these ideas. For example, suppose we compute the DFT of a $2N$-length sequence directly. This results in approximately $2N\log_2(2N)$ complex multiplies. Computation of the DFT of a $2N$-length sequence using Eqs. (5.8.6) and (5.8.7) requires $N \log_2 N$ complex multiplies, plus N complex multiplies in Eq. (5.8.7) for a total of $N + N$

$\log_2(N)$. The ratio of the number of multiplications of this method to the direct method is

$$r = \frac{N + N \log_2 N}{2N \log_2(2N)} = \frac{1}{2}. \tag{5.8.8}$$

Digital Filtering of Real Signals

The FFT algorithm is often used to implement filtering. If $\mathbf{u}$ is a data sequence and $\mathbf{h}$ the unit-pulse response, then the output sequence is $\mathbf{y}$ given by

$$\mathbf{y} = \mathbf{u} * \mathbf{h}. \tag{5.8.9}$$

Assuming we have available an FFT program written for complex input data, we can use the ideas mentioned previously to compute two real convolutions simultaneously.

Thus let $\mathbf{f}$, $\mathbf{g}$, and $\mathbf{h}$ be three real sequences of length N. We wish to compute two cyclic convolutions of $\mathbf{w} = \mathbf{f} * \mathbf{h}$ and $\mathbf{y} = \mathbf{g} * \mathbf{h}$. These two convolutions can be computed simultaneously in the following way. Form a sequence $\mathbf{d}$ where

$$\mathbf{d} = \mathbf{f} + j\mathbf{g}. \tag{5.8.10}$$

Compute the DFTs of $\mathbf{h}$ and $\mathbf{d}$ and form the product $H(n)D(n)$. Now consider the IDFT of this product. It is given by

$$H(n)D(n) \quad \overset{DFT}{\longleftrightarrow} \quad \mathbf{h} * \mathbf{d} = \mathbf{h} * \mathbf{f} + j\mathbf{h} * \mathbf{g}. \tag{5.8.11}$$

Thus we have calculated both $\mathbf{w}$ and $\mathbf{y}$ simultaneously.

There are several simple applications of this result. Suppose one wishes to filter two real signals $\mathbf{f}$ and $\mathbf{g}$ simultaneously with the same filter $\mathbf{h}$. The sequences $\mathbf{f}$ and $\mathbf{g}$ are then N point sections of the input sequences. Or one can filter one real sequence at twice the normal rate by defining $\mathbf{f}$ and $\mathbf{g}$ as adjacent input sections of length N.

The use of the FFT to perform convolutions is often called fast convolution. If the filter's unit-pulse response length is small (say, less than 30 points), then the overhead in the FFT algorithm makes direct computation more efficient than the methods discussed here.

Summary • The use and development of fast algorithms for the DFT and other discrete transforms is a large and ongoing area of application and research. Fast algorithms have led to many diverse applications in such fields as spectral analysis, digital filtering, radar and sonar data processing, biomedical data processing, and seismic processing. Because of space limitations we are able to introduce certain basic principles in the design of fast algorithms only briefly.

The development of the FFT and related algorithms has made the frequency-domain formulation of digital filtering competitive with time-domain approaches. The FFT has also greatly improved the efficiency of correlation and spectral analysis which have application in many fields of science and engineering. Thus the DFT (and fast algorithms for the DFT) are extremely powerful and useful tools in digital signal processing.

PROBLEMS

5.1. One of the techniques for developing fast algorithms is to perform matrix transformations on the DFT matrix **V** that leave **V** invariant. One example is Eq. (5.3.4). Show that the resulting fast algorithm is equivalent to the decimation-in-frequency algorithm.

5.2. Show that bit-reversal permutations can be performed in place.

5.3. Suggest why the SFG of Fig. 5.8.2 is useful compared with other permutations of intermediate variables that leave **X** and **x** in natural order.

5.4. Suppose a data sequence $x(k)$ is given by

$$x(k) = \cos^2\left(\frac{2\pi k}{N}\right) + 2.$$

Find the DFT of $\mathbf{x}$ for $N = 8$.

5.5. Consider the FFT algorithm written in FORTRAN that computes an in-place FFT on the complex data vector $\mathbf{x}$.

a) Is this a decimation-in-time or decimation-in-frequency algorithm?
b) Where is the bit-reversal permutation performed?
c) Where are the butterflys performed?

```
      SUBROUTINE FFT(X,NU,MODE)
C
C  Fortran subroutine to do a base-2 FFT on a complex sequence.
C  Written by Richard T. Behrens, 1986.
C
C  Call list parameters:
C              X - Complex array of length at least N (where
C                  N=2**NU).  Used to pass data to subroutine and
C                  results back to calling program.
C             NU - Integer.  The base-2 log of the number of data
C                  points in X (N=2**NU).
C           MODE - Integer.  Controls the operation performed by
C                  the subroutine as follows:
C                     Ø = FFT
C                     1 = Power spectrum (magnitude squared of FFT)
C                    -1 = Inverse FFT.
      COMPLEX X(1),W,A,B
      N=2**NU
      J=1
      PI2=6.2831853
      DO 1Ø I=1,N-1
         IF (I.LT.J) THEN
            B=X(J)
            X(J)=X(I)
            X(I)=B
         ENDIF
         K=N/2
```

```
  2     IF (K.GE.J) GOTO 10
           J=J-K
           K=K/2
        GOTO 2
 10  J=J+K
     DO 11 L=1,NU
        LE=2**L
        ANG=-PI2/LE
        LE2=LE/2
        A=(1.0,0.0)
        W=CMPLX(COS(ANG),SIN(ANG))
        IF (MODE.EQ.-1) W=CMPLX(COS(ANG),-SIN(ANG))
        DO 11 J=1,LE2
           DO 3 I=J,N,LE
              ID=I+LE2
              B=X(ID)*A
              X(ID)=X(I)-B
  3        X(I)=X(I)+B
 11  A=A*W
     IF (MODE.EQ.1) THEN
        DO 12 I=1,N
 12     X(I)=CMPLX(ABS(X(I))**2,0.0)
     ENDIF
     IF (MODE.EQ.-1) THEN
        DO 14 I=1,N
 14     X(I)=X(I)/CMPLX(FLOAT(N),0.0)
     ENDIF
     RETURN
     END
```

5.6. Suppose **f** is a sequence of length $N = LM$. Derive the expressions needed for computing the DFT of **f** using L DFTs of M points. Derive also the DFT of **f** using M DFTs of L points. Find the number of complex additions and multiplications required to combine these smaller transforms into the length N DFT. Verify these general formulas for the special case $N = 3 \cdot 5 = 15$.

5.7. A complex multiply of $(a_1 + jb_1)(a_2 + jb_2)$ can be implemented with 3 real multiplies and 5 real additions. The usual calculation requires 4 real multiplies and 2 real additions. Construct the algorithm that requires 3 real multiplies. If a real multiply takes 5 times as long to perform as a real addition, which complex multiply algorithm should be used for a 1024-point DFT? What is the time savings for the faster algorithm?

5.8. Prove the following statements:

a) **x** is a real and even periodic sequence if and only if its DFT **X** is a real and even periodic sequence.

b) **x** is a real and odd periodic sequence if and only if its DFT **X** is an imaginary and odd periodic sequence.

5.9. Consider an M-point decimation-in-time FFT. Pad this M-point FFT with zeros so that it becomes an N-point FFT with $N > M$ and $N = 2^\nu$. Because there are zeros in the data sequence **x**, one can alter the SFG or the matrix representation to eliminate operations on zero values in **x**. Suppose $M = 10$ and $N = 16$. Show how to increase the efficiency of this FFT by eliminating nonessential operations.

5.10. Find a mixed radix FFT for $N = 15$. Draw a SFG and a matrix representation of this FFT algorithm.

5.11. The number of multiplications for an FFT with a radix of 2 or 4 is the number of twiddle factor multiplications. There are N of these after each of the M stages where $N = R^M$ and $R = 2$ or 4 for this discussion. Since the radix-2 FFT has twice as many stages as the radix-4 FFT, there are approximately twice as many multiplications in a radix-2 FFT as in a radix-4 FFT. Write a decimation-in-frequency radix-4 FFT.

5.12. One method of reducing computation in a radix-2 FFT is to first compute a table of sines and cosines. The FFT is then calculated by extracting the twiddle factors from the computed table. The following program computes the necessary table. Write a radix-2 FFT using the tabulated sines and cosines in the arrays SN and CS as generated by the program given below.

```
      TWPIN=6.28318/N
      DO10 L=1, N/2
         ARG=(L-1)*TWPIN
         CS(L)=COS(ARG)
10       SN(L)=SIN(ARG)
```

5.13. The Cook-Toom algorithm is an algorithm for linear convolution that computes the convolution by using a polynomial product. The polynomial product is, in turn, computed using the Lagrange interpolation formula. Suppose we have the two sequences $(f(0), f(1))$ and $(g(0), g(1))$. Call the output $(y(0), y(1), y(2))$. The Lagrange interpolation formula states that

$$\begin{aligned}\tilde{y}(z) &= \tilde{f}(z)g(z) = (f(0) + f(1)z)(g(0) + \tilde{g}(1)z)\\ &= \tilde{y}(\alpha_0)L_0(z) + \tilde{y}(\alpha_1)L_1(z) + \tilde{y}(\alpha_2)L_2(z),\end{aligned}$$

where $(\alpha_0, \alpha_1, \alpha_2)$ are constants we are free to choose and

$$L_0(z) = -z^2 + 1, \quad L_1(z) = \tfrac{1}{2}(z^2 + z), \quad L_2(z) = \tfrac{1}{2}(z^2 - z)$$

are the Lagrange interpolation polynomials. Let $\alpha_0 = 0$, $\alpha_1 = 1$, $\alpha_2 = -1$. Write out the computation for $\tilde{y}(z)$ and obtain the linear convolution of the sequences **f** and **g**. Show that it can be expressed in the form

$$\begin{bmatrix} y(0)\\ y(1)\\ y(2)\end{bmatrix} = \begin{bmatrix} 1 & 0 & 0\\ 0 & 1 & -1\\ -1 & 1 & 1\end{bmatrix}\begin{bmatrix} F_0 & 0 & 0\\ 0 & F_1 & 0\\ 0 & 0 & F_2\end{bmatrix}\begin{bmatrix} 1 & 0\\ 1 & 1\\ 1 & -1\end{bmatrix}\begin{bmatrix} g(0)\\ g(1)\end{bmatrix},$$

where

$$F_0 = f(0), \quad F_1 = \tfrac{1}{2}(f(0) + f(1)), \quad F_2 = \tfrac{1}{2}(f(0) - f(1)).$$

5.14. Show that another Cook-Toom algorithm can be written in the form

$$\begin{bmatrix} y(0)\\ y(1)\\ y(2)\end{bmatrix} = \begin{bmatrix} 1 & 0 & 0\\ 1 & -1 & 1\\ 0 & 0 & 1\end{bmatrix}\begin{bmatrix} f(0) & 0 & 0\\ 0 & f(0) - f(1) & 0\\ 0 & 0 & f(1)\end{bmatrix}\begin{bmatrix} 1 & 0\\ 1 & -1\\ 0 & 1\end{bmatrix}\begin{bmatrix} g(0)\\ g(1)\end{bmatrix}.$$

How many multiplications and additions are used in this algorithm?

5.15. Consider the complex multiplication $\alpha + j\beta = (a + jb)(c + jd)$. This multiplication can be performed using three real multiplies and five real additions by using the algorithm

$$\alpha = (a - b)d + a(c - d),$$
$$\beta = (a - b)d + b(c + d).$$

Represent this algorithm in terms of a matrix product as in Problems (5.13) and (5.14) of the form

$$\begin{bmatrix} \alpha \\ \beta \end{bmatrix} = \mathbf{PMQ} \begin{bmatrix} a \\ b \end{bmatrix}$$

where **M** is diagonal, **P** and **Q** contain only ± 1's and 0's.

5.16. The linear convolution of two sequences can be interpreted as passing one of the sequences through an FIR filter whose tap weights are the other filter. Use the Cook-Toom algorithm to construct an algorithm to find the output of a three-tap FIR filter and a length-four data sequence.

5.17. The SFG for Eq. (5.3.3) should be a left to right mirror image of Fig. 5.2.4. The SFG in Fig. 5.3.1 deviates from this in some important respects. Explain.

5.18. True or false: In any power-of-2 FFT signal flow graph, there is exactly one path connecting a given input node to a given output node.

CHAPTER 6

The Approximation Problem for Digital Filters

6.1 INTRODUCTION

The goal of this text is to present a theory of design for linear, shift-invariant digital filters. This design process encompasses many steps, as outlined in Fig. 6.1.1. From a set of specifications, we first obtain an input-output characterization of the filter generally in the form of a transfer function $H(z)$. This part of the design is called the *approximation problem* and is the subject of this and the following chapter. For most applications, this is only the first step in the design process. Given the input-output characterization $H(z)$, there are (theoretically) an infinite number of realizations for a given transfer function. We can use this freedom in the design process to optimize criteria connected with the actual computation such as the effects of finite register lengths or the number of multipliers in the realization. This *realization problem* is less important when filtering is accomplished in software on a large general purpose digital computer. This problem is discussed in Chapters 8, 9, and 10. The *implementation problem* is concerned with hardware implementation of a particular realization. There are many ways to implement a given realization of a digital filter. As technology advances, implementations once considered too costly or too slow actually may prove to be the most efficient. The advent of *Very Large Scale Integration* (VLSI) for example, encourages large amounts of parallelism and a reduction in the connections between logic elements in the design. These considerations suggest that sequential hardware designs are not efficient for some applications and some technologies. Chapter 10 contains a brief discussion of some of the concepts involved in hardware implementation.

In this chapter and the next, we consider the *approximation problem.* We assume that certain specifications are given for a filtering problem. Typically, in this chapter these specifications are expressed in frequency response terms. For example, we may be asked to design a low-pass filter with cutoff frequency of 1 kHz with a given attenuation in the stop-band, and a specification on the

Start the Design

↓

Specifications for a digital filter. e.g., a low-pass filter with cutoff frequency f_c and passband ripple of 1 dB or less.

↓

The Approximation Problem:
Obtain an input-output characterization of the filter [such as $H(z)$] that satisfies the specifications.

↓

The Realization Problem:
obtain a realization that defines the internal structure of the filter that has transfer function $H(z)$. The realization is chosen to optimize criteria associated with the actual computation.

↓

The Implementation Problem:
Determine the hardware to implement the realization so that complexity is minimized and data throughput is optimized.

FIGURE 6.1.1 **The steps in the design of a digital filter.**

sharpness of the transition of the response from the pass-band to the stop-band. The first decision the designer has to make is the sampling time t_0, which determines the foldover frequency $1/2t_0$. The foldover frequency determines the fundamental band of frequencies that will be processed. All other frequencies of interest are best expressed as a fraction of $1/2t_0$ Hz. Furthermore, in applications where the original signal is analog, an anti-aliasing filter with cutoff frequency equal to $1/2t_0$ Hz must be used before the sampling process. We shall often use a normalized frequency $\theta = \omega t_0$.

In Chapter 2, several alternative models for digital filters were discussed. We show there that a transfer function $H(z)$ of the form

$$H(z) = \frac{b_0 + b_1 z^{-1} + \cdots + b_n z^{-n}}{1 + a_1 z^{-1} + \cdots + a_n z^{-n}} \tag{6.1.1}$$

is equivalent to the standard difference equation

$$y(k) = \sum_{i=0}^{n} b_i u(k-i) - \sum_{i=1}^{n} a_i y(k-i). \tag{6.1.2}$$

The goal of the approximation problem is to obtain the coefficients a_i, b_i, for $0 \leqslant i \leqslant n$, so that $H(z)$ meets the specifications for the filter.

There are two primary classes of digital filters (which we have already introduced). If in Eqs. (6.1.1) and (6.1.2) $a_i = 0$, $i = 1, 2, \ldots, n$, the filter is a *finite-impulse response* (FIR) filter. Otherwise it is an *infinite-impulse response* (IIR) filter. We shall divide our approximation methods into two classes: those for FIR filters and those for IIR filters.

Many of the IIR approximation schemes are based on transforming the design of an analog filter by mapping the complex s-plane into the z-plane. These methods benefit from a great deal of analog approximation theory (see Appendix 6A). A number of approximation methods are formulated as multiparameter optimization problems and use a computer to solve the resulting equations. These computer programs are very general and powerful methods of solving the approximation problem. We have space to discuss only a few of the many approximation methods. We begin with approximation methods for FIR filters.

6.2 FINITE-IMPULSE RESPONSE OR NONRECURSIVE FILTER DESIGN

An FIR filter has a transfer function $H(z)$ of the form

$$H(z) = \sum_{k=0}^{N-1} h(k)z^{-k}. \tag{6.2.1}$$

FIR filters have the following properties:

1. A linear phase response is easily obtained. Linear phase filters are called *phase distortionless*. The linear phase response implies a pure time delay. These filters are useful in applications where frequency dispersion effects caused by nonlinear phase response must be minimized, such as in data transmission.

2. Stability of the filter is guaranteed. Thus these structures are often used as adaptive filters in which the coefficients of the filter are modified in accordance with the incoming data.
3. Finite register effects are inherently simpler to analyze and of less consequence than in IIR filters.
4. For sharp cutoff filters, the number of filter taps N is large, implying a large computational burden if realized via Eq. (6.2.1). There are efficient methods of calculation that reduce the computational burden loosely called "fast convolution" methods, which are based on using the FFT to implement the FIR filter.
5. The window design method can be used as a reasonable approximation if efficiency of the design is not critical.

Linear Phase FIR Filters

The length N FIR filter whose transfer function is given by Eq. (6.2.1) will have frequency response function

$$H(e^{j\theta}) = \sum_{k=0}^{N-1} h(k)e^{-jk\theta}. \tag{6.2.2}$$

As a function of θ, the frequency response is periodic with period 2π. It is often desirable to achieve the *linear phase property*, which corresponds to a time delay. The transfer function $H(z)$ is said to have *linear phase* if

$$H(e^{j\theta}) = H_1(\theta)e^{-j(\alpha\theta+\beta)}, \tag{6.2.3}$$

where $H_1(\theta)$ is a real and even function of θ. The magnitude of H is the absolute value of H_1. The phase of H has the form

$$arg\, H(e^{j\theta}) = -\alpha\theta - \beta \tag{6.2.4}$$

when $H_1(\theta)$ is positive, and

$$arg\, H(e^{j\theta}) = -\alpha\theta - \beta - \pi \tag{6.2.5}$$

when $H_1(\theta)$ is negative. The term $(-\alpha\theta)$ corresponds to a time delay of α sampling periods.

There are four ways to achieve the linear phase property with FIR filters. One can have even or odd length, and even or odd symmetry. This is defined as follows:

$$h(N-1-k) = h(k), \quad \text{for all } k \text{ is } \textit{even symmetry}. \tag{6.2.6}$$

$$h(N-1-k) = -h(k), \quad \text{for all } k \text{ is } \textit{odd symmetry}. \tag{6.2.7}$$

For example, if $N = 2M + 1$, and we have even symmetry, then $h(M + k) = h(M - k)$ and

$$\begin{aligned} H(e^{j\theta}) &= \sum_{k=-M}^{M} h(M+k)e^{-j(k+M)\theta} \\ &= e^{-jM\theta}[h(M) + 2\sum_{k=1}^{M} h(M+k)\cos(k\theta)]. \end{aligned} \tag{6.2.8}$$

This satisfies Eq. (6.2.3) with the following identifications:

$$\begin{cases} N = 2M+1 \\ \textit{even symmetry} \end{cases} \Rightarrow \begin{cases} \alpha = M, \quad \beta = 0, \\ H_1(\theta) = h(M) + 2\sum_{k=1}^{M} h(M+k)\cos(k\theta). \end{cases} \tag{6.2.9}$$

The other three possibilities are summarized below.

$$\begin{cases} N = 2M \\ \textit{even symmetry} \end{cases} \Rightarrow \begin{cases} \alpha = M - \frac{1}{2}, \quad \beta = 0, \\ H_1(\theta) = 2\sum_{k=1}^{M} h(M+k-1)\cos[(k-\frac{1}{2})\theta]. \end{cases} \tag{6.2.10}$$

$$\begin{cases} N = 2M + 1 \\ \textit{odd symmetry} \end{cases} \Rightarrow \begin{cases} \alpha = M, \quad \beta = \pi/2, \\ H_1(\theta) = 2 \sum_{k=1}^{M} h(M + k) \sin(k\theta). \end{cases} \quad (6.2.11)$$

$$\begin{cases} N = 2M \\ \textit{odd symmetry} \end{cases} \Rightarrow \begin{cases} \alpha = M - \frac{1}{2}, \quad \beta = \pi/2, \\ H_1(\theta) = 2 \sum_{k=1}^{M} h(M + k - 1) \sin[(k - \frac{1}{2})\theta]. \end{cases} \quad (6.2.12)$$

6.3 WINDOW DESIGN OF FIR FILTERS

Assuming our designs are restricted to linear phase filters, we are now led to consider the design of the amplitude response function. Suppose $H_d(\theta)$ is some desired response function. Because the frequency response of a digital filter is periodic of period 2π, we can represent $H_d(\theta)$ as a DTFT. That is,

$$H_d(\theta) = \sum_{k=-\infty}^{\infty} h_d(k)e^{-jk\theta}, \quad (6.3.1)$$

where

$$h_d(k) = \frac{1}{2\pi} \int_{-\pi}^{\pi} H_d(\theta)e^{jk\theta}\, d\theta. \quad (6.3.2)$$

The sequence $\{h_d(k)\}$ is the unit-pulse response of the desired filter. In applications we impose the condition that the coefficients $h_d(k)$ are real. From Eq. (6.3.1), this implies that

$$H_d(-\theta) = H_d^*(\theta). \quad (6.3.3)$$

Often $H_d(\theta)$ is pure real or pure imaginary. For example, an ideal low-pass filter has the specification

$$H_d(\theta) = \begin{cases} 1, & |\theta| \leq \theta_c \\ 0, & \text{otherwise} \end{cases}. \quad (6.3.4)$$

An ideal differentiator is pure imaginary with response

$$H_d(\theta) = j\theta, \qquad 0 \leq |\theta| \leq \pi. \quad (6.3.5)$$

If $H_d(\theta)$ is real, then $h_d(k) = h_d(-k)$. Similarly, if $H_d(\theta)$ is imaginary, it follows that $h_d(k) = -h_d(-k)$. In this latter case, the unit-pulse response has odd symmetry and $h_d(0) = 0$.

Our goal is to choose an FIR filter with unit-pulse response **h** of length N which is close (in some sense) to $H_d(\theta)$. One criterion is to choose the actual response

$$H_1(e^{j\theta}) = \sum_{k=k_1}^{k_2} h_1(k)e^{-jk\theta}, \quad (6.3.6)$$

so that mean-square error ε^2 between $H_d(\theta)$ and $H_1(e^{j\theta})$ on $[-\pi, \pi]$ is minimized. This is

$$\varepsilon^2 = \frac{1}{2\pi}\int_{-\pi}^{\pi} |H_d(\theta) - H_1(e^{j\theta})|^2\, d\theta. \tag{6.3.7}$$

Using Parseval's relation, we can express Eq. (6.3.7) in the form

$$\varepsilon^2 = \sum_{k=-\infty}^{\infty} |h_d(k) - h_1(k)|^2. \tag{6.3.8}$$

From the theory of Fourier series, we can minimize ε^2 by choosing $h_1(k)$ to be the Fourier coefficients of $H_d(\theta)$ for $k_1 \leqslant k \leqslant k_2$:

$$h_1(k) = h_d(k) = \frac{1}{2\pi}\int_{-\pi}^{\pi} H_d(\theta)e^{jk\theta}\, d\theta. \tag{6.3.9}$$

The resulting minimum mean-square error is

$$\varepsilon^2_{\min} = \|\mathbf{h}_d\|^2 - \|\mathbf{h}_1\|^2 = \frac{1}{2\pi}\int_{-\pi}^{\pi} |H_d(\theta)|^2\, d\theta - \sum_{k=k_1}^{k_2} |h_d(k)|^2. \tag{6.3.10}$$

If $H_d(\theta)$ is real and even, and we take $k_2 = M$, $k_1 = -M$, then $H_1(e^{j\theta})$ will also be real and even, but not causal. A causal filter having the same magnitude response and linear phase can be obtained by simply shifting $\mathbf{h}_1$ in time. Let

$$h(k) = h_1(k - M) \quad \overset{DTFT}{\longleftrightarrow} \quad H(e^{j\theta}) = e^{-jM\theta} H_1(e^{j\theta}). \tag{6.3.11}$$

The filter $H(z)$ is FIR of order $N = 2M + 1$, and has linear phase. We shall assume this relation between $H_1(z)$ and $H(z)$ in the remainder of this section.

EXAMPLE 6.3.1

Design of a low-pass FIR filter.

Consider the design of an FIR filter of length $N = 13$ that approximates an ideal low-pass filter with a cutoff frequency of $f_c = 100$ Hz for a sampling frequency of $f_s = 1000$ Hz. In terms of normalized frequency θ, the cutoff frequency is $\theta_c = 2\pi \cdot (f_c/f_s) = \pi/5$. The desired response is thus

$$H_d(\theta) = \begin{cases} 1, & |\theta| \leqslant \dfrac{\pi}{5} \\ 0, & \dfrac{\pi}{5} < |\theta| < \pi \end{cases}. \tag{6.3.12}$$

Therefore

$$h_d(k) = \frac{1}{2\pi}\int_{-\pi/5}^{\pi/5} 1 \cdot e^{jk\theta}\, d\theta = \tfrac{1}{5}\,\mathrm{sinc}\left(\frac{k\pi}{5}\right). \tag{6.3.13}$$

For an FIR of length $N = 13$, we choose $h_1(k) = h_d(k)$ for $k = 0, \pm 1, \ldots, \pm 6$.

Thus the filter tap weights are

$$\mathbf{h} = \left\{\tfrac{1}{5}\operatorname{sinc}\left(-\frac{6\pi}{5}\right), \tfrac{1}{5}\operatorname{sinc}(-\pi), \ldots, \tfrac{1}{5}\operatorname{sinc}\left(\frac{6\pi}{5}\right)\right\}.$$

The resulting frequency response function is

$$H(e^{j\theta}) = e^{-j6\theta}\left[\sum_{k=-6}^{6} \tfrac{1}{5}\operatorname{sinc}\left(\frac{k\pi}{5}\right)e^{-jk\theta}\right]. \tag{6.3.14}$$

This response is shown in Fig. 6.3.1 along with the corresponding pulse response sequence and an SFG of the filter. The error $\varepsilon_{\min}^2$ of the approximation is, using Eq. (6.3.10),

$$\varepsilon_{\min}^2 = \|\mathbf{h}_d\|^2 - \|\mathbf{h}_1\|^2 = 0.2 - 0.1825 = 0.0175. \tag{6.3.15}$$

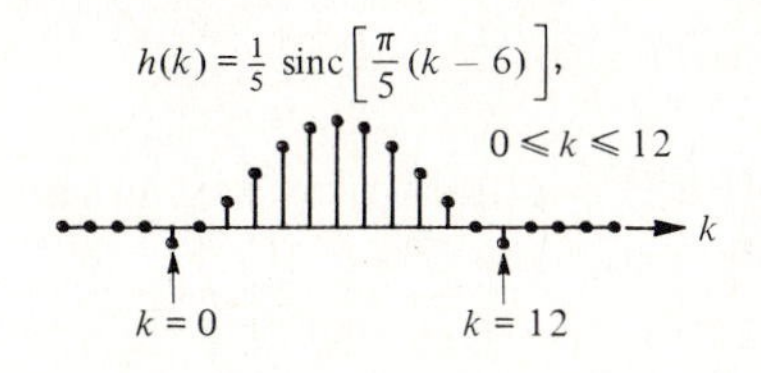

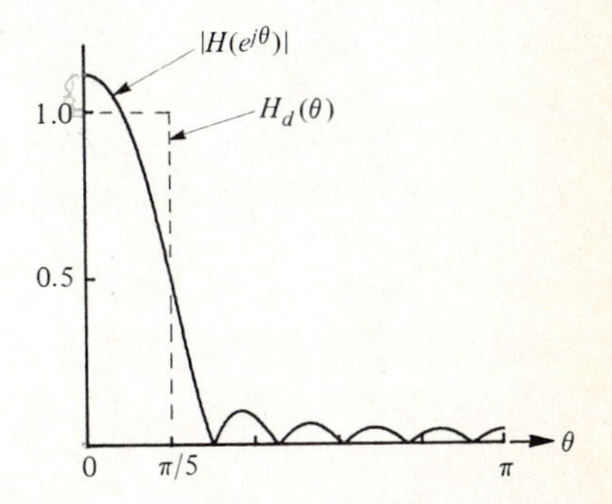

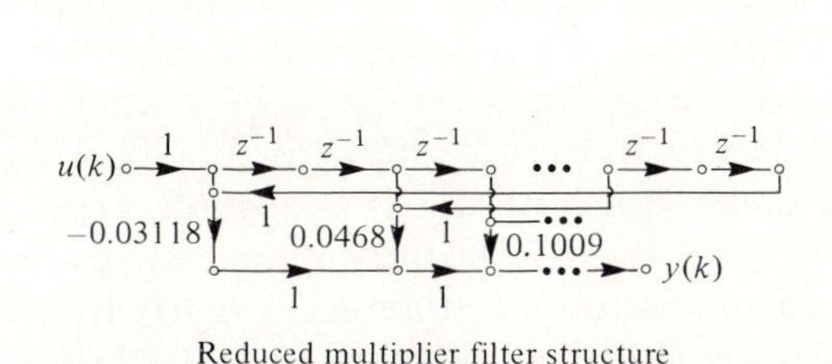

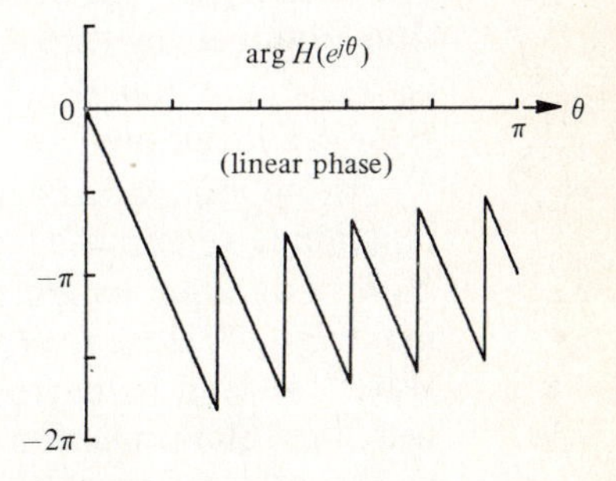

FIGURE 6.3.1 The low-pass FIR filter of Example 6.3.1.

This method of approximation (in which we essentially truncate the Fourier series representation of the desired frequency response) is guaranteed to minimize the mean-square error between the desired response function $H_d(\theta)$ and the approximation $H_1(e^{j\theta})$. There is one major defect in this method of design. At points of discontinuity in $H_d(\theta)$, because of the nonuniform convergence of the approximating sum, a characteristic overshoot in the approximation design $H_1(e^{j\theta})$ always occurs. This oscillation has been studied in connection with Fourier analysis and is known as the *Gibbs phenomenon* (Bracewell 1965). And so, even though we have minimized the mean-square error between $H_d(\theta)$ and $H_1(e^{j\theta})$, because of oscillations of $H_1(e^{j\theta})$ at discontinuities in the desired response, this method of design is not really satisfactory for most applications. As we shall show, even increasing the length of the filter N *will not* reduce the amplitude of these oscillations.

To understand the reason behind the Gibbs phenomenon, consider an infinite (desired) pulse response sequence $\mathbf{h}_d$, which we truncate to length N by multiplying $\mathbf{h}_d$ by a sequence $\mathbf{w}$ of length N. The $\mathbf{w}$ sequence is called a *window sequence*. Thus

$$h_1(k) = h_d(k)w(k), \tag{6.3.16}$$

where we have used a *rectangular window* sequence

$$w(k) = \begin{cases} 1, & |k| \leq (N-1)/2 \\ 0, & \text{otherwise} \end{cases}. \tag{6.3.17}$$

The resulting frequency response of $H(e^{j\theta})$ is obtained by taking the DTFT of both sides of Eq. (6.3.16). We obtain

$$H_1(e^{j\theta}) = \text{DTFT}\{h_d(k)w(k)\}$$

$$= \frac{1}{2\pi}\int_{-\pi}^{\pi} H_d(\phi)W_R(e^{j(\theta-\phi)})\,d\phi. \tag{6.3.18}$$

Equation (6.3.18) is an alternative representation of the approximating frequency response function written as the periodic convolution of the desired response $H_d(\phi)$ and the transform of the window sequence $W_R(e^{j\phi})$. $H_1(e^{j\theta})$ is thus obtained by "smoothing" the desired response $H_d(\phi)$ with the window response function $W_R(e^{j\phi})$. Figure 6.3.2 depicts this smoothing operation for $H_d(\phi)$ an ideal low-pass filter and $W_R(e^{j\phi})$ as defined by Eq. (6.3.17).

From Eq. (6.3.18) it is clear that the most desirable window response function is an impulse function $\delta(\theta)$. In this case, the approximation $H_1(e^{j\theta}) = H_d(\theta)$. For finite length windows sequences, this is impossible. How can we make $H_1(e^{j\theta})$ close to $H_d(\theta)$? From Eq. (6.3.18), the window response function $W_R(e^{j\theta})$ should be narrow compared to the variations in $H_d(\phi)$. If this is not true, then the transition edges in $H_d(\phi)$ will be smeared. $W_R(e^{j\theta})$ should also be smooth with no side lobes beyond the main lobe. The side lobes in $W_R(e^{j\theta})$ cause the ripples in the approximation $H_1(e^{j\theta})$. In general, as we shall show, these are conflicting requirements on the window $W(e^{j\theta})$.

In the case of the rectangular window, we have from Example 4.2.4.

$$W_R(e^{j\theta}) = \left[\frac{\sin\left(\frac{N\theta}{2}\right)}{\sin\left(\frac{\theta}{2}\right)}\right].$$

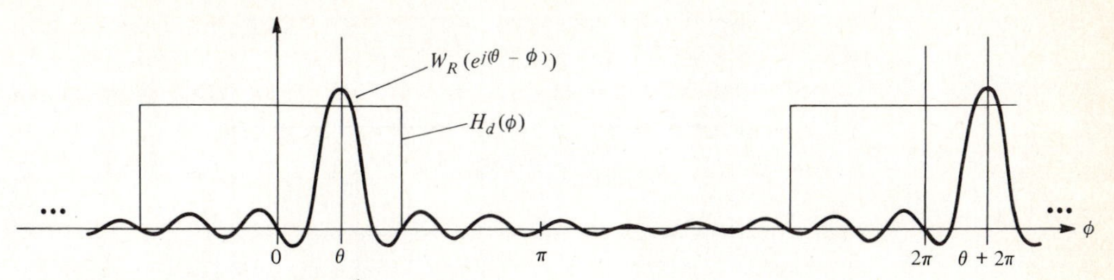

(a) Convolution of $H_d(\phi)$ and $W_R(e^{j\phi})$ at frequency θ.

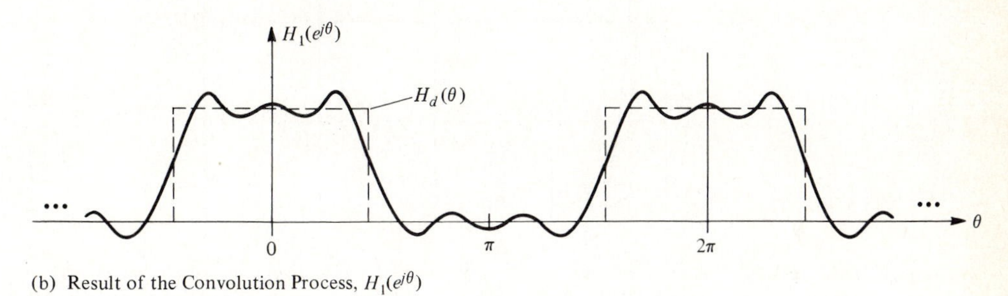

(b) Result of the Convolution Process, $H_1(e^{j\theta})$

FIGURE 6.3.2 The convolution of $W_R(e^{j\theta})$ and $H_d(\phi)$.

A plot of $W_R(e^{j\theta})/N$ is shown in Fig. 6.3.3 for several values of N. We see from this plot that as N increases:

1. The width of the main lobe decreases,
2. The height of the first side lobe approaches a constant value independent of N.

These are the two primary properties of windows that are of interest in FIR filter design. The first is called the *resolution* of the window function and is characterized by the width of the main lobe. This width translates directly into the "smearing" that occurs at jumps in $H_d(\theta)$, resulting in a certain *transition width* whenever a band edge occurs in $H_d(\theta)$. The second property is the side lobe structure of the window function sometimes called *window leakage*. The side lobe structure of the window creates the ripple in the pass- and stop-bands of the FIR filter. It is often measured by the ratio

$$R = \frac{|\text{main lobe amplitude}|}{|\text{first side lobe amplitude}|}. \tag{6.3.19}$$

For the rectangular window, suppose we measure main lobe width by the first zero crossing in $W_R(e^{j\theta})$. The first zero crossing, θ_1, occurs at

$$\frac{N\theta_1}{2} = \pi \quad \Rightarrow \quad \theta_1 = \frac{2\pi}{N} \text{ radians} \tag{6.3.20}$$

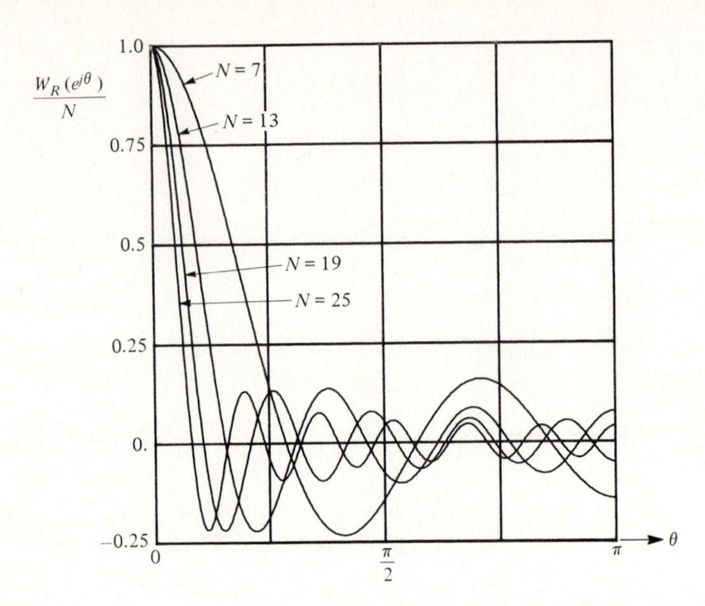

FIGURE 6.3.3 **The DTFT of the rectangular windows of length $N = 7, 13, 19, 25$.**

or since $\theta = \omega t_0 = 2\pi f t_0$, then,

$$f_1 = \frac{1}{N t_0} \text{ Hz.} \tag{6.3.21}$$

Equation (6.3.21) implies that resolution varies inversely as $N t_0$, the total duration of the FIR filter or *span* of the filter. The side lobe ripple is characterized by the ratio R in Eq. (6.3.19). The main lobe amplitude is

$$\lim_{\theta \to 0} \left[\frac{\sin\left(\frac{N\theta}{2}\right)}{\sin\left(\frac{\theta}{2}\right)} \right] = N.$$

The first side lobe occurs (for large) N at approximately $N\theta/2 = 3\pi/2$ or $\theta = 3\pi/N$. Thus R is given by

$$R = \frac{N}{\left[\dfrac{1}{\sin\left(\frac{3\pi}{2N}\right)}\right]} = N \sin\left(\frac{3\pi}{2N}\right) \cong \frac{3\pi}{2}, \qquad N \gg 1. \tag{6.3.22}$$

Thus as N increases, R approaches $3\pi/2$. This implies that the first side lobe never decreases below 21% or -13.56 dB of the main lobe amplitude. This is a manifestation of the Gibbs phenomenon.

Other Window Functions

Thus we see that even though we have minimized the mean square-squared error between $H_d(\theta)$ and $H_1(e^{j\theta})$, the resulting approximation has ripples in the pass- and stop-bands that are generally unacceptable. We can reduce the ripple in $H_1(e^{j\theta})$ by using windows other than the rectangular window. These windows all have the characteristic that the sequences taper smoothly toward zero at the ends of the sequence. The main lobe width of the DTFT of window sequence is broadened as the side lobe ripple is reduced. Some commonly used windows are given in Table 6.3.1 (Rabiner 1975).

The generalized Hanning window can be interpreted as a class of windows obtained as a weighted sum of the rectangular window and shifted versions of the rectangular window. The shifted versions add together to cancel the side lobe structure at the expense of a broader main lobe. Figure 6.3.4 depicts the situation for $\alpha = 0.50$, which results in the Hanning window

$$W_H(e^{j\theta}) = \alpha W_R(e^{j\theta}) + \left(\frac{1-\alpha}{2}\right)\{W_R(e^{j(\theta - 2\pi/N)}) + W_R(e^{j(\theta + 2\pi/N)})\}. \tag{6.3.23}$$

The corresponding sequence is obtained by taking the inverse DTFT of Eq.

TABLE 6.3.1 **Window sequences used in FIR filter design.**

Window Type	Window Sequence Values, $w(k)$; $N = 2M + 1$
1. Rectangular	$w(k) = \begin{cases} 1, & \lvert k\rvert \leqslant M \\ 0, & \text{otherwise} \end{cases}$
2. Generalized Hanning	$w_H(k) = w(k)\left[\alpha + (1-\alpha)\cos\left(\frac{2\pi}{N}k\right)\right], \quad 0 < \alpha < 1$ $\alpha = 0.54$, Hamming window $\alpha = 0.50$, Hanning window
3. Bartlett	$w_B(k) = w(k)\left[1 - \frac{\lvert k\rvert}{M+1}\right]$
4. Kaiser	$w_K(k) = w(k)\, I_0\left(\alpha\sqrt{1 - \left(\frac{k}{M}\right)^2}\right)\Big/ I_0(\alpha)$
5. Chebyshev	See Eqs. (6B.20), (6B.21), (6B.22), and Problem 6.23.
6. Gaussian	$w_G(k) = \begin{cases} \exp\left[-\frac{1}{2}k^2 \tan^2\left(\frac{\theta_0}{2}\right)\right], & \lvert k\rvert < M \\ w_G(M-1)\Big/\left[2M\sin^2\left(\frac{\theta_0}{2}\right)\right], & \lvert k\rvert = M \\ 0, & \lvert k\rvert > M \end{cases}$

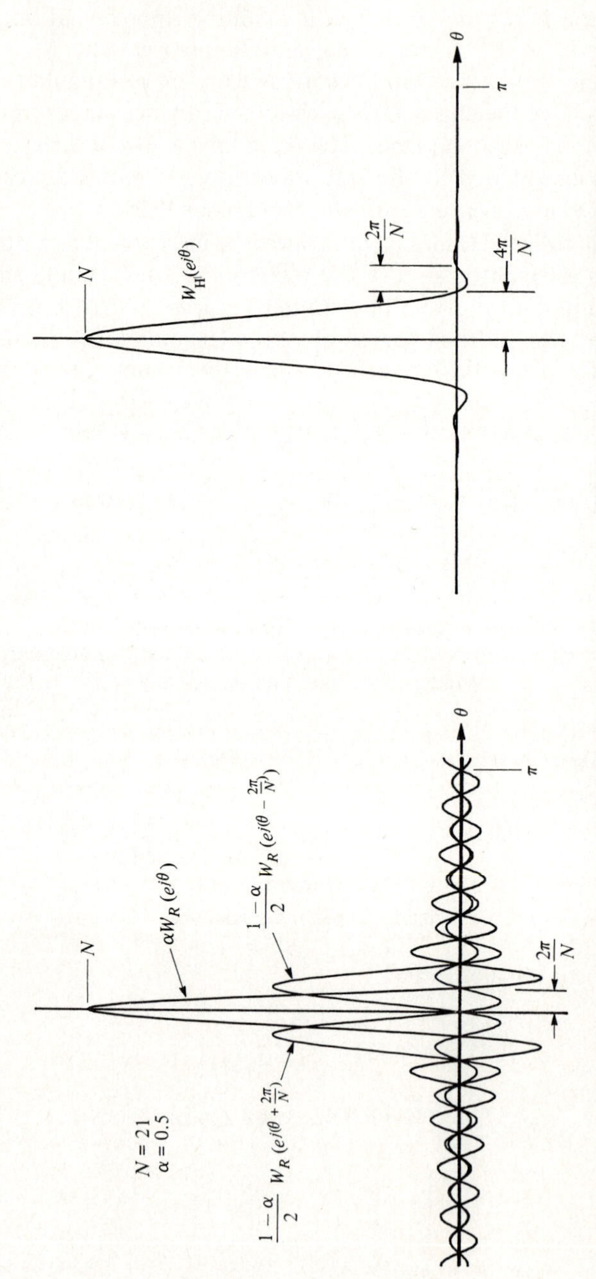

FIGURE 6.3.4 **The Hanning window. On the left are the three rectangular window components. The sum of these is depicted on the right. There is a marked decrease in sidelobe amplitude, resulting from the out-of-phase sidelobes of the three components. The width of the main lobe, however, has been increased.**

(6.2.23). Recall that the frequency shift property states that

$$w(k)e^{j(2\pi k/N)} \quad \overset{DTFT}{\longleftrightarrow} \quad W_R(e^{j(\theta - 2\pi/N)}).$$

Thus we have

$$\left(\frac{1-\alpha}{2}\right)\{W_R(e^{j(\theta-2\pi/N)}) + W_R(e^{j(\theta+2\pi/N)})\}$$

$$\overset{DTFT}{\longleftrightarrow} \quad (1-\alpha)w(k)\left(\frac{e^{j(2\pi/N)k} + e^{-j(2\pi/N)k}}{2}\right)$$

$$= w(k)\left[(1-\alpha)\cos\left(\frac{2\pi}{N}k\right)\right]$$

So that

$$W_H(e^{j\theta}) \quad \overset{DTFT}{\longleftrightarrow} \quad \left[\alpha + (1-\alpha)\cos\left(\frac{2\pi}{N}k\right)\right]w(k), \qquad |\alpha| < 1. \tag{6.3.24}$$

The generalized Hanning windows and the triangular window (and, in fact, all windows) are used to reduce the side lobes of the window from that of the rectangular window at the price of increasing the main lobe width. The reductions in side lobe amplitude are summarized in Table 6.3.2. In these windows, one can choose the side lobe attenuation needed and then increase N to achieve the desired transition width at points of discontinuity in $H_d(\theta)$.

There are three important design parameters in window design for FIR filters. These are N the number of tap weights, $2\theta_0$ the width of the main lobe, and R the ratio of the main lobe amplitude to the first side lobe amplitude. The first three windows in Table 6.3.1 fix the attenuation of the side lobe structure and allow the designer to vary the width of the main lobe by changing N. The last three windows in Table 6.3.1 possess more flexibility in that one can trade among all three of these design parameters. In the Kaiser window (Kaiser 1966), the parameter α allows one to trade main lobe width for side lobe amplitude. And by varying N one can in some sense optimize the choice of one of the three parameters.

TABLE 6.3.2 **Characteristics of some popular windows.**

Window Type	**Peak Side Lobe Amplitude (dB)**	**Main Lobe Width**
Rectangular	−13	$4\pi/N$
Triangular (Bartlett)	−25	$8\pi/N$
Hamming ($\alpha = 0.54$)	−41	$8\pi/N$
Hanning ($\alpha = 0.50$)	−31	$8\pi/N$

In the Chebyshev window, these three parameters are related by

$$\tan\left(\frac{\theta_0}{2}\right) = \sinh\left[\frac{1}{2M}\cosh^{-1}(2R)\right], \qquad N = 2M + 1. \tag{6.3.25}$$

For R large, θ_0 small, and $2M \gg \ln(2R)$, we can approximate Eq. (6.3.25) by the equation

$$\theta_0 \cong \frac{\ln(2R)}{M}. \tag{6.3.26}$$

The Chebyshev window has the desirable properties that for a given R and θ_0, N is minimized; also for fixed θ_0 and N, R is maximized and finally, for fixed R and N, θ_0 is minimized. The Gaussian window is an approximation for the Chebyshev window and so Eq. (6.3.26) applies also as a design formula. The Gaussian window is easy to use and represents a good compromise in terms of ease of design and computational complexity. [See Heyliger (1970).]

EXAMPLE 6.3.2

Design of low-pass FIR filters using Gaussian windows.

To demonstrate the use of these more flexible windows consider the design of a low-pass FIR filter. We shall use a Gaussian window sequence to reduce the side lobe structure of the resulting filter design. The half-lobe width θ_0 is chosen nominally as 15° or 0.2618 radians. To compare how the ratio R varies with N, the number of tap weights, we shall compare three designs for $N = 11, 21, 31$. Assume the cutoff frequency of the low-pass filter is $\pi/2$ radians.

The desired frequency response function $H_d(\theta)$ is given by

$$H_d(\theta) = \sum_{k=-\infty}^{\infty} h_d(k)e^{-jk\theta} = \begin{cases} 1, & |\theta| \leqslant \dfrac{\pi}{2} \\ 0, & \dfrac{\pi}{2} < |\theta| < \pi \end{cases},$$

where the tap weights are given by

$$h_d(k) = \frac{1}{2\pi}\int_{-\pi/2}^{\pi/2} e^{jk\theta}\,d\theta = \tfrac{1}{2}\operatorname{sinc}\left(\frac{k\pi}{2}\right). \tag{6.3.27}$$

This unit-pulse response is shown in Fig. 6.3.5. To design the low-pass filter, we shall truncate the sequence $\mathbf{h}_d$ and multiply by the window sequence $\mathbf{w}_G$ to obtain the design unit-pulse response $\mathbf{h}_1$. The window sequence is obtained using the formula found in Table 6.3.1 for the Gaussian window:

$$h_1(k) = h_d(k)w_G(k), \qquad h(k) = h_1(k - M) \tag{6.3.28}$$

The results are shown in Fig. 6.3.6.

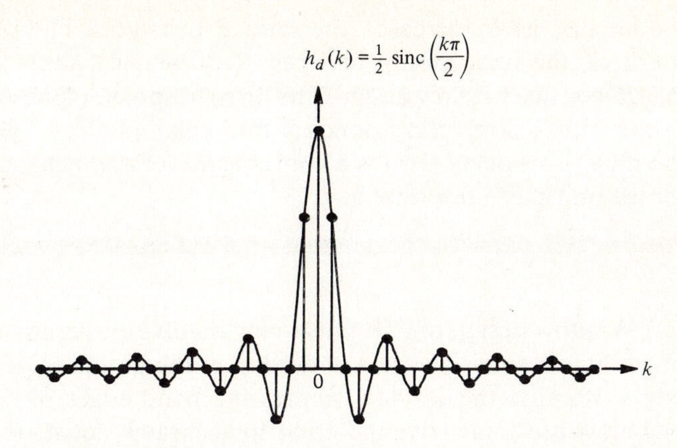

FIGURE 6.3.5 The unit-pulse response h_d for an ideal low-pass filter with cutoff frequency $\pi/2$ radians per second.

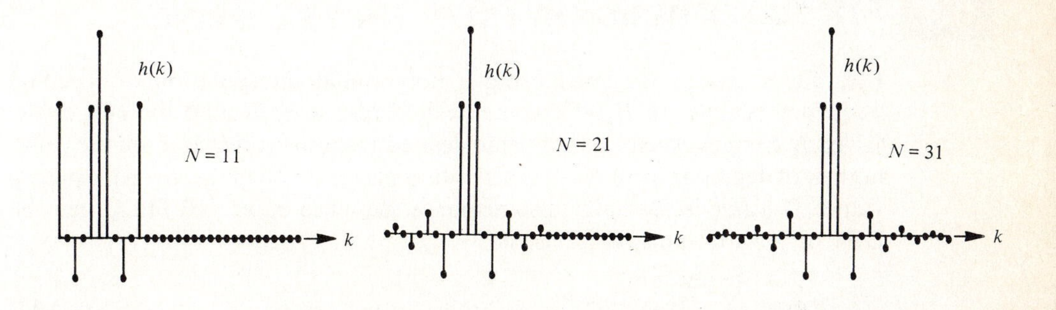

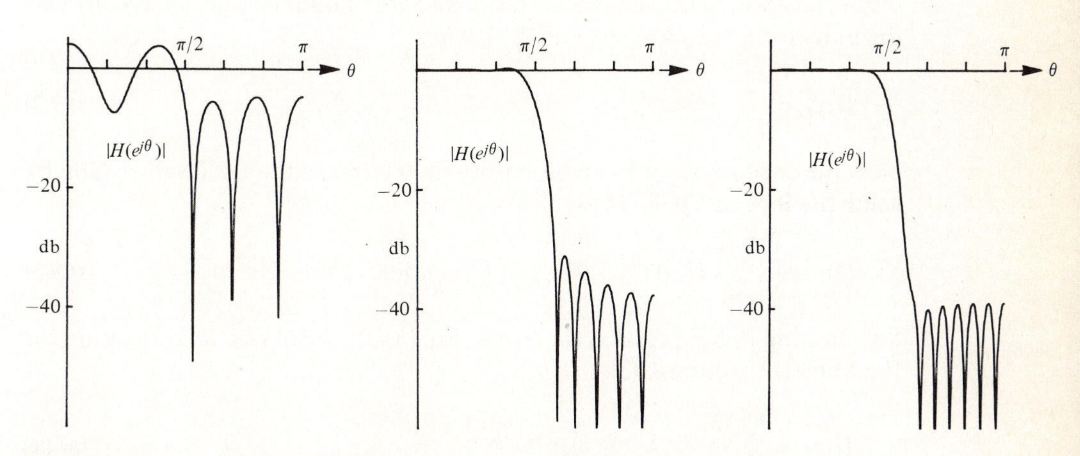

FIGURE 6.3.6 Three low-pass FIR filter designs using the Gaussian window.

Notice that as N increases, the ratio R decreases. The reader is encouraged to check the relationship between R, N, and θ_0 given in Eqs. (6.3.25) and (6.3.26) for this window design. The phase response of these filters are piecewise linear with a slope corresponding to a delay of $(N - 1)/2$, half the length of the filter. This delay term is a consequence of imposing a causality condition on the unit-pulse response $\mathbf{h}$.

Window design of FIR filters may involve an iterative process to produce acceptable results. Iteration is needed in the design to accurately place band edges. Because the window smears the band edges of the desired response, one must use some trial and error to accurately locate the band edges of the filter. Furthermore, in order to use window designs, one must be able to find a closed form expression in Eq. (6.3.2). If the desired frequency response function $H_d(\theta)$ is very complicated, evaluation of $\mathbf{h}_d$ can be difficult.

6.4 FIR FILTER DESIGN BY FREQUENCY SAMPLING

FIR filters can be designed using a polynomial interpolation of specified frequency samples of $H_d(e^{j\theta})$ (some desired response) around the unit circle. Given N frequency samples of some desired response, there is a unique polynomial of degree at most $N - 1$ in z^{-1} that passes through the given frequency values. If $\{h(k)\}$ is the unit-pulse response sequence of an FIR filter, then the DFT of this unit-pulse response is

$$H(n) = \sum_{k=0}^{N-1} h(k)e^{-j2\pi kn/N}. \tag{6.4.1}$$

From our discussion of the DFT, the DFT values $H(n)$ can be interpreted as the z-transform of the sequence $\mathbf{h}$ evaluated at N points equally spaced around the unit circle, i.e., $H(n) = \tilde{H}(e^{j2\pi n/N})$, where

$$\tilde{H}(z) = \sum_{k=0}^{N-1} h(k)z^{-k}. \tag{6.4.2}$$

Now the coefficients of $\mathbf{h}$ can be expressed in terms of the DFT values $H(n)$ by using the inverse DFT. Thus

$$h(k) = \frac{1}{N}\sum_{n=0}^{N-1} H(n)e^{j(2\pi/N)nk}, \qquad k = 0, 1, \ldots, N - 1. \tag{6.4.3}$$

Substituting this expression for $h(k)$ in Eq. (6.4.2), we obtain an expression for the transfer function $\tilde{H}(z)$.

$$\tilde{H}(z) = \sum_{k=0}^{N-1}\left[\frac{1}{N}\sum_{n=0}^{N-1} H(n)e^{j(2\pi/N)nk}\right]z^{-k}. \tag{6.4.4}$$

Now interchange the summations and sum over k to obtain

$$\tilde{H}(z) = \frac{1}{N}\sum_{n=0}^{N-1} H(n)\left[\frac{1 - z^{-N}}{1 - e^{j(2\pi/N)n}z^{-1}}\right]. \qquad (6.4.5)$$

Equation (6.4.5) expresses the transfer function $\tilde{H}(z)$ in terms of N frequency samples $H(n)$ multiplied by an $(N - 1)$ order polynomial in z^{-1}. This polynomial acts to interpolate between the N equally spaced frequency samples. If we specify the frequency samples of a desired response equally spaced around $|z| = 1$, then Eq. (6.4.5) gives us the response at all frequencies $z = e^{j\theta}$. For N large and smooth desired response, the error between the desired response and the actual response given in Eq. (6.4.5) is small. The filter's unit-pulse response is obtained using Eq. (6.4.3).

The disadvantage of this design procedure is that frequency sampling filters have excessive ripple or overshoot at points of discontinuity in the desired response. As in the window design, this problem can be solved by increasing the transition width at frequencies where jumps occur. Thus instead of specifying $H(n_0) = 1$ and $H(n_0 + 1) = 0$, one increases the transition band by setting $H(n_0) = 1$, $H(n_0 + 1) = \frac{1}{2}$, and $H(n_0 + 2) = 0$. The value $H(n_0 + 1)$ can also be optimized, in some sense, to reduce the ripple. That is, if this transition value is left unconstrained and chosen to minimize some measure of the error between the desired and actual response, a smaller overshoot can be obtained. This can be done quickly by trial and error or, alternatively, one can write an optimization routine to solve for the unknown frequency value $H(n_0 + 1)$. Further reduction of the overshoot or ripple can be obtained by using more than one unconstrained frequency sample in the transition band. For example, if $H(n_0) = 1$, $H(n_0 + 1)$ and $H(n_0 + 2)$ are optimized, and $H(n_0 + 3) = 0$, then an even smaller ripple can be obtained.

In this design, we specify the desired frequency samples. Generally, we wish to obtain a real pulse response sequence **h**. If a real sequence **h** is required, the samples $H(n)$ cannot be specified arbitrarily. We have

$$\tilde{H}(z) = \sum_{k=0}^{N-1} h(k)z^{-k}. \qquad (6.4.6)$$

For **h** real, Eq. (6.4.6) implies that $\tilde{H}^*(e^{j\theta}) = \tilde{H}(e^{-j\theta})$, which means $\tilde{H}^*(e^{jn2\pi/N}) = \tilde{H}(e^{-jn2\pi/N})$. Since $H(n)$ is periodic with period N, we must have

$$H(N - n) = H^*(n), \qquad n = 0, 1, \cdots. \qquad (6.4.7)$$

The converse is also true. If the frequency samples satisfy Eq. (6.4.7), then **h** is real.

EXAMPLE 6.4.1

Design of a low-pass FIR filter using frequency sampling.

To illustrate frequency sampling design, consider the design of an ideal low-pass filter. The desired response is

$$H_d(\theta) = \begin{cases} 1, & |\theta| \leq \theta_c \\ 0, & \theta_c < |\theta| \leq \pi. \end{cases}$$

Choose $N = 15$ and $\theta_c = (4.5/7.5)\pi$ so that the desired frequency samples are as shown in Fig. 6.4.1. The unit-pulse response of the filter is given by the IDFT of the frequency samples in Fig. 6.4.1.

$$h(k) = \frac{1}{15} \sum_{n=0}^{14} H(n)e^{j(2\pi/15)nk}, \qquad k = 0, 1, \ldots, 14.$$

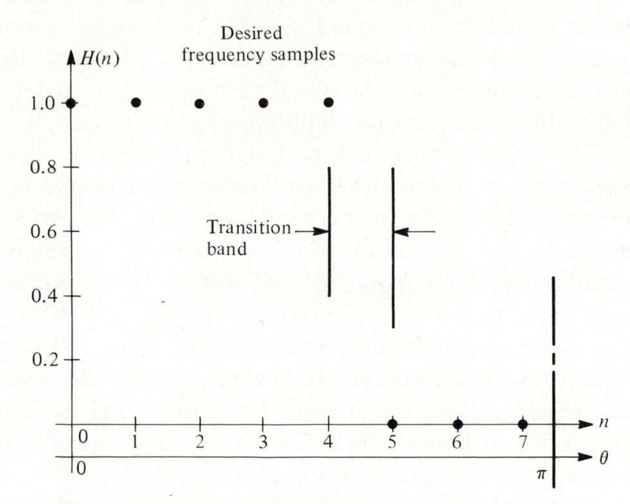

FIGURE 6.4.1 **The samples of the desired frequency response.**

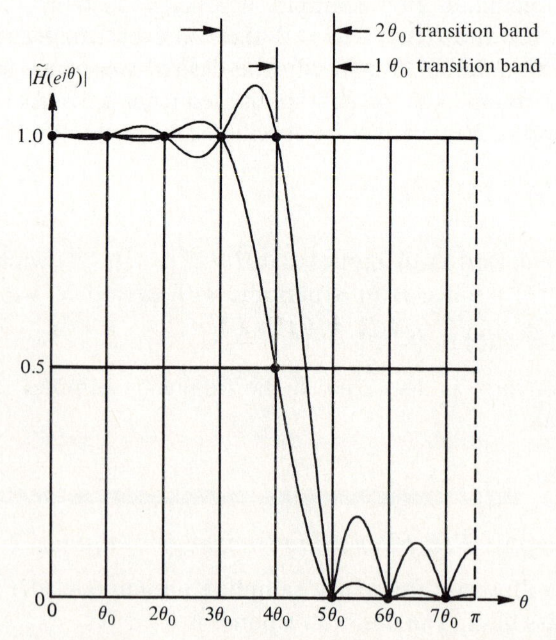

FIGURE 6.4.2 **Frequency response of two frequency sampling FIR designs. $N = 15$, and $\theta_0 = 2\pi/N$.**

The actual response for this unit-pulse sequence is given by Eq. (6.4.5). Thus

$$\tilde{H}(e^{j\theta}) = \frac{1}{15}\sum_{n=0}^{14} H(n)\cdot\left[\frac{1 - e^{-j15\theta}}{1 - e^{j(2\pi n/15 - \theta)}}\right]$$

or

$$\tilde{H}(e^{j\theta}) = \sum_{k=0}^{14} h(k)e^{-jk\theta}.$$

Figure 6.4.2 depicts the actual response. Also shown is the response when the transition width is increased by choosing $H(4)$ to be $\frac{1}{2}$ [$H(11)$ is also set equal to $\frac{1}{2}$]. Notice the decrease in ripple in the filter with the broader transition band.

To summarize, the frequency design method begins by specifying the frequency values of a desired response at N frequencies around the unit circle. At points of discontinuity in the desired response, one or more DFT coefficients (the frequency values) are left unspecified. They are chosen to reduce the approximation error in the vicinity of the jump in order to trade transition width for a smaller ripple in the actual response. The optimization can also be formulated as a linear programming problem (Rabiner 1972).

6.5 MINIMAX DESIGN OF FIR FILTERS

In window design, one minimizes the energy in the error $H_d(\theta) - H_1(e^{j\theta})$. For some applications it is more appropriate to minimize the maximum error:

$$\|H_d - H_1\|_\infty = \max_\theta |H_d(\theta) - H_1(e^{j\theta})|. \tag{6.5.1}$$

The problem of finding

$$H_1(z) = \sum_{k=-M}^{M} h_1(k)z^{-k} \tag{6.5.2}$$

to minimize the error norm in Eq. (6.5.1) is called the *minimax design problem*

The window design admitted the easy solution of Eq. (6.3.16), but obtaining a minimax solution is much more difficult. In general, it must be found numerically using specialized search procedures. Fortunately, the minimax problem is formally equivalent to a classical problem called the *Chebyshev problem* which involves uniform approximation by polynomials. Suppose that $\mathbf{h}_1$ has even symmetry. Then the frequency response becomes

$$H_1(e^{j\theta}) = h_1(0) + 2\sum_{k=1}^{M} h_1(k)\cos(k\theta). \tag{6.5.3}$$

For each positive integer k there is a polynomial $T_k(x)$ of degree k for which

$$\cos(k\theta) = T_k(\cos\theta). \tag{6.5.4}$$

These polynomials are called *Chebyshev polynomials* and are discussed in Appendix 6B. It follows that

$$H_1(e^{j\theta}) = h_1(0) + 2\sum_{k=1}^{M} h_1(k)T_k(\cos\theta) = p(\cos\theta), \tag{6.5.5}$$

where $p(x)$ is a polynomial of degree at most M, with real coefficients. Now the desired response $H_d(\theta)$ is real and 2π-periodic and therefore has a cosine series. This means that there is some real function $d(x)$ for which

$$H_d(\theta) = d(\cos\theta). \tag{6.5.6}$$

The error norm in Eq. (6.5.1) is therefore equivalent to

$$\|d - p\|_\infty = \max_{-1 \leqslant x \leqslant 1} |d(x) - p(x)|, \tag{6.5.7}$$

since the range of $x = \cos(\theta)$ is $-1 \leqslant x \leqslant 1$. The Chebyshev problem is to find a polynomial p of degree at most M to minimize $\|d - p\|_\infty$. If we can find the solution $p(x)$ to the Chebyshev problem, then we can can find the FIR impulse response $\mathbf{h}_1$ using Eq. (6.5.5), since the Chebyshev polynomials are linearly independent. The essential things to know about the Chebyshev problem are covered in the following theorem, which can be found in Cheney (1966).

THEOREM (Chebyshev). Let d be a real valued function which is continuous on $-1 \leqslant x \leqslant 1$.

1. There is unique polynomial p of degree at most M which minimizes $\|d - p\|_\infty$.
2. $p(x)$ is that polynomial if and only if there exist points

$$-1 \leqslant x_0 < x_1 < \cdots < x_{n+1} \leqslant 1 \tag{6.5.8}$$

for which

$$d(x_i) - p(x_i) = (-1)^i \varepsilon, \qquad i = 0, 1, \ldots, n+1 \tag{6.5.9}$$

where

$$|\varepsilon| = \|d - p\|_\infty. \tag{6.5.10}$$

The $n + 2$ points in the ordered list (6.5.8) are the points at which the error $d(x) - p(x)$ takes on its extreme values. If $d(x)$ is differentiable, then the derivative will vanish at the interior points, and at x_0 and x_{n+1} when these differ from -1 and 1. This can be expressed as

$$(x_i^2 - 1)(d'(x_i) - p'(x_i)) = 0, \qquad i = 0, 1, \ldots, n+1. \tag{6.5.11}$$

The fact that the error takes on extreme values at $n + 2$ points leads to the so-called *equal ripple* error phenomenon. A graph of $d(x) - p(x)$ typically resembles a sinusoid which is distorted in the horizontal (but not vertical) axis.

Equations (6.5.9) and (6.5.11), together with the conditions (6.5.8) and (6.5.10) amount to $2n + 4$ equations in $2n + 4$ unknowns; $n + 2$ points,

$n + 1$ polynomial coefficients, and ε. Most techniques for searching for solutions hinge on the fact that the equations are linear in the coefficients of $p(x)$. In fact, if the correct choices for x_0 through x_{n+1} were known, then finding $p(x)$ would be easy, and would be equivalent to the Lagrange interpolation problem.

There are a number of good numerical methods for obtaining minimax FIR designs. In (Hermann 1970a, 1970b) one can find ways to guarantee equal ripple behavior in the pass- and stop-bands of generalized bandpass filters. In (Hofstetter 1972) there is a technique for designing equal ripple filters that extends the Hermann-Schussler formulation. In these papers, tolerances on pass- and stop-band ripple and filter order are specified. One limitation of these methods is that the band edges cannot be accurately determined beforehand. In (McClellan 1973b) and (Parks 1972) the minimax FIR design problem is generalized to all four types of linear phase filters, and a more general weighted norm is allowed. Using the Remez algorithm, this allows for the design of a large class of filter types, including generalized bandpass filters, differentiators, and Hilbert transformers of large order. The IEEE Professional Group on Acoustics, Speech, and Signal Processing has made available a book (and a magnetic tape) that contains several useful design programs at a modest cost (IEEE 1979).

6.6 THE APPROXIMATION PROBLEM FOR IIR DIGITAL FILTERS

IIR digital filters have transfer functions of the form (6.6.1)

$$H(z) = \frac{\sum_{i=0}^{m} b_i z^{-i}}{1 + \sum_{i=1}^{n} a_i z^{-i}}. \tag{6.6.1}$$

We assume that $n > 0$ and $m \leqslant n$ for purposes of this discussion. The approximation problem for IIR filters is to find the $\{a_i\}$ and the $\{b_i\}$ in Eq. (6.6.1) to approximate a desired response function $H_d(\theta)$.

As in the FIR approximation problem, there are many approximation methods for IIR digital filters. One of the most popular formulations is to use the large body of knowledge for the approximation of continuous-time or *analog* filters. From a set of given filter specifications we obtain a continuous-time filter approximation in the form of a transfer function $\hat{H}(s)$. $\hat{H}(s)$ approximates some desired response $\hat{H}_d(s)$. The analog design $\hat{H}(s)$ is then transformed in some manner to obtain the digital filter $H(z)$. The transformation of $\hat{H}(s)$ can often be constructed using a mapping of the s-plane into the z-plane, say $z = \psi(s)$. This is formally accomplished by merely substituting for s in $\hat{H}(s)$, the function $s = \psi^{-1}(z)$. The choice of the mapping is generally based on preserving some property between the two domains or on preserving some invariance between the analog and the corresponding digital filter.

Impulse Invariance

The essential idea in the design of impulse invariant digital filters is that we should choose the unit-pulse response $\{h(k)\}$ of the digital filter to equal sampled values of the impulse response $\hat{h}(t)$ of some desired analog filter. That is, we seek a digital filter with unit-pulse response **h** so that

$$h(k) = t_0 \hat{h}(kt_0), \tag{6.6.2}$$

where the gain t_0 is included to ensure that the digital filter gain is independent of the sampling period. Suppose $\hat{H}(s)$, the Laplace transform of the analog impulse response $\hat{h}(t)$, approximates some desired analog response $\hat{H}_d(s)$. $\hat{H}(s)$ might represent, for example, a Butterworth filter approximation to an ideal filter. $\hat{H}(s)$ is obtained by standard analog network synthesis. We require $\hat{H}(s)$ be rational, that is, the ratio of polynomials in s with numerator order less than the denominator order. Thus let $\hat{H}(s)$ be of the form

$$\hat{H}(s) = \frac{1 + \beta_1 s + \cdots + \beta_m s^m}{\alpha_0 + \alpha_1 s + \cdots + \alpha_n s^n}, \qquad n > m. \tag{6.6.3}$$

Expand $\hat{H}(s)$ in partial fractions assuming the denominator has no repeated roots. Then we obtain

$$\hat{H}(s) = \frac{c_1}{s - p_1} + \frac{c_2}{s - p_2} + \cdots + \frac{c_n}{s - p_n}, \tag{6.6.4}$$

where p_i, $i = 1, 2, \ldots, n$ are the poles of $\hat{H}(s)$. From Eq. (6.6.4), we can immediately obtain the impulse response of the analog approximation by taking the inverse Laplace transform of $\hat{H}(s)$.

$$\hat{h}(t) = \begin{cases} c_1 e^{p_1 t} + c_2 e^{p_2 t} + \cdots + c_n e^{p_n t}, & t \geqslant 0 \\ 0, & t < 0 \end{cases}. \tag{6.6.5}$$

Now write **h** as a sum of n sequences. That is,

$$h(k) = \begin{cases} h_1(k) + h_2(k) + \cdots + h_n(k), & k \geqslant 0 \\ 0, & k < 0 \end{cases}. \tag{6.6.6}$$

Set the ith term of Eq. (6.6.6) equal to the sampled version of the ith term of Eq. (6.6.5). Thus

$$h_i(k) = \begin{cases} t_0 c_i e^{p_i k t_0}, & k \geqslant 0 \\ 0, & k < 0 \end{cases}. \tag{6.6.7}$$

The digital filter transfer function is then the z-transform of Eq. (6.6.6). The ith term is given by

$$H_i(z) = \frac{t_0 c_i}{1 - e^{p_i t_0} z^{-1}} \tag{6.6.8}$$

The transfer function of the digital filter is therefore

$$H(z) = \sum_{i=1}^{n} H_i(z). \tag{6.6.9}$$

Each term in the partial fraction expansion of $\hat{H}(s)$ in Eq. (6.6.4) has been replaced by a term of the form Eq. (6.6.8). Thus each simple pole in $\hat{H}(s)$ at p_i is now a simple pole of $H(z)$ at $e^{p_i t_0}$. Notice that if $\text{Re}\{p_i\} < 0$, then $|e^{p_i t_0}| < 1$. Hence poles in the left half s-plane map into poles inside the unit circle in the z-plane, as we require for stable, causal filters. The impulse invariant design for simple poles is summarized as (for the case of repeated roots, see Problem 6.21):

For

$$\hat{H}(s) = \sum_{i=1}^{n} \frac{c_i}{s - p_i}, \tag{6.6.10}$$

then

$$H(z) = \sum_{i=1}^{n} \frac{t_0 c_i}{1 - e^{p_i t_0} z^{-1}}. \tag{6.6.11}$$

How good is the resulting approximation? To answer this question consider the frequency response function of the resulting approximation. The frequency response of the digital filter is

$$H(e^{j\omega t_0}) = \sum_{k=0}^{\infty} h(k) e^{-jk\omega t_0} = \sum_{k=0}^{\infty} t_0 \hat{h}(kt_0) e^{-j\omega k t_0}. \tag{6.6.12}$$

To understand how good this frequency response function is we need to relate $H(e^{j\theta})$ to the analog approximation $\hat{H}(j\omega)$. Recall that in Section 4.6 we obtained the frequency response of a sampled time function in Eq. (4.6.6). Using this,

$$\begin{aligned} H(e^{j\theta}) = H(e^{j\omega t_0}) &= \sum_{n=-\infty}^{\infty} \hat{H}(j\omega + jn\omega_0) \\ &= \sum_{n=-\infty}^{\infty} \hat{H}\left(j\omega + jn\frac{2\pi}{t_0}\right) = \sum_{n=-\infty}^{\infty} \hat{H}\left(j\frac{\theta}{t_0} + j\frac{n2\pi}{t_0}\right). \end{aligned} \tag{6.6.13}$$

The conclusion we reach is that $H(e^{j\theta})$ is an aliased version of $\hat{H}[j(\theta/t_0)]$ rather than $\hat{H}[j(\theta/t_0)]$ itself. Notice also from Eq. (6.6.13) that the aliasing

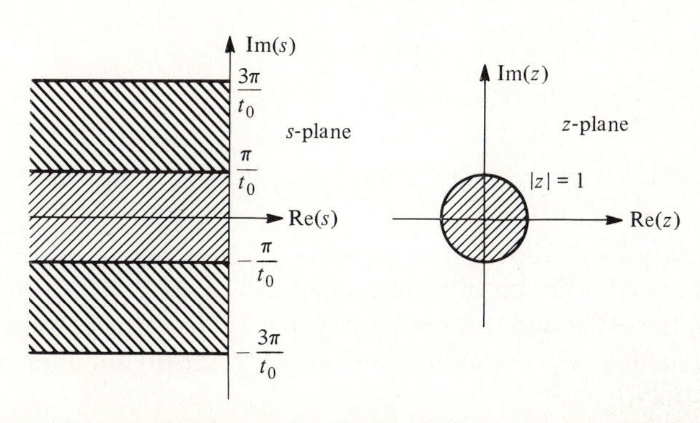

FIGURE 6.6.1 A representation of the aliasing that occurs in impulse invariant design.

interval is $2\pi/t_0$ since it is repeated versions of $\hat{H}[j(\theta/t_0)+j(n2\pi/t_0)]$ that are added together to form $H(e^{j\theta})$. Thus strips of width $2\pi/t_0$ in the s-plane map into the entire z-plane as depicted in Fig. 6.6.1. Because of the aliasing that occurs in the impulse invariant design, this design should be restricted to filters that are narrowband and whose response dies off quickly at the foldover frequency. This implies that the denominator polynomial of $\hat{H}(s)$ should be of considerably higher degree than the numerator polynomial.

EXAMPLE 6.6.1

Design of a third-order Butterworth digital filter.

Suppose we wish to design a low-pass digital filter using the digital equivalent of a three-pole Butterworth filter. The analog filter (see Appendix 6A) with unity dc gain has the transfer function

$$\hat{H}(s) = \frac{-p_1p_2p_3}{(s-p_1)(s-p_2)(s-p_3)} = \frac{\Omega_c^3}{(s+\Omega_c)(s^2+s\Omega_c+\Omega_c^2)}$$

with $p_1 = -\Omega_c$, $p_2 = p_3^* = -\Omega_c[(1-j\sqrt{3})/2]$. Ω_c is the cutoff frequency of the analog low-pass filter. Expanding $\hat{H}(s)$ in partial fractions, we obtain

$$\hat{H}(s) = \frac{c_1}{s-p_1} + \frac{c_2}{s-p_2} + \frac{c_3}{s-p_3}$$

with $c_1 = \Omega_c$, $c_2 = 2\Omega_c/(-3+j\sqrt{3})$, $c_3 = c_2^*$. The impulse invariant design is obtained using Eq. (6.6.11). Thus

$$H(z) = \frac{\mu}{1-z^{-1}e^{-\mu/2}} - \frac{\mu - z^{-1}\mu e^{-\mu/2}[\cos(a)+\sin(a)/\sqrt{3}]}{1-2z^{-1}e^{-\mu/2}\cos(a)+z^{-2}e^{-\mu}},$$

where $\mu = \Omega_c t_0$, $a = \mu\sqrt{3}/2$. If we choose Ω_c to be one-half the half sampling frequency, then

$$\Omega_c = \frac{\pi}{2t_0}, \qquad a = \frac{\sqrt{3}\pi}{4}.$$

In this case, $H(z)$ reduces to

$$H(z) \cong \frac{1.57}{1-0.21z^{-1}} - \left[\frac{1.57-0.55z^{-1}}{1-0.19z^{-1}+0.21z^{-2}}\right].$$

As expressed above, $H(z)$ could be realized as the sum of a first- and second-order system in parallel, as shown in Fig. 6.6.2. The corresponding frequency response function is shown in Fig. 6.6.3. The error between the response $\hat{H}(j\omega)$, the analog approximation, and $H(e^{j\omega t_0})$, the digital filter response, is caused by aliasing.

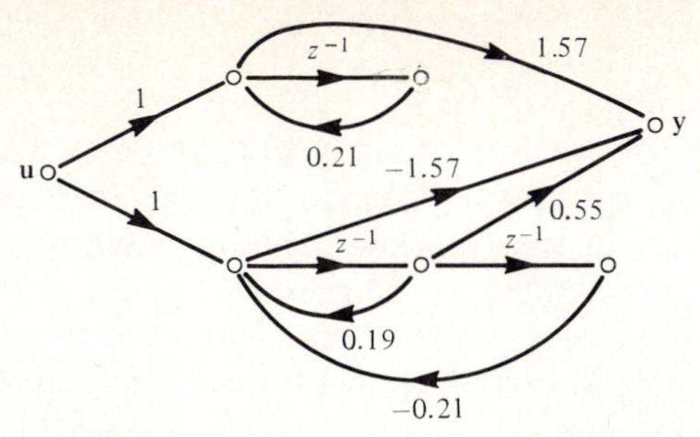

FIGURE 6.6.2 A parallel realization of a third-order low-pass filter based on a Butterworth approximation.

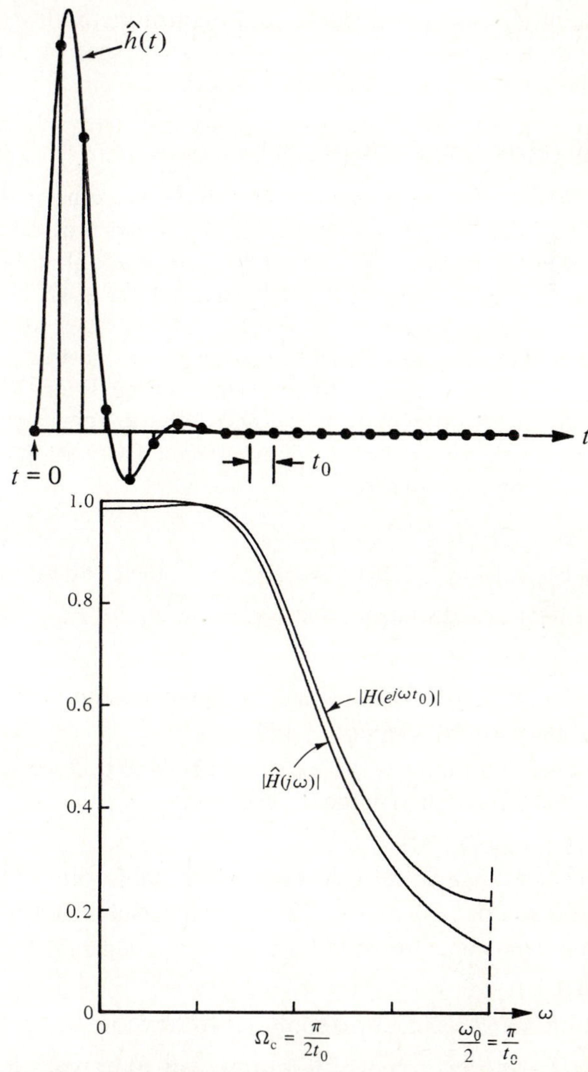

FIGURE 6.6.3 The time and frequency response for the analog Butterworth filter $\hat{H}(s)$, and the impulse invariant but aliased, discrete-time filter $H(z)$.

The impulse invariant design is motivated by a desire to match the unit-pulse response of a digital filter to the impulse response of an analog filter that meets certain specifications. The matching of time response characteristics could be extended to other functions such as the step response or the covariance functions of the filters. The ideas in matching other functions follow in a straightforward manner from the procedure presented for matching the impulse response.

We have seen that the impulse invariance method results in a distortion in the frequency response of the digital filter because of aliasing. This distortion is difficult to ascertain beforehand. The next method we discuss, the bilinear transform method, eliminates the distortion caused by aliasing but introduces a nonlinear warping of the frequency axis.

The Bilinear Transform Method

Suppose that for each analog filter $\hat{H}(s)$, we can construct a digital filter $H(z)$ for which $H(e^{j\omega t_0}) = \hat{H}(j\Omega(\omega))$. This yields a direct relation between the two frequency response functions; the response of the digital filter at $\theta = \omega t_0$ will agree with the response of the analog filter at Ω. The nonlinear function $\Omega(\omega)$ is called a *frequency warping*. If Ω increases from $-\infty$ to ∞ as ωt_0 increases from $-\pi$ to π, then there can be no aliasing. We have simply compacted the frequency response of the analog filter. This property can be achieved by a transformation $z = \psi(s)$ of the complex s-plane into the complex z-plane which meets certain requirements. The analog and digital filters are connected by the equation

$$\hat{H}(s) = H(\psi(s)), \qquad \textbf{(6.6.14)}$$

where the transformation $z = \psi(s)$ should have the following properties.

1. The $j\Omega$ axis should map into the unit circle;

   $\psi(j\Omega) = e^{j\omega t_0}$

   in order that the frequency response functions can be related by a frequency warping $\Omega(\omega)$.

2. An inverse map $s = \psi^{-1}(z)$ should exist so that we can produce the digital filter H from the analog filter $\hat{H}$ via

   $H(z) = \hat{H}(\psi^{-1}(z))$.

3. If $\hat{H}(s)$ is stable, then $H(z)$ should be stable. Now if λ is a pole of $\hat{H}$, then $\psi(\lambda)$ will be a pole of H. Therefore it is sufficient that ψ take the open left half plane into the open unit disc, i.e.,

   $\text{Re}(s) < 0 \quad \Rightarrow \quad |\psi(s)| < 1.$

4. If the zero-frequency response is to stay the same, then $\psi(0) = 1$.

A transformation which satisfies all four requirements is the *bilinear transform*

$$z = \psi(s) = \frac{1 + s}{1 - s}. \qquad \textbf{(6.6.15)}$$

The inverse transformation is

$$s = \psi^{-1}(z) = \frac{z-1}{z+1}. \tag{6.6.16}$$

The frequency warping function is obtained by setting $z = e^{j\omega t_0}$ in Eq. (6.6.16).

$$j\Omega(\omega) = \frac{e^{j\omega t_0} - 1}{e^{j\omega t_0} + 1} = \frac{e^{j\omega t_0/2} - e^{-j\omega t_0/2}}{e^{j\omega t_0/2} + e^{-j\omega t_0/2}} = j\,\frac{\sin(\omega t_0/2)}{\cos(\omega t_0/2)}.$$

Thus

$$\Omega(\omega) = \tan(\omega t_0/2). \tag{6.6.17}$$

This frequency warping is displayed in Fig. 6.6.4. Notice that Ω increases continuously from $-\infty$ to $+\infty$ as ω increases from $-\omega_0/2$ to $\omega_0/2$ where $\omega_0 = 2\pi/t_0$ is the sampling frequency. The relation $\Omega(\omega_0/4) = 1$ scales the bandwidth of the frequency warp.

Use of the bilinear transformation eliminates the aliasing associated with matching samples of the impulse response function $\hat{h}(t)$, but introduces a nonlinear frequency warping. However, it is possible to compensate for this by using a process called "prewarping."

Suppose we wish to design a low-pass filter with a desired cutoff frequency $\theta_c = \omega_c t_0$. θ_c is called a critical frequency since we wish to locate this band edge accurately. In general, there may be several critical frequencies depending on the type of filter one wishes to design. We first locate these critical frequencies on the unit circle. They are then mapped into the s-plane using Eq.

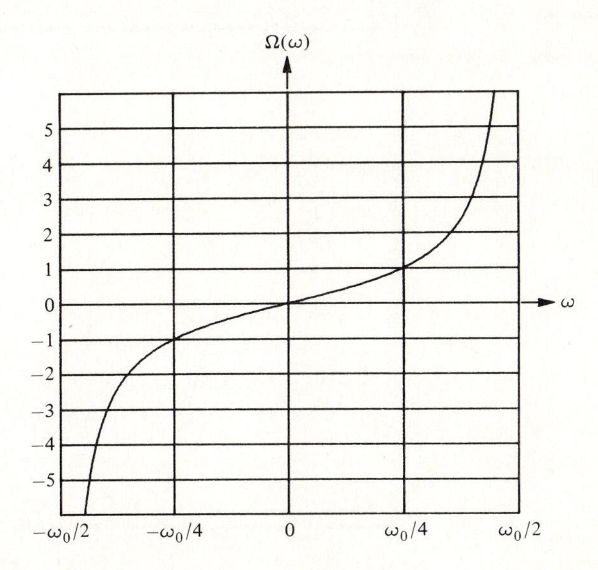

FIGURE 6.6.4 The nonlinear frequency warping $\Omega(\omega) = \tan(\omega t_0/2)$ associated with the bilinear transformation.

(6.6.17). A continuous-time $\hat{H}(s)$ is now constructed using standard analog approximation techniques. This transfer function is then transformed into the digital filter $H(z)$ using Eq. (6.6.14). The resulting transfer function is simply

$$H(z) = \hat{H}\left(\frac{z-1}{z+1}\right). \tag{6.6.18}$$

The digital filter's frequency response function is

$$H(e^{j\theta}) = \hat{H}\left(j \tan\left(\frac{\theta}{2}\right)\right). \tag{6.6.19}$$

Notice that absolutely no aliasing is introduced. The only distortion is that caused by the warping of the frequency axis. This distortion has been anticipated and is reduced by the so-called prewarping process.

EXAMPLE 6.6.2

Design of a third-order Butterworth digital filter via the bilinear transform.

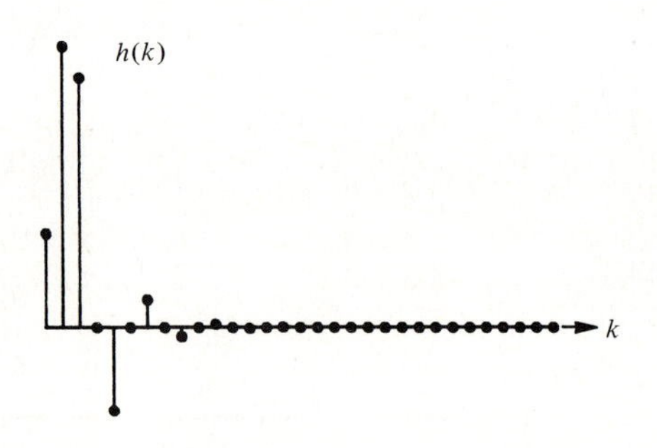

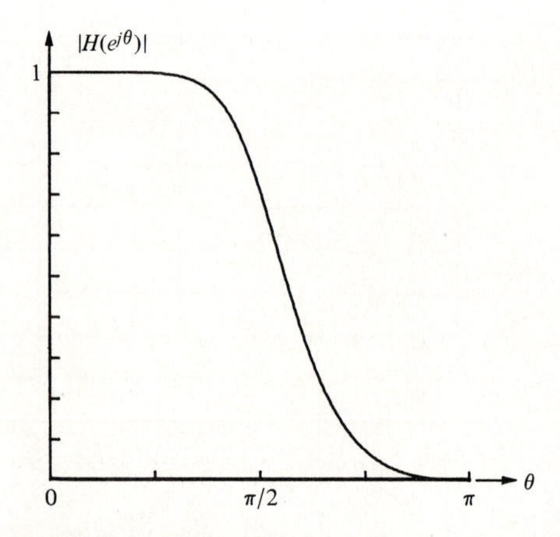

FIGURE 6.6.5 Unit-pulse response and magnitude response for a bilinear design of a third-order Butterworth low-pass filter.

Suppose we use a bilinear design for the third-order, low-pass, Butterworth filter of Example 6.6.1. The analog filter has transfer function

$$\hat{H}(s) = \frac{\Omega_c^3}{(s + \Omega_c)(s^2 + s\Omega_c + \Omega_c^2)}.$$

Suppose we want the cutoff frequency for the discrete-time filter $H(z)$ to be $\theta_c = \omega_c t_0 = \pi/2$. Then because of the frequency warp, we must choose the cutoff frequency for the analog filter $\hat{H}(s)$ to be

$$\Omega_c = \Omega(\omega_c) = \tan(\omega_c t_0/2) = \tan(\pi/4) = 1.$$

Therefore

$$\hat{H}(s) = \frac{1}{(s + 1)(s^2 + s + 1)}.$$

The corresponding digital filter is

$$H(z) = \hat{H}\left(\frac{z - 1}{z + 1}\right) = \frac{(z + 1)^3}{2z(3z^2 + 1)} = \frac{(1 + z^{-1})^3}{2(3 + z^{-2})}.$$

The frequency response function of the digital filter is

$$H(e^{j\theta}) = \frac{(1 + e^{-j\theta})^3}{2(3 + e^{-2j\theta})},$$

as depicted in Fig. 6.6.5.

The bilinear design process can be summarized as follows:

1. Specify the set of critical frequencies $\{\theta_k\}$ for the desired digital filter. There can be any number of critical frequencies.
2. Map these critical frequencies using $\Omega_k = \tan(\theta_k/2)$ to obtain the corresponding analog critical frequencies $\{\Omega_k\}$.
3. Design an analog filter $\hat{H}(s)$ using the critical frequencies in (2).
4. Construct $H(z) = \hat{H}(\psi^{-1}(z))$.

Because of the properties of the bilinear transformation stable, a causal $\hat{H}(s)$ will transform to a stable, causal $H(z)$ with no aliasing in the frequency response function:

$$H(e^{j\omega t_0}) = \hat{H}(j \tan(\omega t_0/2)). \quad \textbf{(6.6.20)}$$

The bilinear transformation filter design procedure is easily implemented, especially in cases where the desired frequency response function takes on values 0 or 1 along the frequency axis, e.g., low-pass, band-pass, band-reject filters. Because there is such an extensive body of theory for the design of analog filters, the bilinear method of design is an excellent procedure for many applications involving digital filters.

The Matched z-Transformation Design

The matched z-transformation is similar to the impulse invariant design. If $\hat{H}(s)$ is an analog response in factored form, then the analog filter is transformed by replacing both poles and zeros by

$$\begin{aligned} \text{zeros:} \quad & s - q_j \leftarrow 1 - z^{-1} e^{q_j t_0} \\ \text{poles:} \quad & s - p_j \leftarrow 1 - z^{-1} e^{p_j t_0}. \end{aligned} \tag{6.6.21}$$

The poles of the digital filter using Eq. (6.6.21) are the same as obtained by the impulse invariant method. The zeros, however, are mapped using Eq. (6.6.21), the same mapping as applied to the poles. This mapping applied to the zeros can sometimes cause high-frequency zeros to map to low-frequency zeros in the digital filter's response. This method of design is thus not often used.

A Minimum Mean Squared Error Design

The next chapter discusses in more depth how to design transfer functions $H(z)$ to minimize a mean squared error criterion. Suppose we assume an IIR filter constructed of a cascade of second-order sections so that the transfer function $H(z)$ is

$$H(z) = g \prod_{i=1}^{N} \left(\frac{1 + b_{1i} z^{-1} + b_{2i} z^{-2}}{1 + a_{1i} z^{-1} + a_{2i} z^{-2}} \right).$$

Let $H_d(e^{j\theta})$ be the desired magnitude response function and consider the squared error function

$$E = \sum_{i=1}^{M} |H_d(\theta_i) - H(e^{j\theta_i})|^2, \tag{6.6.22}$$

where θ_i, $i = 1, 2, \ldots, M$ are a discrete set of frequencies at which the desired response is specified. The squared error function depends on the unknown parameters $\{g, a_{i1}, a_{2i}, b_{i1}, b_{i2}\}$, $i = 1, 2, \ldots, N$. To minimize E, a necessary condition is to find the partial derivatives of E with respect to these unknown parameters and set these derivatives to zero. This generates $4N + 1$ equation in $4N + 1$ unknowns. This formulation is due to Steiglitz (1970).

These equations can be solved computationally using a procedure such as the *Fletcher-Powell method.* Since the parameter values are not constrained, poles and zeros outside $|z| = 1$ are possible. Steiglitz suggests replacing a pole (or zero) at polar coordinates (ζ, ϕ) by a pole (or zero) at $(1/\zeta, \phi)$ since this replacement does not change the magnitude response (to within a constant).

6.7 FREQUENCY TRANSFORMATIONS

A *generalized band-pass filter* has a frequency response function $H(e^{j\theta})$ which is (ideally) zero in each *stop-band* and one in each *pass-band.* A standard approach to the design of such filters is to transform a prototype low-pass filter using *frequency transformations.* Two possible routes from an analog low-pass filter to a digital band-pass filter are shown in the figure below.

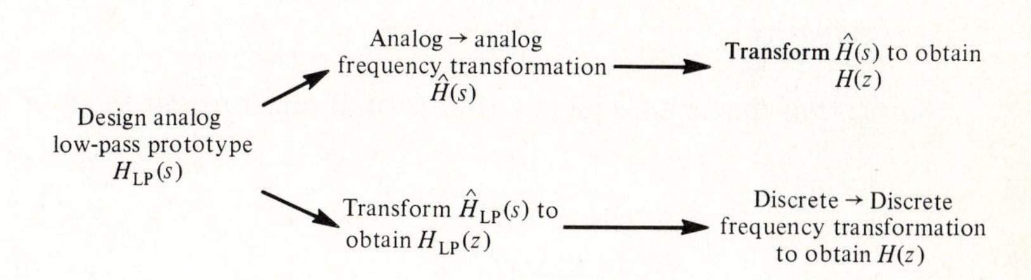

The top branch uses an *analog* frequency transformation followed (typically) by a bilinear transform to obtain $H(z)$. The lower branch uses the bilinear transform to obtain a prototype low-pass $H_{LP}(z)$ followed by a *discrete-time* frequency transformation to obtain $H(z)$. We shall focus on this approach.

The problem is this. Given an acceptable prototype low-pass filter $H(z)$, design a frequency transformation $F(z)$ so that the composition

$$G(z) = H(F(z)) \tag{6.7.1}$$

is a filter with the required pass-bands. The frequency transformation $F(z)$ must have the following properties.

1. $F(z)$ should map the unit circle into itself, i.e.,

   $$F(e^{j\phi}) = e^{j\theta(\phi)}, \tag{6.7.2}$$

   in order that the frequency response of $G(z)$ should derive from that of $H(z)$ via

   $$G(e^{j\phi}) = H(e^{j\theta(\phi)}). \tag{6.7.3}$$

2. If $H(z)$ is stable and minimum phase, then $G(z)$ should have the same properties. Now if λ is a pole (zero) of G, then $F(\lambda)$ is a pole (zero) of H. Therefore if $|\lambda| < 1$ implies $|F(\lambda)| < 1$, then these properties will be preserved.

This leads to the following definition. The complex function $F(z)$ is a *frequency transformation* provided that

$$\begin{aligned} |z| > 1 &\iff |F(z)| > 1, \\ |z| = 1 &\iff |F(z)| = 1, \\ |z| < 1 &\iff |F(z)| < 1. \end{aligned} \tag{6.7.4}$$

It is not hard to see that products $F_1(z)F_2(z)$ of frequency transformations are also frequency transformations, and that compositions $F_2(F_1(z))$ of frequency transformations are frequency transformations. Notice that if $F(z)$ is a frequency transformation, then $1/F(z)$ is a stable all-pass filter. Therefore in the composition $G(z) = H(F(z))$, we are replacing the unit delay z^{-1} with the all-pass filter $[F(z)]^{-1}$.

For IIR filter design, it is mandatory that $G(z)$ must be rational and therefore $F(z)$ must be a rational function. What do rational frequency transformations look like? All first-order frequency transformations have the form

$$F(z) = \pm \frac{z - \alpha}{1 - \alpha^* z}, \qquad |\alpha|^2 < 1. \tag{6.7.5}$$

To see that this is a frequency transformation, consider the following.

$$|F(z)|^2 = \frac{(z - \alpha)(z^* - \alpha^*)}{(1 - \alpha^* z)(1 - \alpha z^*)} = 1 + \frac{zz^* - 1 + \alpha\alpha^* - \alpha\alpha^* zz^*}{1 - \alpha z^* - \alpha^* z + \alpha\alpha^* zz^*}$$

$$= 1 + \frac{[|z|^2 - 1][1 - |\alpha|^2]}{|1 - \alpha^* z|^2}. \tag{6.7.6}$$

This is readily seen to have the property (6.7.4). Order n frequency transformations are products of these, having the form (known as a *Blaschke product*)

$$F(z) = \pm \prod_{k=1}^{n} \frac{z - \alpha_k}{1 - \alpha_k^* z} = \pm \frac{p(z)}{\tilde{p}(z)}, \tag{6.7.7}$$

where the polynomial

$$p(z) = p_0 + p_1 z^{-1} + \cdots + p_n z^{-n} = p_0 z^{-n} \prod_{k=1}^{n} (z - \alpha_k) \tag{6.7.8}$$

has roots satisfying $|\alpha_k| < 1$, and

$$\tilde{p}(z) = p_n + p_{n-1} z^{-1} + \cdots + p_0 z^{-n} = z^{-n} p(z^{-1}). \tag{6.7.9}$$

Now let us return to the problem of obtaining $F(z)$ so that $G(z) = H(F(z))$ will have the required pass bands. Let $H(z)$ have cutoff frequency $\theta = \pi/2$ as shown in Fig. 6.7.1. (These plots are idealized, of course. The perfect corners are not possible.) $G(z)$ will have specified pass bands as in the figure. The problem is to construct $F(z)$ having a frequency warp $\theta(\phi)$ as shown in Fig. 6.7.2, so that the band edges are appropriately mapped. In general, the order of $F(z)$ will be the number of pass bands on the circle (in this case, 6). There are two cases:

$$\textit{low-pass:} \begin{cases} F(1) = 1, & \text{[positive sign in Eq. (6.7.7)]} \\ \theta(0) = 0, & \\ \theta_k = \theta(\phi_k) = (k - \tfrac{1}{2})\pi. & \end{cases} \tag{6.7.10}$$

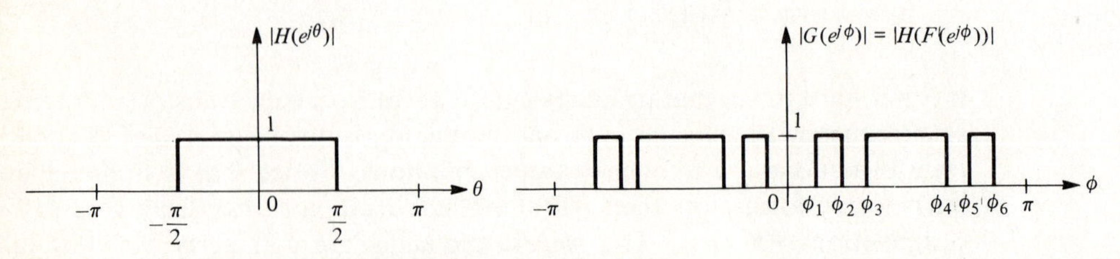

FIGURE 6.7.1 **A frequency transformation example.**

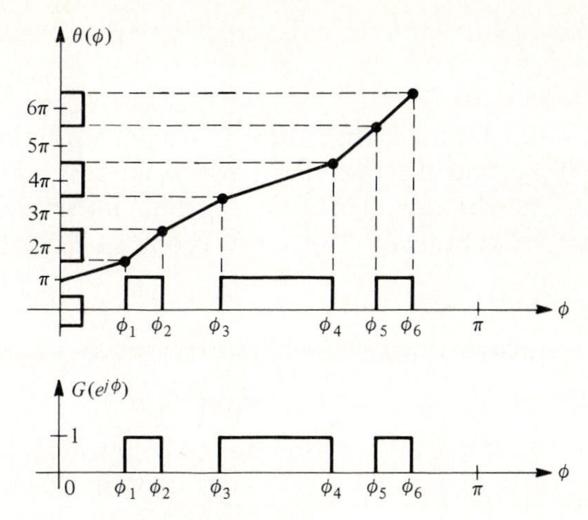

FIGURE 6.7.2 **The frequency warp θ(φ).**

$$\textit{low-stop:}\begin{cases} F(1) = -1, & \text{[negative sign in Eq. (6.7.7)]} \\ \theta(0) = \pi, \\ \theta_k = \theta(\phi_k) = (k + \tfrac{1}{2})\pi. \end{cases} \tag{6.7.11}$$

(In the example of Fig. 6.7.1, $G(z)$ is low-stop.) Clearly, the only difference between these two cases is the addition of π radians to each specified point in the low-pass case, which is equivalent to replacing $F(z)$ with $-F(z)$. Thus we can restrict our attention to the low-pass case.

It turns out that the phase interpolation conditions (6.7.10) generate linear equations for the coefficients of $p(z)$ in Eq. (6.7.8). The development of these equations involves a "symmetrization," as follows:

$$F(e^{j\phi_k}) = e^{j\theta_k}$$

if and only if [using Eq. (6.7.7)]

$$p(e^{j\phi_k}) = e^{j\theta_k}\tilde{p}(e^{j\phi_k})$$

if and only if [using Eq. (6.7.9)]

$$p(e^{j\phi_k}) = e^{j\theta_k}e^{-jn\phi_k}p(e^{-j\phi_k}).$$

The symmetric version of this equation is

$$\begin{aligned} 0 &= [e^{-j(\theta_k - n\phi_k)/2}p(e^{j\phi_k}) - e^{j(\theta_k - n\phi_k)/2}p(e^{-j\phi_k})]/[2j] \\ &= \sum_{i=0}^{n} p_i \sin\left[\frac{n\phi_k}{2} - \frac{\theta_k}{2} - i\phi_k\right]. \end{aligned} \tag{6.7.12}$$

These form n real equations in the $n + 1$ real unknowns $p_0, p_1, \ldots, p_n$. For the low-pass case,

$$\theta_k = (k - \tfrac{1}{2})\pi, \qquad 1 \leqslant k \leqslant n. \tag{6.7.13}$$

It can be shown that if the band-edge angles are ordered via

$$0 < \phi_1 < \phi_2 < \cdots < \phi_n < \pi, \tag{6.7.14}$$

and Eqs. (6.7.13) hold, then there is a unique solution to Eqs. (6.7.12) for a specified p_0, and the resulting polynomial $p(z)$ must have roots satisfying $|\alpha_k| < 1$. There is also an efficient algorithm for solving these equations, given in Fig. 6.7.3 (Franchitti 1985). See problems 6.24 through 6.26 for further elaboration.

EXAMPLE 6.7.1

A low-pass to low-pass frequency transformation.

The cutoff frequency of the prototype low-pass filter $H(z)$ can be moved from $\pi/2$ to ϕ_1 by using the frequency transformation

$$F(z) = \frac{z - \alpha}{1 - \alpha z}, \tag{6.7.15}$$

where α is to be determined. Using Eq. (6.7.12), with $\theta_1 = \pi/2$,

$$p_0 \sin\left[\frac{\phi_1}{2} - \frac{\pi}{4}\right] + p_1 \sin\left[\frac{\phi_1}{2} - \frac{\pi}{4} - \phi_1\right] = 0,$$

and therefore

$$\alpha = -\frac{p_1}{p_0} = \frac{\sin\left[\frac{\phi_1}{2} - \frac{\pi}{4}\right]}{\sin\left[-\frac{\phi_1}{2} - \frac{\pi}{4}\right]} = -\tan\left[\frac{\phi_1}{2} - \frac{\pi}{4}\right] = \frac{1 - \tan[\phi_1/2]}{1 + \tan[\phi_1/2]}. \tag{6.7.16}$$

The frequency warping induced by the transformation $F(z)$ in Eq. (6.7.15) is

$$\theta(\phi) = 2 \arg[e^{j\phi/2} - \alpha e^{-j\phi/2}], \tag{6.7.17}$$

which satisfies the equation

$$\tan\left[\frac{\theta(\phi)}{2}\right] = \frac{1 + \alpha}{1 - \alpha} \tan\left[\frac{\phi}{2}\right] \tag{6.7.18}$$

Setting $\phi = \phi_1$ and $\theta = \pi/2$ and solving for α, one obtains Eq. (6.7.16).

Suppose that $F(z)$ is order n and $H(z)$ is order m. Then as ϕ increases from $-\pi$ to π, $\theta(\phi)$ will increase from $-n\pi$ to $n\pi$, making n complete circles and passing through the pass-band of $H(z)$ n times. Thus $G(z)$ will have n pass-bands on $[-\pi, \pi]$. The price that must be paid for this accomplishment is that $G(z)$ will have order $n \cdot m$. If λ is a pole of $H(z)$ then $G(z)$ will have n associated poles satisfying

$$F(\mu) = \lambda, \qquad \text{or } p(\mu) \pm \lambda \tilde{p}(\mu) = 0. \tag{6.7.19}$$

This is an nth degree polynomial in μ, with complex coefficients, unless λ is real. There will be n poles of $G(z)$ conjugate to these, associated with λ^*.

Given n, $\phi[1, n]$

To compute $p[0, n]$ satisfying $F(e^{j\phi_k}) = e^{j(k-\frac{1}{2})\pi}$, where

$$F(z) = p(z)/\tilde{p}(z)$$
$$p(z) = p_0 + p_1 z^{-1} + \cdots + p_n z^{-n}$$
$$\tilde{p}(z) = z^{-n} p(z^{-1})$$

Initialization
$v \leftarrow \frac{1}{2}$
$p_0 \leftarrow 1$

Body

For $k = 1$ to n, Do
  $v \leftarrow -v$
  $\phi' \leftarrow (\phi_k - \pi)v$
  For $j = 0$ to k, Do
    $\alpha \leftarrow 0$
    $\beta \leftarrow 0$
    If $j > 0$ then
      $\alpha \leftarrow \alpha + p_{j-1}$
      $\beta \leftarrow \beta - p_{k-j}$
    (end if)
    If $j < k$ then
      $\alpha \leftarrow \alpha + p_j$
      $\beta \leftarrow \beta + p_{k-j-1}$
    (end if)
    $q_j \leftarrow \alpha \cos(\phi') + \beta \sin(\phi')$
  (end loop on j)
  For $j = 0$ to k, Do
    $p_j \leftarrow q_j$
  (end loop on j)
(end loop on k)

FIGURE 6.7.3 **Algorithm for the computation of parameters of the frequency transformation $F(z)$.**

6.8 A COMPARISON OF FIR AND IIR DIGITAL FILTERS

The choice of whether to use an FIR or an IIR filter depends on the particular application and the resources available for implementing the filter. FIR filters realized in nonrecursive structures are inherently simpler, and they are always stable. Thus they are ideally suited for use in adaptive or time-varying applications because stability does not have to be imposed as another constraint. FIR designs are essentially all computer aided designs. Although hand calculations can produce reasonable designs, to obtain more efficient designs, one must iterate the design process. FIR filters are ideally suited to so-called multirate digital processing problems. These are applications that involve more than one sampling rate, e.g., the synchronization of two data sequences sampled at different rates that are not commensurate. Linear phase response is easily obtainable with FIR filters.

IIR filters have simple, hand calculation design methods, such as the bilinear transform method, which produce excellent designs. IIR filters can produce much sharper transitions in amplitude response with significantly fewer coefficients than nonrecursive FIR filters. However, if the FIR filter is realized using fast convolution algorithms based on the FFT or some other fast algorithm, then the computational burden is essentially equal for comparable designs.

FIR filters have inherently simpler finite register effects than do IIR filters. It is easier to analyze these effects in FIR filters.

Both types of filters are useful. The application and the resources available for implementation are generally the deciding factors in the choice of a particular design. Generally, the choice is not clear-cut and one must evaluate several factors before choosing a particular type of filter.

Summary • This chapter has presented several design methods for FIR and IIR digital filters. The word "design" used in this chapter is really a misnomer. Our goal here is to obtain the transfer function $H(z)$ from a given set of specifications. This is really only the first step in the design procedure. It is often the simplest part of the design process. From a given $H(z)$ we must next determine a realization or algorithm to compute $H(z)$. (This is discussed in Chapters 8, 9, and 10.) After designing the realization, we must still determine an implementation in software or hardware or a combination of the two.

The next chapter continues the discussion of obtaining the transfer function $H(z)$ from a given set of specifications. The specifications we assumed in this chapter are not expressed in terms of the characteristics of the input sequence (although the designer has certainly taken this into account, at least implicitly). The specifications here generally assume a circuit synthesis point of view and result in low-pass, high-pass, bandpass, etc., approximations. In the next chapter the specifications for $H(z)$ are defined in terms of criteria that involve the characteristics of the input sequence, e.g., obtain $H(z)$ to filter a signal from a mixture of signal and noise. Again the problem is to find the transfer function (or some other I/O characterization).

PROBLEMS

6.1. We wish to design an FIR bandpass filter by using a low-pass prototype. The desired response is shown in Fig. 6P.1. Consider a low-pass prototype with frequency response $H_{d,\mathrm{LP}}(\theta)$ given by

$$H_{d,\mathrm{LP}}(\theta) = \begin{cases} 1, & |\theta| \leq \dfrac{\pi}{8} \\ 0, & \dfrac{\pi}{8} < |\theta| \leq \pi \end{cases}.$$

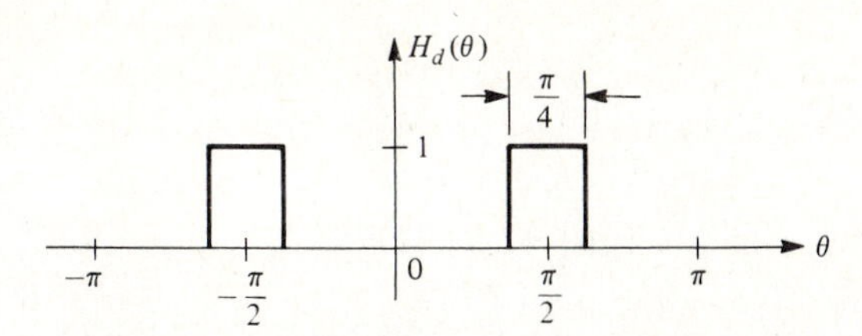

FIGURE 6P.1 The desired frequency response function of a bandpass filter.

a) Find the filter tap weights for $N = 17$ for $H_{d,\mathrm{LP}}(\theta)$.

b) To obtain a bandpass design centered on $\theta = \pi/2$, multiply the filter tap weights in a) by $\cos(k\pi/2)$, i.e., use as filter tap weights

$$h_{\mathrm{BP}}(k) = \begin{cases} h_{d,\mathrm{LP}}(k)\cos\left(\dfrac{k\pi}{2}\right), & |k| \leqslant 8 \\ 0, & \text{otherwise.} \end{cases}$$

c) Sketch the frequency response of the window function that corresponds to the bandpass window, namely,

$$w_{\mathrm{BP}}(k) = w(k)\cos\left(\frac{k\pi}{2}\right),$$

where

$$w(k) = \begin{cases} 1, & |k| \leqslant 8 \\ 0, & \text{otherwise} \end{cases}.$$

6.2. Let $W(e^{j\theta})$ be the frequency response function

$$W(e^{j\theta}) = \sum_{k=-M}^{M} w(k)e^{-jk\theta}$$

with $w(k) = w(-k)$.

a) Show that multiplying $w(k)$ by $\cos(k\theta_0)$, $0 \leqslant \theta_0 \leqslant \pi$, results in a window frequency function

$$W_E(e^{j\theta};\theta_0) = \tfrac{1}{2}\{W(e^{j(\theta-\theta_0)}) + W(e^{j(\theta+\theta_0)})\}.$$

b) Suppose $w(k) = -w(-k)$. Show that multiplying $w(k)$ by $\sin(k\theta_0)$, $0 \leqslant \theta_0 \leqslant \pi$, results in a window frequency function

$$W_O(e^{j\theta};\theta_0) = \frac{j}{2}\{W(e^{j(\theta-\theta_0)}) - W(e^{j(\theta+\theta_0)})\}.$$

c) How can these results be used in FIR filter design?

6.3. The Chebyshev window is based on the Chebyshev polynomials $T_n(x)$. These polynomials satisfy:

$$T_n(x) = \begin{cases} \cos(n\cos^{-1}x), & |x| \leqslant 1 \\ \cosh(n\cosh^{-1}x), & x > 1 \end{cases}$$

with

$$T_n(x) = 2xT_{n-1}(x) - T_{n-2}(x)$$

and

$$T_0(x) = 1, \qquad T_1(x) = x.$$

a) Plot $T_6(x)$ on the interval [0, 2].
b) Define x_0 by $T_n(x_0) = R > 1$. Show $x_0 > 1$.
c) Change variables from x to θ via the relation $x = x_0 \cos(\theta/2)$. The Chebyshev window frequency response function is

$$W(e^{j\theta}) = \sum_{k=-M}^{M} w(k)e^{-jk\theta} = T_{2M}\left(x_0 \cos\left(\frac{\theta}{2}\right)\right).$$

Plot the function for $2M = 6$ on the interval $[0, \pi]$.
d) Show that the point where $T_{2M}(x_0 \cos(\theta/2)) = 1$ is given by $\theta_0 = 2\cos^{-1}(1/x_0)$.

6.4. To find a relationship between $\{N, R, \theta_0\}$ for the Chebyshev window, we know that $T_{2M}(x_0) = R$ from Problem 6.3 and that $\theta_0 = 2\cos^{-1}(1/x_0)$.

a) From the equation $T_{2M}(x_0) = R$, show that

$$R = \cosh[2M \cosh^{-1}(x_0)], \qquad N = 2M + 1$$

b) Using $R = \cosh[2M \cosh^{-1}(x_0)]$ and $\theta_0 = 2\cos^{-1}(1/x_0)$, eliminate the variable x_0 and thus obtain the relationship

$$\tan\left(\frac{\theta_0}{2}\right) = \sinh\left(\frac{1}{2M}\cosh^{-1}(R)\right).$$

c) Show that the preceding formula can be approximated by

$$\theta_0 \cong \frac{\ln(2R)}{M}.$$

This approximate relationship can be used in FIR filter design to obtain a reasonable starting point for a Chebyshev window design.

6.5. Design a low-pass FIR filter with cutoff frequency $\theta_c = \pi/2$. Choose N equal to 13, 17, and 21, θ_0 equal to 15°, and use a Gaussian window. Compare the values of R as N varies. How do these results compare with the design formula relating N, R, and θ_0 given in Problem 6.4b?

6.6. One application of FIR filters are so-called interpolation filters whose transfer functions are designed to approximate an ideal delay, $\exp(-j2\pi f\tau)$, of τ seconds (Gabel 1980). These filters find application in data reconstruction, time-delay beamforming, waveform shaping, and other signal processing operations. Let $d = \tau/t_0$ be a normalized delay and $\theta = 2\pi f\tau$ so that the desired transfer function is $H(e^{j\theta}) = e^{-j\theta d}$. If the delay d is a multiple of $\frac{1}{2}$, then a linear phase FIR filter as designed by the McCellan-Parks algorithm is sufficient. However, if d is not a multiple of $\frac{1}{2}$, then there are no linear phase filters. One approach to solve this problem is to define an appropriate error measure in terms of the filter coefficients and then apply some nonlinear optimization technique to find the coefficients that minimize the criterion function. Let the transfer function of FIR, N tap, filter be

$$H(e^{j\theta}) = \sum_{k=0}^{N-1} h(k)e^{-jk\theta + jk_0\theta},$$

where k_0 is an arbitrary shift of the unit-pulse response to make the filter causal. Define the amplitude and phase errors as

$$E_1(\theta) = |H_d(\theta)| - |H(e^{j\theta})|, \qquad E_2(\theta) = \arg H_d(\theta) - \arg H(e^{j\theta}).$$

We can combine these two errors to form a weighted average error as

$$E(\theta) = (E_1(\theta)w_1)^l + (E_2(\theta)w_2)^l,$$

where l is a positive even integer and w_i, $i = 1, 2$, are weight functions.

a) Find the gradient function $\partial E/\partial h(k)$. Show that $\partial E_1(\theta)/\partial h(k)$ can be expressed in the form

$$\frac{\sum_{n=1}^{N-1} h(k)\cos[(n-k)\theta]}{|H(e^{j\theta})|}.$$

Show also that

$$\frac{\partial \arg H(e^{j\theta})}{\partial h(l)} = \frac{\mathrm{Re}[H(e^{j\theta})]\sin[(l-k_0)\theta] - \mathrm{Im}[H(e^{j\theta})]\cos[(l-k_0)\theta]}{H(e^{j\theta})}.$$

6.7. An ideal differentiator has a frequency response function

$$H_d(\theta) = j\theta, \qquad 0 \leqslant |\theta| \leqslant \pi.$$

a) Find the Fourier coefficients for $H_d(\theta)$.
b) Calculate the tap weights using a window design and a Hanning window for $N = 19$.
c) Plot the amplitude and phase response of the actual filter response in b).
d) Repeat b) and c) using a Gaussian window. Choose θ_0 or R to "improve" the response in c).

6.8. A Hilbert transform filter has a frequency response

$$H_d(\theta) = -j\,\mathrm{sign}(\theta), \qquad 0 \leqslant \theta \leqslant \pi.$$

Design an FIR filter to approximate this frequency function using $N = 17$ taps. Use a design method of your choice.

6.9. The frequency response function of a causal FIR filter is linear phase provided

$$H(e^{j\omega t_0}) = A(e^{j\omega t_0})\exp[-j(\alpha\omega t_0 + \beta)],$$

where $A(e^{j\omega t_0})$ is real and α and β are constants. Show that α and β must satisfy the equations

$$\beta = 0 \text{ or } \frac{\pi}{2}, \qquad \alpha = \frac{N-1}{2}.$$

6.10. Let $G(z)$ be the transfer function of a causal filter with real unit pulse response sequence **g**. Let $H(n) = G(e^{j2\pi n/N})$, and let $\tilde{H}(z)$ be the length N FIR frequency sampling filter defined in Eq. (6.4.5).

a) Express $h(k)$ in terms of **g**.
b) Show that $h(k)$ is real.
c) Express $\tilde{H}(e^{j\theta})$ in terms of $G(e^{j\theta})$.

6.11. Suppose an FIR filter has unit-pulse response

$$
\begin{aligned}
h(k) &= k + 1, && k = 0, 1, 2, 3, 4 \\
h(8 - k) &= h(k), && k = 5, 6, 7, 8 \\
h(k) &= 0, && \text{otherwise.}
\end{aligned}
$$

a) How would you classify this filter (low-pass, high-pass, etc.)?
b) Write the transfer function $H(z)$ as both a nonrecursive **and** as a recursive filter.

6.12. Although we have emphasized the linear phase property of FIR filters, IIR filters realized in nonreal time can also be linear phase. Two methods are shown in Fig. P6.12. Show the resulting transfer function between input and output is linear phase. The time reversal box has an I/O characteristic of the form $y(k) = u(-k)$. What is the actual phase response in both methods?

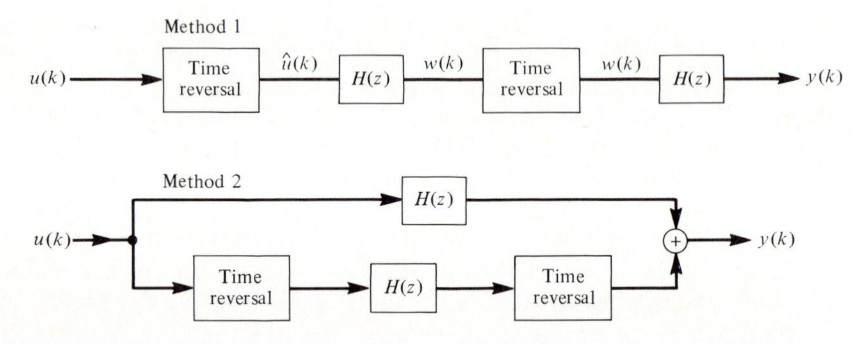

FIGURE 6P.12 **Two theoretical methods for obtaining linear phase IIR filters.**

6.13. An IIR filter has a transfer function $H(z)$ given by

$$H(z) = \frac{(1 + z^{-1})^2}{1 - \frac{1}{2}z^{-2}}.$$

a) Construct a pole-zero diagram for $H(z)$.
b) Find the frequency response function and plot the amplitude and phase responses.
c) Let $G(n) = H(e^{j2\pi n/N})$, with $N = 13$. Let $\tilde{G}(z)$ be the FIR filter derived from these samples via Eq. (6.4.4). Plot magnitude and phase of $\tilde{G}(e^{j\theta})$.

6.14. Suppose the transfer function of an approximating analog filter is $\hat{H}(s)$ given by

$$\hat{H}(s) = \frac{s^3}{(s+1)(s^2+s+1)}.$$

This is a third-order high-pass Butterworth filter with a cutoff frequency normalized to 1 radian/sec. Obtain the corresponding digital filter $H(z)$ using:

a) the impulse invariance method.
b) the bilinear transform method.
c) plot the amplitude response of the digital filters in a) and b). Which response do you prefer?

6.15. A low-pass digital filter is required to have a maximally flat amplitude response function with a cutoff frequency of 1 kHz. The attenuation at 1.5 kHz must be greater than 20 dB. The sampling frequency is 8 kHz. Find a digital filter transfer function $H(z)$ that meets these specifications.

6.16. Consider the design of a low-pass digital filter meeting the following specifications. The cutoff frequency is 500 Hz, the stopband attenuation is at least 30 dB at 1500 Hz, the passband is equiripple with no more than 1 dB ripple, and the sampling frequency is 8 kHz. Find a digital filter transfer function that meets these specifications.

6.17. Consider a normalized fourth-order analog Butterworth filter $\hat{H}(s)$ given by

$$\hat{H}(s) = \frac{1}{(s^2+0.765s+1)(s^2+1.848s+1)}.$$

If the sampling frequency is 1 kHz, find the digital filter transfer function using:

a) the impulse invariance method.
b) the bilinear transform method.
c) Plot the amplitude response for each of the digital filters obtained in a) and b).

6.18. If $F(z)$ is a frequency transformation, then a frequency transformed filter based on the prototype filter $H(z)$ is $G(z) = H(F(z))$. Suppose we assume our prototype is a low-pass Butterworth filter with cutoff frequency is $w_c = \pi/2$ of order n. Find the frequency transformation needed to transform the prototype low-pass filter to a bandpass filter with center frequency $\pi/2$ and bandwidth $\pi/10$.

6.19. In the frequency transformation of prototype low-pass filters, the essential calculation is the determination of the coefficients of the polynomial $p(z)$, which in turn defines the frequency mapping $F(z) = \pm p(z)/\tilde{p}(z)$. Show that the coefficients of $p(z)$, p_l, are the n solutions to the equation [assuming the sign of $F(z)$ is positive]

$$\sum_{l=1}^{n} \sin\left\{(n-l)\phi_k - \left(\frac{\theta_k + n\phi_k}{2}\right)\right\} p_l = \sin\left(\frac{\theta_k - n\phi_k}{2}\right), \qquad k = 1, 2, \ldots, n.$$

6.20. The low-cost of analog-to-digital and digital-to-analog converters has made the digitization of analog circuits very attractive in many applications. Suppose, for example, we wish to filter a low-pass voice signal in the frequency range 300 to 3800 Hz from broad band noise of uniform power from 0 to 10 kHz. Design a digital filter of no more than eighth order that will enhance the voice signal. Assuming a constant noise power in the frequency range 0 to 10 kHz, what is the increase in signal-to-noise ratio obtainable by this filter?

6.21. To use the impulse invariance method in the case of repeated roots in $\hat{H}(s)$, we must find the unit-pulse response that results from terms of the form $c/(s - p)^r$. Find the inverse Laplace transform of the form $c/(s - p)^r$. Sample this continuous-time impulse response $\hat{h}_i(t)$ at $t = kt_0, k = 0, 1, 2, \ldots$ to obtain the corresponding unit-pulse response sequence $\{h_i(k)\}$.

6.22. Let $d(x) = x^{M+1}$. Find the solution $p(x)$ to the Chebyshev problem (Section 6.5) where p has degree at most M.

6.23. Develop a subroutine to do Chebyshev FIR filter design.

Input: $R, M, h_d(0), \cdots, h_d(M)$.
Output: $h(k) = w(k - M)h_d(k - M), \quad 0 \leqslant k \leqslant 2M$.
Compute x_0 using Eq. (6B.13). Then compute the Chebyshev window sequence of order M using Eqs. (6B.20) through (6B.22). Finally, compute the unit pulse response. Assume that $\mathbf{h}_d$ is even so that $\mathbf{h}$ will have linear phase.

6.24. (This and the following two problems develop computational tools that can be used to produce $H(z)$ for any Butterworth band-pass filter, with one or more pass-bands.) The prototype discrete-time Butterworth low-pass filter of order L is defined by

$$H(z) = \hat{H}\left(\frac{z - 1}{z + 1}\right) = \frac{\beta(z)}{\alpha(z)},$$

where $\hat{H}(s)$ satisfies Eq. (6A.7) and

$$\alpha(z) = \sum_{k=0}^{L} \alpha_k z^{-k}, \quad \beta(z) = \sum_{k=0}^{L} \beta_k z^{-k}.$$

1) Show that

$$|H(e^{j\theta})|^2 = \frac{1}{1 + \left[\dfrac{1 - \cos(\theta)}{1 + \cos(\theta)}\right]^L}.$$

What is the cutoff frequency?

2) Show that $H(z)$ has a zero of order L at $z = -1$.

3) Show that the poles of $H(z)$ are

$$\lambda_k = j \sin(\theta_k)/[1 - \cos(\theta_k)], \quad \text{where}$$

$$\theta_k = \frac{\pi}{2}\left[1 + \frac{2k - 1}{L}\right], \quad 1 \leqslant k \leqslant L.$$

4) Develop a subroutine with

Input: L.
Output: $\alpha_0, \cdots, \alpha_L, \beta_0, \cdots, \beta_L$.

6.25. [Frequency transformation to produce a generalized band-pass filter.] Let

$$p(z) = \sum_{k=0}^{M} p_k z^{-k}, \quad \tilde{p}(z) = z^{-M} p(z^{-1}),$$

$$F(z) = p(z)/\tilde{p}(z).$$

Using the algorithm in Fig. 6.7.3, develop a subroutine with

Input: $M, \phi_1, \cdots, \phi_M$,
Output: $p_0, \cdots, p_M$,

for which $F(e^{j\phi_k}) = e^{j(k-1/2)\pi}$.

6.26. Given the coefficients of $\alpha(z)$, $\beta(z)$, $p(z)$ from the previous two problems, we require the coefficients of $a(z)$ and $b(z)$ in

$$G(z) = H(\sigma F(z)) = \frac{b(z)}{a(z)},$$

where $\sigma = 1$ for low-pass or $\sigma = -1$ for low-stop.

1) Show that if we do not require $a_0 = 1$, we can take

$$a(z) = \sum_{k=0}^{LM} a_k z^{-k} = \sum_{k=0}^{L} \alpha_k [\sigma \tilde{p}(z)]^k [p(z)]^{L-k},$$

$$b(z) = \sum_{k=0}^{LM} b_k z^{-k} = \sum_{k=0}^{L} \beta_k [\sigma \tilde{p}(z)]^k [p(z)]^{L-k}.$$

2) Show that the following steps will produce the coefficients of $a(z)$ and $b(z)$. Take $N \geqslant LM + 1$.

$$[p_0, \cdots, p_M, 0, \cdots, 0] \quad \xleftrightarrow{DFT;N} \quad [P(0), P(1), \cdots, P(N-1)]$$

$$A(n) = \alpha(e^{j\theta_n})[P(n)]^L, \quad B(n) = \beta(e^{j\theta_n})[P(n)]^L,$$

$$\text{where } \theta_n = 2\arg[P(n)] + \left[\frac{1-\sigma}{2} + \frac{2nM}{N}\right]\pi.$$

$$[a_0, \cdots, a_{LM}, 0, \cdots, 0] \quad \xleftrightarrow{DFT;N} \quad [A(0), \cdots, A(N-1)]$$

$$[b_0, \cdots, b_{LM}, 0, \cdots, 0] \quad \xleftrightarrow{DFT;N} \quad [B(0), \cdots, B(N-1)].$$

3) Develop a subroutine with

Input: $L, M, \alpha_0, \cdots, \alpha_L, \beta_0, \cdots, \beta_L, p_0, \cdots, p_M, \sigma$.

Output: $a_0, \cdots, a_{LM}, b_0, \cdots, b_{LM}$.

(The approach in part 2 was suggested by G. Feyh.)

Design of Analog Low-Pass Filter Prototypes

Using the bilinear transform (Section 6.5) followed by a frequency transformation (Section 6.7), one can generate the transfer function $H(z)$ for a wide class of generalized band-pass filters. But one must start this process with a prototype analog low-pass filter $\hat{H}(s)$, with cutoff frequency $\omega_c = 1$. In this appendix, we will summarize four classical low-pass filter types:

1. *Butterworth filters* have a magnitude response which is maximally flat near $\omega = 0$, and declines monotonically for $\omega > 0$.
2. *Chebyshev filters* have a magnitude response which exhibits equal ripple for $0 < \omega < 1$, and declines monotonically for $\omega > 1$.
3. *Inverse Chebyshev filters* have a magnitude response which is monotone decreasing for $0 < \omega < 1$ and exhibits equal ripple for $\omega > 1$.
4. *Elliptic* or *Cauer filters* exhibit equal ripple in the pass-band $0 < \omega < 1$, and equal ripple in the stop-band $\omega > 1$.

Each of these filter types is characterized by the functional form of the square of its magnitude response. Using the equation

$$|\hat{H}(j\omega)|^2 = \hat{H}(s)\hat{H}(-s)|_{s=j\omega}, \tag{6A.1}$$

one must extend this definition of $\hat{H}(s)\hat{H}(-s)$ from the $j\omega$-axis to the entire complex plane, as a rational function of s. One then extracts the left half plane poles and zeros. These are the poles and zeros of $\hat{H}(s)$. The right half plane poles and zeros belong to $\hat{H}(-s)$.

The ideal low-pass filter has magnitude response

$$|\hat{H}(j\omega)|^2 = \begin{cases} 1, & |\omega| < 1 \\ 0, & |\omega| > 1 \end{cases}. \tag{6A.2}$$

Approximations must be judged by the degree to which they approach this response. This gives rise to a set of *specifications* for closeness of fit. One divides the frequency range $0 \leqslant \omega < \infty$ into three intervals, and requires that $|\hat{H}(j\omega)|^2$ meet conditions in each interval as follows:

Pass-band $(0 \leqslant \omega \leqslant \omega_l)$:

$$1 - \varepsilon \leqslant |\hat{H}(j\omega)|^2 \leqslant 1. \tag{6A.3}$$

Transition band $(\omega_l \leqslant \omega \leqslant \omega_u)$:

$$|\hat{H}(j\omega)| \text{ is monotone decreasing.} \tag{6A.4}$$

Stop-band $(\omega_u \leqslant \omega < \infty)$:

$$|\hat{H}(j\omega)|^2 \leqslant \delta. \tag{6A.5}$$

Figures 6A.2 through 6A.4 exhibit three filter types in relation to these specifications. The parameters of the specification are ε, δ, ω_l, ω_u where

$$\varepsilon > 0, \quad \delta > 0, \quad \omega_l \leqslant 1, \quad \omega_u \geqslant 1. \tag{6A.6}$$

As these parameters approach their limits, the specifications approach the conditions for an ideal response.

For each class of analog low-pass filter, there will be equations which constrain the possible values of ε, δ, ω_l, and ω_u. These equations will also involve the filter order n. For a fixed order n, these *design equations* characterize the possibilities for tradeoffs. For example, one can usually decrease ε if he is willing to increase the width of the transition band. As n is increased, however, all four of these specifications can be improved.

Butterworth Filters

The order n Butterworth filter is characterized by

$$|\hat{H}(j\omega)|^2 = \frac{1}{1 + \omega^{2n}}. \tag{6A.7}$$

This response is plotted in Fig. 6A.1 for various filter orders. For $\omega^2 < 1$, the Taylor series for $|\hat{H}|^2$ is

$$|\hat{H}(j\omega)|^2 = 1 - \omega^{2n} + \omega^{4n} \cdots. \tag{6A.8}$$

It follows that the derivatives of $|\hat{H}|^2$ with respect to ω at $\omega = 0$ vanish for all orders up to $2n - 1$. This is what is meant by "maximally flat," and one can see from Fig. 6A.1 that the effect is pronounced. The Butterworth squared magnitude response is monotone decreasing for $\omega > 0$, is equal to 1 for $\omega = 0$, and to $1/2$ at the cutoff frequency $\omega = 1$, and approaches 0 as ω approaches infinity.

The design equations for Butterworth filters involve no more than the definition of $|\hat{H}|^2$ in Eq. (6A.7) and the specifications, as depicted in Fig. 6A.2.

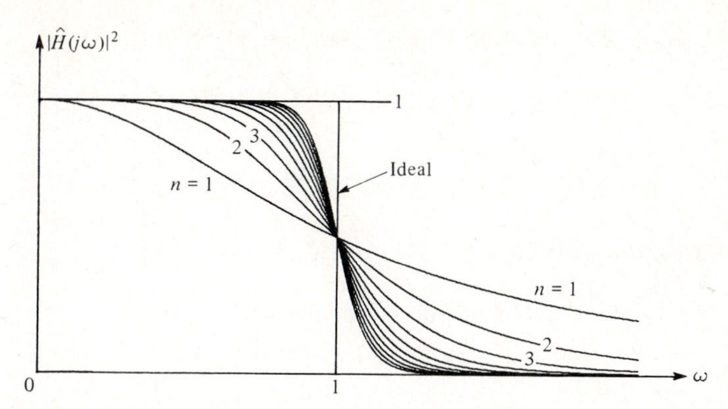

FIGURE 6A.1. **The magnitude response function for Butterworth filters as a function of filter order *n*.**

We have

$$|\hat{H}(j\omega_l)|^2 = 1 - \varepsilon \quad \Rightarrow \quad \omega_l = \left[\frac{\varepsilon}{1-\varepsilon}\right]^{1/(2n)}, \tag{6A.9}$$

$$|\hat{H}(j\omega_u)|^2 = \delta \quad \Rightarrow \quad \omega_u = \left[\frac{1-\delta}{\delta}\right]^{1/(2n)}. \tag{6A.10}$$

Notice that ω_l and ω_u both approach 1 as n approaches infinity, provided ε and δ are between 0 and $\frac{1}{2}$.

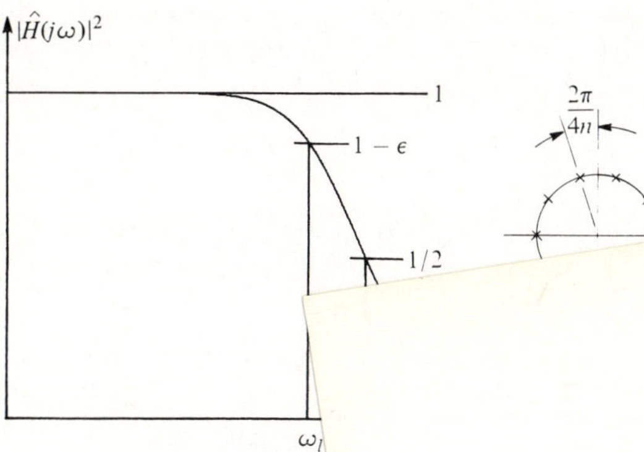

FIGURE 6A.2. **Parameters for low-pas**

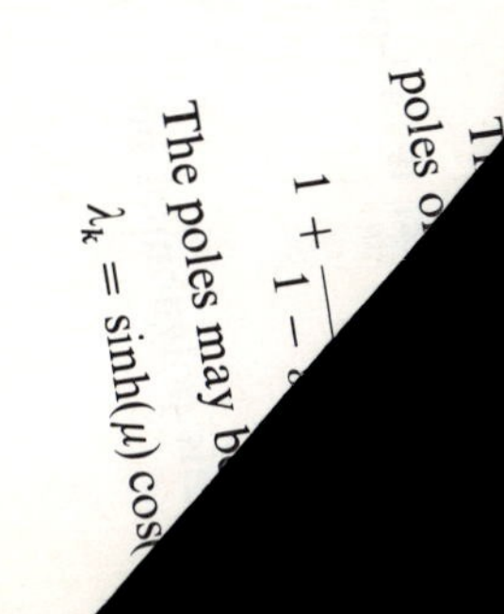

The zeros of the Butterworth filter
The poles must satisfy the equation

$$1 + (-js)^{2n} = 0,$$

obtained by setting $\omega = s/j = -js$.

unit circle as shown in Fig. 6A.1 and are given by

$$\lambda_k = e^{j\theta_k}, \quad \theta_k = \frac{\pi}{2}\left(1 + \frac{2k-1}{n}\right), \quad 1 \leqslant k \leqslant n. \tag{6A.12}$$

Only the left half plane roots are kept.

Chebyshev Filters

The order n Chebyshev filters are characterized by

$$|\hat{H}(j\omega)|^2 = \frac{1}{1 + \dfrac{\varepsilon}{1-\varepsilon} T_n(\omega)^2}. \tag{6A.13}$$

This family involves the parameter ε as well as the filter order n, and makes use of the Chebyshev polynomials $T_n(\omega)$ which are discussed in Appendix 6B. The response is depicted in Fig. 6A.3. The equal ripple behavior in the pass-band is due to the property

$$T_n(\cos\theta) = \cos n\theta, \tag{6A.14}$$

which is limited, of course, to arguments in the range $\omega^2 < 1$. The monotone decrease in the transition- and stop-bands is due to the property

$$T_n(\cosh(\theta)) = \cosh(n\theta). \tag{6A.15}$$

Since $T_n(1) = 1$, we have $|\hat{H}(j)|^2 = 1 - \varepsilon$, and we may therefore set $\omega_l = 1$. The relation between ω_u, δ, and ε is more subtle, and uses Eq. (6A.15). The design equations for Chebyshev filters are

$$|\hat{H}(j\omega_l)|^2 = 1 - \varepsilon \quad \Rightarrow \quad \omega_l = 1 \tag{6A.16}$$

$$|\hat{H}(j\omega_u)|^2 = \delta \quad \Rightarrow \quad T_n(\omega_u)^2 = \frac{(1-\varepsilon)(1-\delta)}{\varepsilon\delta}$$

$$\Rightarrow \qquad n\cosh^{-1}(\omega_u) = \cosh^{-1}\left[\sqrt{\frac{(1-\varepsilon)(1-\delta)}{\varepsilon\delta}}\right]. \tag{6A.17}$$

The usual tradeoff is between ripple magnitude ε and transition bandwidth ... $- 1$.

...he Chebyshev filters are also all-pole and have zeros at infinity. The ...f $\hat{H}(s)$ satisfy the polynomial equation

$$\ldots \varepsilon \ldots T_n(-js)^2 = 0. \tag{6A.18}$$

...e shown to be

$$\ldots \theta_k) + j\cosh(\mu)\sin(\theta_k), \tag{6A.19}$$

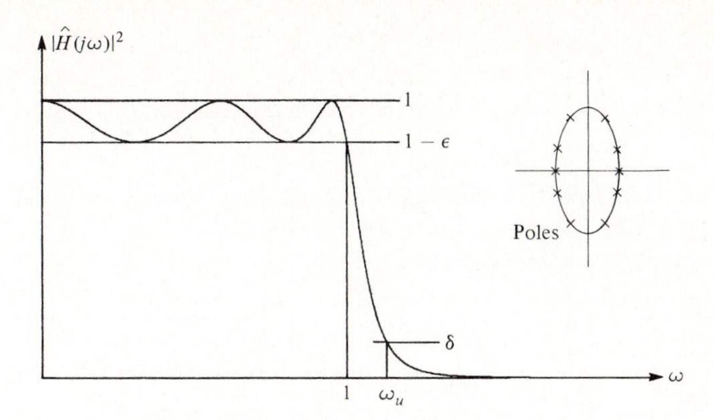

FIGURE 6A.3 **Magnitude response function for a Chebyshev filter.**

where $1 \leqslant k \leqslant n$,

$$\theta_k = \frac{\pi}{2}\left(1 + \frac{2k-1}{n}\right), \tag{6A.20}$$

and

$$\mu = \frac{1}{n}\sinh^{-1}\left(\sqrt{\frac{1-\varepsilon}{\varepsilon}}\right). \tag{6A.21}$$

[See Guillemin (1957) and Van Valkenburg (1982).] These poles lie on an ellipse, as shown in Fig. 6A.3.

Inverse Chebyshev Filters

Order n inverse Chebyshev filters are characterized by

$$|\hat{H}(j\omega)|^2 = \frac{\dfrac{\delta}{1-\delta}\, T_n(\omega^{-1})^2}{1 + \dfrac{\delta}{1-\delta}\, T_n(\omega^{-1})^2}. \tag{6A.22}$$

The response is shown in Fig. 6A.4. These filters have equal ripple in the stop-band, and nontrivial zeros on the $j\omega$-axis. Since $T_n(1) = 1$, the squared response at the cutoff frequency is δ, which is the height of the ripples.

The design equations for inverse Chebyshev filters are

$$|\hat{H}(j\omega_l)|^2 = 1 - \varepsilon \quad \Rightarrow \quad T_n\left(\frac{1}{\omega_l}\right)^2 = \frac{(1-\varepsilon)(1-\delta)}{\varepsilon\delta}$$
$$\Rightarrow \quad n\cosh^{-1}\left(\frac{1}{\omega_l}\right) = \cosh^{-1}\left[\sqrt{\frac{(1-\varepsilon)(1-\delta)}{\varepsilon\delta}}\right], \tag{6A.23}$$

$$|\hat{H}(j\omega_u)|^2 = \delta \quad \Rightarrow \quad \omega_u = 1. \tag{6A.24}$$

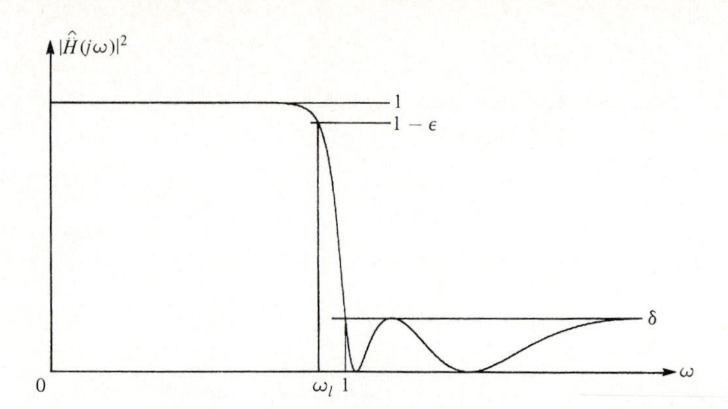

FIGURE 6A.4 **The magnitude response function for an inverse Chebyshev filter.**

The zeros of $\hat{H}(s)$ satisfy

$$T_n\left(\frac{j}{s}\right) = 0. \tag{6A.25}$$

In view of Eq. (6A.14), however, the zeros of $T_n(x)$ are at $\cos[(2k-1)\pi/2n]$. Therefore the zeros of $\hat{H}(s)$ are at

$$j\omega_k = \frac{j}{\cos\left[\dfrac{(2k-1)\pi}{2n}\right]}. \tag{6A.26}$$

(If n is odd, then one zero is at infinity.) The poles of $\hat{H}(s)$ satisfy

$$1 + \frac{\delta}{1-\delta} T_n\left(\frac{j}{s}\right)^2 = 0. \tag{6A.27}$$

These may be obtained by replacing ε with δ in Eq. (6A.21) and then taking the reciprocals of the complex numbers in Eqs. (6A.19).

Elliptic Filters

An even order ($n = 2m$) elliptic filter is characterized by

$$|\hat{H}(j\omega)|^2 = \frac{1}{1 + \mu^2 E^2(\omega)} \tag{6A.28}$$

where

$$E(\omega) = \prod_{k=1}^{m} \frac{\omega_k^2 - \omega^2}{1 - \omega^2\omega_k^2}, \tag{6A.29}$$

and the parameters ω_k all satisfy $\omega_k^2 < 1$. For $n = 2m + 1$, $E(\omega)$ has the form of Eq. (6A.29), but is multiplied by ω. Notice that $E(\omega)$ has the symmetry property

$$E(\omega^{-1}) = 1/E(\omega). \tag{6A.30}$$

All the zeros of E are real and lie in the interval $-1 < \omega < 1$. The poles are real and are the reciprocals of the zeros. The parameters ω_1 through ω_m are constrained so that $E(\omega)$ has a Chebyshev-like equal ripple; the local peaks and valleys of $E(\omega)$ for which the derivative vanishes all have the same absolute value. In other words

$$\omega^2 < 1, \quad E'(\omega) = 0 \quad \Rightarrow \quad |E(\omega)| = E_0. \tag{6A.31}$$

Because of the symmetry in Eq. (6A.30), the local peaks and valleys of $E(\omega)$ for $|\omega| > 1$ (between the poles) will satisfy

$$\omega^2 > 1, \quad E'(\omega) = 0 \quad \Rightarrow \quad |E(\omega)| = E_0^{-1}. \tag{6A.32}$$

In the pass-band, the elliptic filters will exhibit equal ripple with

$$E(\omega_k) = 0 \quad \Rightarrow \quad |\hat{H}(j\omega_k)|^2 = 1. \tag{6A.33}$$

$$E'(\omega) = 0 \quad \Rightarrow \quad |\hat{H}(j\omega)|^2 = 1 - \varepsilon = \frac{1}{1 + \mu^2 E_0^2}. \tag{6A.34}$$

In the stop-band, they will also exhibit equal ripple with

$$E(\omega_k^{-1}) = \infty \quad \Rightarrow \quad |\hat{H}(j\omega_k^{-1})|^2 = 0. \tag{6A.35}$$

$$E'(\omega) = 0 \quad \Rightarrow \quad |\hat{H}(j\omega)|^2 = \delta = \frac{1}{1 + \mu^2 E_0^{-2}}. \tag{6A.36}$$

Because of the nonlinear equations (6A.31) which must be satisfied by the parameters ω_1 through ω_m, the design of elliptic filters and the calculation of the resulting poles is quite complicated. See Van Valkenburg (1982) and Lam (1979). A program for designing discrete time elliptic filters is given in Gray and Markel (1976).

APPENDIX

6B

The Chebyshev Window Function

Using Chebyshev polynomials, one can obtain the two parameter family of *Chebyshev windows.* The design equations for these windows allow one to control the main lobe width and the ratio of the main lobe height to the heights of the side lobes.

The Chebyshev polynomial of order n is defined as follows:

$$T_n(x) = [1 \quad 0]\begin{bmatrix} 0 & 1 \\ -1 & 2x \end{bmatrix}^n \begin{bmatrix} 1 \\ x \end{bmatrix}. \tag{6B.1}$$

This sequence of polynomials satisfies the difference equation

$$T_{n+2}(x) = 2xT_{n+1}(x) - T_n(x), \quad n \geqslant 0 \tag{6B.2}$$

where

$$T_0(x) = 1, \quad T_1(x) = x. \tag{6B.3}$$

Using the techniques for solving difference equations developed in Chapter 2, one can further characterize these polynomials for specific ranges of x.

$$x < -1, \qquad T_n(-\cosh(u)) = (-1)^n \cosh(nu) \tag{6B.4}$$

$$-1 \leqslant x \leqslant 1, \qquad T_n(\cos\theta) = \cos(n\theta) \tag{6B.5}$$

$$x > 1, \qquad T_n(\cosh(u)) = \cosh(nu) \tag{6B.6}$$

The first nine polynomials are displayed in Fig. 6B.1. For $|x| < 1$ the graph resembles a horizontally distorted sinusoid. This is the "equal ripple" region of the polynomial. For $x > 1$ the polynomials are monotone increasing (except T_0) and the rate of increase is that of x^n. The zeros of $T_n(x)$ are all real and in the range $-1 < x < 1$. From Eq. (6B.5) these are

$$x = \cos\left[\frac{(2k-1)\pi}{2n}\right], \qquad k \text{ an integer.} \tag{6B.7}$$

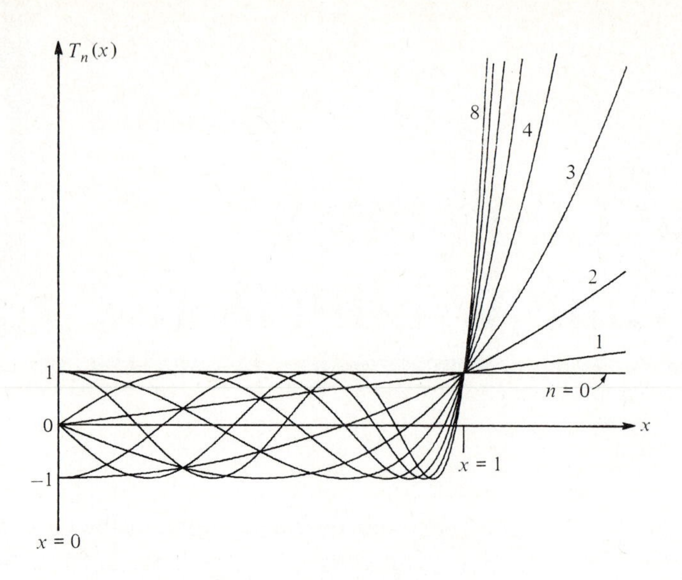

FIGURE 6B.1 The Chebyshev polynomials of orders 0 to 8.

The *Chebyshev window function* (of length $N = 2M + 1$, with parameter $x_0 > 1$) is defined implicitly by

$$W(e^{j\theta}) = \sum_{k=-M}^{M} w(k)e^{-jk\theta} = T_{2M}(x_0 \cos(\theta/2)). \tag{6B.8}$$

In light of the presence of $\cos(\theta/2)$ in the argument, it may not be clear that this identity is possible. But Chebyshev polynomials satisfy the identity

$$T_{2M}(x) = T_M(2x^2 - 1), \tag{6B.9}$$

and therefore

$$T_{2M}(x_0 \cos(\theta/2)) = T_M(x_0^2(1 + \cos\theta) - 1). \tag{6B.10}$$

This is clearly a polynomial of order M in $\cos\theta$, and admits the expansion of Eq. (6B.8).

The height of the main lobe of W is

$$R = W(1) = T_{2M}(x_0). \tag{6B.11}$$

Define θ_0 to be the first crossing of the level 1, as shown in Fig. 6B.2. Since $T_{2M}(1) = 1$, we have

$$1 = x_0 \cos(\theta_0/2). \tag{6B.12}$$

Then $2\theta_0$ is the width of the main lobe of W. From Eq. (6B.11) and Eq. (6B.6),

we have

$$\cosh^{-1}(R) = 2M \cosh^{-1}(x_0). \tag{6B.13}$$

One can eliminate x_0 from this relation, using Eq. (6B.12), thereby producing the design equation relating the height and width of the main lobe of W.

$$\cosh^{-1}(R) = 2M \cosh^{-1}\left[\frac{1}{\cos(\theta_0/2)}\right]. \tag{6B.14}$$

All side lobes have height 1, since the window is derived from a Chebyshev polynomial. For large R, the relation (6B.14) is well approximated by the simpler rule

$$\theta_0 \simeq \frac{\ln(2R)}{M}. \tag{6B.15}$$

It is not obvious from the definition (6B.8) how the coefficients of the Chebyshev window are to be obtained, but for numerical computation the definition leads to a difference equation. This difference equation involves

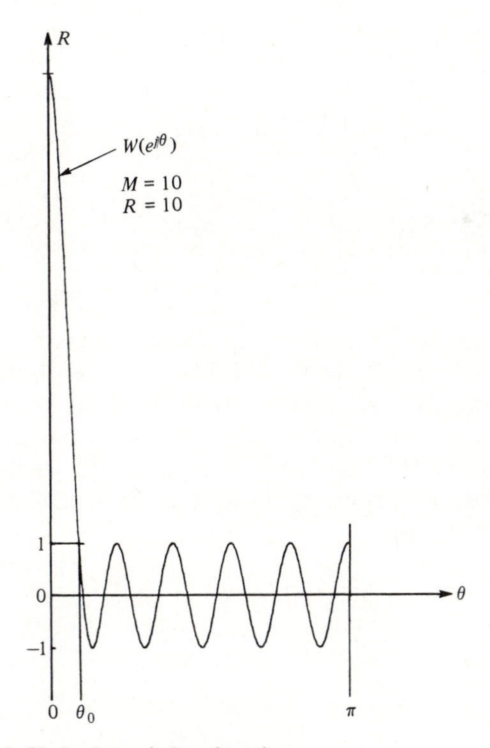

FIGURE 6B.2 **A Chebyshev window function.**

both the window order M and the coefficient index k, but allows for efficient computation of the coefficients. Use Eqs. (6B.8) and (6B.10), and substitute $(z + z^{-1})/2$ for $\cos\theta$. The result is

$$W_M(z) = \sum_{k=-M}^{M} w_M(k)z^{-k} = T_M\left(x_0^2\left[1 + \frac{z + z^{-1}}{2}\right] - 1\right). \tag{6B.16}$$

The initial conditions for the Chebyshev polynomials of Eqs. (6B.3) provide initial conditions for the windows.

$$W_0(z) = 1. \tag{6B.17}$$

$$W_1(z) = x_0^2\left[1 + \frac{z + z^{-1}}{2}\right] - 1. \tag{6B.18}$$

The Chebyshev difference equation (6B.2) also forces a window difference equation.

$$W_M(z) = [x_0^2(2 + z + z^{-1}) - 2]W_{M-1}(z) - W_{M-2}(z), \quad M > 1. \tag{6B.19}$$

Taking the inverse z-transform of Eqs. (6B.17) through (6B.19) yields

$$w_0(k) = \begin{cases} 1, & k = 0 \\ 0, & \text{otherwise} \end{cases}, \tag{6B.20}$$

$$w_1(k) = \begin{cases} x_0^2 - 1, & k = 0 \\ x_0^2/2, & |k| = 1, \\ 0, & \text{otherwise} \end{cases} \tag{6B.21}$$

and for $M > 1$,

$$w_M(k) = 2(x_0^2 - 1)w_{M-1}(k) + x_0^2[w_{M-1}(k-1) + w_{M-1}(k+1)] - w_{M-2}(k). \tag{6B.22}$$

This set of equations allows for the efficient computation of window coefficients for increasing order, for a fixed value of x_0. But there are two caveats. The right hand side of Eq. (6B.22) will involve window coefficients that are out of range, i.e., $|k|$ is greater than the order of the window. One should be careful to set these to zero before they are used. Second, the resulting window is not normalized. One should multiply all coefficients by the appropriate scalar to achieve unity energy, or the desired DC gain, or to make some other condition valid.

CHAPTER 7

Least-Squares Filter Design

7.1 Introduction

In Chapter 6 we outlined the steps in the design and implementation of digital filters. The first step in this process, the approximation problem, consists of determining the input/output characteristics of the filter from a given set of specifications. In Chapter 6 these specifications are expressed in terms that are not explicitly dependent on the characteristics of the input signal. For example, we might be asked to find the transfer function of a low-pass filter with a given cutoff frequency, transition bandwidth, and attenuation in the stop-band. In this chapter we consider the approximation problem for specifications that include characteristics of the input signal. For example, we may wish to design a filter to "clean up" a signal in added noise. Our goal is again to determine the input/output characteristics of the filter in the form of the transfer function $H(z)$ or the unit-pulse response sequence $\mathbf{h}$.

In our discussions in Chapter 6 we often seek $H(z)$ to approximate some ideal characteristic. For example, we cannot obtain an FIR filter that is an ideal low-pass filter. In a similar manner we can often only approximate filters to extract wanted signals in a mixture of signal and noise. If $\mathbf{u}$ is the sum of a wanted signal $\mathbf{s}$ and a noise $\mathbf{n}$, it is usually not possible to design a filter $\mathbf{h}$ for which the output $\mathbf{y}$ is exactly $\mathbf{s}$. However, it is possible to approximate such performance if the statistical properties of the signal and noise are known (and are different). Furthermore, the more we know concerning the statistical characteristics of $\mathbf{s}$ and $\mathbf{n}$, the better we can design $\mathbf{h}$.

In approximation problems we must always have a measure of the quality of the approximation. This measure is a key ingredient in the problem formulation. A classic example occurs in the Fourier series approximation to an arbitrary function. One measure that has found wide application and leads to simple and practical implementations is the mean square error (MSE). Often this criterion results in optimum or near optimum performance over all possible criteria that could have reasonably been chosen. We shall call such problems *least-squares problems* since the goal is to find a filter that minimizes the squared error. The error is the difference between the desired output and the actual output. For example, suppose we wish to select a signal $\mathbf{x}$ from a set $\mathbf{X}$ to best approximate some other signal $\mathbf{y}$ not in the set $\mathbf{X}$. One way to choose $\mathbf{x}$ in $\mathbf{X}$ which is "closest" to $\mathbf{y}$ is to find $\mathbf{x}$ to minimize the Euclidean distance $d(\mathbf{x}, \mathbf{y})$ where

$$d^2(\mathbf{x}, \mathbf{y}) = \sum_k [x(k) - y(k)]^2. \tag{7.1.1}$$

This is equivalent to minimizing the scalar function $V(\mathbf{x}) = d^2(\mathbf{x}, \mathbf{y})$. That is,

$$V(\mathbf{x}) = \sum_k [x^2(k) - 2x(k)y(k) + y^2(k)]. \tag{7.1.2}$$

For least-squares designs, the problem is always equivalent to finding a filter $\mathbf{h}$ to minimize a scalar function $V(\mathbf{h})$ where $V(\mathbf{h})$ is a quadratic form as in Eq. (7.1.2). $V(\mathbf{h})$ consists of a term quadratic in $\mathbf{h}$, a term linear in $\mathbf{h}$, and a constant. Thus we can express $V(\mathbf{h})$ in the form

$$\begin{aligned} V(\mathbf{h}) &= \rho - 2\mathbf{q}^T\mathbf{h} + \mathbf{h}^T\mathbf{R}\mathbf{h} \\ &= \rho - 2\sum_k q(k)h(k) + \sum_{k,l}\sum R(k, l)h(k)h(l). \end{aligned} \tag{7.1.3}$$

In the example of Eq. (7.1.2), the parameters corresponding to the general form of Eq. (7.1.3) are

$$\rho = \sum_k y^2(k), \qquad \mathbf{q} = \mathbf{y}, \qquad \mathbf{R} = \mathbf{I}. \tag{7.1.4}$$

In this chapter we formulate several least-squares problems to determine a filter $\mathbf{h}$ or $H(z)$. The quadratic form obtained for $\mathbf{h}$ may arise from deterministic or statistical criteria. We discuss these two cases in Sections 7.2 and 7.4, respectively. In both cases, however, the formulation results in the minimization of a quadratic form $V(\mathbf{h})$ as in Eq. (7.1.3). It is only the definition of the parameters ρ, $\mathbf{q}$, and $\mathbf{R}$ that distinguishes the two problems. Once $V(\mathbf{h})$ is defined, the mathematical procedure for minimizing V is the same in each case.

It may occur in practice that explicit measurement of the parameters ρ, $\mathbf{q}$, and $\mathbf{R}$ is not possible. The only measurements possible occur on signals that are indirectly related to the parameters in the quadratic form $V(\mathbf{h})$. In these cases it may be possible to design an approximation to the least-squares filter in real time. These so-called *adaptive* filters are discussed briefly in Section 7.5.

7.2 DETERMINISTIC LEAST SQUARES

Least-squares problems can be classified as *deterministic* or *statistical* depending on the nature of the signals and filters involved. In statistical problems, one must deal with random signals. The goal might be to separate a desired signal from background noise, or to alter the frequency content of a signal in some way. In this context, the parameters of the quadratic form $V(\mathbf{h})$ are correlations between signal samples. In deterministic problems, there may be random signals, but our interest is in filters present in the system that have unsatisfactory transfer functions. For example, one might be required to transmit a signal across a communication channel which is linear but has a frequency response which attenuates some frequency bands. This channel cannot be replaced, but it can be corrected by cascading it with an *equalization filter*, whose frequency response approximates the inverse of the channel response. In this section, we shall consider a general version of this problem.

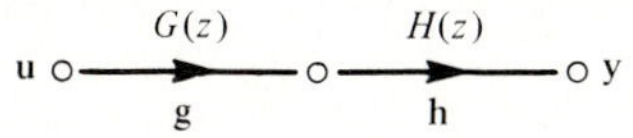

FIGURE 7.2.1.

Consider the cascade of filters in Fig. 7.2.1. The transfer function $G(z)$ represents a communication path which must be used. The filter $H(z)$ (in this case at the output) is present to change the overall transfer function from $G(z)$ to $G(z)H(z)$. Suppose that for some reason, we would like the overall response to be $F(z)$. In the channel equalization problem, for instance, we would set $F(z) = 1$. Ideally, then, we would construct $H(z)$ so that

$$\mathbf{f} = \mathbf{g} * \mathbf{h} \quad \overset{z}{\longleftrightarrow} \quad F(z) = G(z)H(z). \tag{7.2.1}$$

This would mean of course that $H(z) = F(z)/G(z)$. This filter is, in typical situations, inadmissible for one of three reasons. First, if $G(z)$ is not minimum phase, then $1/G(z)$ cannot be causal and stable, and we must require these properties of $H(z)$. Second, $F(z)$ may not be causal. Finally it may be necessary to restrict $H(z)$ to be FIR of order n.

If we cannot meet Eq. (7.2.1) exactly, then we must resort to some sort of approximation. In particular, minimizing the energy of the error $\mathbf{f} - \mathbf{g} * \mathbf{h}$ leads to a least-squares problem.

Given $\mathbf{f} \in l_2$, $\quad \mathbf{g} \in l_2$ with $\mathbf{g}$ causal,

Minimize

$$V(\mathbf{h}) = \|\mathbf{f} - \mathbf{g} * \mathbf{h}\|^2 = \sum_{k=-\infty}^{\infty} [f(k) - (\mathbf{g} * \mathbf{h})(k)]^2$$

$$= \frac{1}{2\pi} \int_{-\pi}^{\pi} |F(e^{j\theta}) - G(e^{j\theta})H(e^{j\theta})|^2 d\theta \tag{7.2.2}$$

Subject to the constraints

$$h(k) = 0 \quad \text{for} \quad k < 0 \quad \text{or} \quad k > n. \tag{7.2.3}$$

The constraints make $H(z)$ an FIR filter. Note that the Parseval relation allows either time or frequency domain expressions for the error energy $V(\mathbf{h})$.

The function $V(\mathbf{h})$ in Eq. (7.2.2) is quadratic but is not in the standard form of Eq. (7.1.3). Let us proceed to put it into this form and then derive the equations that characterize the filter $H(z)$ which minimizes V. These equations are called the *normal equations* for the problem. From Eq. (7.2.2),

$$V(\mathbf{h}) = \sum_{k=-\infty}^{\infty} \left[f(k) - \sum_{m=0}^{n} h(m)g(k-m) \right]^2$$
$$= \left[\sum_{k=-\infty}^{\infty} f^2(k) \right] - 2 \sum_{m=0}^{n} h(m) \left[\sum_{k=-\infty}^{\infty} f(k)g(k-m) \right]$$
$$+ \sum_{l=0}^{n} \sum_{m=0}^{n} h(m)h(l) \left[\sum_{k=-\infty}^{\infty} g(k-m)g(k-l) \right]. \tag{7.2.4}$$

Comparing this with Eq. (7.1.3), we can identify the parameters of the quadratic form. For the constant term,

$$\rho = \sum_{k=-\infty}^{\infty} f(k)^2 = \|\mathbf{f}\|^2. \tag{7.2.5}$$

For the linear term,

$$q(m) = \sum_{k=-\infty}^{\infty} f(k)g(k-m)$$
$$= \sum_{k=0}^{\infty} f(k+m)g(k) = \frac{1}{2\pi} \int_{-\pi}^{\pi} F(e^{j\theta})G^*(e^{j\theta})e^{jm\theta} d\theta. \tag{7.2.6}$$

Finally, for the quadratic term,

$$R(m, l) = \sum_{k=-\infty}^{\infty} g(k-m)g(k-l)$$
$$= \sum_{k=0}^{\infty} g(k+m-l)g(k) = \frac{1}{2\pi} \int_{-\pi}^{\pi} |G(e^{j\theta})|^2 e^{j(m-l)\theta} d\theta,$$
$$= r(m-l). \tag{7.2.7}$$

Since the elements of the matrix $\mathbf{R}$ depend only on the difference between indices, all the elements on any diagonal will be the same. Such a matrix is called a *Toeplitz* matrix. (The sequence $\mathbf{r}$ is called the *deterministic autocorrelation sequence* for $\mathbf{g}$, and was introduced in Problem 3.24.).

Thus

$$V(\mathbf{h}) = [1 \quad \mathbf{h}^T] \begin{bmatrix} \rho & -\mathbf{q}^T \\ -\mathbf{q} & \mathbf{R} \end{bmatrix} \begin{bmatrix} 1 \\ \mathbf{h} \end{bmatrix}$$
$$= \rho - 2\mathbf{q}^T\mathbf{h} + \mathbf{h}^T\mathbf{R}\mathbf{h}, \tag{7.2.8}$$

where

$$\mathbf{q} = \begin{bmatrix} q(0) \\ \vdots \\ q(n) \end{bmatrix}, \tag{7.2.9}$$

$$\mathbf{h} = \begin{bmatrix} h(0) \\ \vdots \\ h(n) \end{bmatrix}, \tag{7.2.10}$$

$$\mathbf{R} = \begin{bmatrix} r(0) & r(1) & \cdots & r(n) \\ r(1) & \ddots & \ddots & \vdots \\ \vdots & \ddots & \ddots & r(1) \\ r(n) & \cdots & r(1) & r(0) \end{bmatrix}. \tag{7.2.11}$$

Setting the gradient of V to zero yields the necessary conditions for a minimum. The result is called the *normal equations*:

$$\sum_{m=0}^{n} r(k-m)h(m) = q(k), \quad 0 \leqslant k \leqslant n,$$

or

$$\mathbf{Rh} = \mathbf{q}. \tag{7.2.12}$$

If $\mathbf{h}$ satisfies Eq. (7.2.12), then

$$V(\mathbf{h}) = V_{\min} = \rho - \mathbf{q}^T\mathbf{R}^{-1}\mathbf{q} = \rho - \mathbf{q}^T\mathbf{h}. \tag{7.2.13}$$

FIGURE 7.2.2 **Algorithm for the solution of a Toeplitz system of equations.**

Given: n, $q[0, n]$, $r[0, n]$

To compute: $h[0, n]$ satisfying Eq. (7.2.12)

Initialization:

$\alpha \leftarrow r(0)$
$a_0(0) = 1$
$h(0) \leftarrow q(0)/r(0)$

Body:

For $k = 1$ to n, Do
  $a_k(k-1) = 0$
  $a_0(k) = 1$
  $\beta \leftarrow \left[\sum_{j=0}^{k-1} r(k-j)a_j(k-1)\right] \Big/ \alpha$
  For $j = 1$ to k, Do
    $a_j(k) = a_j(k-1) - \beta a_{k-j}(k-1)$
  (end loop on j)
  $\alpha \leftarrow \alpha(1-\beta^2)$
  $h(k) \leftarrow \left[q(k) - \sum_{j=0}^{k-1} r(k-j)h(j)\right] \Big/ \alpha$
  For $j = 0$ to $k-1$, Do
    $h(j) \leftarrow h(j) + a_{k-j}(k)h(k)$
  (end loop on j)
(end loop on k)

Most of the computational effort to find the least squares filter is in the infinite sums of Eqs. (7.2.6) and (7.2.7). Once the parameters $\mathbf{q}[0, n]$ and $\mathbf{r}[0, n]$ are known, the normal equations can be solved using an efficient algorithm which generalizes the Levinson-Durbin algorithm (see Chapter 11). The algorithm is described in Fig. 7.2.2.

EXAMPLE 7.2.1

A least-squares inverse for a nonminimum phase filter.

If $F(z) = 1$, then $V(\mathbf{h}) = \|\boldsymbol{\delta} - \mathbf{g} * \mathbf{h}\|^2$, and the goal is to equalize $G(z)$ by $H(z) \approx 1/G(z)$. Such problems are variously referred to as *equalization*, or *deconvolution*, or *least-squares inverse* problems. Suppose we take

$$G(z) = 1 - \mu z^{-1}, \qquad |\mu| > 1. \tag{7.2.14}$$

Notice that since G has a zero at μ that is outside the unit circle, then G is nonminimum phase, and $1/G(z)$ cannot be both stable and causal. What, if anything, can be done to equalize such a filter? With $F(z) = 1$, and $G(z)$ given by Eq. (7.2.14), we have

$$r(k) = \begin{cases} 1 + \mu^2, & k = 0, \\ -\mu, & k = \pm 1, \\ 0, & \text{otherwise}, \end{cases} \tag{7.2.15}$$

$$q(k) = \begin{cases} 1, & k = 0 \\ 0, & k > 0 \end{cases}. \tag{7.2.16}$$

The normal equations (7.2.12) become

$$\begin{bmatrix} 1 + \mu^2 & -\mu & & 0 \\ -\mu & \ddots & \ddots & \\ & \ddots & \ddots & -\mu \\ 0 & & -\mu & 1 + \mu^2 \end{bmatrix} \begin{bmatrix} h(0) \\ h(1) \\ \vdots \\ h(n) \end{bmatrix} = \begin{bmatrix} 1 \\ 0 \\ \vdots \\ 0 \end{bmatrix}. \tag{7.2.17}$$

For specific choices of μ and n, the algorithm of Fig. 7.2.2 could be used to determine $\mathbf{h}$. In the interests of exposing the asymptotic behavior, however, let us attack head on. For $1 \leqslant k \leqslant n - 1$, we have

$$-\mu h(k - 1) + (1 + \mu^2)h(k) - \mu h(k + 1) = 0. \tag{7.2.18}$$

This homogeneous difference equation has solutions of the form

$$h(k) = c_1 \mu^k + c_2 \mu^{-k}. \tag{7.2.19}$$

To determine the coefficients, we must use the equations for

$$\begin{aligned} k = 0: &\qquad (1 + \mu^2)h(0) - \mu h(1) = 1 \\ k = n: &\qquad -\mu h(n - 1) + (1 + \mu^2)h(n) = 0. \end{aligned} \tag{7.2.20}$$

Solving for c_1 and c_2 and substituting into Eq. (7.2.19) leads to

$$h_n(k) = \begin{cases} \dfrac{\mu^{n+1-k} - \mu^{-(n+1-k)}}{\mu^{n+3} - \mu^{-(n+1)}}, & 0 \leqslant k \leqslant n \\ 0, & \text{otherwise.} \end{cases} \tag{7.2.21}$$

What happens as $n \to \infty$? Since $|\mu| > 1$,

$$\lim_{n\to\infty} h_n(k) = \frac{\mu^{n+1-k}}{\mu^{n+3}} = \mu^{-2}\mu^{-k}, \quad k \geqslant 0. \tag{7.2.22}$$

Thus

$$\lim_{n\to\infty} H_n(z) = \mu^{-2}\frac{1}{1-\mu^{-1}z^{-1}}. \tag{7.2.23}$$

Clearly, this is no inverse of $G(z)$. However,

$$\lim_{n\to\infty} H_n(z)G(z) = \frac{1}{\mu} E(z), \tag{7.2.24}$$

where

$$E(z) = \frac{1-\mu z^{-1}}{\mu - z^{-1}} \tag{7.2.25}$$

is an all-pass filter! A short computation yields

$$\lim_{n\to\infty} \|\boldsymbol{\delta} - \mathbf{g} * \mathbf{h}_n\|^2 = 1 - \frac{1}{\mu^2}. \tag{7.2.26}$$

Therefore, the error energy is small if the zero of $G(z)$ is only slightly outside the unit circle, but approaches one as $\mu \to \infty$.

EXAMPLE 7.2.2

A deconvolution filter with delay.

A communication channel has transfer function

$$G(z) = \frac{1}{4} - z^{-2},$$

which is nonminimum phase. Suppose that we wish to equalize $G(z)$ but will tolerate some delay. That is, we can set $F(z) = z^{-L}$ rather than $F(z) = 1$. This device can lead to a dramatic decrease in mean square error. In particular, let us set

$$F(z) = z^{-12}$$

and let the filter order be $n = 10$. We have

$$r(k) = \begin{cases} 1.0625, & k = 0 \\ -0.25, & k = \pm 2 \\ 0, & \text{otherwise.} \end{cases}$$

$$q(k) = \begin{cases} 0, & 0 \leqslant k \leqslant 9, \\ -1, & k = 10. \end{cases}$$

Applying the algorithm to these input parameters yields the following results:

$$\mathbf{h} = \{-9.155 \times 10^{-4}, 0, -3.891 \times 10^{-3}, 0, -1.562 \times 10^{-2}, 0, -6.25 \times 10^{-2}, 0, -.25, 0, -1.0\}$$

Now $\mathbf{h} * \mathbf{g}$ should approximate the sequence $\{\delta(k - 12)\}$. This turns out to be

$$\mathbf{g} * \mathbf{h} = \{-2.289 \times 10^{-4}, 0, -5.722 \times 10^{-5}, 0, -1.431 \times 10^{-5}, 0, -3.576 \times 10^{-6}, 0, -8.941 \times 10^{-7}, 0, -2.236 \times 10^{-7}, 0, 1.\}$$

The mean square error is 5.588×10^{-8}, which indicates that the approximation is excellent. Deconvolution filters with delay are sometimes called *spiking* filters.

The FIR Whitening Problem

The whitening problem which we are about to set up is interesting in itself, but turns out to be an integral part of least-squares problems in general, in that it yields a decomposition which simplifies the general problem.

Given $\mathbf{g} \in l_2$, causal,

Minimize

$$V_0(\mathbf{a}) = \|\mathbf{g} * \mathbf{a}\|^2 = \frac{1}{2\pi} \int_{-\pi}^{\pi} |G(e^{j\theta})A(e^{j\theta})|^2 d\theta \quad \textbf{(7.2.27)}$$

Subject to the constraints

$$a(0) = 1,$$
$$a(k) = 0 \quad \text{for } k < 0, k > n. \quad \textbf{(7.2.28)}$$

The constraint that $a(0) = 1$ is important, for without it the solution would be $A(z) = 0$. The reason that $A(z)$ is called a *whitening* filter is that (provided it minimizes V_0) the product $A(z)G(z)$ will approximate an all-pass filter, that is

$$|G(e^{j\theta})A(e^{j\theta})| \approx \text{constant.}$$

This result is not obvious based on the problem statement. Statistical versions of the whitening filter problem form the basis for *autoregressive power spectral estimates.* These are discussed in Chapter 11.

To obtain the normal equations for the problem of minimizing V_0, we will use the technique of "completing the square." In matrix form

$$V_0(\mathbf{a}) = \mathbf{a}^T\mathbf{R}\mathbf{a} \tag{7.2.29}$$

where $\mathbf{R}$ is given by Eq. (7.2.11) and

$$\mathbf{a} = \begin{bmatrix} 1 \\ a(1) \\ \vdots \\ a(n) \end{bmatrix} \tag{7.2.30}$$

Consider a perturbation vector

$$\boldsymbol{\Delta} = \begin{bmatrix} 0 \\ \Delta(1) \\ \vdots \\ \Delta(n) \end{bmatrix} \tag{7.2.31}$$

which satisfies $\Delta(0) = 0$ so that the constraint Eq. (7.2.28) holds, i.e., $(\mathbf{a} + \boldsymbol{\Delta})(0) = 1$. Then

$$V_0(\mathbf{a} + \boldsymbol{\Delta}) = \mathbf{a}^T\mathbf{R}\mathbf{a} + \mathbf{a}^T\mathbf{R}\boldsymbol{\Delta} + \boldsymbol{\Delta}^T\mathbf{R}\mathbf{a} + \boldsymbol{\Delta}^T\mathbf{R}\boldsymbol{\Delta}. \tag{7.2.32}$$

The goal is to choose $\mathbf{a}$ so that the cross terms vanish for all perturbations. This leads to the *normal equations*

$$\mathbf{R}\mathbf{a} = \begin{bmatrix} \alpha \\ 0 \\ \vdots \\ 0 \end{bmatrix} \tag{7.2.33}$$

(Since $\boldsymbol{\Delta}(0)$ is always zero, the first element of $\mathbf{Ra}$ can be nonzero. The role of this parameter α will shortly become clear.) If Eq. (7.2.33) holds, then $\mathbf{a}^T\mathbf{R}\boldsymbol{\Delta} = 0$, and

$$\begin{aligned} V_0(\mathbf{a} + \boldsymbol{\Delta}) &= V_0(\mathbf{a}) + V_0(\boldsymbol{\Delta}) \\ &= \|\mathbf{g} * \mathbf{a}\|^2 + \|\mathbf{g} * \boldsymbol{\Delta}\|^2 \geqslant V_0(\mathbf{a}). \end{aligned} \tag{7.2.34}$$

In other words, any perturbation increases $V_0(\mathbf{a})$. The parameter α in Eq. (7.2.33) turns out to be the minimum value of V_0 since if $\mathbf{a}$ satisfies the normal equations (7.2.33), then

$$V_0(\mathbf{a}) = \mathbf{a}^T \begin{bmatrix} \alpha \\ 0 \\ \vdots \\ 0 \end{bmatrix} = \alpha, \tag{7.2.35}$$

since $a(0) = 1$.

The whitening problem has an interesting property. The solution filter $A(z)$ depends on the magnitude of $G(z)$ only. This is because the normal equations (7.2.33) involve $G(z)$ only through the matrix $\mathbf{R}$, and from Eqs. (7.2.7), the elements of $\mathbf{R}$ depend only on $|G|^2$. If we were to substitute $G(z)E(z)$ for $G(z)$, where $E(z)$ is an all-pass filter, the resulting $A(z)$ would be the same.

WHITENING FILTER DECOMPOSITION. The collection of whitening filters of all orders can be used to obtain a triangular decomposition of the matrix **R** in Eq. (7.2.11) (as is done in Appendix 7A). Let

$$A_m(z) = \sum_{k=0}^{m} a_m(k) z^{-k} \tag{7.2.36}$$

be the solution to the order m whitening filter problem. Construct two $(n+1) \times (n+1)$ matrices

$$\mathbf{T} = \begin{bmatrix} 1 & a_1(1) & \cdots & a_n(n) \\ & \ddots & & \vdots \\ & 0 & 1 & a_n(1) \\ & & & 1 \end{bmatrix}, \quad T_{ij} = a_j(j-i) \tag{7.2.37}$$

$$\mathbf{D} = \text{Diag}\{\alpha_0, \ldots, \alpha_n\}, \tag{7.2.38}$$

where α_m is the parameter appearing in Eq. (7.2.35); i.e., the minimum value of V_0 for whitening filters of order m. The indices i and j of these matrices range from 0 to n.

Using the normal equations (7.2.33), and the fact that **R** is Toeplitz, one can show in a matter analogous to the development of Eq. (7A.36) that

$$\mathbf{R}^{-1} = \mathbf{T}\mathbf{D}^{-1}\mathbf{T}^T. \tag{7.2.39}$$

(See Problem 7.7.) This is a triangular decomposition of the inverse of **R**. Now consider the original problem of minimizing $\|\mathbf{f} - \mathbf{g} * \mathbf{h}\|^2$ where $H_n(z)$ is an order n FIR filter. Combining Eqs. (7.2.39) and (7.2.12), we have

$$\mathbf{h} = \mathbf{R}^{-1}\mathbf{q} = \mathbf{T}\mathbf{D}^{-1}\mathbf{T}^T\mathbf{q}. \tag{7.2.40}$$

This equation yields a representation for $H_n(z)$ in terms of the whitening filters of order 0 to n, together with the elements of the vector **q**. The result (which is developed in Problem 7.7) is

$$H_n(z) = \sum_{m=0}^{n} \frac{c_m}{\alpha_m} \tilde{A}_m(z), \tag{7.2.41}$$

where

$$\tilde{A}_m(z) = z^{-m} A_m(z^{-1}) = \sum_{k=0}^{m} a_m(m-k) z^{-k} \tag{7.2.42}$$

and

$$c_m = \sum_{k=0}^{m} a_m(m-k) q(k). \tag{7.2.43}$$

This representation for $H_n(z)$ leads to a lattice signal flow graph with application to filter structures. (See Problem 7.11.)

The sequence of whitening filters is best computed using the Levinson-Durbin algorithm (see Chapter 11). This procedure forms a part of the algorithm in Fig. 7.2.2.

TABLE 7.2.1 Whitening filter examples.

$G(z)$		$r(k)$	$A_n(z)$	α_n
$\frac{1}{1-\lambda z^{-1}}$, (*single pole*)	$\lvert\lambda\rvert < 1$	$\frac{\lambda^{\lvert k\rvert}}{1-\lambda^2}$	$1-\lambda z^{-1}, \quad n > 0$	$1, \quad n > 0$
$1-\mu z^{-1}$, (*single zero*)	$\mu \neq 1$	$\begin{cases} 1+\mu^2, & k=0 \\ -\mu, & \lvert k\rvert = 1 \\ 0, & \text{otherwise} \end{cases}$	$\sum_{k=0}^{n}\left[\frac{\mu^{n+1-k}-\mu^{-(n+1-k)}}{\mu^{n+1}-\mu^{-(n+1)}} z^{-k}\right]$	$\mu\frac{\mu^{n+2}-\mu^{-(n+2)}}{\mu^{n+1}-\mu^{-(n+1)}}$
$1-z^{-1}$		$\binom{2}{k+1}(-1)^k$	$\sum_{k=0}^{n}\left(\frac{n+1-k}{n+1}\right) z^{-k}$	$\frac{n+2}{n+1}$
$(1-z^{-1})^2$		$\binom{4}{\mu+2}(-1)^k$	$\sum_{k=0}^{n}\frac{(n+1-k)(n+2-k)(k+1)}{(n+1)(n+2)} z^{-k}$	$\frac{(n+3)(n+4)}{(n+1)(n+2)}$
$\frac{\lambda-z^{-1}}{1-\lambda z^{-1}}$, (*all-pass*)	$\lvert\lambda\rvert < 1$	$\delta(k)$	1	1

EXAMPLE 7.2.3

Whitening Filters.

Table 7.2.1 gives the whitening filters $A_n(z)$ (of all orders) for some simple choices of $G(z)$. In the first line, $G(z)$ has a simple pole and no zeros. The filters $A_n(z)$ for any $n > 0$ are the same; they simply cancel the pole and result in $G(z)A_n(z) = 1$. This is perfect whitening since $|G(e^{j\theta})A_n(e^{j\theta})|$ is constant. In the next three lines, $G(z)$ is FIR, and has only zeros. Thus $A_n(z)$ cannot perfectly whiten since it would have to have poles to cancel the zeros of $G(z)$; an impossibility for FIR filters. In these cases we get a sequence of ever larger $A_n(z)$ and decreasing α_n.

In the last line of the table, $G(z)$ is itself all-pass; $|G(e^{j\theta})| = 1$. From Eq. (7.2.7) it follows that $\mathbf{R}$ is the identity matrix and therefore $A_n(z) = 1$ for all n. An all-pass filter cannot be further whitened. The whitening filter can correct magnitude distortion, but it cannot correct phase distortion for a nonminimum phase $G(z)$.

Asymptotic Behavior ($n \to \infty$)

What happens as the FIR filter order n is allowed to approach infinity? Does error energy approach zero? For example, suppose that $F(z)/G(z)$ is causal and has finite energy. Then we would expect the filters $H_n(z)$ which minimize $\|\mathbf{f} - \mathbf{g} * \mathbf{h}\|^2$ to approach $F(z)/G(z)$, resulting in zero error energy. But if G is nonminimum phase, then $1/G(z)$ is not causal and stable. And if $F(z)$ is not causal, then the error cannot vanish for $k < 0$, since

$$(\mathbf{f} - \mathbf{g} * \mathbf{h})(k) = f(k) \quad \text{for } k < 0. \tag{7.2.44}$$

Either of these situations can result in positive error energy, even as $n \to \infty$.

The examples listed in Table 7.2.1 can be used to demonstrate what happens as $n \to \infty$ when $G(z)$ is nonminimum phase. The results are listed in Table 7.2.2. The last column in this table contains the error energy $\|\boldsymbol{\delta} - \mathbf{g} * \mathbf{h}\|^2$ for the least-squares inverse problem. This turns out to be related to the whitening filter error energy through the equation

$$V_{\min} = 1 - \frac{g(0)^2}{\alpha_n} \tag{7.2.45}$$

which is derived in Problem 7.6. In all cases listed in Table 7.2.2 the limiting error for the least-squares inverse problem is zero when $G(z)$ is minimum phase, and positive when $G(z)$ is nonminimum phase. Likewise

$$G(z)A_n(z) \to g(0) \tag{7.2.46}$$

when $G(z)$ is minimum phase, else $G(z)A_n(z)$ will approach an all-pass filter. Now recall that both α_n and $A_n(z)$ depend only on the magnitude of G. A result of

TABLE 7.2.2 Asymptotic behavior for the whitening filters of Table 7.2.1. The last column gives the asymptotic error energy for the associated least-squares inverse problems.

$G(z)$		$g(0)$	$\lim_{n\to\infty} A_n(z)$	$\lim_{n\to\infty} G(z)A_n(z)$	$\lim_{n\to\infty} \alpha_n$	$\lim_{n\to\infty} \left[1 - \frac{g^2(0)}{\alpha_n}\right]$
$\frac{1}{1-\lambda z^{-1}}$, (*single pole*)	$\lvert\lambda\rvert < 1$	1	$1 - \lambda z^{-1}$	1	1	0
$1 - \mu z^{-1}$, (*single zero*)	$\lvert\mu\rvert \neq 1$	1	$\begin{cases} \frac{1}{1-\mu z^{-1}}, & \lvert\mu\rvert < 1 \\ \frac{1}{1-\mu^{-1}z^{-1}}, & \lvert\mu\rvert > 1 \end{cases}$	$\begin{cases} 1, & \lvert\mu\rvert < 1 \\ \mu\frac{\mu^{-1}-z^{-1}}{1-\mu^{-1}z^{-1}}, & \lvert\mu\rvert > 1 \end{cases}$	$\begin{cases} 1, & \lvert\mu\rvert < 1 \\ \mu^2, & \lvert\mu\rvert > 1 \end{cases}$	$\begin{cases} 0, & \lvert\mu\rvert < 1 \\ 1 - \frac{1}{\mu^2}, & \lvert\mu\rvert > 1 \end{cases}$
$1 - z^{-1}$		1	$\frac{1}{1-z^{-1}}$	1	1	0
$(1 - z^{-1})^2$		1	$\left(\frac{1}{1-z^{-1}}\right)^2$	1	1	0
$\frac{\lambda - z^{-1}}{1-\lambda z^{-1}}$, (*all-pass*)	$\lvert\lambda\rvert < 1$	λ	1	$\frac{\lambda - z^{-1}}{1-\lambda z^{-1}}$	1	$1 - \lambda^2$

Szëgo (1958) is that

$$\lim_{n\to\infty} \alpha_n = \exp\left[\frac{1}{2\pi}\int_{-\pi}^{\pi} \ln|G(e^{j\theta})|^2 d\theta\right]. \tag{7.2.47}$$

Thus, a usable minimum phase condition for $G(z)$ is

$$g(0)^2 = \exp\left[\frac{1}{2\pi}\int_{-\pi}^{\pi} \ln|G(e^{j\theta})|^2 d\theta\right]. \tag{7.2.48}$$

When this holds, we can expect

$$\lim A_n(z) = \frac{g(0)}{G(z)}, \tag{7.2.49}$$

and the error energy in Eq. (7.2.45) to approach zero.

Now let us consider the case where $G(z)$ is nonminimum phase and $F(z)$ is not causal. Let

$$G(z) = G_0(z)E(z) \tag{7.2.50}$$

where $G_0(z)$ meets the minimum phase condition (7.2.48), but $E(z)$ is all-pass with magnitude one. Let $n = \infty$ (IIR filters) and set

$$H(z) = H_0(z)/G_0(z), \tag{7.2.51}$$

where $H_0(z)$ remains to be determined. Then $\mathbf{g} * \mathbf{h} = \mathbf{e} * \mathbf{h}_0$ and therefore

$$V(\mathbf{h}) = \|\mathbf{f} - \mathbf{g} * \mathbf{h}\|^2 = \|\mathbf{f} - \mathbf{e} * \mathbf{h}_0\|^2. \tag{7.2.52}$$

The determination of $\mathbf{h}_0$ to minimize the right-hand side of Eq. (7.2.52) is greatly simplified by the fact that the $\mathbf{R}$ matrix for the sequence $\mathbf{e}$ is the identity. Thus the normal equations become

$$h_0(m) = \sum_{k=0}^{\infty} f(k+m)e(k), \quad m \geqslant 0. \tag{7.2.53}$$

The question of interest to us, however, is "what is the minimum of $V(\mathbf{h})$?" Using Problem 3.23, one can show that

$$V_{\min} = \left[\sum_{k=-\infty}^{-1} f(k)^2\right] + \sum_{m=1}^{\infty}\left[\sum_{k=0}^{\infty} f(k)e(k+m)\right]^2. \tag{7.2.54}$$

The first term in this sum is zero if and only if $F(z)$ is causal. The second term is zero if $G(z)$ is minimum phase (i.e., $\mathbf{e} = \boldsymbol{\delta}$). For example, for the least-squares inverse problem, $F(z) = 1$, which is causal. In this case Eq. (7.2.54) reduces to

$$V_{\min} = \sum_{m=1}^{\infty} e(m)^2 = 1 - e(0)^2, \tag{7.2.55}$$

which is consistent with the right-hand column of Table 7.2.2. For the least-squares inverse with delay L, $F(z) = z^{-L}$. The error energy in Eq. (7.2.54) is then

$$V_{\min} = 1 - \sum_{k=0}^{L} e^2(k), \tag{7.2.56}$$

which decreases as the delay is increased.

7.3 WIDE SENSE STATIONARY SIGNALS

The unit-pulse sequence $\boldsymbol{\delta}$ and complex sinusoids $\{e^{jk\theta}\}$ are standard test inputs for linear systems. The response of a linear filter to $\boldsymbol{\delta}$ is the unit-pulse response sequence $\mathbf{h}$ that characterizes the input/output relation $\mathbf{y} = \mathbf{h} * \mathbf{u}$. Likewise, the frequency response function can be measured using sinusoidal inputs. Despite their value for analysis, however, one does not expect to find these signals in actual applications. In fact, we can rarely be sure what to expect. This suggests that we should consider some sort of analysis based on probability theory. But this would be a formidable exercise if done in generality. We need a compromise. In words, we need a statistical description for signals that is simple enough to be computationally useful.

A "random" signal is actually an infinite number of scalar random variables. A complete probabilistic description (joint density functions, moments of all orders, etc.) is enormously detailed. Suppose that we were to throw out all but a small part of that detail. What should we keep? In the context of this chapter, we could say "let's keep the bare minimum to be able to do least-squares problems." In statistical problems, one typically seeks to minimize the "*mean squared error*," which is to say the expected value of the square of the error. Now if the "error" is a linear function of the signal, then this would have the form

$$E\left(\sum_k c_k u(k)\right)^2 = \sum_k \sum_l c_k c_l E[u(k)u(l)]. \tag{7.3.1}$$

Thus in order to evaluate this quantity, we need *second moments*, i.e., the quantities

$$R_{uu}(k,l) = E[u(k)u(l)]. \tag{7.3.2}$$

In a little more general case, we would also need the *first moments* $E[u(k)]$. This amount of detail from the complete statistical description of the random signal $\mathbf{u}$ begins to look manageable.

For most applications, however, a further simplification is justified. Just as it was reasonable to consider shift-invariant, linear filters, it is also reasonable to consider "mean shift-invariant random signals." Such a signal possesses moments invariant under time shifts. Thus

$$\begin{aligned} E[u(k+l)] &= E[u(k)], \\ E[u(k+l)u(j+l)] &= E[u(k)u(j)] \quad \text{for all } l. \end{aligned} \tag{7.3.3}$$

This property is called *weak stationarity*. If $\mathbf{u}$ is a weakly stationary sequence, then we need only a single number μ for the first moments, instead of a sequence.

$$\mu = E[u(k)], \quad \text{(for all } k). \tag{7.3.4}$$

Furthermore, we need only a sequence of numbers $\{r_{uu}(l)\}$ to characterize all second moments, instead of the two-dimensional array $\{R_{uu}(k, l)\}$.

$$r_{uu}(l) = E[u(k+l)u(k)], \quad \text{(for all } k). \tag{7.3.5}$$

(We shall employ subscripts when discussing more than one signal. Otherwise they are optional.) Any random sequence that satisfies Eqs. (7.3.4) and (7.3.5) and the further condition

$$r_{uu}(0) = E[u^2(k)] < \infty, \tag{7.3.6}$$

is called a *wide sense stationary* (WSS) random sequence. (It's "wide sense" because we have required stationarity only for first- and second-order moments.) To make matters even simpler, we shall also take μ to be zero.

$$\mu = E[u(k)] = 0, \tag{7.3.7}$$

(i.e., $\mathbf{u}$ is a "zero-mean" random sequence). Thus all that is left of our abbreviated statistical description of $\mathbf{u}$ is the single sequence $\{r_{uu}(l)\}$ in Eq. (7.3.5). This is called the *autocorrelation sequence* of the zero-mean WSS signal $\mathbf{u}$.

Not every sequence can be an autocorrelation sequence. It must have the following two properties:

$$\text{Symmetry:} \quad r_{uu}(k) = r_{uu}(-k) \quad \text{for all } k, \tag{7.3.8}$$

$$\text{Positivity:} \quad \det \begin{bmatrix} r(0) & r(1) & \dots & r(n) \\ r(1) & r(0) & \dots & r(n-1) \\ & & \vdots & \\ r(n) & & & r(0) \end{bmatrix} \geqslant 0 \quad \text{for all } n \geqslant 0. \tag{7.3.9}$$

The second condition requires the Toeplitz matrix constructed from the autocorrelation sequence to be positive semidefinite for all $n \geqslant 0$ (see Appendix 7A and Problems 7.12 and 7.15).

The discrete time Fourier transform of the autocorrelation sequence (which is an equivalent amount of information) is also useful. By definition, the *power spectral density function* $S(\theta)$ for the zero mean WSS signal $\mathbf{u}$ is given by

$$\{r_{uu}(k)\} \xleftrightarrow{DTFT} S_{uu}(\theta) = \sum_{k=-\infty}^{\infty} r_{uu}(k) e^{-jk\theta}. \tag{7.3.10}$$

[It is customary to write $S(\theta)$ instead of $S(e^{j\theta})$ for power spectra.]

The Wiener-Khintchine theorem, which we shall discuss shortly, provides some intuition about what $S(\theta)$ measures. The symmetry and positivity conditions (7.3.8) and (7.3.9) have simple counterparts in the frequency domain. They are

$$\text{Symmetry:} \quad S(\theta) = S(-\theta), \tag{7.3.11}$$

$$\text{Positivity:} \quad S(\theta) \geqslant 0 \quad \text{for all } \theta. \tag{7.3.12}$$

White Noise

A WSS signal $\mathbf{v}$ that satisfies (7.3.13) is called a *white noise signal.*

$$\begin{aligned} &E[v(k)] = 0, \\ &r_{vv}(k) = \delta(k) \xleftrightarrow{DTFT} S_{vv}(\theta) = 1. \end{aligned} \tag{7.3.13}$$

It is called "white" because its spectral density function is constant, which is the meaning Sir Isaac Newton gave to white light. It clearly satisfies the symmetry and positivity conditions. We shall see that a white noise signal is the "canonical" test input for statistical models. So called "pseudo-random number generator" computer functions attempt to produce a sequence that is uncorrelated, although perhaps not zero mean and unit variance. Thus a random number generator together with a simple linear scaling operation can be used to simulate a white noise signal.

PASSING WHITE NOISE THROUGH LINEAR FILTERS. A key question, which we will shortly address, is "what happens when a WSS signal is passed through a linear filter $H(z)$?" It turns out that the output signal is also WSS. It is not difficult to show that if the input is zero mean, then so is the output. Therefore, the interesting problem becomes the computation of second moments, given the autocorrelation sequence of the input and the filter unit-pulse response. The situation is simplified when the input is a white noise signal, and we shall start with this case. Probably the most fundamental property of white noise inputs deals with *two* filters, however.

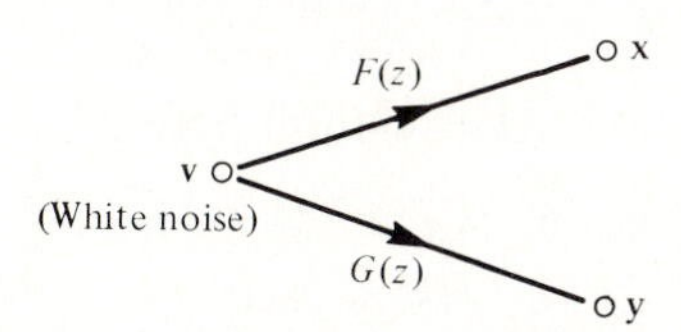

FIGURE 7.3.1 **Two linear filters driven by the same white noise source.**

Consider two linear filters driven by the same white noise input as shown in Fig. 7.3.1. How are the random variables $x(k)$ and $y(k)$ correlated? Let **f** and **g** be the unit pulse response sequences for the filters $F(z)$ and $G(z)$, respectively. We then have

$$\begin{aligned} E[x(k)y(k)] &= E\left[\sum_l \sum_j f(l)g(j)v(k-l)v(k-j)\right] \\ &= \sum_l \sum_j f(l)g(j)E[v(k-l)v(k-j)] \\ &= \sum_l \sum_j f(l)g(j)\delta(l-j) \\ &= \sum_j f(j)g(j). \end{aligned} \tag{7.3.14}$$

Grouping this together with the Parseval identity (4.2.13) gives a very useful set of relations:

$$E\,[x(k)y(k)] = \sum_l f(l)g(l) = \frac{1}{2\pi}\int_{-\pi}^{\pi} F(e^{j\theta})G^*(e^{j\theta})d\theta. \tag{7.3.15}$$

These three expressions represent three kinds of inner products, but since they are all equal we can regard them as three possible interpretations of the same thing. The first involves WSS signals; the second, time domain sequences; and the third, frequency domain functions. If, in statistical least-squares problems, we can construct a model for which two WSS signals of interest derive from the same white noise input, then we can study the relation of the two signals in three ways.

As an application of the construction in Fig. 7.3.1, let us compute all the second moments for the case of a single filter driven by white noise, as shown in Fig. 7.3.2. Since there are two WSS signals, we have three second moments to compute. These are

$$\begin{aligned} &\text{Input autocorrelation:} && r_{vv}(l) = E[v(k+l)v(k)], \\ &\text{Cross correlation between input and output:} && r_{sv}(l) = E[s(k+l)v(k)], \\ &\text{Output autocorrelation:} && r_{ss}(l) = E[s(k+l)s(k)]. \end{aligned} \tag{7.3.16}$$

By assumption, $\mathbf{v}$ is white and therefore

$$r_{vv}(l) = \delta(l) \quad \xleftrightarrow{DTFT} \quad S_{vv}(\theta) = 1. \tag{7.3.17}$$

To compute the cross correlation between $\mathbf{s}$ and $\mathbf{v}$, use Fig. 7.3.1 with

$$\begin{aligned} F(z) &= z^l H(z) && \Rightarrow && x(k) = s(k+l), \\ G(z) &= 1 && \Rightarrow && y(k) = v(k). \end{aligned}$$

Then Eq. (7.3.15) gives

$$r_{sv}(l) = E[s(k+l)v(k)] = \sum_k f(k)g(k) = \sum_k h(k+l)\delta(k) = h(l)$$

so that

$$r_{sv}(l) = h(l) \quad \xleftrightarrow{DTFT} \quad S_{sv}(\theta) = H(e^{j\theta}). \tag{7.3.18}$$

Thus, for white noise inputs the cross correlation between the input and output is the unit-pulse response sequence. This result has certain practical applications (see Problems 7.17 and 7.37).

To compute the output autocorrelation function, use Fig. 7.3.1 with

$$\begin{aligned} F(z) &= z^l H(z) && \Rightarrow && x(k) = s(k+l), \\ G(z) &= H(z) && \Rightarrow && y(k) = s(k). \end{aligned}$$

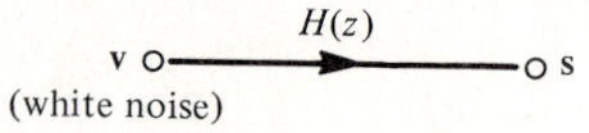

FIGURE 7.3.2 **A linear filter driven by white noise.**

Then Eq. (7.3.15) gives

$$r_{ss}(l) = E[s(k + l)s(k)] = \sum_k f(k)g(k) = \sum_k h(k + l)h(k). \tag{7.3.19}$$

Note that since the right-hand side of Eq. (7.3.19) is independent of k, the output signal is wide sense stationary. Using the second relation in Eq. (7.3.15), we have

$$\begin{aligned} r_{ss}(l) &= \frac{1}{2\pi}\int_{-\pi}^{\pi} [e^{j\theta l}H(e^{j\theta})][H^*(e^{j\theta})]\,d\theta \\ &= \frac{1}{2\pi}\int_{-\pi}^{\pi} |H(e^{j\theta})|^2 e^{j\theta l}\,d\theta. \end{aligned} \tag{7.3.20}$$

Since this is an obvious inverse DTFT, we have

$$r_{ss}(l) \quad \overset{DTFT}{\longleftrightarrow} \quad S_{ss}(\theta) = |H(e^{j\theta})|^2. \tag{7.3.21}$$

The output power spectral density function is the square of the magnitude of the frequency response function (for a white noise input).

EXAMPLE 7.3.1

The output power spectral density of a linear filter driven by white noise.

Suppose a white noise sequence **v** is used to excite the digital filter shown in Fig. 7.3.3. What is the output power spectral density function of the sequence **y**? From Eq. (7.3.21) we have $S_{yy}(\theta) = |H(e^{j\theta})|^2$. Thus we need only find $H(z)$ for this filter. From the SFG, we have

$$y(k) = v(k) - a_1 y(k-1) - a_2 y(k-2)$$

and so

$$H(z) = \frac{1}{1 + a_1 z^{-1} + a_2 z^{-2}}.$$

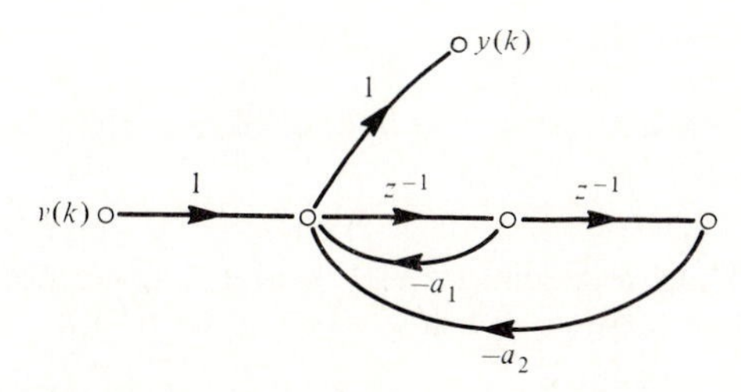

FIGURE 7.3.3 **A second-order digital filter.**

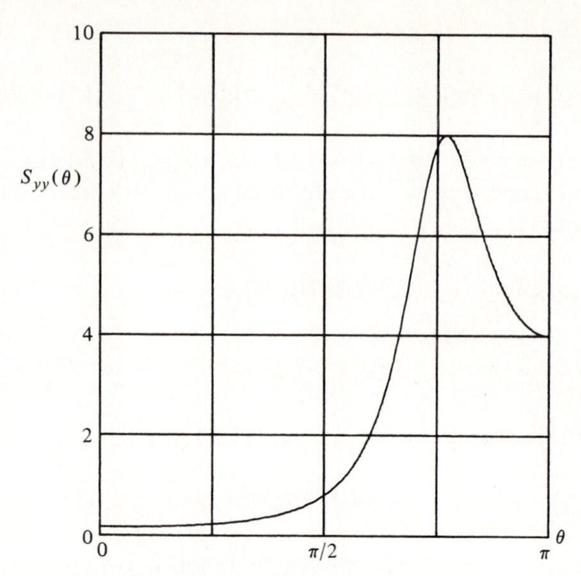

FIGURE 7.3.4 **The spectral density function $S_{yy}(\theta)$.**

The power spectral density function of **y** is thus

$$S_{yy}(\theta) = |H(e^{j\theta})|^2 = \left|\frac{1}{1 + a_1 e^{-j\theta} + a_2 e^{-2j\theta}}\right|^2.$$

Given the values of a_1 and a_2, we can use $S_{yy}(\theta)$ to find the distribution of power in **y** as a function of frequency.

For example, if $a_1 = 1$ and $a_2 = \frac{1}{2}$, then

$$S_{yy}(\theta) = \left|\frac{1}{1 + e^{-j\theta} + \dfrac{e^{-2j\theta}}{2}}\right|^2.$$

A plot of this function is shown in Fig. 7.3.4.

The Wiener-Khintchine Theorem

Suppose **u** is WSS, and **y** is the output of a linear filter **h**:

$$\mathbf{y} = \mathbf{h} * \mathbf{u}. \tag{7.3.22}$$

What is $\mathbf{S}_{yy}(\theta)$? As a special case, consider **u** generated by passing **v**, a white noise sequence, through a linear system with unit-pulse response **g**. Thus $\mathbf{u} = \mathbf{g} * \mathbf{v}$ and so we have

$$\mathbf{y} = \mathbf{h} * \mathbf{u} = \mathbf{h} * \mathbf{g} * \mathbf{v}. \tag{7.3.23}$$

This case is special because of the mechanism for generating **u**. Using Eq. (7.3.21) twice, we have

$$S_{uu}(\theta) = |G(e^{j\theta})|^2$$
$$S_{yy}(\theta) = |G(e^{j\theta})H(e^{j\theta})|^2,$$

which can be written as

$$S_{yy}(\theta) = S_{uu}(\theta)|H(e^{j\theta})|^2. \tag{7.3.24}$$

This relation can be derived without resorting to the construction in Eq. (7.3.23). In the general case, with **u** a WSS input signal, it is known as the *Wiener-Khintchine theorem* and provides some insight into what it is that the power spectral density actually measures. The variance of the WSS signal **y** is

$$r_{yy}(0) = E[y^2(k)] = \frac{1}{2\pi}\int_{-\pi}^{\pi} S_{yy}(\theta)d\theta \tag{7.3.25}$$

(using the inverse DTFT at $k = 0$). For *ergodic* sequences, one can estimate *expected values* by computing *time averages*. Thus the variance of an ergodic WSS signal is its *average power* (with probability one):

$$r_{yy}(0) = E[y^2(k)] = \lim_{N\to\infty} \frac{1}{2N+1} \sum_{k=-N}^{N} y^2(k). \tag{7.3.26}$$

Consider then the Wiener-Khintchine relation (7.3.24) in the case that $H(z)$ is a very narrow bandpass filter with bandwidth B. We conclude that the total output power is that portion of the input power that lies in the band B. But from Eqs. (7.3.24) and (7.3.25),

$$\text{Average power in bandwidth } B = \frac{1}{2\pi}\int_B S_{uu}(\theta)d\theta. \tag{7.3.27}$$

Thus the power spectral density function measures the average power content of the signal (as a function of frequency).

PASSING COLORED NOISE THROUGH LINEAR SYSTEMS. If one passes the white noise signal **v** through two filters $F(z)$ and $G(z)$, then the WSS outputs **x** and **y** will satisfy the basic inner product relations (7.3.15). Since we will need the results for Section 7.4, let us generalize this to the case of a colored input signal.

Let the input signal **u** in Fig. 7.3.5 be colored with autocorrelation

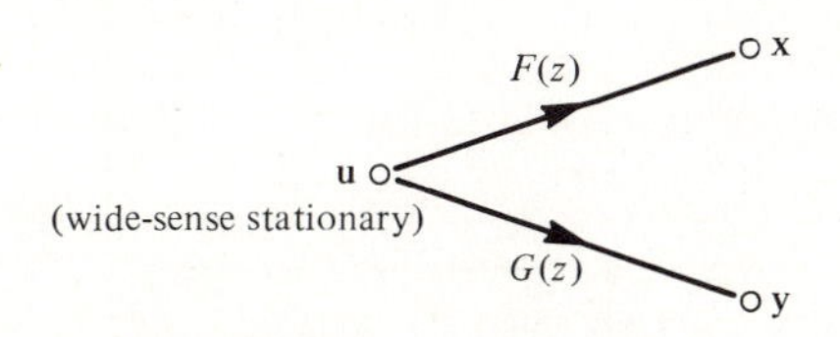

FIGURE 7.3.5 **A WSS signal driving two linear filters.**

sequence $\mathbf{r}_{uu}$. Since $\mathbf{x} = \mathbf{f} * \mathbf{u}$ and $\mathbf{y} = \mathbf{g} * \mathbf{u}$, we have

$$\begin{aligned} r_{xy}(l) = E[x(k+l)y(k)] &= \sum_m \sum_n f(m)g(n)E[u(k+l-m)u(k-n)] \\ &= \sum_m \sum_n f(m)g(n)r_{uu}(l-m+n). \end{aligned} \tag{7.3.28}$$

The DTFT of this cross correlation sequence is

$$\begin{aligned} S_{xy}(\theta) &= \sum_l r_{xy}(l)e^{-jl\theta} \\ &= \sum_l \sum_m \sum_n [f(m)e^{-jm\theta}][g(n)e^{jn\theta}][r_{uu}(l-m+n)e^{-j(l-m+n)\theta}] \\ &= F(e^{j\theta})[G(e^{j\theta})]^* S_{uu}(\theta). \end{aligned} \tag{7.3.29}$$

Note that if $F = G$, then $\mathbf{x} = \mathbf{y}$ and the result is the Wiener-Khintchine relation (7.3.24). Now set $l = 0$ in Eq. (7.3.28) to get the generalization of Eq. (7.3.15):

$$\begin{aligned} E[x(k)y(k)] &= \sum_m \sum_n f(m)g(n)r_{uu}(n-m) \\ &= \frac{1}{2\pi}\int_{-\pi}^{\pi} F(e^{j\theta})[G(e^{j\theta})]^* S_{uu}(\theta)d\theta. \end{aligned} \tag{7.3.30}$$

The Signal Modeling Problem (Spectral Factorization)

Let $\mathbf{y}$ be a WSS signal with power spectrum $S_{yy}(\theta)$. Can this signal be modeled as the output of a linear filter driven by white noise? That is, can we find $G(z)$ as depicted in Fig. 7.3.6? The frequency response function must satisfy $S_{yy}(\theta) = |G(e^{j\theta})|^2$. Now in one sense this is a trivial problem, since we can set

$$G(e^{j\theta}) = [S_{yy}(\theta)]^{1/2}. \tag{7.3.31}$$

This is possible provided only that $\mathbf{y}$ has finite total power

$$\frac{1}{2\pi}\int_{-\pi}^{\pi} S_{yy}(\theta)d\theta < \infty, \tag{7.3.32}$$

so that $\mathbf{g}$ has finite energy, i.e.,

$$\sum_{k=-\infty}^{\infty} |g(k)|^2 = \frac{1}{2\pi}\int_{-\pi}^{\pi} |G(e^{j\theta})|^2 d\theta < \infty. \tag{7.3.33}$$

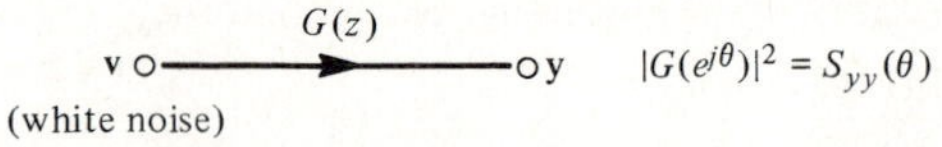

FIGURE 7.3.6.

One must be aware, however, that **g** may not be *causal*. Since $G(e^{j\theta})$ is real and even [if defined by Eq. (7.3.31)], then so is **g**. The answer is also clearly nonunique, since if $F(z)$ is any all-pass filter so that

$$|F(e^{j\theta})| = 1, \tag{7.3.34}$$

then

$$|G(e^{j\theta})|^2 = |G(e^{j\theta})F(e^{j\theta})|^2. \tag{7.3.35}$$

The question is made much more interesting if we require $G(z)$ to be causal. This question turns out to be critical for the solution of a great many least-squares problems. The answer was given by Szëgo (Grenander 1958).

THE SZËGO CONDITION. Suppose $S_{yy}(\theta)$ is real, even, nonnegative, and satisfies Eq. (7.3.32) and the condition

$$\exp\left[\frac{1}{2\pi}\int_{-\pi}^{\pi} \ln[S_{yy}(\theta)]\,d\theta\right] > 0. \tag{7.3.36}$$

Then there exists a causal finite energy sequence **g** with transform

$$G(z) = \sum_{k=0}^{\infty} g(k)z^{-k}, \tag{7.3.37}$$

satisfying $|G(e^{j\theta})|^2 = S_{yy}(\theta)$.

Obtaining such a $G(z)$ given $S_{yy}(\theta)$ is called the *spectral factorization problem.* If $S(\theta)$ is a rational function of $\cos(\theta)$, then a solution may be obtained by factoring the numerator and denominator polynomials after first replacing $\cos(n\theta)$ with

$$\cos(n\theta) = \left(\frac{z^n + z^{-n}}{2}\right)\bigg|_{z=e^{j\theta}}. \tag{7.3.38}$$

EXAMPLE 7.3.2

Spectral factorization.

Suppose $S(\theta)$ is given by

$$S(\theta) = \frac{5 + 4\cos 2\theta}{5 - 4\cos 2\theta} = \frac{5 + 2(z^2 + z^{-2})}{5 - 2(z^2 + z^{-2})}$$
$$= \frac{(2z^2 + 1)(2z^{-2} + 1)}{(2z^2 - 1)(2z^{-2} - 1)}. \tag{7.3.39}$$

We want to factor this as $G(z)G(z^{-1})$, where $G(z)$ has poles inside the circle. This will give us causality. Usually we also want $G(z)$ to be minimum phase, which means that we should also have the zeros inside the unit circle. For Eq. (7.3.39), take

$$G(z) = \frac{2z^2 + 1}{2z^2 - 1}. \tag{7.3.40}$$

7.4 STATISTICAL LEAST SQUARES

Statistical least-squares filter design problems arise when random signals must be considered. Typically the input $\mathbf{u}$ to the filter $H(z)$ that we must design is not the signal we would like to have. There is another signal $\mathbf{w}$ which is desired, but is not available for observation. For example, $\mathbf{u}$ might consist of *signal plus noise* and $\mathbf{w}$ would be the *signal only*. The idea is to design $H(z)$ so that the output $\hat{\mathbf{w}} = \mathbf{h} * \mathbf{u}$ approximates $\mathbf{w}$. This is reminiscent of the deterministic least-squares problem where the filter $G(z)$ was available but $F(z)$ was desired. In that problem, the goal was to design $H(z)$ so that $H(z)G(z)$ approximated $F(z)$. Despite the different goals of the statistical and deterministic problems, we will find that there is a formal equivalence between the two. In fact, for the FIR design problem, the normal equations are the same although the parameters $\mathbf{R}$ and $\mathbf{q}$ are given different meanings.

We will assume that the observed signal $\mathbf{u}$ and the desired signal $\mathbf{w}$ are jointly wide sense stationary, and have zero means. For a given filter $H(z)$, we form the WSS error signal $\mathbf{e}$ as follows:

$$\mathbf{e} = \mathbf{w} - \hat{\mathbf{w}} = \mathbf{w} - \mathbf{h} * \mathbf{u}. \tag{7.4.1}$$

The degree to which $\hat{\mathbf{w}}$ approximates $\mathbf{w}$ is measured by the variance of $e(k)$. This is the *mean squared error* $E[e^2(k)]$. This quantity is a quadratic function of the elements of the unit pulse response sequence $\mathbf{h}$, and gives rise to a least-squares problem. The parameters of the quadratic form are elements of the following cross- and autocorrelation sequences, which are assumed to be known.

$$q(l) \triangleq r_{wu}(l) = E[w(k + l)u(k)] \tag{7.4.2}$$

$$r(l) \triangleq r_{uu}(l) = E[u(k + l)u(k)] \tag{7.4.3}$$

The autocorrelation sequence $r_{ww}(k)$ may or may not be known. It is not needed to characterize the least-squares filter $H(z)$. We need the variance of $w(k)$, however, to compute the mean squared error. This is

$$\rho \triangleq \sigma_w^2 = E[w^2(k)]. \tag{7.4.4}$$

There are constraints on the unit pulse response sequence $\mathbf{h}$ which are related to the amount of data that is available at time k. We may be in possession of all or only part of the signal **u** at the time $w(k)$ is to be computed. The pulse response is constrained to be zero for those values which would correspond to unknown values of $u(k)$.

The estimate equation is

$$\hat{w}(k) = \sum_l h(l)u(k-l). \tag{7.4.5}$$

We shall treat three cases:

Noncausal case: $u(k-l)$ is known for all l, therefore no constraints on **h**.

FIR case: $u(k-l)$ is known for $0 \leqslant l \leqslant n$, so $h(l) = 0$ for $l < 0$ or $l > n$.

Causal IIR case: $u(k-l)$ is known for $l \geqslant 0$, so $h(l) = 0$ for $l < 0$.

We can now state the *statistical least-squares design problem:*

Given $\rho, \{r(k), q(k)\}$ of Eqs. (7.4.2), (7.4.3), (7.4.4),

Minimize

$$\begin{aligned} V(\mathbf{h}) &= E[w(k) - \hat{w}(k)]^2 \\ &= E\left[w(k) - \sum_l h(l)u(k-l)\right]^2 \end{aligned} \tag{7.4.6}$$

Subject to one of the following sets of constraints:

$$\|\mathbf{h}\|^2 < \infty \quad \text{(noncausal case)}, \tag{7.4.7}$$

$$h(l) = 0 \text{ for } l < 0 \text{ and } l > n \quad \text{(FIR case)}, \tag{7.4.8}$$

$$\|\mathbf{h}\|^2 < \infty \text{ and } h(l) = 0 \text{ for } l < 0 \quad \text{(causal IIR case)}. \tag{7.4.9}$$

The function $V(\mathbf{h}) = E[e^2(k)]$ is quadratic in the elements of the sequence **h**, and can therefore be put into the standard form of Eq. (7.1.3). Using the error definition Eq. (7.4.1),

$$\begin{aligned} V(\mathbf{h}) = E[e^2(k)] &= E[w(k) - \sum_m h(m)u(k-m)]^2 \\ &= E[w^2(k)] - 2\sum_m h(m)\, E[w(k)u(k-m)] \\ &\quad + \sum_m \sum_l h(l)h(m)\, E[u(k-l)u(k-m)]. \end{aligned}$$

This is quadratic with constant term

$$E[w^2(k)] = \rho, \tag{7.4.10}$$

linear term coefficients

$$E[w(k)u(k-m)] = E[w(k+m)u(k)] = q(m), \tag{7.4.11}$$

and quadratic term coefficients

$$E[u(k-l)u(k-m)] = r(m-l). \tag{7.4.12}$$

Thus the known quantities in Eqs. (7.4.2), (7.4.3), and (7.4.4) parameterize the quadratic form $V(\mathbf{h})$. Notice that the matrix is Toeplitz, as it was in the deterministic least-squares problem.

It is helpful to use the Wiener-Khintchine relation and its generalizations in Eqs. (7.3.28) through (7.3.30) to provide a frequency domain formula for the error function V. This is summarized in Table 7.4.1. The goal is to express everything in terms of the sequences $\boldsymbol{\rho}$, $\mathbf{q}$, $\mathbf{r}$, or their discrete time Fourier transforms. This is done step by step. The bottom line contains the autocorrelation sequence and power spectrum of the error signal $\mathbf{e}$. The mean square error, of course, is $V(\mathbf{h}) = r_{ee}(0)$. The error power spectrum $S_{ee}(\theta)$ is quadratic in $H(e^{j\theta})$. Since this is a scalar quantity (for each value of θ), one can easily complete the square in this quadratic form and obtain the representation on the last line. It is tempting, while gazing at this relation, to conclude that the solution to the problem of minimizing V is to set

$$H(e^{j\theta}) = \begin{cases} \dfrac{S_{wu}(\theta)}{S_{uu}(\theta)}, & S_{uu}(\theta) > 0 \\ 0, & \text{otherwise.} \end{cases} \tag{7.4.13}$$

This choice of $H(z)$ minimizes the error spectrum pointwise, i.e., at each value of θ. Therefore, it will also minimize V, if there are no constraints on H. But

TABLE 7.4.1 A summary of relations for the statistical least-squares problem, leading to the autocorrelation sequence and power spectrum of the error signal.

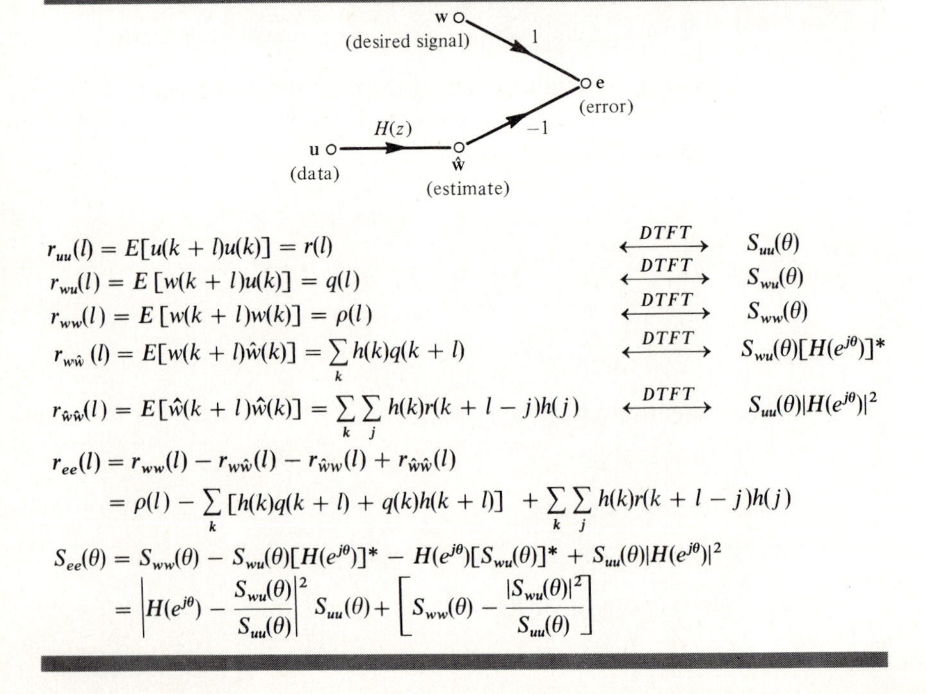

$$r_{uu}(l) = E[u(k+l)u(k)] = r(l) \quad \xleftrightarrow{DTFT} \quad S_{uu}(\theta)$$

$$r_{wu}(l) = E[w(k+l)u(k)] = q(l) \quad \xleftrightarrow{DTFT} \quad S_{wu}(\theta)$$

$$r_{ww}(l) = E[w(k+l)w(k)] = \rho(l) \quad \xleftrightarrow{DTFT} \quad S_{ww}(\theta)$$

$$r_{w\hat{w}}(l) = E[w(k+l)\hat{w}(k)] = \sum_k h(k)q(k+l) \quad \xleftrightarrow{DTFT} \quad S_{wu}(\theta)[H(e^{j\theta})]^*$$

$$r_{\hat{w}\hat{w}}(l) = E[\hat{w}(k+l)\hat{w}(k)] = \sum_k \sum_j h(k)r(k+l-j)h(j) \quad \xleftrightarrow{DTFT} \quad S_{uu}(\theta)|H(e^{j\theta})|^2$$

$$\begin{aligned} r_{ee}(l) &= r_{ww}(l) - r_{w\hat{w}}(l) - r_{\hat{w}w}(l) + r_{\hat{w}\hat{w}}(l) \\ &= \rho(l) - \sum_k [h(k)q(k+l) + q(k)h(k+l)] + \sum_k \sum_j h(k)r(k+l-j)h(j) \end{aligned}$$

$$\begin{aligned} S_{ee}(\theta) &= S_{ww}(\theta) - S_{wu}(\theta)[H(e^{j\theta})]^* - H(e^{j\theta})[S_{wu}(\theta)]^* + S_{uu}(\theta)|H(e^{j\theta})|^2 \\ &= \left| H(e^{j\theta}) - \frac{S_{wu}(\theta)}{S_{uu}(\theta)} \right|^2 S_{uu}(\theta) + \left[S_{ww}(\theta) - \frac{|S_{wu}(\theta)|^2}{S_{uu}(\theta)} \right] \end{aligned}$$

there's the rub. The nontrivial constraints in the FIR and causal IIR problems do not allow the filter in Eq. (7.4.13) in general. For the noncausal case, however, that is the solution. Notice that the bracketed term in the error power spectrum is independent of $H(z)$. Therefore, this part of the error cannot be reduced. The integral of this term will be a lower bound on the error variance.

Another approach is to attack the FIR problem (which is a finite quadratic form). Since $V(\mathbf{h})$ has exactly the same form as the deterministic version of Eq. (7.2.8), the normal equations (7.2.12) characterize the solution. The causal IIR case can then be viewed as the limiting case of a sequence of FIR problems of increasing order. We will use the method of spectral factorization to build a model for the signals involved in the least-squares problem, which will lead to an explicit solution.

The Noncausal Case

When all of the data $\mathbf{u}$ is available, one can use a noncausal filter $\mathbf{h}$. In this case $H(z)$ in Eq. (7.4.13) minimizes $V(\mathbf{h})$ provided it exists as a finite energy function. A classic example is the filtering of a signal in additive noise. Let $\mathbf{s}$ and $\mathbf{n}$ be WSS signals with known autocorrelation functions called *signal* and *noise*, respectively. Assume the two signals are uncorrelated so that

$$E[s(k+l)n(k)] = 0 \quad \text{for all } k, l. \tag{7.4.14}$$

The observation or *data* is $\mathbf{u}$, which is the sum of $\mathbf{s}$ and $\mathbf{n}$. The desired output from $\mathbf{h}$ is $\mathbf{w} = \mathbf{s}$. Thus we have

$$\text{Data:} \qquad \mathbf{u} = \mathbf{s} + \mathbf{n}. \tag{7.4.15}$$

$$\text{Desired output:} \qquad \mathbf{w} = \mathbf{s}. \tag{7.4.16}$$

The parameters of the quadratic form $V(\mathbf{h})$ for these choices of $\mathbf{w}$ and $\mathbf{u}$ are:

$$\begin{aligned} q(l) = r_{wu}(l) &= E[w(k+l)u(k)] = E[s(k+l)[s(k)+n(k)]] \\ &= r_{ss}(l) \quad \xleftrightarrow{DTFT} \quad S_{wu}(\theta) = S_{ss}(\theta), \end{aligned} \tag{7.4.17}$$

$$\begin{aligned} r(l) = r_{uu}(l) &= E[[s(k+l)+n(k+l)][s(k)+n(k)]] \\ &= r_{ss}(l) + r_{nn}(l) \quad \xleftrightarrow{DTFT} \quad S_{uu}(\theta) = S_{ss}(\theta) + S_{nn}(\theta). \end{aligned} \tag{7.4.18}$$

From Eq. (7.4.13), the least mean square error filter is

$$H(e^{j\theta}) = \frac{S_{wu}(\theta)}{S_{uu}(\theta)} = \frac{S_{ss}(\theta)}{S_{ss}(\theta)+S_{nn}(\theta)} = \frac{1}{1+\left(\dfrac{S_{ss}(\theta)}{S_{nn}(\theta)}\right)^{-1}}. \tag{7.4.19}$$

Since this function is real and even, the corresponding unit pulse response $\mathbf{h}$ is also real and even (and thus noncausal). The form of the answer is intuitively satisfying. The frequency response function is close to unity when $S_{ss}(\theta)/S_{nn}(\theta)$ is large and is close to zero when this signal-to-noise ratio is small.

The minimum mean squared error is obtained by integrating the resulting error power spectrum, from Table 7.4.1.

$$\begin{aligned} V_{\min} &= \frac{1}{2\pi}\int_{-\pi}^{\pi}\left[S_{ww}(\theta) - \frac{|S_{wu}(\theta)|^2}{S_{uu}(\theta)}\right]d\theta \\ &= \frac{1}{2\pi}\int_{-\pi}^{\pi}\left[S_{ss}(\theta) - \frac{S_{ss}(\theta)^2}{S_{ss}(\theta)+S_{nn}(\theta)}\right]d\theta \\ &= \frac{1}{2\pi}\int_{-\pi}^{\pi}\left[\frac{S_{ss}(\theta)S_{nn}(\theta)}{S_{ss}(\theta)+S_{nn}(\theta)}\right]d\theta. \end{aligned} \tag{7.4.20}$$

The FIR Case

In the FIR case, we process the data $\mathbf{u}$ with an nth-order FIR filter. We seek $\mathbf{h}$ to minimize

$$V(\mathbf{h}) = E[e^2(k)] = \rho - 2\sum_{l=0}^{n} h(l)q(l) + \sum_{l=0}^{n}\sum_{j=0}^{n} h(l)h(j)r(l-j) \tag{7.4.21}$$

with $q(l) = r_{wu}(l)$. This quadratic form is identical to Eq. (7.2.8) for the deterministic problem. Thus the normal equations (7.2.12) characterize the minimizing $\mathbf{h}$ [or $H(z)$]. We rewrite the normal equations here as

$$q(l) = r_{wu}(l) = \sum_{j=0}^{n} r(l-j)h(j), \qquad 0 \leqslant l \leqslant n. \tag{7.4.22}$$

The algorithm given in Fig. 7.2.2 can be used to compute the solution to these equations.

If $\mathbf{h}$ minimizes error variance, then the normal equations must hold. In the context of the statistical least-squares problem, some valuable intuition can be gained by expressing them in terms of the random signals involved. This may be done in a few lines as follows:

$$\begin{aligned} 0 &= q(l) - \sum_{j=0}^{n} r(l-j)h(j) \\ &= E\left\{\left[w(k) - \sum_{j=0}^{n} h(j)u(k-j)\right]u(k-l)\right\} \\ &= E\{[w(k) - \hat{w}(k)]u(k-l)\} \\ &= E[e(k)u(k-l)], \qquad 0 \leqslant l \leqslant n. \end{aligned} \tag{7.4.23}$$

The error $e(k)$ is uncorrelated with each of the random variables $u(k)$, $u(k-1), \ldots, u(k-n)$ which are used to form the estimate

$$\hat{w}(k) = \sum_{j=0}^{n} h(j)u(k-j). \tag{7.4.24}$$

Calling this set of input values the "data," we may interpret Eq. (7.4.23) as requiring that the *error be uncorrelated with the data.* This is sometimes called the *orthogonality principle*. The filter $H(z)$ must combine the data in such a way that this is true, or else the error variance has not been minimized.

EXAMPLE 7.4.1

An FIR filter for LMS estimation of a signal in noise.

Consider an FIR version of the signal in additive noise problem. We are given the autocorrelation functions of the signal **s** and the noise **n** and the fact that **s** and **n** are uncorrelated, i.e., $r_{sn}(l) = 0$ for all l. The input is **u** and the desired output is **w** where

$$\text{Data:} \qquad u(k) = s(k) + n(k) \tag{7.4.25}$$

$$\text{Desired output:} \quad w(k) = s(k - L) \tag{7.4.26}$$

and $n = 2L$. The filter must form its estimate of **s** on $2L + 1$ values of **u**. Notice that we have intentionally specified a delay in the desired output **w**. This places the desired output in the center of the data interval and leads to a linear phase filter $H(z)$.

To set up the normal equations, we need the parameters $r_{uu}(l)$ and $q(l) = r_{wu}(l)$.

$$r(l) = r_{uu}(l) = r_{ss}(l) + r_{nn}(l), \tag{7.4.27}$$

$$\begin{aligned} q(l) = r_{wu}(l) &= E[w(k + l)u(k)] = E[s(k + l - L)][s(k) + n(k)] \\ &= r_{ss}(l - L). \end{aligned} \tag{7.4.28}$$

The matrix in Eq. (7.4.27) is the sum of two Toeplitz matrices. Using the notation

$$\mathbf{R}_{xx}(n) \triangleq \begin{bmatrix} r_{xx}(0) & \cdots & r_{xx}(n) \\ \vdots & & \vdots \\ r_{xx}(n) & \cdots & r_{xx}(0) \end{bmatrix},$$

we can express the normal equations (7.4.22) in the form

$$[\mathbf{R}_{ss}(2L) + \mathbf{R}_{nn}(2L)] \begin{bmatrix} h(0) \\ \vdots \\ h(L) \\ \vdots \\ h(2L) \end{bmatrix} = \mathbf{R}_{ss}(2L) \begin{bmatrix} 0 \\ \vdots \\ 0 \\ 1 \\ 0 \\ \vdots \\ 0 \end{bmatrix} \begin{matrix} \left.\begin{matrix} \\ \\ \\ \end{matrix}\right\} L \text{ zeros} \\ \\ \left.\begin{matrix} \\ \\ \\ \end{matrix}\right\} L \text{ zeros} \end{matrix} . \tag{7.4.29}$$

This equation has a form similar to the noncausal filter Eq. (7.4.19), which is repeated below:

$$[S_{ss}(\theta) + S_{nn}(\theta)]H(e^{j\theta}) = S_{ss}(\theta). \tag{7.4.30}$$

The solution to Eq. (7.4.29) is a linear phase FIR filter. To see this, we express the linear phase condition $h(k) = h(2L - k)$, $k = 0, 1, \ldots, 2L$, in the form

$$\mathbf{h} = \mathbf{J}\mathbf{h}, \tag{7.4.31}$$

where $\mathbf{J}$ is the matrix

$$\mathbf{J} = \begin{bmatrix} \bigcirc & & 1 \\ & \diagup & \\ 1 & & \bigcirc \end{bmatrix}. \tag{7.4.32}$$

Now let $\mathbf{p}$ be the column vector on the right-hand side of Eq. (7.4.29). By inspection

$$\mathbf{p} = \mathbf{J}\,\mathbf{p}. \tag{7.4.33}$$

If $\mathbf{R}$ is Toeplitz, then it is not difficult to show that $\mathbf{J}$ commutes with $\mathbf{R}$, *i.e.*,

$$\mathbf{J}\mathbf{R} = \mathbf{R}\mathbf{J}. \tag{7.4.34}$$

Now suppose that

$$(\mathbf{R}_{ss} + \mathbf{R}_{nn})\mathbf{h} = \mathbf{R}_{ss}\mathbf{p}. \tag{7.4.35}$$

Then

$$\begin{aligned} (\mathbf{R}_{ss} + \mathbf{R}_{nn})\mathbf{J}\mathbf{h} &= [\mathbf{J}\mathbf{R}_{ss} + \mathbf{J}\mathbf{R}_{nn}]\mathbf{h} \\ &= \mathbf{J}\,\mathbf{R}_{ss}\mathbf{p} \\ &= \mathbf{R}_{ss}\mathbf{J}\,\mathbf{p} = \mathbf{R}_{ss}\mathbf{p}. \end{aligned} \tag{7.4.36}$$

Thus if $\mathbf{h}$ satisfies the normal equations (7.4.29), so does $\mathbf{J}\mathbf{h}$. Since the solution to the normal equations is unique when $\mathbf{R}_{ss} + \mathbf{R}_{nn}$ is positive definite, it follows that $\mathbf{h} = \mathbf{J}\mathbf{h}$ and $H(z)$ is linear phase, i.e.,

$$z^{L}H(z) = z^{-L}H(z^{-1}). \tag{7.4.37}$$

In Farden (1974) a method of designing high-order, linear phase, FIR filters with specified stop- and pass-bands is presented that uses this approach. One "designs" the signal and noise spectra, placing the signal power in the pass-band and the noise power in the stop-band.

EXAMPLE 7.4.2

A correspondence between deterministic and statistical least-squares problems.

One can generate an exact analog to the deterministic least-squares problem (7.2.2) using Eq. (7.3.15). Suppose we generate the signals $\mathbf{u}$ and $\mathbf{w}$ as in Fig. 7.4.1. The signal $\mathbf{v}$ is white noise. The filter $G(z)$ is causal but $F(z)$ need not be. Using Eq. (7.3.16), we have direct analogs between the meanings of the parameters $\mathbf{r}$ and $\mathbf{q}$, which appear in the two sets of normal equations (7.2.12) and (7.4.22), namely,

$$q(l) = r_{wu}(l) = E[w(k+l)u(k)] = \sum_{k=0}^{\infty} f(k+l)g(k), \tag{7.4.38}$$

$$r(l) = r_{uu}(l) = E[u(k+l)u(k)] = \sum_{k=0}^{\infty} g(k+l)g(k), \tag{7.4.39}$$

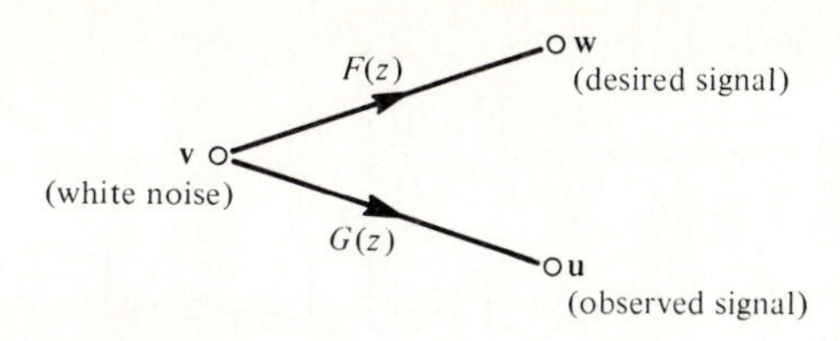

FIGURE 7.4.1 **Generation of u and w.**

$$\rho = r_{ww}(0) = E[w^2(k)] = \sum_{k=-\infty}^{\infty} f^2(k). \tag{7.4.40}$$

Therefore, the quadratic forms for the two problems are identical.

$$V(\mathbf{h}) = \|\mathbf{f} - \mathbf{g} * \mathbf{h}\|^2 = E[(w(k) - \hat{w}(k))^2]. \tag{7.4.41}$$

Using the construction in Fig. 7.4.1, one can generate a statistical least-squares problem from a deterministic problem. The problem descriptions are related as follows:

$$\begin{array}{lcl} F(z),\ G(z) & \rightarrow & \begin{array}{l} S_{wu}(\theta) = F(e^{j\theta})G^*(e^{j\theta}) \\ S_{uu}(\theta) = |G(e^{j\theta})|^2 \end{array} \\ \textit{Deterministic problem} & & \textit{Statistical problem} \end{array} \tag{7.4.42}$$

It is more useful (and more difficult) to reverse this situation. Can we find $F(z)$ and $G(z)$ given the power density functions $S_{wu}(\theta)$ and $S_{uu}(\theta)$? This is called the *modeling* problem and is the basis for the Wiener-Hopf theory.

EXAMPLE 7.4.3

A one-step FIR prediction filter.

A least-squares problem which is very important in the context of parametric power spectrum estimation is the *one-step prediction* problem. Let $\mathbf{y}$ be an observable WSS signal with autocorrelation $\mathbf{r}_{yy}$ and spectral density $S_{yy}(\theta)$. The problem is to design an FIR filter to predict $y(k+1)$ given the present and n previous values of $\mathbf{y}$. Thus the data is $\{y(k), y(k-1), \ldots y(k-n)\}$ and the desired signal is $y(k+1)$:

$$\begin{array}{ll} \text{Desired output:} & w(k) = y(k+1). \\ \text{Observed input:} & u(k) = y(k). \end{array} \tag{7.4.43}$$

The parameters of the normal equations are

$$r(l) = r_{uu}(l) = E[y(k+l)y(k)] = r_{yy}(l), \tag{7.4.44}$$

$$q(l) = r_{wu}(l) = E[y(k+l+1)y(k)] = r_{yy}(l+1). \tag{7.4.45}$$

Thus the normal equations for the one-step FIR predictor are

$$\begin{bmatrix} r(0) & r(1) & \cdots & r(n) \\ r(1) & r(0) & \cdots & r(n-1) \\ & \vdots & & \\ r(n) & & \cdots & r(0) \end{bmatrix} \begin{bmatrix} h(0) \\ h(1) \\ \vdots \\ h(n) \end{bmatrix} = \begin{bmatrix} r(1) \\ r(2) \\ \vdots \\ r(n+1) \end{bmatrix}. \tag{7.4.46}$$

We can build a model for this prediction problem also. Suppose $\mathbf{y}$ is generated by passing white noise $\mathbf{v}$ through a minimum phase filter $G(z)$. Then we have the situation depicted in Fig. 7.4.2. The choice of $F(z)$ is governed by the relation of $y(k+1)$ to $y(k)$. Notice that even when $G(z)$ is causal, $F(z)$ is not causal because $f(-1) = g(0) \neq 0$. The squared error is given by

$$V(\mathbf{h}) = \|\mathbf{f} - \mathbf{g} * \mathbf{h}\|^2. \tag{7.4.47}$$

Thus we want $F(z) - G(z)H(z)$ to approximate zero. For this problem

$$F(z) - G(z)H(z) = G(z)[z - H(z)]. \tag{7.4.48}$$

If $V(\mathbf{h}) = 0$, then we would have $H(z) = z$, which is the noncausal prediction filter and is not, of course, permitted.

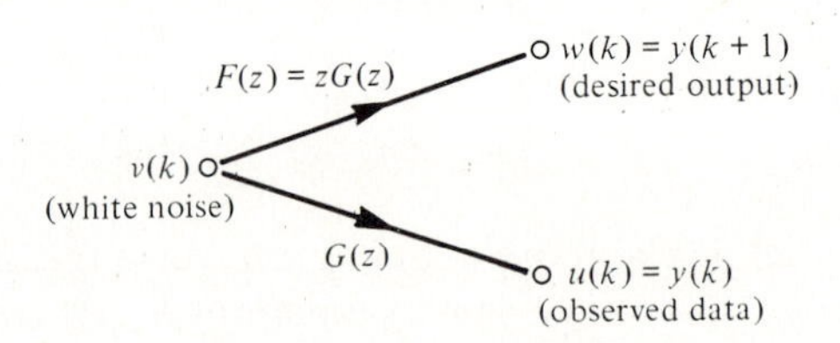

FIGURE 7.4.2 **A model for one-step prediction.**

In Section 7.2 we related the whitening filter and the deterministic least-squares filter. (See also Problems 7.4 and 7.23.) There is also a close connection between the one-step prediction and whitening problems, which can be seen by comparing the normal equations (7.2.33) and (7.4.46). From these two sets of normal equations we make the correspondence

$$A(z) = 1 - z^{-1}H(z). \tag{7.4.49}$$

Then

$$A(z)G(z) = z^{-1}[F(z) - G(z)H(z)], \tag{7.4.50}$$

with $F(z) = zG(z)$. Since the z^{-1} on the right-hand side is merely a unit delay operator, we have

$$\|\mathbf{a} * \mathbf{g}\|^2 = \|\mathbf{f} - \mathbf{g} * \mathbf{h}\|^2. \tag{7.4.51}$$

Therefore, solutions to the FIR prediction problem $\mathbf{h}$ and the whitening problem **a** are related by the following pair of equations:

$$\begin{aligned} A_{n+1}(z) &= 1 - z^{-1}H_n(z), \\ H_n(z) &= z[1 - A_{n+1}(z)]. \end{aligned} \tag{7.4.52}$$

The subscripts in Eq. (7.4.52) denote the order of the FIR filter in each case.

EXAMPLE 7.4.4

An nth-order FIR prediction filter.

Suppose we take as an example the case of a sequence **y** generated by passing white noise **v** through a one-pole filter $G(z) = 1/(1 - \lambda z^{-1})$. Then we have the power spectral density

$$S_{yy}(\theta) = |G(e^{j\theta})|^2 = \frac{1}{1 - 2\lambda \cos\theta + \lambda^2}.$$

The corresponding nth-order, FIR whitening filter (from Table 7.2.1) is

$$A_n(z) = 1 - \lambda z^{-1}, \qquad n > 0.$$

And so the least-squares prediction nth-order filter is

$$H_n(z) = z[1 - A_{n+1}(z)] = \lambda, \qquad n \geqslant 0.$$

In this case, the prediction filter is no more than a constant. Is this a reasonable answer?

Refer back to the model. The generating process is white noise through the single pole filter $G(z)$. Therefore

$$y(k+1) = \lambda y(k) + v(k).$$

The prediction filter equation is merely

$$\hat{w}(k) = \hat{y}(k+1) = \lambda y(k).$$

Thus the error between the true value and the estimate is

$$e(k) = y(k+1) - \hat{y}(k+1) = v(k).$$

The error sequence is the white noise input sequence. Since white noise is unpredictable, the simple predictor $\hat{y}(k+1) = \lambda y(k)$ is, in fact, reasonable.

The Causal IIR Case

Consider now the least-squares problem with $H(z)$ required to be causal but allowed to have an infinite unit-pulse response. This problem may be regarded as the limiting case of the FIR problem as the order n is allowed to approach infinity.

We therefore hypothesize that the necessary conditions for the IIR case are the normal equations of the form

$$\sum_{k=0}^{\infty} r_{uu}(l-k)h(k) = r_{wu}(l) = q(l), \qquad 0 \leqslant l < \infty \tag{7.4.53}$$

or, equivalently,

$$E[e(k)u(k-l)] = 0, \qquad 0 \leqslant l < \infty. \tag{7.4.54}$$

In this context Eq. (7.4.53) is called the *discrete-time Wiener-Hopf* (*WH*) *equation*. Equation (7.4.54) is called the *orthogonality condition*, since it requires the data $\{u(j)\}$, $j \leqslant k$, to be orthogonal (uncorrelated) to the error $e(k) = w(k) - \hat{w}(k)$.

The sufficiency of the normal equations (7.4.53) is established by "completing the square." This can be viewed geometrically. In Fig. 7.4.3, the space spanned by the data is depicted as a plane. Each point in the plane represents a particular linear combination of the data, or equivalently, a filter output (since all filter outputs are merely linear combinations of the data). The desired output $\mathbf{w}$ cannot, in general, be obtained as a linear combination of the data, and thus is a point off the plane. Consider two filters, $\mathbf{h}$ and $\mathbf{h}'$, and assume that $\mathbf{h}$ satisfies the orthogonality condition (7.4.54). The error vector $e(k)$ is then orthogonal to the data plane. The Pythogorean theorem for the right triangle with sides $e(k)$, $e'(k)$, and $x(k) = e'(k) - e(k)$ yields

$$E[e'(k)^2] = E[e^2(k)] + E[x^2(k)] \geqslant E[e^2(k)]. \tag{7.4.55}$$

Thus the error variance for the filter $\mathbf{h}'$ (which is arbitrary) is greater than or equal to the variance for $\mathbf{h}$ [which satisfies Eq. (7.4.54)]. Thus any filter satisfying Eq. (7.4.53) or (7.4.54) has minimum error variance. Equation (7.4.55) follows directly from the fact that

$$E[e'(k)^2] = E[e(k) + x(k)]^2 = E[e^2(k)] + 2E[e(k)x(k)] + E[x^2(k)].$$

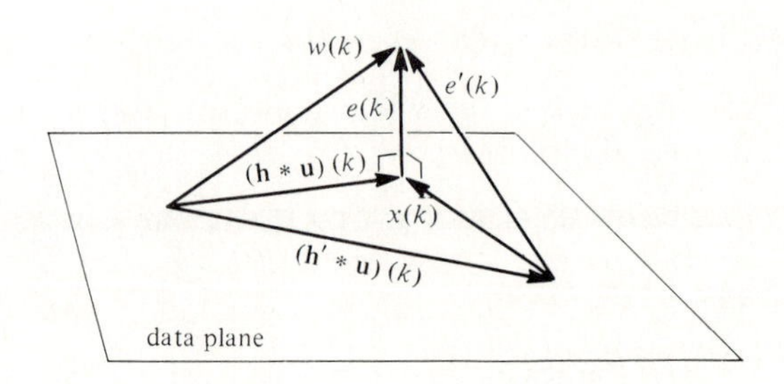

FIGURE 7.4.3 **A geometric analogy for the statistical least-squares problem.**

But $\mathbf{x}$ is a linear combination of the data since it lies in the data plane. Thus $E[e(k)x(k)] = 0$ by Eq. (7.4.54).

The problem is to find a solution for the WH equations (7.4.53). It is tempting to identify the left-hand side as the convolution of two sequences and apply the DTFT. This, however, assumes we have an equality between sequences and requires that Eq. (7.4.53) hold for all values of l, both positive and negative. Since this is not the case, we cannot proceed in this way.

One method of solving the WH equations begins by building a model for the generation of the signals $\mathbf{u}$ and $\mathbf{w}$, as was suggested in Example 7.4.2, as shown in Fig. 7.4.4. This device will leave us with an equivalent deterministic least-squares problem. The key step in the construction is the use of spectral factorization.

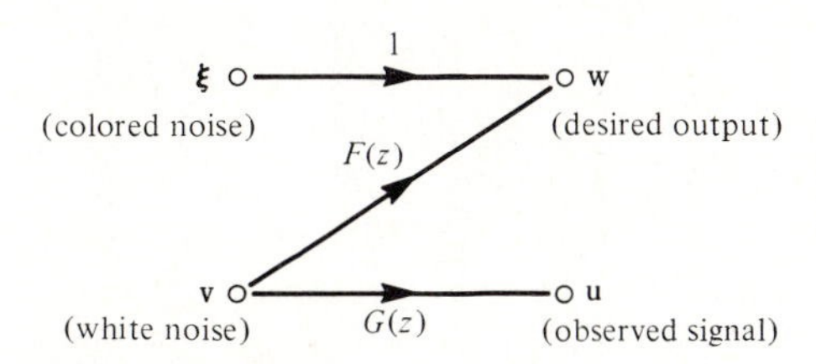

FIGURE 7.4.4 **A model for the IIR case.**

We are given $S_{uu}(\theta)$, $S_{wu}(\theta)$, and $S_{ww}(\theta)$. We seek $F(z)$, $G(z)$, and $S_{\xi\xi}(\theta)$ so that the model is consistent with those power spectra. The noise input spectra satisfy

$$\begin{aligned} S_{vv}(\theta) &= 1 \qquad \text{(white noise)}, \\ S_{v\xi}(\theta) &= 0 \qquad (v \text{ and } \xi \text{ are uncorrelated}). \end{aligned} \tag{7.4.56}$$

The *consistency conditions* are

$$S_{uu}(\theta) = |G(e^{j\theta})|^2, \tag{7.4.57}$$

$$S_{wu}(\theta) = F(e^{j\theta})[G(e^{j\theta})]^*, \tag{7.4.58}$$

$$S_{ww}(\theta) = |F(e^{j\theta})|^2 + S_{\xi\xi}(\theta). \tag{7.4.59}$$

The first two conditions are the same as in Example 7.4.2. The third amounts to a definition of the power spectrum for ξ, which accounts for that component of the desired signal $\mathbf{w}$ which is uncorrelated with the observed signal $\mathbf{u}$.

To construct the model, we must obtain $F(z)$, $G(z)$, and $S_{\xi\xi}(\theta)$. If $S_{uu}(\theta)$ satisfies the Szëgo condition (7.3.36), then we can find a causal, minimum phase, and finite energy $\mathbf{g}$ which satisfies Eq. (7.4.57). (We will discover shortly the importance of the minimum phase condition.) Once $G(z)$ has been obtained, we may use Eq. (7.4.58) to construct $F(z)$:

$$F(e^{j\theta}) = \frac{S_{wu}(\theta)}{[G(e^{j\theta})]^*}. \tag{7.4.60}$$

And from this it follows that

$$
\begin{aligned}
S_{\xi\xi}(\theta) &= S_{ww}(\theta) - |F(e^{j\theta})|^2 \\
&= S_{ww}(\theta) - \frac{|S_{wu}(\theta)|^2}{S_{uu}(\theta)}.
\end{aligned} \tag{7.4.61}
$$

This power spectrum is precisely that component of the error spectrum which is independent of $H(z)$ (see Table 7.4.1).

By virtue of the consistency conditions, we can write the error spectrum as

$$
S_{ee}(\theta) = |F(e^{j\theta}) - G(e^{j\theta})H(e^{j\theta})|^2 + S_{\xi\xi}(\theta). \tag{7.4.62}
$$

This is validated by using the relations (7.4.57), (7.4.60), and (7.4.61) to reduce $S_{ee}(\theta)$ to the form given in Table 7.4.1. Now integrate the error spectrum to obtain

$$
\begin{aligned}
V(\mathbf{h}) = E[e^2(k)] &= \frac{1}{2\pi}\int_{-\pi}^{\pi} S_{ee}(\theta)d\theta \\
&= \|\mathbf{f} - \mathbf{g} * \mathbf{h}\|^2 + V_0,
\end{aligned} \tag{7.4.63}
$$

where

$$
V_0 = E[\xi^2(k)]. \tag{7.4.64}
$$

We see that the minimization of $V(\mathbf{h})$ is equivalent to the minimization of $\|\mathbf{f} - \mathbf{g} * \mathbf{h}\|^2$. Thus we have constructed an equivalent deterministic least-squares problem.

We recall that if $G(z)$ is not minimum phase, then there will be a term [the second term in Eq. (7.2.54)] which contributes to $V_{\min}$. This is to be avoided. If $G(z)$ is minimum phase and causal and

$$
|G(e^{j\theta})|^2 \geqslant \sigma^2 \qquad \text{for all } \theta, \tag{7.4.65}
$$

(as will be the case if $\mathbf{u}$ has a white noise component with variance σ^2), then $1/G(z)$ is minimum phase and causal and

$$
|G(e^{j\theta})|^{-2} \leqslant 1/\sigma^2 \quad \Rightarrow \quad \mathbf{g}^{-1} \in l_2.
$$

We will make this assumption. Now referring back to Fig. 7.4.4, we can regard $\mathbf{u}$ and $\mathbf{v}$ as equivalent in the sense that one can obtain either signal from the other, with a stable causal filter [either $G(z)$ or $G^{-1}(z)$]. The signal $g(o)\mathbf{v}$ is called the *innovations* part of $\mathbf{u}$ (Kailath 1974). It is that part of $\mathbf{u}$ which is unpredictable from knowledge of the past history of $\mathbf{u}$. If we had chosen a nonminimum phase $G(z)$, then we could not have recovered $\mathbf{v}$ from $\mathbf{u}$.

The problem can now be solved in a manner similar to the one that was used in Section 7.2. Suppose we set

$$
H(z) = H_0(z)/G(z). \tag{7.4.66}
$$

Then the mean squared error in Eq. (7.4.63) becomes

$$\begin{aligned} V(\mathbf{h}) &= \|\mathbf{f} - \mathbf{h}_0\|^2 + V_0 \\ &= V_0 + \sum_k [f(k) - h_0(k)]^2. \end{aligned} \tag{7.4.67}$$

The solution to this problem is trivial. We set

$$h_0(k) = f_+(k) \triangleq \begin{cases} f(k), & k \geqslant 0. \\ 0, & k < 0 \end{cases}. \tag{7.4.68}$$

In other words, since **h** must be causal, we must set

$$H_0(z) = F_+(z), \tag{7.4.69}$$

the "causal part" of $F(z)$. The result is depicted in Fig. 7.4.5. Let us summarize this method of solving the Wiener-Hopf equations:

Given $S_{uu}(\theta)$, find the causal and minimum phase $G(z)$ satisfying

$$|G(e^{j\theta})|^2 = S_{uu}(\theta).$$

Set $F(e^{j\theta}) = S_{wu}(\theta)/[G(e^{j\theta})]^*$.

Then $H(z) = F_+(z)/G(z)$.

The mean squared error is [using Eq. (7.4.67)]

$$V_{\min} = V_0 + \sum_{k=-\infty}^{-1} f^2(k). \tag{7.4.70}$$

The error signal itself can be written

$$e(k) = w(k) - \hat{w}(k) = \xi(k) + \sum_{l=-\infty}^{-1} f(l)v(k-l). \tag{7.4.71}$$

The first part of the error is $\xi(k)$ having variance V_0. This is that part of the desired signal **w** which is uncorrelated with the observed signal **u** (*both past and future*). The second term is that part of **w** which is uncorrelated with the past history of **u**, but could be determined if the future of **u** were available. This component of the error corresponds to the noncausal part of $F(z)$. The second term in $V_{\min}$ [Eq. (7.4.70)] is the difference between the minimum mean square errors of the noncausal and causal IIR least-squares problems.

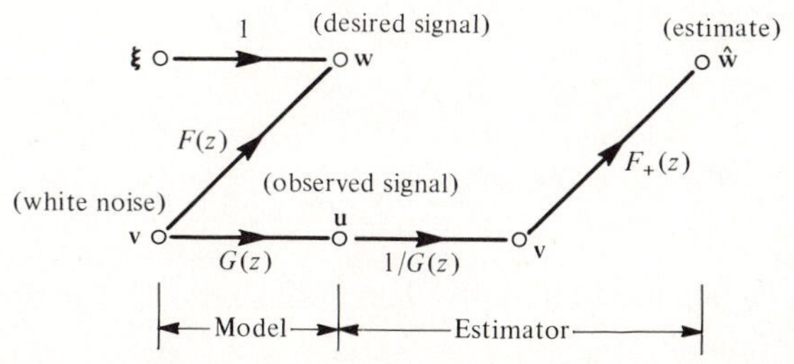

FIGURE 7.4.5 **The solution to the IIR least-squares causal filter problem.**

EXAMPLE 7.4.5

An IIR prediction filter.

Let **y** be a WSS signal with power spectral density function $S_{yy}(\theta)$. Consider the problem of predicting $y(k+m)$ based on $\{y(j)\}$, $j \leq k$. Let

$$\begin{aligned} &\text{Observed signal:} \quad u(k) = y(k). \\ &\text{Desired signal:} \quad w(k) = y(k+m). \end{aligned} \tag{7.4.72}$$

Suppose a causal, minimum phase filter $G(z)$ can be found for which

$$S_{yy}(\theta) = |G(e^{j\theta})|^2. \tag{7.4.73}$$

(This is the spectral factorization problem.) Then $w(k) = y(k+m)$ forces

$$F(z) = z^m G(z) \tag{7.4.74}$$

(see Fig. 7.4.5). The causal part of $F(z)$ is

$$\begin{aligned} F_+(z) &= \sum_{k=0}^{\infty} g(k+m) z^{-k} \\ &= z^m \left[G(z) - \sum_{k=0}^{m-1} g(k) z^{-k} \right]. \end{aligned} \tag{7.4.75}$$

Suppose, for example, that

$$S_{yy}(\theta) = \frac{5 + 4\cos 2\theta}{5 - 4\cos 2\theta}.$$

We must choose $G(z)$ having both poles and zeros inside $|z| = 1$ and satisfying

$$G(z)G(z^{-1}) = \frac{5 + 4\left(\dfrac{z^2 + z^{-2}}{2}\right)}{5 - 4\left(\dfrac{z^2 + z^{-2}}{2}\right)}.$$

The solution is

$$G(z) = \frac{z^2 + \frac{1}{2}}{z^2 - \frac{1}{2}},$$

and

$$H(z) = \frac{F_+(z)}{G(z)} = z^m \left[G(z) - \sum_{k=0}^{m-1} g(k) z^{-k} \right] \Big/ G(z).$$

For any m-step prediction problem of this kind, the prediction error variance is

$$E[y(k+m) - \hat{y}(k+m)]^2 = \sum_{k=0}^{m-1} g^2(k). \tag{7.4.76}$$

This is the variance of the second term in Eq. (7.4.71). For this example $S_{\xi\xi}(\theta) = 0$.

7.5 ADAPTIVE FILTERS

In many real-time applications, the second-order statistics for the signals **u** and **w** are not known. For example, when a telephone call is initiated, the transfer function of the telephone channel is unknown, and the connection may involve signal reflections that produce undesirable echoes. This problem can be corrected with a linear filter once the transfer function is measured. The measurement can be made, however, only after the system is running. Therefore, the correction filter must be designed by the system itself, and in a reasonably short time. These requirements impose new constraints and change the goal of the filter design problem.

The situation is depicted in Fig. 7.5.1. Let us reconsider the FIR statistical least-squares problem, with two changes. We do not know the second-order statistics

$$r(l) = E[u(k + l)u(k)], \tag{7.5.1}$$

$$q(l) = E[w(k + l)u(k)], \tag{7.5.2}$$

but we are allowed to measure the desired signal **w** (with a possible delay). The filter $H(z)$ is changed while the error

$$e(k) = w(k) - \hat{w}(k) \tag{7.5.3}$$

is not zero, in such a way as to reduce future error variance. Since the filter itself is changing with the data, it is actually nonlinear. If the adaptation is turned off, however, then the result is a linear, shift invariant filter whose input is **u** and whose output is $\hat{\mathbf{w}}$.

Assuming that the signals **u** and **w** are WSS, there is a minimum mean square error FIR filter. We cannot expect that the adaptation will find the least-squares filter, but must accept some degradation in performance as a consequence of the necessity to allow the system to adapt. Our goal is no longer to design the filter $H(z)$, it is to design the adaptation algorithm in hopes of reasonable performance. Furthermore, because the adaptation must be

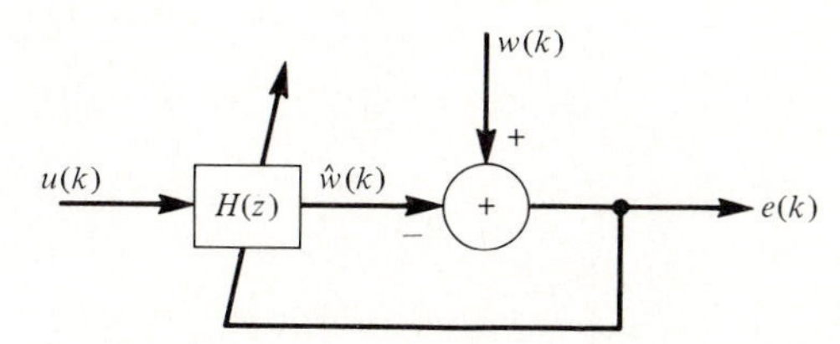

FIGURE 7.5.1 **An adaptive filter. (The transfer function is modified while the error is not zero.)**

done in real time, the computation involved in updating $H(z)$ must be kept to an absolute minimum.

The least-squares filter would minimize the quadratic form

$$V(\mathbf{h}) = \rho - 2\mathbf{q}^T\mathbf{h} + \mathbf{h}^T\mathbf{R}\,\mathbf{h} \tag{7.5.4}$$

and would satisfy the normal equations

$$(\tfrac{1}{2}\nabla V)(\mathbf{h}) = \mathbf{R}\,\mathbf{h} - \mathbf{q} = \mathbf{0}, \tag{7.5.5}$$

where ∇V is the *gradient* of V. Now if the second-order statistics $\mathbf{r}$, $\mathbf{q}$ are not known *a priori*, then one does not know the coefficients in the normal equation. It is possible to first estimate them and then solve the normal equations, but this is usually not feasible because of the time and computation involved.

The Widrow Algorithm

The Widrow, or "noisy gradient" algorithm (Widrow 1985) eliminates the estimation of the second-order statistics and the matrix inversion in the normal equation, and uses the data directly. It may be developed in two steps. First, rather than working with the normal equations (7.5.5), one applies a gradient-following procedure to the quadratic form (7.5.4). Second, the gradient is estimated in a very primitive fashion, using the current error and the previous n inputs. This estimate is called the "noisy gradient."

The level curves of a positive definite quadratic form are ellipsoidal, as shown in Fig. 7.5.2 (with $n = 1$). The gradient of $V(\mathbf{h})$ is the vector

$$\nabla V(\mathbf{h}) = \left[\frac{\partial V}{\partial h(0)}, \frac{\partial V}{\partial h(1)}, \ldots, \frac{\partial V}{\partial h(n)}\right]^T = 2(\mathbf{R}\,\mathbf{h} - \mathbf{q}). \tag{7.5.6}$$

This vector will always point in the direction of greatest increase for the function V and is an outward normal to a line (or hyperplane) tangent to the level surface. The gradient search procedure

$$\mathbf{h}' = \mathbf{h} - \frac{\mu}{2}\nabla V(\mathbf{h}) = \mathbf{h} - \mu(\mathbf{R}\,\mathbf{h} - \mathbf{q}) \tag{7.5.7}$$

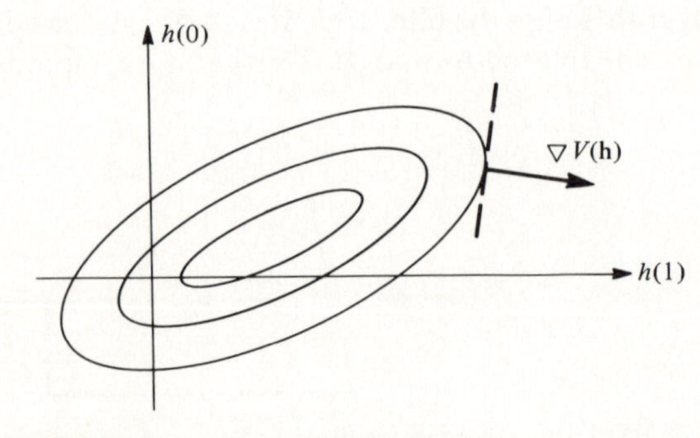

FIGURE 7.5.2 **A contour plot of the quadratic form $V(h)$. The level curves are ellipsoidal.**

moves the current choice of **h** in the direction of the negative gradient, which is the direction of steepest descent. The parameter μ is called the *step size.* Iteration of this update leads to the vector difference equation

$$\mathbf{h}(k+1) = (\mathbf{I} - \mu\mathbf{R})\mathbf{h}(k) + \mu\mathbf{q}, \tag{7.5.8}$$

which converges to

$$[\mathbf{I} - (\mathbf{I} - \mu\mathbf{R})]^{-1}\mu\mathbf{q} = \mathbf{R}^{-1}\mathbf{q}, \tag{7.5.9}$$

provided that all eigenvalues of $\mathbf{I} - \mu\mathbf{R}$ satisfy

$$|\lambda| = |1 - \mu\lambda'| < 1, \qquad (\lambda' \text{ an eigenvalue of } \mathbf{R}). \tag{7.5.10}$$

The matrix **R** is positive semidefinite and symmetric; consequently, all of its eigenvalues are nonnegative real. The requirement (7.5.10) imposes a stability condition on the step size, namely,

$$0 < \mu < \frac{2}{\lambda_{\max}}, \tag{7.5.11}$$

where $\lambda_{\max}$ is the largest eigenvalue of **R**.

The second step in the development of the Widrow algorithm is to estimate the gradient directly from the data, since the values of **r** and **q** are unknown. Let us expand the components of the vector equation (7.5.8).

$$\begin{aligned} h(k+1, l) &= h(k, l) - \mu\left(\sum_{j=0}^{n} r(l-j)h(k, j) - q(l)\right) \\ &= h(k, l) - \mu E\left[\sum_{j=0}^{n} [u(k-j)h(k, j) - w(k)]u(k-l)\right] \\ &= h(k, l) - \mu E[(\hat{w}(k) - w(k))u(k-l)] \\ &= h(k, l) + \mu E[e(k)u(k-l)]. \end{aligned} \tag{7.5.12}$$

The final expression in this development is simpler than we might have expected, and is also intuitively appealing. The orthogonality condition version of the normal equations says that the error must be uncorrelated with the data. If this is true, then the coefficients of the filter are left unchanged:

$$E[e(k)u(k-l)] = 0 \qquad \Rightarrow \qquad h(k+1, l) = h(k, l). \tag{7.5.13}$$

Equation (7.5.12) is the exact equivalent of the true gradient follower (7.5.7). It is made into the "noisy gradient" follower by the simple expedient of removing the expectation operator. The result is the *noisy gradient follower*:

$$h(k+1, l) = h(k, l) + \mu e(k)u(k-l). \tag{7.5.14}$$

This amount of computation is often feasible in real time. It uses only the current error $e(k)$, the current values of the unit-pulse response [$h(0)$ through $h(n)$], and the current data "in the filter" [$u(k)$ through $u(k-n)$].

How well does it work? More to the point, does it even work at all? Hard analytical answers to these questions have, to our knowledge, not been given.

However, a wealth of experimental evidence shows that it sometimes works very well. It seems to be the nature of this approach that one must work experimentally rather than theoretically. However, some intuition and rules of thumb can be obtained and we shall conclude this section with a short discussion of the issues of selection of step size and convergence.

The notion of "convergence" for adaptive filters has only intuitive meaning, except in special cases. If the input sequence is arbitrary, then the error can be driven to zero and held there only if

$$w(k) = \sum_{j=0}^{n} g(j)u(k-j), \tag{7.5.15}$$

which is to say that the desired signal $w(k)$ can be produced from the observable signal $u(k)$ using an nth-order FIR filter. If this is not the case, then even the least-squares filter will produce an error signal whose variance is greater than zero. Calling this minimum variance σ^2, the adaptive filter must generate an error sequence whose variance is greater than σ^2, no matter how much time is allowed for the algorithm to settle. And as long as the error is not zero, then the Widrow algorithm will continue to change the filter coefficients. Therefore, there can be no strict convergence. Nevertheless the error can often be made "small enough," with the filter coefficients moving randomly in the vicinity of those of the least-squares filter.

Even though strict convergence is not possible in general, the algorithm can diverge spectacularly. This occurs for too large a step size. Let us consider a very simple example.

EXAMPLE 7.5.1

A simple adaptive filter.

Let

$$w(k) = u(k-2) + \xi(k), \tag{7.5.16}$$

where ξ is a white noise signal with variance

$$E[\xi^2(k)] = 0.01. \tag{7.5.17}$$

For the moment, let $\mathbf{u}$ be another white noise signal with

$$E[u^2(k)] = 1. \tag{7.5.18}$$

The least-squares filter is easily shown to be

$$H(z) = z^{-2}, \tag{7.5.19}$$

which is FIR of order 2 and produces

$$\hat{w}(k) = u(k-2). \tag{7.5.20}$$

The error is then ξ, and the error variance is 0.01.

Figure 7.5.3 exhibits the results of a simulation of the Widrow algorithm for this example. The step size for this simulation was $\mu = 0.2$, and the order

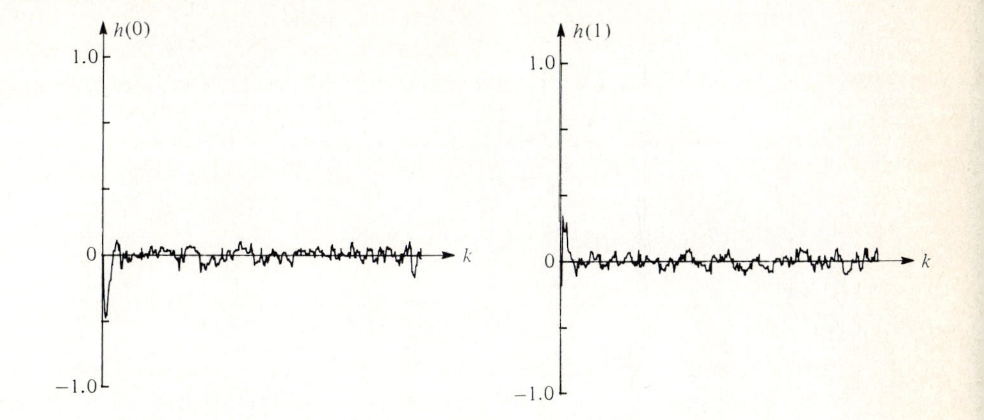

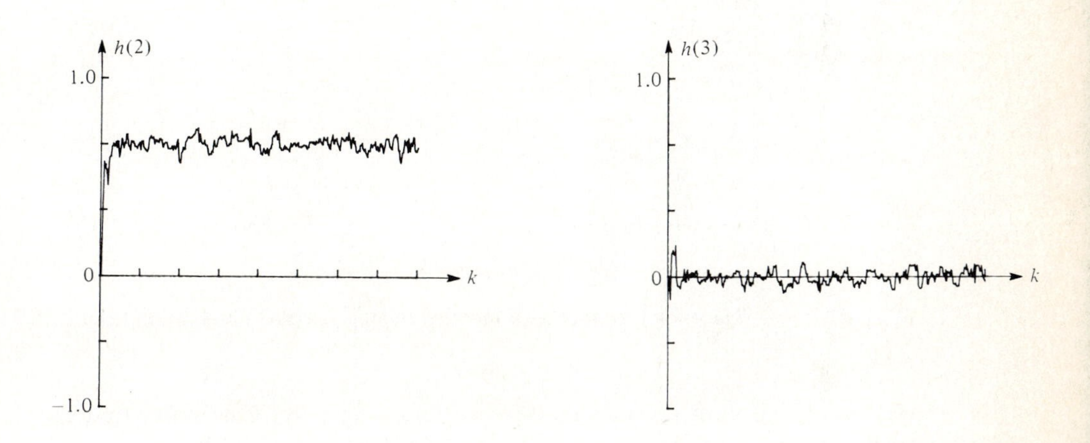

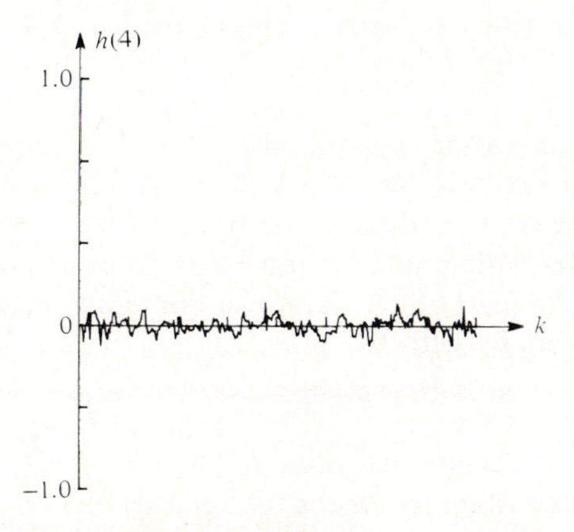

FIGURE 7.5.3 **A simulation of a simple adaptive filter.**

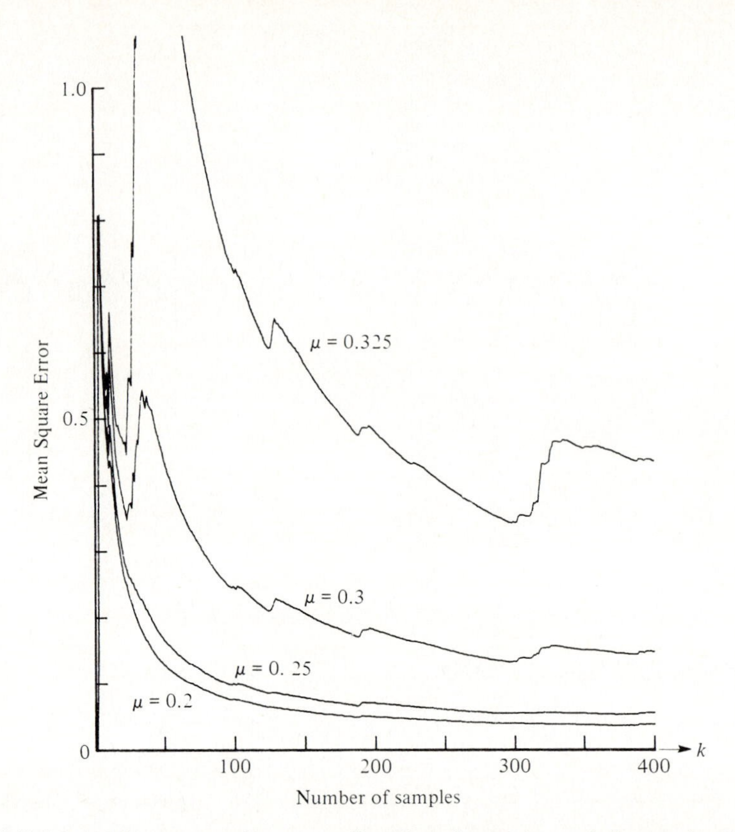

FIGURE 7.5.4 **The error variance as a function of step size and observation length.**

was $n = 4$. All filter coefficients were initialized to zero. One notes that the coefficients tend toward those of the least-squares filter (7.5.19).

Now let us consider the behavior of the algorithm as the step size μ is varied. In Fig. 7.5.4, we have shown the approximate final error variance as a function of the step size μ and elapsed time k. With the trivial filter

$$H(z) = 0, \qquad \hat{w}(k) = 0, \tag{7.5.21}$$

the error variance would be 1.01. The least-squares filter (7.5.19) would produce error variance 0.01. A useful filter would therefore produce an error variance between these two extremes. For μ less than roughly 0.4, the variance decreases with k, and for most choices of μ actually nears the performance of the least-squares filter. However, for values of μ greater than 0.4, the variance grows rapidly with k.

How should one choose the step size? The inequality (7.5.11) which guarantees stability for the true gradient follower is inappropriate in the noisy gradient situation, even if we knew the autocorrelation sequence (and could therefore estimate λ_{max}). For Example 7.5.1, the inequality would be $\mu < 2.0$ when in fact the algorithm does not behave for μ greater than about 0.4.

An approach to the selection of step size which requires only an estimate of $E[u^2(k)]$ is to set

$$\mu \approx [(n+1)E[u^2(k)]]^{-1}. \tag{7.5.22}$$

This value can be explained in the following way. The error that enters into the Widrow algorithm is

$$\begin{aligned} e(k) &= w(k) - \hat{w}(k) \\ &= w(k) - \sum_{l=0}^{n} h(l)u(k-l). \end{aligned}$$

Using this error, the filter coefficients are modified using Eq. (7.5.14). If a new error were then computed, it would be

$$\begin{aligned} e'(k) &= w(k) - \sum_{l=0}^{n} h'(l)u(k-l) \\ &= e(k)\left[1 - \mu \sum_{l=0}^{n} u^2(k-l)\right]. \end{aligned} \tag{7.5.23}$$

The step size μ which would force $e'(k) = 0$ is

$$\mu = \left[\sum_{l=0}^{n} u^2(k-l)\right]^{-1}. \tag{7.5.24}$$

This suggests the estimate (7.5.22). Using Eq. (7.5.24) to produce a time-varying step size would require more computation, but it has some interesting properties. (See Problem 7.32.)

A more subtle phenomenon related to the "convergence" issue involves the "coloration" of the input signal $\mathbf{u}$. If $\mathbf{u}$ is a white sequence then, $\mathbf{R} = \sigma^2\mathbf{I}$, and the level surfaces in Fig. 7.5.2 will be spherical. In this case the negative gradient at any point will lie on a line that goes through the minimum of the quadratic function $V(\mathbf{h})$. Thus the gradient follower will head directly to the minimum. However, if $\mathbf{u}$ is colored (that is to say if $S_{uu}(\theta)$ has significant variation), then $\mathbf{R}$ will have significant off-diagonal elements. In this case the level surfaces of Fig. 7.5.2 will be elliptical. If they are highly eccentric, then the negative gradient can lie on a line which misses the minimum by a great deal. This situation is typical in practice. The gradient follower will then (even in the noise-free case) perform a series of direction changes, and follow a jagged trajectory toward the minimum. A measure of ill-conditioning here is the *condition number* of $\mathbf{R}$; involving the minimum and maximum eigenvalues:

$$K(\mathbf{R}) = \frac{\lambda_{max}}{\lambda_{min}}. \tag{7.5.25}$$

(See Problem 7.30.)

Summary • In Chapters 6 and 7, we discussed the problem of obtaining the transfer function $H(z)$ of a digital filter to meet certain specifications. In Chapter 6, the specifications are given in terms of classical filter parameters such as one might experience in the design of traditional analog filters. In this chapter, the specifications are given in terms that explicitly involve parameters of the input data. In both cases, the goal is to obtain a digital filter input/output characterization that approximates the given specifications.

In this chapter, the specifications are expressed in terms that lead to so-called "least-squares" filters. We have considered two cases: deterministic least-squares designs and statistical least-squares designs. In both cases we are led to the minimization of a quadratic form. The difference between these two cases lies in the definitions of the parameters ρ, $\mathbf{q}$, $\mathbf{R}$ of the quadratic form.

In both the deterministic and statistical least-squares problems, the quadratic form is Toeplitz. Thus, from an abstract point of view, to study least-squares problems with stationarity is to study Toeplitz quadratic forms. There are three cases, depending on the range of indices in the summations. These are the finite case (FIR filter design), the semi-infinite case (causal IIR filter design), and the doubly infinite case (noncausal filter design). One can use Fourier transforms to write down the solution only in the last case. The finite case is, of course, the easiest, and we have given an algorithm in Fig. 7.2.2 which solves Toeplitz systems of equations. The semi-infinite case is usually solved using the method of spectral factorization.

This chapter marks an end to our concern with the approximation problem in filter design. The remainder of this text deals with more detailed descriptions of a digital filter. We shall concern ourselves not only with the input/output characterization, but also with precisely how this I/O characteristic is computed. For this we shall require descriptions that incorporate how computations are actually performed. The next chapter introduces the descriptions we shall use to study the internal structure of a digital filter.

PROBLEMS

7.1 Using the algorithm for evaluation of polynomials on the unit circle given in Chapter 2, plot

$$|G(e^{j\theta})|^{-1} \quad \text{and} \quad |A_n(e^{j\theta})| \quad \text{for } n = 1, 2, 3,$$

where $G(z) = 1 - z^{-1}/2$. (See Table 7.2.1.)

7.2 Let $G(z) = 1 - \beta z^{-m}$ for some integer $m > 0$. Minimize

$$V(\mathbf{h}) = \|\boldsymbol{\delta} - \mathbf{g} * \mathbf{h}\|^2$$

where

$$H(z) = \sum_{k=0}^{n} h(k)z^{-k}$$

and $n > 0$ is fixed. Compute $H(z)$ and $V_{\min}$ as a function of n and m.

7.3 An FIR whitening problem with $n = 4$ has solution

$$A(z) = 1 - z^{-1} + \frac{\sqrt{2}}{4} z^{-2}$$

and least-squares error $\alpha = 1$. Compute $r(k)$ for $k = 0, 1, 2, 3, 4$.

7.4 Suppose $\mathbf{f}$ is a causal, finite energy sequence. Set $G(z) = z^{-1}F(z)$ and consider the following deterministic least-squares problems:

$$\textit{Minimize } V_0(\mathbf{a}) = \|\mathbf{a} * \mathbf{g}\|^2, \qquad a(0) = 1, \qquad A(z) \text{ of order } n + 1.$$

$$\textit{Minimize } V(\mathbf{h}) = \|\mathbf{f} - \mathbf{g} * \mathbf{h}\|^2, \qquad H(z) \text{ of order } n.$$

Show that the solutions satisfy $H_n(z) = z[1 - A_{n+1}(z)]$, and that $\min V_0 = \min V$.

7.5 Let $F(z) = (1 + 0.4\,z^{-1})$ and $G(z) = (1 - 0.4\,z^{-1})^{-1}$. Find the FIR filter $H(z)$ of order $n = 4$ which minimizes $\|\mathbf{f} - \mathbf{g} * \mathbf{h}\|^2$, and compute the error energy $V_{\min}$.

7.6 Suppose that $\mathbf{g}$ is a causal, finite energy sequence. Consider the following two deterministic least-squares problems:

$$\textit{Minimize } V(\mathbf{h}) = \|\boldsymbol{\delta} - \mathbf{g} * \mathbf{h}\|^2, \quad H(z) \text{ of order } n.$$

$$\textit{Minimize } V_0(\mathbf{a}) = \|\mathbf{g} * \mathbf{a}\|^2, \qquad a(0) = 1, \quad A(z) \text{ of order } n.$$

Show that the solutions satisfy $H_n(z) = h(0)A_n(z)$, $h(0)\alpha_n = g(0)$, and that

$$\min V(\mathbf{h}) = 1 - h(0)g(0) = 1 - g^2(0)/\alpha_n \quad \text{where}$$

$$\min V_0(\mathbf{a}) = \alpha_n.$$

7.7 Show that the normal equations (7.2.33) for the deterministic whitening filter problem can also be written in the form

$$\mathbf{R}\begin{bmatrix} a(n) \\ \vdots \\ a(1) \\ 1 \end{bmatrix} = \begin{bmatrix} 0 \\ \vdots \\ 0 \\ \alpha \end{bmatrix}.$$

Then, using the development of the triangular decomposition in Appendix 7A, establish Eqs. (7.2.39) and (7.2.41).

7.8 Derive Eq. (7.2.54).

7.9 In Eq. (7.2.52), suppose that $F(z) = D(z)E(z)$ where

$$D(z) = \sum_{k=0}^{n} d(k)z^{-k}.$$

Show that the right-hand side of Eq. (7.2.54) is zero, using Problem 3.23.

7.10 In communications problems, it is often the case that a reasonable amount of delay is tolerable. Consider a deterministic least-squares problem with

$$F(z) = z^{-L}, \qquad G(z) = \frac{\frac{1}{2} - z^{-1}}{1 + \frac{1}{2}z^{-1}}.$$

Find the least-squares IIR $H(z)$ with $L = 0$. Then compute, as a function of L, the minimum error energy using Eq. (7.2.54).

7.11 The FIR filters $A_m(z)$, $\tilde{A}_m(z)$, and $H_m(z)$ of Eqs. (7.2.36), (7.2.41), and (7.2.42) satisfy the recursion

$$\begin{bmatrix} H_m(z) \\ A_m(z) \\ \tilde{A}_m(z) \end{bmatrix} = \begin{bmatrix} 1 & \dfrac{-c_m\beta_m}{\alpha_m} & \dfrac{c_m}{\alpha_m}z^{-1} \\ 0 & 1 & -z^{-1}\beta_m \\ 0 & -\beta_m & z^{-1} \end{bmatrix} \begin{bmatrix} H_{m-1}(z) \\ A_{m-1}(z) \\ \tilde{A}_{m-1}(z) \end{bmatrix}.$$

Develop a lattice FIR signal flow graph based on this recursion. There should be one input node and at least $3n$ other nodes whose transfer functions (relative to the input node) are these three FIR filters, with orders $1 \leqslant m \leqslant n$.

7.12 Let $\mathbf{u}$ be a WSS signal with autocorrelation sequence $\mathbf{r}_{uu}$. Prove Eq. (7.3.8). Then prove the inequality (7.3.9) by considering

$$E\left[\sum_{l=0}^{n} c_l u(k-l)\right]^2$$

as a quadratic form in c_0 through c_n.

7.13 Assuming $\mathbf{u}$ is a WSS signal and $E[u(k)^2] > 0$, compute the probability that $\mathbf{u}$ has finite energy.

7.14 Let $\mathbf{y}$ be a WSS signal, and let

$$x(k) = \sum_{j=0}^{n-1} y(nk+j).$$

Express $r_{xx}(k)$ in terms of $r_{yy}(k)$, and $S_{xx}(\theta)$ in terms of $S_{yy}(\theta)$.

7.15 Let $\mathbf{u}$ be WSS and $\mathbf{r}_{uu}$ its autocorrelation sequence.

a) Show that $r(l)^2 \leqslant r(0)^2$ for all l.

b) Suppose that $r(3)^2 = r(0)^2$. What can you say about the rest of the autocorrelation sequence? What can you say about the signal $\mathbf{u}$?

7.16 What happens to the matrix in Eq. (7.3.9) if a white noise signal of variance σ^2 (which is uncorrelated with $\mathbf{u}$) is added to the signal $\mathbf{u}$?

7.17 Let $H(z)$ be a linear digital filter. We wish to measure the unit-pulse response by using a white noise input sequence $\mathbf{v}$. Show that the cross correlation of the input and the output yields $\mathbf{h}$, the unit-pulse response.

7.18 Find $S_{uu}(\theta)$ if $r_{uu}(l) = e^{-|l|}\cos(\theta_0 l)$ for all l.

7.19 Prove the spectral factorization theorem: Let

$$S(\theta) = \frac{N(\cos\theta)}{D(\cos\theta)}$$

where $N(x)$ and $D(x)$ are polynomials with real coefficients, and $S(\theta) > 0$ for all θ. Then S has the representation

$$S(\theta) = |G(e^{j\theta})|^2,$$

where

$$G(z) = \frac{\sum_{k=0}^{m} b(k)z^{-k}}{\sum_{k=0}^{n} a(k)z^{-k}}$$

has real coefficients and poles and zeros satisfying $|\lambda| < 1$.

7.20 Find the spectral factorization of $S_{uu}(\theta) = \dfrac{1-\alpha^2}{1+\alpha^2 - 2\alpha\cos(3\theta)}$.

7.21 Assume that you have a random number generator that produces a sequence of real numbers $\mathbf{w}$ which are uncorrelated and uniformly distributed between zero and one. How can you use this sequence to produce a zero mean WSS signal $\mathbf{y}$ with power spectral density function $S_{yy}(\theta)$ given by

$$S_{yy}(\theta) = \frac{2(1-\cos(\theta))}{1+a^2+2a\cos(2\theta)}.$$

7.22 The spectral density function for an observed data sequence is

$$S_{uu}(\theta) = \frac{(1-0.2e^{-j\theta})(1-0.2e^{j\theta})}{(1-0.9e^{-j\theta})(1-0.9e^{j\theta})}.$$

Find the filter $G(z)$ which when driven by white noise $\mathbf{v}$ has an output $\mathbf{u}$ with the given spectral density function. What is the autocorrelation function for $\mathbf{u}$?

7.23 A statistical version of the whitening problem of Section 7.2 can be set up for any WSS signal $\mathbf{y}$. Let $\mathbf{e} = \mathbf{a} * \mathbf{y}$, where

$$A(z) = 1 + a(1)z^{-1} + \cdots + a(n)z^{-n}.$$

The problem is to minimize

$$V(\mathbf{a}) = E[e^2(k)], \qquad \text{subject to } a(0) = 1.$$

a) Show that if $S_{yy}\theta = |G(e^{j\theta})|^2$, then

$$V(\mathbf{a}) = \|\mathbf{a} * \mathbf{g}\|^2.$$

b) If $\mathbf{a}$ is the solution to this problem, must it follow that $r_{ee}(k) = 0$ for $1 \leqslant k \leqslant n$?

7.24 Let $\mathbf{y}$ be a WSS signal with power spectral density function

$$S_{yy}(\theta) = \frac{25}{9} \frac{5 + 4\cos(\theta)}{17 + 8\cos(2\theta)}.$$

Find the causal IIR minimum mean square error m-step predictor for $\mathbf{y}$ (as in Example 7.4.5), and compute the error variance.

7.25 A discrete time signal $\mathbf{s}$ is combined with white noise $\mathbf{n}$ to produce an observed signal $\mathbf{u} = \mathbf{s} + \mathbf{n}$. The signal and noise have spectral density functions

$$S_{ss}(\theta) = \frac{0.51}{1.49 - 1.4\cos\theta}, \qquad S_{nn}(\theta) = 1,$$

and are uncorrelated. Find the minimum error variance noncausal filter for $\mathbf{w} = \mathbf{s}$, and the mean square error. Then find the minimum error variance causal IIR filter for $\mathbf{w} = \mathbf{s}$, and the corresponding mean square error.

7.26 Repeat Problem 7.25 for a signal $\mathbf{s}$ with autocorrelation sequence $r_{ss}(l) = \alpha^{|l|}$, $|\alpha| < 1$ and noise $\mathbf{n}$ with autocorrelation $\mathbf{r}_{nn} = \boldsymbol{\delta}$.

7.27 Find a causal IIR m-step predictor for a signal $\mathbf{s}$ with autocorrelation sequence

$$r_{ss}(k) = \frac{4}{3}\left(\frac{1}{2}\right)^{|k|}.$$

7.28 Let $\mathbf{v}$ be unit variance white noise, and let $\mathbf{w} = \mathbf{a} * \mathbf{v}$, $\mathbf{u} = \mathbf{b} * \mathbf{w}$, where

$$A(z) = 1 + 2z^{-1}, \qquad B(z) = \frac{z + 1.1}{z - 0.9}.$$

Find the minimum mean square error causal IIR filter $H(z)$ to estimate $\mathbf{w}$ given $\mathbf{u}$.

7.29 Let $G(z) = G_0(z)E(z)$ where $G_0(z)$ has a finite energy causal inverse, but $E(z)$ is all-pass, so that $|E(e^{j\theta})| = 1$. Let $\mathbf{w}$ be unit variance white noise and let $\mathbf{u} = \mathbf{g} * \mathbf{w}$. Find the minimum mean square error causal IIR filter $H(z)$ to estimate $\mathbf{w}$ given $\mathbf{u}$, and compute the error variance in terms of $G_0(z)$ and $E(z)$.

7.30 Consider the selection of step size for the true gradient follower.

a) Using Eq. (7.5.8), show that

$$[\mathbf{h}(k) - \mathbf{R}^{-1}\mathbf{q}] = [\mathbf{I} - \mu\mathbf{R}]^k[\mathbf{h}(0) - \mathbf{R}^{-1}\mathbf{q}]$$

Therefore the convergence rate is determined by

$$\Lambda(\mu) \triangleq \max_i |1 - \mu\lambda_i|,$$

i.e., the eigenvalue of $\mathbf{I} - \mu\mathbf{R}$ with greatest absolute value (the eigenvalues of $\mathbf{R}$ being $\lambda_0, \ldots, \lambda_n$).

b) Find μ to minimize $\Lambda(\mu)$, and express min Λ as a function of the condition number of **R**, defined in Eq. (7.5.25).

7.31 Let the matrix **R** in Eq. (7.5.4) have eigenvalues $\lambda_0, \ldots, \lambda_n$ (all positive). The step size inequality (7.5.11) which is necessary for the convergence of the true gradient follower is

$$\mu < 2/\lambda_{\max}.$$

Show that the choice of step size dictated by Eq. (7.5.22) is

$$\mu = \left[\sum_{i=0}^{n} \lambda_i\right]^{-1} < \frac{1}{\lambda_{\max}}.$$

7.32 This problem examines the time-varying step size of Eq. (7.5.24) for use with the noisy gradient follower Eq. (7.5.14). Let

$$\mathbf{h} = [h(0), \ldots, h(n)]^T$$
$$\mathbf{v} = [u(k), \ldots, u(k-n)]^T$$
$$e = e(k) = w(k) - \hat{w}(k) = w - \mathbf{h}^T\mathbf{v}.$$

We want to change **h** to $\mathbf{h} + \mathbf{d}$, using the error measurement e, where **d** is chosen to minimize

$$\alpha\|\mathbf{d}\|^2 + [e']^2 = \alpha\|\mathbf{d}\|^2 + [e - \mathbf{d}^T\mathbf{v}]^2.$$

(This attempts to drive the error to zero, but penalizes large changes in the coefficient vector.)

a) Show that the solution is

$$\mathbf{d} = \mu e\mathbf{v}, \quad \text{with } \mu = \frac{1}{\alpha + \|\mathbf{v}\|^2}.$$

[This yields Eq. (7.5.24) as $\alpha \to 0$.]
Now suppose that Eq. (7.5.15) holds, i.e., $w = \mathbf{g}^T\mathbf{v}$.
Let $\mathbf{\Delta} = \mathbf{g} - \mathbf{h}$ be the coefficient error vector.
Then $e = \mathbf{\Delta}^T\mathbf{v}$ is the measured error. Using the step size

$\mu = \|\mathbf{v}\|^{-2}$ (with $\alpha = 0$), set $\mathbf{\Delta}' = \mathbf{g} - \mathbf{h}'$, where $\mathbf{h}' = \mathbf{h} + \mu e\mathbf{v} = \mathbf{h} + \mathbf{d}$.

b) Show that $\mathbf{\Delta}' = \mathbf{E}\mathbf{\Delta}$, where

$$\mathbf{E} = \mathbf{E}^2 = \mathbf{E}^T = \mathbf{I} - \mathbf{v}(\mathbf{v}^T\mathbf{v})^{-1}\mathbf{v}^T$$

is an orthogonal projection.

c) Then show that $\|\mathbf{\Delta}'\|^2 \leqslant \|\mathbf{\Delta}\|^2$ with equality if and only if $e = 0$. [Thus the coefficient error vector can only decrease. Note, however, that there is no additive noise in Eq. (7.5.15).]

7.33 Suppose the signals **w** and **u** of Fig. 7.5.1 satisfy Eq. (7.5.15), i.e., $\mathbf{w} = \mathbf{g} * \mathbf{u}$, where both **g** and **u** are causal, and $u(0) \neq 0$. Then the adaptive filter should attempt to match the filter $G(z)$. Suppose we modify the Widrow algorithm in

Eq. (7.5.14) as follows:

$$h(k+1, l) = h(k, l) + \delta(k, l)\,\mu e(k)u(k-l).$$

Show that with the appropriate choice of step size μ; at time k we will have

$$h(k, l) = g(l) \text{ for } 0 \leqslant l < k,$$

no matter what the initial values $h(0, l)$ were.

7.34 Let $\mathbf{R}$ be an $n \times n$ symmetric and positive definite matrix.

a) Show that $\mathbf{R}^{-1}$ exists.
b) Show that $\mathbf{R}^{-1} = (\mathbf{R}^{-1})^T > \mathbf{0}$.

7.35 Prove Eq. (7A.37).

7.36 "Duals" to Eqs. (7A.19), (7A.20), (7A.23), and the inequalities (7A.24) may be obtained by reversing the roles of $\mathbf{x}$ and $\mathbf{y}$. In other words, using the same quadratic form $V(\mathbf{x}, \mathbf{y})$, we minimize V with respect to $\mathbf{y}$ holding $\mathbf{x}$ fixed. The resulting $\mathbf{T}$ has zeros above the diagonal. Find these duals.

7.37 a) The deterministic autocorrelation sequence $\mathbf{r}_g$ for a causal filter $G(z)$ was defined in Problem 3.24. Show that if $\mathbf{v}$ is unit variance white noise and $\mathbf{u} = \mathbf{g} * \mathbf{v}$, then $r_{uu}(k) = r_g(k)$. This relates a deterministic autocorrelation sequence and the autocorrelation sequence for a WSS signal.

b) The unit pulse response of a filter may be measured [using Eq. (7.3.18)] by employing a white noise input signal. There is a deterministic analog to this as well. Let $E(z)$ be causal and all-pass, and let $\mathbf{y} = \mathbf{h} * \mathbf{e}$. Compute

$$\sum_{k=0}^{\infty} y(k+l)e(k).$$

c) For the application in part b), it is often desirable to limit the amplitude of the input signal to avoid nonlinear "saturation" effects. This leads to the following interesting problem. For a fixed order m, find the all-pass filter of the form given in Problem 3.19 which has least l_∞ norm $\|\mathbf{e}\|_\infty = \max_k |e(k)|$. Solve this problem for the case $m = 1$.

APPENDIX

7A

Quadratic Forms

Quadratic forms are the type of function that implicitly understood when one speaks of least-squares problems. In this appendix we shall give a brief account of the properties of quadratic forms including the triangular or "Cholesky" decomposition.

A function V of a scalar variable x is *quadratic* if it has the form

$$V(x) = \rho - 2qx + rx^2, \tag{7A.1}$$

where ρ, q, and r are constants. The graph of such a function is in general parabolic, as in Fig. 7A.1. The graph is shaped like a valley if $r > 0$, in which case V has a strict minimum at

$$x_0 = \frac{q}{r}. \tag{7A.2}$$

The minimum value is

$$V(x_0) = \rho - \frac{q^2}{r}. \tag{7A.3}$$

Another way of writing the quadratic form is

$$V(x) = V(x_0) + r(x - x_0)^2, \tag{7A.4}$$

and is the result of "completing the square." From this representation, we may conclude immediately that if $r > 0$, then

$$V(x) \geqslant V(x_0) \tag{7A.5}$$

with equality *only* at the unique point x_0.

This approach is easily extended to the vector case. Let

$$\mathbf{x} = \begin{bmatrix} x(1) \\ x(2) \\ \vdots \\ x(n) \end{bmatrix} \tag{7A.6}$$

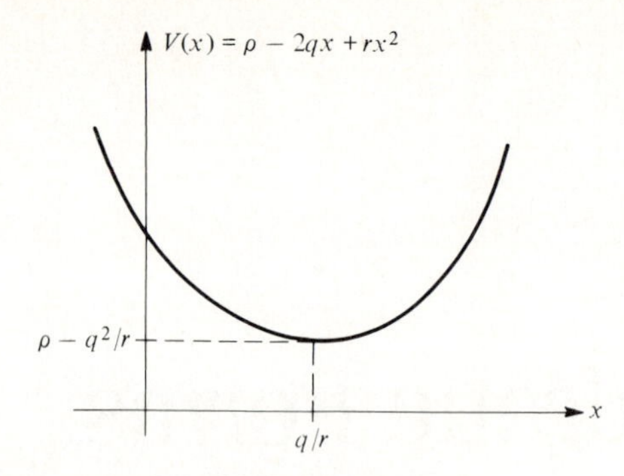

FIGURE 7.A.1 **The graph of a quadratic function.**

be an n-dimensional vector, and let

$$
\begin{aligned}
V(\mathbf{x}) &= \rho - 2\sum_{i=1}^{n} q(i)x(i) + \sum_{i=1}^{n}\sum_{j=1}^{n} x(i)R(i,j)x(j) \\
&= \rho - 2\mathbf{q}^T\mathbf{x} + \mathbf{x}^T\mathbf{R}\mathbf{x} \\
&= [1 \quad \mathbf{x}^T]\begin{bmatrix} \rho & -\mathbf{q}^T \\ -\mathbf{q} & \mathbf{R} \end{bmatrix}\begin{bmatrix} 1 \\ \mathbf{x} \end{bmatrix}.
\end{aligned} \tag{7A.7}
$$

Suppose that the $n \times n$ matrix $\mathbf{R}$ is nonsingular (has an inverse). Then the process of completing the square can be applied to $V(\mathbf{x})$ in the form

$$
\begin{aligned}
V(\mathbf{x}) &= [\mathbf{x} - \mathbf{R}^{-1}\mathbf{q}]^T\mathbf{R}[\mathbf{x} - \mathbf{R}^{-1}\mathbf{q}] + (\rho - \mathbf{q}^T\mathbf{R}^{-1}\mathbf{q}) \\
&= V(\mathbf{x}_0) + (\mathbf{x} - \mathbf{x}_0)^T\mathbf{R}(\mathbf{x} - \mathbf{x}_0),
\end{aligned} \tag{7A.8}
$$

where

$$\mathbf{x}_0 = \mathbf{R}^{-1}\mathbf{q}, \tag{7A.9}$$

[which is an n-dimensional version of Eq. (7A.2)].

Now in order to generalize the inequality (7A.5), we need a positivity condition for the matrix $\mathbf{R}$, similar to the scalar condition $r > 0$. The symmetric real matrix $\mathbf{R}$ is *positive definite* if for real vectors $\mathbf{y}$,

$$\mathbf{y} \neq \mathbf{0} \quad \Rightarrow \quad \mathbf{y}^T\mathbf{R}\mathbf{y} \geqslant 0. \tag{7A.10}$$

We will use the shorthand

$$\mathbf{R} > \mathbf{0} \tag{7A.11}$$

to indicate this situation. The matrix $\mathbf{R}$ is *positive semidefinite* if the strict inequality is replaced with

$$\mathbf{y}^T\mathbf{R}\mathbf{y} \geqslant 0.$$

We will use the shorthand

$$\mathbf{R} \geqslant \mathbf{0}$$

if $\mathbf{R}$ is positive semidefinite. We have

$$\mathbf{R} > \mathbf{0} \quad \Leftrightarrow \quad V(\mathbf{x}) > V(\mathbf{x}_0) \quad \text{when} \quad \mathbf{x} \neq \mathbf{x}_0. \tag{7A.12}$$

(If $\mathbf{R} > \mathbf{0}$, then $\mathbf{R}$ is nonsingular. See Problem 7.36.)

All the FIR least-squares problems we have studied in Chapter 7 involve the quadratic form $V(\mathbf{x})$ in Eqs. (7A.7) and (7A.8), with $\mathbf{R} > \mathbf{0}$. The equations that one must solve to find the minimizing vector $\mathbf{x}_0$ are called the *normal equations* and are of the form

$$\mathbf{R}\,\mathbf{x}_0 = \mathbf{q}, \tag{7A.13}$$

which is consistent with Eq. (7A.9).

Decomposition of Quadratic Forms

Consider a symmetric matrix $\mathbf{W} > \mathbf{0}$ which is partitioned as

$$\mathbf{W} = \begin{bmatrix} \mathbf{P} & -\mathbf{Q}^T \\ -\mathbf{Q} & \mathbf{R} \end{bmatrix} \tag{7A.14}$$

with $\mathbf{P}$ and $\mathbf{R}$ both square matrices. Then

$$\begin{aligned} V(\mathbf{x}, \mathbf{y}) &= [\mathbf{x}^T \quad \mathbf{y}^T]\,\mathbf{W}\begin{bmatrix} \mathbf{x} \\ \mathbf{y} \end{bmatrix} \\ &= \mathbf{x}^T\mathbf{P}\mathbf{x} - \mathbf{x}^T\mathbf{Q}^T\mathbf{y} - \mathbf{y}^T\mathbf{Q}\mathbf{x} + \mathbf{y}^T\mathbf{R}\mathbf{y} \end{aligned} \tag{7A.15}$$

must be positive as long as the vectors $\mathbf{x}$ and $\mathbf{y}$ are not both equal to zero. Since

$$\mathbf{x} \neq \mathbf{0} \quad \Rightarrow \quad V(\mathbf{x}, \mathbf{0}) = \mathbf{x}^T\mathbf{P}\,\mathbf{x} > 0,$$

and

$$\mathbf{y} \neq \mathbf{0} \quad \Rightarrow \quad V(\mathbf{0}, \mathbf{y}) = \mathbf{y}^T\mathbf{R}\,\mathbf{y} > 0,$$

the diagonal blocks of a positive definite matrix must also be positive definite. A sharper condition is obtained by minimizing V with respect to $\mathbf{x}$, holding $\mathbf{y}$ fixed. The minimum occurs at

$$\mathbf{x} = \mathbf{P}^{-1}\mathbf{Q}^T\mathbf{y} \tag{7A.16}$$

and

$$V(\mathbf{P}^{-1}\mathbf{Q}^T\mathbf{y}, \mathbf{y}) = \mathbf{y}^T[\mathbf{R} - \mathbf{Q}\,\mathbf{P}^{-1}\mathbf{Q}^T]\mathbf{y}. \tag{7A.17}$$

This must likewise be positive for $\mathbf{y}$ not equal to zero, and therefore

$$\mathbf{R} - \mathbf{Q}\,\mathbf{P}^{-1}\mathbf{Q}^T > \mathbf{0}. \tag{7A.18}$$

A decomposition of the matrix $\mathbf{W}$ which exhibits the foregoing piecemeal minimization is

$$\mathbf{W} = \begin{bmatrix} \mathbf{P} & -\mathbf{Q} \\ -\mathbf{Q} & \mathbf{R} \end{bmatrix} = \mathbf{T}^T\begin{bmatrix} \mathbf{P} & \mathbf{0} \\ \mathbf{0} & \mathbf{R} - \mathbf{Q}\,\mathbf{P}^{-1}\mathbf{Q}^T \end{bmatrix}\mathbf{T}, \tag{7A.19}$$

where

$$\mathbf{T} = \begin{bmatrix} \mathbf{I} & -\mathbf{P}^{-1}\mathbf{Q}^T \\ \mathbf{0} & \mathbf{I} \end{bmatrix}. \tag{7A.20}$$

The factorization (7A.19) is called a *block triangular decomposition* of $\mathbf{W}$. The matrix $\mathbf{T}$ is block diagonal and nonsingular. In fact

$$\mathbf{T}^{-1} = \begin{bmatrix} \mathbf{I} & \mathbf{P}^{-1}\mathbf{Q}^T \\ \mathbf{0} & \mathbf{I} \end{bmatrix}. \tag{7A.21}$$

This decomposition is yet another way to describe the technique of completing the square. We have

$$\mathbf{T}\begin{bmatrix} \mathbf{x} \\ \mathbf{y} \end{bmatrix} = \begin{bmatrix} \mathbf{x} - \mathbf{P}^{-1}\mathbf{Q}^T\mathbf{y} \\ \mathbf{y} \end{bmatrix}, \tag{7A.22}$$

and therefore, using Eq. (7A.19) we have

$$\begin{aligned} V(\mathbf{x}, \mathbf{y}) &= [\mathbf{x}^T \quad \mathbf{y}^T]\mathbf{W}\begin{bmatrix} \mathbf{x} \\ \mathbf{y} \end{bmatrix} \\ &= [\mathbf{x} - \mathbf{P}^{-1}\mathbf{Q}^T\mathbf{y}]^T\mathbf{P}[\mathbf{x} - \mathbf{P}^{-1}\mathbf{Q}^T\mathbf{y}] + \mathbf{y}^T[\mathbf{R} - \mathbf{Q}\mathbf{P}^{-1}\mathbf{Q}^T]\mathbf{y}. \end{aligned} \tag{7A.23}$$

The right-hand side is positive for all choices of $\mathbf{x}$ and $\mathbf{y}$ except both zero if and only if

$$\mathbf{P} > \mathbf{0} \quad \text{and} \quad \mathbf{R} - \mathbf{Q}\mathbf{P}^{-1}\mathbf{Q}^T > \mathbf{0}. \tag{7A.24}$$

These are necessary and sufficient conditions for $\mathbf{W} > \mathbf{0}$.

The block triangular decomposition (7A.19) may be further refined by decomposing the diagonal blocks $\mathbf{P}$ and $\mathbf{R} - \mathbf{Q}\mathbf{P}^{-1}\mathbf{Q}^T$ in the same way. The limiting form would be a factorization of the form

$$\mathbf{W} = \mathbf{T}^T\mathbf{D}\,\mathbf{T} \tag{7A.25}$$

where

$$\mathbf{T} = \begin{bmatrix} 1 & & X \\ & \ddots & \\ 0 & & 1 \end{bmatrix} \tag{7A.26}$$

is upper triangular with ones on the diagonal, and

$$\mathbf{D} = \text{Diag}\{\alpha_1, \alpha_2, \ldots, \alpha_n\} \triangleq \begin{bmatrix} \alpha_1 & & & 0 \\ & \alpha_2 & & \\ & & \ddots & \\ 0 & & & \alpha_n \end{bmatrix} \tag{7A.27}$$

is diagonal. This form is called a "triangular" or "Cholesky" decomposition of the matrix $\mathbf{W}$. If we let

$$\mathbf{x} = \begin{bmatrix} x(1) \\ x(2) \\ \vdots \\ x(n) \end{bmatrix} = \mathbf{T}^{-1}\mathbf{v} = \mathbf{T}^{-1}\begin{bmatrix} v(1) \\ v(2) \\ \vdots \\ v(n) \end{bmatrix}, \tag{7A.28}$$

then

$$\mathbf{x}^T\mathbf{W}\mathbf{x} = \mathbf{v}^T\mathbf{D}\mathbf{v} = \sum_{k=1}^{n} \alpha_k v^2(k). \tag{7A.29}$$

The quadratic form has been decomposed into n scalar quadratic terms. It is apparent from this that

$$\mathbf{W} > \mathbf{0} \quad \Leftrightarrow \quad \alpha_k > 0 \quad \text{for} \quad 1 \leqslant k \leqslant n. \tag{7A.30}$$

The matrix $\mathbf{T}$ and the scalars α_k which parameterize the triangular decomposition (7A.25) are obtained in the following way. Let $\mathbf{W}(k)$ be the upper left $k \times k$ block of the $n \times n$ matrix $\mathbf{W}$. This submatrix must be positive definite if $\mathbf{W}$ is. The equation

$$\mathbf{W}(k) \begin{bmatrix} a_{1k} \\ a_{2k} \\ \vdots \\ a_{k-1,k} \\ 1 \end{bmatrix} = \begin{bmatrix} 0 \\ 0 \\ \vdots \\ 0 \\ \alpha_k \end{bmatrix} \tag{7A.31}$$

must therefore have a unique solution. Since $\mathbf{W}(k)$ is an upper left block of $\mathbf{W}$, we have

$$\mathbf{W} \begin{bmatrix} a_{1k} \\ \vdots \\ a_{k-1,k} \\ 1 \\ 0 \\ \vdots \\ 0 \end{bmatrix} = \begin{bmatrix} 0 \\ \vdots \\ 0 \\ \alpha_k \\ \times \\ \vdots \\ \times \end{bmatrix}. \tag{7A.32}$$

Let

$$\mathbf{T} = \begin{bmatrix} 1 & a_{12} & a_{13} & \cdots & a_{1n} \\ & \ddots & & & \vdots \\ & & \ddots & & a_{n-1,n} \\ & 0 & & & 1 \end{bmatrix}^{-1}. \tag{7A.33}$$

Then

$$\mathbf{W}\mathbf{T}^{-1} = \begin{bmatrix} \alpha_1 & & 0 \\ & \ddots & \\ \times & & \alpha_n \end{bmatrix} \tag{7A.34}$$

and

$$(\mathbf{T}^{-1})^T\mathbf{W}\mathbf{T}^{-1} = (\mathbf{W}\mathbf{T}^{-1})^T\mathbf{T}^{-1} = \begin{bmatrix} \alpha_1 & & \times \\ & \ddots & \\ 0 & & \alpha_n \end{bmatrix} \mathbf{T}^{-1}. \tag{7A.35}$$

But the right-hand side is the product of upper triangular matrices, and is therefore upper triangular. However, the left-hand side is symmetric. Therefore the right-hand side must also be symmetric and consequently diagonal. The diagonal elements are simply the α_k since the diagonal elements of $\mathbf{T}^{-1}$ are all ones. Thus

$$(\mathbf{T}^{-1})^T \mathbf{W} \mathbf{T}^{-1} = \mathbf{D}, \tag{7A.36}$$

which is quickly transformed into Eq. (7A.25).

In the problems, the reader is asked to show that

$$\det \mathbf{W}(k) = \prod_{j=1}^{k} \alpha_j, \qquad 1 \leqslant j \leqslant n. \tag{7A.37}$$

From this it follows that

$$\alpha_k = \frac{\det \mathbf{W}(k)}{\det \mathbf{W}(k-1)}. \tag{7A.38}$$

Using Eqs. (7A.28) and (7A.29), this diagonal element has the interpretation

$$\alpha_k = \min \mathbf{x}^T \mathbf{W} \mathbf{x}$$

subject to the constraints $x(k) = 1, \qquad x(j) = 0 \qquad \text{for } j > k.$ **(7A.39)**

The formula (7A.38) for the minimum value in Eq. (7A.39) is known as *Gram's theorem*.

Furthermore, the matrix **W** in Eq. (7A.25) satisfies

$$\begin{aligned} \mathbf{W} > \mathbf{0} \quad &\Leftrightarrow \quad \alpha_k > 0, \qquad 1 \leqslant k \leqslant n \\ &\Leftrightarrow \quad \det \mathbf{W}(k) > 0, \qquad 1 \leqslant k \leqslant n. \end{aligned}$$

This test for positivity is called *Sylvester's test*.

CHAPTER 8

Internal Descriptions for Digital Filters

8.1 Introduction

Until now we have been concerned only with low-level "input-output" descriptions of digital filters. The relation

$$\mathbf{y} = \mathbf{h} * \mathbf{u} \tag{8.1.1}$$

between the input, output, and unit-pulse response sequences demonstrates that the sequence $\mathbf{h}$ is sufficient to characterize input-output properties, and knowing the transfer function $H(z)$ is the same as knowing $\mathbf{h}$, since they form a z-transform pair:

$$\mathbf{h} \quad \overset{z}{\leftrightarrow} \quad H(z). \tag{8.1.2}$$

Therefore, the first problem in digital filter design is to find $H(z)$ to meet input-output specifications, subject only to order constraints for the allowable functional forms:

$$\textbf{FIR}: \quad H(z) = \sum_{k=0}^{N-1} h(k)z^{-k} \tag{8.1.3}$$

$$\textbf{IIR}: \quad H(z) = \frac{b_0 + b_1 z^{-1} + \cdots + b_n z^{-n}}{1 + a_1 z^{-1} + \cdots + a_n z^{-n}} \tag{8.1.4}$$

We have treated this problem in Chapters 6 and 7.

But there is more to filter design than the determination of the transfer function. The second problem is to obtain an *algorithm* or *realization* which describes exactly how the computation is to be performed. The first reason for developing more detailed filter descriptions is to be able to formulate the *realization problem*, which is

Given a rational transfer function $H(z)$, produce a list of primitive operations for the processor to execute.

Any processor under consideration for use in a digital filtering application will have a set of "primitive" operations that can be performed. Digital filters do not require a great variety of operations, but do need the ability to add, multiply, and move data. Consider, for example, the following loop body:

$$\begin{aligned}&\text{Read } u\\ &\quad y \leftarrow c \cdot x + d \cdot u\\ &\text{Write } y\\ &\quad x \leftarrow a \cdot x + b \cdot u\end{aligned} \tag{8.1.5}$$

Repeated execution of these statements will realize a digital filter whose transfer function is

$$H(z) = \frac{c \cdot b}{z - a} + d \tag{8.1.6}$$

as one can see by examining the difference equations equivalent to the statements in (8.1.5):

$$y(k) = cx(k) + du(k), \qquad x(k+1) = ax(k) + bu(k). \tag{8.1.7}$$

We shall show that any IIR or FIR filter [with rational $H(z)$] can be implemented as an infinite loop containing statements like those in Eq. (8.1.5). Let us examine them in more detail.

The typical statement involves a sum of products. Each operand in the expression must be read either from memory or from an input port, and results are written either to memory or to an output port. Notice that each multiplication involves one constant and one variable [we exhibit time explicitly when writing the difference equations as in Eq. (8.1.7)]. Multiplication of two variables would lead to a nonlinear filter, and multiplication of two constants would be pointless. Because of this dichotomy, we may use separate memories for the constants and the variables. And because constants don't change, we can store them in read-only memory (ROM). Let us write a sum of products statement as

$$v(m_0) = c(l_1)v(m_1) + \cdots + c(l_n)v(m_n). \tag{8.1.8}$$

This statement is parameterized by indices $\{l_k, m_k\}$ which are memory addresses, with l an address in read-only memory, and m an address in read-write memory (or RAM). It simplifies things to think of the input and output variables (which might in fact require special hardware) as residents of RAM, but with specific addresses. (If the processor is designed this way, then it is said to use "memory-mapped I/O.")

One solution of the realization problem is to obtain a list of statements in the form (8.1.8), which, when executed as an infinite loop, will realize a given transfer function $H(z)$. From such a list one can generally produce an implementation using digital signal processors that are designed to execute statements like Eq. (8.1.8) efficiently. These processors usually exhibit some degree of parallelism, such as the ability to do a multiplication, an accumulation, and a memory access concurrently. A simple example is given below.

EXAMPLE 8.1.1

A digital filter realization.

The read-only memory in Figure 8.1.1 has stored in it all constant coefficients, as well as the variable addresses and control information necessary to execute a fixed sequence of operations of the form (8.1.8). The beginning of a loop is initiated by the RESET signal going active, which forces the ROM address to all zeros. At the beginning of a sum of products operation the accumulator is cleared, and at the end of an operation the most significant bits of the result is written to memory by activating the WRITE signal. In between, a product of a constant (from ROM) and a variable (from RAM) is being formed, and a running sum is kept in the accumulator. (Typically, the accumulator is double precision, to reduce roundoff error.)

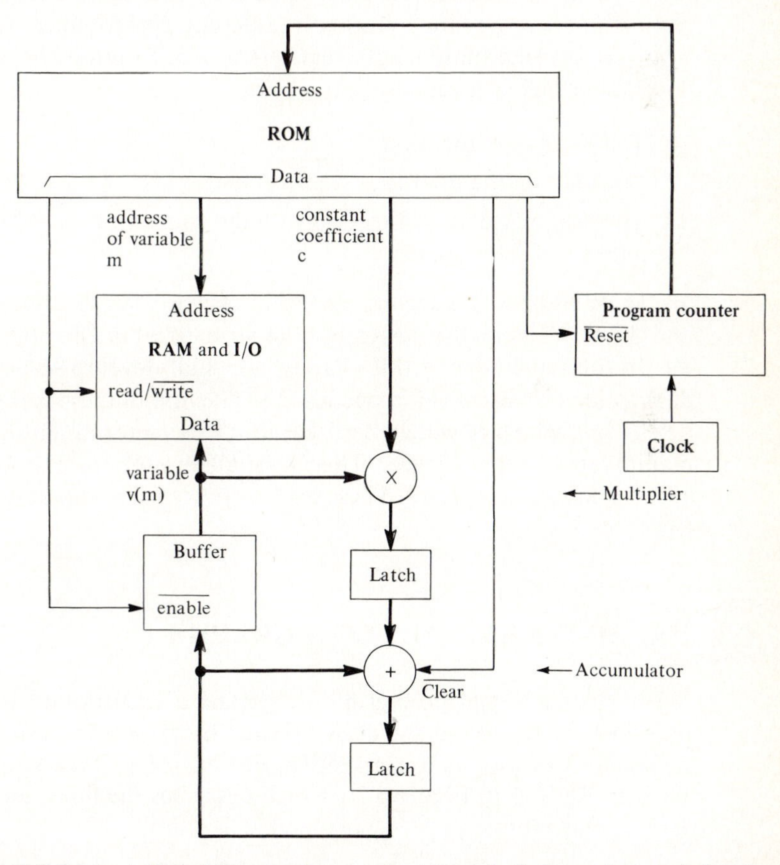

FIGURE 8.1.1 A digital filter implementation. Certain timing and control signals are not shown.

The Reasons for More Detailed Filter Descriptions

We have seen that in order to bridge the gap between a knowledge of $H(z)$ and an actual hardware implementation, we must study the realization problem. The realization itself [a body of statements like Eq. (8.1.8)] amounts to a more detailed description. Once formulated, we should ask some questions about the realization problem. Does it always have a solution? Does it have more than one solution? In fact, as we shall show the realization problem has an infinite number of solutions, and for this reason there is a need for efficient descriptions.

If there is more than one possible realization, which one should be chosen? This question must be answered on the basis of other issues than $H(z)$, because by construction they all generate the same transfer function. One issue is the cost of the realization, which might depend, for example, on the number of multiplications required. Another issue is output signal quality, since different realizations can produce markedly different performance due to the errors involved in using finite length registers in realizations. The main reasons for developing higher-level descriptions are:

1. To specify a realization.
2. To evaluate the quality of performance.
3. To manipulate the realization with the goal of increasing efficiency and/or quality of performance.

We shall address the second issue (finite word length effects) in Chapter 9, and the third issue (the design of filter structures) in Chapter 10.

In the remainder of this chapter, we shall develop several more detailed descriptions as discussed in Section 2.8. The first of these is the *primitive signal flow graph*, which is widely used because it is pictorial and because it can be manipulated to some degree. Then we shall develop *state variable* and *factored state variable descriptions*. These lead to powerful computational and analytical tools.

8.2 PRIMITIVE SIGNAL FLOW GRAPHS

When does a signal flow graph determine a realization? In answering this question we are led to the class of *primitive signal flow graphs* (PSFG). Not all signal flow graphs are primitive. For example, if we know $H(z)$, then we have an SFG as in Figure 8.2.1 which describes the filter, but does not determine a realization.

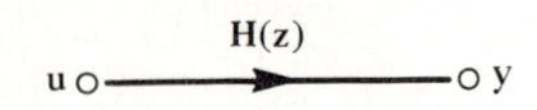

FIGURE 8.2.1.

If a signal flow graph determines a realization, then every node in the graph is a variable [the $v(m)$ in Eq. (8.1.8)] and every variable must have its own node in the graph. Some of the variables are available at the "top of the loop." From these variables alone, we must be able to compute all the rest using sums of products operations like Eq. (8.1.8). These available variables are either inputs, or variables that were updated during the previous pass through the loop and are in memory. Such variables can be represented in a signal flow graph as being outputs of unit delays. (The unit delay is a model for memory storage.) Since the node equations for all other variables must look like Eq. (8.1.8), the branches that do not correspond to unit delays must have constant gains. These branches represent the coefficients $c(l)$ in Eq. (8.1.8). From these considerations, and one other, we are led to the following definition.

Definition

A signal flow graph is *primitive* if

1. All branches are "primitive," that is, they have branch gains which are either constant, or z^{-1} (a unit delay).
2. There are no delay free loops in the graph.
3. The number of nodes and branches is finite.

Primitive signal flow graphs generate realizations (loop bodies), but somewhat indirectly. One must determine the order in which the node variables are updated, since the variables on the right-hand side of the equation must be valid before the variable on the left can be computed. The ordering is often easily done by inspection. However, it is worthwhile to consider the steps in some detail. One should keep the following observation in mind. At the top of the loop one has available the input variables and the stored variables, which are outputs of unit delays. All other variables must be computed from these.

Node Ordering Procedure

STEP 1. Examine each unit delay. If there is an incoming branch at the output, i.e.,

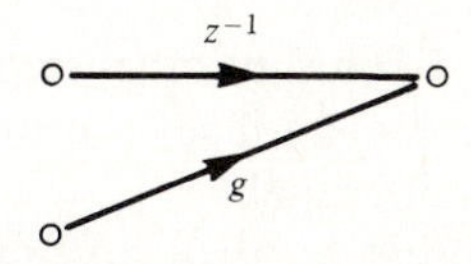

then isolate the output by inserting a new node and unity branch as in the following diagram.

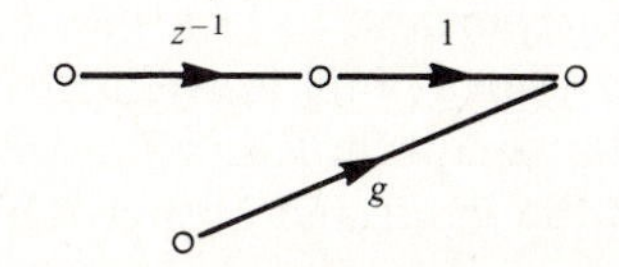

(The unit delays model stored variables. Since each such variable must be available at the beginning of each pass, it must be assigned a node.)

STEP 2. Label all input nodes, and all of the z^{-1} output nodes with the index $k = 0$. (These nodes represent variables that are available at the beginning of the loop, and need not be computed.) Set the index k to 1.

STEP 3. Examine all unlabeled nodes. Find all nodes that can be computed from labeled nodes and then label these with the index k. (All branches into an unlabeled node must be constant multipliers because of Step 1. A node can be computed if all nodes at the input side of these branches have been labeled.)

STEP 4. If there remain any unlabeled nodes in the graph, then increment the index k, and go back to Step 3.

How do we know that all nodes will be labeled by this procedure? Since the outputs of the z^{-1} branches are considered known, we may remove these branches. These variables are now input variables. By hypothesis, there are no loops in the graph that remain. Consequently, all paths have finite length. In fact, the longest path can involve no more than $m - 1$ branches for a PSFG with m nodes. Let S_k be the set of all nodes with index k. All nodes in S_1 are connected to S_0 nodes via exactly one branch. Similarly, all nodes in S_k are connected to input nodes by paths of at length at most k branches, with at least one path having length equal to k. If L is the length of the longest delay free path in the graph, then S_L is not empty, but S_{L+1} is empty. Since every node in the graph is connected to S_0 nodes by paths of length at most L, every node is labeled with index less than or equal to L.

EXAMPLE 8.2.1

The node sets for a lattice filter.

A "lattice" structure is shown in Fig. 8.2.2. The reader should check that this graph satisfies the requirements in the definition of a PSFG. Each node has been indexed according to the node ordering procedure. Note that there is a delay-free path along the bottom of the graph of length 4 connecting the input variable u to the single variable v_4 in S_4. This is the longest such path, and, therefore, $L = 4$. This number represents the computation delay inherent in this structure, even if several processors were available to compute separate statements in parallel.

The variable subsets S_k force an order of computation. No variable in S_k may be computed until *all* the variables in S_{k-1} have been computed. However, if there are several variables in S_k, then the order in which they are computed is not important. In fact, they could be computed concurrently if more than one arithmetic unit is available.

The signal flow graph in Fig. 8.2.2(a) would appear to have a total of 11 nodes, including the input and output. This would suggest that 11 memory

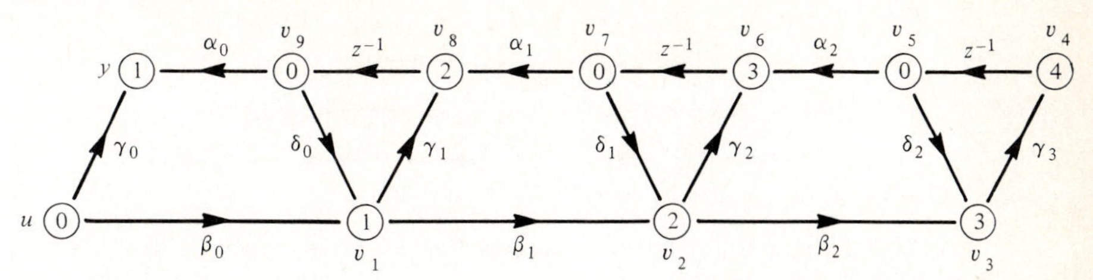

(a) Primitive signal flow graph

S_0: u, v_9, v_7, v_5

S_1: $y = \alpha_0 v_9 + \gamma_0 u$
$v_1 = \delta_0 v_9 + \beta_0 u$

S_2: $v_8 = \alpha_1 v_7 + \gamma_1 v_1$
$v_2 = \delta_1 v_7 + \beta_1 v_1$

S_3: $v_6 = \alpha_2 v_5 + \gamma_2 v_2$
$v_3 = \delta_2 v_5 + \beta_2 v_2$

S_4: $v_4 = \gamma_3 v_3$

Update: $v_9 \leftarrow v_8$
$v_7 \leftarrow v_6$
$v_5 \leftarrow v_4$

(b) Statement list for realization

FIGURE 8.2.2 Lattice Example.

locations are needed. But this is not really the case. Note that variables v_9 and v_8 could share the same memory location (thereby eliminating the "update" data move) since v_9 is used only in S_1 but v_8 is not computed until S_2. Thus we can regard $\{v_8, v_9\}$ as an equivalence class, corresponding to a single storage element. Similarly $\{v_6, v_7\}$, $\{v_4, v_5\}$, and $\{v_1, v_2, v_3\}$ are equivalence classes. It is profitable to identify these so that the hardware resources (both time and memory) are used efficiently. The signal flow graph description does not, however, make them obvious. We shall return to this point in Section 8.4.

There are systematic methods of generating primitive signal flow graphs, starting with $H(z)$ alone. Some of these were discussed in Section 3.7. The PSFG in Fig. 8.2.3 can be used for any second-order transfer function and is "canonical" in the sense that the branch gains are coefficients of the transfer function. This can be used as a building block in two different ways. If we factor $H(z)$ as

$$H(z) = \prod_{k=1}^{m} H_k(z), \tag{8.2.1}$$

where each $H_k(z)$ is order two with real coefficients, then we may obtain a PSFG for $H(z)$ by "cascading" second-order sections of the form in Fig. 8.2.3. Alternately, a partial fraction expansion of $H(z)$ followed by pairing complex conjugate poles leads to a sum

$$H(z) = \sum_{k=1}^{m} H_k(z) \tag{8.2.2}$$

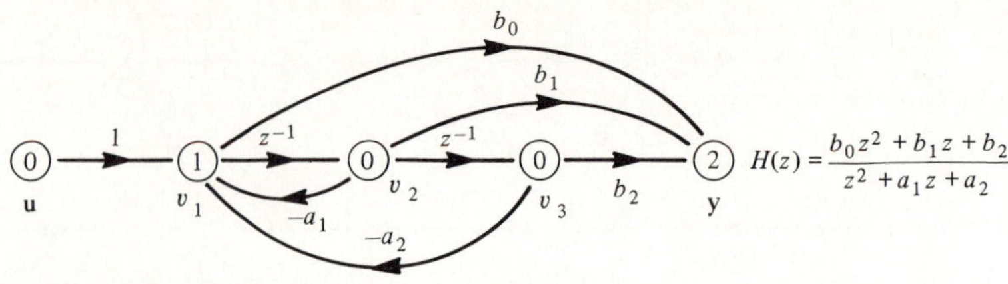

(a) Primitive signal flow graph

S_0: u, v_2, v_3

S_1: $v_1 = u - a_1 v_2 - a_2 v_3$

S_2: $y = b_0 v_1 + b_1 v_2 + b_2 v_3$

Updates: $v_3 \leftarrow v_2$
$v_2 \leftarrow v_1$

(b) Statement list for realization

FIGURE 8.2.3 Direct form, second-order section.

of second-order terms. This leads to a "parallel" combination of direct form second-order sections. Lattice realizations may be obtained by continued fraction expansions of $H(z)$. Problem 8.36 is an example of this.

EXAMPLE 8.2.2

A method of generating direct forms recursively.

A procedure for generating a direct form of arbitrary order is shown in Fig. 8.2.4. Part (a) of this figure shows the basic section, which contains a single unit delay and two nontrivial coefficient branches. If we connect this section to a one-input, two-output graph as shown by the dotted lines, we produce another one-input, two-output graph. The construction is terminated

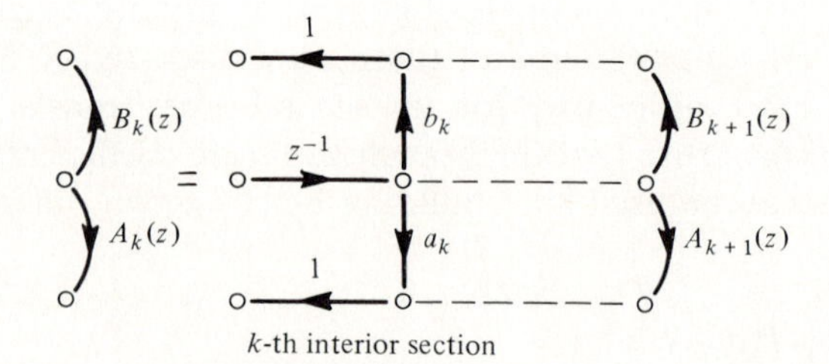

k-th interior section

$$B_k(z) = z^{-1}\,[b_k + B_{k+1}(z)]$$
$$A_k(z) = z^{-1}\,[a_k + A_{k+1}(z)]$$

(a) "Growing" the signal flow graph

Right side termination: $B_{n+1}(z) = A_{n+1}(z) = 0$

Then $B_1(z) = z^{-1}b_1 + z^{-2}b_2 + \cdots + z^{-n}b_n$

$A_1(z) = z^{-1}a_1 + z^{-2}b_2 + \cdots z^{-n}a_n$

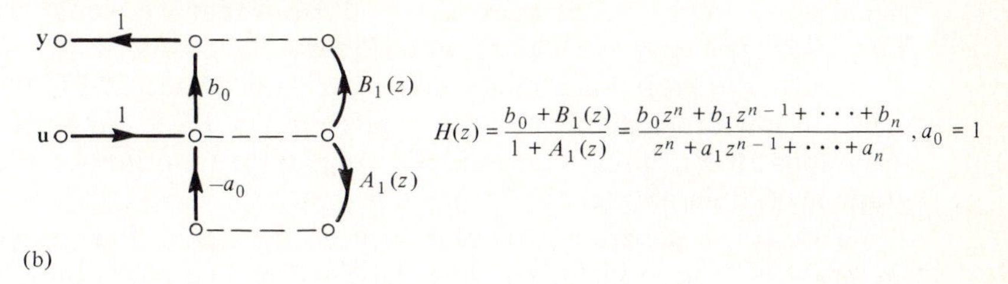

(b)

(c) Resulting direct form, for $n = 5$.

FIGURE 8.2.4 **High-order direct forms.**

as shown in part (b) of the figure, and the resulting transfer function has coefficients in one-to-one correspondence with the branch gains. (Note the reversal of the direction of the branch labeled a_0, which introduces loops for the first time.) The result of this construction for the case $n = 5$ is shown in part (c) and the nodes have been ordered. The statements in the realization that results from this construction are of the form

$$x \leftarrow y + c \cdot x, \qquad \textbf{(8.2.3)}$$

involving one multiplication and one addition. Direct forms of order greater than two are seldom used because of their poor finite word length performance.

EXAMPLE 8.2.3

Delay free loops in PSFGs.

Signal flow graphs that are otherwise primitive, but contain delay-free loops are excluded from membership by the definition of a PSFG. To see why this is so, let us consider an example. In Section 6.7, the design of arbitrary bandpass filters via frequency transformations is discussed. From a given low-pass filter $H(z)$, a new filter

$$G(z) = H(F(z)) \qquad \textbf{(8.2.4)}$$

is constructed, where $F(z)$ is a frequency transformation. Operating on signal flow graphs in this manner would mean that we would replace z^{-1} with $1/F(z)$, i.e., replace each z^{-1} with $(F(z))^{-1}$.

In order to keep the graph primitive, we should use a PSFG that realizes $1/F(z)$. A low-order example of this is given in Fig. 8.2.5. Note that although the graphs for $H(z)$ in part (a) and $1/F(z)$ in part (b) are primitive, the resulting graph for $G(z)$ in part (c) has a delay-free loop.

If we try to generate a realization from the signal flow graph for $G(z)$, the problem is soon identified. After labeling the two nodes in S_0 as shown, no other nodes can be labeled. Neither of the two nodes in the delay-free loop can be labeled until the other is. A delay-free loop is like a circular argument; every node variable in the loop depends on the one preceding it. This does not mean that we cannot realize $G(z)$, of course. But we must remove the delay-free loop nodes, e.g., by signal flow graph reduction techniques, to obtain a PSFG.

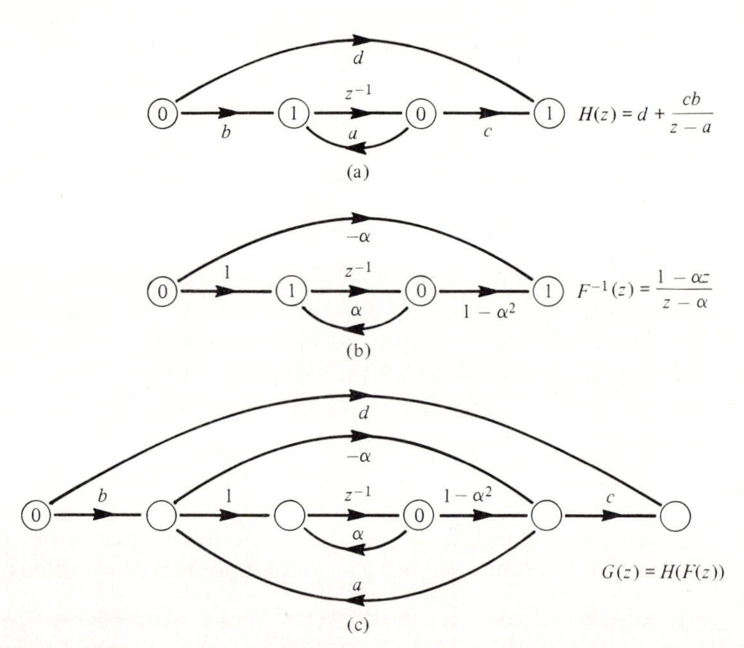

FIGURE 8.2.5 **Frequency transformations in signal flow graphs can lead to delay-free loops.**

How well do primitive signal flow graphs describe the "structure" of a digital filter? We have seen that from a PSFG, one can generate a realization or, in other words, an ordered list of operations. But the realization is obtained somewhat indirectly, through the node ordering procedure. The graph is pictorial, however, and suggests hardware if one equates z^{-1} branches with registers, nodes with accumulators, and constant branches with multipliers.

However, this can be misleading. It is very expensive to construct a filter in this manner. Actual digital filters will look more like the system of Fig. 8.1.1. (Digital signal processing devices usually have only one multiplier.)

The interpretation of z^{-1} branches as delay is also misleading. There is delay throughout the entire computational process, which is not modeled. We have seen that delay-free paths of length greater than one correspond to unavoidable delay, as some node variables cannot be updated until all the variables that precede it in the path have been updated. In an attempt to model these computational delays, in a very large scale integration (VLSI) device environment where parallel processing is feasible, generalizations of primitive signal flow graphs containing "fractional" delays (like $z^{-1/2}$) have been introduced. We shall see examples of these in Section 8.4.

8.3 STATE VARIABLE DESCRIPTIONS

State variable descriptions are a specialization of PSFG descriptions. For every PSFG, there is a unique state variable description (SVD), but the reverse is not true. The SVD can be thought of as the reduced PSFG, which is *canonical* with respect to the unit delay branches. Why are the unit delay variables selected to be preserved? It is because, at a given time k, the *independent* variables in the PSFG are the inputs and the stored variables, or (outputs of the unit delays). All other variables may be computed from these.

There are important problems involving digital filter structures that are quite naturally formulated using the SVD. With this matrix description, and linear algebra, one can apply powerful analytical methods. Furthermore, when it comes time to compute, the state variable approach is invaluable.

We shall define the SVD by construction, starting from the PSFG.

State Variable Description from a PSFG

STEP 1. Replace each unit delay with the equivalent path and label the input and output variables as shown.

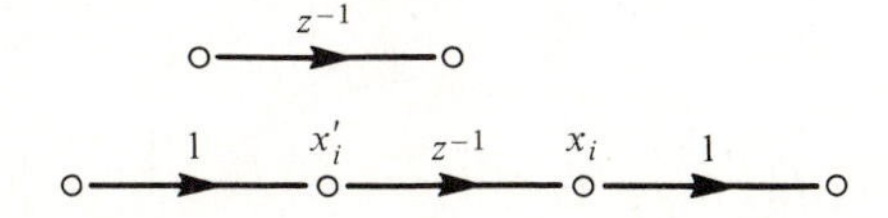

STEP 2. Remove the unit delays, thereby creating new outputs x_i' and inputs x_i. Note $x_i'(k) \triangleq x_i(k+1)$.

STEP 3. Completely reduce the graph. (Eliminate all nodes that are not inputs or outputs.)

STEP 4. Replace the unit delays.

After Step 3, the only remaining variables or nodes are the original input(s), the original output(s), and the delay variables x_i, x_i' for $i = 1, \ldots, n$. The equations (without the delays) have the matrix form

$$\mathbf{X}' = \mathbf{AX} + \mathbf{BU}, \tag{8.3.1}$$

$$\mathbf{Y} = \mathbf{CX} + \mathbf{DU}. \tag{8.3.2}$$

These represent equations for the two kinds of outputs $\mathbf{X}'$ and $\mathbf{Y}$ in terms of the two kinds of inputs $\mathbf{X}$ and $\mathbf{U}$. The coefficient matrices ($\mathbf{A}$, $\mathbf{B}$, $\mathbf{C}$, $\mathbf{D}$) must be constant, since they cannot involve the variable z. These matrices are called the *state variable parameterization*, and have the following dimensions:

$$\begin{array}{c} \\ n \\ \\ l \end{array} \overset{\quad n \quad m}{\left[\begin{array}{c|c} \mathbf{A} & \mathbf{B} \\ \hline \mathbf{C} & \mathbf{D} \end{array}\right]}, \tag{8.3.3}$$

where n is the number of unit delays (the dimension of $\mathbf{X}$ and $\mathbf{X}'$), m is the number of original input nodes (the dimension of $\mathbf{U}$), and l the number of original output nodes (the dimension of $\mathbf{Y}$).

Replacing the unit delays merely adds the equation

$$\mathbf{X} = z^{-1}\mathbf{X}' \tag{8.3.4}$$

to Eqs. (8.3.1) and (8.3.2). Equations (8.3.1), (8.3.2), and (8.3.4) are the z-transform version of the *state variable description* for the original PSFG.

EXAMPLE 8.3.1

A state variable description of a digital filter.

An example is given in Fig. 8.3.1. With practice, it is not difficult to write down a state variable parameterization by inspection. However, care should be exercised. A common error for this example is to overlook the fact that there are two paths to the output node from the nodes x_1 and x_2, which produce the two terms for each component of the vector $\mathbf{C}$.

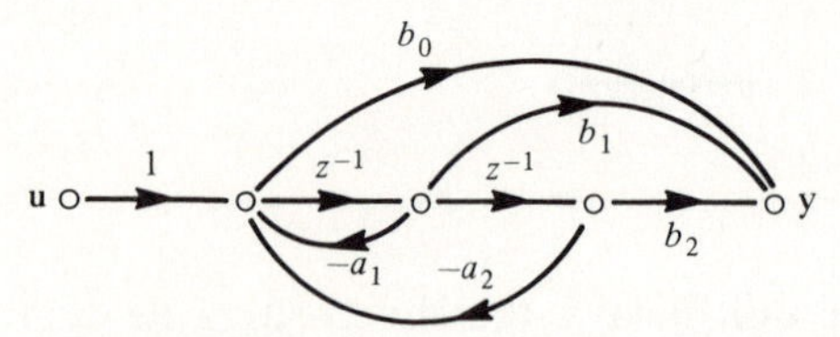

(a) Original PSFG

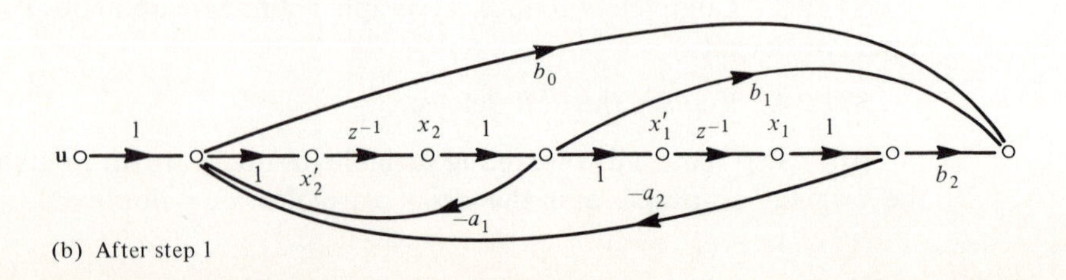

(b) After step 1

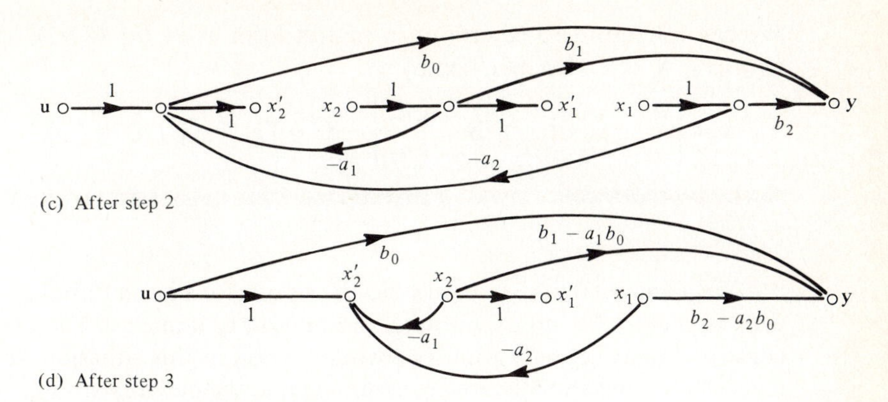

State variable description:

$$A = \begin{bmatrix} 0 & 1 \\ -a_2 & -a_1 \end{bmatrix}, \quad B = \begin{bmatrix} 0 \\ 1 \end{bmatrix}, \quad C = [b_2 - a_2 b_0, b_1 - a_1 b_0], \quad D = b_0$$

FIGURE 8.3.1 An example of a state variable description.

EXAMPLE 8.3.2
A state variable description.

Another example of obtaining the SVD from an SFG is shown here. We identify the outputs of the unit delays as $x_1(k)$ and $x_2(k)$. The inputs are $x'_1(k) = x_1(k + 1)$ and $x'_2(k) = x_2(k + 1)$. We first write equations for $x_1(k + 1)$ and $x_2(k + 1)$ in terms of the input $u(k)$ and the state variables $x_1(k)$ and $x_2(k)$:

$$x'_1(k) = x_1(k + 1) = a_{11}x_1(k) + a_{12}x_2(k) + b_1 u(k),$$
$$x'_2(k) = x_2(k + 1) = a_{21}x_1(k) + a_{22}x_2(k) + b_2 u(k).$$

We next write an equation for $y(k)$ in terms of the state $x^T = (x_1, x_2)$ and the input $u(k)$. Thus

$$y(k) = c_1 x_1(k) + c_2 x_2(k) + du(k).$$

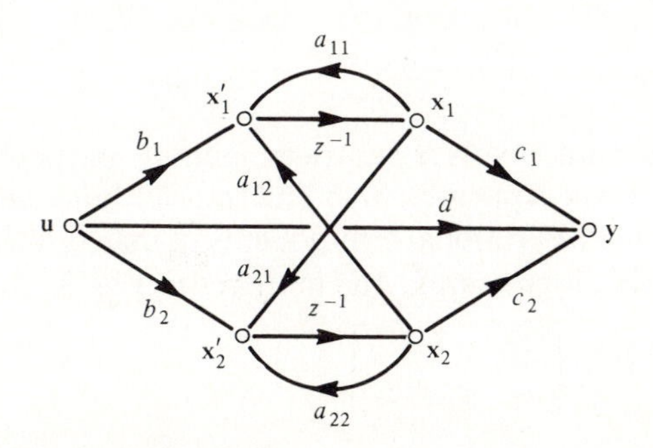

FIGURE 8.3.2 A state variable SFG.

We can rewrite these equations in matrix form as in Eq. (8.3.1), where the matrices (**A**, **B**, **C**, **D**) are given by

$$\mathbf{A} = \begin{bmatrix} a_{11} & a_{12} \\ a_{21} & a_{22} \end{bmatrix}, \quad \mathbf{B} = \begin{bmatrix} b_1 \\ b_2 \end{bmatrix}, \quad \mathbf{C} = [c_1 \quad c_2], \quad \mathbf{D} = [d].$$

We point out that a state variable description for a given PSFG is *canonical* with respect to the inputs, outputs, and unit delay branches. The steps in the construction are exactly what one would expect in this situation. Intuitively then, when using an SVD, we are assuming that the memory storage primitive elements (unit delays) are all that really matter about the internal structure of the filter. One should keep this in mind when considering how much detail is required.

An SVD is equivalent to a PSFG with one extra property. In a state variable PSFG, the longest delay-free path always has length $L = 1$. Thus one can consider state variable PSFGs as the subclass of all PSFGs for which $L = 1$. [Note that in the original PSFG of Figure 8.3.1, we have $L = 2$.]

We turn now to the development of some standard applications of the SVD. To begin with, let us develop some of the links shown in Fig. 2.8.1. The goal is to generate lower-level descriptions from the state variable parameterization (**A**, **B**, **C**, **D**).

Transfer Function from the State Variable Description

To obtain the transfer function, one simply eliminates the internal variables **X** and **X**′ from the set of Eqs. (8.3.1)–(8.3.4). Elimination of **X**′ from Eqs. (8.3.1) and (8.3.4) leads to

$$\mathbf{X}(z) = [z\mathbf{I} - \mathbf{A}]^{-1}\mathbf{B}\mathbf{U}(z). \tag{8.3.5}$$

Substitution of this term into Eq. (8.3.2) yields

$$\mathbf{Y}(z) = \mathbf{C}(z\mathbf{I} - \mathbf{A})^{-1}\mathbf{B}\mathbf{U}(z) + \mathbf{D}\mathbf{U}(z). \tag{8.3.6}$$

We therefore identify the transfer function as

$$\mathbf{H}(z) = \mathbf{D} + \mathbf{C}(z\mathbf{I} - \mathbf{A})^{-1}\mathbf{B}. \tag{8.3.7}$$

This expression involves matrices, and in particular the matrix resolvant $(z\mathbf{I} - \mathbf{A})^{-1}$. (See Appendix 8A.) The transfer function $\mathbf{H}(z)$ given by Eq. (8.3.7) is *rational* (or has rational components, if matrix-valued.)

Consider, for example, the matrices in Fig. 8.3.1.

$$(z\mathbf{I} - \mathbf{A}) = \begin{bmatrix} z & -1 \\ a_2 & z + a_1 \end{bmatrix}. \tag{8.3.8}$$

Thus

$$(z\mathbf{I} - \mathbf{A})^{-1}\mathbf{B} = \frac{1}{(z^2 + a_1 z + a_2)}\begin{bmatrix} z + a_1 & 1 \\ -a_2 & z \end{bmatrix}\begin{bmatrix} 0 \\ 1 \end{bmatrix}. \qquad \textbf{(8.3.9)}$$

And so

$$H(z) = D + \mathbf{C}(z\mathbf{I} - \mathbf{A})^{-1}\mathbf{B} = \frac{b_0 z^2 + b_1 z + b_2}{z^2 + a_1 z + a_2}. \qquad \textbf{(8.3.10)}$$

Standard Difference Equation from the State Variable Description

We have seen that a PFSG with n unit delays determines a SVD whose state vector has dimension n. From the parameterization (**A**, **B**, **C**, **D**), we generate the transfer function $H(z)$ in Eq. (8.3.7). For a single-input, single-output filter, this function is the ratio of two nth-degree polynomials. This same transfer function can be obtained from a system satisfying a standard difference equation given in Section 2.6. Since $H(z)$ completely determines the I/O map of the filter, we have shown that every single-input, single-output, finite-order filter must produce output sequences that satisfy an SDE. The coefficients in the difference equation are obtained from the numerator and denominator polynomials of $H(z)$. An example is given in Fig. 8.3.3.

Time Domain State Variable Descriptions

State variable descriptions are usually given in the time domain. Let us therefore convert Eqs. (8.3.1), (8.3.2), and (8.3.4) to the time domain. Equation (8.3.4) becomes

$$\mathbf{x}(k) = \mathbf{x}'(k - 1)$$

or

$$\mathbf{x}'(k) = \mathbf{x}(k + 1). \qquad \textbf{(8.3.11)}$$

Using this, Eqs. (8.3.1) and (8.3.2) become

$$\mathbf{x}(k + 1) = \mathbf{A}\mathbf{x}(k) + \mathbf{B}\mathbf{u}(k), \qquad \textbf{(8.3.12)}$$

$$\mathbf{y}(k) = \mathbf{C}\mathbf{x}(k) + \mathbf{D}\mathbf{u}(k). \qquad \textbf{(8.3.13)}$$

These equations are generally what is meant by a state variable description. The vector $\mathbf{x}(k)$, whose dimension is the number of unit delays, is called the "state" of the filter. Equation (8.3.12) is called the "next state" equation, and Eq. (8.3.13) is called the "output" equation.

For a given input sequence, the sequence of states is called a "trajectory." We turn now to the question of how the trajectory is determined, using Eq. (8.3.12). We seek, in other words, a closed form solution. Using Eq. (8.3.12)

EXAMPLE 8.3.3
Several descriptions of a digital filter.

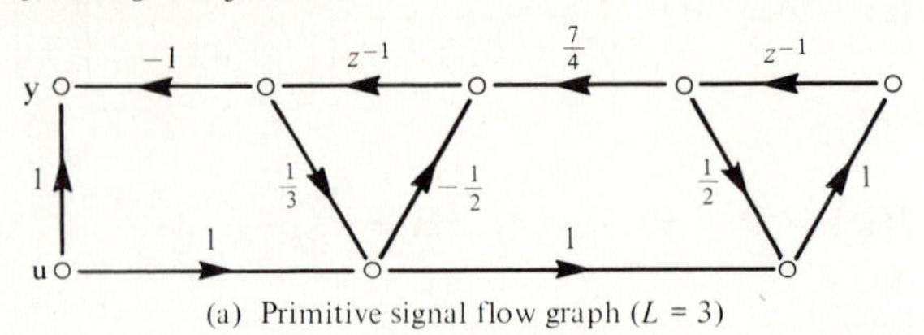

(a) Primitive signal flow graph ($L = 3$)

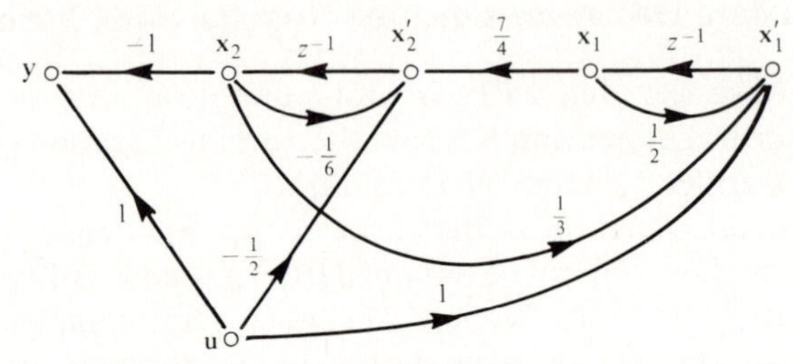

(b) Resulting state variable primitive signal flow graph ($L = 1$)

$$\mathbf{A} = \begin{bmatrix} \frac{1}{2} & \frac{1}{3} \\ \frac{7}{4} & -\frac{1}{6} \end{bmatrix} \cdot \quad \mathbf{B} = \begin{bmatrix} 1 \\ -\frac{1}{2} \end{bmatrix}$$

$$\mathbf{C} = [0 \quad -1] \ , \quad \mathbf{D} = [1]$$

(c) State variable parameterization

$$H(z) = D + \mathbf{C}(z\mathbf{I} - \mathbf{A})^{-1}\mathbf{B}$$

$$= \frac{z^2 + z/6 - 8/3}{z^2 - z/3 - 2/3}$$

(d) Transfer function

$$y(k) - \frac{1}{3}y(k-1) - \frac{2}{3}y(k-2) = u(k) + \frac{1}{6}u(k-1) - \frac{8}{3}u(k-2)$$

(e) Standard difference equation

FIGURE 8.3.3 **An example of several descriptions for a second-order digital filter.**

twice, we find that

$$\begin{aligned} \mathbf{x}(k+2) &= \mathbf{A}(\mathbf{A}\mathbf{x}(k) + \mathbf{B}\mathbf{u}(k)) + \mathbf{B}\mathbf{u}(k+1), \\ &= \mathbf{A}^2\mathbf{x}(k) + \mathbf{A}\mathbf{B}\mathbf{u}(k) + \mathbf{B}\mathbf{u}(k+1). \end{aligned} \tag{8.3.14}$$

The generalization of this is the following, which can be verified by direct substitution:

$$\mathbf{x}(k) = \mathbf{A}^{k-k_0}\mathbf{x}(k_0) + \sum_{l=1}^{k-k_0} \mathbf{A}^{l-1}\mathbf{B}\mathbf{u}(k-l), \qquad k > k_0. \tag{8.3.15}$$

Equation (8.3.15) gives the value of the state of the system at time k in terms of (1) the state of some previous time k_0 and (2) the input sequence in the time interval k_0 to $k - 1$. Note that it involves powers of the matrix $\mathbf{A}$. (We discuss powers of a matrix in Appendix 8A.) Equation (8.3.15) has some interesting consequences.

The state $\mathbf{x}(k_0)$ summarizes the entire past history of the input sequence [$\mathbf{u}(k)$ for $k < k_0$]. In other words, the trajectory for all time $k > k_0$ depends only on $\mathbf{x}(k_0)$ and the subsequent input sequence. For this reason, it is often said that the state of a system embodies its memory of the past. It represents all that is necessary to know about the past in order to predict the future.

Suppose that the input sequence is identically zero. Then Eq. (8.3.15) reduces to

$$\mathbf{x}(k) = \mathbf{A}^{k-k_0}\mathbf{x}(k_0). \tag{8.3.16}$$

This is the solution to a vector version of the *homogeneous initial value problem*:

$$\begin{aligned} &\mathbf{x}(k_0) \quad \text{given} \\ &\mathbf{x}(k+1) = \mathbf{A}\mathbf{x}(k), \qquad k \geqslant k_0. \end{aligned} \tag{8.3.17}$$

What happens in the long run as k approaches infinity? We show in Appendix 8A that the matrix powers $\mathbf{A}^k$ tend to the zero matrix as k approaches infinity if and only if the eigenvalues λ_k of the matrix $\mathbf{A}$ satisfy the inequality

$$|\lambda_k| < 1, \qquad k = 1, 2, \ldots, n. \tag{8.3.18}$$

This is equivalent to the stability condition given in Section 2.6, for the eigenvalues of $\mathbf{A}$ are the roots of the polynomial

$$\hat{a}(z) = \det(z\mathbf{I} - \mathbf{A}),$$

which (by Cramer's Rule) is the denominator of the transfer function $H(z)$ as expressed in Eq. (8.3.7). In other words, the poles of the transfer function $H(z)$ are the eigenvalues of the matrix $\mathbf{A}$.

If we now assume that the filter is stable, in the sense that all eigenvalues have modulus less than one, then we may rewrite Eq. (8.3.15) with $k_0 = -\infty$, and drop the first term. The result is

$$\mathbf{x}(k) = \sum_{l=1}^{\infty} \mathbf{A}^{l-1}\mathbf{B}\mathbf{u}(k - l). \tag{8.3.19}$$

Unit-Pulse Response from the State Variable Description

We know that the input and output sequences are related by the equation

$$\mathbf{y} = \mathbf{h} * \mathbf{u}.$$

In order to identify the unit-pulse response sequence in terms of the SVD, let us compute $y(k)$ using Eq. (8.3.19) (which assumes stability) and the output equation (for a scalar input and output)

$$\begin{aligned} y(k) &= \mathbf{C}\mathbf{x}(k) + \mathbf{D}u(k) \\ &= \sum_{l=1}^{\infty} \mathbf{C}\mathbf{A}^{l-1}\mathbf{B}u(k-l) + \mathbf{D}u(k) \\ &= \sum_{l=-\infty}^{\infty} h(l)u(k-l). \end{aligned} \tag{8.3.20}$$

In order for the last two lines to agree, we must have

$$h(k) = \begin{cases} 0, & k < 0 \\ \mathbf{D}, & k = 0 \\ \mathbf{C}\mathbf{A}^{k-1}\mathbf{B}, & k > 0 \end{cases} \tag{8.3.21}$$

which is the desired relationship.

The unit-pulse response has three separate parts, in agreement with Eq. (2.6.20) (which assumes causality). The elements of $\mathbf{h}$ are zero for $k < 0$ (causality), equal to $\mathbf{D}$ for $k = 0$, and a linear combination of the powers of the matrix $\mathbf{A}$ for $k > 0$. We show in Appendix 8A that the powers of an $n \times n$ matrix $\mathbf{A}$ satisfy an nth-order homogeneous difference equation. It is this third term that must take the form (2.6.20). Recall that the elements $h(1)$ through $h(n)$ were used as initial data for the homogeneous SDE (2.6.19). These elements come from the powers of the matrix $\mathbf{A}$. The element $h(0) = \mathbf{D}$, which does not contain powers of $\mathbf{A}$, represents the "straight through" path from $\mathbf{u}$ to $\mathbf{y}$ in the system.

Coordinate Transformations

An input/output relation like $\mathbf{y} = \mathbf{h} * \mathbf{u}$ is sometimes called an "external" description because it gives no information about the internal structure. The state $\mathbf{x}(k)$ in Eqs. (8.3.12) and (8.3.13) represents n internal variables, if n is the dimension of the vector $\mathbf{x}(k)$. One can change the coordinate system for $\mathbf{x}(k)$ without changing the *external* description (the unit-pulse response sequence). Such a transformation will, however, change the *internal* description $(\mathbf{A}, \mathbf{B}, \mathbf{C}, \mathbf{D})$.

Let $\mathbf{T}$ be an $n \times n$ nonsingular matrix, and let

$$\mathbf{q}(k) = \mathbf{T}^{-1}\mathbf{x}(k). \tag{8.3.22}$$

Substitution of this transformation into the state variable equations (8.3.12) and (8.3.13) yields

$$\begin{aligned} \mathbf{q}(k+1) &= \mathbf{T}^{-1}[\mathbf{A}\mathbf{x}(k) + \mathbf{B}u(k)] \\ &= \mathbf{T}^{-1}\mathbf{A}\mathbf{T}\mathbf{q}(k) + \mathbf{T}^{-1}\mathbf{B}u(k) \end{aligned} \tag{8.3.23}$$

$$y(k) = \mathbf{C}\mathbf{T}\mathbf{q}(k) + \mathbf{D}u(k). \tag{8.3.24}$$

Equations (8.3.23) and (8.3.24) have the same form as the original, except that the parameterization has undergone the following transformation:

$$(\mathbf{A}, \mathbf{B}, \mathbf{C}, \mathbf{D}) \quad \leftarrow \quad (\mathbf{T}^{-1}\mathbf{A}\mathbf{T}, \mathbf{T}^{-1}\mathbf{B}, \mathbf{C}\mathbf{T}, \mathbf{D}). \tag{8.3.25}$$

This transformation allows one to study an infinite set of externally equivalent SVDs. It is a very useful method of modifying realizations in the digital filter synthesis problem. One can easily demonstrate that the external descriptions are *invariant under coordinate transformation.*

Transfer Function Invariance:

$$\begin{aligned}\mathbf{H}'(z) &= \mathbf{D}' + \mathbf{C}'(z\mathbf{I} - \mathbf{A}')^{-1}\mathbf{B}' = \mathbf{D} + \mathbf{C}\mathbf{T}(z\mathbf{I} - \mathbf{T}^{-1}\mathbf{A}\mathbf{T})^{-1}\mathbf{T}^{-1}\mathbf{B} \\ &= \mathbf{D} + \mathbf{C}\mathbf{T}(\mathbf{T}^{-1}(z\mathbf{I} - \mathbf{A})\mathbf{T})^{-1}\mathbf{T}^{-1}\mathbf{B} = \mathbf{D} + \mathbf{C}\mathbf{T}\mathbf{T}^{-1}(z\mathbf{I} - \mathbf{A})^{-1}\mathbf{T}\mathbf{T}^{-1}\mathbf{B} \\ &= \mathbf{D} + \mathbf{C}(z\mathbf{I} - \mathbf{A})^{-1}\mathbf{B} \equiv \mathbf{H}(z).\end{aligned} \tag{8.3.26}$$

Unit-Pulse Response Sequence Invariance:

$$\begin{aligned}\mathbf{h}'(k) &= \begin{cases} 0 \\ \mathbf{D}' \\ \mathbf{C}'\mathbf{A}'^{k-1}\mathbf{B}' \end{cases} = \begin{cases} 0 \\ \mathbf{D} \\ \mathbf{C}\mathbf{T}(\mathbf{T}^{-1}\mathbf{A}\mathbf{T})\cdots(\mathbf{T}^{-1}\mathbf{A}\mathbf{T})\mathbf{T}^{-1}\mathbf{B} \end{cases} \\ &= \begin{cases} 0, & k < 0 \\ \mathbf{D}, & k = 0. \\ \mathbf{C}\mathbf{A}^{k-1}\mathbf{B}, & k > 0 \end{cases}\end{aligned} \tag{8.3.27}$$

The state variable PSFGs which correspond to identical external descriptions may look markedly different, however, as Fig. (8.3.4) demonstrates.

Poles and Zeros and the State Variable Description

All finite-order filters have rational transfer functions. In other words, $H(z)$ is the ratio of two polynomials. Let

$$\hat{a}(z) = z^n + a_1 z^{n-1} + \cdots + a_n, \tag{8.3.28}$$

$$\hat{b}(z) = b_0 z^n + b_1 z^{n-1} + \cdots + b_n, \tag{8.3.29}$$

be the denominator and numerator of such a (single-input, single-output) filter having transfer function

$$H(z) = \frac{\hat{b}(z)}{\hat{a}(z)}. \tag{8.3.30}$$

We say that the SVD $(\mathbf{A}, \mathbf{B}, \mathbf{C}, \mathbf{D})$ *realizes* $H(z)$ if

$$H(z) = \mathbf{D} + \mathbf{C}(z\mathbf{I} - \mathbf{A})^{-1}\mathbf{B}. \tag{8.3.31}$$

We have not yet shown that a realization always exists.

Consider the relation of $(\mathbf{A}, \mathbf{B}, \mathbf{C}, \mathbf{D})$ to the poles and zeros of $H(z)$, which are the roots of the polynomials $\hat{a}(z)$ and $\hat{b}(z)$, respectively. We know that the

EXAMPLE 8.3.4

Externally equivalent digital filters.

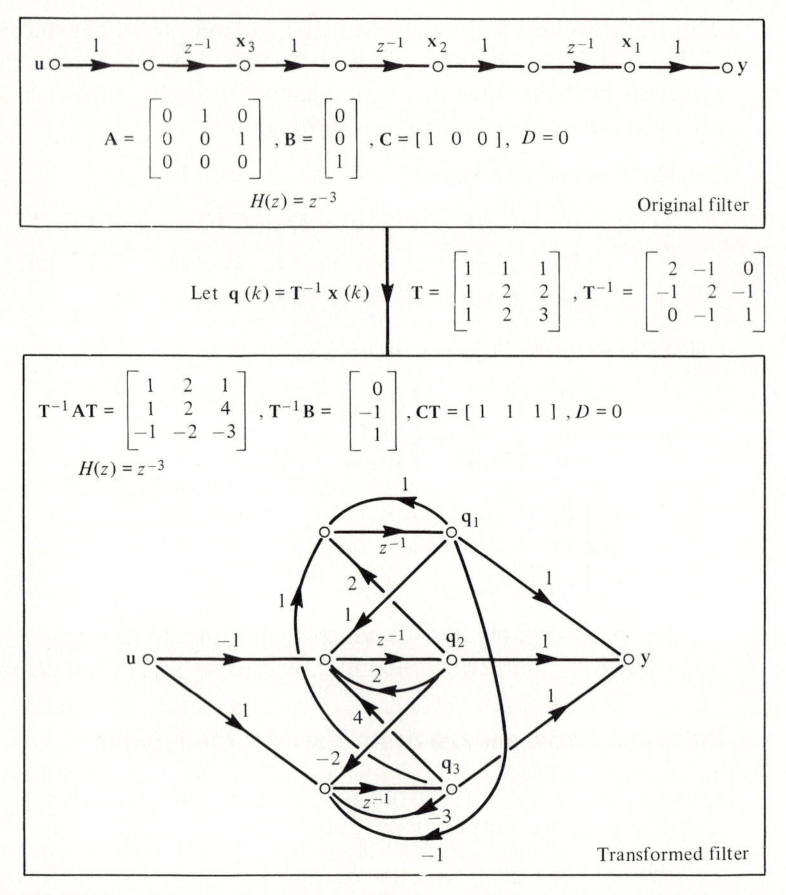

FIGURE 8.3.4 **An example of two externally equivalent filters. Both have transfer function $H(z) = z^{-3}$.**

poles of the filter are the eigenvalues of **A**. This can be seen from the relation

$$\hat{a}(z) = \det(z\mathbf{I} - \mathbf{A}), \tag{8.3.32}$$

which holds by Cramer's Rule for matrix inversion. The polynomial $\hat{a}(z)$ is called the *characteristic polynomial of* **A**. Thus the poles of the filter depend only on the matrix **A**.

What about the zeros, or to be precise, the numerator polynomial $\hat{b}(z)$ (which includes the gain constant b_0)? There are $n + 1$ coefficients in this polynomial. These coefficients are related to the $n + 1$ components in **C** and **D** via linear equations. This system of equations is invertible if and only if the matrices **A** and **B** form a controllable pair, which means that the $n \times n$ "controllability" matrix

$$\mathcal{B} = [\mathbf{B}, \mathbf{AB}, \mathbf{A}^2\mathbf{B}, \ldots, \mathbf{A}^{n-1}\mathbf{B}], \tag{8.3.33}$$

is nonsingular. Although there are a number of ways to write these equations, let us note that

$$h(0) = \mathbf{D} \tag{8.3.34}$$

$$[h(1), h(2), \ldots, h(n)] = \mathbf{C}[\mathbf{B}, \mathbf{AB}, \ldots, \mathbf{A}^{n-1}\mathbf{B}]. \tag{8.3.35}$$

We may conclude from these equations that if the controllability matrix is nonsingular for a given pair (**A**, **B**), then we may choose the first $n + 1$ elements of the unit-impulse response sequence arbitrarily. To complete our original discussion, we need only note that the numbers $h(0)$ through $h(n)$ and the numbers b_0 through b_n are related by a triangular system of equations which we can write as

$$\begin{bmatrix} 1 & 0 & 0 & \cdots & 0 \\ a_1 & 1 & 0 & \cdots & 0 \\ & & \vdots & & \\ a_n & a_{n-1} & & \cdots & 1 \end{bmatrix} \begin{bmatrix} h(0) \\ h(1) \\ \vdots \\ h(n) \end{bmatrix} = \begin{bmatrix} b_0 \\ b_1 \\ \vdots \\ b_n \end{bmatrix}. \tag{8.3.36}$$

(Note that the $b_0, b_1, \ldots, b_n$ used here are not to be confused with the elements of the matrix **B**.)

8.4 FACTORED STATE VARIABLE DESCRIPTIONS

The goal of a realization loop body is to compute the next-state variables and output(s) from the current-state variables and input(s). A state variable description eliminates consideration of any variables but these, and in so doing may combine several arithmetic operations. A *factored state variable description* combines the detail of a primitive signal flow graph with the matrix description of state variables. Such a description is useful in complex but "regular" structures that are desirable in VLSI implementation. It is called a factored description because it amounts to a matrix factorization of the state variable description:

$$\begin{bmatrix} \mathbf{A} & \mathbf{B} \\ \mathbf{C} & \mathbf{D} \end{bmatrix} = \mathbf{F}_L \cdot \mathbf{F}_{L-1} \cdots \mathbf{F}_1. \tag{8.4.1}$$

The number of factors is the length of the longest delay-free path in a PSFG description. We shall define the FSVD by construction, starting with a PSFG.

Construction of the Factored State Variable Description

Given a primitive signal flow graph, use the node ordering procedure of Section 8.2 to get the sets $S_0, S_1, \ldots, S_L$ of node variables. Let $\mathbf{v}_k$ denote a vector whose components are the variables in S_k. In particular, let

$$\mathbf{v}_0 = \begin{bmatrix} x_1 \\ x_2 \\ \vdots \\ x_n \\ u \end{bmatrix}, \tag{8.4.2}$$

where the x_k are the outputs of unit delays. Each variable in S_k is a linear combination of variables in $S_0 \cup S_1 \cdots \cup S_{k-1}$. Therefore, there is a coefficient matrix $\mathbf{G}_k$ for which

$$\mathbf{v}_k = \mathbf{G}_k \begin{bmatrix} \mathbf{v}_{k-1} \\ \mathbf{v}_{k-2} \\ \vdots \\ \mathbf{v}_0 \end{bmatrix}, \qquad 1 \leqslant k \leqslant L. \tag{8.4.3}$$

Finally, let $\mathbf{G}_0$ be a matrix containing zeros and ones that picks out the next-state and output(s):

$$\begin{bmatrix} x_1' \\ x_2' \\ \vdots \\ x_n' \\ y \end{bmatrix} = \mathbf{G}_0 \begin{bmatrix} \mathbf{v}_L \\ \mathbf{v}_{L-1} \\ \vdots \\ \mathbf{v}_0 \end{bmatrix}. \tag{8.4.4}$$

Taken together, Eqs. (8.4.3) and (8.4.4) become

$$\begin{bmatrix} \mathbf{x}' \\ y \end{bmatrix} = \mathbf{G}_0 \begin{bmatrix} \mathbf{G}_L \\ \mathbf{I} \end{bmatrix} \begin{bmatrix} \mathbf{G}_{L-1} \\ \mathbf{I} \end{bmatrix} \cdots \begin{bmatrix} \mathbf{G}_1 \\ \mathbf{I} \end{bmatrix} \begin{bmatrix} \mathbf{x} \\ u \end{bmatrix} \tag{8.4.5}$$

These matrices will increase in size from right to left, with the exception of $\mathbf{G}_0$. Since $\mathbf{G}_0$ does not contain any actual multipliers, it may be combined with the matrix to its right, leaving L factors. Now, reduce the size of these matrices by eliminating any column containing only zeros, together with the corresponding row of the matrix to its right. The identity matrices in Eq. (8.4.5) preserve the previously computed node variables. These identity matrices grow in size from right to left. Several examples follow to illustrate this construction.

Examples of FSVDs

Figure 8.4.1 contains the development of a FSVD for a second-order direct form structure. At the top of the figure, we find the PSFG with the node variables ordered. We then find the sequence of updates, denoted by left arrows. For each left arrow there is a matrix factor, appearing on the line below. Note that the leftmost matrix $\mathbf{G}_0$ is trivial in that each row contains one 1 and the rest all zeros. After combining the two matrices on the left,

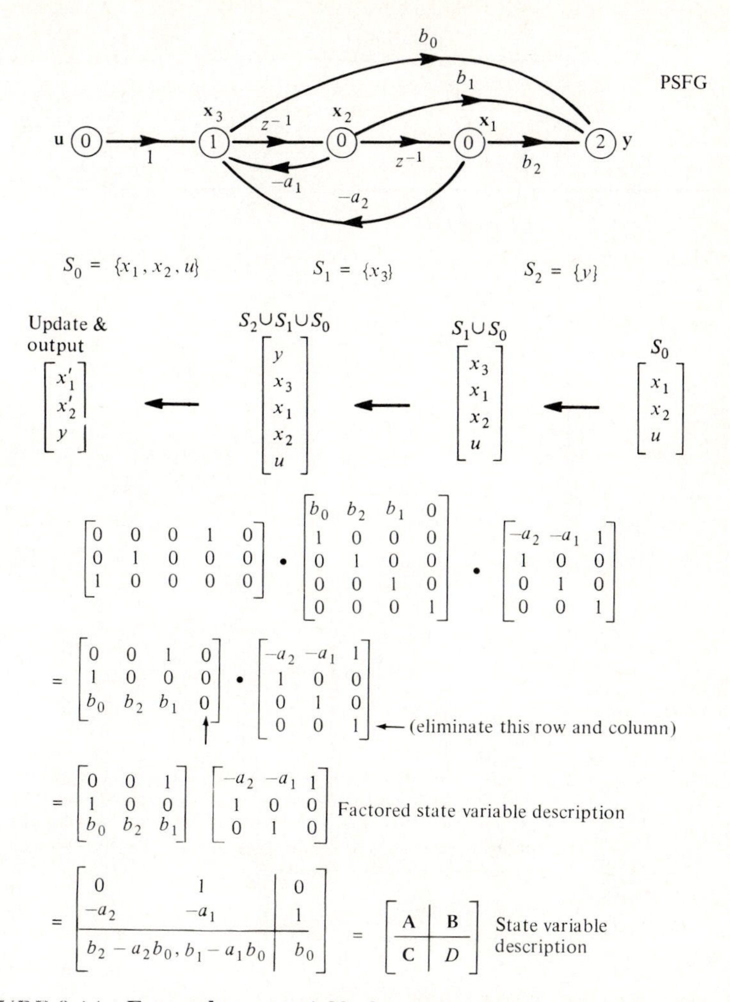

FIGURE 8.4.1 Factored state variable description for second-order direct form.

and eliminating one inessential row and column, we have the factored state variable description. There are $L = 2$ factors, corresponding to the length of the longest delay-free path. These matrices yield the state variable description (shown on the last line) when multiplied together.

Figure 8.4.2 contains a more complicated example, a "lattice" structure. In this case several rows and columns are eliminated before yielding a rather simple FSVD. We have $L = 3$ factors, each of which contains a 2×2 submatrix of nontrivial coefficients. Note that the state variable description involves more nontrivial multiplications (13) than does the FSVD (12). The disparity would increase as the order of the lattice is increased.

It is often helpful to simplify flow graphs by gathering several node variables into a single *vector*-valued node. If nodes are vectors, then branches

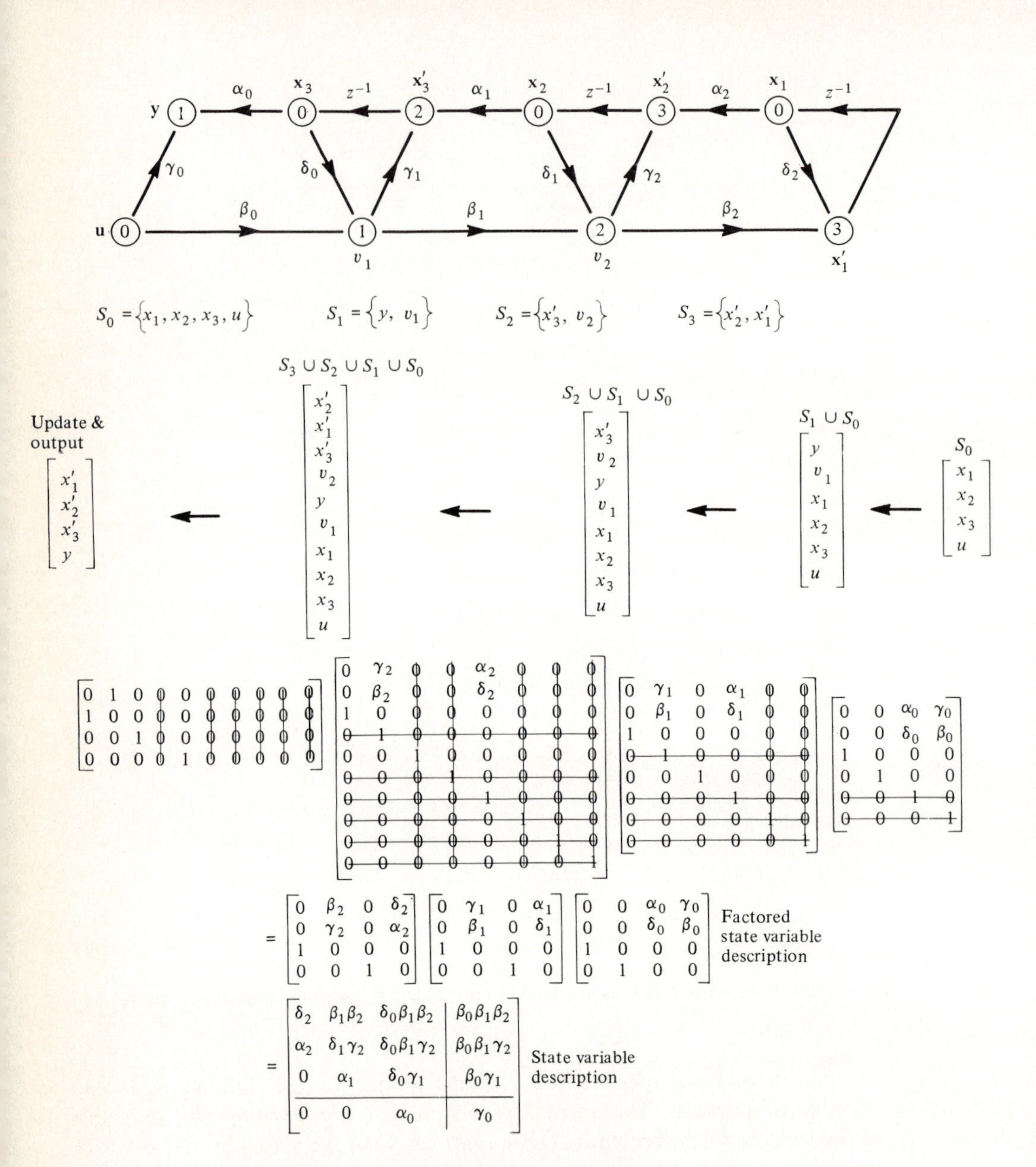

FIGURE 8.4.2 **Factored state variable description for a lattice.**

would become matrices. Figure 8.4.3 is an example of such a graph and represents the cascade connection of two state variable PSFGs. In a state variable graph, L must be one, but in the cascade L is two, since we have produced delay-free paths of that length. The same development can be followed as before to produce a factored state variable description, but the matrices contain matrix blocks, rather than scalars. The last line represents the rule

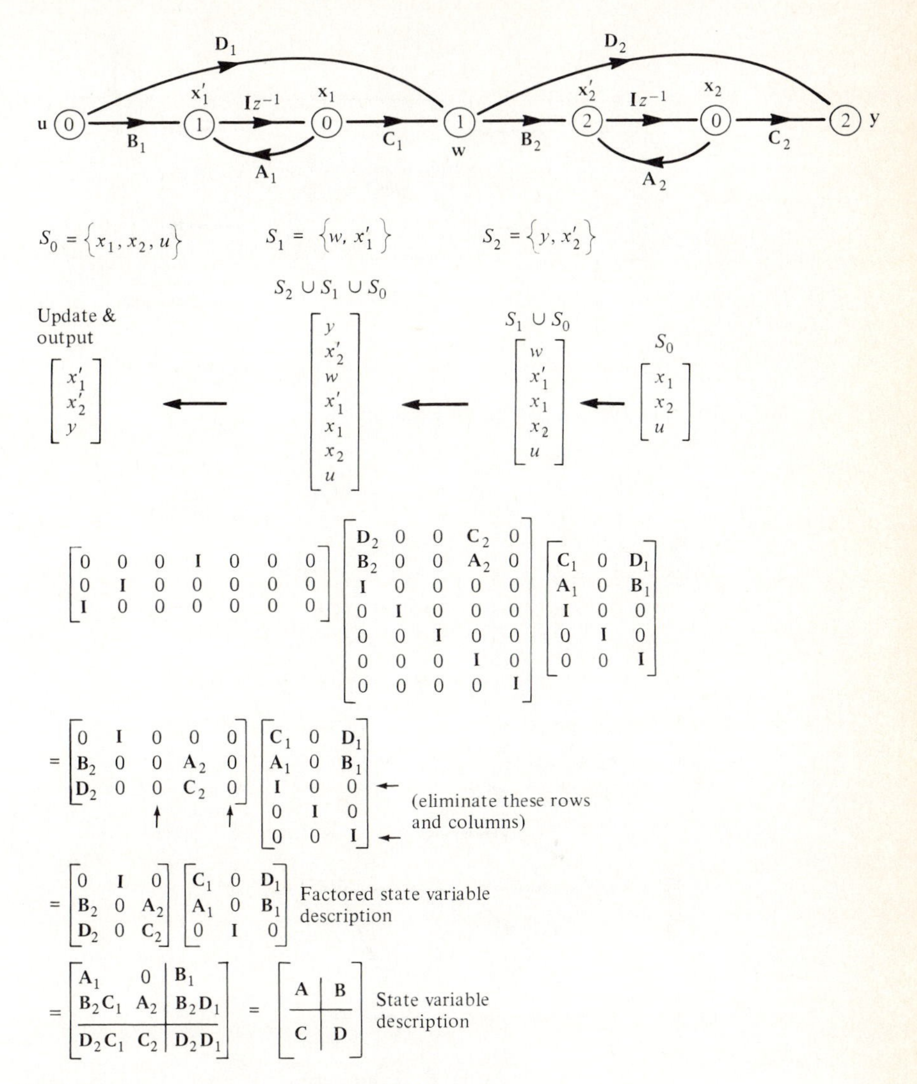

FIGURE 8.4.3 **Cascade of state variable structures.**

for finding the state variable description for the cascade of two state variable structures.

Another scheme which may be described using a FSVD is that of "block I/O" processing. Instead of updating at every sample time, an input buffer of size m is filled with m consecutive input samples. Then the state is updated every m units of time and m consecutive output samples are placed in an output buffer and presumably shifted out one at a time. There results a system having the form of Fig. 8.4.4. The development of a state variable description for the linear system between the buffers is obtained by combining the state variable equations for m consecutive updates. Each update involves the product of a matrix and a vector. The case of a block size of $m = 2$ is shown in Fig. 8.4.5.

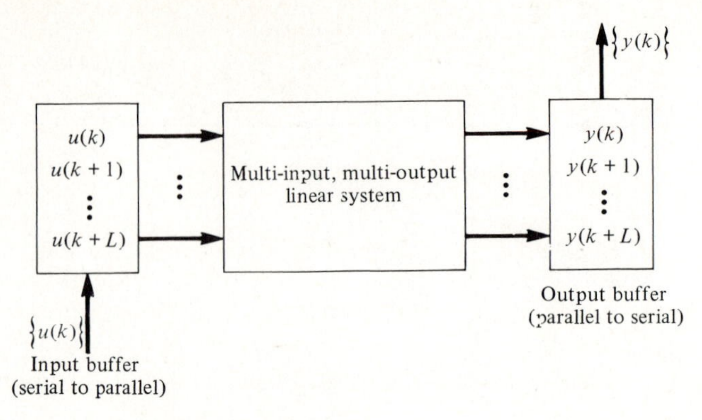

FIGURE 8.4.4 **Block I/O filtering scheme.**

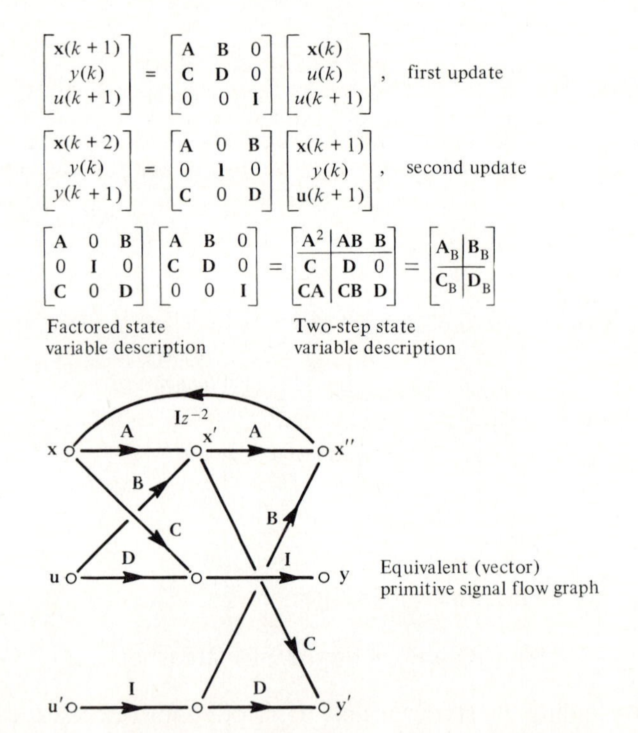

FIGURE 8.4.5 **Block input/output state variable structure, with block size of two.**

The factored state variable description satisfies

$$\mathbf{F}_L \cdot \mathbf{F}_{L-1} \cdots \mathbf{F}_1 = \begin{bmatrix} \mathbf{A} & \mathbf{B} \\ \mathbf{C} & \mathbf{D} \end{bmatrix}. \tag{8.4.6}$$

Therefore, we may "factor" the state variable equations

$$\begin{bmatrix} \mathbf{x}(k+1) \\ y(k) \end{bmatrix} = \begin{bmatrix} \mathbf{A} & \mathbf{B} \\ \mathbf{C} & \mathbf{D} \end{bmatrix} \begin{bmatrix} \mathbf{x}(k) \\ u(k) \end{bmatrix} \tag{8.4.7}$$

to get

$$\mathbf{q}_l(k) = \mathbf{F}_l \mathbf{q}_{l-1}(k), \tag{8.4.8}$$

where

$$\mathbf{q}_0(k) = \begin{bmatrix} \mathbf{x}(k) \\ u(k) \end{bmatrix}, \qquad \mathbf{q}_L(k) = \begin{bmatrix} \mathbf{x}(k+1) \\ y(k) \end{bmatrix}. \tag{8.4.9}$$

The factored state variable description preserves all the node variable computations and the order in which they are computed. If v is a variable in S_k, then the node equation for v is represented by a row of the matrix $\mathbf{F}_k$ and v can be identified with a component of the vector $\mathbf{q}_k$ in Eq. (8.4.8). But this identification may cease to be valid as we proceed to further updates. The act of throwing out rows and columns is equivalent to discarding node variables that have ceased to be useful.

Primitive Signal Flow Graphs with Fractional Delays

There are L updates in the factored state variable equations (8.4.8). A primitive signal flow graph constructed from these equations would have the form shown in Fig. 8.4.6, part (a). The nodes in that graph are vector-valued and the branch gains are matrix-valued. Notice that splitting the vectors $\mathbf{q}_0$ and $\mathbf{q}_L$ into components $\mathbf{x}$, u, and y corresponds to partitions of the matrices $\mathbf{F}_1$ and $\mathbf{F}_L$. All the delay is lumped into one z^{-1} (matrix) branch and joins $\mathbf{x}'$ to $\mathbf{x}$.

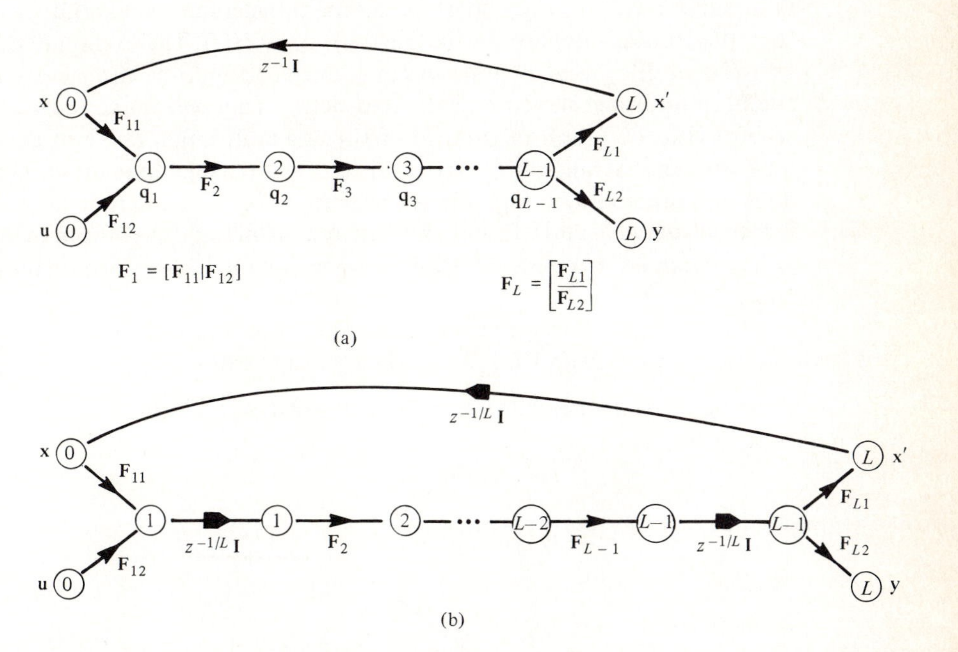

FIGURE 8.4.6 **Signal flow graphs for FSV equations.**

In order to better describe the computational delay inherent in the updates of Eq. (8.4.8), some authors have used fractional delays, distributed uniformly throughout the cycle of computation as shown in part (b) of Fig. 8.4.6. If we draw the graph in this way, it is clear that the input/output properties are changed only in that the output y is delayed by $(L - 1)$ of the fractional delays. This is realistic if the processor is essentially computation bound rather than I/O bound. In any event, the possibility of using fractional delay signal flow graphs to better describe the timing relations between filter variables has been exposed.

Fractional delay PSFGs may be created in other ways. Consider, for example, the lattice shown in part (a) of Fig. 8.4.7. This graph has maximal delay-free path length of $L = 5$, leading to a factored state variable description with five factors. Except for the fifth one of these, each factor contains one 2×2 nontrivial submatrix, which begins at the lower right-hand corner of $\mathbf{F}_1$ and works its way up, one step at a time. Now, one would like to decrease the value of L without increasing the number of multiplications (as would happen if we were to multiply the $\mathbf{F}_k$'s together). Intuitively, this can be done by the simple device of "decoupling" the graph by distributing the delays equally between the upper and lower tiers of the lattice. If we move half of each delay to the bottom (and splitting the node variables v_1 through v_5) we obtain the graph shown in part (b) of the figure. Two questions arise. First, "What happens to the transfer function?" Second, "How are we to interpret the resulting signal flow graph (i.e., what filter realization does it suggest)?"

One can compute the transfer function for lattice (b) in the normal way, although it is tedious to do so. If we call the transfer function for lattice (a) $H(z)$, then the transfer function for lattice (b) is $z^{-1/2}H(z)$. The extra half delay lies between nodes y and y' and was introduced to enforce the rule that every computation must have an associated delay. This will make lattice (b) consistent with the graph of part (b) of Fig. 8.4.6 although it is not at all clear how we can rearrange it to make this so. The reasons that the modification does not otherwise change the transfer function is (first) that no loop gain has been changed, and (second) that any path from input to output in lattice (a) will correspond to a path in lattice (b) having an identical path gain (multiplied by $z^{-1/2}$).

But how are we to interpret the result? Let us consider the general case of a signal flow graph with the following properties.

1. The graph is the connection of subgraphs of the form

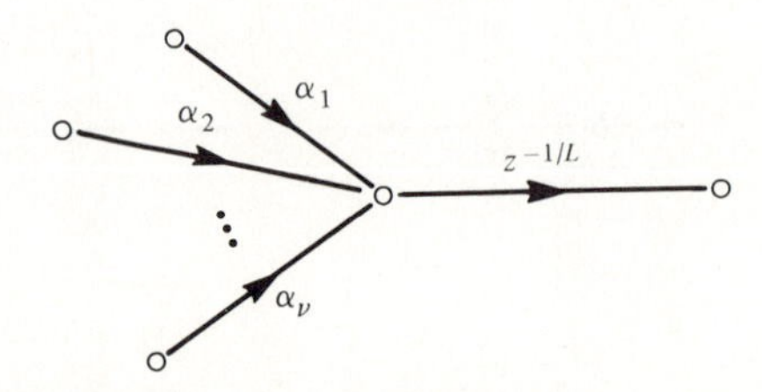

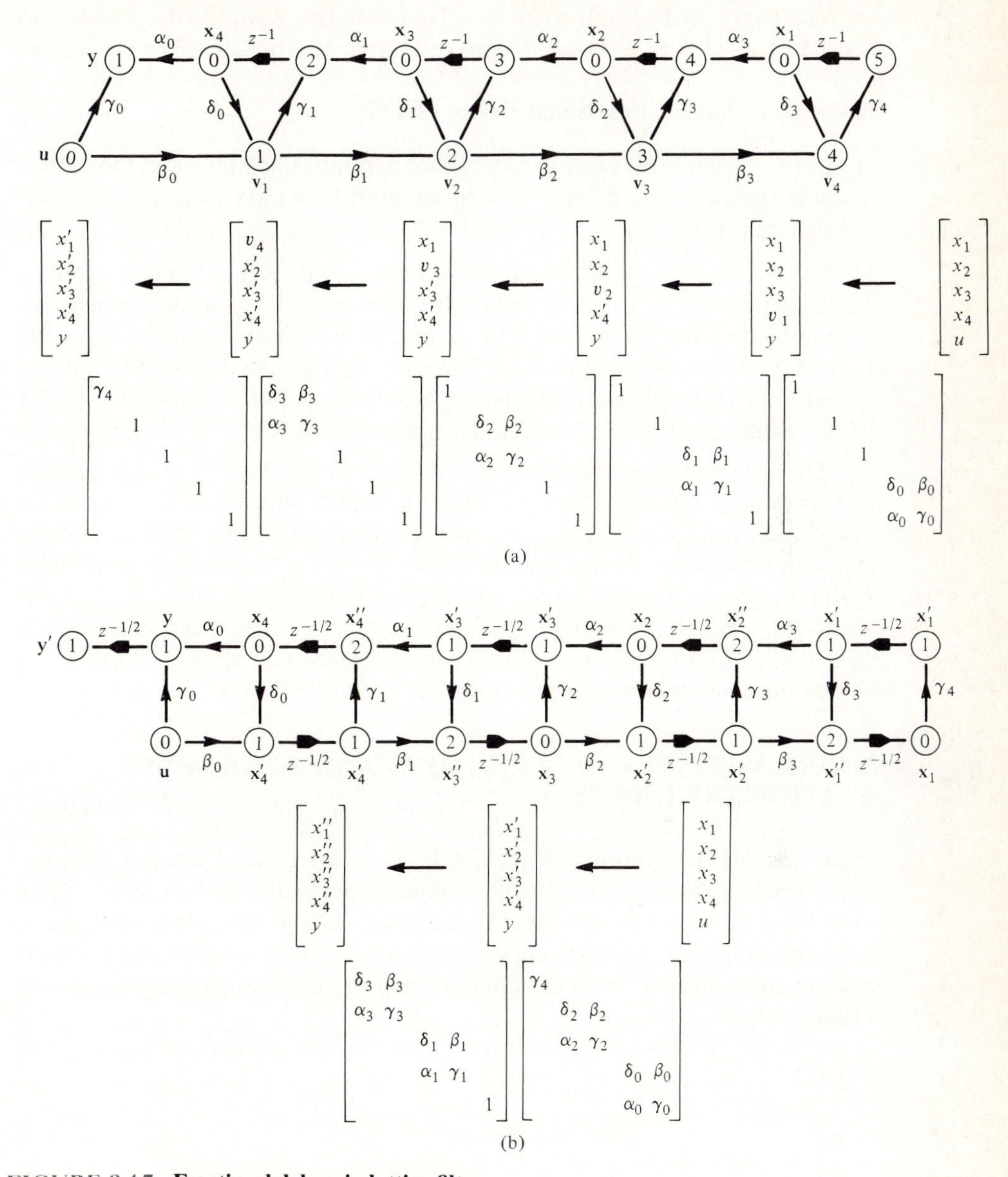

FIGURE 8.4.7 Fractional delays in lattice filters.

where every connection joins a delay output to one or more constant branch inputs. (This models the basic "sum of products" operation. It follows that no delay-free paths having length greater than one is allowed.)

2. All loop gains must contain a positive integer delay. (In other words, every loop involves L, $2L$, $3L$, etc., fractional delays.)
3. Any two paths joining two given nodes can differ only by an integer delay.

For every such graph there is a factored state variable description. In order to construct it, however, we must modify the node ordering procedure.

Node Ordering for Fractional Delay PSFGs

1. Let S_0 be the set of node variables that contains the inputs and the output of every fractional delay joined to an input by a path with integer delay. [See part (b) of Fig. 8.4.7.]
2. For $k = 1, \ldots, L$, let S_k be the set of node variables that are connected to variables in S_{k-1} by one constant branch, or by one constant branch followed by a fractional delay. [This makes the variables on the two ends of a fractional delay equivalent. The only exception to this is that the delay outputs in S_L have already been placed in S_0. For example, the nodes marked (2) in Fig. 8.4.7 feed nodes marked (0).]

The node ordering procedure determines a factored state variable description by simply writing the matrix equation that produces the S_{k+1} variables from the S_k variables. We can conclude that fractional delay PSFGs generate factored state variable descriptions and may be interpreted in that light. Notice that several nodes in such a graph can actually share the same physical hardware (time-multiplexing of the same hardware) because the variables are needed at different minor cycle times between successive inputs. This may not be obvious from the graph itself, however.

8.5 STATE VARIABLE DESCRIPTIONS AND WIDE SENSE STATIONARY INPUTS

State variable descriptions of linear filters are most useful when it becomes necessary to compute quantities that depend on the internal structure of the filter, and then to ask how these quantities change when the structure is modified. Some of the more important quantities that need to be computed are statistical, arising when the filter is being driven by wide sense stationary random inputs.

Let us begin with a causal filter $H(z)$ driven by white noise as shown in Fig. 8.5.1.

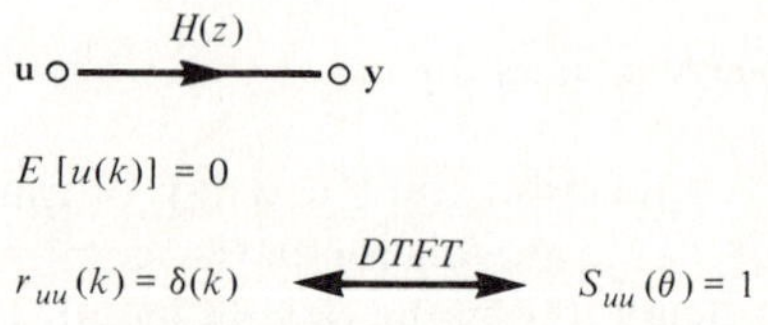

$H(z)$

u o——►——o y

$E[u(k)] = 0$

$r_{uu}(k) = \delta(k) \xleftrightarrow{DTFT} S_{uu}(\theta) = 1$

FIGURE 8.5.1 **A filter with a white noise input.**

Since the input is white, the unit pulse response is a cross correlation of the input and output.

$$h(k) = E[y(k + l)u(l)]. \tag{8.5.1}$$

The autocorrelation sequence for the output is

$$r_{yy}(k) = E[y(k + l)y(l)] = \sum_{l=0}^{\infty} h(k + l)h(l). \tag{8.5.2}$$

Using a state variable description, we have

$$h(k) = \begin{cases} 0, & k < 0 \\ \mathbf{D}, & k = 0 \\ \mathbf{C}\mathbf{A}^{k-1}\mathbf{B}, & k > 0 \end{cases} \quad \overset{DTFT}{\longleftrightarrow} \quad H(e^{j\theta}) = \mathbf{D} + \mathbf{C}(e^{j\theta}\mathbf{I} - \mathbf{A})^{-1}\mathbf{B}. \tag{8.5.3}$$

Since

$$r_{yy}(k) \quad \overset{DTFT}{\longleftrightarrow} \quad S_{yy}(\theta) = 1 \cdot |H(e^{j\theta})|^2, \tag{8.5.4}$$

we can express the autocorrelation sequence directly from (**A**, **B**, **C**, **D**). The key to such an expression is the covariance matrix for the filter state, defined by

$$\mathbf{K} = E(\mathbf{x}(k)\mathbf{x}^T(k)) \tag{8.5.5}$$

with components

$$K_{ij} = E(x_i(k)x_j(k)). \tag{8.5.6}$$

This matrix depends only on **A**, **B** and the input autocorrelation sequence, because

$$\mathbf{x}(k) = \sum_{l=0}^{\infty} \mathbf{A}^l\mathbf{B}u(k - l - 1). \tag{8.5.7}$$

Substitution of this equation into the definition (8.5.5) leads to

$$\mathbf{K} = \sum_{l=0}^{\infty} \sum_{m=0}^{\infty} \mathbf{A}^l\mathbf{B}r_{uu}(l - m)(\mathbf{A}^m\mathbf{B})^T, \tag{8.5.8}$$

which is a doubly infinite sum. For white inputs, the cross terms are zero resulting in the equation

$$\mathbf{K} = \sum_{l=0}^{\infty} (\mathbf{A}^l\mathbf{B})(\mathbf{A}^l\mathbf{B})^T. \tag{8.5.9}$$

These equations have frequency domain counterparts obtained by using the Parseval identity for the discrete time Fourier transform. Consider the z-transform of $\mathbf{A}^l\mathbf{B}$. We have

$$\sum_{l=0}^{\infty} \mathbf{A}^l\mathbf{B}z^{-l} = \sum_{l=0}^{\infty} \left(\frac{\mathbf{A}}{z}\right)^l \mathbf{B} = (\mathbf{I} - z^{-1}\mathbf{A})^{-1}\mathbf{B}. \tag{8.5.10}$$

In Eq. (8.5.10), we have made use of the identity $\sum_{l=0}^{\infty} \mathbf{A}^l = (\mathbf{I} - \mathbf{A})^{-1}$ if **A** has eigenvalues λ_i such that $|\lambda_i| < 1$, $i = 1, 2, \ldots, n$. Thus let

$$\Psi(z) = (z\mathbf{I} - \mathbf{A})^{-1}\mathbf{B}. \tag{8.5.11}$$

Then the counterpart of Eq. (8.5.8) is

$$\mathbf{K} = \frac{1}{2\pi}\int_{-\pi}^{\pi} \Psi(e^{j\theta})S_{uu}(\theta)\Psi^*(e^{j\theta})\,d\theta, \tag{8.5.12}$$

which is simplified for white noise inputs to

$$\mathbf{K} = \frac{1}{2\pi}\int_{-\pi}^{\pi} \Psi(e^{j\theta})\Psi^*(e^{j\theta})\,d\theta. \tag{8.5.13}$$

Neither of the two equations (8.5.8) or (8.5.14) provide much help in the computation of **K**, however.

The right-hand side of Eq. (8.5.9) is a sort of matrix-valued geometric series. It is possible to indirectly sum the series by equating the covariance of the two sides of the state equation

$$\begin{aligned}\mathbf{K} &= E(\mathbf{x}(k+1)\mathbf{x}^T(k+1)) = E(\mathbf{A}\mathbf{x}(k) + \mathbf{B}u(k))(\mathbf{A}\mathbf{x}(k) + \mathbf{B}u(k))^T \\ &= E[\mathbf{A}\mathbf{x}(k)\mathbf{x}^T(k)\mathbf{A}^T + \mathbf{B}u(k)u^T(k)\mathbf{B}^T].\end{aligned} \tag{8.5.14}$$

Thus

$$\mathbf{K} = \mathbf{AKA}^T + \mathbf{BB}^T. \tag{8.5.15}$$

In Eq. (8.5.14), we have used the fact that $\mathbf{x}(k)$ and $u(k)$ are uncorrelated for white inputs. It is a simple matter to verify that the right-hand side of Eq. (8.5.9) satisfies Eq. (8.5.15). Although Eq. (8.5.15) generates **K** only implicitly, it can be done by solving linear equations involving the components of **K**. In this way, one can avoid infinite sums.

EXAMPLE 8.5.1

Finding the covariance matrix of a direct form filter.

Let

$$\mathbf{A} = \begin{bmatrix} 0 & 1 \\ -a_2 & -a_1 \end{bmatrix}, \quad \mathbf{B} = \begin{bmatrix} 0 \\ 1 \end{bmatrix},$$

a second-order, direct form filter.

Using Eq. (8.5.15), we obtain

$$\begin{aligned}\mathbf{K} = \begin{bmatrix} k_1 & k_2 \\ k_2 & k_3 \end{bmatrix} &= \begin{bmatrix} 0 & 1 \\ -\bar{a}_2 & -a_1 \end{bmatrix}\begin{bmatrix} k_1 & k_2 \\ k_2 & k_3 \end{bmatrix}\begin{bmatrix} 0 & -a_2 \\ 1 & -a_1 \end{bmatrix} + \begin{bmatrix} 0 \\ 1 \end{bmatrix}[0 \quad 1] \\ &= \begin{bmatrix} k_3 & -a_2k_2 - a_1k_3 \\ -a_2k_2 - a_1k_3 & a_2^2k_1 + 2a_1a_2k_2 + a_1^2k_3 \end{bmatrix} + \begin{bmatrix} 0 & 0 \\ 0 & 1 \end{bmatrix}.\end{aligned}$$

The components of Eq. (8.5.1) become

$$\begin{aligned}k_1 &= k_3, \\ k_2 &= -a_2k_2 - a_1k_3, \\ k_3 &= a_2^2k_1 + 2a_1a_2k_2 + a_1^2k_3 + 1.\end{aligned}$$

Solving these equations yields

$$\mathbf{K} = \frac{1}{(1 - a_2)(1 - a_1 + a_2)(1 + a_1 + a_2)} \begin{bmatrix} 1 + a_2 & -a_1 \\ -a_1 & 1 + a_2 \end{bmatrix}. \tag{8.5.16}$$

For a white input sequence, the present state, which depends only on past inputs, is correlated with past inputs but uncorrelated with present and future inputs. Thus the cross correlation of the state and a white noise input is

$$\begin{aligned} E[\mathbf{x}(k + l)u(l)] &= E\left(\sum_{j=0}^{\infty} \mathbf{A}^j\mathbf{B}u(k + l - j - 1)u(l)\right) \\ &= \sum_{j=0}^{\infty} \mathbf{A}^j\mathbf{B}\delta(k - j - 1) \\ &= \begin{cases} \mathbf{A}^{k-1}\mathbf{B}, & k > 0 \\ \mathbf{0}, & k \leqslant 0 \end{cases}. \end{aligned} \tag{8.5.17}$$

(This is also the unit-pulse response from the input to internal states.)

Calculation of the Output Autocorrelation Function in Terms of (A, B, C, D)

The output autocorrelation function is given by

$$\begin{aligned} r_{yy}(k) = E[y(k + l)y^T(l)] &= E[(\mathbf{C}\mathbf{x}(k + l) + Du(k + l)(\mathbf{C}\mathbf{x}(l) + Du(l))^T] \\ &= E\{\mathbf{C}\mathbf{x}(k + l)\mathbf{x}^T(l)\mathbf{C}^T + Du(k + l)\mathbf{x}^T(l)\mathbf{C}^T + \mathbf{C}\mathbf{x}(k + l)u^T(l)D^T \\ &\quad + Du(k + l)u^T(l)D^T\}. \end{aligned} \tag{8.5.18}$$

In this expression, the first term is evaluated as follows

$$E(\mathbf{x}(k+l)\mathbf{x}^T(l)) = E\left[\sum_{m=0}^{\infty} \mathbf{A}^m\mathbf{B}u(k+l-m-1)\left(\sum_{m'=0}^{\infty} \mathbf{A}^{m'}\mathbf{B}u(l-m'-1)\right)^T\right]. \tag{8.5.19}$$

Now, taking the expected value inside the double sum, we obtain

$$\begin{aligned} E(\mathbf{x}(k + l)\mathbf{x}^T(l)) &= \sum_{m,m'=0}^{\infty}\sum (\mathbf{A}^m\mathbf{B})(\mathbf{A}^{m'}\mathbf{B})^T r_{uu}(k - m + m') \\ &= \sum_{m=0}^{\infty} \mathbf{A}^k\mathbf{A}^m\mathbf{B}(\mathbf{A}^m\mathbf{B})^T \\ &= \mathbf{A}^k\mathbf{K}, \qquad k \geqslant 0 \end{aligned} \tag{8.5.20}$$

In Eq. (8.5.20), we have used the fact that $r_{uu}(k - m + m') = \delta(k - m + m')$. Since the present state is uncorrelated with future inputs, the term in Eq.

(8.5.18) containing $u(k + l)\mathbf{x}^T(k)$ is zero. Thus we find that

$$\begin{aligned} r_{yy}(k) &= \mathbf{CA}^k\mathbf{KC}^T + \mathbf{CA}^{k-1}\mathbf{BD}^T \\ &= \mathbf{CA}^k\mathbf{KC}^T + h(k)\mathbf{D}, \qquad k > 0. \end{aligned} \tag{8.5.21}$$

An important application of this computation is the evaluation of the energy in the filter unit-pulse response and frequency response. This is obtained by setting $k = 0$ in Eq. (8.5.21) to get

$$\begin{aligned} r_{yy}(0) = E(y^2(k)) &= \sum_{l=0}^{\infty} h^2(l) \\ &= \frac{1}{2\pi}\int_{-\pi}^{\pi} |H(e^{j\theta})|^2\, d\theta \\ &= \mathbf{D}^2 + \mathbf{CKC}^T. \end{aligned} \tag{8.5.22}$$

To compute this quantity from the state variable description (**A**, **B**, **C**, **D**), one computes **K** from (**A**, **B**) using Eq. (8.5.15) and then combines it with (**C**, **D**) using Eq. (8.5.22).

The calculation of the autocorrelation function or the second moment of the output using Eqs. (8.5.21) and (8.5.22) eliminates any infinite sums. The calculation involves solving linear equations and matrix multiplication. We shall use the covariance matrix **K** in other calculations also. For example, in Chapter 9 we use **K** to "scale" digital filters.

The Matrix K and Liapunov Stability

The state covariance matrix satisfies the equation

$$\mathbf{K} = \mathbf{AKA}^T + \mathbf{BB}^T, \tag{8.5.23}$$

which amounts to a linear system of n^2 equations.

Suppose we write these n^2 equations for the unknown components of **K** in the following form:

$$\hat{\mathbf{A}}\hat{\mathbf{K}} = \hat{\mathbf{B}}, \tag{8.5.24}$$

where $\hat{\mathbf{K}}$ is a column vector consisting of the n^2 unknowns in **K**.

A solution exists and is unique if the determinant of $\hat{\mathbf{A}}$ is not zero. It can be shown that

$$\det \hat{\mathbf{A}} = \prod_{k=1}^{n}\prod_{j=1}^{n} (1 - \lambda_k\lambda_j^*), \qquad \{\lambda_i\} \text{ are eigenvalues of } \mathbf{A}. \tag{8.5.25}$$

Therefore, a solution exists if the system is stable ($|\lambda_i| < 1$). But covariance matrices must also be sign definite. Using the sum formula (8.5.9), we see that for any vector **v**,

$$\mathbf{v}^T\mathbf{K}\mathbf{v} = \sum_{k=0}^{\infty} (\mathbf{v}^T\mathbf{A}^k\mathbf{B})^2 \geqslant 0. \tag{8.5.26}$$

If the system is controllable, then

$$\mathbf{v}^T\mathbf{K}\mathbf{v} = 0 \quad \Rightarrow \quad \mathbf{v}^T[\mathbf{B}, \mathbf{AB}, \mathbf{A}^2\mathbf{B}, \ldots, \mathbf{A}^{n-1}\mathbf{B}] = \mathbf{0} \quad \Rightarrow \quad \mathbf{v} = \mathbf{0}. \tag{8.5.27}$$

Thus if (**A**, **B**) is stable and controllable, then **K** is positive definite. It is also true that if (**A**, **B**) is controllable and the solution to Eq. (8.5.23) is positive definite, then **A** must be stable. (This is actually a special case of Liapunov stability theory.) To see this, let λ be any eigenvalue of **A**. Then there exists a row eigenvector:

$$\boldsymbol{\rho}\mathbf{A} = \lambda\boldsymbol{\rho}, \qquad \boldsymbol{\rho} \neq \mathbf{0}. \tag{8.5.28}$$

Using Eq. (8.5.23),

$$\begin{aligned}\boldsymbol{\rho}\mathbf{K}\boldsymbol{\rho}^* &= (\boldsymbol{\rho}\mathbf{A})\mathbf{K}(\boldsymbol{\rho}\mathbf{A})^* + (\boldsymbol{\rho}\mathbf{B})(\boldsymbol{\rho}\mathbf{B})^* \\ &= \lambda\boldsymbol{\rho}\mathbf{K}\boldsymbol{\rho}^*\lambda^* + |\boldsymbol{\rho}\mathbf{B}|^2,\end{aligned}$$

which can be rearranged to give

$$1 - |\lambda|^2 = \frac{|\boldsymbol{\rho}\mathbf{B}|^2}{\boldsymbol{\rho}\mathbf{K}\boldsymbol{\rho}^*}. \tag{8.5.29}$$

(We are assuming that **K** is positive definite.)

Now

$$\begin{aligned}&1 - |\lambda|^2 = 0 \quad \text{if and only if } \boldsymbol{\rho}\mathbf{B} = 0 \\ &\Leftrightarrow \boldsymbol{\rho}[\mathbf{B}, \mathbf{AB}, \mathbf{A}^2\mathbf{B}, \ldots, \mathbf{A}^{n-1}\mathbf{B}] = \boldsymbol{\rho}\mathbf{B}[1, \lambda, \ldots, \lambda^{n-1}] = \mathbf{0} \\ &\Leftrightarrow \boldsymbol{\rho} = \mathbf{0}, \quad \text{(controllability)}.\end{aligned}$$

Therefore, since the system is controllable, $1 - |\lambda|^2 > 0$, which is to say, $|\lambda| < 1$.

EXAMPLE 8.5.2

Stability of a digital filter using **K**.

The matrix **K** in Example 8.5.1 has principal minors

$$K_{11} = \frac{1 + a_2}{(1 - a_2)(1 + a_1 + a_2)(1 - a_1 + a_2)}, \tag{8.5.30}$$

$$\det \mathbf{K} = \frac{1}{(1 - a_2)^2(1 + a_1 + a_2)(1 - a_1 + a_2)}. \tag{8.5.31}$$

One can show that both of these are positive and thus **K** is positive definite if and only if

$$\left.\begin{aligned}(1 - a_2) > 0 \\ (1 + a_2 - a_1) > 0 \\ (1 + a_1 + a_2) > 0\end{aligned}\right\}. \tag{8.5.32}$$

The set of parameters (a_1, a_2) satisfying Eq. (8.5.32) is called the stability triangle for second-order filters. These inequalities are necessary and sufficient conditions for the two roots of the polynomial $\hat{a}(z) = z^2 + a_1 z + a_2 = (z - \lambda_1)(z - \lambda_2)$ to satisfy $|\lambda_1| < 1$ and $|\lambda_2| < 1$.

Factored State Variable Equations

Computing the variance of any variable in a realization is not difficult if the factored state variable description is used. Let

$$\mathbf{q}_0(k) = \begin{bmatrix} \mathbf{x}(k) \\ u(k) \end{bmatrix}. \tag{8.5.33}$$

Then the covariance matrix for $\mathbf{q}_0(k)$ is

$$\begin{aligned} E(\mathbf{q}_0(k)\mathbf{q}_0^T(k)) &= \begin{bmatrix} E(\mathbf{x}(k)\mathbf{x}^T(k)) & E(\mathbf{x}(k)u(k)) \\ E(u(k)\mathbf{x}^T(k)) & E(u^2(k)) \end{bmatrix} \\ &= \left[\begin{array}{c|c} \mathbf{K} & \mathbf{0} \\ \hline \mathbf{0} & 1 \end{array}\right] \triangleq \tilde{\mathbf{K}}_0. \end{aligned} \tag{8.5.34}$$

The factored state variable equations

$$\mathbf{q}_{l+1}^{(k)} = \mathbf{F}_{l+1}\mathbf{q}_l^{(k)}, \qquad 0 \leqslant l \leqslant L - 1 \tag{8.5.35}$$

lead to equations for $\{\tilde{\mathbf{K}}_l\}$.

$$\tilde{\mathbf{K}}_{l+1} = E[\mathbf{q}_{l+1}^{(k)}\mathbf{q}_{l+1}^T(k)] = \mathbf{F}_{l+1}\tilde{\mathbf{K}}_l\mathbf{F}_{l+1}^T. \tag{8.5.36}$$

Thus each covariance $\mathbf{K}_l$ can be computed by iterating this matrix difference equation. Since

$$\begin{bmatrix} \mathbf{A} & \mathbf{B} \\ \mathbf{C} & \mathbf{D} \end{bmatrix} = \mathbf{F}_L \cdot \mathbf{F}_{L-1} \cdots \mathbf{F}_2 \cdot \mathbf{F}_1, \tag{8.5.37}$$

we have

$$\begin{aligned} \tilde{\mathbf{K}}_L &= E(\mathbf{q}_L(k)\mathbf{q}_L^T(k)) = E\left\{\begin{bmatrix} \mathbf{x}(k+1) \\ y(k) \end{bmatrix} [\mathbf{x}^T(k+1)\ y(k)]\right\} \\ &= \begin{bmatrix} \mathbf{A} & \mathbf{B} \\ \mathbf{C} & \mathbf{D} \end{bmatrix} \tilde{\mathbf{K}}_0 \begin{bmatrix} \mathbf{A}^T & \mathbf{C}^T \\ \mathbf{B}^T & \mathbf{D}^T \end{bmatrix} = \begin{bmatrix} \mathbf{AKA}^T + \mathbf{BB}^T & \mathbf{AKC}^T + \mathbf{BD}^T \\ \mathbf{CKA}^T + \mathbf{DB}^T & \mathbf{CKC}^T + \mathbf{D}^2 \end{bmatrix} \\ &= \begin{bmatrix} \mathbf{K} & \mathbf{AKC}^T + \mathbf{BD}^T \\ \mathbf{CKA}^T + \mathbf{DB}^T & r_{yy}(0) \end{bmatrix}. \end{aligned} \tag{8.5.38}$$

Summary • This chapter has presented several internal descriptions for digital filters. We have also discussed how they are used in various calculations. In the design of digital filters and other problems in digital signal processing it is often necessary to use internal descriptions in order to optimize certain aspects of the computation. Thus in Chapter 5 we used internal descriptions of the DFT calculation in order to reduce the number of computations. In the following chapters, we shall use these descriptions to optimize other criteria associated with the computation.

State variable and factored state variable descriptions have been emphasized in this discussion because these internal descriptions are algebraic. They provide a mathematical formulation for studying various structures that is not feasible with any other description. This is not to say that SVD or FSVD should always be employed to describe digital filters. One should employ the simplest description in any given situation. However, many applications in digital signal processing require the use of more than I/O descriptions.

PROBLEMS

8.1. Find a PSFG which has the following characteristics (simultaneously):

a) Has only integer multipliers.
b) Has loops with nonzero multipliers.
c) Realizes the transfer function $H(z) = 4z^{-3} + z^{-1}$.

8.2. Discuss and *illustrate by example* the most efficient method to obtain:

a) $(\mathbf{A}, \mathbf{B}, \mathbf{C}, \mathbf{D})$ from a unit-pulse response $\mathbf{h}$.
b) An SDE from a PSFG.
c) A software program from the unit-pulse response $\mathbf{h}$.

8.3. Find $(\mathbf{A}, \mathbf{B}, \mathbf{C}, \mathbf{D})$, transfer function $H(z)$, and the unit-pulse response $\mathbf{h}$ for the PSFG shown in Fig. 8P.3.

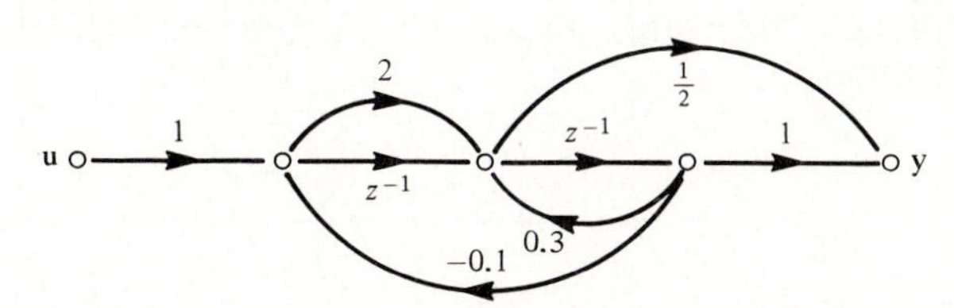

FIGURE 8P.3.

8.4. Find an SVD, $(\mathbf{A}, \mathbf{B}, \mathbf{C}, \mathbf{D})$, for Fig. 8P.4.

8.5. Apply a nonsingular transformation $\mathbf{T}$ to the state vector in Fig. 8P.4 so that $\mathbf{x}' = \mathbf{T}^{-1}\mathbf{x}$ results in a matrix $\mathbf{A}$ which is diagonal. Sketch the new SFG of the transformed structure. (Assume that $ea > 0$.)

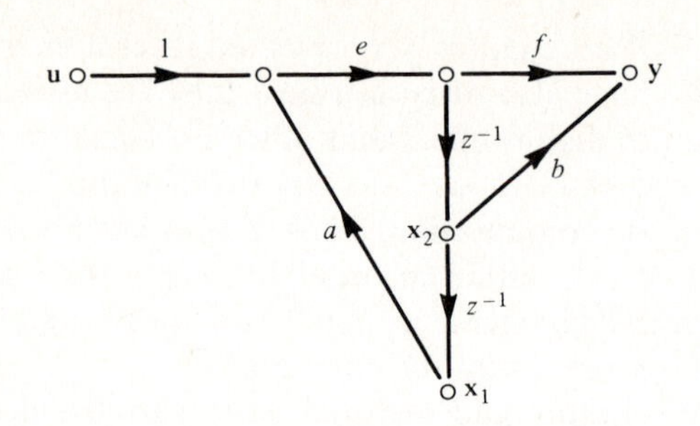

FIGURE 8P.4.

8.6. For the PSFG in Fig. 8P.6 find the following:

a) The factored state variable description (FSVD).
b) The state variable description (SVD).
c) An SFG for the SVD of part b).
d) A software program.
e) The length of the longest delay-free path.

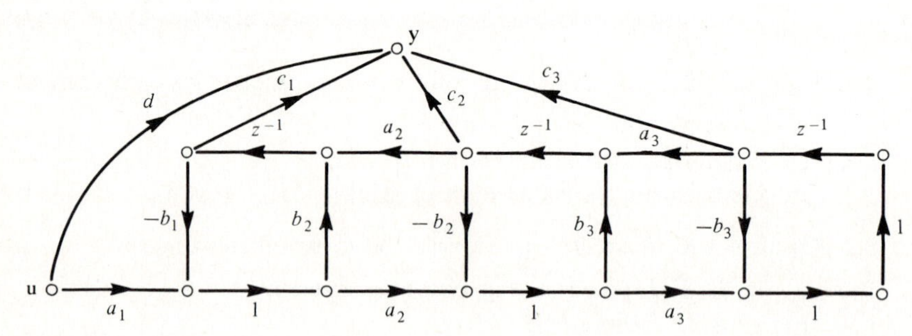

FIGURE 8P.6.

8.7. Find a factored state variable description for a parallel connection of two state variable descriptions, $(\mathbf{A}_1, \mathbf{B}_1, \mathbf{C}_1, \mathbf{D}_1)$ and $(\mathbf{A}_2, \mathbf{B}_2, \mathbf{C}_2, \mathbf{D}_2)$. System 1 and System 2 are assumed to have m inputs and n outputs.

8.8. Find a factored state variable description for the following nonrecursive lattice filter:

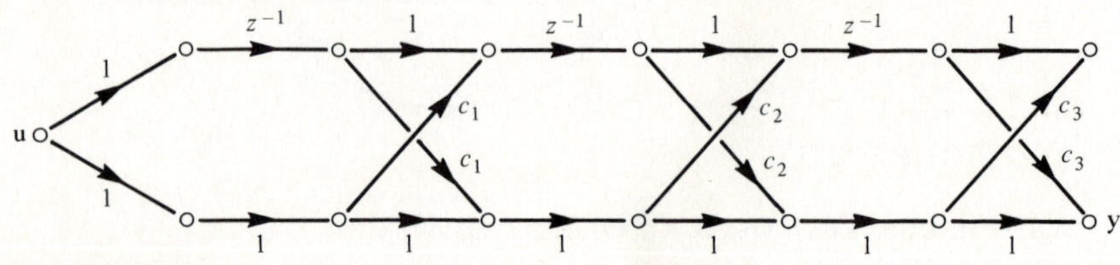

FIGURE 8P.8.

8.9. Find a state variable description for a feedback connection of two two-input, two-output systems as shown in Fig. 8P.9. (Beware of delay-free loops when the feedback connection is made.)

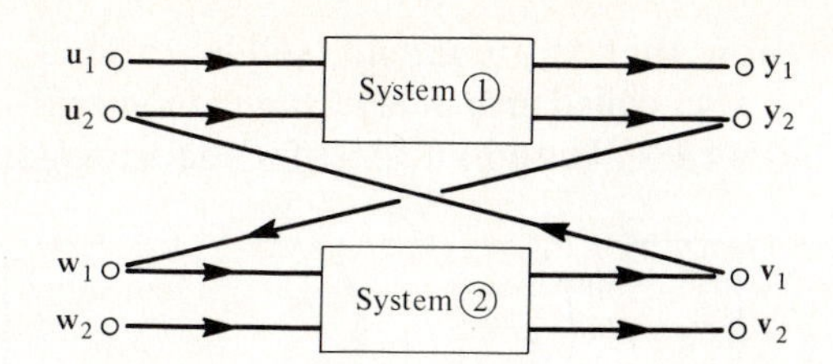

FIGURE 8P.9.

8.10. Find a factored state variable description for a block processing filter with a block length of $m = 5$.

8.11. Answer the following demands for the PSFG in Fig. 8P.11.

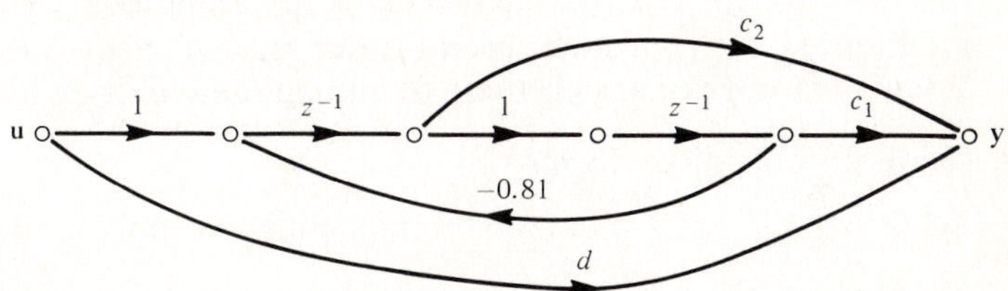

FIGURE 8P.11.

a) Choose c_1, c_2, and d so that the unit-pulse response begins with $\{1, 1, 1\}$.
b) Choose c_1, c_2, and d so that the steady-state response to a unit-step input is $y(k) = 1$ and is $y(k) = 0$ for an input $u(k) = \cos(k\pi/2)$.
c) Choose c_1, c_2, and d so that the numerator polynomial in $H(z)$ is $\hat{b}(z) = z^2 + z + 1$.
d) Choose c_1, c_2, and d so that the frequency response function has magnitude identically one, i.e., $|H(e^{j\theta})| \equiv 1$.

8.12. In Problem 8.11, choose c_1, c_2, and d to maximize $\sum_{k=0}^{\infty} kh^2(k)$ subject to the constraint $\sum_{k=0}^{\infty} h^2(k) = 1$.

8.13. Show that if a PSFG has no loops, then the unit-pulse response has finite duration.

8.14. Suppose a lattice filter with a factored state variable description as in portion (a) of Fig. 8.4.7 is given. How can the FSVD in part (b) be obtained analytically?

8.15. Consider the PSFG shown in Fig. 8P.15. Find

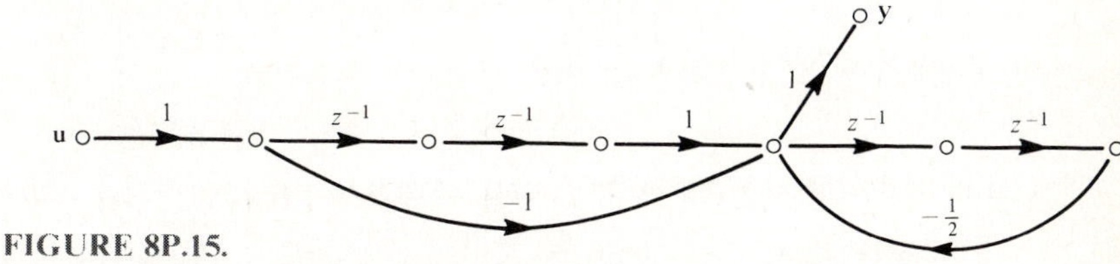

FIGURE 8P.15.

a) The filter's transfer function $H(z)$.
b) An SVD for the filter.
c) Use b) to check your result in a).

8.16. One can show that any minimal, stable, rational transfer function can be realized by a so-called *tridiagonal realization* with PSFG as shown in Fig. 8P.16. Find an FSVD and an SVD for the tridiagonal realization.

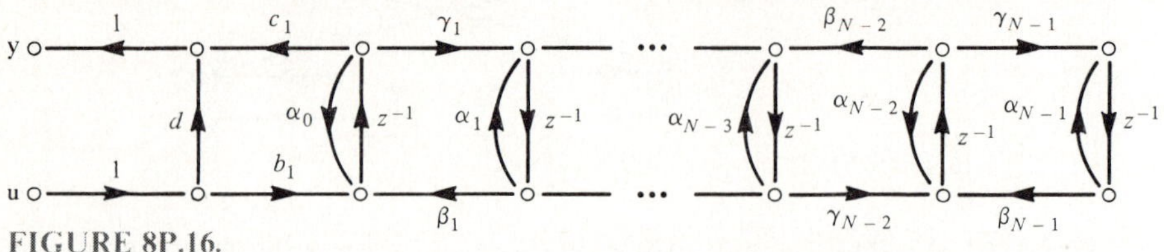

FIGURE 8P.16.

8.17. Consider a cascade of several second-order digital filters of the form $(b_0 + b_1 z^{-1} + b_2 z^{-2})/(1 + a_1 z^{-1} + a_2 z^{-2})$. Assuming the ith section has an SVD $(\mathbf{A}_i, \mathbf{B}_i, \mathbf{C}_i, \mathbf{D}_i)$, find the FSVD for N sections in cascade. If unit-delays are inserted between each second-order section, what is the resulting FSVD. What advantage (if any) is there to adding these unit-delays?

8.18. (State variable direct forms.)

Let $\psi^T(z) = [1, z, z^2, \ldots, z^{n-1}]$ and define two polynomials:

$$\hat{a}(z) = z^n + a_1 z^{n-1} + \cdots + a_n,$$
$$\hat{b}(z) = b_0 z^n + b_1 z^{n-1} + \cdots + b_n.$$

a) Show the identity $(z\mathbf{I} - \mathbf{A})\psi(z) = \hat{a}(z)\mathbf{B}$ or its equivalent $(z\mathbf{I} - \mathbf{A})^{-1}\mathbf{B} = [1/\hat{a}(z)]\psi(z)$ uniquely defines a matrix $\mathbf{A}$ and a vector $\mathbf{B}$. Find these matrices.

b) Determine $\mathbf{C}$ and $\mathbf{D}$ for which

$$\mathbf{D} + \mathbf{C}(z\mathbf{I} - \mathbf{A})^{-1}\mathbf{B} = \frac{\hat{b}(z)}{\hat{a}(z)}.$$

c) Draw a PSFG equivalent to the SVD $(\mathbf{A}, \mathbf{B}, \mathbf{C}, \mathbf{D})$.

d) If $\hat{a}(\lambda) = 0$, then λ is an eigenvalue of $\mathbf{A}$. What is the eigenvector?

8.19. Let $\mathbf{A}, \mathbf{B}, \mathbf{C}$ be the solutions to Problem 8.18.

a) Show that $$[\mathbf{B}, \mathbf{AB}, \mathbf{A}^2\mathbf{B}, \ldots, \mathbf{A}^{n-1}\mathbf{B}] = \begin{bmatrix} a_{n-1} & \cdots & a_1 & 1 \\ \vdots & & ⋰ & \\ a_1 & ⋰ & & \\ 1 & & 0 & \end{bmatrix}^{-1}.$$

b) Show that $$\hat{b}(\mathbf{A}) = b_0\mathbf{A}^n + b_1\mathbf{A}^{n-1} + \cdots + b_n\mathbf{I} = \begin{bmatrix} \mathbf{C} \\ \mathbf{CA} \\ \vdots \\ \mathbf{CA}^{n-1} \end{bmatrix}.$$

c) Let $\hat{b}(z)/\hat{a}(z) = \sum_{k=0}^{\infty} h(k)z^{-k}$, and let $\hat{a}(z) = \prod_{k=1}^{n} (z - \lambda_k)$. Show that

$$\det \begin{bmatrix} h(1) & h(2) & \cdots & h(n) \\ h(2) & h(3) & \cdots & h(n+1) \\ & & \vdots & \\ h(n) & & \cdots & h(2n-1) \end{bmatrix} = \pm \prod_{k=1}^{n} \hat{b}(\lambda_k).$$

8.20. (An algorithm to find a transformation to direct form.) Let **A**, **B** satisfy $\det[\mathbf{B}, \mathbf{AB}, \ldots, \mathbf{A}^{n-1}\mathbf{B}] \neq 0$ [the pair $(\mathbf{A}, \mathbf{B})$ is said to *controllable* if this is so]. Let $\det(z\mathbf{I} - \mathbf{A}) = a_0 z^n + a_1 z^{n-1} + \cdots + a_n$.

$$\text{Let } \begin{cases} \mathbf{x}(0) = \mathbf{0} \\ \mathbf{x}(k+1) = \mathbf{A}\mathbf{x}(k) + \mathbf{B}a_k \end{cases}.$$

(This is a trajectory that begins at the origin and has for input the coefficients of the characteristic polynomial.)

a) Show that $\mathbf{x}(n+1) = \mathbf{0}$.
b) Let $\mathbf{T} = [\mathbf{x}(n), \mathbf{x}(n-1), \ldots, \mathbf{x}(1)]$.

Show that

$$\mathbf{T}^{-1}\mathbf{A}\mathbf{T} = \left[\begin{array}{c|ccc} 0 & 1 & & 0 \\ \vdots & & \ddots & \\ 0 & 0 & & 1 \\ \hline -a_n & -a_{n-1} & \cdots & -a_1 \end{array}\right], \quad \mathbf{T}^{-1}\mathbf{B} = \begin{bmatrix} 0 \\ \vdots \\ 0 \\ 0 \\ 1 \end{bmatrix}.$$

8.21. Show that the system

$$\mathbf{x}(0) = \mathbf{0}$$
$$\mathbf{x}(k+1) = \mathbf{A}\mathbf{x}(k) + \mathbf{B}u(k)$$

can be driven to any specified vector $\mathbf{x}(n)$ if and only if the pair $(\mathbf{A}, \mathbf{B})$ is controllable (see Problem 8.20).

8.22. Suppose that the pair $(\mathbf{A}, \mathbf{B})$ is not controllable. Show that there must be a pole-zero cancellation in

$$H(z) = \mathbf{D} + \mathbf{C}(z\mathbf{I} - \mathbf{A})^{-1}\mathbf{B}$$

no matter what choice of **C** and **D** is made.

8.23. The pair (**A**,**C**) is said to be *observable* if the matrix $[\mathbf{C}^T, (\mathbf{CA})^T, \ldots, (\mathbf{CA}^{n-1})^T]^T$ is invertible (for a single output variable). Show that if the pair $(\mathbf{A}, \mathbf{C})$ is observable, then the state $\mathbf{x}(0)$ can be uniquely determined from $(u(k), y(k))$ for $0 \leqslant k < n$.

8.24. (A least-squares problem.) Let **g** be a given finite energy sequence and let $(\mathbf{A}, \mathbf{B})$ be given. Assume that the eigenvalues of **A** satisfy $|\lambda_i| < 1$, $i = 1, 2, \ldots, n$. Choose **C** and **D** to minimize $\|\mathbf{h} - \mathbf{g}\|^2$ where

$$h(k) = \begin{cases} \mathbf{D}, & k = 0 \\ \mathbf{CA}^{k-1}\mathbf{B}, & k > 0 \end{cases}.$$

a) Show that $\|\mathbf{h} - \mathbf{g}\|^2 = \|\mathbf{g}\|^2 - 2g_0\mathbf{D} - 2\mathbf{C}\mathbf{v} + \mathbf{D}^2 + \mathbf{CKC}^T$. Find the vector **v** and the matrix **K**.
b) Show that the solution is given by

$$\mathbf{D} = g(0),$$
$$\mathbf{C} = \mathbf{v}^T\mathbf{K}^{-1}.$$

8.25. Given $\mathbf{g}$ and $(\mathbf{A}, \mathbf{B})$ as in Problem 8.24, choose $\mathbf{C}$ and $\mathbf{D}$ to minimize $\|\delta - \mathbf{g} * \mathbf{h}\|^2$.

8.26. Suppose that $(\mathbf{A}, \mathbf{B})$ are given (stable and controllable). Define a matrix $\mathbf{K}$ implicitly by

$$\mathbf{K} = \mathbf{AKA}^T + \mathbf{BB}^T.$$

Choose $\mathbf{C}$, $\mathbf{D}$ so that $\mathbf{AKC}^T + \mathbf{BD} = \mathbf{0}$ and $\mathbf{CKC}^T + \mathbf{DD}^T = \mathbf{I}$. Show that in this case $\mathbf{H}(z)^T\mathbf{H}(z^{-1}) = \mathbf{I}$.

8.27. Let $(\mathbf{A}, \mathbf{B}, \mathbf{C}, \mathbf{D})$ generate $H(z)$. Show that if z is a complex number for which $z = H(z)$, then z is an eigenvalue of the matrix $\begin{bmatrix} \mathbf{A} & \mathbf{B} \\ \mathbf{C} & \mathbf{D} \end{bmatrix}$. Hint. Construct an eigenvector of the form

$$\begin{bmatrix} \psi(z) \\ 1 \end{bmatrix}$$

8.28. Suppose $H(z) = \hat{b}(z)/\hat{a}(z) = \mathbf{D} + \mathbf{C}(z\mathbf{I} - \mathbf{A})^{-1}\mathbf{B}$ and $\mathbf{D} \neq 0$. Show that

$$H^{-1}(z) = \frac{\hat{a}(z)}{\hat{b}(z)} = \mathbf{D}^{-1} - \mathbf{D}^{-1}\mathbf{C}(z\mathbf{I} - \mathbf{A} + \mathbf{BD}^{-1}\mathbf{C})^{-1}\mathbf{BD}^{-1}$$

8.29. Suppose $\mathbf{D} \neq 0$ and $H(z)$ is as defined in Problem 8.28. Let $\Psi(z) = (z\mathbf{I} - \mathbf{A})^{-1}\mathbf{B}$ and $\Psi_2(z) = (z\mathbf{I} - \mathbf{A} + \mathbf{BD}^{-1}\mathbf{C})^{-1}\mathbf{BD}^{-1}$. Show that

$$\Psi_2(z) = \Psi(z)\frac{\hat{a}(z)}{\hat{b}(z)}$$

8.30. Let $H(z) = \hat{b}(z)/\hat{a}(z)$ with $\hat{b}(z) = b_0 z^n + b_1 z^{n-1} + \cdots + b_n$ and $\hat{a}(z) = z^n + a_1 z^{n-1} + \cdots + a_n$.

a) Under what conditions will $\mathbf{x}(k) = [y(k-1) \ldots y(k-n)]^T$ be a valid state vector?
b) Under what conditions will $\mathbf{x}(k) = [u(k-1) \ldots u(k-n)]^T$ be a valid state vector? [Give your answers in terms of conditions involving $\hat{a}(z)$, $\hat{b}(z)$. Find $(\mathbf{A}, \mathbf{B}, \mathbf{C}, \mathbf{D})$ for each case.]

8.31. With $H(z)$ as in the previous problem, show that

$$\mathbf{x}(k) = \begin{bmatrix} y(k) \\ y(k+1) \\ \vdots \\ y(k+n-1) \end{bmatrix} - \begin{bmatrix} h(0) & & & \bigcirc \\ h(1) & \ddots & & \\ & \ddots & \ddots & \\ h(n-1) & & h(1) & h(0) \end{bmatrix} \begin{bmatrix} u(k) \\ u(k+1) \\ \vdots \\ u(k+n-1) \end{bmatrix}$$

is a valid state vector. Determine $(\mathbf{A}, \mathbf{B}, \mathbf{C}, \mathbf{D})$. Compute the state covariance $\mathbf{K}$ in terms of $\mathbf{r}_{yy}$, $\mathbf{h}$.

8.32. Suppose $(\mathbf{A}, \mathbf{B}, \mathbf{C}, \mathbf{D})$ has a unit-pulse sequence $\mathbf{h}$ and transfer function $H(z)$. Construct a new state variable description

$$\begin{bmatrix} \mathbf{A}\cos\omega_0 & \mathbf{A}\sin\omega_0 \\ -\mathbf{A}\sin\omega_0 & \mathbf{A}\cos\omega_0 \end{bmatrix}, \begin{bmatrix} \mathbf{B} \\ \mathbf{0} \end{bmatrix}, [\mathbf{C}\cos\omega_0, \mathbf{C}\sin\omega_0], \mathbf{D}$$

which produces $G(z)$. Express $\mathbf{g}$ in terms of $\mathbf{h}$. Express $G(z)$ in terms of $H(z)$.

8.33. Suppose $u(k + n) = u(k)$ (input is n-periodic).

a) Show that the state vector is also n-periodic, i.e., $\mathbf{x}(k + n) = \mathbf{x}(k)$ where $\mathbf{x}(k + 1) = \mathbf{A}\mathbf{x}(k) + \mathbf{B}u(k)$.

b) Construct the matrix $\mathbf{G}$ for which

$$\begin{bmatrix} \mathbf{x}(0) \\ \mathbf{x}(1) \\ \vdots \\ \mathbf{x}(n-1) \end{bmatrix} = \mathbf{G} \begin{bmatrix} \mathbf{B}u(0) \\ \mathbf{B}u(1) \\ \vdots \\ \mathbf{B}u(n-1) \end{bmatrix}.$$

Show that $\mathbf{G}$ is block circulant.

8.34. Let $H(z) = (z + 1)^3/[6z(z^2 + \frac{1}{3})]$. Compute $\sum_{k=0}^{\infty} h^2(k)$, by first finding an SVD, and then using Eq. (8.5.22). (This is a third-order Butterworth low-pass filter.)

8.35. Given $(\mathbf{A}, \mathbf{B}, \mathbf{C}, \mathbf{D})$ and the matrices $\mathbf{K}$ and $\mathbf{W}$ defined implicitly as follows:

$$\mathbf{K} = \mathbf{A}\mathbf{K}\mathbf{A}^T + \mathbf{B}\mathbf{B}^T,$$
$$\mathbf{W} = \mathbf{A}^T\mathbf{W}\mathbf{A} + \mathbf{C}^T\mathbf{C}.$$

a) If $\mathbf{A}' = \mathbf{T}^{-1}\mathbf{A}\mathbf{T}$, $\quad \mathbf{B}' = \mathbf{T}^{-1}\mathbf{B}$, $\quad \mathbf{C}' = \mathbf{C}\mathbf{T}$, $\quad$ then find $\mathbf{K}'$ and $\mathbf{W}'$.

b) Show that the eigenvalues of the matrix $(\mathbf{K}\mathbf{W})$ are coordinate free, i.e., are identical to the eigenvalues of $(\mathbf{K}'\mathbf{W}')$.

c) Show that $\operatorname{Tr}(\mathbf{K}\mathbf{W}) = \sum_{k=0}^{\infty} kh^2(k)$. [In general, $\operatorname{Tr}(\mathbf{K}\mathbf{W}) = \sum_{k=0}^{\infty} k \operatorname{Tr}(\mathbf{h}(k)\mathbf{h}^T(k))$.]

8.36. There are several possible "lattice structure" realizations for a given transfer function $H(z)$. This problem provides the means for constructing one. Let $H_n(z) = H(z)$. We wish to construct a sequence of transfer functions of decreasing order:

$$H_k(z) = \frac{\hat{b}_k(z)}{\hat{a}_k(z)}, \qquad 0 \leqslant k \leqslant n$$

where

$$\hat{b}_k(z) = \sum_{i=0}^{k} b_{ki} z^{-i}, \qquad \hat{a}_k(z) = \sum_{i=0}^{k} a_{ki} z^{-i}, \qquad a_{k0} = 1.$$

a) Suppose that $H_k(z)$ is given. Choose scalars α_k and β_k so that with

$$H_k(z) = \beta_k + z^{-1} \cfrac{1}{z^{-1}\alpha_k + \cfrac{1}{H_{k-1}(z)}},$$

the function $H_{k-1}(z)$ will be of order $k - 1$. (The expansion of this equation which expresses $H_n(z)$ in terms of $H_0(z) = b_{00}$, is called a *continued fraction expansion*.)

b) Using the results of part a), find a matrix function

$$\mathbf{Q} = \mathbf{Q}(\alpha, \beta, z)$$

for which

$$\begin{bmatrix} \hat{b}_k(z) \\ \hat{a}_k(z) \end{bmatrix} = \mathbf{Q}(\alpha_k, \beta_k, z) \begin{bmatrix} \hat{b}_{k-1}(z) \\ \hat{a}_{k-1}(z) \end{bmatrix}$$

Also, compute det $\mathbf{Q}$, and $\mathbf{Q}^{-1}$.

c) Now construct a signal flow graph meeting the following conditions: It has a single input and a single output and the transfer function from input to output is $H_k(z)$. All branches are primitive except one, whose path gain is $H_{k-1}(z)$. (The expansion of this graph, obtained by replacing $H_{k-1}(z)$ with its own graph and so on for decreasing k, is called a *lattice*.)

d) Is there a correspondence between the signal flow graph in part c) and the matrix in part b)?

e) Can *any* transfer function be decomposed in this way?

Spectral Representation and Functions of a Matrix

In our treatment of state variable descriptions, we have seen the matrix expressions

$$\mathbf{A}^k \quad \text{(powers)}$$
$$(z\mathbf{I} - \mathbf{A})^{-1} \quad \text{(matrix resolvent)}$$

play a central role in time and frequency domain analyses. These expressions involve the $n \times n$ matrix $\mathbf{A}$ and are therefore *matrix functions of the matrix* $\mathbf{A}$. If $\mathbf{A}$ were a scalar variable, we would write these functions as

$$f(a) = a^k, \tag{8A.1}$$

$$g(a) = \frac{1}{z - a}. \tag{8A.2}$$

The primary goal of this appendix is to characterize functions of the *matrix* $\mathbf{A}$ that derive from functions of a *scalar* variable. There is some fascinating algebraic structure beneath this characterization which has an amazing breadth of application, both for analysis and for computation.

The material in this section is supplementary to our discussion of system descriptions. Most results are given without proof, and the discussion is not meant to be a formal treatise. Rather, we have chosen a shortest path to the central issue, which is how to deal with functions of a square matrix.

One can superficially divide matrix algebra into two kinds of problems. The first group of problems involves characterizing or computing solutions to a system of linear equations like

$$\mathbf{Ax} = \mathbf{b}, \tag{8A.3}$$

where $\mathbf{A}$ is a known matrix and $\mathbf{b}$ a known vector. One might call these problems the "easy" problems, since finite procedures are known that characterize the set of all solution vectors $\mathbf{x}$, if there are any. The other group of problem come under the heading of "spectral theory of matrices" and involve such things as eigenvalues, eigenvectors, and functions of a matrix. Historically, the "easy" problems were studied first and are very old. The spectral theory was not studied until late in the nineteenth century. The bridge from the first group to the second is provided by the *theory of determinants*, first studied for their own sake by Vandermonde in the eighteenth century (see Kline 1972).

Determinants were invented (evidently by Maclaurin in 1729) to characterize solutions to Eq. (8A.3). In this context, there are three essential results, which we shall briefly indicate.

1. If $\mathbf{A}$ is an $n \times n$ matrix, then there exists a nonzero solution $\mathbf{x}$ to the *homogeneous equation*

$$\mathbf{Ax} = \mathbf{0} \quad \text{iff } \det \mathbf{A} = 0. \tag{8A.4}$$

2. If $\mathbf{A}$ is an $n \times n$ matrix, then there exists a unique solution $\mathbf{x}$ to Eq. (8A.3) if and only if $det(\mathbf{A}) \neq 0$, and if this is so, then the solution can be expressed in terms of determinants by Cramer's rule (1750).
3. If $\mathbf{A}$ and $\mathbf{B}$ are two $n \times n$ matrices, then

$$\det \mathbf{AB} = \det \mathbf{A} \det \mathbf{B} \tag{8A.5}$$

(We assume that the reader is familiar with the determinant of a matrix, and with Cramer's rule.)

The spectral theory begins with three definitions involving a determinant. Let us assume, for the remainder of this section, that $\mathbf{A}$ is an $n \times n$ matrix. Then the *characteristic polynomial of* $\mathbf{A}$ *is*

$$p(z) = \det(z\mathbf{I} - \mathbf{A}). \tag{8A.6}$$

The *eigenvalues* of $\mathbf{A}$ are the n (possibly repeated) roots of this polynomial. Finally, for each eigenvalue λ there is a nonzero *eigenvector* ψ which satisfies

$$\mathbf{A}\psi = \lambda\psi, \qquad \psi \neq \mathbf{0}. \tag{8A.7}$$

It is important to realize that there are some deep mathematical results that support these so called "definitions." First, the fact that the function of z in Eq. (8A.6) is in fact a *polynomial*. That is,

$$p(z) = \det(z\mathbf{I} - \mathbf{A}) = z^n + a_1 z^{n-1} + \cdots + a_n \tag{8A.8}$$

follows from the properties of determinants. That polynomials have roots (and therefore that $\mathbf{A}$ has eigenvalues) was first proven by Gauss in 1799 and is a result so important that it is called the "fundamental theorem of algebra." That an eigenvector must exist for each eigenvalue follows from Th. (8A.4), since we are searching for a nonzero solution to the homogeneous equation, $(\lambda\mathbf{I} - \mathbf{A})\psi = \mathbf{0}$.

The most important result in the spectral theory of $n \times n$ matrices is the Cayley-Hamilton theorem, which we shall approach by studying the *matrix resolvent* $(z\mathbf{I} - \mathbf{A})^{-1}$. Using Cramer's rule, one establishes a representation for this function.

4. The matrix resolvent must have the form

$$(z\mathbf{I} - \mathbf{A})^{-1} = \frac{1}{p(z)}\{z^{n-1}\mathbf{I} + z^{n-2}\mathbf{B}_2 + \cdots + \mathbf{B}_n\} \tag{8A.9}$$

where $p(z)$ is given by Eq. (8A.8) and where each $\mathbf{B}_k$ is an $n \times n$ matrix.
Rewrite Eq. (8A.9) as follows:

$$(z^n + a_1 z^{n-1} + \cdots + a_n)\mathbf{I} = (z\mathbf{I} - \mathbf{A})(z^{n-1}\mathbf{I} + z^{n-2}\mathbf{B}_2 + \cdots + \mathbf{B}_n). \tag{8A.10}$$

Using the properties of determinants and some extended use of the chain rule for derivatives, one proves the identity

$$\frac{d}{dz}p(z) = \mathrm{Tr}(z^{n-1}\mathbf{I} + z^{n-2}\mathbf{B}_2 + \cdots + \mathbf{B}_n\}. \tag{8A.11}$$

[The notation Tr(**A**) stands for the *trace*, or sum of diagonal elements of the matrix **A**.] Now by Eq. (8A.9), the matrix resolvant is fully parameterized by the a_k's and $\mathbf{B}_k$'s. If one equates coefficients of like powers of z in Eqs. (8A.10) and (8A.11), one can easily derive the following algorithm for computing these parameters.

LEVERRIER'S ALGORITHM

Initialization:
$$\begin{cases} a_0 = 1 \\ \mathbf{B}_1 = \mathbf{I} \end{cases}$$

Recursion:
$$\begin{cases} a_k = -\dfrac{1}{k}\mathrm{Tr}(\mathbf{A}\mathbf{B}_k) \\ \mathbf{B}_{k+1} = \mathbf{A}\mathbf{B}_k + a_k\mathbf{I} \end{cases} \tag{8A.12}$$

This algorithm requires about n^4 multiplications.

5. (Cayley-Hamilton theorem.) Let **A** have characteristic polynomial $p(z)$. Then

$$p(\mathbf{A}) = \mathbf{A}^n + a_1\mathbf{A}^{n-1} + \cdots + a_n\mathbf{I} = \mathbf{0}. \tag{8A.13}$$

This result is often stated "every square matrix satisfies its own characteristic equation." Recall that $p(z)$ is an nth-degree polynomial with n roots. Yet there are actually n^2 scalar equations in Eq. (8A.13), more than we have a right to expect. How does it follow from our previous results? Set $z = 0$ in Eq. (8A.10) to obtain

$$a_n\mathbf{I} = -\mathbf{A}\mathbf{B}_n. \tag{8A.14}$$

Now observe that the sequence of matrices $\{\mathbf{B}_k\}$ is exactly the sequence one obtains when using Horner's rule to compute $p(\mathbf{A})$. In other words,

$$\mathbf{B}_{k+1} = \mathbf{A}^k + a_1\mathbf{A}^{k-1} + \cdots + a_k\mathbf{I}. \tag{8A.15}$$

It follows that

$$\mathbf{AB}_n + a_n\mathbf{I} = \mathbf{A}^n + a_1\mathbf{A}^{n-1} + \cdots + a_n\mathbf{I} \tag{8A.16}$$

or $p(\mathbf{A})$. But, by Eq. (8A.14), this is the zero matrix. This proves the Cayley-Hamilton theorem.

Why is this theorem so significant? It is because of the tight structure it imposes on all functions of the matrix **A**. Put informally, it will reduce the computation of $f(\mathbf{A})$ to the computation of $f(\lambda_1)$ through $f(\lambda_n)$ (and possible derivatives) where the λ_k are the eigenvalues of **A**. It is this structure that we seek to develop. To begin with, however, we point out a surprising consequence of the Cayley-Hamilton theorem. Consider the infinite sequence of powers $\{\mathbf{I}, \mathbf{A}, \mathbf{A}^2, \ldots\}$. Each of these matrices is an n^2-dimensional object. Thus we know that no more than n^2 of them can be linearly independent. The Cayley-Hamilton theorem guarantees, however, that for any nonnegative integer k, the matrix $\mathbf{A}^k$ can be expressed as a linear combination of the first n powers, $\mathbf{A}^0$ through $\mathbf{A}^{n-1}$. Therefore there are at most n linearly independent matrices in the sequence! This is much less than the n^2 we might expect. The proof of this is a corollary of the following two results.

6. Let $p(z) = \det(z\mathbf{I} - \mathbf{A})$, and let $f(z)$ and $g(z)$ be two arbitrary polynomials. Then

$$f(z) = g(z) \bmod p(z) \Rightarrow f(\mathbf{A}) = g(\mathbf{A}). \tag{8A.17}$$

7. Let $f(z)$ be any polynomial of arbitrary degree. Then there exists a polynomial $r(z)$ of degree less than n (where **A** is $n \times n$) for which

$$f(\mathbf{A}) = r(\mathbf{A}). \tag{8A.18}$$

The notation $\{f = g \bmod p\}$ means that $f(z) - g(z) = p(z)q(z)$ for some polynomial $q(z)$. Every polynomial has a unique representation

$$f(z) = q(z)p(z) + r(z), \tag{8A.19}$$

where $r(z)$ (the "remainder" polynomial) has degree less than n. Using this and the Cayley-Hamilton theorem, one establishes Eq. (8A.18).

From these two results, we see that the set of all polynomial functions of the $n \times n$ matrix **A** forms at most an n-parameter family [the coefficients of $r(z)$]. This follows from the Cayley-Hamilton theorem and is the key structural result. One now asks for an efficient and elegant representation for this n-parameter family. This motivates the so-called *spectral representation* of the matrix **A**. We will approach this via the classical Lagrange interpolation problem.

Let us begin with the simple case, wherein all n eigenvalues of the matrix **A** are distinct. Let

$$p(z) = \det(z\mathbf{I} - \mathbf{A}) = \prod_{k=1}^{n} (z - \lambda_k). \tag{8A.20}$$

For this case, we have the following fact.

8. If all roots of $p(z)$ in Eq. (8A.20) are distinct, then

$$f(z) = g(z) \bmod p(z)$$

if and only if

$$f(\lambda_k) = g(\lambda_k) \quad \text{for } 1 \leqslant k \leqslant n. \tag{8A.21}$$

It follows that if $f(z)$ agrees with $g(z)$ on the eigenvalues of **A**, then $f(\mathbf{A}) = g(\mathbf{A})$. Now the Lagrange interpolation problem is to find the unique polynomial $r(z)$ of degree less than n for which

$$r(\lambda_k) = f(\lambda_k), \qquad 1 \leqslant k \leqslant n. \tag{8A.22}$$

The solution is found using the polynomials

$$e_k(z) = \prod_{\substack{j=1 \\ j \neq k}}^{n} \frac{(z - \lambda_j)}{(\lambda_k - \lambda_j)}, \tag{8A.23}$$

which are chosen to satisfy

$$e_k(\lambda_j) = \begin{cases} 1, & j = k \\ 0, & \text{otherwise.} \end{cases} \tag{8A.24}$$

The interpolating polynomial is then

$$r(z) = \sum_{k=1}^{n} f(\lambda_k) e_k(z). \tag{8A.25}$$

Using the results (8A.17) and (8A.21), we have

$$f(\mathbf{A}) = r(\mathbf{A}) = \sum_{k=1}^{n} f(\lambda_k) \mathbf{E}_k, \tag{8A.26}$$

where

$$\mathbf{E}_k = e_k(\mathbf{A}). \tag{8A.27}$$

Equation (8A.26) allows us to compute any polynomial function of the matrix **A** provided we first compute the eigenvalues λ_1 through λ_n, and matrices $\mathbf{E}_1$ through $\mathbf{E}_n$. This equation actually holds for *all functions* (not just polynomials) that are analytic on little discs centered on the eigenvalues. An example will expose the surprising power of this result.

EXAMPLE 8A.1

Computing functions of a matrix.

Let

$$\mathbf{A} = \begin{bmatrix} \frac{1}{2} & \frac{1}{3} \\ \frac{7}{4} & -\frac{1}{6} \end{bmatrix}.$$

First, compute the eigenvalues of **A**,

$$\det(z\mathbf{I} - \mathbf{A}) = z^2 - \tfrac{1}{3}z - \tfrac{2}{3} = (z - 1)(z + \tfrac{2}{3}).$$

Thus $\lambda_1 = 1$ and $\lambda_2 = -\frac{2}{3}$.

Use Eqs. (8A.23) and (8A.27) to obtain

$$\mathbf{E}_1 = \left.\frac{z + \frac{2}{3}}{1 + \frac{2}{3}}\right|_{z=\mathbf{A}} = \tfrac{3}{5}\mathbf{A} + \tfrac{2}{5}\mathbf{I} = \begin{bmatrix} \frac{7}{10} & \frac{1}{5} \\ \frac{21}{20} & \frac{3}{10} \end{bmatrix}.$$

Similarly,

$$\mathbf{E}_2 = \left.\frac{z - 1}{-1 - \frac{2}{3}}\right|_{z=\mathbf{A}} = \begin{bmatrix} \frac{3}{10} & -\frac{1}{5} \\ -\frac{21}{20} & \frac{7}{10} \end{bmatrix} = \mathbf{I} - \mathbf{E}_1.$$

We now have λ_1, λ_2, $\mathbf{E}_1$, and $\mathbf{E}_2$. From these we can compute functions of **A**. For example,

$$f(a) = 1 \Rightarrow f(\mathbf{A}) = \mathbf{E}_1 + \mathbf{E}_2 = \mathbf{I},$$

$$f(a) = a \Rightarrow f(\mathbf{A}) = \lambda_1\mathbf{E}_1 + \lambda_2\mathbf{E}_2 = \mathbf{A},$$

$$f(a) = \frac{1}{a} \Rightarrow f(\mathbf{A}) = \mathbf{A}^{-1} = \frac{1}{\lambda_1}\mathbf{E}_1 + \frac{1}{\lambda_2}\mathbf{E}_2 = \begin{bmatrix} \frac{1}{4} & \frac{1}{2} \\ \frac{21}{8} & -\frac{3}{4} \end{bmatrix},$$

$$f(a) = a^k \Rightarrow f(\mathbf{A}) = \mathbf{A}^k = \lambda_1^k\mathbf{E}_1 + \lambda_2^k\mathbf{E}_2 = \mathbf{E}_1 + (-\tfrac{2}{3})^k\mathbf{E}_2$$

$$= \begin{bmatrix} \frac{7}{10} + \frac{3}{10}(-\frac{2}{3})^k & \frac{1}{5}(1 - (-\frac{2}{3})^k) \\ \frac{21}{20}(1 - (-\frac{2}{3})^k) & \frac{3}{10} + \frac{7}{10}(-\frac{2}{3})^k \end{bmatrix},$$

$$f(a) = \frac{1}{z - a} \Rightarrow f(\mathbf{A}) = (z\mathbf{I} - \mathbf{A})^{-1} = \frac{1}{z - 1}\mathbf{E}_1 + \frac{1}{z + \frac{2}{3}}\mathbf{E}_2.$$

The computationally difficult part of the problem of finding the λ_k's and $\mathbf{E}_k$'s is the problem of factoring the polynomial $p(z)$. The matrices $\mathbf{E}_k$ have a number of properties which can be exploited. These are enumerated in the following, which covers the general case (with possibly repeated roots).

9. Let **A** be an $n \times n$ matrix and let

$$p(z) = \det(z\mathbf{I} - \mathbf{A}) = \prod_{k=1}^{m} (z - \lambda_k)^{m_k}, \tag{8A.28}$$

where the m_k are the multiplicities of the distinct roots λ_1 through λ_m. Then

$$f(z) = g(z) \bmod p(z)$$

if and only if

$$f^{(j)}(\lambda_k) = g^{(j)}(\lambda_k), \qquad 1 \leqslant k \leqslant m, \qquad 0 \leqslant j \leqslant m_k - 1. \tag{8A.29}$$

There exist matrices $\mathbf{E}_1$ through $\mathbf{E}_m$ and (possibly zero) matrices $\mathbf{N}_1$ through $\mathbf{N}_m$ for which

$$\begin{aligned}
&\mathbf{E}_k^2 = \mathbf{E}_k \quad (\mathbf{E}_k \text{ is } \textit{idempotent} \text{ or a } \textit{projection})\\
&\mathbf{E}_k\mathbf{E}_j = \mathbf{0} \quad \text{when } k \neq j\\
&\sum_{k=1}^{m} \mathbf{E}_k = \mathbf{I} \quad \text{rank}(\mathbf{E}_k) = \text{Tr}(\mathbf{E}_k) = m_k\\
&\mathbf{N}_k^{m_k} = \mathbf{0} \quad (\mathbf{N}_k \text{ is } \textit{nilpotent})\\
&\mathbf{N}_k\mathbf{E}_k = \mathbf{E}_k\mathbf{N}_k = \mathbf{N}_k\\
&\mathbf{N}_k\mathbf{E}_j = \mathbf{E}_j\mathbf{N}_k = \mathbf{0} \quad \text{when } k \neq j\\
&\mathbf{N}_k\mathbf{N}_j = \mathbf{0} \quad \text{when } k \neq j
\end{aligned} \tag{8A.30}$$

and

$$f(\mathbf{A}) = \sum_{k=1}^{m} f(\lambda_k)\mathbf{E}_k + \sum_{k=1}^{m} \sum_{j=1}^{m_k - 1} \frac{f^{(j)}(\lambda_k)\mathbf{N}_k^j}{j!}. \tag{8A.31}$$

The extra terms involving the nilpotent matrices all vanish when the eigenvalues of **A** are distinct. Three important special cases of Eq. (8A.31) are the following:

$$f(a) = a \quad \Rightarrow \quad \mathbf{A} = \sum_{k=1}^{m} \lambda_k\mathbf{E}_k + \mathbf{N}_k, \tag{8A.32}$$

$$f(a) = \frac{1}{z - a} \quad \Rightarrow \quad (z\mathbf{I} - \mathbf{A})^{-1} = \sum_{k=1}^{m} \left\{\frac{1}{z - \lambda_k}\mathbf{E}_k + \sum_{j=1}^{m_k - 1} \frac{1}{(z - \lambda_k)^{j+1}}\mathbf{N}_k^j\right\}, \tag{8A.33}$$

$$f(a) = a^l \quad \Rightarrow \quad \mathbf{A}^l = \sum_{k=1}^{m} \left\{\lambda_k^l\mathbf{E}_k + \sum_{j=1}^{m_k - 1} \binom{l}{j} \lambda_k^{l-j}\mathbf{N}_k^j\right\}. \tag{8A.34}$$

Equation (8A.32) is called the *spectral representation of* **A**. It involves the scalars λ_1 through λ_m and the matrices $\{\mathbf{E}_k\}$ and $\{\mathbf{N}_k\}$. These matrices may be computed in various ways. We shall discuss two possibilities.

The generalization of the Lagrange interpolation method involves Eq. (8A.31). Since $\sum_{k=1}^{m} m_k = n$, there are precisely n separate terms in the double sum (8A.31), each of which involves an evaluation of f or one of its derivatives. If one constructs a polynomial for which all of these terms *except one* vanishes, then one can pick out the respective coefficient matrix by evaluating this polynomial at **A**.

Another method involves a simple partial fraction expansion of the matrix resolvent $(z\mathbf{I} - \mathbf{A})^{-1}$ given by Eq. (8A.9). Each of the n^2 components of this matrix is a rational function of z, and the denominator for all of them is $p(z)$. Thus they can all be expanded in a partial fraction expansion. The result is precisely Eq. (8A.33). One need only identify the coefficient matrices properly to obtain the $\mathbf{E}_k$'s and $\mathbf{N}_k$'s.

EXAMPLE 8A.2

Computing the spectral representation of a matrix.

Let

$$\mathbf{A} = \begin{bmatrix} 6 & 9 & 0 \\ -4 & -6 & 1 \\ 0 & 0 & 2 \end{bmatrix}$$

The matrix resolvent for **A** is

$$(z\mathbf{I} - \mathbf{A})^{-1} = \begin{bmatrix} z-6 & -9 & 0 \\ 4 & z+6 & -1 \\ 0 & 0 & z-2 \end{bmatrix}^{-1}$$

$$= \frac{1}{z^3 + a_1 z^2 + a_2 z + a_3}\{z^2\mathbf{I} + z\mathbf{B}_2 + \mathbf{B}_3\}. \tag{8A.35}$$

Using Leverrier's algorithm, we obtain the matrices $\mathbf{B}_2$ and and the coefficients a_1, a_2, and a_3 as follows:

$$a_1 = -\mathrm{Tr}\{\mathbf{A}\} = -2,$$

$$\mathbf{B}_2 = \mathbf{AB}_1 + a_1\mathbf{I} = \mathbf{A} - 2\mathbf{I} = \begin{bmatrix} 4 & 9 & 0 \\ -4 & -8 & 1 \\ 0 & 0 & 0 \end{bmatrix},$$

$$a_2 = -\tfrac{1}{2}\mathrm{Tr}\{\mathbf{AB}_2\} = 0,$$

$$\mathbf{B}_3 = \mathbf{AB}_2 + a_2\mathbf{I} = \begin{bmatrix} -12 & -18 & 9 \\ 8 & 12 & -6 \\ 0 & 0 & 0 \end{bmatrix},$$

$$a_3 = -\tfrac{1}{3}\mathrm{Tr}(\mathbf{AB}_3) = 0.$$

Thus we have

$$(z\mathbf{I} - \mathbf{A})^{-1} = \frac{1}{z^2(z-2)}\begin{bmatrix} z^2 + 4z - 12 & 9z - 18 & 9 \\ -4z + 8 & z^2 - 8z + 12 & z - 6 \\ 0 & 0 & z^2 \end{bmatrix}.$$

A partial fraction expansion of $(z\mathbf{I} - \mathbf{A})^{-1}$ is

$$\left[\frac{1}{z}\mathbf{E}_1 + \frac{1}{z^2}\mathbf{N}_1 + \frac{1}{z-2}\mathbf{E}_2\right],$$

where

$$\mathbf{E}_1 = \begin{bmatrix} 1 & 0 & -\frac{9}{4} \\ 0 & 1 & 1 \\ 0 & 0 & 0 \end{bmatrix}, \quad \mathbf{N}_1 = \begin{bmatrix} 6 & 9 & -\frac{9}{2} \\ -4 & -6 & 3 \\ 0 & 0 & 0 \end{bmatrix}, \quad \mathbf{E}_2 = \begin{bmatrix} 0 & 0 & \frac{9}{4} \\ 0 & 0 & -1 \\ 0 & 0 & 1 \end{bmatrix}.$$

One can use the relations (8A.30) to check our results. For example, $\mathbf{E}_1^2 = \mathbf{E}_1$, $\mathbf{E}_2^2 = \mathbf{E}_2$, $\mathbf{E}_1 + \mathbf{E}_2 = \mathbf{I}$, $\mathbf{N}_1^2 = \mathbf{0}$.

Eigenvectors

The spectral representation for the matrix **A** in Eq. (8A.32) obviously involves the eigen*values* of **A**. What about the associated eigen*vectors*? The answer is that they are superceded by the matrices $\{\mathbf{E}_k\}$.

Consider the roles played by the eigenvectors. If $\mathbf{A}\psi = \lambda\psi$, then on the one-dimensional subspace spanned by the vector ψ, the $n \times n$ matrix **A** acts like multiplication by the scalar λ. A complicated thing has been reduced to a simple thing. Now consider the case where the matrix **A** has spectral representation

$$\mathbf{A} = \sum_{k=1}^{n} \lambda_k \mathbf{E}_k$$

with no nontrivial nilpotent matrices. Using the properties in Eq. (8A.30), we see that

$$\mathbf{A}\mathbf{E}_k = \mathbf{E}_k\mathbf{A} = \lambda_k \mathbf{E}_k.$$

We may infer from this that the matrix $\mathbf{E}_k$ is both a left and right "eigenmatrix" in the sense that **A** acts on it like multiplication by a scalar. In fact, the "projection" matrix $\mathbf{E}_k$ projects all vectors onto a space (the kth eigenspace) of eigenvectors, since

$$\mathbf{A}(\mathbf{Ex}) = \lambda(\mathbf{Ex}).$$

In the general case, where there are repeated roots in $p(z)$ and nontrivial nilpotents, the situation is not quite so simple. However, every eigenvalue λ_k must have at least one eigenvector ψ_k, and this vector must lie in the range of $\mathbf{E}_k$. It is not necessarily the case, however, that all vectors in the range of $\mathbf{E}_k$ must be eigenvectors.

Transformations

A change in the coordinate system used to represent vectors produces the transformation on matrix operators

$$\mathbf{A} \quad \leftarrow \quad \mathbf{T}^{-1}\mathbf{A}\mathbf{T}. \tag{8A.36}$$

What does this transformation do to the spectral representation of **A**? This question has been of great interest to mathematicians, since one can make the case that any "intrinsic" property of the matrix **A** as a linear operator should not depend on the coordinate system. In other words, it should be invariant under transformation. Among the invariants are the eigenvalues of **A**. To show this, we need two applications of Eq. (8A.5). First,

$$1 = \det \mathbf{I} = \det \mathbf{T}^{-1}\mathbf{T} = \det \mathbf{T}^{-1} \det \mathbf{T} \Rightarrow \det \mathbf{T}^{-1} = \frac{1}{\det \mathbf{T}}.$$

Second,

$$\begin{aligned}\det(z\mathbf{I} - \mathbf{T}^{-1}\mathbf{A}\mathbf{T}) &= \det(\mathbf{T}^{-1}(z\mathbf{I} - \mathbf{A})\mathbf{T}) \\ &= \det \mathbf{T}^{-1} \det(z\mathbf{I} - \mathbf{A}) \det \mathbf{T} \\ &= \det(z\mathbf{I} - \mathbf{A}).\end{aligned}$$

The matrices $\mathbf{E}_k$ and $\mathbf{N}_k$ are not invariant, but transform as

$$\begin{aligned}\mathbf{E}_k &\leftarrow \mathbf{T}^{-1}\mathbf{E}_k\mathbf{T}, \\ \mathbf{N}_k &\leftarrow \mathbf{T}^{-1}\mathbf{N}_k\mathbf{T}.\end{aligned}$$

One can check that all the relations in the list (8A.30) remain true.

Normal Matrices

An $n \times n$ matrix **A** is *normal* if

$$\mathbf{A}^*\mathbf{A} = \mathbf{A}\mathbf{A}^*, \qquad \textbf{(8A.37)}$$

where $\mathbf{A}^*$ is the transpose conjugate of **A**, defined by

$$[\mathbf{A}^*]_{ij} = (a_{ji})^*. \qquad \textbf{(8A.38)}$$

This extra property involves studying a *Euclidean* vector space (with the extra notion of orthogonality via an inner product) rather than an abstract vector space. None of the results so far stated in this section involved this concept. Important special cases of normal matrices are the

---

Hermitian: $\mathbf{A}^* = \mathbf{A}$, **(8A.39)**

Anti-Hermitian: $\mathbf{A}^* = -\mathbf{A}$, **(8A.40)**

Unitary: $\mathbf{A}^*\mathbf{A} = \mathbf{I}$. **(8A.41)**

The spectral theory for normal matrices is much more restrictive. We shall state the principal results. Keep in mind the fact that all the previously stated results for general $n \times n$ matrices must also hold for normal matrices.

10. Let **A** be an $n \times n$ normal matrix and let

$$p(z) = \det(z\mathbf{I} - \mathbf{A}) = \prod_{k=1}^{m} (z - \lambda_k)^{m_k},$$

where the m_k are the multiplicities of the distinct roots. Then **A** has the spectral representation

$$\mathbf{A} = \sum_{k=1}^{m} \lambda_k \mathbf{E}_k, \qquad \textbf{(8A.42)}$$

where

$$\mathbf{E}_k = \mathbf{E}_k^* = \mathbf{E}_k^2$$

$$\operatorname{rank}(\mathbf{E}_k) = \operatorname{Tr}(\mathbf{E}_k) = m_k$$

$$\sum_{k=1}^{m} \mathbf{E}_k = \mathbf{I}; \qquad \mathbf{E}_k\mathbf{E}_j = 0, \quad k \neq j.$$

Thus, if **A** is normal, it cannot have nontrivial nilpotents, and its $\mathbf{E}_k$ must be Hermitian. The following result says that the three previously mentioned special normal matrices are characterized by two properties: normality and location of eigenvalues.

11. If **A** is normal, and

- $\lambda_k^* = \lambda_k$ for each k, then **A** is Hermitian,
- $-\lambda_k^* = \lambda_k$ for each k, then **A** is anti-Hermitian,
- $\lambda_k^*\lambda_k = 1$ for each k, then **A** is unitary.

CHAPTER 9

Finite Length Register Effects in Fixed Point Digital Filters

9.1 INTRODUCTION

In Chapters 6 and 7 we considered the design of an input/output representation for a digital filter from a set of specifications. Chapter 6 discusses the deterministic approximation problem in which we design an $H(z)$ which is low-pass, band-pass, high-pass, etc. In Chapter 7, we obtain $H(z)$ as the result of an optimization process in order to filter out the effects of additive noise or, in general, to optimize some statistical performance measure. In either case the result of the approximation problem is an I/O characterization such as the transfer function $H(z)$ of a digital filter.

In Chapter 8 we showed that there are an infinite number of realizations for a given transfer function $H(z)$. How should we use this freedom in choosing a particular realization of the filter? The answer to this question depends on several factors. This freedom is generally used to optimize some criterion associated with the actual algorithm or realization used and on the manner in which we perform the computations. For example, if $H(z)$ is realized as a FORTRAN program on a large general purpose computer, then we might seek the realization that has the fewest number of multiplies per output sample. On the other hand, if we realize $H(z)$ as a hardware structure using fixed point arithmetic, then the internal noise caused by the use of finite length registers may be the most important parameter to optimize.

In general we use the freedom we have in selecting a realization for $H(z)$ to optimize criteria that are tied to the actual computation and, in some cases, to the hardware implementation of $H(z)$. Implementation considerations can influence the form of an algorithm or realization, and, vice versa, the nature of the algorithms often suggests a form of the hardware. In this chapter we shall present a theory of design that deals with the adverse effects of finite-length registers. We assume that fixed point arithmetic is used and that our starting point is an input/output description of the filter in the form of $H(z)$.

There are three primary finite-length register effects in fixed point digital filters. These effects are:

1. Roundoff noise caused by the rounding of products within the realization;
2. Limit cycles due to nonlinear quantization effects;
3. Changes in the input/output description of the filter due to approximating real parameters with a finite binary representation.

In addition, there is another error caused whenever analog-to-digital conversion is used to convert a continuous-time signal into a sequence of numbers that the digital filter can process. Strictly speaking, this error occurs outside the filter (at the input) and is therefore independent of the filter realization. The filter designer cannot change this quantization noise. However, it is important to understand. We shall find that many of the same concepts in A/D conversion also occur in the study of roundoff noise.

Of the three finite-length register effects, roundoff noise is the most important. The third effect, loosely called coefficient sensitivity, is a deterministic effect and can be bounded in some sense by the roundoff noise. The limit cycle properties of a realization can also be minimized by structures that fortuitously have low-roundoff noise.

9.2 ANALOG-TO-DIGITAL CONVERSION AND QUANTIZATION NOISE

In most applications of digital filtering, the original data or signal to be filtered is analog in nature. Thus an A/D converter is often an integral part of a digital filtering system. Analog-to-digital conversion consists of the dual process of sampling a continuous-time signal and converting the samples (real numbers) into a sequence of finite-length binary numbers. We assume the sampling process occurs at a rate that is at least twice the highest frequency contained in the signal. The error that occurs in A/D conversion is caused by the quantization of the real numbers.

The binary number generated by the A/D converter is assumed to be a fixed point number. Several formats such as signed magnitude, one's complement, or two's complement are used. Most often two's complement is chosen, because it is easily implemented. A two's complement representation of the real number r is

$$r = \Delta\left(-b_0 + \sum_{n=1}^{\infty} b_n 2^{-n}\right), \qquad -\Delta \leqslant r \leqslant \Delta \tag{9.2.1}$$

where b_n, $n \geqslant 0$ is one or zero. The first bit b_0 is the sign bit ($b_0 = 1$ if $r < 0$, $b_0 = 0$ if $r > 0$). The value of Δ is arbitrary.

When we use a finite-length register of $B + 1$ bits, the actual number stored is quantized to $[r]_Q$, where

$$[r]_Q = \Delta\left(-b_0 + \sum_{n=1}^{B} b_n 2^{-n}\right). \tag{9.2.2}$$

The quantized representation of r, $[r]_Q$, must be an integral multiple of the smallest quantum q. This smallest quantum is

$$q = \Delta 2^{-B}. \tag{9.2.3}$$

The number q is also called the *quantization step size* and is the finest separation between the 2^{B+1} numbers we can represent with $B + 1$ bits. We can arbitrarily define q by the choice of Δ.

The error between the real number r and its finite binary representation is

$$e = r - [r]_Q. \tag{9.2.4}$$

We shall assume that in forming $[r]_Q$ we round r to the nearest integer multiple of q. In this case the quantizer that defines the relationship between r and $[r]_Q$ has a characteristic as shown in Fig. 9.2.1. If a number r is larger than the largest number representable (or smaller than the smallest number representable), an overflow occurs. Overflows generally create large errors and thus must be avoided. For example, in two's complement if a large positive number, say, 011 ... 11 overflows, it becomes a number like 100 ... 00, which is a large negative number. Thus the error incurred in this overflow is approximately 2Δ. Overflows can be minimized by increasing the range of possible representations. This is done by increasing the value of Δ in Eq. (9.2.3). However, if the number of levels remains the same, the quantization step size q must also increase ($q = \Delta 2^{-B}$). Thus a decrease in the possibility of overflow is obtained only by increasing the error due to quantization.

The method used to "represent" an overflowed number will determine the error caused by overflows. After a number has overflowed, we can transform the overflowed number in a variety of ways. In two's complement representations, if nothing is done after an overflow, the overflow transformation or characteristic is as shown in Fig. 9.2.2(a). The overflow characteristic is

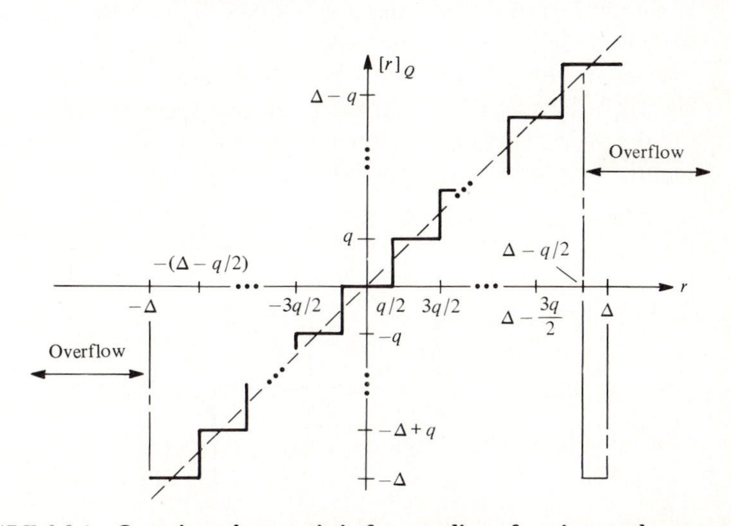

FIGURE 9.2.1 **Quantizer characteristic for rounding of two's complement numbers.**

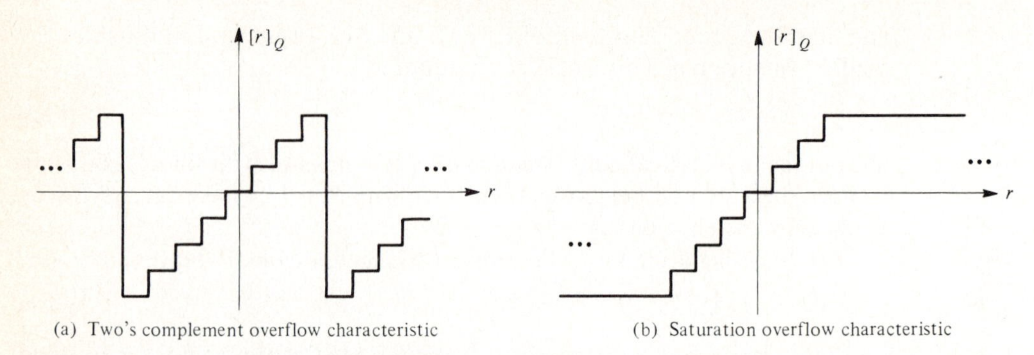

FIGURE 9.2.2 **Overflow Characteristics.**

periodic and a small overflow causes an error of approximately 2Δ, as explained previously.

Another method of handling overflows is to use a so-called saturation characteristic, as shown in Fig. 9.2.2(b). In this case the overflowed register is reset to the largest (or smallest) number representable. A saturation characteristic produces, on the average, a smaller overflow error than does the two's complement overflow characteristic. However, it is not as easy to implement as the two's complement characteristic. The two's complement characteristic also has the desirable property that if several quantized variables are added together, no overflow will occur if the sum is in-range *even though individual terms of the sum are out of range.*

From Fig. 9.2.2, we see that the error due to quantization for each sample $e(k)$ lies between $-q/2$ and $q/2$, i.e.,

$$-q/2 \leqslant e(k) \leqslant q/2. \tag{9.2.5}$$

Theoretical studies and numerical simulations have shown (Bennett 1948; Widrow 1956) that $e(k)$ can be approximated by a white noise sequence, uncorrelated with $u(k)$ and uniformly distributed on $[-q/2, q/2]$, as shown in Fig. 9.2.3. This assumes that the frequency spectrum of $\mathbf{u}$ is reasonably wide and the probability density function for $\mathbf{u}$ is broad relative to q so that several quantization levels are usually crossed between samples of $\mathbf{u}$. The variance or

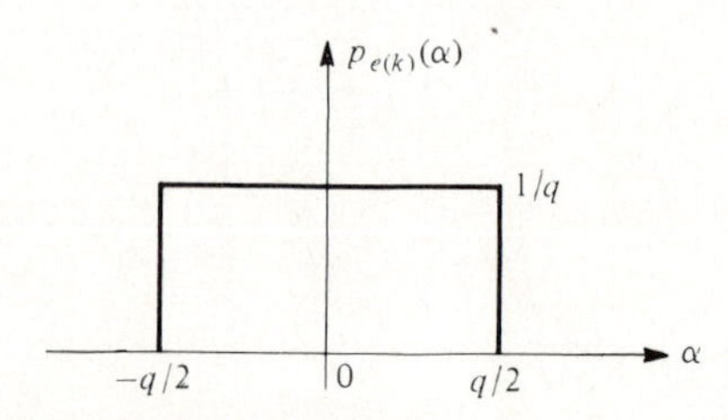

FIGURE 9.2.3 **The probability density function of the quantizer error.**

noise power associated with this error distribution is given by

$$\sigma_e^2 = E[e(k)^2] = \int_{-q/2}^{q/2} \frac{\alpha^2}{q}\, d\alpha = \frac{q^2}{12}. \tag{9.2.6}$$

[The mean value of $e(k)$ is zero.]

EXAMPLE 9.2.1

The signal-to-noise ratio for a uniform quantizer.

Suppose a sampled signal $\{u(k)\}$ limited in amplitude to the range $[-1, 1]$ is quantized to B bits. How many bits are needed to ensure the output signal-to-noise ratio is greater than some given value, say SN_Q?

Since $u(k)$ is limited to values between ± 1, we can choose $\Delta = 1$ and with B bits the quantization step size is

$$q = 1 \cdot 2^{-B+1}.$$

The output of the quantizer is

$$[u(k)]_Q = u(k) + e(k).$$

Define the signal-to-quantization noise ratio as

$$SN_Q = E[u(k)^2]/E[e(k)^2].$$

We shall assume $\{e(k)\}$ is a zero-mean, white noise sequence with variance $\sigma_e^2 = q^2/12$. The ratio SN_Q depends on both the statistics of $e(k)$ and $u(k)$. Suppose we assume $u(k)$ is uniformly distributed over the interval $[-1, 1]$. Then

$$\sigma_u^2 = E[u(k)^2] = \int_{-1}^{1} \frac{\alpha^2}{2}\, d\alpha = \frac{1}{3}.$$

In this case,

$$SN_Q = \frac{\frac{1}{3}}{q^2/12} = \frac{4}{q^2} = \frac{4}{2^{-2B+2}} = 2^{2B}.$$

Thus the required number of bits B to obtain the signal-to-noise ratio SN_Q is

$$B = \frac{\log_2(SN_Q)}{2}, \tag{9.2.7}$$

where B is rounded to the smallest integer greater than or equal to the right-hand side of Eq. (9.2.7).

In summary, the effect of quantizing a real number can be modeled by simply adding a zero-mean, white-noise source of variance $q^2/12$ to the original unquantized variable, as shown schematically in Fig. 9.2.4. This additive

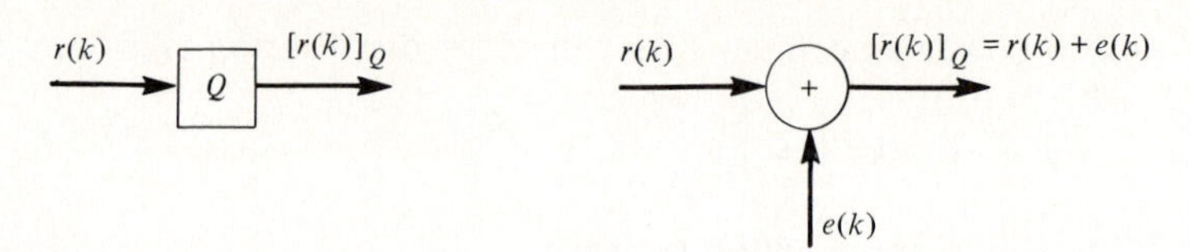

FIGURE 9.2.4 **A nonlinear quantizer can be modeled as a linear element with additive noise.**

noise model accounts *only* for the roundoff error caused by the quantizer. Overflow errors are *not accounted for in this model, and are assumed to be negligible.* This latter assumption is met by scaling the quantizer for a class of input signals. The method used to restore an overflowed number is determined by implementation.

9.3 LIMIT CYCLES—OVERFLOW OSCILLATIONS

We shall assume in the sequel that numbers are represented as signed, two's complement numbers and that we shall employ the two's complement overflow characteristic shown in Fig. 9.2.2. (In other words, we do not explicitly detect overflows and correct for them in some manner.) Thus depending on how well we have scaled a digital filter, there is the possibility that internal registers will overflow. The question we wish to answer in this section is what happens to the digital filter immediately following an overflow?

The answer to this question depends on several factors, which include:

1. The filter realization or structure,
2. The number representation used in the filter,
3. The nature of the input,
4. The overflow characteristic used in the filter.

The earliest studies of overflow effects considered the second-order direct form filter shown in Fig. 9.3.1. This realization has the advantage of having the minimum number of multipliers for a second-order filter. However, with a two's complement overflow characteristic, a disastrous effect can occur after an internal overflow—so-called *overflow oscillations.* For a stable direct form filter, the output of the filter after an internal overflow can (depending on the

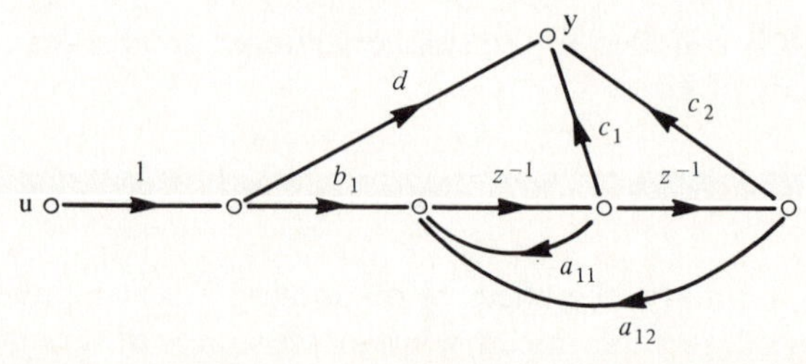

FIGURE 9.3.1 **A signal flow graph for a second-order direct form filter.**

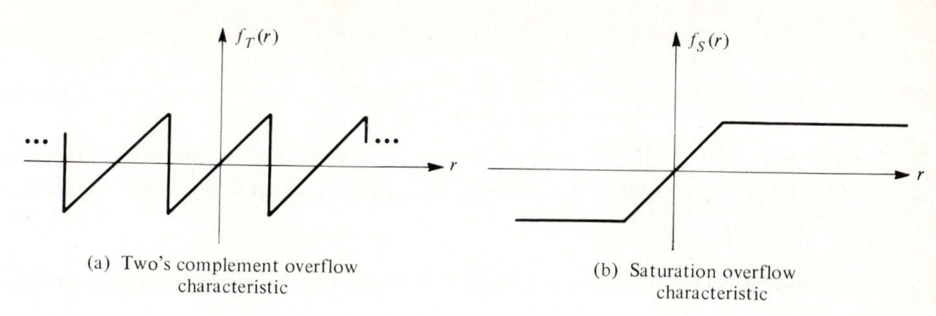

(a) Two's complement overflow characteristic

(b) Saturation overflow characteristic

FIGURE 9.3.2.

poles of the filter) become independent of the input sequence. This condition is called an overflow oscillation and is caused by the overflow nonlinearity characteristic.

For purposes of this discussion, we can disregard the roundoff error caused by the quantizer and focus only on the overflow nonlinearity characteristic. Thus the two's complement and saturation overflow characteristics are as shown in Fig. 9.3.2.

An overflow function f describes how an out-of-range number is to be represented within the number system. If we assume the range of the quantizer is $(-1, 1)$, then f must satisfy

$$f(r) = r, \qquad |r| < 1 \tag{9.3.1}$$

$$|f(r)| \leq 1. \tag{9.3.2}$$

Equation (9.3.1) states that in-range numbers are unaffected. Equation (9.3.2) states that the result of applying the overflow characteristic is always an in-range number. Together Eqs. (9.3.1) and (9.3.2) imply that the magnitude of $f(r)$ never exceeds the magnitude of r, i.e.,

$$|f(r)| \leq |r|. \tag{9.3.3}$$

This is the essential property we shall use in characterizing overflow functions f.

EXAMPLE 9.3.1

A property of the two's complement overflow characteristic.

One well-known and useful property of the two's complement overflow function f_T is the cancellation of intermediate overflows in an accumulator whenever the resultant sum is in range. This fact is easy to see if one regards the function f_T as plotted around a circle creating the periodicity seen in Fig. 9.3.2(a). Stated analytically, we have

$$f_T(f_T(f_T \cdots f_T(f_T(r_1) + r_2) \cdots) + r_n) = f_T(r_1 + r_2 \cdots + r_n). \tag{9.3.4}$$

Consider the case of two values summed together. We have

$$f_T(f_T(r_1) + r_2) = \begin{cases} f_T(r_1 + r_2), & |r_1| < 1 \\ f_T(r_1 + r_2 + 2n), & |r_1| > 1 \end{cases}. \tag{9.3.5}$$

However, $f_T(r + 2n) = f_T(r)$, because f_T is periodic of period 2. Thus the term $f_T(r_1 + r_2 + 2n) = f_T(r_1 + r_2)$ and since $(r_1 + r_2)$ is in range, the result is $r_1 + r_2$.

A State Variable Model for Overflow

Our study of overflow oscillations will use the state variable description of a digital filter. Recall from Chapter 8 that we can represent a filter in the form

$$\mathbf{x}(k + 1) = \mathbf{A}\mathbf{x}(k) + \mathbf{B}u(k), \qquad y(k) = \mathbf{C}\mathbf{x}(k) + Du(k). \tag{9.3.6}$$

Suppose for this discussion, we simplify Eq. (9.3.6) by setting the input to zero. Consider the evolution of the idealized filter state described by the equation

$$\mathbf{x}(k + 1) = \mathbf{A}\mathbf{x}(k). \tag{9.3.7}$$

In the case of finite-length registers, Eq. (9.3.7) becomes

$$\mathbf{x}(k + 1) = f(\mathbf{A}\mathbf{x}(k)), \tag{9.3.8}$$

where f is the overflow characteristic. The overflow function f applied to a vector operates on each component; i.e., $[f(\mathbf{x})]_i \triangleq f(x_i)$. If the filter of Eq. (9.3.6) is stable, then for any initial state $\mathbf{x}(0)$, $\mathbf{x}(k)$ will approach zero as $k \to \infty$. This may not occur for a system described by Eq. (9.3.8), depending on $\mathbf{A}$ and f.

Consider, for example, a second-order digital filter with states $x_1(k)$ and $x_2(k)$. If each state component is stored in a register of B bits, then each register can represent 2^B different numbers. The filter state $(x_1, x_2)^T$ thus can take on 2^{2B} values (numbers). Schematically we can represent all these possible states as a grid of points as shown in Fig. 9.3.3. Each intersection point represents a possible value of the state vector $(x_1, x_2)^T$.

This schematic of the 2^{2B} possible states a filter can assume is called a *state-space grid.* We can use the grid to describe roundoff and overflow. Suppose we have some state $\mathbf{x}(k)$ as shown in Fig. 9.3.3. An idealized filter produces a new state $\mathbf{x}(k + 1) = \mathbf{A}\mathbf{x}(k)$. If this new state $\mathbf{A}\mathbf{x}(k)$ falls outside of the representable numbers as shown in Fig. 9.3.3, then overflow occurs. The overflowed state is transformed into a representable number by the overflow nonlinearity and the result is $f(\mathbf{A}\mathbf{x}(k))$. Figure 9.3.3 indicates how the overflowed state $\mathbf{A}\mathbf{x}(k)$ is transformed under two's complement. The overflow error is

$$e_{\text{overflow}} = \mathbf{A}\mathbf{x}(k) - f(\mathbf{A}\mathbf{x}(k)).$$

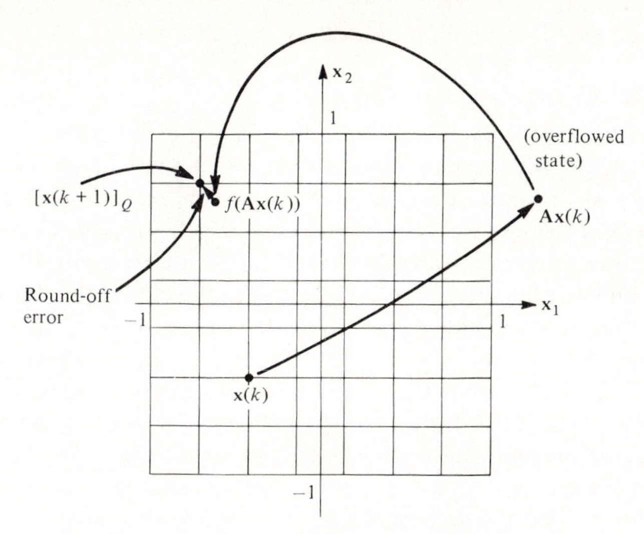

FIGURE 9.3.3 **The state-space of a second-order digital filter.**

In this schematic a *roundoff error* is the error incurred in transforming a point inside the $(-1, 1)$ square to a nearby grid point. (See Fig. 9.3.3.)

Notice again the tradeoff between the dynamic range of the states defined by the size or extent of the overall grid and the roundoff error determined by the size of the cells within the grid boundaries. For a fixed word length B, the number of grid points or cells is fixed. If we expand the boundaries of the $(-1, 1)$ square to $(-\Delta, \Delta)$, $\Delta > 1$, then we reduce the possibility of overflow. However, the cells within the grid must also increase in size, thereby increasing roundoff error. This is discussed in more detail in Section 9.8.

To demonstrate the phenomenon of overflow oscillation, consider the direct form filter shown in Fig. 9.3.4. The state equation for this filter is

$$\mathbf{x}(k+1) = \begin{bmatrix} 0 & 1 \\ -\frac{1}{2} & -1 \end{bmatrix} \mathbf{x}(k) + \begin{bmatrix} 0 \\ 1 \end{bmatrix} u(k). \tag{9.3.9}$$

The eigenvalues of $\mathbf{A}$ [which are the poles of $H(z)$] are given by the roots of the characteristic equation $g(\lambda) = 0$.

$$g(\lambda) = \det(\lambda\mathbf{I} - \mathbf{A}) = \lambda^2 + \lambda + \tfrac{1}{2} = 0. \tag{9.3.10}$$

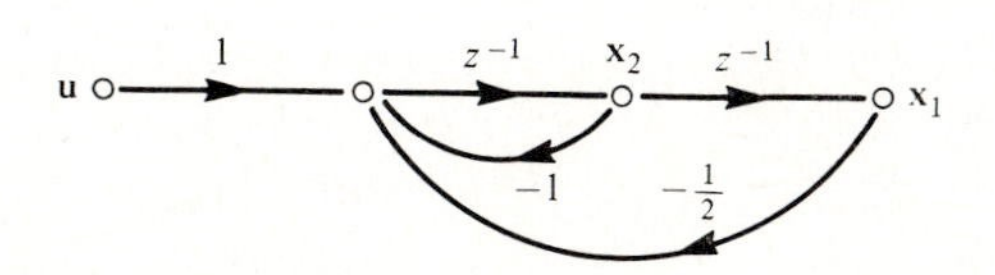

FIGURE 9.3.4 **A second-order direct form filter with poles at $-1/2 \pm j/2$.**

Solving for the roots, we obtain

$$\lambda_1, \lambda_2 = -\tfrac{1}{2} \pm \tfrac{1}{2}j. \tag{9.3.11}$$

Since $|\lambda_1|$ and $|\lambda_2|$ are less than unity, the filter is stable.

Figure 9.3.5 depicts the boundaries of the state-space grid of this filter. (Since we are not concerned with roundoff error in this discussion, we can disregard the internal cells in the state-space.) Suppose we ask how the state of this filter evolves for an initial state $\mathbf{x}(0)$ with no input, i.e., $u(k) = 0$ for all k. For an ideal filter $\mathbf{x}(k) = \mathbf{A}^k\mathbf{x}(0)$ and since the eigenvalues of $\mathbf{A}$ have magnitude less than unity (stable filter), $\mathbf{x}(k) \to \mathbf{0}$ as $k \to \infty$.

In the case of finite length registers, we must calculate the trajectory of $\mathbf{x}(k)$ and determine if this trajectory ever falls outside the state space grid of Fig. 9.3.5(a). If the trajectory leaves the $(-1, 1)$ square, then the path of the trajectory is changed by the overflow nonlinearity. We can determine if the state leaves the $(-1, 1)$ square by transforming the boundaries of the $(-1, 1)$ square by the linear transformation defined by the system matrix $\mathbf{A}$ of Eq. (9.3.9). For example, the point $[1, 1]^T$ is transformed into $\mathbf{A}[1, 1]^T = [1, -1.5]^T$. Similarly, we can calculate how all states are transformed by $\mathbf{A}$.

The result is shown in Fig. 9.3.5(b). What we find is that the matrix $\mathbf{A}$ maps two triangles $\overline{AEG}$ and $\overline{CFH}$ *outside* of the original state space boundaries. Any state contained within these two triangles overflows at the next transition with *zero input*. The two's complement overflow function restores the overflowed states by "wrapping" the overflow regions into the original state space, as shown in Fig. 9.3.6. (The restored values are denoted by two primes.)

A comparison of Figs. 9.3.5(a) and 9.3.6 shows that the polygon $\overline{DGEBHF}$

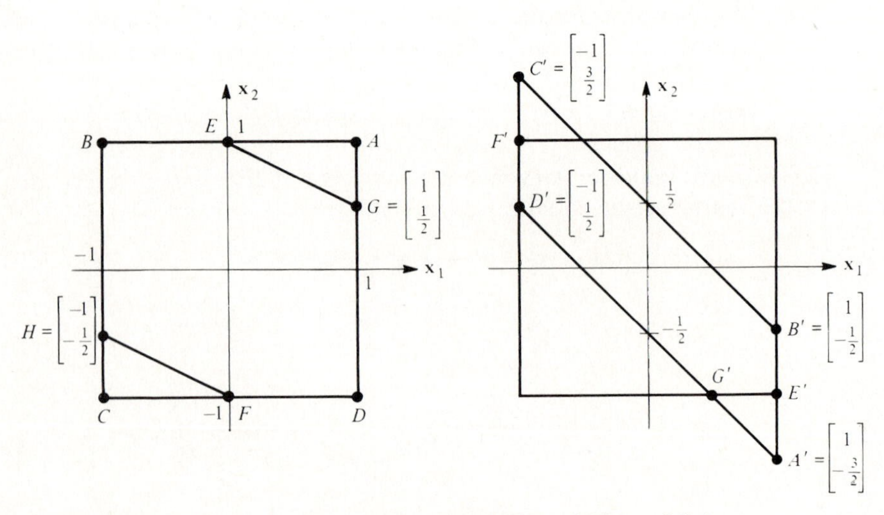

FIGURE 9.3.5 **The state space for the digital filter of Figure 9.3.4.**

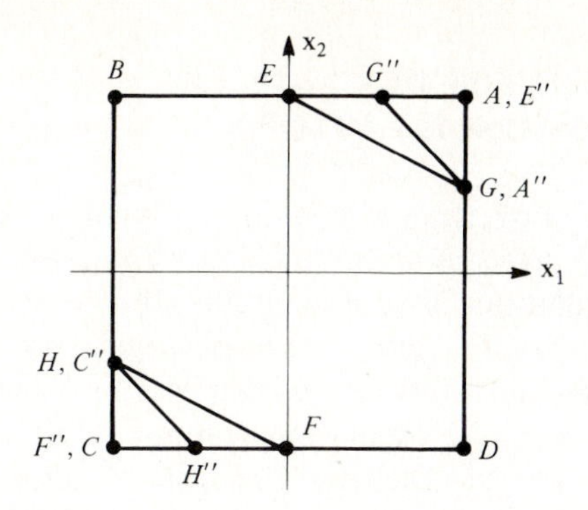

FIGURE 9.3.6 **The effect of the two's complement overflow nonlinearity applied to Figure 9.3.5(b).**

has mapped entirely into its own interior. Thus states that originate inside $\overline{DGEBHF}$ will approach the origin as $k \to \infty$. On the other hand, the triangles $\overline{AEG}$ and $\overline{CFH}$ map inside themselves via overflow. Successive applications of $f_T(\mathbf{Ax}(k))$ map $\overline{AEG}$ and $\overline{CFH}$ into smaller and smaller regions which converge, in this filter, to the fixed points (0.8, 0.8) and (−0.8, −0.8), respectively. By changing the sign of the feedback coefficients in the filter so that the eigenvalues are $\frac{1}{2} \pm j/2$, the filter output can be made to oscillate between (0.8, 0.8) and (−0.8, −0.8) for initial states in $\overline{AEG}$ and $\overline{CFH}$.

Thus for certain initial states *the filter output does not converge to zero for a zero input*. This is called *overflow oscillation.* If the input is small and an overflow occurs, this phenomenon can take over and the filter's output becomes independent of the input. How can we avoid this catastrophic effect? If we use a direct form and two's complement arithmetic, we must guarantee that no overflows occur. Thus we must scale the filter very conservatively to ensure that filter states never enter the northeast and southwest corners of the state space. This means we increase the boundaries of the state space grid and thus increase the cell sizes which, finally, results in more roundoff noise.

Another solution is to use a saturation overflow characteristic. Suppose the filter of Fig. 9.3.4 is implemented using a saturation nonlinearity. The matrix $\mathbf{A}$ is unchanged, and thus Fig. 9.3.5 is valid. Under a saturation overflow characteristic, the triangle $\overline{A'E'G'}$ is mapped into the line segment $\overline{E'G'}$. Similarly for the triangle $\overline{G'F'H'}$. Thus if a saturation nonlinearity is used, every state trajectory approaches the origin. The filter is free of overflow oscillation. The reason this is not a popular solution is because it is more costly to implement. The result of every multiplication must be checked for overflow in order to apply the saturation nonlinearity. Whether this is done in software or hardware, the result is more cost and slower throughput for a given implementation.

9.4 FILTER STRUCTURES FREE OF OVERFLOW OSCILLATIONS

The analysis of overflow oscillations using state space methods as described in Section 9.3 provides an insight that we can use to characterize the causes of overflow oscillations. Overflow oscillations occur because some state vectors *increase in length* after multiplication by the system matrix **A**. Since overflow functions always decrease or do not change the length of the state vector, repeated oscillations cannot occur unless the process of multiplying by **A** increases the length of a state vector. Roughly speaking, then, a necessary condition for the occurrence of overflow oscillations is that the length of **Ax** is greater than the length of **x** for some **x** in the state space.

We know from our discussion of Chapter 8 that for a given transfer function $H(z)$, we can find infinitely many realizations, i.e., there are infinitely many system matrices **A** that realize $H(z)$. Intuition suggests that by changing the system matrices $\{\mathbf{A}, \mathbf{B}, \mathbf{C}, \mathbf{D}\}$, we could find structures that do not increase the length of **x** upon multiplication by **A**. How do we characterize these structures?

We need a method of measuring the "gain" of a matrix **A**. We wish to compare various system matrices **A** in terms of their "amplifying power" on a vector **x**. There are several definitions we can use in defining the gain of a matrix. These definitions depend on how the length of a vector in the space is defined. Most often we define the length or *norm* of a vector as

$$\|\mathbf{x}\| = (\mathbf{x}^T\mathbf{x})^{1/2} = \left(\sum_{i=1}^{n} x_i^2\right)^{1/2}. \tag{9.4.1}$$

With this definition of length, the "gain" or "amplifying power" or *norm* of a matrix **A** can be defined as

$$\|\mathbf{A}\| = \max_{\mathbf{x}\neq 0} \frac{\|\mathbf{Ax}\|}{\|\mathbf{x}\|} = \max_{\mathbf{x}\neq 0}\left(\frac{\mathbf{x}^T\mathbf{A}^T\mathbf{Ax}}{\mathbf{x}^T\mathbf{x}}\right)^{1/2}. \tag{9.4.2}$$

In other words, $\|\mathbf{A}\|$ is the maximum increase in the length of the vector. If $\|\mathbf{A}\| < 1$, then all vectors **x** decrease in length under multiplication by **A**. Thus Eq. (9.4.2) is a test for the occurrence of overflow oscillations in digital filters. If we can find a structure with a system matrix **A** such that $\|\mathbf{A}\| < 1$, then zero-input overflow oscillations cannot occur. The next example presents a structure that is free of zero-input overflow oscillations for any transfer function $H(z)$.

EXAMPLE 9.4.1

The normal filter structure.

Consider the direct form filter of Fig. 9.3.4. This filter has state variable equations given by

$$\mathbf{x}(k+1) = \mathbf{A}\mathbf{x}(k) + \mathbf{B}u(k) = \begin{bmatrix} 0 & 1 \\ -\frac{1}{2} & -1 \end{bmatrix} \mathbf{x}(k) + \begin{bmatrix} 0 \\ 1 \end{bmatrix} u(k),$$

$$y(k) = \mathbf{C}\mathbf{x}(k) + Du(k) = [-\tfrac{1}{2} \quad -1]\mathbf{x}(k) + [1]u(k). \tag{9.4.3}$$

As shown in Section 9.3, this filter realization suffers from zero-input overflow oscillations for initial states near the corners $(-1, -1)$ and $(1, 1)$. Consider a change of coordinates using the nonsingular transformation $\mathbf{T}^{-1}$ defined in Eq. (9.4.4).

$$\hat{\mathbf{x}}(k) = \mathbf{T}^{-1}\mathbf{x}(k) = \begin{bmatrix} -\beta & 0 \\ -\alpha & 1 \end{bmatrix} \mathbf{x}(k), \tag{9.4.4}$$

where α and β are the real and imaginary parts of the complex eigenvalues of $\mathbf{A}$. That is, the poles of the system are

$$\lambda_1, \lambda_2 = \alpha \pm j\beta.$$

In this case, the eigenvalues of $\mathbf{A}$ are $-1/2 \pm j/2$ so that α and β are

$$\alpha = -\beta = -\tfrac{1}{2}.$$

For this nonsingular transformation of the state, the new state variable matrices are $\{\hat{\mathbf{A}}, \hat{\mathbf{B}}, \hat{\mathbf{C}}, \mathbf{D}\}$, where

$$\begin{aligned} \hat{\mathbf{A}} &= \mathbf{T}^{-1}\mathbf{A}\mathbf{T} = \begin{bmatrix} -\beta & 0 \\ -\alpha & 1 \end{bmatrix} \begin{bmatrix} 0 & 1 \\ -\frac{1}{2} & -1 \end{bmatrix} \begin{bmatrix} -1/\beta & 0 \\ -\alpha/\beta & 1 \end{bmatrix} \\ &= \begin{bmatrix} \alpha & -\beta \\ \beta & \alpha \end{bmatrix} = \begin{bmatrix} -\frac{1}{2} & -\frac{1}{2} \\ \frac{1}{2} & -\frac{1}{2} \end{bmatrix}, \\ \hat{\mathbf{B}} &= \mathbf{T}^{-1}\mathbf{B} = \begin{bmatrix} -\beta & 0 \\ -\alpha & 1 \end{bmatrix} \begin{bmatrix} 0 \\ 1 \end{bmatrix} = \begin{bmatrix} 0 \\ 1 \end{bmatrix}, \\ \hat{\mathbf{C}} &= \mathbf{C}\mathbf{T} = [-\tfrac{1}{2} \quad -1] \begin{bmatrix} -1/\beta & 0 \\ -\alpha/\beta & 1 \end{bmatrix} = [0, \ -1], \\ \hat{\mathbf{D}} &= \mathbf{D} = [1]. \end{aligned} \tag{9.4.5}$$

The structure $\{\hat{\mathbf{A}}, \hat{\mathbf{B}}, \hat{\mathbf{C}}, \mathbf{D}\}$ defined by Eq. (9.4.5) has a system matrix $\hat{\mathbf{A}}$ that satisfies the matrix equation

$$\hat{\mathbf{A}}^T\hat{\mathbf{A}} = \hat{\mathbf{A}}\hat{\mathbf{A}}^T. \tag{9.4.6}$$

Matrices satisfying Eq. (9.4.6) are known as *normal matrices.* For this reason, we term any filter with a system matrix $\mathbf{A}$ satisfying Eq. (9.4.6) a *normal filter.* (Note that this property depends only on $\mathbf{A}$.) An SFG for the normal filter defined by Eq. (9.4.5) is shown in Fig. 9.4.1. The distinguishing characteristic is the symmetric form of the structure. These realizations are also known as *uniform grid* or *coupled resonator filters* (Radar 1967).

To determine the overflow oscillation properties of this realization, we can transform the state space using the system matrix $\hat{\mathbf{A}}$ of Eq. (9.4.5). The

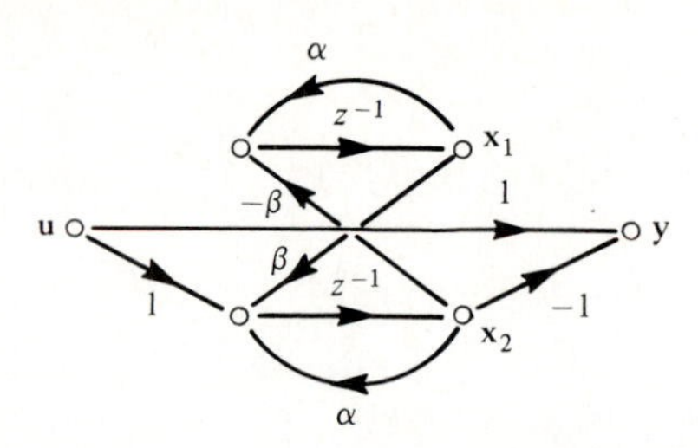

FIGURE 9.4.1 **A SFG of a normal digital filter defined by (9.4.5).**

result is shown in Fig. 9.4.2. Referring to Fig. 9.4.2, we see that the original state space $\overline{ABCD}$ is mapped into $\overline{A'B'C'D'}$ with half the original area ($\det \hat{\mathbf{A}} = \frac{1}{2}$). More importantly, the image of $\overline{ABCD}$ under the linear transformation defined by $\hat{\mathbf{A}}$ is contained completely within itself. The length of all state vectors $\mathbf{x}(k)$ are reduced upon multiplication by $\hat{\mathbf{A}}$. Contrast this with the direct form realization of Fig. 9.3.4. In this filter, dimensions of the original state space $\overline{ABCD}$ are *not* reduced in all directions. In particular, the diagonal $\overline{A'C'}$ is longer than the original diagonal $\overline{AC}$. Overflows oscillations in the direct form filter are caused by the fact that points in the corners near A and C are mapped by the matrix $\mathbf{A}$ of Eq. (9.39) into points that are farther from the origin than the original points.

We can also demonstrate analytically that normal digital filters are free of zero-input overflows by calculating the matrix norm $\|\mathbf{A}\|$ of Eq. (9.4.2). For complex eigenvalues $\lambda_1, \lambda_2 = \alpha \pm j\beta$, the system matrix is $\hat{\mathbf{A}}$ given by

$$\hat{\mathbf{A}} = \begin{bmatrix} \alpha & -\beta \\ \beta & \alpha \end{bmatrix}.$$

From Eq. (9.4.2), we have

$$\|\hat{\mathbf{A}}\| = \max_{\mathbf{x} \neq 0} \left(\frac{\mathbf{x}^T \hat{\mathbf{A}}^T \hat{\mathbf{A}} \mathbf{x}}{\mathbf{x}^T \mathbf{x}} \right)^{1/2}.$$

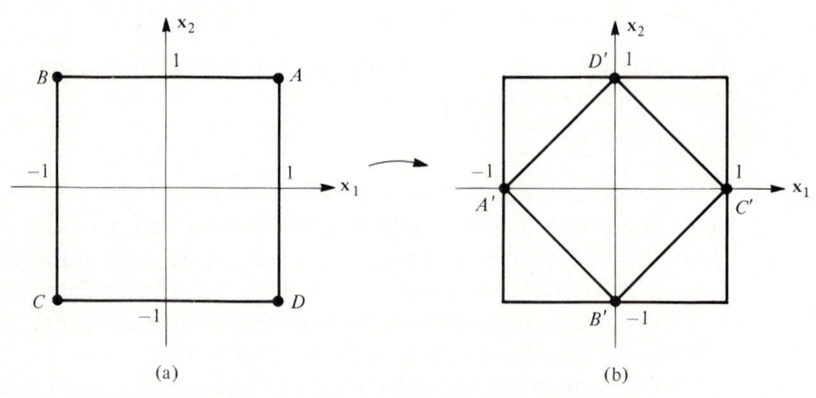

FIGURE 9.4.2 **State-space transformation under a normal matrix $\hat{\mathbf{A}}$.**

Now,

$$\hat{\mathbf{A}}^T\hat{\mathbf{A}} = \begin{bmatrix} \alpha & \beta \\ -\beta & \alpha \end{bmatrix}\begin{bmatrix} \alpha & -\beta \\ \beta & \alpha \end{bmatrix} = (\alpha^2 + \beta^2)\mathbf{I}.$$

Thus

$$\|\hat{\mathbf{A}}\| = \max_{\mathbf{x} \neq 0} \left((\alpha^2 + \beta^2) \frac{\mathbf{x}^T\mathbf{I}\mathbf{x}}{\mathbf{x}^T\mathbf{x}} \right)^{1/2} = (\alpha^2 + \beta^2)^{1/2}. \tag{9.4.7}$$

Equation (9.4.7) proves that for stable filters the matrix norm is less than unity. Thus no zero-input overflows can occur for normal digital filters.

We can generalize these results and obtain sufficient conditions for characterizing the absence of overflows in a given digital filter structure (Mills 1981). The generalization occurs in redefining the length of a vector as

$$\|\mathbf{x}\| = (\mathbf{x}^T\hat{\mathbf{D}}\mathbf{x})^{1/2}, \tag{9.4.8}$$

where $\hat{\mathbf{D}}$ is a positive diagonal matrix, i.e.,

$$\hat{\mathbf{D}} = \mathrm{diag}(d_1, d_2, \ldots d_n), \qquad d_i > 0 \quad \text{for } i = 1, 2, \ldots, n.$$

In this case the length of each component is weighted by the diagonal values of $\hat{\mathbf{D}}$. The matrix norm of $\mathbf{A}$ becomes

$$\|\mathbf{A}\| = \max_{\mathbf{x} \neq 0} \left(\frac{\mathbf{x}^T\mathbf{A}^T\hat{\mathbf{D}}\mathbf{A}\mathbf{x}}{\mathbf{x}^T\hat{\mathbf{D}}\mathbf{x}} \right)^{1/2}. \tag{9.4.9}$$

To prevent zero-input overflow oscillations, it is sufficient that

$$\|\mathbf{A}\mathbf{x}\| \leqslant r\|\mathbf{x}\|, \qquad 0 < r < 1. \tag{9.4.10}$$

If the inequality (9.4.10) is to hold, then it must be true that

$$0 \leqslant r^2\|\mathbf{x}\|^2 - \|\mathbf{A}\mathbf{x}\|^2 = r^2\mathbf{x}^T\hat{\mathbf{D}}\mathbf{x} - \mathbf{x}^T\mathbf{A}^T\hat{\mathbf{D}}\mathbf{A}\mathbf{x} = \mathbf{x}^T(r^2\hat{\mathbf{D}} - \mathbf{A}^T\hat{\mathbf{D}}\mathbf{A})\mathbf{x}. \tag{9.4.11}$$

Equation (9.4.11) states that Eq. (9.4.10) is true for all $\mathbf{x}$ if and only if the matrix $\mathbf{Q} = r^2\hat{\mathbf{D}} - \mathbf{A}^T\hat{\mathbf{D}}\mathbf{A}$ is a positive definite matrix. Thus we have a *sufficient* condition for determining whether a given filter is free of zero-input overflow oscillations. The problem has been reduced to finding a positive diagonal matrix $\hat{\mathbf{D}}$ for which $\mathbf{Q}$ is positive definite where

$$\mathbf{Q} = r^2\hat{\mathbf{D}} - \mathbf{A}^T\hat{\mathbf{D}}\mathbf{A}. \tag{9.4.12}$$

This result does not tell us how to explicitly choose the matrix $\hat{\mathbf{D}}$.

From this characterization (9.4.12) we can conclude, for example, that if $\mathbf{A}$ satisfies Eq. (9.4.12), then any diagonal (nonsingular) transformation of $\mathbf{A}$, $\mathbf{T}^{-1}\mathbf{A}\mathbf{T}$ also satisfies Eq. (9.4.12). (See Problem 9.7.) For the important case of second-order filters, this result reduces to the following (see Problem 9.8).

If the eigenvalues of $\mathbf{A}$ have magnitude less than unity, $\hat{\mathbf{D}}$ exists for which $\mathbf{Q}$ is positive definite if and only if

$$a_{12}a_{21} \geqslant 0$$

or **(9.4.13)**

$$\text{if } a_{12}a_{21} < 0, \qquad \text{then} \qquad |a_{11} - a_{22}| + \det \mathbf{A} < 1,$$

where

$$\mathbf{A} = \begin{bmatrix} a_{11} & a_{12} \\ a_{21} & a_{22} \end{bmatrix}.$$

EXAMPLE 9.4.2

Testing for overflow oscillations in a second-order direct form filter.

Consider the direct form filter shown in Fig. 9.4.3.

The system matrix for this filter is

$$\mathbf{A} = \begin{bmatrix} 0 & 1 \\ b & a \end{bmatrix}.$$

The poles of the filter are inside $|z| = 1$ provided

$$|a| + b < 1, \qquad -b < 1. \tag{9.4.14}$$

Equations (9.4.13) imply that overflow oscillations do not occur provided

$$b \geqslant 0 \tag{9.4.15}$$

or

$$b < 0 \quad \text{and} \quad |a| - b < 1.$$

Combining Eqs. (9.4.14) and (9.4.15) yields

$$|a| + |b| < 1. \tag{9.4.16}$$

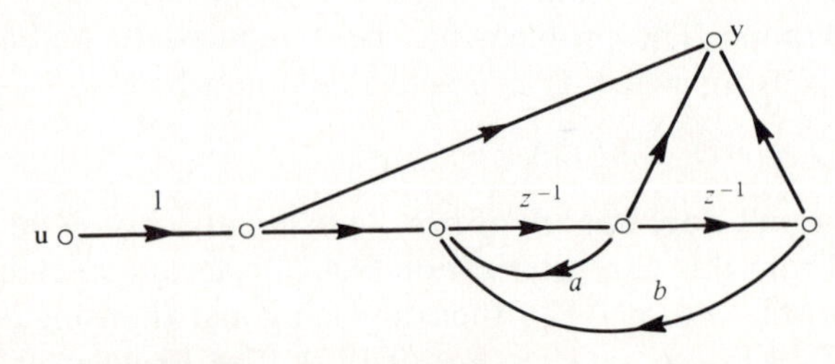

FIGURE 9.4.3 A second-order direct form filter.

This inequality is exactly the necessary and sufficient condition found by Ebert et al. (1969) for the absence of overflow oscillations in direct form filters. Notice that it depends on the location of the poles of the filter.

If a filter is stable, then the system matrix **A** satisfies $\det \mathbf{A} < 1$. From Eq. (9.4.13), it is clear that a second-order filter is free of overflow oscillations if the diagonal elements of **A** are equal.

Recall that for second-order normal filters with complex poles $\alpha \pm j\beta$, the system matrix **A** is of the form

$$\mathbf{A} = \begin{bmatrix} \alpha & -\beta \\ \beta & \alpha \end{bmatrix}$$

Thus all second-order normal filters satisfy this condition. As we have seen previously, they are free of overflow oscillations.

The sufficient conditions given here for the absence of overflow oscillations are a generalization of so-called minimum norm digital filter realizations (Barnes 1977). A stable filter is defined to be minimum norm provided

$$\|\mathbf{A}\| = \max_i \{|\lambda_i|\}, \qquad \{\lambda_i\} \text{ are eigenvalues of } \mathbf{A}. \tag{9.4.17}$$

Equation (9.4.17) can be shown to be equivalent to $\mathbf{Q} > \mathbf{0}$ with $\hat{\mathbf{D}} = \mathbf{I}$.

Summary on Overflow Limit Cycles

The analysis of the detailed behavior of a digital filter is greatly complicated by the occurrence of overflows. The simplicity of a linear system is replaced by the complexity of a nonlinear system. Fortunately, the concept of a filter state can be used to analyze filter behavior in this nonlinear mode of operation.

We have assumed the overflow nonlinearity satisfies Eq. (9.3.3). Under this assumption, we can decompose the operation of the filter into two mappings on the filter state. The first is a linear transformation $\mathbf{x}(k) \to \mathbf{A}\mathbf{x}(k)$. The second is a nonlinear transformation $\mathbf{A}\mathbf{x}(k) \to f(\mathbf{A}\mathbf{x}(k))$. These two operations determine whether overflows will lead to oscillation.

The main result of our discussion is that if the matrix $\mathbf{Q} = \hat{\mathbf{D}} - \mathbf{A}^T\hat{\mathbf{D}}\mathbf{A}$ is positive definite, the filter cannot sustain zero-input overflow oscillations. Filters satisfying this condition can also be expected to behave in a satisfactory manner with nonzero inputs.

Among filters that satisfy the condition $\mathbf{Q} > \mathbf{0}$ are normal filters, minimum roundoff noise filters, ladder or lattice filters, and a multitude of others. The often-used direct form filter is not, unfortunately, free of overflow oscillations. The ability to oscillate depends on the location of the poles of the direct form filter.

9.5 ROUNDOFF NOISE IN IIR DIGITAL FILTERS

Any realization of an IIR digital filter employs three basic operations: multiplication by constants (the filter parameters), accumulation of the products, and storage into memory. The results of accumulations inside the filter must eventually be quantized because the multiplications always increase the number of bits required to represent the products. If two B bit numbers are multiplied together, the product is $2B$ bits long. This product must eventually be quantized to B bits. We shall assume the quantization is accomplished by rounding. The error that occurs as this quantization of internal accumulation is called *roundoff noise.*

The model we shall use for roundoff noise is the same as used in our discussion of A/D conversion noise and depicted in Fig. 9.5.1. The quantizer is replaced by an additive white noise source of noise power (variance) $q^2/12$ where q is the quantization step size. The roundoff noise is assumed to be uniformly distributed on $[-q/2, q/2]$ with zero mean. This model also assumes that (a) noise from different accumulators is uncorrelated, and (b) each noise source is uncorrelated with the input.

The validity of this model has been established by numerous simulations by many investigators over a period of several years. As in the case of A/D conversion, the primary requirement is that the input changes sufficiently from sample to sample that several quantization levels are crossed. In terms of our state-space grid of Section 9.3 several cells are crossed from sample to sample.

The problem we shall discuss in this and the following sections is: given an I/O description of the filter such as the transfer function $H(z)$ and the roundoff noise quantization model just discussed, what state variable realization minimizes the output roundoff noise due to the quantization of internal products? Notice that the problem statement does not explicitly mention overflow errors. We must assume overflow errors are negligible. This implies the filter is *correctly scaled so that overflows rarely occur.* Roundoff errors can always be reduced by simply compressing the allowable range of representable numbers, i.e., by compressing the boundaries of the state space grid of Fig. 9.3.3. This reduces q which determines the (local) noise power of noise sources ($q^2/12$). This is not an acceptable solution because then overflow errors dominate the output error. Since they are *not* included in the analysis of roundoff error, we must ensure they are negligible by first scaling the filter.

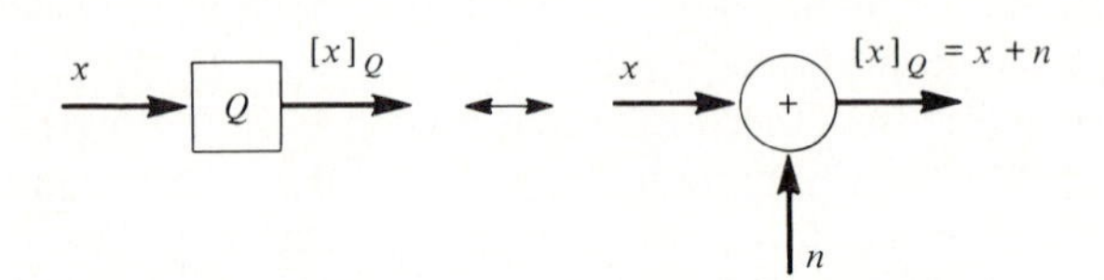

FIGURE 9.5.1 A linear model of internal quantization of products.

Computation of Output Roundoff Noise

Before discussing how to scale a fixed point digital filter, we shall introduce the use of the quantization noise model of Fig. 9.5.1. We can demonstrate the essential calculation in finding the output roundoff noise of a filter by means of the SFG shown in Fig. 9.5.2.

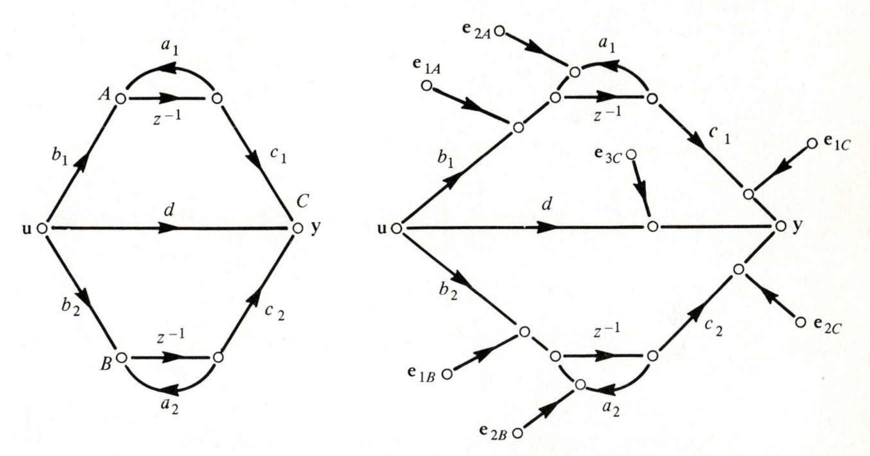

FIGURE 9.5.2 **A roundoff noise model for a second order digital filter.**

Referring to Fig. 9.5.2, we find there are three sources of roundoff noise in the filter, which occur at the nodes labeled *A*, *B*, and *C*. Nodes *A* and *B* have two noise sources and node *C*, the output register, has three contributing noise sources. Each product in the original SFG is replaced by its equivalent noise model. This is shown in Fig. 9.5.2(b). The total output noise power due to internal quantizations of multiplies is obtained by summing the output noise powers associated with all of the internal noise sources.

Using Fig. 9.5.2 as an example, consider the output noise due to quantization at node *A*. Let $\mathbf{g}_A$ be the unit-pulse response sequence from node *A* to the output node. Then, as shown in Chapter 7, the output noise variance due to the white noise sources e_{1A} and e_{2A} is

$$\sigma_A^2 = \sigma_{e_{1A}}^2 \sum_{k=0}^{\infty} g_A^2(k) + \sigma_{e_{2A}}^2 \sum_{k=0}^{\infty} g_A^2(k) = 2\left(\frac{q^2}{12}\right) \sum_{k=0}^{\infty} g_A^2(k)$$

$$= 2\left(\frac{q^2}{12}\right) \|\mathbf{g}_A\|_2^2. \tag{9.5.1}$$

Similarly, if $\mathbf{g}_B$ is the unit-pulse response sequence from node *B* to the output node, the output noise variance due to e_{1B} and e_{2B} is

$$\sigma_B^2 = 2\left(\frac{q^2}{12}\right) \|\mathbf{g}_B\|_2^2. \tag{9.5.2}$$

The noise caused by quantization at the output node C is

$$\sigma_C^2 = 3\left(\frac{q^2}{12}\right). \tag{9.5.3}$$

The total output roundoff noise is the sum of σ_A^2, σ_B^2, and σ_C^2 (since we have assumed these noise sources are uncorrelated). Thus the total output roundoff noise is

$$\sigma_{\text{total}}^2 = \frac{q^2}{12}\left\{3 + 2\left(\|\mathbf{g}_A\|^2 + \|\mathbf{g}_B\|^2\right)\right\}. \tag{9.5.4}$$

By Parseval's equality, we have

$$\|\mathbf{g}\|^2 = \sum_{k=0}^{\infty} g^2(k) = \frac{1}{2\pi}\int_{-\pi}^{\pi} |G(e^{j\theta})|^2\, d\theta = \frac{1}{2\pi j}\oint_{|z|=1} G(z)G(z^{-1})\frac{dz}{z}, \tag{9.5.5}$$

where $G(z)$ is the z-transform of the sequence $\mathbf{g}$. Thus Eq. (9.5.4) can be expressed as

$$\sigma_{\text{total}}^2 = \frac{q^2}{12}\left\{\frac{4}{2\pi j}\oint_{|z|=1}\left[G_A(z)G_A(z^{-1}) + G_B(z)G_B(z^{-1})\right]\frac{dz}{z} + 3\right\}. \tag{9.5.6}$$

In general for a nth-order state variable SFG there are $(n + 1)$ nodes that generate roundoff noise. The total output noise power is thus of the form

$$\sigma_{\text{total}}^2 = \frac{q^2}{12}\left\{\sum_{i=1}^{n} v_i\left[\sum_{k=0}^{\infty} g_i^2(k)\right] + v_{n+1}\right\}, \tag{9.5.7}$$

where v_i is the number of multipliers entering node i. The sequence $\mathbf{g}_i$ is the unit-pulse response from node i to the output node. ($i = n + 1$ corresponds to the output node with $\mathbf{g}_{n+1} = \boldsymbol{\delta}$.)

Given any SFG of a filter with transfer function $H(z)$, we can evaluate the output roundoff noise of the realization using Eq. (9.5.7). The process is straightforward, but often tedious. The essential calculation is Eq. (9.5.5).

We remind the reader once more that we have only calculated errors due to quantization. Overflow errors are not included. Scaling must be performed to ensure that overflows rarely occur.

In deriving Eqs. (9.5.4), (9.5.6), and (9.5.7), we have assumed that each product is quantized *before* it is accumulated at a node. This results in a multiplication by v_i, the number of multipliers entering a node. A better practice is to use a double-length register to accumulate the products and then perform *one* quantization at each node. In this case, Eq. (9.5.7) can be written

$$\sigma_{\text{total}}^2 = \left(\frac{q^2}{12}\right)\left\{\sum_{i=1}^{n}\left[\sum_{k=0}^{\infty} g_i^2(k)\right] + 1\right\} = \left(\frac{q^2}{12}\right)\left\{\sum_{i=1}^{n}\|\mathbf{g}_i\|^2 + 1\right\}. \tag{9.5.8}$$

In the sequel, we will assume that double-length accumulators are used.

EXAMPLE 9.5.1

Calculation of output roundoff noise.

To illustrate the calculation of output roundoff noise, consider the following second-order direct form filter shown in Fig. 9.5.3. There are two sources of quantization noise, at node A and the output node B. If $\mathbf{g}_A$ is the unit-pulse response from node A to the output, then the output roundoff noise is given by

$$\sigma_{\text{total}}^2 = \frac{q^2}{12}\{\|\mathbf{g}_A\|^2 + 1\}.$$

To find the output roundoff noise we first find $\mathbf{g}_A$, which in this case is identical to the filter's unit-pulse response $\mathbf{h}$. Using the techniques of Chapter 2, we find that

$$g_A(k) = h(k) = \begin{cases} 0, & k < 0 \\ \frac{3}{2}(-\frac{1}{4})^k - \frac{1}{2}(\frac{1}{4})^k, & k \geqslant 0 \end{cases}$$

Thus

$$\begin{aligned} \|\mathbf{g}_A\|^2 &= \sum_{k=0}^{\infty} [\tfrac{3}{2}(-\tfrac{1}{4})^k - \tfrac{1}{2}(\tfrac{1}{4})^k]^2 \\ &= \frac{9}{4}\frac{1}{1-\frac{1}{16}} - \frac{3}{2}\frac{1}{1+\frac{1}{16}} + \frac{1}{4}\frac{1}{1-\frac{1}{16}} \\ &= \frac{64}{51}. \end{aligned}$$

The output roundoff noise is therefore

$$\sigma_{\text{total}}^2 = \frac{q^2}{12}\{\tfrac{64}{51} + 1\}.$$

The number $(\frac{64}{51})$ is often called the *noise gain* of the filter. An alternate calculation of $\|\mathbf{g}_A\|^2$ can be obtained using residue theory. The essential formula is

$$\|\mathbf{g}_A\|^2 = \frac{1}{2\pi j}\oint_{|z|=1} G_A(z)G_A(z^{-1})\frac{dz}{z} = \frac{1}{2\pi}\int_{-\pi}^{\pi} |G_A(e^{j\theta})|^2\, d\theta. \qquad \textbf{(9.5.9)}$$

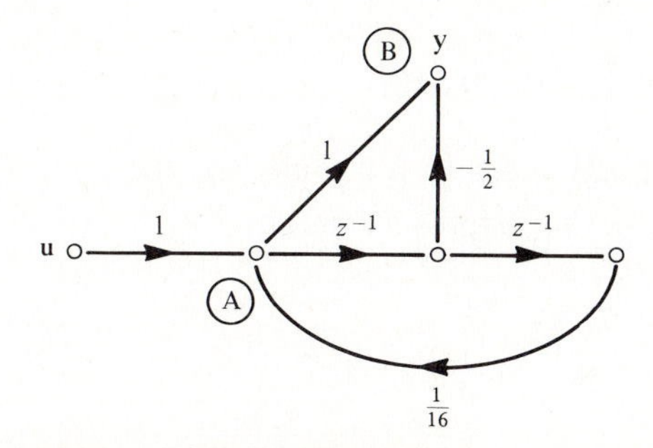

FIGURE 9.5.3 **A second-order direct form filter.**

In any case, the evaluation of Eq. (9.5.9) is somewhat tedious. We shall use the theory of state variables in Section 9.7 to simplify the calculation of output roundoff noise. We shall show that roundoff noise variance can be found as the solution to a set of simultaneous linear equations.

The calculation of roundoff noise as presented here is meaningless if the filter is not correctly scaled. The next section discusses this important topic.

9.6 SCALING FIXED POINT DIGITAL FILTERS TO PREVENT OVERFLOW

As we have previously stated, roundoff noise is the dominant component of output error in a digital filter *only when overflows in internal registers are negligible.* Internal overflows cause large errors, and must be prevented. This is accomplished by properly scaling the realization. Scaling constrains the numerical values of internal filter variables to remain in a range appropriate to the hardware. Because of finite length registers, the range of a filter variable is necessarily limited. For fixed point number representations, we can express this as a bound on internal variables $v(k)$ such that

$$|v(k)| \leqslant \Delta. \tag{9.6.1}$$

Δ is arbitrary and is related to the quantization step size q by Eq. (9.2.3). We shall assume Δ is unity. The variable $v(k)$ ideally satisfies the equation

$$v(k) = (\mathbf{f} * \mathbf{u})(k). \tag{9.6.2}$$

In Eq. (9.6.2), $\mathbf{f}$ is the unit-pulse response from the input to the variable $\mathbf{v}$. The range of values of $\mathbf{v}$ determined by this equation depends on the nature of the input and on the sequence $\mathbf{f}$.

For example, if the input is bounded $|u(k)| \leqslant 1$, then

$$\begin{aligned} |v(k)| &= \left| \sum_l f(l)u(k-l) \right| \leqslant \sum_l |f(l)|\,|u(k-l)| \\ &= \sum_l |f(l)| = \|\mathbf{f}\|_1. \end{aligned} \tag{9.6.3}$$

Similarly if $u(k) = \cos k\theta$ for some θ, then

$$v(k) = \operatorname{Re}[F(e^{j\theta})e^{jk\theta}].$$

Thus

$$|v(k)| \leqslant \max_\theta |F(e^{j\theta})|. \tag{9.6.4}$$

Five cases are summarized in Table 9.6.1.

TABLE 9.6.1 The range of the variable $v(k) = (f*u)(k)$ for various classes of inputs u.

Input	Range of v
1. *Sinusoidal inputs* $u(k) = \cos(k\theta)$	$\lvert v(k)\rvert \leqslant \max_{\theta} \lvert F(e^{j\theta})\rvert$
2. *Bounded inputs* $\lvert u(k)\rvert \leqslant 1$	$\lvert v(k)\rvert \leqslant \sum_{l} \lvert f(l)\rvert = \lVert f\rVert_1$
3. *Finite energy inputs* $\sum_{l \leqslant k} u^2(l) \leqslant 1$	$\lvert v(k)\rvert \leqslant \left[\sum_{l} f^2(l)\right]^{1/2} = \lVert f\rVert_2$
4. *Wide sense stationary inputs* $r_{uu}(k) \leftrightarrow S_{uu}(\theta)$	$[E[v^2(k)]]^{1/2} = \left[\frac{1}{2\pi}\int_{-\pi}^{\pi} S_{uu}(\theta)\lvert F(e^{j\theta})\rvert^2\, d\theta\right]^{1/2}$
5. *White inputs* $S_{uu}(\theta) = 1$	$[E[v^2(k)]]^{1/2} = \lVert f\rVert_2$

The first three cases represent true bounds on the range of the variable $\mathbf{v}$, for the given classes of input sequences. The last two give standard deviations of random variables, and can, of course, be exceeded. The most conservative bound is the second. One can show that

$$\lVert \mathbf{f}\rVert_2 \leqslant \max_{\theta} \lvert F(e^{j\theta})\rvert \leqslant \lVert \mathbf{f}\rVert_1. \tag{9.6.5}$$

In order to meet the bound $\lVert \mathbf{f}\rVert_1$, in Eq. (9.6.3), the input sequence will have to satisfy $\lvert u(k)\rvert = 1$ for as many values of k as there are nonzero elements of $\mathbf{f}$, and the signs must be chosen carefully. (In automatic control theory, these are called "bang-bang" controller inputs.) Thus the second bound is usually far too conservative.

Scaling Rules

In order to reconcile bounds like those in Table 9.6.1 with the constraint on $v(k)$ in Eq. (9.6.1), we must constrain the "gain" or norm of the unit-pulse response sequences for internal variables. Two such rules are

$$l_1 \text{ scaling:} \quad \lVert \mathbf{f}\rVert_1 = \sum_{l=0}^{\infty} \lvert f(l)\rvert = 1, \tag{9.6.6}$$

$$l_2 \text{ scaling:} \quad \delta\lVert \mathbf{f}\rVert_2 = \delta\left[\sum_{l=0}^{\infty} f^2(l)\right]^{1/2} = 1. \tag{9.6.7}$$

The parameter δ is subjectively chosen. It can be interpreted to represent the number of standard deviations representable in the register containing $v(k)$ if the input is unit-variance white noise. (A value of δ greater than 5 is considered conservative.)

The Scaling Operation

If a digital filter realization is not scaled (i.e., does not satisfy the scaling rule), then it must be changed in such a way that the rule is satisfied but the transfer function from input to output is not changed. Choose any internal variable **v**. By eliminating all variables except **u**, **v**, and **y** the signal flow graph of the filter becomes the one shown in Fig. 9.6.1.

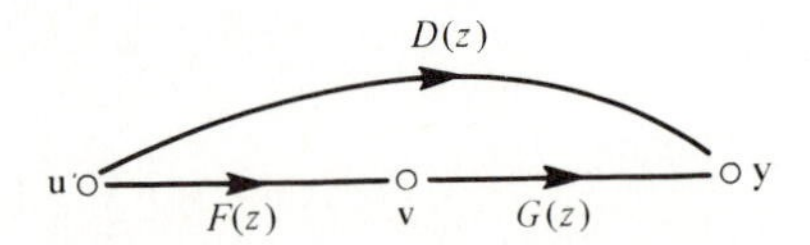

FIGURE 9.6.1 **A SFG depicting an unscaled node variable v.**

This graph might be called "canonical" with respect to the three variables **u**, **v**, and **y** since it is the smallest graph that preserves their relationships.

One can scale the variable **v** up or down without changing the transfer function

$$H(z) = D(z) + F(z)G(z) \tag{9.6.8}$$

by changing $F(z)$ and $G(z)$ as shown in Fig. 9.6.2. We merely divide $F(z)$ by some number β and multiply $G(z)$ by the same number. The product $F(z)/\beta$ and $\beta G(z)$ is, of course, $F(z)G(z)$ and so the original transfer function does not change.

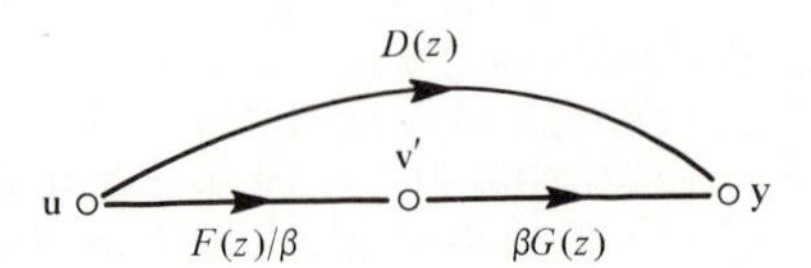

FIGURE 9.6.2 **A SFG depicting a scaled node variable v′.**

The parameter β is then chosen to meet the scaling rule. For example,

$$l_1 \text{ scaling:} \quad \beta = \|\mathbf{f}\|_1 = \sum_l |f(l)| \tag{9.6.9}$$

$$l_2 \text{ scaling:} \quad \beta = \delta\|\mathbf{f}\|_2 = \delta\left[\sum_{l=0}^{\infty} f^2(l)\right]^{1/2}. \tag{9.6.10}$$

In the original primitive signal flow graph, the modification indicated in Fig. 9.6.2 may be accomplished by including the factor $1/\beta$ in all branches entering the node **v** and including the factor β in all branches leaving **v**.

Scaling State Variable Descriptions

To scale a state variable description, we need only apply an appropriate diagonal transformation to the state vector. The computation involved in doing this is appropriate only for the l_2 scaling rule, however. To see how this is done, let us begin with the expression for the present state in terms of the past input sequence. We have from Eq. (8.3.19) that

$$\mathbf{x}(k) = \sum_{l=0}^{\infty} \mathbf{A}^l \mathbf{B} u(k-1-l). \tag{9.6.11}$$

The response to a unit-pulse input $u(k) = \delta(k)$ is therefore

$$\mathbf{f}(k) = \begin{cases} \mathbf{0}, & k \leqslant 0 \\ \mathbf{A}^{k-1}\mathbf{B}, & k > 0 \end{cases} \quad \overset{z}{\longleftrightarrow} \quad (z\mathbf{I} - \mathbf{A})^{-1}\mathbf{B}. \tag{9.6.12}$$

Here $\mathbf{f}(k)$ is an $n \times 1$ vector. In Chapter 8 we introduced the state covariance matrix $\mathbf{K}$, which is intimately related to this sequence. In fact, from Eq. (8.5.9) and (9.6.12) we have

$$\mathbf{K} = \sum_{k=0}^{\infty} \mathbf{f}(k)\mathbf{f}^T(k). \tag{9.6.13}$$

The elements of this matrix are inner products involving components of the vector sequence $\mathbf{f}(k)$:

$$K_{ij} = (\mathbf{f}_i, \mathbf{f}_j) = \sum_{k=0}^{\infty} f_i(k) f_j(k), \tag{9.6.14}$$

in particular,

$$K_{ii} = (\mathbf{f}_i, \mathbf{f}_i) = \sum_{k=0}^{\infty} f_i^2(k) = \|\mathbf{f}_i\|_2^2. \tag{9.6.15}$$

Therefore, the l_2 scaling constraint (9.6.7) applied to the ith component of the state vector is

$$\delta\sqrt{K_{ii}} = 1, \qquad i = 1, 2, \ldots, n. \tag{9.6.16}$$

How can we change the state variable description $(\mathbf{A}, \mathbf{B}, \mathbf{C}, \mathbf{D})$ to satisfy these constraints?

First of all, the infinite sum in Eq. (9.6.13) can be computed by solving the linear system of equations

$$\mathbf{K} = \mathbf{A}\mathbf{K}\mathbf{A}^T + \mathbf{B}\mathbf{B}^T. \tag{9.6.17}$$

Recall that if we apply a transformation to the state $\mathbf{x}$ to obtain $\mathbf{x}'$,

$$\mathbf{x}' = \mathbf{T}^{-1}\mathbf{x}, \tag{9.6.18}$$

then

$$(\mathbf{A}, \mathbf{B}, \mathbf{C}, \mathbf{D}) \quad \leftarrow \quad (\mathbf{T}^{-1}\mathbf{A}\mathbf{T}, \mathbf{T}^{-1}\mathbf{B}, \mathbf{C}\mathbf{T}, \mathbf{D}) \tag{9.6.19}$$

and

$$\mathbf{K} \quad \leftarrow \quad \mathbf{T}^{-1}\mathbf{K}(\mathbf{T}^{-1})^T. \tag{9.6.20}$$

[This follows from Eq. (9.6.17).] In particular, if

$$\mathbf{T} = \text{Diag}\{t_1, t_2, \ldots, t_n\} \tag{9.6.21}$$

then

$$\mathbf{T}^{-1} = \text{Diag}\left\{\frac{1}{t_1}, \frac{1}{t_2}, \ldots, \frac{1}{t_n}\right\},$$

and so

$$(\mathbf{T}^{-1}\mathbf{K}\mathbf{T}^{-T})_{ij} = \frac{K_{ij}}{t_i t_j}. \tag{9.6.22}$$

Therefore the scaling rule (9.6.16) can be achieved by using a diagonal transformation with diagonal elements

$$t_i = \delta\sqrt{K_{ii}}, \qquad i = 1, 2, \ldots, n. \tag{9.6.23}$$

We emphasize that using this state variable formulation we need only solve a set of linear equations, Eq. (9.6.17), to find **K**. We then use Eq. (9.6.23) to find **T**, and finally Eq. (9.6.20) to obtain the l_2 scaled realization. Contrast this computation with the calculation of $\|\mathbf{f}_i\|_2$, $i = 1, 2, \ldots, n$ for each state variable x_i. In fact, given any (**A, B, C, D**), the process described here for scaling can be easily programmed for any state variable filter description. The next two examples demonstrate the computations implied in this development.

EXAMPLE 9.6.1

Scaling a second-order direct form.

Consider the unscaled second-order direct form filter shown in Fig. 9.6.3. The state variable matrices (**A**, **B**) are given by

$$\mathbf{A} = \begin{bmatrix} 0 & 1 \\ -a_2 & -a_1 \end{bmatrix}, \qquad \mathbf{B} = \begin{bmatrix} 0 \\ 1 \end{bmatrix}.$$

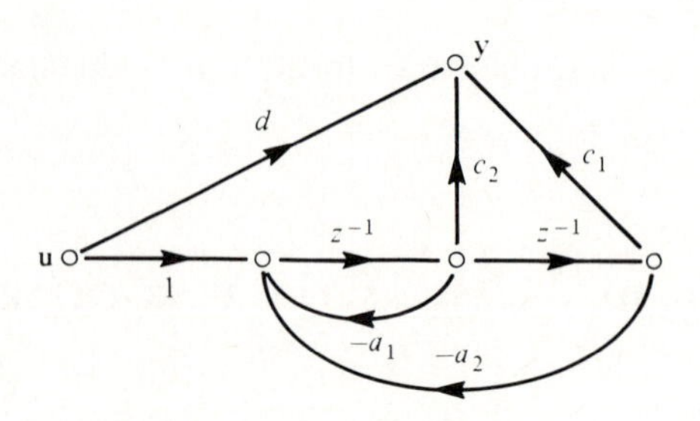

FIGURE 9.6.3 **A second-order direct form filter.**

From Example 8.5.1,

$$\mathbf{K} = \mathbf{AKA}^T + \mathbf{BB}^T$$

$$= \frac{1}{(1 - a_2)(1 + a_1 + a_2)(1 - a_1 + a_2)} \begin{bmatrix} 1 + a_2 & -a_1 \\ -a_1 & 1 + a_2 \end{bmatrix}$$

$$= \frac{1}{\gamma(a_1, a_2)} \begin{bmatrix} 1+a_2 & -a_1 \\ -a_1 & 1+a_2 \end{bmatrix}, \quad \gamma(a_1, a_2) = (1 - a_2)(1 - a_1 + a_2)(1 + a_1 + a_2).$$

The scaling rule is

$$\delta\sqrt{K_{ii}} = 1, \qquad i = 1, 2.$$

And for $\delta = 1$, $\mathbf{T}$ is given by Eqs. (9.6.21) and (9.6.23). Thus

$$\mathbf{T} = \sqrt{\frac{1 + a_2}{\gamma(a_1, a_2)}}\,\mathbf{I}, \qquad \mathbf{T}^{-1} = \sqrt{\frac{\gamma(a_1, a_2)}{1 + a_2}}\,\mathbf{I}.$$

Thus the scaled filter is given by

$$\mathbf{T}^{-1}\mathbf{AT} = \mathbf{A} = \begin{bmatrix} 0 & 1 \\ -a_2 & -a_1 \end{bmatrix}, \qquad \mathbf{T}^{-1}\mathbf{B} = \begin{bmatrix} 0 \\ \sqrt{\dfrac{\gamma(a_1, a_2)}{1 + a_2}} \end{bmatrix},$$

$$\mathbf{CT} = \sqrt{\frac{1 + a_2}{\gamma(a_1, a_2)}}\,[c_1, c_2]$$

with covariance matrix

$$\mathbf{T}^{-1}\mathbf{KT}^{-T} = \begin{bmatrix} 1 & \dfrac{-a_1}{1 + a_2} \\ \dfrac{-a_1}{1 + a_2} & 1 \end{bmatrix}.$$

The transformed SFG is as shown in Fig. 9.6.4. (In this example, **A** is unchanged only because $K_{11} = K_{22}$ in the original filter.)

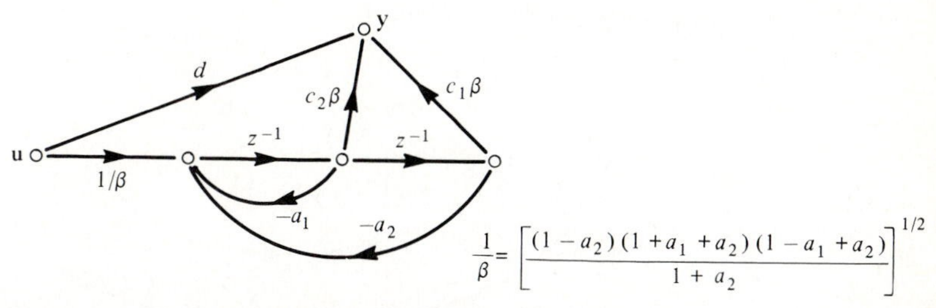

FIGURE 9.6.4 **An l_2-scaled direct form filter.**

EXAMPLE 9.6.2

Scaling a second-order direct form filter.

Consider the direct form second-order filter given in Fig. 9.6.5.

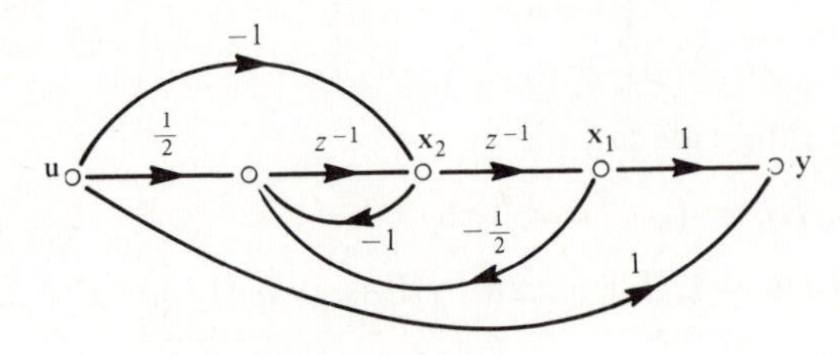

FIGURE 9.6.5 **A second-order direct form filter.**

The state variable matrices (**A**, **B**) are given by

$$\mathbf{A} = \begin{bmatrix} 0 & 1 \\ -\frac{1}{2} & -1 \end{bmatrix}, \qquad \mathbf{B} = \begin{bmatrix} -1 \\ \frac{1}{2} \end{bmatrix}.$$

The equations for the components of **K** from Eq. (9.6.17) are

$$\begin{aligned} k_{12} &= k_{21}, \\ k_{11} &= k_{22} + 1, \\ k_{12} &= -\tfrac{1}{2}k_{12} - k_{22} - \tfrac{1}{2}, \\ k_{22} &= k_{22} + k_{12} + \tfrac{1}{4}k_{11} + \tfrac{1}{4}. \end{aligned}$$

The solution to these equations is

$$\mathbf{K} = \begin{bmatrix} k_{11} & k_{12} \\ k_{21} & k_{22} \end{bmatrix} = \frac{1}{5}\begin{bmatrix} 7 & -3 \\ -3 & 2 \end{bmatrix}.$$

For scaling, we choose **T** so that $t_i = \delta\sqrt{K_{ii}}$, which implies that

$$\mathbf{T} = \delta\begin{bmatrix} \sqrt{\frac{7}{5}} & 0 \\ 0 & \sqrt{\frac{2}{5}} \end{bmatrix}, \qquad \mathbf{T}^{-1} = \frac{1}{\delta}\begin{bmatrix} \sqrt{\frac{5}{7}} & 0 \\ 0 & \sqrt{\frac{5}{2}} \end{bmatrix}.$$

Thus the scaled direct form is given by

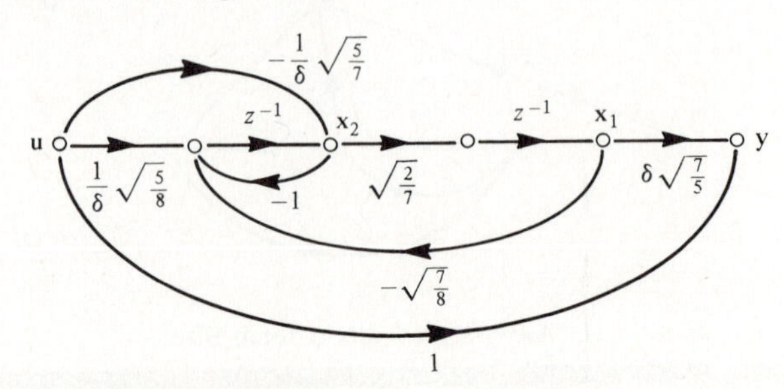

FIGURE 9.6.6 **An l_2 scaled second-order direct form filter.**

$$\mathbf{T}^{-1}\mathbf{A}\mathbf{T} = \begin{bmatrix} 0 & \sqrt{\frac{2}{7}} \\ -\sqrt{\frac{7}{8}} & -1 \end{bmatrix}, \mathbf{T}^{-1}\mathbf{B} = \frac{1}{\delta}\begin{bmatrix} -\sqrt{\frac{5}{7}} \\ \sqrt{\frac{5}{8}} \end{bmatrix}, \quad \mathbf{C}\mathbf{T} = \delta[\sqrt{\tfrac{7}{5}}, 0], \quad D = [1].$$

The l_2 scaled direct-form filter is shown in Fig. 9.6.6.

From the two calculations presented in Examples 9.6.1 and 9.6.2, it is clear that scaling changes the coefficients in the filter. Thus the output roundoff noise is affected by the scaling operation. The next section discusses this interaction between scaling and output roundoff noise and also presents an alternative method for calculating the output roundoff noise that requires solving a set of linear equations.

9.7 THE TRADEOFF BETWEEN OVERFLOW AND ROUNDOFF NOISE

The choice of the parameter δ in the scaling rule

$$\delta\sqrt{K_{ii}} = 1, \qquad i = 1, 2, \ldots, n \tag{9.7.1}$$

for state variable filters may seem subjective. How should it be chosen to produce the best results? In order to answer this question, one could simply set up the experiment shown in Fig. 9.7.1. The outputs of a "perfect" (double precision floating point) and an actual (fixed point arithmetic) digital filter are compared. The two filters have the same nominal transfer function. The fixed point filter has been scaled with $K_{ii} = 1$ (or $\delta = 1$). The scaling parameter may be changed using the scheme shown in the figure to satisfy the rule (9.7.1) for any choice of δ. The filter is then simulated and the output error variance is measured for each value of δ.

The results are shown in Fig. 9.7.2. For large values of δ, the internal variables are scaled so conservatively that no overflow occurs.

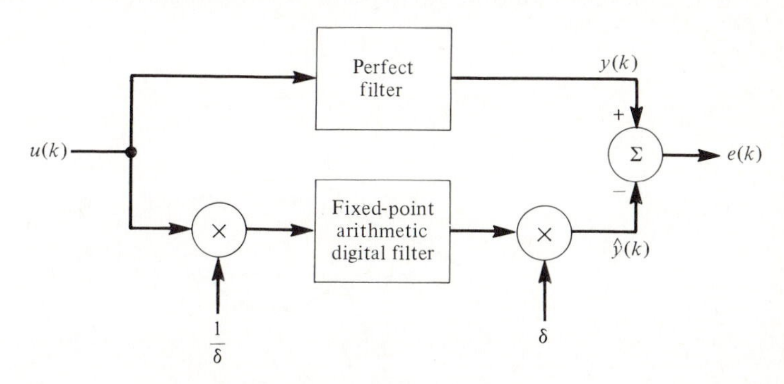

FIGURE 9.7.1 **Comparing the outputs of a perfect filter and a fixed point arithmetic filter, with provision for changing the scaling parameter δ.**

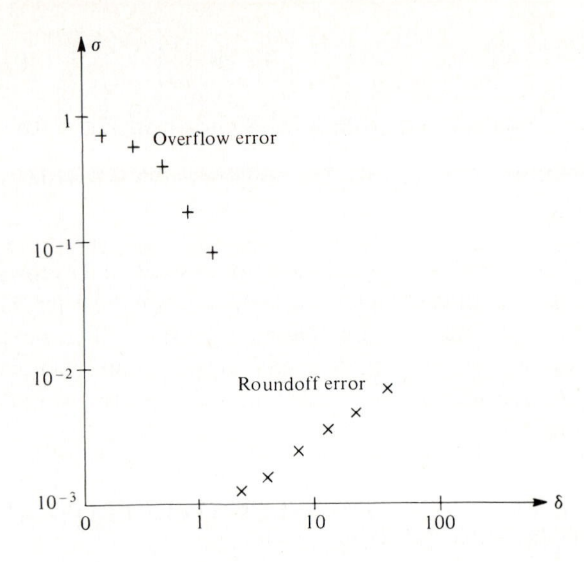

FIGURE 9.7.2 **Results of simulation. σ is the standard deviation of the error *e*(*k*).**

But as δ is decreased there is a sudden and dramatic increase in output error. This is due to the introduction for the first time of overflows, which produce large errors. As δ is decreased further, the filter is in a continuous state of overflow and the output ceases to have any relation to the input.

Notice the situation for large δ. Here the output error is much lower, but increases roughly proportionally to δ. Since these errors cannot be due to overflow, they must be due to roundoff in the accumulation of products inside the filter. But why the increase with δ? We shall explain this phenomenon in the remainder of the section. It represents the tradeoff between overflow (large scale) and quantization (small scale) sources of error. Overflow errors decrease with δ, but roundoff noise increases with δ.

Consider a state variable system, with roundoff modeled as individual noise sources—one at each point of accumulation. The equations for this

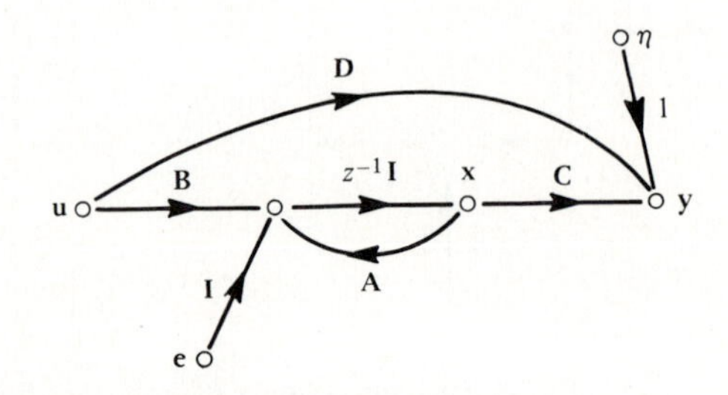

FIGURE 9.7.3 **A SFG for Equation (9.7.2).**

model are

$$\begin{aligned}\mathbf{x}(k+1) &= \mathbf{A}\mathbf{x}(k) + \mathbf{B}u(k) + \mathbf{e}(k),\\ y(k) &= \mathbf{C}\mathbf{x}(k) + Du(k) + \eta(k).\end{aligned} \tag{9.7.2}$$

As we have seen in previous sections, scaling rules involve transfer functions from input to internal state, while roundoff noise involves transfer functions from internal state to the output. In order to visualize this difference, consider the signal flow graph for Eq. (9.7.2) in Fig. 9.7.3. For the consideration of output error, the graph should be reduced to obtain Fig. 9.7.4.

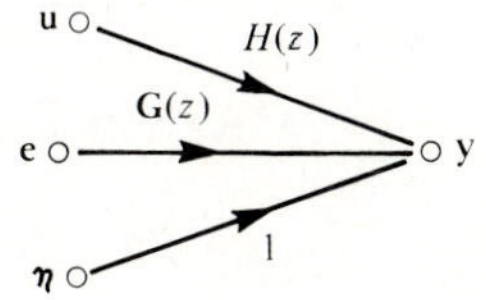

$$H(z) = D + \mathbf{C}(z\mathbf{I} - \mathbf{A})^{-1}\mathbf{B}. \tag{9.7.3}$$

$$\mathbf{G}(z) = \mathbf{C}(z\mathbf{I} - \mathbf{A})^{-1}. \tag{9.7.4}$$

FIGURE 9.7.4 **A SFG for output error.**

Recall that the output roundoff noise variance due to $e_i(k)$ is given by $\sigma_e^2 \sum_{k=0}^{\infty} g_i^2(k)$, where $G_i(z) \leftrightarrow \mathbf{g}_i$. But for the purpose of scaling, one is interested only in the relation of the state to the inputs, as in Fig. 9.7.5.

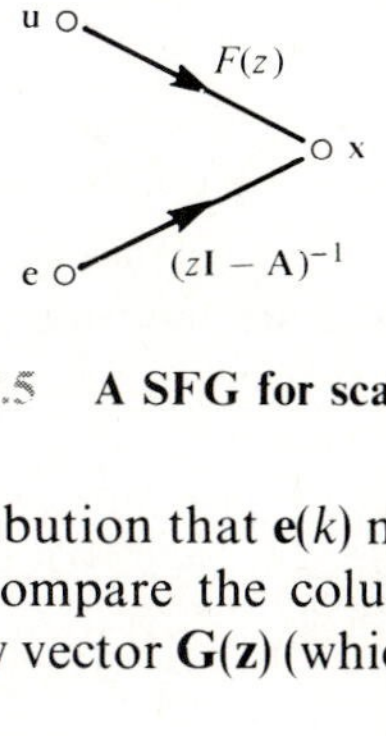

$$\mathbf{F}(z) = (z\mathbf{I} - \mathbf{A})^{-1}\mathbf{B}. \tag{9.7.5}$$

FIGURE 9.7.5 **A SFG for scaling x.**

[The contribution that $\mathbf{e}(k)$ makes to the range of $\mathbf{x}(k)$ is usually negligible.]

If we compare the column vector $\mathbf{F}(z)$ (which is important for scaling) and the row vector $\mathbf{G}(\mathbf{z})$ (which is important for output noise), we see a striking duality:

$$\mathbf{f}(k) = \begin{cases}\mathbf{0}, & k \leq 0\\ \mathbf{A}^{k-1}\mathbf{B}, & k > 0\end{cases} \quad \overset{z}{\leftrightarrow} \quad \mathbf{F}(z) = (z\mathbf{I} - \mathbf{A})^{-1}\mathbf{B}, \tag{9.7.6}$$

$$\mathbf{g}(k) = \begin{cases}\mathbf{0}, & k \leq 0\\ \mathbf{C}\mathbf{A}^{k-1}, & k > 0\end{cases} \quad \overset{z}{\leftrightarrow} \quad \mathbf{G}(z) = \mathbf{C}(z\mathbf{I} - \mathbf{A})^{-1}. \tag{9.7.7}$$

In Eqs. (9.7.6) and (9.7.7), $\mathbf{f}(k)$ and $\mathbf{g}(k)$ are n-dimensional vector sequences. The ith components, $\mathbf{f}_i$ and $\mathbf{g}_i$, are the unit-pulse responses from the input to x_i and from x_i to the output, respectively. The important thing about these sequences is their energy. The scaling rule is

$$\delta^2 \sum_{k=1}^{\infty} f_i^2(k) = 1. \tag{9.7.8}$$

If we call σ_i^2 the variance of that part of the output which is the response to $e_i(k)$, then

$$\sigma_i^2 = \sigma_e^2 \sum_{k=1}^{\infty} g_i^2(k) = \left(\frac{q^2}{12}\right) \|\mathbf{g}_i\|^2, \tag{9.7.9}$$

where

$$\sigma_e^2 = \frac{q^2}{12} = E[e_i^2(k)]. \tag{9.7.10}$$

Now the energy in the sequence $\mathbf{f}_i$ is the ith diagonal element of a matrix, namely,

$$\mathbf{K} = \sum_{k=1}^{\infty} \mathbf{f}(k)\mathbf{f}^T(k) = \mathbf{AKA}^T + \mathbf{BB}^T = \sum_{k=0}^{\infty} (\mathbf{A}^k\mathbf{B})(\mathbf{A}^k\mathbf{B})^T. \tag{9.7.11}$$

Because of the duality between Eqs. (9.7.6) and (9.7.7), we expect a similar characterization for the energy in the sequence $\mathbf{g}_i$. Let

$$\mathbf{W} = \sum_{k=1}^{\infty} \mathbf{g}(k)^T\mathbf{g}(k) = \sum_{k=0}^{\infty} (\mathbf{CA}^k)^T(\mathbf{CA}^k)$$
$$= \mathbf{A}^T\mathbf{WA} + \mathbf{C}^T\mathbf{C}. \tag{9.7.12}$$

Then

$$K_{ij} = (\mathbf{f}_i, \mathbf{f}_j) = \sum_{k=1}^{\infty} f_i(k) f_j(k), \tag{9.7.13}$$

$$W_{ij} = (\mathbf{g}_i, \mathbf{g}_j) = \sum_{k=1}^{\infty} g_i(k) g_j(k). \tag{9.7.14}$$

Therefore, the diagonal elements are energies, and we can rewrite Eqs. (9.7.8) and (9.7.9) as

$$\text{Scaling:} \quad \delta^2 K_{ii} = 1, \qquad i = 1, 2, \ldots, n \tag{9.7.15}$$

$$\text{Output roundoff noise from } i\text{th state:} \quad \sigma_i^2 = \sigma_e^2 W_{ii}, \quad i = 1, 2, \ldots, n \tag{9.7.16}$$

The tradeoff between overflow and roundoff errors is implicit in these two equations. To see it, we must discover how the matrices **K** and **W** change under coordinate transformation, since that is the tool for scaling (and also for minimizing output roundoff noise, as we shall see in the next section).

TRANSFORMATION RULES:

If

$$(\mathbf{A}, \mathbf{B}, \mathbf{C}) \leftarrow (\mathbf{T}^{-1}\mathbf{AT}, \mathbf{T}^{-1}\mathbf{B}, \mathbf{CT}),$$

then

$$(\mathbf{K}, \mathbf{W}) \leftarrow (\mathbf{T}^{-1}\mathbf{KT}^{-T}, \mathbf{T}^T\mathbf{WT}). \tag{9.7.17}$$

(These rules are easily verified by direct substitution.)

Suppose that the original structure, having parameters (**A**, **B**, **C**, **K**, **W**) does not satisfy the scaling rule (9.7.16). We must scale by applying the diagonal transformation

$$\mathbf{T} = \text{Diag}\{\delta\sqrt{K_{11}}, \delta\sqrt{K_{22}}, \ldots, \delta\sqrt{K_{nn}}\}. \tag{9.7.18}$$

Then

$$K'_{ij} = (\mathbf{T}^{-1}\mathbf{K}\mathbf{T}^{-T})_{ij} = \frac{K_{ij}}{\delta^2\sqrt{K_{ii}K_{jj}}} \tag{9.7.19}$$

and so

$$K'_{ii} = \frac{1}{\delta^2}. \tag{9.7.20}$$

But what has happened to the matrix $\mathbf{W}$ under the transformation $\mathbf{T}$ in Eq. (9.7.18)?

Since $\mathbf{W}' = \mathbf{T}^T\mathbf{W}\mathbf{T}$, we find that

$$W'_{ij} = W_{ij}\delta^2\sqrt{K_{ii}K_{jj}} \tag{9.7.21}$$

and so

$$W'_{ii} = \delta^2 W_{ii}K_{ii}. \tag{9.7.22}$$

The total output error due to e_1 through e_n is therefore

$$\sigma^2_{\text{total}} = \sum_{i=1}^{n} \sigma_i^2 = \sigma_e^2 \sum_{i=1}^{n} W'_{ii} = \delta^2\left(\frac{q^2}{12}\right)\sum_{i=1}^{n} W_{ii}K_{ii}. \tag{9.7.23}$$

Equation (9.7.23) corresponds to Eq. (9.5.8) for *scaled digital filters.* [We have omitted the roundoff noise due to the output node in Eq. (9.7.23).] This formula gives the output roundoff noise power in terms of the *unscaled* filter parameters. Note that the product $K_{ii}W_{ii}$ is invariant under diagonal transformation (i.e., is unaffected by scaling), and therefore combines scaling and error power.

We now see from Eq. (9.7.23) the explanation for the right-hand side of Fig. (9.7.2). The output variance is proportional to the square of the scaling parameter δ. Thus δ should be chosen large enough to prevent overflow, with perhaps some margin of safety. However, it should not be chosen much larger than the critical value.

EXAMPLE 9.7.1

Calculation of output roundoff noise using state variables.

Consider the calculation of output roundoff noise for the second-order direct form of Example 9.5.1. The SFG of the unscaled filter is shown in Fig. 9.7.6. Suppose we choose $\delta = 4$. To scale the filter, we first find the state variable description $(\mathbf{A}, \mathbf{B}, \mathbf{C}, D)$. We obtain

$$\mathbf{A} = \begin{bmatrix} 0 & 1 \\ \frac{1}{16} & 0 \end{bmatrix}, \quad \mathbf{B} = \begin{bmatrix} 0 \\ 1 \end{bmatrix}, \quad \mathbf{C} = [\tfrac{1}{16}, -\tfrac{1}{2}], \quad D = [1].$$

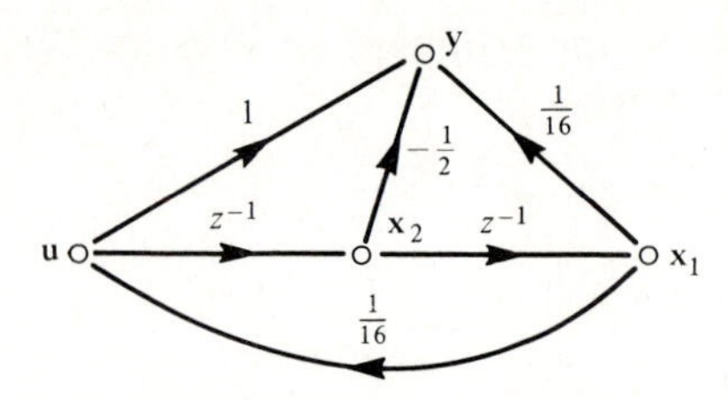

FIGURE 9.7.6 **A SFG of an unscaled second-order filter.**

To scale the filter, we need **K**. From Example 8.5.1,

$$\mathbf{K} = \frac{1}{(1 - a_2)(1 - a_1 + a_2)(1 + a_1 + a_2)} \begin{bmatrix} 1 + a_2 & -a_1 \\ -a_1 & 1 + a_2 \end{bmatrix}, \tag{9.7.24}$$

where, in our case, $a_1 = 0$ and $a_2 = -\frac{1}{16}$. Thus

$$\mathbf{K} = \frac{16^2}{(15)(17)} \mathbf{I}.$$

The scaling rule requires that $\delta\sqrt{K_{ii}} = 1$, which in this case means

$$4\sqrt{K_{ii}} = 1, \qquad i = 1, 2.$$

The transformation **T** required is thus given by Eq. (9.7.18) as

$$\mathbf{T} = \text{Diag}\left\{\frac{64}{\sqrt{255}}, \frac{64}{\sqrt{255}}\right\} = 4.0078\,\mathbf{I}.$$

Thus the scaled filter becomes

$$\mathbf{T}^{-1}\mathbf{A}\mathbf{T} = \begin{bmatrix} 0 & 1 \\ \frac{1}{16} & 0 \end{bmatrix}, \mathbf{T}^{-1}\mathbf{B} = 0.2495 \begin{bmatrix} 0 \\ 1 \end{bmatrix}, \mathbf{C}\mathbf{T} = [0.25049 \quad -2.0039]$$

with an SFG as shown in Fig. 9.7.7.

To find the output roundoff noise for the l_2 scaled filter, we calculate **W** from Eq. (9.7.12). Thus, using the unscaled filter, we have

$$\mathbf{W} = \begin{bmatrix} w_1 & w_2 \\ w_2 & w_3 \end{bmatrix} = \begin{bmatrix} 0 & -a_2 \\ 1 & 0 \end{bmatrix} \begin{bmatrix} w_1 & w_2 \\ w_2 & w_3 \end{bmatrix} \begin{bmatrix} 0 & 1 \\ -a_2 & 0 \end{bmatrix} + \begin{bmatrix} c_1 \\ c_2 \end{bmatrix} [c_1 \quad c_2].$$

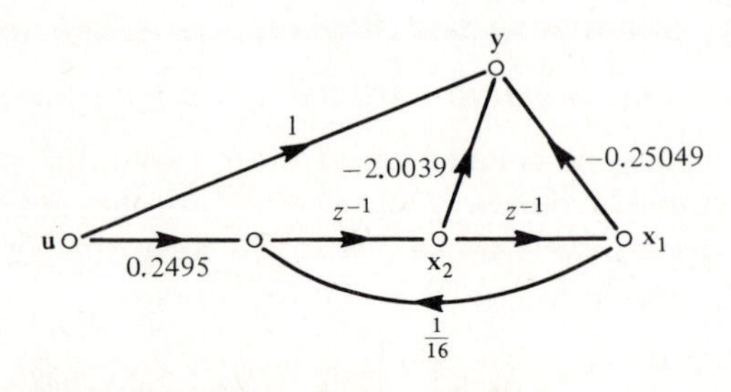

FIGURE 9.7.7 **A SFG of a scaled second-order filter.**

The components of this matrix equation are

$$w_1 = a_2^2 w_3 + c_1^2,$$
$$w_2 = -a_2 w_2 + c_1 c_2,$$
$$w_3 = w_1 + c_2^2.$$

Thus we have three linear equations in three unknowns. The solution is

$$w_1 = \frac{c_1^2 + a_2^2 c_2^2}{1 - a_2^2},$$
$$w_2 = \frac{c_1 c_2}{1 + a_2},$$
$$w_3 = \frac{c_1^2 + c_2^2}{1 - a_2^2}. \qquad \textbf{(9.7.25)}$$

From Eq. (9.7.23), output roundoff noise (not including the output register) can be written as

$$\sigma_{\text{total}}^2 = \frac{q^2}{12}\delta^2 \sum_{i=1}^{2} W_{ii} K_{ii}, \qquad \textbf{(9.7.26)}$$

where the diagonal elements are obtained from Eqs. (9.7.24) and (9.7.25). Thus we have

$$\begin{aligned}\sigma_{\text{total}}^2 &= \frac{q^2}{12}\delta^2[K_{11}W_{11} + K_{22}W_{22}] \\ &= \frac{q^2}{12}(4)^2\left[\left(\frac{256}{255}\right)\left(\frac{1}{204}\right) + \left(\frac{256}{255}\right)\left(\frac{13}{51}\right)\right] \\ &= 0.348q^2. \end{aligned} \qquad \textbf{(9.7.27)}$$

We again remind the reader that these calculations require only the solving of linear equations. Contrast this with the more classical solution as given in Example 9.5.1 (for the unscaled filter).

9.8 REACHABLE SETS: A GEOMETRICAL INTERPRETATION OF SCALING

Scaling fixed point filters has a geometrical interpretation in terms of so-called reachable sets. This interpretation is useful in understanding why certain structures are more efficient than others in terms of finite register effects. Consider the range of values of the state **x** that can be "reached" for some class of inputs **u**. This set of values is called the reachable set for that class of inputs. It is *not*, in general, an n-dimensional cube. In a two-dimensional case, for example, a reachable set for bounded inputs might be as shown in

Fig. 9.8.1. If the filter is scaled, then the reachable set should lie inside the n-dimensional cube of representable states. But to use this cube efficiently, the volume of the reachable set should be as large as possible. If the eccentricity of the reachable set is large, the efficiency of the filter structure is poor.

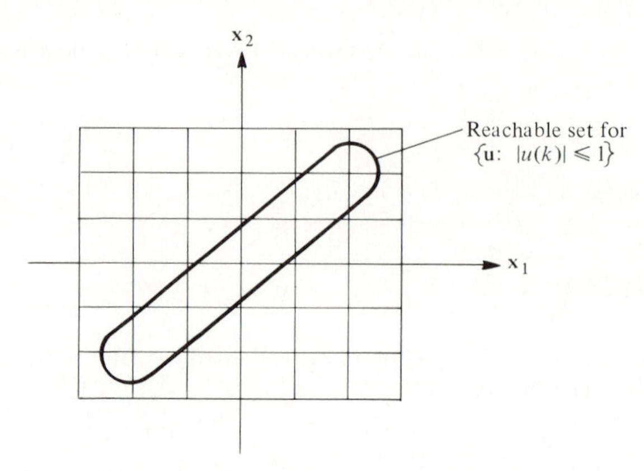

FIGURE 9.8.1 A two-dimensional example of a reachable set for a bounded input.

We shall treat two cases, bounded inputs and finite energy inputs corresponding to l_1 and l_2 scaling, respectively. Let us begin by rewriting Eq. (9.6.11) in more efficient notation. Define

$$\mathbf{x} = \mathbf{x}(k), \tag{9.8.1}$$

$$\mathscr{F} = [\mathbf{f}(1), \mathbf{f}(2), \cdots] = [\mathbf{B}, \mathbf{AB}, \mathbf{A}^2\mathbf{B}, \cdots], \tag{9.8.2}$$

$$\hat{\mathbf{u}} = [u(k-1), u(k-2), u(k-3), \cdots]^T. \tag{9.8.3}$$

Equation (9.6.11), which gives the present state in terms of the past history of the input, becomes

$$\mathbf{x} = [\mathbf{B}, \mathbf{AB}, \mathbf{A}^2\mathbf{B}, \cdots] \begin{bmatrix} u(k-1) \\ u(k-2) \\ \vdots \end{bmatrix} = \mathscr{F}\hat{\mathbf{u}}. \tag{9.8.4}$$

Suppose $|u(k)| \leq 1$ for all k. Then for any direction in n-dimensional space defined by the row vector $\boldsymbol{\rho}$ we have

$$\boldsymbol{\rho}\mathbf{x} = \boldsymbol{\rho}\mathscr{F}\hat{\mathbf{u}} \leq \|\boldsymbol{\rho}\mathscr{F}\|_1 = \sum_{k=0}^{\infty} |\boldsymbol{\rho}\mathbf{A}^k\mathbf{B}|. \tag{9.8.5}$$

The bound is obtained by setting the input equal to the sign of $(\boldsymbol{\rho}\mathbf{A}^k\mathbf{B})$, i.e.,

$$u(k-j) = \operatorname{sign}[\boldsymbol{\rho}\mathbf{A}^{j-1}\mathbf{B}] = \pm 1. \tag{9.8.6}$$

The set of all vectors $\mathbf{x}$ satisfying inequality (9.8.5) is a "half-space" whose boundary is the plane

$$\boldsymbol{\rho}\mathbf{x} = \|\boldsymbol{\rho}\mathscr{F}\|_1. \tag{9.8.7}$$

The reachable set is the intersection of all such half-spaces (one for each vector $\boldsymbol{\rho}$). This set is difficult to calculate.

The bounded energy input case is simpler. Consider an input sequence $\mathbf{u}$ such that

$$\|\mathbf{u}\|_2 = \left[\sum_{j>0} u^2(k-j) \right]^{1/2} \leqslant 1. \tag{9.8.8}$$

In this case, the reachable set is characterized by an ellipsoid.

$$\|\mathbf{u}\|_2 \leqslant 1 \quad \Rightarrow \quad \mathbf{x}^T\mathbf{K}^{-1}\mathbf{x} \leqslant 1. \tag{9.8.9}$$

Here $\mathbf{K}$ is the covariance matrix of the state and can be written as

$$\mathbf{K} = [\mathbf{B}, \mathbf{AB}, \cdots] \begin{bmatrix} (\mathbf{B})^T \\ (\mathbf{AB})^T \\ \vdots \end{bmatrix} = \mathscr{F}\mathscr{F}^T. \tag{9.8.10}$$

To show the reachable set is the ellipsoid $\mathbf{x}^T\mathbf{K}^{-1}\mathbf{x} \leqslant 1$, consider the following argument. Given any vector $\mathbf{x}$, we can reach $\mathbf{x}$ with an input $\mathbf{u}_x$ given by

$$\mathbf{u}_x = \mathscr{F}^T\mathbf{K}^{-1}\mathbf{x}. \tag{9.8.11}$$

This follows from Eqs. (9.8.10) and (9.8.11) since

$$\mathscr{F}\mathbf{u}_x = \mathscr{F}\mathscr{F}^T\mathbf{K}^{-1}\mathbf{x} = \mathbf{K}\mathbf{K}^{-1}\mathbf{x} = \mathbf{x}, \tag{9.8.12}$$

This input has values

$$u_x(k-j) = \mathbf{f}^T(j)\mathbf{K}^{-1}\mathbf{x}, \tag{9.8.13}$$

and has minimum energy among all inputs that reach $\mathbf{x}$. Its energy is

$$\|\mathbf{u}_x\|_2^2 = \mathbf{u}_x^T\mathbf{u}_x = \mathbf{x}^T\mathbf{K}^{-1}\mathscr{F}\mathscr{F}^T\mathbf{K}^{-1}\mathbf{x} = \mathbf{x}^T\mathbf{K}^{-1}\mathbf{x}. \tag{9.8.14}$$

Suppose $\mathbf{u}$ is some other input for which $\mathbf{x} = \mathscr{F}\mathbf{u}$. Then

$$\mathbf{u} = \mathbf{u}_x + (\mathbf{u} - \mathbf{u}_x).$$

But

$$\begin{aligned} \mathbf{u}_x^T(\mathbf{u} - \mathbf{u}_x) &= \mathbf{x}^T\mathbf{K}^{-1}\mathscr{F}(\mathbf{u} - \mathbf{u}_x) \\ &= \mathbf{x}^T\mathbf{K}^{-1}(\mathbf{x} - \mathbf{x}) = 0, \end{aligned}$$

and so

$$\|\mathbf{u}\|_2^2 = \|\mathbf{u}_x\|_2^2 + \|\mathbf{u} - \mathbf{u}_x\|_2^2 \geqslant \|\mathbf{u}_x\|_2^2. \tag{9.8.15}$$

Therefore,

$$\mathscr{F}\mathbf{u} = \mathscr{F}\mathbf{u}_x \quad \Rightarrow \quad \|\mathbf{u}\|_2^2 \geqslant \|\mathbf{u}_x\|_2^2 = \mathbf{x}^T\mathbf{K}^{-1}\mathbf{x}. \tag{9.8.16}$$

This is the minimum energy property. Put another way,

$$\mathbf{x} = \mathscr{F}\mathbf{u} \quad \Rightarrow \quad \mathbf{x}^T\mathbf{K}^{-1}\mathbf{x} \leqslant \|\mathbf{u}\|_2^2. \tag{9.8.17}$$

This is what we set out to show.

We can relate this discussion to white noise inputs. If $\mathbf{u}$ is Gaussian, wide sense stationary with $S_{uu}(\theta) = 1$, then the set

$$\{\mathbf{x}: \quad \mathbf{x}^T\mathbf{K}^{-1}\mathbf{x} \leqslant 1\} \tag{9.8.18}$$

is the "one standard deviation error ellipsoid." The level curves of the probability density function of the state vector $\mathbf{x}$ are the level curves of $\mathbf{x}^T\mathbf{K}^{-1}\mathbf{x}$.

EXAMPLE 9.8.1

Reachable sets for a second-order direct form.

Consider a second-order filter whose transfer function is

$$H(z) = \frac{b_0 z^2 + b_1 z + b_2}{(z - 0.9e^{j\pi/8})(z - 0.9e^{-j\pi/8})}. \tag{9.8.19}$$

An l_2 scaled, direct form state variable description for Eq. (9.8.19) is

$$\mathbf{A} = \begin{bmatrix} 0 & 1 \\ -a_2 & -a_1 \end{bmatrix}, \quad \mathbf{B} = \begin{bmatrix} 0 \\ B_2 \end{bmatrix}, \quad \mathbf{K} = \begin{bmatrix} 1 & k \\ k & 1 \end{bmatrix} \tag{9.8.20}$$

with

$$a_1 = -1.8\cos\left(\frac{\pi}{8}\right), \qquad a_2 = 0.81$$

$$B_2 = \left[\left(\frac{1 - a_2}{1 + a_2}\right)(1 - a_1 + a_2)(1 + a_1 + a_2)\right]^{1/2} = 0.23151,$$

$$k = -\frac{a_1}{1 + a_2} = 0.91878.$$

The set of all vectors in the x_1, x_2 plane reachable with inputs satisfying $|u(k)| \leqslant 1$ is

$$\left\{\mathbf{x}: \quad \boldsymbol{\rho}\mathbf{x} \leqslant \sum_{k=0}^{\infty} |\boldsymbol{\rho}\mathbf{A}^k\mathbf{B}| \quad \text{for all row vectors } \boldsymbol{\rho}\right\}. \tag{9.8.21}$$

The set of all vectors reachable with inputs satisfying $\sum_k u^2(k) \leqslant 1$ is the ellipsoid

$$\{\mathbf{x}: \quad \mathbf{x}^T\mathbf{K}^{-1}\mathbf{x} \leqslant 1\}. \tag{9.8.22}$$

These two reachable sets are shown in Fig. 9.8.2(a). The boundary of the set in Eq. (9.8.21) is the envelope of the family of straight lines, one line for each vector $\boldsymbol{\rho}$. The boundary of the set (9.8.22) is the small ellipse. Note how similar in shape these two sets are, despite the differences in their construction.

Sinusoidal inputs $u(k) = \cos(k\theta)$ lead to elliptical trajectories. Since these inputs are bounded, the state trajectories must lie inside the reachable set (9.8.21). In Fig. 9.8.2(b) several of these ellipses are shown, for different choices of θ. The actual trajectories would, of course, be discrete points on an ellipse.

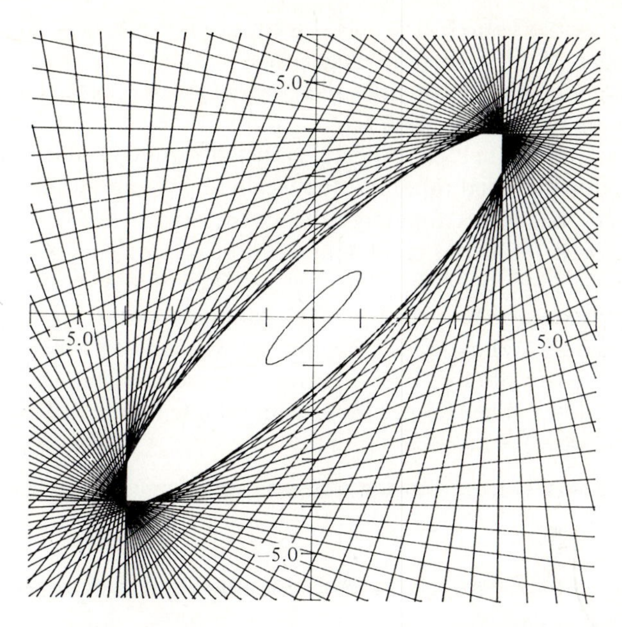

a) The reachable set for bounded energy inputs, $\|u\|_2 \leqslant 1$ is the interior of the ellipse. The reachable set for bounded inputs $|u(k)| \leqslant 1$ is the large area inside the tangent lines. The structure is a direct form.

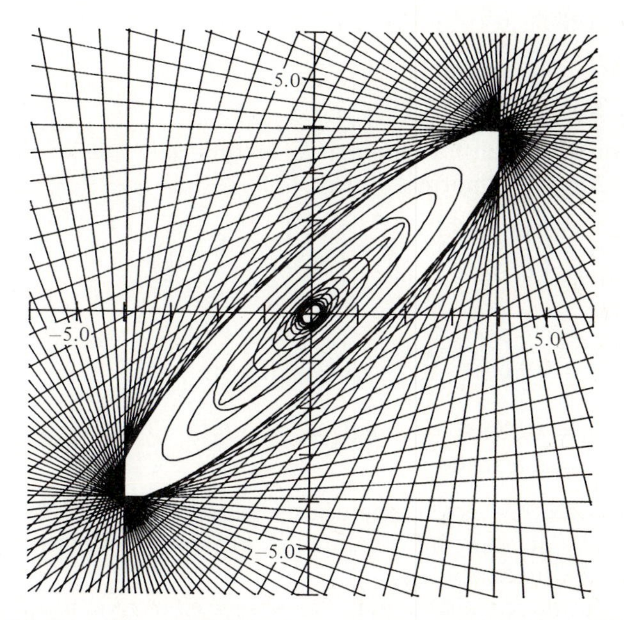

b) Sinusoidal inputs $u(k) = \cos(k\theta)$ lead to trajectories which lie on ellipses. Since the inputs satisfy $|u(k)| \leqslant 1$, the ellipses will lie inside the reachable set for bounded inputs. The largest ellipse is for $\theta = \pi/8$, which is the angle of the system poles.

FIGURE 9.8.2

For some values of θ, the ellipses are quite small. The large ellipses are for values of θ for which $e^{j\theta}$ is near a system pole. Notice again, however, that the family of ellipses for sinusoidal inputs gives a set that is remarkably similar in shape to the reachable set.

The error ellipsoid (9.8.22) is usually quite similar in shape to the reachable set for bounded inputs. If it is scaled up by a few standard deviations, it will then be approximately the same size. If we consider an ellipsoid $\{\mathbf{x}: \mathbf{x}^T\mathbf{K}^{-1}\mathbf{x} \leqslant \delta^2\}$ consistent with the scaling rule (9.7.15), then we see that it grows linearly with δ. In Fig. 9.8.2(a), if $\delta = 4$, the ellipse would closely approximate the larger set. For scaling purposes, it is important to answer the question "Where does the state vector spend its time?" Exact answers can be difficult to obtain. The virtue of ellipsoidal estimates are that they are easily computed. They do not involve an infinite number of infinite sums, as does the set (9.8.21).

EXAMPLE 9.8.2

Reachable sets and coordinate transformations.

Another virtue of ellipsoidal reachable sets is that coordinate transformations are easily accommodated. Recall that if

$$\mathbf{A}, \mathbf{B} \quad \leftarrow \quad \mathbf{T}^{-1}\mathbf{A}\mathbf{T}, \mathbf{T}^{-1}\mathbf{B},$$

then

$$\mathbf{K} \quad \leftarrow \quad \mathbf{T}^{-1}\mathbf{K}\mathbf{T}^{-T}.$$

A suitable transformation (see Section 9.9) will bring the matrices in Eq. (9.8.20) to the "normal" form:

$$\hat{\mathbf{A}} = \begin{bmatrix} \alpha & -\beta \\ \beta & \alpha \end{bmatrix} \quad \hat{\mathbf{B}} = \begin{bmatrix} B_1 \\ B_2 \end{bmatrix}, \quad \hat{\mathbf{K}} = \begin{bmatrix} 1 & k \\ k & 1 \end{bmatrix}, \tag{9.8.23}$$

where

$$\alpha = 0.9 \cos\left(\frac{\pi}{8}\right), \qquad \beta = -0.9 \sin\left(\frac{\pi}{8}\right)$$

$$B_1 = 0.58506, \qquad B_2 = -0.19417$$

$$k = -0.26590.$$

This filter has exactly the transfer function given by Eq. (9.8.19). Figure 9.8.3 exhibits the same information for the new coordinate system as Fig. 9.8.2 did for the original system.

Intuitively, the digital filter is more efficient in the new coordinate system. The actual state can occupy only discrete points, which are uniformly distributed in a cube. If the system is properly scaled according to some rule, then that cube will be the smallest one that contains the reachable set appropriate to that rule. In Fig. 9.8.3 many more of the discrete points will be reachable

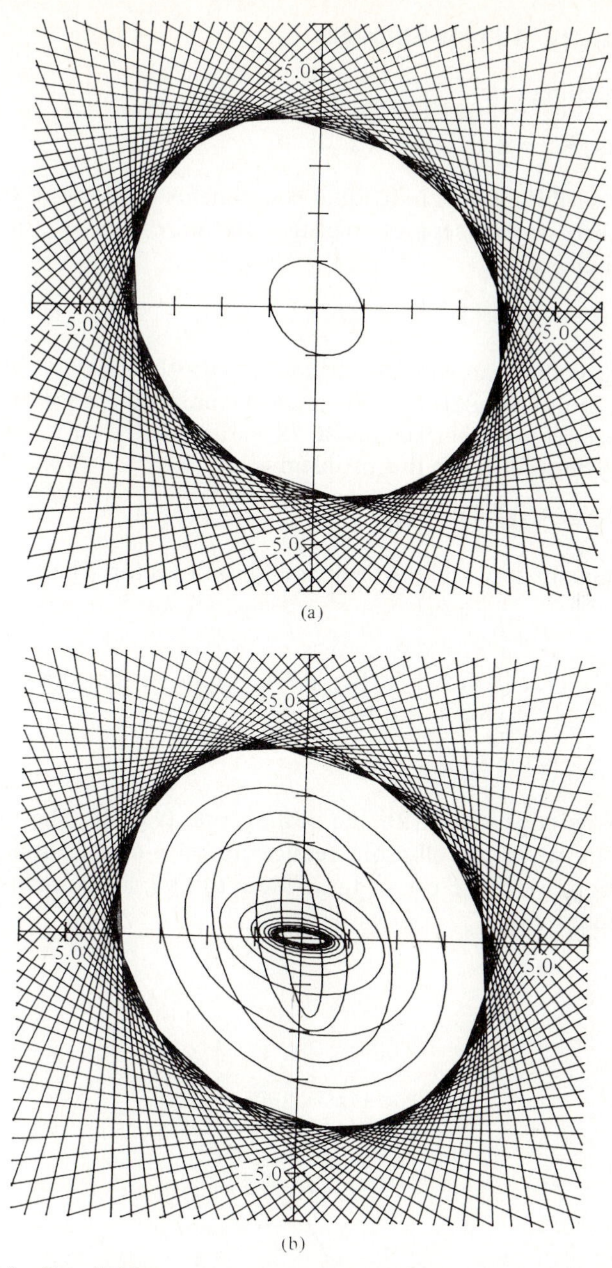

FIGURE 9.8.3 **Reachable sets for a normal form filter structure with transfer function (9.8.19).**

because of the smaller eccentricity of the reachable set. We shall see that this increase in efficiency does indeed provide a higher quality of performance with respect to finite length registers.

Geometry of Error Ellipsoids

The scaling rule

$$\delta\sqrt{K_{ii}} = 1, \qquad i = 1, 2, \ldots, n \tag{9.8.24}$$

necessarily involves individual components of the state vector, because these numbers occupy separate registers. But they must also be consistent with the error ellipsoid:

$$E = \{\mathbf{x}\colon\ \mathbf{x}^T\mathbf{K}^{-1}\mathbf{x} \leqslant \delta^2\}. \tag{9.8.25}$$

What is the vector in the ellipsoid E with the greatest value of x_i. To answer this, let us formulate a constrained maximization problem: Find $\mathbf{x}$ to maximize $\boldsymbol{\rho}\mathbf{x}$, subject to the constraint $\mathbf{x}^T\mathbf{K}^{-1}\mathbf{x} = \delta^2$.

The solution to this problem is

$$\mathbf{x} = \delta(\boldsymbol{\rho}\mathbf{K}\boldsymbol{\rho}^T)^{-1/2}\mathbf{K}\boldsymbol{\rho}^T. \tag{9.8.26}$$

To maximize x_i, take $\boldsymbol{\rho}$ to be the ith unit vector. Then

$$x_i = \delta\sqrt{K_{ii}}. \tag{9.8.27}$$

This vector is shown in Fig. 9.8.4. It follows that the smallest rectangle that contains the ellipsoid E would be

$$\{\mathbf{x}\colon\ |x_i| \leqslant \delta\sqrt{K_{ii}}\,. \tag{9.8.28}$$

This is consistent with the scaling rule (9.8.24). The most efficient scaling occurs when the ellipsoid reduces to an n-dimensional sphere. In the next section, we make use of these ideas to find structures that minimize output roundoff noise.

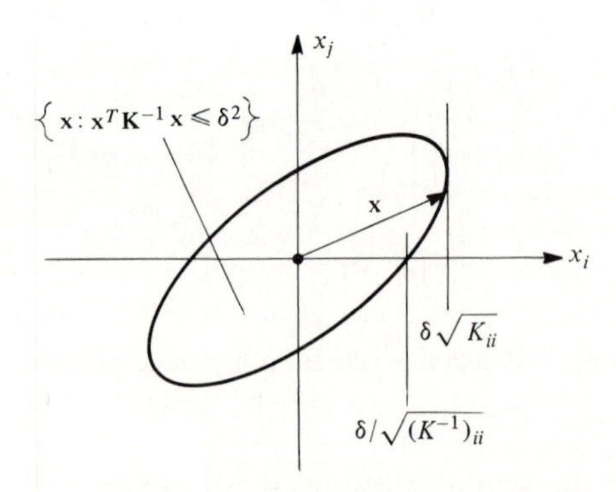

FIGURE 9.8.4 Geometry of error ellipsoids. The vector x has the greatest component in the i-th coordinate.

9.9 THE SYNTHESIS OF MINIMUM ROUNDOFF NOISE STATE VARIABLE FILTERS

We have discussed in some detail the process of scaling fixed point filters because the calculation of roundoff noise (and thus its minimization) is meaningless unless the filter is correctly scaled. We now have available the tools to synthesize structures that minimize output roundoff noise. The key formula is Eq. (9.7.23), repeated here.

$$\sigma_{\text{total}}^2 = \frac{\delta^2 q^2}{12} \sum_{i=1}^{n} W_{ii} K_{ii} \tag{9.9.1}$$

Equation (9.9.1) expresses the output roundoff noise of the filter in terms of diagonal elements of **K** and **W**, which in turn are given in terms of the state variable description of the filter (**A**, **B**, **C**), namely,

$$\begin{aligned} \mathbf{K} &= \mathbf{AKA}^T + \mathbf{BB}^T = \sum_{k=0}^{\infty} (\mathbf{A}^k\mathbf{B})(\mathbf{A}^k\mathbf{B})^T, \\ \mathbf{W} &= \mathbf{A}^T\mathbf{WA} + \mathbf{C}^T\mathbf{C} = \sum_{k=0}^{\infty} (\mathbf{CA}^k)^T(\mathbf{CA}^k). \end{aligned} \tag{9.9.2}$$

Thus Eq. (9.9.1) expresses the output roundoff noise in terms of the internal structure of the filter. The problem is to find a nonsingular transformation of the state $\mathbf{x}$, where $\mathbf{x}' = \mathbf{T}^{-1}\mathbf{x}$ such that σ_{total}^2 is minimized. This transformation changes (**A**, **B**, **C**, **K**, **W**) as in Eq. (9.7.17), repeated below for convenience:

$$(\mathbf{A}, \mathbf{B}, \mathbf{C}, \mathbf{D}, \mathbf{K}, \mathbf{W}) \leftarrow (\mathbf{T}^{-1}\mathbf{AT}, \mathbf{T}^{-1}\mathbf{B}, \mathbf{CT}, \mathbf{D}, \mathbf{T}^{-1}\mathbf{KT}^{-T}, \mathbf{T}^T\mathbf{WT}). \tag{9.9.3}$$

Notice that the product matrix (**KW**) undergoes a similarity transformation, i.e.,

$$\mathbf{KW} \leftarrow \mathbf{T}^{-1}(\mathbf{KW})\mathbf{T}. \tag{9.9.4}$$

This implies that the eigenvalues of **KW** are invariant with respect to nonsingular transformations of the state $\mathbf{x}$. These eigenvalues shall prove to be important in our theory of minimum roundoff noise filters.

There are two cases of interest in the minimization of Eq. (9.9.1). In the first case, we shall choose the word lengths of the n state variable registers optimally subject to the constraint the average word length is B bits. In other words, we shall distribute the total number of bits, nB, among n registers optimally. If the ith register has B_i bits, then we have

$$\sum_{i=1}^{n} B_i = nB. \tag{9.9.5}$$

In the second case, we shall assume the word lengths of the n state variable register are equal.

The Optimal (Unequal) Word Length (OWL) Case

For B_i bits in the ith register and a quantization step size of q, we have

$$q2^{B_i} = 2, \tag{9.9.6}$$

which implies that

$$q = 2^{-B_i+1}. \tag{9.9.7}$$

Substituting Eq. (9.9.7) into Eq. (9.9.1) gives

$$\sigma_{\text{total}}^2 = \frac{\delta^2}{3} \sum_{i=1}^{n} \frac{W_{ii}K_{ii}}{(2^{B_i})^2}. \tag{9.9.8}$$

We wish now to minimize Eq. (9.9.8) by the correct choice of the B_i, subject to the constraint (9.9.5). The key idea is the so-called arithmetic-geometric mean inequality (Noble 1969). It states that for n real numbers r_i,

$$\frac{1}{n}\sum_{i=1}^{n} r_i \geqslant \left[\prod_{i=1}^{n} r_i\right]^{1/n} \tag{9.9.9}$$

with equality if and only if $r_1 = r_2 = \cdots = r_n$. [The term on the right-hand side of inequality (9.9.9) is called the geometric mean of the numbers r_i, $i = 1, 2, \ldots, n$.]

Thus by inequality (9.9.9),

$$\frac{1}{n}\sum_{i=1}^{n} \frac{W_{ii}K_{ii}}{2^{2B_i}} \geqslant \left[\prod_{i=1}^{n} \frac{W_{ii}K_{ii}}{2^{2B_i}}\right]^{1/n}. \tag{9.9.10}$$

To optimize the choice of the B_i in inequality (9.9.10), we choose the B_i so that all the terms in the sum are equal. Thus take

$$\frac{K_{ii}W_{ii}}{2^{2B_i}} = c, \qquad i = 1, 2, \ldots, n \tag{9.9.11}$$

where c is chosen so that

$$\sum_{i=1}^{n} B_i = nB. \tag{9.9.12}$$

Eliminating c between Eqs. (9.9.11) and (9.9.12) yields the required result.

$$B_i = B + \frac{1}{2}\log_2(K_{ii}W_{ii}) - \frac{1}{2n}\sum_{j=1}^{n} \log_2(K_{jj}W_{jj}). \tag{9.9.13}$$

Substituting this value of B_i into Eq. (9.9.8) gives

$$\sigma_{\text{total,OWL}}^2 = \left[\frac{n}{3}\left(\frac{\delta}{2^B}\right)^2\right]\left[\prod_{i=1}^{n} K_{ii}W_{ii}\right]^{1/n}. \tag{9.9.14}$$

Equation (9.9.14) neglects that fact that B_i must be an integer. The right-hand side of Eq. (9.9.14) is now to be minimized by the correct choice of a nonsingular transformation $\mathbf{T}$.

The Equal Word Length (EWL) Case

The equal word length case follows directly from Eq. (9.9.8) with $B_i = B$. The expression for output roundoff noise in this case is

$$\sigma^2_{\text{total,EWL}} = \left[\frac{n}{3}\left(\frac{\delta}{2^B}\right)^2\right]\left[\frac{1}{n}\sum_{i=1}^{n} K_{ii}W_{ii}\right]. \tag{9.9.15}$$

Comparing Eqs. (9.9.14) and (9.9.15), we see that for the optimal word length case it is the geometric mean of $K_{ii}W_{ii}$, $i = 1, 2, \ldots, n$ that is important. In the equal word length case, it is the arithmetic mean of $K_{ii}W_{ii}$, $i = 1, 2, \ldots, n$.

Minimization of the Geometric Mean of $\mathbf{K}_{ii}\mathbf{W}_{ii}$

Minimization of Eq. (9.9.14) is based on *Hadamard's inequality* (Noble 1969), which states: If $\mathbf{P}$ is a positive definite, symmetric matrix, then the scalar $e(\mathbf{P})$ defined by

$$e(\mathbf{P}) = \left[\frac{\det \mathbf{P}}{\prod_{i=1}^{n} P_{ii}}\right]^{1/2} \tag{9.9.16}$$

lies between 0 and 1, $0 < e(\mathbf{P}) \leqslant 1$, and obtains the value unity if and only if $\mathbf{P}$ is diagonal.

To use Eq. (9.9.16), we rewrite the product of the diagonal elements of $\mathbf{K}$ and $\mathbf{W}$ as follows

$$\begin{aligned}\prod_{i=1}^{n} K_{ii}W_{ii} &= \prod_{i=1}^{n} K_{ii}W_{ii}\frac{\det \mathbf{KW}}{\det \mathbf{KW}} = \prod_{i=1}^{n} \frac{K_{ii}}{\det \mathbf{K}}\frac{W_{ii}}{\det \mathbf{W}}\det \mathbf{KW} \\ &= \prod_{i=1}^{n}\frac{K_{ii}}{\det \mathbf{K}}\prod_{i=1}^{n}\frac{W_{ii}}{\det \mathbf{W}}\det \mathbf{KW} = \left[\frac{1}{e(\mathbf{K})}\right]^2\left[\frac{1}{e(\mathbf{W})}\right]^2 \det \mathbf{KW}.\end{aligned} \tag{9.9.17}$$

If μ_i^2, $i = 1, 2, \ldots, n$ are the eigenvalues of $\mathbf{KW}$, then the determinant of $\mathbf{KW}$ is $\prod_{i=1}^{n} \mu_i^2$. Thus we can write

$$[\det \mathbf{KW}]^{1/2n} = \left[\prod_{i=1}^{n} \mu_i^2\right]^{1/2n} = \left[\prod_{i=1}^{n} \mu_i\right]^{1/n} = M_g, \tag{9.9.18}$$

where M_g is the geometric mean of the square roots of the eigenvalues of $\mathbf{KW}$. We shall call these square roots μ_i, $i = 1, 2, \ldots, n$ the *second-order modes* of the filter. Substituting Eqs. (9.9.18) and (9.9.17) into Eq. (9.9.14), we obtain the total output noise as

$$\sigma^2_{\text{total,OWL}} = \left[\frac{n}{3}\left(\frac{\delta}{2^B}\right)^2\right]\frac{M_g^2}{[e(\mathbf{K})e(\mathbf{W})]^{2/n}}. \tag{9.9.19}$$

Under nonsingular transformations of the state M_g is invariant. Thus Eq. (9.9.19) is minimized when $e(\mathbf{K}) = e(\mathbf{W}) = 1$! By Hadamard's inequality, this occurs if and only if $\mathbf{K}$ and $\mathbf{W}$ are simultaneously diagonal, in which case

Eq. (9.9.19) reduces to

$$\min \sigma^2_{\text{total,OWL}} = \left[\frac{n}{3}\left(\frac{\delta}{2^B}\right)^2\right] M_g^2. \tag{9.9.20}$$

The minimum output roundoff noise is proportional to the square of the geometric mean of the filter's second-order modes. M_g^2 is the noise gain of the optimal filter in this case. The optimizing transformation **T** simultaneously diagonalizes the two positive definite, symmetric matrices **K** and **W**. The simultaneous diagonalization of two such matrices is a well-known problem that occurs, for example, in the analysis of Hamiltonian systems and can be found in many linear algebra texts (Noble 1969). We shall be primarily concerned with the solution in the equal word length case.

Minimization of the Arithmetic Mean of $\mathbf{K}_{ii}\mathbf{W}_{ii}$

The equal word length case requires that we minimize the arithmetic mean of $K_{ii}W_{ii}$, $i = 1, 2, \ldots, n$. This case is, in some sense, less intuitive than the optimal word length case. The solution is embodied in the following result (Mullis 1976a).

If **K** and **W** are positive definite, symmetric real matrices, then

$$\frac{1}{n}\sum_{i=1}^{n} K_{ii}W_{ii} \geqslant M_a^2, \qquad M_a = \frac{1}{n}\sum_{i=1}^{n} \mu_i \tag{9.9.21}$$

with equality if and only if

$$\begin{aligned} &\mathbf{K} = \mathbf{DWD}, \qquad \mathbf{D} = \text{diag}\{d_1^2, \ldots, d_n^2\}, \\ &K_{ii}W_{ii} = K_{jj}W_{jj}, \qquad i, j = 1, 2, \ldots, n. \end{aligned} \tag{9.9.22}$$

These conditions imply that

$$e(\mathbf{K}) = e(\mathbf{W}) = \left(\frac{M_g}{M_a}\right)^n,$$

so that

$$\min \sigma^2_{\text{total,EWL}} = \left[\frac{n}{3}\left(\frac{\delta}{2^B}\right)^2\right] M_a^2. \tag{9.9.23}$$

The noise gain of the optimal EWL filter is M_a^2, the arithmetic mean squared of the second-order modes. The simplicity of the noise gains in Eqs. (9.9.20) and (9.9.23) is striking. The ratio M_g^2/M_a^2 is the advantage one can expect to gain by distributing bits optimally among the registers in the filter. The distribution is made in accordance with the values of the second-order modes μ_i, $i = 1, 2, \ldots, n$. Large values of the μ_i receive more bits than registers corresponding to small values of μ_i. Notice that using optimal word lengths is advantageous only when there is a large variation among the second-order modes.

The transformation that obtains the minimum EWL filter, i.e., transforms **K** and **W** so that they satisfy the optimality conditions (9.9.22) is described in (Mullis 1976a). For practical reasons that will become apparent in the next section, we shall limit our discussion of this theory applied to second-order sections.

9.10 PROPERTIES OF MINIMUM ROUNDOFF NOISE STRUCTURES

The transformations that optimize state variable structures to minimize either Eq. (9.9.14) or (9.9.15), result in filter structures (**A**, **B**, **C**, **D**), which contain, in general, nontrivial numbers in every component. For an nth-order single-input, single-output filter this implies that there are $(n + 1)^2$ multiplies for each output sample. This is a large increase in the number of multiplies as compared to more traditional structures. It is questionable whether this large number of multiplies can be justified in most applications.

The traditional method of reducing roundoff noise in digital filters is to decompose an nth-order filter into a cascade or parallel connections of second (and possibly first)-order sections. The idea is to separate closely spaced poles and zeros into different sections of the filter. These second-order sections are most often realized as direct forms. This results in $(2.5n + 1)$ multiplies per output sample.

One way of reducing the number of multiplies in minimum roundoff noise realizations is to apply the theory to smaller-order subfilters. In direct form realizations, decomposing an nth-order filter into lower-order subfilters is done to reduce the roundoff noise. In the case of minimum noise filters decomposition is used to reduce the number of multiplies from $(n + 1)^2$ to $(4n + 1)$ (for second-order sections in cascade). The decomposed realization is no longer minimum noise, however.

If we decompose $H(z)$ into r subfilters which are then connected in some way, we have r separate minimization problems involving the filters $\mathbf{A}^{(j)}$, $\mathbf{B}^{(j)}$, $\mathbf{C}^{(j)}$, $\mathbf{D}^{(j)}$, $j = 1, 2, \ldots, r$. This optimization must be formulated in such a way so as to preserve the global connection of the r subfilters. This is accomplished by restricting the nonsingular transformations **T** so that a particular global connection is preserved.

The **K** and **W** matrices of the overall filter satisfy the equations

$$\mathbf{K} = \mathbf{A}\mathbf{K}\mathbf{A}^T + \mathbf{B}\mathbf{B}^T, \tag{9.10.1}$$

$$\mathbf{W} = \mathbf{A}^T\mathbf{W}\mathbf{A} + \mathbf{C}^T\mathbf{C}, \tag{9.10.2}$$

where (**A**, **B**, **C**) describe the overall filter. The matrices $\mathbf{K}^{(j)}$, $\mathbf{W}^{(j)}$ do not satisfy these equations. To obtain the individual $\mathbf{K}^{(j)}\mathbf{W}^{(j)}$ matrices, one must compute the **K** and **W** matrices for the overall filter. The individual $\mathbf{K}^{(j)}$, $\mathbf{W}^{(j)}$ are then extracted from **K** and **W** for the complete filter. If the optimizing transformations **T** preserve the global structure, then the eigenvalues of each product $\mathbf{K}^{(j)}\mathbf{W}^{(j)}$ are invariant. Thus the noise gain of the ith subfilter is $[M_a^{(i)}]^2$, where $M_a^{(i)}$ is the arithmetic mean of the second-order modes of the ith subfilter. These *block second-order modes* are not the same as the second-order modes of the nth-order filter (and so, the noise gain of the block structure is not the same as the noise gain of the global minimum noise filter). Furthermore, the block second-order modes depend on the arrangement of the subfilters. For example, if cascaded sections are rearranged, the block second-order modes change. Thus the problem of pairing and ordering of poles and zeros in the subfilters of a cascade connection is not addressed by this theory.

We shall call minimum noise filters that preserve some global connection of subfilters a *block optimal* structure. We shall call a connection of subfilters that have been optimized in isolation *sectional optimal* structures. For example, parallel connections of sectional optimal structures are also block optimal. However, a cascade of sectional optimal forms is *not* block optimal. As we shall see, the simplicity of design for isolated filter sections is an attractive feature in their use even though their performance is not as good as block optimal realizations. In Section 9.12, we present a simplified theory of design for both block optimal and isolated second-order sections.

Frequency Transformations and Minimum Roundoff Noise Filters

One of the most remarkable properties of minimum roundoff noise filters is the invariance of the second-order modes with respect to frequency transformations. If $H(z)$ is a prototype filter and $F(z)$ is a frequency transformation (see Section 6.7), then $G(z) = H(F(z))$ is a frequency transformed filter. The second-order modes of $G(z)$ are m copies of the second-order modes of $H(z)$ where m is the parameter in the frequency transformation $F(z)$. That is,

$$F(z) = \pm \prod_{i=1}^{m} \left(\frac{z - \alpha_i^*}{1 - \alpha_i z} \right), \qquad |\alpha_i| < 1. \tag{9.10.3}$$

For a proof of this result, see Mullis (1976b).

This remarkable property has many significant consequences. Since the second-order modes of a filter define its minimum noise performance, the optimal noise performance of any frequency transformed filter is immediately known from the prototype filter. Suppose, for example, we use a low-pass to low-pass transformation for Eq. (9.10.3). In this case, the appropriate frequency transformation is $F(z) = (z - \alpha)/(1 - \alpha z)$. As $\alpha \to 1$, the frequency transformed filters bandwidth $\to 0$. In this case $m = 1$ in Eq. (9.10.3). Thus the second-order modes of the transformed filter are identical to the original filter. This means the output noise of optimal realizations (optimum word length, equal word length, or block optimal filters) are *independent of the filter's bandwidth*.

In contrast, if we calculate the noise gain of an nth-order, direct form realization, we find a noise gain given by Mullis (1976b):

$$\text{Noise gain (direct form)} = \frac{P(\alpha)}{(1 - \alpha^2)^{2n-2}}, \tag{9.10.4}$$

where $P(\alpha)$ is a polynomial of order $4(n - 1)$ and is strictly positive for real α. Thus the family of nth-order direct forms with transfer function $H(F(z))$ and $F(z) = (z - \alpha)/(1 - \alpha z)$ has a state variable noise gain given by Eq. (9.10.4). Notice that as the filter bandwidth $\rightarrow 0 (\alpha \rightarrow 1)$, the noise gain in Eq. (9.10.4) approaches infinity. For a low-pass prototype $H(z)$, this corresponds to a very narrow passband in $G(z)$ with poles clustered close to $z = 1$ or $z = -1$.

This result suggests that the theory of low-roundoff noise filters is most profitably used in the case of narrowband low- or high-pass filters. The roundoff noise performance of direct forms deteriorates without limit as the poles of the filter cluster together. The process of decomposing an nth-order filter into second-order subfilters is a way to isolate closely spaced poles. However, if all the poles cluster at $z = 1$ or $z = -1$, this decomposition is no longer effective in separating closely spaced pole pairs.

State Variable and Nonstate Variable Nodes

The development presented thus far assumes the filter realization is described by an SVD (**A**, **B**, **C**, **D**). There are, as discussed in Chapter 8, useful realizations that are not describable by a state variable description. These realizations (like a lattice or a cascade of subfilters) contain nodes that are not connected to unit-delay branches. These nodes or variables thus do not appear in a state variable description. However, they do produce roundoff noise.

One method of including the noise from nonstate nodes in our calculation of output roundoff noise is to revert to first principles. Calculate the l_2 norms of the unit-pulse responses $\mathbf{f}_i$ and $\mathbf{g}_i$ corresponding to these nodes ($\mathbf{f}_i$ is the unit-pulse response from the input to node i and $\mathbf{g}_i$ is the unit-pulse response from node i to the output). The total output roundoff noise from n state variable nodes and m nonstate nodes is

$$\sigma_{\text{total}}^2 = \frac{\delta^2 q^2}{12}\left\{\sum_{i=1}^{n} K_{ii} W_{ii} + \sum_{j=1}^{m} \|\mathbf{f}_j\|^2 \|\mathbf{g}_j\|^2\right\}. \tag{9.10.5}$$

Another method of including noise caused by nonstate nodes is to use a description that includes these nodes, the factored state variable description. In Chapter 8, we showed how to compute $(\tilde{\mathbf{K}}_l, \tilde{\mathbf{W}}_l)$, $l = 1, 2, \ldots, L$ for a factored state variable description:

$$\begin{bmatrix} \mathbf{A} & \mathbf{B} \\ \mathbf{C} & \mathbf{D} \end{bmatrix} = \mathbf{F}_L \mathbf{F}_{L-1} \cdots \mathbf{F}_2 \mathbf{F}_1. \tag{9.10.6}$$

To compute the individual $\tilde{\mathbf{K}}_l$, $l = 1, 2, \ldots, L$, we use Eq. (8.5.36), repeated below,

$$\tilde{\mathbf{K}}_{l+1} = \mathbf{F}_{l+1} \tilde{\mathbf{K}}_l \mathbf{F}_{l+1}^T, \qquad l = 0, 1, 2, \ldots, L-1 \tag{9.10.7}$$

with an initialization given by Eq. (8.5.34):

$$\tilde{\mathbf{K}}_0 = \begin{bmatrix} \mathbf{K} & 0 \\ 0 & \mathbf{1} \end{bmatrix}, \tag{9.10.8}$$

where $\mathbf{K} = \mathbf{A}\mathbf{K}\mathbf{A}^T + \mathbf{B}\mathbf{B}^T$. Similarly, the individual $\mathbf{W}_l$, $l = 1, 2, \ldots, L$ are given by

$$\mathbf{W}_l = \mathbf{F}_l^T \mathbf{W}_{l+1} \mathbf{F}_l, \qquad l = 1, 2, \ldots, L \tag{9.10.9}$$

where the initialization is

$$\mathbf{W}_{L+1} = \begin{bmatrix} \mathbf{W} & 0 \\ 0 & \mathbf{1} \end{bmatrix}, \tag{9.10.10}$$

where $\mathbf{W} = \mathbf{A}^T\mathbf{W}\mathbf{A} + \mathbf{C}^T\mathbf{C}$. The output roundoff noise for any FSVD is contained in the matrices $(\mathbf{K}_l, \mathbf{W}_l)$, $l = 1, 2, \ldots, L$ and is given by

$$\sigma_{\text{total}}^2 = \frac{q^2\delta^2}{12} \sum_{l=1}^{L} \left[\sum_{i=1}^{n} \gamma_{i,l} K_{ii,l} W_{ii,l} \right]. \tag{9.10.11}$$

In Eq. (9.10.11), $\gamma_{il} = 0$ if the ith row of $\mathbf{F}_l$ contains only zeros and ones; otherwise it is unity (this assumes double length accumulators are used at nodes of accumulation). The $(\mathbf{A}, \mathbf{B}, \mathbf{C})$ used to find $\mathbf{K}_0$ and $\mathbf{W}_{L+1}$ are obtained from Eq. (9.10.6).

We can summarize the steps for finding the output roundoff noise for a FSVD as follows:

1. Compute $\mathbf{K}$ and $\mathbf{W}$ using $(\mathbf{A}, \mathbf{B}, \mathbf{C})$ in Eq. (9.10.6).
2. Compute $\mathbf{K}_0$ and $\mathbf{W}_{L+1}$ using Eqs. (9.10.8) and (9.10.10), respectively.
3. Compute $(\mathbf{K}_l, \mathbf{W}_l)$, $l = 1, 2, \ldots, L$ using Eqs. (9.10.7) and (9.10.9), respectively.
4. Compute the output roundoff noise from Eq. (9.10.11).

9.11 COMPUTATION OF K AND W

One of the important practical computations of the theory we have presented is the computation of the matrices $\mathbf{K}$ and $\mathbf{W}$ given $(\mathbf{A}, \mathbf{B}, \mathbf{C})$. These matrices can be found using Eqs. (9.10.1) and (9.10.2). The computation for computing $\mathbf{K}$ also computes $\mathbf{W}$ by substituting $\mathbf{A}^T$ for $\mathbf{A}$ and $\mathbf{C}^T$ for $\mathbf{B}$. We shall present two algorithms.

One simple and yet very effective algorithm is the following:

Initialize: $\mathbf{F} \leftarrow \mathbf{A}$; $\mathbf{K} \leftarrow \mathbf{B}\mathbf{B}^T$.

Loop: $\mathbf{K} \leftarrow \mathbf{F}\mathbf{K}\mathbf{F}^T + \mathbf{K}$

$\mathbf{F} \leftarrow \mathbf{F}^2$.

Continue until $\mathbf{F} = \mathbf{0}$.

We can see the results of this algorithm by merely computing each step as indicated. It is a quadratic type of convergence. Each step doubles the number of terms in the sum

$$\mathbf{K} = \sum_{l=0}^{\infty} (\mathbf{A}^l\mathbf{B})(\mathbf{A}^l\mathbf{B})^T. \tag{9.11.1}$$

Another algorithm for **K** proceeds as follows:

---

1. Given (**A**, **B**), compute the coefficients $\{a_k\}$ of the characteristic polynomial $\det(\lambda\mathbf{I} - \mathbf{A}) = 0$.
2. Set $\mathbf{x}(0) = \mathbf{0}$ and compute: $\mathbf{x}(k + 1) = \mathbf{A}\mathbf{x}(k) + \mathbf{B}a_k$, for $0 < k < n$. Set

$$\mathbf{T} = [\mathbf{x}(n)\cdots\mathbf{x}(1)]. \tag{9.11.2}$$

3. Given the polynomial coefficients, compute $r(0)$ through $r(n)$ satisfying Eq. (11.3.41). The inverse Levinson algorithm developed in Section 11.3 may be used for this purpose. Construct an n by n matrix **R** having elements

$$R_{ij} = r(|i - j|). \tag{9.11.3}$$

4. Compute **K** from **T** and **R** via: $\mathbf{K} = \mathbf{TRT}^T$.

This algorithm essentially finds an intermediate direct form, computes the covariance of the direct form, and then transforms the covariance matrix of the direct form back to that of the original (**A**, **B**).

9.12 SECTIONAL OPTIMAL STRUCTURES

As discussed previously, the number of multiplies per output sample for an nth-order, minimum noise filter is $(n + 1)^2$. If we apply the theory to second-order subfilters, we increase the minimum noise somewhat but reduce the number of multiplies from $(n + 1)^2$ to approximately $4n$ per output sample. Furthermore, by applying the theory to second-order subfilters, we can simplify the design process. In this section we present a theory for the design of isolated, minimum noise, second-order filters and a design of so-called block optimal cascaded sections. If second-order filters optimized in isolation are connected, we call the overall filter *sectional optimal*. In general, sectional optimal structures are not block optimal. We shall restrict our discussion to the equal word length case.

The necessary and sufficient conditions for minimum roundoff noise filters are given by Eq. (9.9.22) in terms of the matrices **K** and **W**. In the case of second-order filters (in isolation) Jackson (1979) has shown these conditions are equivalent to:

$$a_{11} = a_{22}, \tag{9.12.1}$$

$$b_1c_1 = b_2c_2, \tag{9.12.2}$$

the filter is l_2 scaled. (9.12.3)

The first two conditions are invariant with respect to diagonal transformations of the state. Thus l_2 scaling can be applied after Eqs. (9.12.1) and (9.12.2) are satisfied to obtain a minimum noise filter.

There are many ways to derive structures satisfying the three optimality conditions (9.12.1)–(9.12.3). One method is algebraic in nature. [See Bomar (1985).] Suppose the transfer function of a second-order filter is

$$H(z) = D + \mathbf{C}(z\mathbf{I} - \mathbf{A})^{-1}\mathbf{B} = d + \frac{q_1 z^{-1} + q_2 z^{-2}}{1 + p_1 z^{-1} + p_2 z^{-2}}. \tag{9.12.4}$$

Now equating coefficients of like powers of z^{-1}, we obtain the following four equations:

$$\begin{aligned} q_1 &= c_1 b_1 + c_2 b_2, \\ q_2 &= c_1 b_2 a_{12} + c_2 b_1 a_{21} - c_1 b_1 a_{22} - c_2 b_2 a_{11}, \\ p_1 &= -(a_{11} + a_{22}), \\ p_2 &= a_{11}a_{22} - a_{12}a_{21}. \end{aligned} \tag{9.12.5}$$

These four equations place four constraints on the eight coefficients of (**A**, **B**, **C**). There are also four possible permutations of the signs of the coefficients on the right-hand sides of Eq. (9.12.5). The first two optimality conditions (9.12.1) and (9.12.2) introduce two more constraints. The final two constraints are obtained from the scaling conditions

$$K_{11} = K_{22} = 1, \tag{9.12.6}$$

using $\mathbf{K} = \mathbf{AKA}^T + \mathbf{BB}^T$,

$$\begin{bmatrix} a_{11}^2 - 1 & 2a_{11}a_{12} & a_{12}^2 \\ a_{11}a_{21} & a_{12}a_{21} + a_{11}a_{22} - 1 & a_{12}a_{22} \\ a_{21}^2 & 2a_{21}a_{22} & a_{22}^2 - 1 \end{bmatrix} \begin{bmatrix} K_{11} \\ K_{12} \\ K_{22} \end{bmatrix} = \begin{bmatrix} -b_1^2 \\ -b_1 b_2 \\ -b_2^2 \end{bmatrix}. \tag{9.12.7}$$

Thus we have a total of eight constraints on eight coefficients in (**A**, **B**, **C**). It is just a matter of a good deal of algebra to write the coefficients in (**A**, **B**, **C**) in terms of the parameters in $H(z)$, namely, (q_1, q_2, p_1, p_2). The result is given in Fig. 9.12.1 for a l_2-scaled, second-order, minimum roundoff noise filter.

This purely algebraic design process is certainly straightforward, but it is not very insightful. Perhaps a clearer explanation of the design process can be obtained by breaking the optimization process into a series of steps. These steps are outlined in Fig. 9.12.2. One of the advantages of this design process is that the calculation of **K** and **W** is a natural consequence. Thus one not only obtains the sectional optimal structure, but also the performance of this structure.

To detail the design process of Fig. 9.12.2 assume that the transfer function of the filter is given by Eq. (9.12.4). We shall assume the poles of the filter are complex conjugates given by

$$\lambda = \alpha + j\beta, \qquad \lambda^* = \alpha - j\beta. \tag{9.12.8}$$

Compute: $v_1, v_2, \ldots, v_8$

$$v_1 = \left(\frac{q_2}{q_1}\right), \quad v_2 = (v_1^2 - p_1 v_1 + p_2)^{1/2}, \quad v_3 = v_1 - v_2, \quad v_4 = v_1 + v_2$$

$$v_5 = p_2 - 1, \quad v_6 = p_2 + 1, \quad v_7 = v_5[v_6^2 - p_1^2], \quad v_8 = \left(\frac{p_1}{2}\right)^2 - p_2$$

$$a_{11} = a_{22} = -p_1/2$$

$$b_1 = [v_7/(2p_1 v_3 - v_6(1 - v_3^2)]^{1/2}$$

$$b_2 = [v_7/(2p_1 v_4 - v_6(1 + v_4^2)]^{1/2}$$

$$a_{21} = [(p_2^2 + v_5)v_8/(p_1^2 + v_5)]^{1/2}$$

$$a_{12} = v_8/a_{21}$$

$$c_1 = q_1/2b_1$$

$$c_2 = q_1/2b_2$$

FIGURE 9.12.1 **Design equations for a l_2 scaled, second-order, minimum roundoff noise filter.**

$$H(z) = d + \frac{q_1 z^{-1} + q_2 z^{-2}}{1 + p_1 z^{-1} + p_2 z^{-2}}$$

$\downarrow \mathbf{T}_N$

Unscaled normal form: $(\mathbf{A}_0, \mathbf{B}_0, \mathbf{C}_0, D)$; $\quad a_{11} = a_{22}$

$\downarrow \mathbf{R}\left(\frac{\phi}{2}\right) = \mathbf{T}_R$, rotation

Unscaled normal form: $(\mathbf{A}_1, \mathbf{B}_1, \mathbf{C}_1, D)$; $a_{11} = a_{22}$; $\quad b_1 c_1 = b_2 c_2$

$\downarrow \mathbf{T}_S$, diagonal

l_2 scaled minimum noise form: $(\mathbf{A}_2, \mathbf{B}_2, \mathbf{C}_2, D)$

FIGURE 9.12.2 **A design for minimum noise, second-order filters.**

From the transfer function (9.12.4), we can write down a direct form SVD for the filter as

$$\mathbf{A} = \begin{bmatrix} 0 & 1 \\ -p_2 & -p_1 \end{bmatrix} = \begin{bmatrix} 0 & 1 \\ -(\alpha^2 + \beta^2) & 2\alpha \end{bmatrix}, \quad \mathbf{B} = \begin{bmatrix} 0 \\ 1 \end{bmatrix},$$
$$\mathbf{C} = [q_2, \quad q_1], \quad D = d. \tag{9.12.9}$$

The corresponding normal form is obtained using a transformation $\mathbf{T}_N$, where

$$\mathbf{T}_N = \begin{bmatrix} -\dfrac{1}{\beta} & 0 \\ \\ -\dfrac{\alpha}{\beta} & 1 \end{bmatrix}. \tag{9.12.10}$$

Normal Form (Unscaled)

Applying $\mathbf{T}_N$ from Eq. (9.12.10) to (9.12.9), we obtain a normal filter $(\mathbf{A}_0, \mathbf{B}_0, \mathbf{C}_0)$.

$$\mathbf{A}_0 = \mathbf{T}_N^{-1}\mathbf{A}\mathbf{T}_N = \begin{bmatrix} \alpha & -\beta \\ \beta & \alpha \end{bmatrix}, \quad \mathbf{B}_0 = \begin{bmatrix} 0 \\ 1 \end{bmatrix}, \quad \mathbf{C}_0 = \left[\frac{-q_2 - \alpha q_1}{\beta}, \; q_1\right]. \tag{9.12.11}$$

Notice that $a_{11} = a_{22} = \alpha$, so that this structure satisfies the first optimality condition. The second optimality condition is $b_1c_1 = b_2c_2$. This condition can be interpreted geometrically. Rewrite $\mathbf{C}_0$ in the form

$$\mathbf{C}_0 = r[\cos\phi, \quad \sin\phi], \tag{9.12.12}$$

where

$$r^2 = q_1^2 + \left(\frac{q_2 + \alpha q_1}{\beta}\right)^2, \qquad \phi = \tan^{-1}\left(\frac{\beta q_1}{-q_2 - \alpha q_1}\right). \tag{9.12.13}$$

We interpret $\mathbf{B}_0$ and $\mathbf{C}_0$ as two-dimensional vectors as shown in Fig. 9.12.3. The condition $b_1c_1 = b_2c_2$ can be obtained by a rotation of $\mathbf{B}_0$ and $\mathbf{C}_0$ by an amount $\theta = \phi/2$, where ϕ is given by Eq. (9.12.13). (We have sketched only the case for $\mathbf{B}_0$, $\mathbf{C}_0$ in the first quadrant.)

To satisfy the constraint $b_1c_1 = b_2c_2$, we can use a rotational transformation $\mathbf{T}_R$ where

$$\mathbf{T}_R = \mathbf{R}\left(\frac{\phi}{2}\right) = \begin{bmatrix} \cos\dfrac{\phi}{2} & -\sin\dfrac{\phi}{2} \\ \\ \sin\dfrac{\phi}{2} & \cos\dfrac{\phi}{2} \end{bmatrix}, \tag{9.12.14}$$

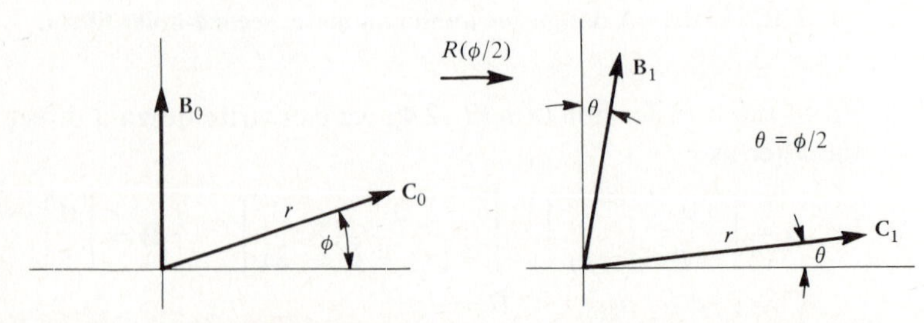

FIGURE 9.12.3 A geometric interpretation of the optimality conditions $b_1c_1 = b_2c_2$.

where we use the notation $\mathbf{R}(\psi)$ to denote a rotation by an angle ψ. Transforming $(\mathbf{A}_0, \mathbf{B}_0, \mathbf{C}_0)$ using $\mathbf{R}(\phi/2)$, we obtain an unscaled normal filter that satisfies the first two optimality conditions.

$$\mathbf{A}_1 = \mathbf{R}\left(-\frac{\phi}{2}\right)\mathbf{A}_0\mathbf{R}\left(\frac{\phi}{2}\right) = \begin{bmatrix} \alpha & -\beta \\ \beta & \alpha \end{bmatrix},$$

$$\mathbf{B}_1 = R\left(-\frac{\phi}{2}\right)\mathbf{B}_0 = \begin{bmatrix} \sin\dfrac{\phi}{2} \\ \cos\dfrac{\phi}{2} \end{bmatrix}, \tag{9.12.15}$$

$$\mathbf{C}_1 = \mathbf{C}_0\mathbf{R}\left(\frac{\phi}{2}\right) = r\left[\cos\frac{\phi}{2},\quad \sin\frac{\phi}{2}\right].$$

$(\mathbf{A}_1, \mathbf{B}_1, \mathbf{C}_1)$ satisfies $a_{11} = a_{22}$ and $b_1c_1 = b_2c_2$ but is not l_2 scaled. To scale $(\mathbf{A}_1, \mathbf{B}_1, \mathbf{C}_1)$, we require an expression for the covariance matrix $\mathbf{K}_1$. To calculate $\mathbf{K}_1$ from the covariance matrix $\mathbf{K}_0$ of the filter $(\mathbf{A}_0, \mathbf{B}_0, \mathbf{C}_0)$ use

$$\mathbf{K}_1 = \mathbf{R}\left(-\frac{\phi}{2}\right)\mathbf{K}_0\mathbf{R}\left(\frac{\phi}{2}\right). \tag{9.12.16}$$

Consider the calculation of $\mathbf{K}_0$. We have

$$\mathbf{K}_0 = \sum_{k=0}^{\infty} (\mathbf{A}_0^k\mathbf{B}_0)(\mathbf{A}_0^k\mathbf{B}_0)^T. \tag{9.12.17}$$

Express $\mathbf{A}_0$ in the form

$$\mathbf{A}_0 = \lambda_1\mathbf{E}_1 + \lambda_2\mathbf{E}_2, \tag{9.12.18}$$

where λ_1 and λ_2 are the poles of the filter given in Eq. (9.12.8). Thus using the techniques of Appendix 8A, we can write

$$\mathbf{A}_0^k = \lambda_1^k\mathbf{E}_1 + \lambda_2^k\mathbf{E}_2. \tag{9.12.19}$$

Thus the covariance matrix $\mathbf{K}_0$ is

$$\mathbf{K}_0 = \sum_{k=0}^{\infty} [(\lambda^k\mathbf{E}_1 + \lambda^{*k}\mathbf{E}_1^*)\mathbf{B}_0][(\lambda^k\mathbf{E}_1 + \lambda^{*k}\mathbf{E}_1^*)\mathbf{B}_0]^T. \tag{9.12.20}$$

Using Appendix 8A, we find

$$\mathbf{E}_1 = \frac{1}{2}\begin{bmatrix} 1 & j \\ -j & 1 \end{bmatrix} \tag{9.12.21}$$

Substituting for $\mathbf{E}_1$ and $\mathbf{E}_1^*$ in Eq. (9.12.20), we obtain

$$\mathbf{K}_0 = \frac{1}{2}\begin{bmatrix} a - \mathrm{Re}(b) & -\mathrm{Im}(b) \\ -\mathrm{Im}(b) & a + \mathrm{Re}(b) \end{bmatrix} = \mathbf{R}(\xi)\mathbf{D}_0\mathbf{R}(-\xi), \tag{9.12.22}$$

where

$$a = \frac{1}{1 - |\lambda|^2}, \qquad b = \frac{1}{1 - \lambda^2}, \qquad \xi = -\tfrac{1}{2}\arg(1 - \lambda^2),$$

$$\mathbf{D}_0 = \begin{bmatrix} \gamma_1 & 0 \\ 0 & \gamma_2 \end{bmatrix}, \quad \begin{aligned} \gamma_1 &= \tfrac{1}{2}(a - |b|) \\ \gamma_2 &= \tfrac{1}{2}(a + |b|) \end{aligned}. \tag{9.12.23}$$

Using the representation (9.12.22), we find an expression for the covariance matrix $\mathbf{K}_1$.

$$\mathbf{K}_1 = \mathbf{T}_R^{-1}\mathbf{R}(\xi)\mathbf{D}_0\mathbf{R}(-\xi)\mathbf{T}_R^{-T} = \mathbf{R}\left(\xi - \frac{\phi}{2}\right)\mathbf{D}_0\mathbf{R}\left(-\xi + \frac{\phi}{2}\right). \tag{9.12.24}$$

In Eq. (9.12.24), we have used the fact that the product of two rotations is another rotation with angle equal to the sum of the angles of the individual rotations. The covariance matrix $\mathbf{K}_1$ is

$$\mathbf{K}_1 = \begin{bmatrix} d_1^2 & k_{12} \\ k_{21} & d_2^2 \end{bmatrix}, \quad \begin{aligned} d_1^2 &= \gamma_1 \cos^2\Delta + \gamma_2 \sin^2\Delta \\ d_2^2 &= \gamma_1 \sin^2\Delta + \gamma_2 \cos^2\Delta \end{aligned},$$

$$\Delta = \xi - \frac{\phi}{2} = -\tfrac{1}{2}\arg(1 - \lambda^2) - \tfrac{1}{2}\tan^{-1}\left(\frac{\beta q_1}{-q_2 - \alpha q_1}\right). \tag{9.12.25}$$

To l_2 scale $(\mathbf{A}_1, \mathbf{B}_1, \mathbf{C}_1)$, we use the diagonal transformation

$$\mathbf{T}_S = \delta\begin{bmatrix} d_1 & 0 \\ 0 & d_2 \end{bmatrix}. \tag{9.12.26}$$

This results in a minimum roundoff noise filter given by

$$\mathbf{A}_2 = \mathbf{T}_S^{-1}\mathbf{A}_1\mathbf{T}_S = \begin{bmatrix} \alpha & -\beta\dfrac{d_2}{d_1} \\ \beta\dfrac{d_1}{d_2} & \alpha \end{bmatrix}, \qquad \mathbf{B}_2 = \mathbf{T}_S^{-1}\mathbf{B}_1 = \frac{1}{\delta}\begin{bmatrix} \dfrac{\sin\phi/2}{d_1} \\ \dfrac{\cos\phi/2}{d_2} \end{bmatrix}$$

$$\mathbf{C}_2 = \mathbf{C}_1\mathbf{T}_S = r\delta\left[d_1 \cos\frac{\phi}{2}, \quad d_2 \sin\frac{\phi}{2}\right]. \tag{9.12.27}$$

The filter $(\mathbf{A}_2, \mathbf{B}_2, \mathbf{C}_2)$ meets all the conditions of optimality (9.12.1)–(9.12.3) and thus is the minimum roundoff noise filter. This derivation uses the matrix $\mathbf{K}_1$ to scale the filter. A similar calculation can be used to obtain $\mathbf{W}_1$, and from $\mathbf{K}_1$ and $\mathbf{W}_1$ we can evaluate the noise performance of the optimal design.

9.13 NOISE GAIN FORMULAS FOR SECOND-ORDER MINIMUM NOISE FILTERS

To obtain the noise performance of minimum noise, second-order filters we proceed as in the calculation $\mathbf{K}_1$. The essential calculation required is that of $\mathbf{W}_0$. We have, in general,

$$\mathbf{W}_0 = \sum_{k=0}^{\infty} (\mathbf{C}_0\mathbf{A}_0^k)^T(\mathbf{C}_0\mathbf{A}_0^k). \tag{9.13.1}$$

We can write $\mathbf{C}_0$ as

$$\mathbf{C}_0 = r\left[\cos\frac{\phi}{2}, \sin\frac{\phi}{2}\right] = r\mathbf{B}_0^T\mathbf{R}\left(\frac{\pi}{2} - \phi\right). \tag{9.13.2}$$

Substituting Eq. (9.13.2) into Eq. (9.13.1) yields

$$\mathbf{W}_0 = r^2 \sum_{k=0}^{\infty} (\mathbf{A}_0^k)^T\mathbf{R}\left(-\frac{\pi}{2} + \phi\right)\mathbf{B}_0\mathbf{B}_0^T\mathbf{R}\left(\frac{\pi}{2} - \phi\right)\mathbf{A}_0^k. \tag{9.13.3}$$

The matrix $\mathbf{A}_0$ can be expressed as a scalar times a rotation matrix. That is, $\mathbf{A}_0$ is of the form

$$\mathbf{A}_0 = v\mathbf{R}(\psi). \tag{9.13.4}$$

This implies that $\mathbf{A}_0$ and $\mathbf{R}(\phi = \pi/2)$ commute. Thus Eq. (9.13.3) can be expressed as

$$\begin{aligned}
\mathbf{W}_0 &= r^2\mathbf{R}\left(\phi - \frac{\pi}{2}\right)\left(\sum_{k=0}^{\infty} (\mathbf{A}_0^k)^T\mathbf{B}_0\mathbf{B}_0^T\mathbf{A}_0^k\right)\mathbf{R}\left(\frac{\pi}{2} - \phi\right) \\
&= r^2\mathbf{R}\left(\phi - \frac{\pi}{2}\right)\mathbf{R}(-\xi)\mathbf{D}_0\mathbf{R}(\xi)\mathbf{R}\left(\frac{\pi}{2} - \phi\right) \\
&= r^2\mathbf{R}(\psi)\mathbf{D}_0\mathbf{R}(-\psi),
\end{aligned} \tag{9.13.5}$$

where

$$\psi = \phi - \frac{\pi}{2} - \xi = \tan^{-1}\left(\frac{\beta q_1}{-q_2 - \alpha q_1}\right) + \tfrac{1}{2}\arg(1 - \lambda^2) - \frac{\pi}{2}. \tag{9.13.6}$$

The noise gain matrix for the minimum noise filter is $\mathbf{W}_2$ given by

$$\mathbf{W}_2 = (\mathbf{T}_S\mathbf{T}_R)^T\mathbf{W}_0(\mathbf{T}_S\mathbf{T}_R) \tag{9.13.7}$$

where (with $\delta = 1$)

$$\mathbf{T}_S = \begin{bmatrix} d_1 & 0 \\ 0 & d_2 \end{bmatrix}, \qquad \mathbf{T}_R = \mathbf{R}\left(\frac{\phi}{2}\right). \tag{9.13.8}$$

Thus the noise gain matrix $\mathbf{W}_2$ is given by

$$\mathbf{W}_2 = r^2\begin{bmatrix} d_1^2d_2^2 & w_{12} \\ w_{21} & d_1^2d_2^2 \end{bmatrix}. \tag{9.13.9}$$

The noise gain of the minimum noise (MN) structure ($\mathbf{A}_2$, $\mathbf{B}_2$, $\mathbf{C}_2$) is given by the trace of $\mathbf{W}_2$:

$$\text{Noise gain (MN)} = \mathrm{Tr}(\mathbf{W}_2) = 2r^2 d_1^2d_2^2. \tag{9.13.10}$$

Using some algebra (see Problem 9.24), we can rewrite Eq. (9.13.10) in the form

$$\text{Noise gain (MN)} = \frac{r^2}{2}\left\{\left(\frac{1}{1 - |\lambda|^2}\right)^2 - h^2\left|\frac{1}{1 - \lambda^2}\right|^2\right\}, \tag{9.13.11}$$

where

$$h^2 = \cos^2(\phi - 2\xi) \tag{9.13.12}$$

with ϕ and ξ as defined in Eqs. (9.12.13) and (9.13.6), respectively. The utility of the noise gain formulas (9.13.10) and (9.13.11) is that they express the noise gain of an isolated second-order minimum noise filter in terms of parameters of the transfer function of the filter. We can thus easily evaluate the noise performance of a given transfer function. The total noise power of these filters is obtained from the noise gain using

$$\sigma_{\text{total}}^2 = \frac{\delta^2 q^2}{12}\,[\text{Noise gain (MN)}]. \tag{9.13.13}$$

9.14 BLOCK OPTIMAL STRUCTURES

Block optimal structures are optimized within the constraint that the block topology of the structure is maintained. We shall concentrate on cascade connections of second-order sections. Why is a connection of second-order sections optimized in isolation not block optimal? A simple explanation is that the downstream filters are not correctly scaled. In isolation, the covariance matrix of each section is used to l_2 scale each filter for a white noise input. In cascade, the covariance matrix, for each downstream section, must be calculated based not on a white noise input but rather on a colored input. In order to correctly scale a block structure, we must calculate the overall **K** and **W** matrices of the overall filter. Only with this information can we optimize the block structure.

To illustrate the ideas involved consider a cascade connection of second-order sections. Assume we know the transfer functions of each section, $H_i(z)$, $i = 1, 2, \ldots, n/2$. The first step is to obtain some SVD of each section $(\mathbf{A}_i, \mathbf{B}_i, \mathbf{C}_i, \mathbf{D}_i)$, for example a direct form realization.

To obtain the overall $(\mathbf{K}, \mathbf{W})$ matrices, we first find the $(\mathbf{A}, \mathbf{B}, \mathbf{C}, \mathbf{D})$ for the overall cascade. This can be done using an FSVD and the techniques of Chapter 8. Once $(\mathbf{A}, \mathbf{B}, \mathbf{C}, \mathbf{D})$ for the overall filter is known, we can calculate $(\mathbf{K}, \mathbf{W})$ since

$$\begin{aligned} \mathbf{K} &= \mathbf{A}\mathbf{K}\mathbf{A}^T + \mathbf{B}\mathbf{B}^T, \\ \mathbf{W} &= \mathbf{A}^T\mathbf{W}\mathbf{A} + \mathbf{C}^T\mathbf{C}. \end{aligned} \tag{9.14.1}$$

From $(\mathbf{K}, \mathbf{W})$ for the overall filter, we extract 2×2 blocks on the diagonals. These individual $(\mathbf{K}_i, \mathbf{W}_i)$, $i = 1, 2, \ldots, n/2$ are the covariance and noise gain matrices for the individual sections. We now use the results for characterizing the optimal **K** and **W** given in Eq. (9.9.22) and apply the necessary transformations to each 2×2 pair $(\mathbf{K}_i, \mathbf{W}_i)$. The necessary transformations needed to bring the individual sections to optimality are summarized in Fig. 9.14.1. Once the necessary transformation $\mathbf{T}_i$ that transforms $(\mathbf{K}_i, \mathbf{W}_i)$ is found so that Eq. (9.9.22) is satisfied, this transformation can be used to find the optimal

Given $\mathbf{K} = \begin{bmatrix} k_{11} & k_{12} \\ k_{21} & k_{22} \end{bmatrix}$, $\mathbf{W} = \begin{bmatrix} w_{11} & w_{12} \\ w_{21} & w_{22} \end{bmatrix}$

1. Transform **K** and **W** using a Cholesky transformation.

$$\mathbf{T}_c = \begin{bmatrix} \sqrt{\dfrac{k_{11}k_{22} - k_{12}^2}{k_{22}}} & \dfrac{k_{12}}{\sqrt{k_{22}}} \\ 0 & \sqrt{k_{22}} \end{bmatrix}.$$

We obtain:

$$\mathbf{K}' = \mathbf{T}_c^{-1}\mathbf{K}\mathbf{T}_c^{-T} = \begin{bmatrix} 1 & 0 \\ 0 & 1 \end{bmatrix}, \qquad \mathbf{W}' = \mathbf{T}_c^T\mathbf{W}\mathbf{T}_c = \begin{bmatrix} w'_{11} & w'_{12} \\ w'_{21} & w'_{22} \end{bmatrix}$$

2. Apply a rotation transformation to $\mathbf{W}'$ so that $w'_{12} = w'_{21} = 0$. Since $\mathbf{K}' = \mathbf{I}$ it is unchanged. The eigenvalues of **KW** are thus the eigenvalues of $\mathbf{W}''$.

$$\mathbf{R}(\theta) = \begin{bmatrix} \cos\theta & -\sin\theta \\ \sin\theta & \cos\theta \end{bmatrix}, \; \theta = \begin{cases} \dfrac{\pi}{4}, & w'_{11} = w'_{22} \\ \tfrac{1}{2}\tan^{-1}\left(\dfrac{2w'_{12}}{w'_{11} - w'_{22}}\right), & w'_{11} \neq w'_{22} \end{cases}$$

Then:

$$\mathbf{K}'' = \mathbf{I}, \qquad \mathbf{W}'' = \mathbf{R}(-\theta)\mathbf{W}'\mathbf{R}(\theta) = \begin{bmatrix} \mu_1^2 & 0 \\ 0 & \mu_2^2 \end{bmatrix}.$$

Now apply a transformation $\mathbf{T} = \dfrac{\delta}{2}\begin{bmatrix} (1+\mu)^{1/2} & (1+\mu)^{1/2} \\ -\left(1 + \dfrac{1}{\mu}\right)^{1/2} & \left(1 + \dfrac{1}{\mu}\right)^{1/2} \end{bmatrix}, \; \mu = \dfrac{\mu_2}{\mu_1}.$

Then

$$\mathbf{K}''' = \frac{1}{\delta^2}\begin{bmatrix} 1 & \dfrac{\mu_1 - \mu_2}{\mu_1 + \mu_2} \\ \dfrac{\mu_1 - \mu_2}{\mu_1 + \mu_2} & 1 \end{bmatrix}, \quad \mathbf{W}''' = \delta^2\begin{bmatrix} \left(\dfrac{\mu_1 + \mu_2}{2}\right)^2 & \dfrac{\mu_1^2 - \mu_2^2}{4} \\ \dfrac{\mu_1^2 - \mu_2^2}{4} & \left(\dfrac{\mu_1 + \mu_2}{2}\right)^2 \end{bmatrix}$$

The optimizing transformation is $\mathbf{T}_c\mathbf{R}(\theta)\mathbf{T}$.

FIGURE 9.14.1 **The optimization of second-order sections using K and W.**

$(\hat{\mathbf{A}}_i, \hat{\mathbf{B}}_i, \hat{\mathbf{C}}_i)$ for the cascade connection via

$$\hat{\mathbf{A}}_i = \mathbf{T}_i^{-1}\mathbf{A}_i\mathbf{T}_i, \quad \hat{\mathbf{B}}_i = \mathbf{T}_i^{-1}\mathbf{B}_i, \quad \hat{\mathbf{C}}_i = \mathbf{C}_i\mathbf{T}_i. \tag{9.14.2}$$

For example, in the cascade of two second-order sections, we obtain an FSVD for the cascade as

$$\begin{bmatrix} \mathbf{A} & \mathbf{B} \\ \mathbf{C} & \mathbf{D} \end{bmatrix} = \begin{bmatrix} \mathbf{I} & 0 & 0 \\ 0 & \mathbf{A}_2 & \mathbf{B}_2 \\ 0 & \mathbf{C}_2 & \mathbf{D}_2 \end{bmatrix}\begin{bmatrix} \mathbf{A}_1 & 0 & \mathbf{B}_1 \\ 0 & \mathbf{I} & 0 \\ \mathbf{C}_1 & 0 & \mathbf{D}_1 \end{bmatrix}. \tag{9.14.3}$$

Now assuming the individual sections are optimized, we obtain two transformations.

$$\begin{bmatrix} \mathbf{T}_1^{-1} & 0 & 0 \\ 0 & \mathbf{T}_2^{-1} & 0 \\ 0 & 0 & 1 \end{bmatrix} \begin{bmatrix} \mathbf{A}_1 & 0 & \mathbf{B}_1 \\ \mathbf{B}_2\mathbf{C}_1 & \mathbf{A}_2 & \mathbf{B}_2\mathbf{D}_1 \\ \mathbf{D}_2\mathbf{C}_1 & \mathbf{C}_2 & \mathbf{D}_2\mathbf{D}_1 \end{bmatrix} \begin{bmatrix} \mathbf{T}_1 & 0 & 0 \\ 0 & \mathbf{T}_2 & 0 \\ 0 & 0 & 1 \end{bmatrix}. \tag{9.14.4}$$

The resulting block optimal structure is

$$\begin{bmatrix} \mathbf{T}_1^{-1}\mathbf{A}_1\mathbf{T}_1 & 0 & \mathbf{T}_1^{-1}\mathbf{B}_1 \\ \mathbf{T}_2^{-1}\mathbf{B}_2\mathbf{C}_1\mathbf{T}_1 & \mathbf{T}_2^{-1}\mathbf{A}_2\mathbf{T}_2 & \mathbf{T}_2^{-1}\mathbf{B}_2\mathbf{D}_1 \\ \mathbf{D}_2\mathbf{C}_1\mathbf{T}_1 & \mathbf{C}_2\mathbf{T}_2 & \mathbf{D}_2\mathbf{D}_1 \end{bmatrix} = \begin{bmatrix} \mathbf{I} & 0 & 0 \\ 0 & \mathbf{T}_2^{-1}\mathbf{A}_2\mathbf{T}_2 & \mathbf{T}_2^{-1}\mathbf{B}_2 \\ 0 & \mathbf{C}_2\mathbf{T}_2 & \mathbf{D}_2 \end{bmatrix}$$

$$\times \begin{bmatrix} \mathbf{T}_1^{-1}\mathbf{A}_1\mathbf{T}_1 & 0 & \mathbf{T}_1^{-1}\mathbf{B}_1 \\ 0 & \mathbf{I} & 0 \\ \mathbf{C}_1\mathbf{T}_1 & 0 & \mathbf{D}_1 \end{bmatrix}. \tag{9.14.5}$$

In Eq. (9.14.4), the individual optimizing transformations are applied in block diagonal form to preserve the block cascade structure. This result is equivalent to applying the optimizing transformations as in Eq. (9.14.2) as shown in Eq. (9.14.5).

We can summarize the steps in this design for a block optimal cascade as follows:

1. Find $(\mathbf{A}, \mathbf{B}, \mathbf{C}, \mathbf{D})$ for the overall structure.
2. Calculate $(\mathbf{K}, \mathbf{W})$ for the overall structure.
3. Extract 2×2 diagonal blocks from $(\mathbf{K}, \mathbf{W})$ in Step 2 and optimize these $(\mathbf{K}_i, \mathbf{W}_i)$, $i = 1, 2, \ldots, n/2$, using the algorithm in Fig. 9.14.1.
4. Apply the individual optimizing transformations in Step 3 to original second-order sections $(\mathbf{A}_i, \mathbf{B}_i, \mathbf{C}_i, \mathbf{D}_i)$, $i = 1, 2, \ldots, n/2$.

9.15 AN EXAMPLE

To illustrate this theory, consider the design of a sixth-order low-pass Butterworth filter. The transfer

$$H(z) = H_1(z)H_2(z)H_3(z), \tag{9.15.1}$$

where

$$H_i(z) = \frac{g_i(z + 1)^2}{z^2 + a_{1i}z + a_{2i}}, \qquad i = 1, 2, 3 \tag{9.15.2}$$

and the parameters of the individual sections are given in Fig. 9.15.1. The individual gains g_i, $i = 1, 2, 3$ were chosen so that the dc gain of each section is unity, i.e.,

$$g_i = (1 + a_{1i} + a_{2i})/4. \tag{9.15.3}$$

From the individual transfer functions, we next obtain direct form state variable realizations as in Eq. (9.12.9). These results are summarized below.

$$\mathbf{A}_i = \begin{bmatrix} 0 & 1 \\ -a_{2i} & -a_{1i} \end{bmatrix}, \quad \mathbf{B}_i = \begin{bmatrix} 0 \\ 1 \end{bmatrix}, \quad \mathbf{C}_i = [g_i - a_{2i}g_i, 2g_i - a_{1i}g_i]$$

$$D_i = g_i; \qquad i = 1, 2, 3.$$

Thus

$$\begin{aligned} \mathbf{C}_1 &= [3.1340 \times 10^{-5} \quad 3.88418 \times 10^{-3}], \\ \mathbf{C}_2 &= [8.03439 \times 10^{-5} \quad 3.696084 \times 10^{-3}], \\ \mathbf{C}_3 &= [1.0665 \times 10^{-4} \quad 3.619872 \times 10^{-3}]. \end{aligned} \tag{9.15.4}$$

To find (**A**, **B**, **C**, **D**) for the overall cascade in terms of ($\mathbf{A}_i$, $\mathbf{B}_i$, $\mathbf{C}_i$, $\mathbf{D}_i$), $i = 1, 2, 3$, we can use the FSVD and the techniques of Chapter 8. The result is

$$\begin{bmatrix} \mathbf{A} & \mathbf{B} \\ \mathbf{C} & \mathbf{D} \end{bmatrix} = \left[\begin{array}{ccc|c} \mathbf{A}_3 & \mathbf{B}_3\mathbf{C}_2 & \mathbf{B}_3\mathbf{D}_2\mathbf{C}_1 & \mathbf{B}_3\mathbf{D}_2\mathbf{D}_1 \\ 0 & \mathbf{A}_2 & \mathbf{B}_2\mathbf{C}_1 & \mathbf{B}_2\mathbf{D}_1 \\ 0 & 0 & \mathbf{A}_1 & \mathbf{B}_1 \\ \hline \mathbf{C}_3 & \mathbf{D}_3\mathbf{C}_2 & \mathbf{D}_3\mathbf{D}_1\mathbf{C}_1 & \mathbf{D}_3\mathbf{D}_2\mathbf{D}_1 \end{array}\right]. \tag{9.15.5}$$

Substituting from Eq. (9.15.4), we obtain the overall (**A**, **B**, **C**, **D**) as

$$\mathbf{A} = \begin{bmatrix} 0 & 1 & 0 & 0 & 0 & 0 \\ -0.88563 & 1.8819 & 8.034 \times 10^{-5} & 3.696 \times 10^{-3} & 2.96 \times 10^{-8} & 3.67 \times 10^{-6} \\ 0 & 0 & 0 & 1 & 0 & 0 \\ 0 & 0 & -0.91498 & 1.9112 & 3.134 \times 10^{-5} & 3.885 \times 10^{-3} \\ 0 & 0 & 0 & 0 & 0 & 1 \\ 0 & 0 & 0 & 0 & -0.96802 & 1.9641 \end{bmatrix},$$

$$\mathbf{B} = \begin{bmatrix} 0 \\ 9.261 \times 10^{-7} \\ 0 \\ 9.81 \times 10^{-4} \\ 0 \\ 1 \end{bmatrix}, \quad \mathbf{C} = \begin{bmatrix} 1.0665 \times 10^{-4} \\ 3.6200 \times 10^{-3} \\ 7.49 \times 10^{-8} \\ 3.45 \times 10^{-6} \\ 2.76 \times 10^{-11} \\ 3.42 \times 10^{-9} \end{bmatrix}, \quad D = 8.636 \times 10^{-10}. \tag{9.15.6}$$

Using these matrices, one now computes **K** and **W** using any one of several possible algorithms. One of the easiest is the following:

$$\begin{aligned} &\text{Initalize:} \quad \mathbf{F} \leftarrow \mathbf{A} \\ &\qquad\qquad\quad \mathbf{K} \leftarrow \mathbf{B}\mathbf{B}^T \\ &\text{Recursion:} \quad \mathbf{K} \leftarrow \mathbf{F}\mathbf{K}\mathbf{F}^T + \mathbf{K} \\ &\qquad\qquad\quad\;\; \mathbf{F} \leftarrow \mathbf{F}^2 \end{aligned} \tag{9.15.7}$$

Section, i	*Gain*, g_i	a_{1i}	a_{2i}
1	9.8×10^{-4}	−1.9641	0.96802
2	9.45×10^{-4}	−1.9112	0.91498
3	9.325×10^{-4}	−1.8819	0.88563

FIGURE 9.15.1 **Filter parameters for Eq. (9.15.2).**

Repeat the recursion until $\mathbf{F} = \mathbf{0}$. In this example it takes about 12 iterations. The same algorithm computes $\mathbf{W}$ by replacing $(\mathbf{A}, \mathbf{B})$ with $(\mathbf{A}^T, \mathbf{C}^T)$. The resulting $(\mathbf{K}, \mathbf{W})$ matrices are approximately:

$$\mathbf{K} = \begin{bmatrix} 1459.5520 & 1458.4893 & 910.3087 & 845.7256 & -653.2163 & -679.7735 \\ 1458.4893 & 1459.5520 & 974.3374 & 910.3087 & -623.1365 & -653.2163 \\ 910.3087 & 974.3374 & 2707.2885 & 2704.1776 & 1090.8679 & 955.9436 \\ 845.7256 & 910.3087 & 2704.1776 & 2707.2885 & 1225.8321 & 1090.8679 \\ -653.2163 & -623.1365 & 1090.8679 & 1225.8321 & 3992.4383 & 3984.4860 \\ -679.7735 & -653.2163 & 955.9436 & 1090.8679 & 3984.4860 & 3992.4383 \end{bmatrix},$$

$$\mathbf{W} = \begin{bmatrix} 0.01278 & -0.01440 & 0.00761 & -0.00859 & 0.00466 & -0.00503 \\ -0.01440 & 0.01629 & -0.00831 & 0.00939 & -0.00503 & 0.00544 \\ 0.00761 & -0.00831 & 0.01074 & -0.01174 & 0.01139 & -0.01199 \\ -0.00859 & 0.00939 & -0.01174 & 0.01283 & -0.01222 & 0.01287 \\ 0.00466 & -0.00503 & 0.01139 & -0.01222 & 0.01899 & -0.01961 \\ -0.00503 & 0.00544 & -0.01199 & 0.01287 & -0.01961 & 0.02027 \end{bmatrix} \tag{9.15.8}$$

To optimize the individual sections, we extract the 2×2 matrices on the diagonals of $\mathbf{K}$ and $\mathbf{W}$ in Eq. (9.15.8), and then perform the series of transformations indicated in Fig. 9.14.1 with $\delta = 4$. The individual optimizing transformations as computed in Fig. 9.14.1 are

$$\mathbf{T}_1 = \begin{bmatrix} 107.05 & 273.91 \\ 90.93 & 273.46 \end{bmatrix}, \quad \mathbf{T}_2 = \begin{bmatrix} -37.15 & -234.62 \\ -22.08 & -224.22 \end{bmatrix},$$
$$\mathbf{T}_3 = \begin{bmatrix} 16.97 & -139.14 \\ 26.34 & -131.29 \end{bmatrix}. \tag{9.15.9}$$

The resulting optimized sections are given by

$$\hat{\mathbf{A}}_i = \mathbf{T}_i^{-1}\mathbf{A}_i\mathbf{T}_i, \quad \hat{\mathbf{B}}_i = \mathbf{T}_i^{-1}\mathbf{B}_i, \quad \hat{\mathbf{C}}_i = \mathbf{C}_i\mathbf{T}_i, \quad \hat{\mathbf{D}}_i = \mathbf{D}_i, \tag{9.15.10}$$

which results in numerical values given below.

$$\hat{\mathbf{A}}_1 = \begin{bmatrix} 0.99179 & 0.06670 \\ -0.05554 & 0.97252 \end{bmatrix}, \quad \hat{\mathbf{B}}_1 = \begin{bmatrix} -0.06272 \\ 0.02451 \end{bmatrix},$$
$$\hat{\mathbf{C}}_1 = \begin{bmatrix} 0.3566 \\ 1.071 \end{bmatrix}^T, \quad \hat{D}_1 = 9.8 \times 10^{-4}$$
$$\hat{\mathbf{A}}_2 = \begin{bmatrix} 0.96053 & 0.03162 \\ -0.05801 & 0.95058 \end{bmatrix}, \quad \hat{\mathbf{B}}_2 = \begin{bmatrix} 0.07449 \\ -0.01180 \end{bmatrix},$$
$$\hat{\mathbf{C}}_2 = \begin{bmatrix} -0.08459 \\ -0.8476 \end{bmatrix}^T, \quad \hat{D}_2 = 9.45 \times 10^{-4}$$
$$\hat{\mathbf{A}}_3 = \begin{bmatrix} 0.93797 & 0.00324 \\ -0.07485 & 0.94379 \end{bmatrix}, \quad \hat{\mathbf{B}}_3 = \begin{bmatrix} 0.09689 \\ 0.01182 \end{bmatrix},$$
$$\hat{\mathbf{C}}_3 = \begin{bmatrix} 0.09716 \\ -0.49008 \end{bmatrix}, \quad \hat{D}_3 = 9.325 \times 10^{-4}. \tag{9.15.11}$$

From these individual matrices we obtain $(\hat{\mathbf{K}}, \hat{\mathbf{W}})$ for the overall block optimal structure using Eq. (9.15.5) and Eq. (9.15.7). The results for $(\hat{\mathbf{K}}, \hat{\mathbf{W}})$ are given below. Notice the diagonal elements of $\hat{\mathbf{K}}$ are approximately $1/\delta^2 = 0.0625$, indicating the filter is l_2 scaled. The noise gain of the block optimal structure is, using $\hat{\mathbf{W}}$, approximately

$$\text{Noise gain (MN)} = \sum_{i=1}^{6} \hat{K}_{ii}\hat{W}_{ii} = 0.93534, \tag{9.15.12}$$

$$\hat{\mathbf{K}} = \begin{bmatrix} 0.0621 & -0.0489 & 0.0144 & -0.0497 & 0.0305 & -0.0147 \\ -0.0489 & 0.0624 & 0.0149 & 0.0197 & -0.0086 & 0.0204 \\ 0.0144 & 0.0149 & 0.0632 & -0.0472 & 0.0273 & 0.0395 \\ -0.0497 & 0.0197 & -0.0472 & 0.0624 & -0.0402 & -0.0090 \\ 0.0305 & -0.0086 & 0.0273 & -0.0402 & 0.0629 & -0.0241 \\ -0.0147 & 0.0204 & 0.0395 & -0.0090 & -0.0241 & 0.0626 \end{bmatrix},$$

$$\hat{\mathbf{W}} = \begin{bmatrix} 2.10641 & -1.65210 & 1.09570 & -1.71945 & -0.46742 & 1.16445 \\ -1.65210 & 2.10595 & -0.37034 & 0.89872 & 0.07363 & -0.36489 \\ 1.09570 & -0.37034 & 1.83411 & -1.37922 & -1.77101 & 2.04162 \\ -1.71945 & 0.89872 & -1.37922 & 1.83331 & 0.49471 & -1.67197 \\ -0.46742 & 0.07363 & -1.77101 & 0.49471 & 3.53816 & -1.35891 \\ 1.16445 & -0.36489 & 2.04162 & -1.67197 & -1.35891 & 3.54889 \end{bmatrix}.$$

Sectional Optimal Design

The sectional optimal designs can be calculated using the design equations in Fig. 9.12.1 or series of transformations in Fig. 9.12.2. In either case, the results (assuming $\delta = 4$) are summarized below.

$$\mathbf{A}_1 = \begin{bmatrix} 0.98205 & 0.0641 \\ -0.0561 & 0.98205 \end{bmatrix}, \qquad \mathbf{B}_1 = \begin{bmatrix} 0.0020 \\ 0.0612 \end{bmatrix},$$

$$\mathbf{C}_1 = \begin{bmatrix} 0.9814 \\ 0.0317 \end{bmatrix}^T, \qquad D_1 = 9.523 \times 10^{-4}$$

$$\mathbf{A}_2 = \begin{bmatrix} 0.95560 & -0.02456 \\ 0.07364 & 0.95560 \end{bmatrix}, \qquad \mathbf{B}_2 = \begin{bmatrix} 0.084158 \\ 0.003169 \end{bmatrix},$$

$$\mathbf{C}_2 = \begin{bmatrix} 0.0220 \\ 0.5831 \end{bmatrix}^T, \qquad D_2 = D_1$$

$$\mathbf{A}_3 = \begin{bmatrix} 0.94095 & 0.0824 \\ -0.0030 & 0.94095 \end{bmatrix}, \qquad \mathbf{B}_3 = \begin{bmatrix} 0.0037 \\ 0.0861 \end{bmatrix},$$

$$\mathbf{C}_3 = \begin{bmatrix} 0.4957 \\ 0.0210 \end{bmatrix}^T, \qquad D_3 = D_1.$$

From these isolated sections, we can check to see if the filters are l_2 scaled for $\delta = 4$ by calculating the individual $(\mathbf{K}_i, \mathbf{W}_i)$, $i = 1, 2, 3$. These are given below. Notice that each $\mathbf{K}_i$, $i = 1, 2, 3$ has diagonal elements of $1/\delta^2 = 0.0625$

indicating each filter is l_2-scaled.

$$\mathbf{K}_1 = \begin{bmatrix} 0.0625 & 0.0156 \\ 0.0156 & 0.0625 \end{bmatrix}, \quad \mathbf{W}_1 = \begin{bmatrix} 16.0707 & 4.008 \\ 4.008 & 16.0707 \end{bmatrix}, \quad \text{Noise gain} = 2.009$$

$$\mathbf{K}_2 = \begin{bmatrix} 0.0625 & 0.0361 \\ 0.0361 & 0.0625 \end{bmatrix}, \mathbf{W}_2 = \begin{bmatrix} 3.0008 & 1.7322 \\ 1.7322 & 3.0008 \end{bmatrix}, \quad \text{Noise gain} = 0.3751$$

$$\mathbf{K}_3 = \begin{bmatrix} 0.0625 & 0.0434 \\ 0.0434 & 0.0625 \end{bmatrix}, \quad \mathbf{W}_3 = \begin{bmatrix} 2.0753 & 1.4406 \\ 1.4406 & 2.0753 \end{bmatrix}, \quad \text{Noise gain} = 0.2594.$$

These matrices establish the optimality of these second-order sections in isolation. When we combine these sections in a cascade connection, however, the result is no longer an optimal structure as we can easily see by calculating the (**K**, **W**) matrices for the overall structures. These are given below.

$$\mathbf{K}_{SO} = \begin{bmatrix} 0.0523 & 0.0540 & 0.0135 & 0.0633 & -0.0275 & -0.0037 \\ 0.0540 & 0.0778 & -0.0143 & 0.0421 & -0.0173 & -0.0189 \\ 0.0135 & -0.0143 & 0.0784 & 0.0627 & -0.0153 & 0.0598 \\ 0.0633 & 0.0421 & 0.0627 & 0.1087 & -0.0451 & 0.0274 \\ -0.0275 & -0.0173 & -0.0153 & -0.0451 & 0.0625 & 0.0156 \\ -0.0037 & -0.0189 & 0.0598 & 0.0274 & 0.0156 & 0.0625 \end{bmatrix},$$

$$\mathbf{W}_{SO} = \begin{bmatrix} 2.07533 & 1.44063 & 1.23291 & 1.23419 & 0.99455 & 0.92791 \\ 1.44063 & 2.07445 & 0.39596 & 0.59102 & 0.18356 & 0.26449 \\ 1.23291 & 0.39596 & 1.73637 & 1.00234 & 2.42562 & 1.30156 \\ 1.23419 & 0.59102 & 1.00234 & 0.88946 & 0.81030 & 0.90295 \\ 0.99455 & 0.18356 & 2.42562 & 0.81030 & 5.32632 & 1.32868 \\ 0.92791 & 0.26449 & 1.30156 & 0.90295 & 1.32868 & 2.15611 \end{bmatrix}.$$

Notice that covariance matrix **K** for the sectional optimal structure no longer represents a correctly l_2-scaled filter ($\delta = 4$). The diagonal elements are not equal to $1/\delta^2 = 0.0625$. In fact, the noise gain of this structure is

$$\text{Noise gain (SO)} = \sum_{i=1}^{6} K_{\text{SO}_{ii}} W_{\text{SO}_{ii}} = 0.97258. \tag{9.15.13}$$

This compares to 0.93534 for the block optimal structure. In this example, the difference between the two structures is small.

To complete this example we should compute the roundoff noise generated by the nodes (registers) between the three cascaded sections. This noise is *independent* of how the individual sections are realized. In that sense it really is an external kind of noise depending only on the ordering of the filters. There are several ways to calculate the l_2 norms shown in Fig. 9.15.2. The noise gains for nodes 1 and 2 are given by $\|\mathbf{h}_1\|^2 \cdot \|\mathbf{h}_2 * \mathbf{h}_3\|^2$ and $\|\mathbf{h}_1 * \mathbf{h}_2\|^2 \cdot \|\mathbf{h}_3\|^2$, respectively. The calculation of these l_2 norms yields

$$\begin{aligned} &\|\mathbf{h}_1\|^2 \cdot \|\mathbf{h}_2 * \mathbf{h}_3\|^2 + \|\mathbf{h}_1 * \mathbf{h}_2\|^2 \cdot \|\mathbf{h}_3\|^2 \\ &= (0.0617)(0.01346) + (0.03975)(0.01692) = 0.00128. \end{aligned} \tag{9.15.14}$$

As compared to either 0.93534 or 0.97258, this contribution to the total noise gain is very small and, for this example, really insignificant.

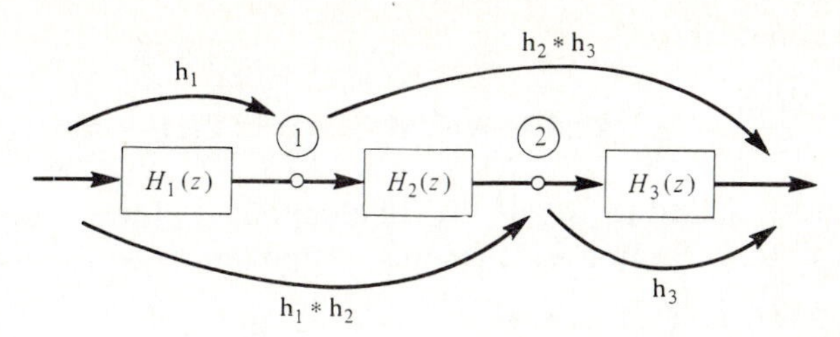

FIGURE 9.15.2.

One final calculation may serve to indicate the magnitude of the reduction of the output roundoff noise using these minimum noise structures vis-à-vis direct form structures. We calculated in Eq. (9.15.8) the (**K**, **W**) matrices for a cascade of direct forms. Thus the noise gain using direct form (DF) structures in cascade is

$$\text{Noise gain (DF)} = \sum_{i=1}^{6} K_{ii}W_{ii} = 262.98. \tag{9.15.15}$$

The ratio of direct form noise gain to the block optimal noise gain is thus

$$\frac{\text{Noise gain (DF)}}{\text{Noise gain (BO)}} \cong 281.3. \tag{9.15.16}$$

This completes the example and serves to illustrate the kinds of calculations involved in designing realizations for good roundoff noise performance. The computations involved are quite straightforward and can usually be performed on a hand calculator.

9.16 COEFFICIENT SENSITIVITY OF DIGITAL FILTERS

Another effect of finite length registers is the quantization of filter parameters. This is manifested by a deterministic change in the I/O characteristic of the filter $H(z)$. In any given design, the effects of coefficient quantization can be calculated. If serious changes occur, one can lengthen coefficient word lengths accordingly. Because this effect is deterministic, it is easier to analyze than roundoff noise and is generally of less consequence.

One way to quantify coefficient sensitivity is to measure the change in the I/O characteristic of the filter with respect to small changes in the parameters of filter. And one way to measure this phenomenon is to calculate the partial derivatives of $H(z)$ (or **h**) with respect to the coefficients in the filter.

Consider a filter described by the usual state variable equations

$$\mathbf{x}(k+1) = \mathbf{A}\mathbf{x}(k) + \mathbf{B}u(k),$$
$$y(k) = \mathbf{C}\mathbf{x}(k) + Du(k).$$

The parameters of the filter are the coefficients in the matrices $\{\mathbf{A}, \mathbf{B}, \mathbf{C}, \mathbf{D}\}$. For example, the partial derivative of $H(z)$ with respect to the coefficients in the **B** matrix is

$$\frac{\partial H(z)}{\partial b_i} = \frac{\partial}{\partial b_i}[\mathbf{C}(z\mathbf{I} - \mathbf{A})^{-1}\mathbf{B} + D] = [\mathbf{C}(z\mathbf{I} - \mathbf{A})^{-1}]_i \tag{9.16.1}$$

where $]_i$ is used to denote the ith component of the $1 \times n$ vector $\mathbf{C}(z\mathbf{I} - \mathbf{A})^{-1}$. Recall from Section 9.7 that this component is equal to $G_i(z)$, where

$$G_i(z) = [\mathbf{C}(z\mathbf{I} - \mathbf{A})^{-1}]_i \quad \overset{z}{\longleftrightarrow} \quad \mathbf{g}_i \tag{9.16.2}$$

and $\mathbf{g}_i$ is the unit-pulse response from the ith state variable register to the output. Similarly,

$$\frac{\partial H(z)}{\partial c_i} = \frac{\partial}{\partial c_i}[\mathbf{C}(z\mathbf{I} - \mathbf{A})^{-1}\mathbf{B} + D] = [(z\mathbf{I} - \mathbf{A})^{-1}\mathbf{B}]_i = F_i(z), \tag{9.16.3}$$

where $F_i(z)$ is the z-transform of $\mathbf{f}_i$, the unit-pulse response from the input to the ith register. Also,

$$\frac{\partial H(z)}{\partial a_{ij}} = \frac{\partial}{\partial a_{ij}}\{\mathbf{C}(z\mathbf{I} - \mathbf{A})^{-1}\mathbf{B} + D\} = G_i(z)F_j(z). \tag{9.16.4}$$

Define the sensitivities in Eq. (9.16.4) as $S_{ij}(z)$ so that

$$S_{ij}(z) = G_i(z)F_j(z). \tag{9.16.5}$$

Using the Cauchy-Schwarz inequality in the form

$$\left[\frac{1}{2\pi}\int_{-\pi}^{\pi} |F(e^{j\theta})G(e^{j\theta})|\, d\theta\right]^2 \leqq \left[\frac{1}{2\pi}\int_{-\pi}^{\pi} |F(e^{j\theta})|^2\, d\theta\right]\left[\frac{1}{2\pi}\int_{-\pi}^{\pi} |G(e^{j\theta})|^2\, d\theta\right], \tag{9.16.6}$$

we can bound the sensitivities in (9.16.5). By Parseval's relation we have

$$K_{jj} = \|F_j(e^{j\theta})\|_2^2 = \frac{1}{2\pi}\int_{-\pi}^{\pi} |F_j(e^{j\theta})|^2\, d\theta, \tag{9.16.7}$$

$$W_{ii} = \|G_i(e^{j\theta})\|_2^2 = \frac{1}{2\pi}\int_{-\pi}^{\pi} |G_i(e^{j\theta})|^2 d\theta.$$

And so substituting in Eq. (9.16.6)

$$\|S_{ij}(\theta)\|_1^2 = \left[\frac{1}{2\pi}\int_{-\pi}^{\pi} |S_{ij}(\theta)|\, d\theta\right]^2 \leqq W_{ii}K_{jj}. \tag{9.16.8}$$

Thus

$$\sum_i \sum_j W_{ii}K_{jj} = \mathrm{Tr}(\mathbf{K})\mathrm{Tr}(\mathbf{W}) \geqq \sum_i \sum_j \|S_{ij}(\theta)\|_1^2. \tag{9.16.9}$$

For minimum roundoff noise filters Tr(**K**) is fixed by scaling and equals $n\delta^2$. Tr(**W**) is minimized. Thus inequality (9.16.9) is a good bound on the sensitivities. From (9.16.8) we can also write

$$\|S_{ij}\|_1 \leqslant \|G_i\|_2 \|F_j\|_2. \tag{9.16.10}$$

For a scaled digital filter, the L_2 norms $\|F_i\|_2, i = 1, 2, \ldots, n$ satisfy the scaling constraints, namely,

$$\delta\|F_i\|_2 = 1, \qquad i = 1, 2, \ldots, n. \tag{9.16.11}$$

Thus (assuming $\delta = 1$) Eq. (9.16.10) can be written as

$$\|S_{ij}\|_1 \leqslant \|G_i\|_2. \tag{9.16.12}$$

Thus the L_2 norms of $G_i(z)$ form an upper bound on the sensitivity norms $\|S_{ij}\|_1$. For the minimum roundoff noise structures, the optimality conditions imply that $\|G_1(z)\|_2 = \|G_2(z)\|_2 = \cdots = \|G_n(z)\|_2$, and so the upper bounds on the sensitivity norms are the same.

Equation (9.16.12) can also be interpreted as providing lower bounds on the roundoff noise in terms of the sensitivities of the coefficient in the scaled filter in the matrix **A**. Experience has shown these bounds to be reasonably tight (Jackson 1976). This means that low roundoff noise and low-coefficient sensitivity generally occur together in digital filters. There are several examples of this including the minimum roundoff noise realizations of Sections 9.9 and 9.10, the so-called normal form or uniform grid structure, and the Markel and Gray normalized lattice realization (Gray 1975).

Thus there is an intimate connection between the coefficient sensitivity functions and the transforms of the unit-pulse responses that define the scaling and noise gain in digital filters. We have used these relationships to bound the coefficient sensitivities in terms of L_2 norms of $G_i(z)$, $i = 1, 2, \ldots, n$.

Coefficient Quantization Effects Measured by Pole-Zero Movement

One popular method of measuring the effects of coefficient quantization is to examine the movement of poles and zeros caused by coefficient quantization (Kaiser 1966). Consider, for example, a direct form realization with transfer function $H(z)$ given by

$$H(z) = \frac{\sum_{k=0}^{m} b_k z^{-k}}{1 + \sum_{k=1}^{n} a_k z^{-k}} = \prod_{i=1}^{n} \frac{(1 - z_i z^{-1})}{(1 - p_i z^{-1})}, \tag{9.16.13}$$

where p_i and z_i are the poles and zeros of the filter, respectively. Changes in the pole locations can be measured by calculating the incremental change of some pole, say p_i, with respect to an incremental change in the coefficients in the direct form realization. Since the direct form realization is defined directly in terms of the coefficients $\{a_k, b_k\}$, consider the calculation of dp_i in terms of $da_1, \ldots, da_n$. Using the rule for a total differential, we can write

$$dp_i = \sum_{k=1}^{n} \left.\frac{\partial p_i}{\partial a_k}\right|_{z = p_i} da_k. \tag{9.16.14}$$

The partial derivatives in Eq. (9.16.14) can be found from

$$\frac{\partial p_i}{\partial a_k} = \frac{\partial \hat{a}/\partial a_k}{\partial \hat{a}/\partial p_i}, \tag{9.16.15}$$

where

$$\hat{a}(z) = 1 + \sum_{k=1}^{n} a_k z^{-k} = \prod_{i=1}^{n} (1 - p_i z^{-1}).$$

Thus for the case of distinct poles Eq. (9.16.14) reduces to

$$dp_i = -\sum_{k=1}^{n} \frac{p_i^{-k+1}}{\prod_{j=1}^{n} (1 - p_j p_i^{-1})} da_k \qquad (j \neq i). \tag{9.16.16}$$

Equation (9.16.16) expresses the incremental change in the pole p_i due to incremental changes in the coefficients $a_1, a_2, \ldots, a_n$. If the poles are close together, then the terms $(1 - p_j p_i^{-1})$ can be very small. The reciprocal values are large. And thus small changes in the coefficients a_i, $i = 1, 2, \ldots, n$ can cause large changes in the poles of the filter. (A similar argument can be carried out for the zeros of the filter.) Thus it behooves the filter designer to separate the poles (and zeros) of a filter. This is the reason for using cascade or parallel connections of smaller-order subfilters. One can use the subfilters to isolate closely packed poles and zeros into separate realizations.

In general, finite register effects become severe as poles of a filter cluster together. For a filter bandwidth, say W Hz, the ratio W/f_s, where f_s is the sampling frequency is a measure of this clustering. When W is a very small percentage of f_s, i.e., W/f_s is small, then finite register effects can be extremely important in the realization chosen to implement a given I/O characteristic. We have seen this in the roundoff noise performance and now again in terms of coefficient sensitivity.

9.17 THE DEADBAND EFFECT: CONSTANT INPUT LIMIT CYCLES

If the input to a fixed point digital filter is a constant, internal rounding errors are not uncorrelated random variables. In fact, these rounding errors are highly correlated. Thus the use of a white noise source to account for rounding errors is not valid in this case. We shall find that for recursive filters, there is no unique steady-state output for a constant input. There is a so-called deadband containing several possible steady-state outputs that depend on where the state first encounters the deadband region. This deadband or constant input limit cycle is caused by rounding the lowest-order bits under constant inputs. The nonlinear mapping that is the result of the quantization process produces the deadband limit cycles.

Deadband limit cycles depend on several parameters, including the filter realization or structure, the location of the filter poles, the quantization step q, the input to the filter, and the filter's initial state. Consider the evolution or trajectory of the state vector $\mathbf{x}(k)$ under a constant input u_0. The state evolves according to

$$\mathbf{x}(k+1) = \mathbf{A}\mathbf{x}(k) + \mathbf{B}u_0. \tag{9.17.1}$$

For a constant input, the solution to Eq. (9.17.1) is

$$\mathbf{x}(k) = \mathbf{A}^k\mathbf{x}(0) + \sum_{n=1}^{k} \mathbf{A}^{n-1}\mathbf{B}u_0. \tag{9.17.2}$$

And so the steady-state (SS) solution is

$$\mathbf{x}_{SS} = \sum_{l=1}^{\infty} \mathbf{A}^{l-1}\mathbf{B}u_0 = (\mathbf{I} - \mathbf{A})^{-1}\mathbf{B}u_0. \tag{9.17.3}$$

Let us assume for this discussion the input u_0 is zero so that $\mathbf{x}_{SS} = 0$. This is the unquantized solution. If the state vector is quantized, then as $\mathbf{x}(k)$ decays toward the origin, we shall find a value of $k = k_0$, say, for which the state enters a limit cycle with some period p. If, for example, $p = 1$, then the difference between $[\mathbf{A}[\mathbf{x}(k)]_Q]$ (the next value unquantized) and $[\mathbf{x}(k)]_Q$ (the present value quantized) is less than one-half the quantization step size in *all* dimensions of the state. That is,

$$|\mathbf{A}[x(k)]_Q - [x(k)]_Q| \leq \frac{q}{2}\mathbf{1}, \qquad k > k_0. \tag{9.17.4}$$

(We are assuming in this discussion that the quantization is the result of rounding to the nearest quantized value of the state.) If Eq. (9.17.4) is satisfied, then $[\mathbf{x}(k)]_Q$ cannot escape a hypercube of size $q/2$ containing $[\mathbf{x}(k)]_Q$. This is a shown schematically in Fig. 9.17.1.

Limit cycles of periods greater than 1 can also occur. Each realization of a filter possesses its own unique unquantized autonomous state vector trajectory. If we were to start two realizations at the same initial state and then plot their autonomous trajectories, we might obtain trajectories of the form shown in Fig. 9.17.2. Realization 1 possesses a state trajectory

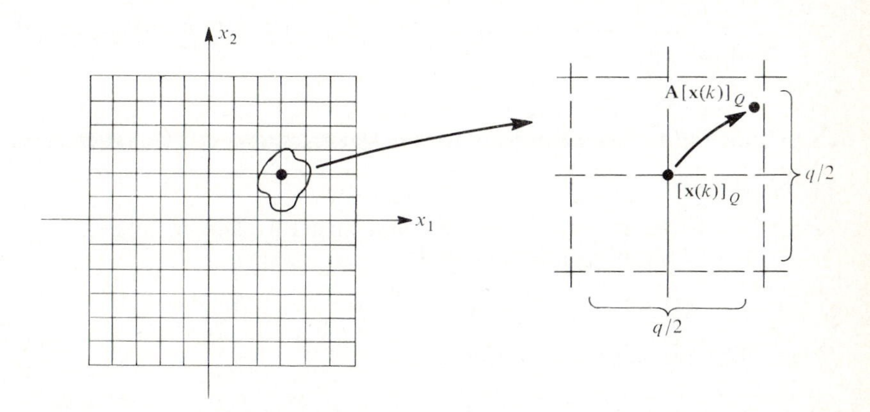

FIGURE 9.17.1 A two-dimensional example for a limit cycle of period $p = 1$.

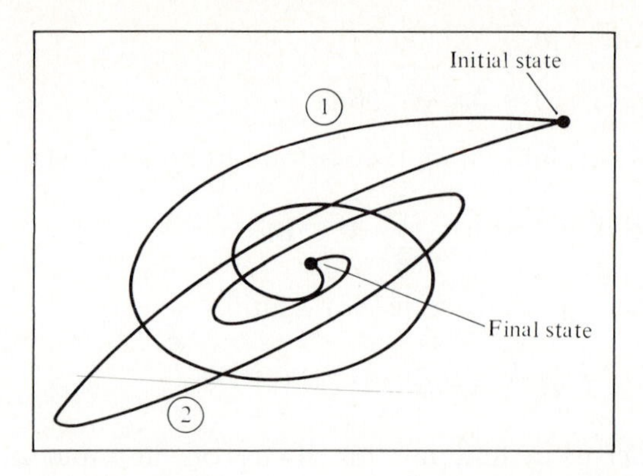

FIGURE 9.17.2 **The state trajectories of two realizations of a given I/O characteristic $H(z)$.**

that is nearly circular dying slowly into the origin. Realization 2, on the other hand, goes quickly toward the original but "overshoots" and then turns quickly back toward the origin. (These trajectories are roughly comparable to a normal form for 1 and a direct form for 2.)

The limit cycle corresponding to realization 1 likely possesses a period $p > 1$. The "horizontal" movement of the trajectory is much larger than the "vertical" movement in one portion of the trajectory. The reverse is true in the other portion of the trajectory. Thus one dimension is "trapped" while the other is not, thereby causing the trajectory to approximately circle the origin with some period p.

The direct calculation of deadband limit cycles is generally not possible by hand calculations except in the most elementary cases. Because of the nonlinear nature of the problem, limit cycles are generally found by direct simulation. Other methods can be used to obtain bounds (Jackson 1979; Long 1973; Parker 1971; Sandberg 1972) on limit cycles for various realizations of the filter.

The next example presents a simple one-dimensional case, that of a first-order IIR filter.

EXAMPLE 9.17.1

Deadband limit cycles for a first-order filter.

Consider the first-order filter structure shown in Fig. 9.17.3. This filter has the state equation given by

$$x(k + 1) = \beta x(k) + u_0.$$

The unquantized solution is

$$x(k) = \beta^k x(0) + u_0 \sum_{l=1}^{k} \beta^{l-1} = \beta^k \left[x(0) - \frac{u_0}{1 - \beta} \right] + \frac{u_0}{1 - \beta}.$$

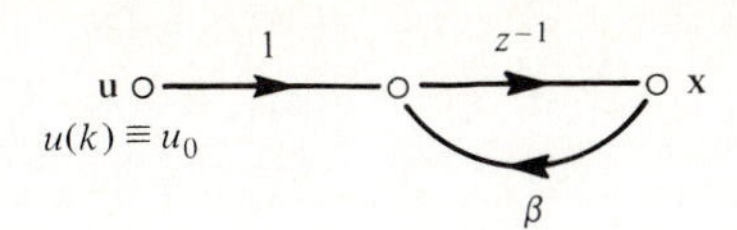

FIGURE 9.17.3.

As $k \to \infty$, $x(k) \to u_0/(1 - \beta)$, the unquantized steady-state solution. If the state $x(k)$ is quantized by rounding, as k increases, eventually the difference

$$\left|\beta[x(k)]_Q + u_0 - [x(k)]_Q\right|$$

becomes less than $q/2$. In this one-dimensional case, the trajectories of the filter's state are along a line. The period of the limit cycle is either 1 or 2, depending on the sign of β. Assume $\beta > 0$, which implies $p = 1$.

For zero input, the deadband region is included in the set of x that satisfies the equation

$$x = \beta x \pm \frac{q}{2}.$$

That is, the set of x such that

$$-\frac{q}{2} < x(1 - \beta) < \frac{q}{2}$$

or, equivalently,

$$\frac{-q}{2(1 - \beta)} < x < \frac{q}{2(1 - \beta)}. \tag{9.17.5}$$

The deadband region for a particular constant input u_0 and initial condition $x(0)$ is contained within the region

$$\left[\frac{-q}{2(1 - \beta)} + x_{ss}, \qquad \frac{q}{2(1 - \beta)} + x_{ss}\right].$$

If $\beta < 0$, then the deadband limits are obtained by replacing β with $|\beta|$; i.e., the limits are unchanged.

Suppose, for example, $u_0 = 5$, $\beta = 0.92$, and $q = 1$. Then the predicted deadband region is contained in the interval [56.25, 68.75]. Suppose the initial state is 72. Then we obtain the following sequence of states.

$x(0) = 72$, initial state

$x(1) = 71.24 \quad \Rightarrow \quad [x(1)]_Q = 71$

$x(2) = 70.32 \quad \Rightarrow \quad [x(2)]_Q = 70$

$$x(3) = 69.40 \quad \Rightarrow \quad [x(3)]_Q = 69$$
$$x(4) = 68.48 \quad \Rightarrow \quad [x(4)]_Q = 68$$
$$x(5) = 67.56 \quad \Rightarrow \quad [x(5)]_Q = 68 \text{ ("sticks" at 68)}$$

The filter's initial state $[x(0) = 72]$ lies above the deadband region. Hence the filter's state decays toward the steady-state solution $u_0/(1 - \beta) = 62.5$ but "sticks" at 68 (period $p = 1$), where the state first encountered the deadband region.

If the input u_0 is changed to some other constant, then the steady-state solution $u_0/(1 - \beta)$ changes. The extent of the possible deadband, however, is unchanged. If the sign of the feedback multiplier β is negative, then the state alternates signs with increasing n. The deadband limit cycle now oscillates with period $p = 2$ about the steady-state solution. The two possible states in this oscillatory mode are again contained with the deadband region

$$\left[\frac{-q}{2(1 - \beta)} + x_{SS}, \qquad \frac{q}{2(1 - \beta)} + x_{SS}\right].$$

Even in this simple one-dimensional example, there are several possible modes of operation for the filter. In higher-order filters, a complete analytical description is most likely not possible. In these cases, one can consider possible bounds for a given structure or one can simulate the filter over the possible range of input parameters and determine the limit cycle behavior of the filter experimentally.

One approach to the reduction of deadband limit cycles is to add a small amount of random noise to the state $\mathbf{x}$ at each iteration. The added noise causes the unquantized state $\mathbf{x}$ to vary (or "dither"). Eventually the added noise will change the unquantized state sufficiently so that rounding causes the state to be further reduced, eventually driving the state toward the steady-state value. Because of the added noise, the true steady state can never be reached. The filter's state will vary about the true state with a variance dependent on the variance of the added noise. This method of reducing the effects of limit cycles has been used with good success for many years in feedback control systems (Truxal 1955).

Another method that has been used to eliminate deadband limit cycles is to quantize using magnitude truncation instead of rounding. Thus, for example, if the unquantized state has a value 67.56 magnitude truncation produces the value 67. While rounding gives 68. as the quantized value, Fig. 9.17.4 depicts the state trajectories for magnitude truncation and rounding on the quantized state for a second-order normal filter with poles at $\lambda_1, \lambda_2 = 0.9 \angle \pm 10°$. The filter's initial state is $[0.75, \quad 75]^T$. Notice that in the case of rounding there is an approximately circular deadband region about x_{SS}, while in the case of magnitude truncation this limit cycle is eliminated.

For the same initial conditions and filter poles a direct form filter has quantized state trajectories, as shown in Fig. 9.17.4. In the case of a direct form, the deadband region under rounding is a line at 45° through the origin. Notice that for these poles (near $z = 1$), the direct form has a larger limit cycle than does the normal form. Also shown in Fig. 9.17.4 is the quantized state trajectory under magnitude truncation. Again we see that magnitude truncation eliminates zero input limit cycles.

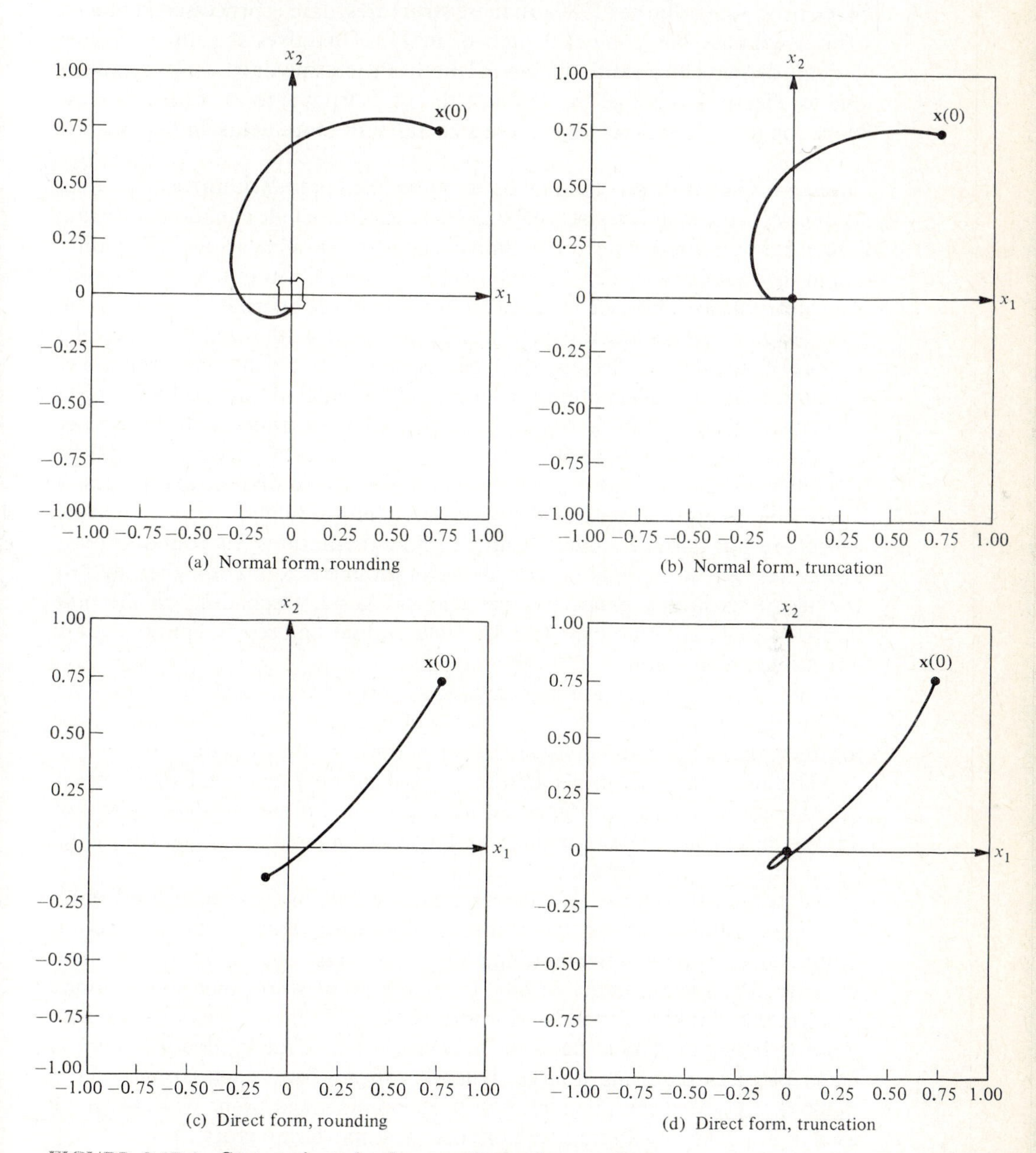

FIGURE 9.17.4 State trajectories for rounding and truncation.

Magnitude truncation does not always eliminate deadband limit cycles. The noise variance caused by truncation of internal products is also larger than for the case of rounding (Rabiner 1975). In addition, one must check each product to ascertain the sign of the number to apply the correct quantization.

Another method of eliminating deadband limit cycles is to use so-called block processing (Barnes 1980). In these structures, data is processed in blocks. (The usual case has a block length of unity.) This gives the filter designer another degree of freedom, the block length. Processing data in blocks allows one to effectively change the poles of the filter from λ to λ^L where L is the block length. We shall discuss these structures in more detail in Chapter 10.

Summary • The synthesis of digital filters using fixed point arithmetic must take account of finite length register effects. These effects include roundoff noise due to internal roundings of products, limit cycles due to nonlinearities in rounding and in the manner in which internal overflows are carried out, and coefficient truncation effects. These effects are all a function of the internal structure of the filter. Given an I/O characteristic $H(z)$, a filter can be synthesized to reduce one or more of these effects. Because these effects depend on the internal structure of the filter, a natural description for their study is the SVD or the FSVD.

The direct form filter has long been a popular realization in both practice and theory. In practical implementations, it possesses the fewest multipliers. In theory, the canonical form is simple and easy to manipulate and possesses many interesting mathematical properties. Unfortunately, its finite register effects are not good. It suffers from overflow oscillations. Its roundoff noise properties can be especially poor for poles grouped closely together, and its coefficient sensitivity properties are also not good. Depending on the pole locations of $H(z)$, some care and foresight should be used in employing the direct form realization.

The theory presented in this chapter suggests there are many alternatives to the direct form. Whenever the poles of a given $H(z)$ are grouped close together, these alternatives should be considered in some detail.

The next chapter on structures (or realizations) of digital filters offers additional alternatives to the direct form and the minimum noise structures derived here. The problems of this and the next chapter also develop additional structures for digital filtering.

We remind the reader that we have limited this discussion to fixed point arithmetic. Many of the problems discussed here are reduced in floating point arithmetic. However, if the floating point word length including the exponent possesses the same number of bits as a fixed point word, then output noise performance ranking depends on several factors. There are cases where fixed point is better than floating point and vice versa. Usually, floating point is better in cases requiring a large dynamic range. However, floating point is significantly more complicated than fixed point. Fixed point will always be used in applications where speed or cost are important factors.

In general, finite register effects become severe as poles of a filter cluster together. For a filter bandwidth, say W Hz, the ratio W/f_s, where f_s is the sampling frequency is a measure of this clustering. When W is a very small percentage of f_s, i.e., W/f_s is small, then finite register effects can be extremely important in the realization chosen to implement a given I/O characteristic. We have seen this in roundoff noise performance and in terms of coefficient sensitivity.

PROBLEMS

9.1 A random variable is passed through a uniform quantizer with an input-output characteristic as shown below. If **u** is uniformly distributed on $[-\Delta, \Delta]$, what is $[E(u - y)^2]$?

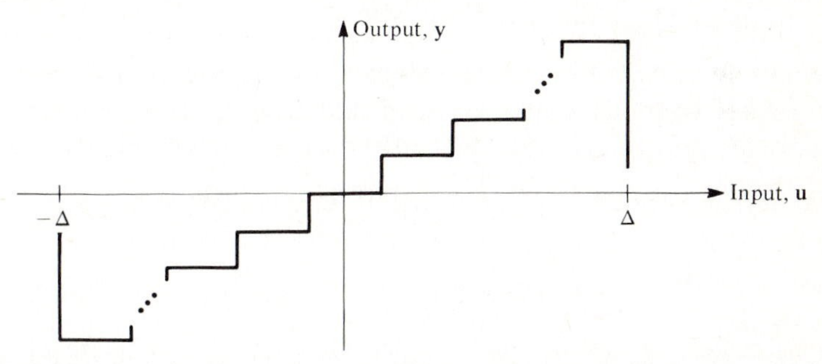

FIGURE 9P.1.

9.2. An input analog signal **u** is Gaussian with zero mean and variance σ^2. Find the mean-square quantization noise if 3 bit quantization is used over a dynamic range of $\pm 3\sigma$. Assume uniform quantization steps with each interval represented by the middle value of the interval.

9.3. The probability density function for the values of an analog signal are as shown below. A 2 bit uniform quantizer is used to quantize the samples of **u**. The middle value in each quantization interval is used to represent that interval. Find the signal to quantization noise power ratio where the signal power is the sum of the squared sampled values (quantized).

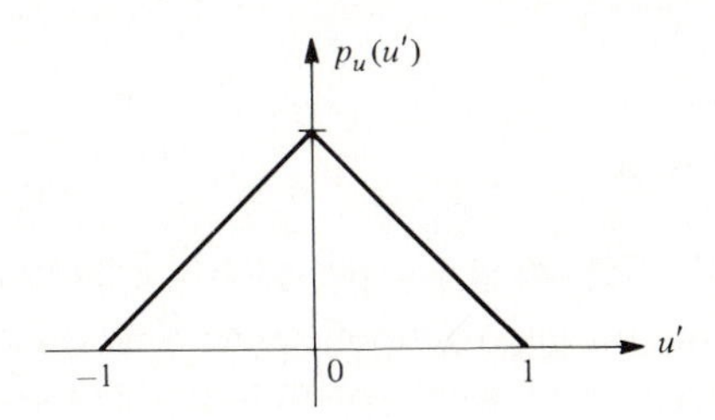

FIGURE 9P.3.

9.4. The input to an analog-to-digital converter is equally likely to fall between 1 and -1 V. Find the number of bits needed to produce a 70 dB signal to quantization noise ratio.

9.5. How many bits/register are needed in a digital audio system to ensure a dynamic range of over 90 dB?

9.6. Consider a second-order direct form filter with poles at $\frac{3}{4} \pm \frac{1}{2}j$. Find the region in the state space grid for which the initial state decays to the origin (for zero input). Find the fixed points in the state space grid to which the initial state converges after an overflow. What periods of overflow oscillation are possible? (Hint: The period can be greater than 1.)

9.7. A digital filter has a system matrix $\mathbf{A}$ that satisfies $\mathbf{Q} = r^2\hat{\mathbf{D}} - \mathbf{A}^T\hat{\mathbf{D}}\mathbf{A} > \mathbf{0}$, where $\hat{\mathbf{D}} = \text{diag}\{d_1, \ldots, d_n\} > 0$. Show that any nonsingular diagonal transformation of the state $\mathbf{x}' = \mathbf{T}^{-1}\mathbf{x}$, results in a matrix $\mathbf{A}'$, which also satisfies $\mathbf{Q}' = r^2\hat{\mathbf{D}} - \mathbf{A}'^T\hat{\mathbf{D}}\mathbf{A}' > \mathbf{0}$.

9.8. Show that if $\mathbf{A}$ has eigenvalues with magnitude less than unity and there exists $\hat{\mathbf{D}}$ as in Eq. (9.4.12) such that $\mathbf{Q} > \mathbf{0}$, this is equivalent in the second-order case to a) $a_{12}a_{21} \geq 0$ or b) if $a_{12}a_{21} < 0$, then $|a_{11} - a_{22}| + \det \mathbf{A} < 1$.

9.9. The *storage efficiency* of a digital filter is defined as

$$e(\mathbf{K}) = \left[\frac{\text{Det } \mathbf{K}}{\prod_{i=1}^{n} K_{ii}}\right]^{1/2}.$$

Roughly speaking, the storage efficiency is a measure of how well a filter realization uses the available 2^{2B} states in the state-space-grid. Find the storage efficiency for:

a) A second-order direct form as shown below.
b) An FIR filter as shown below. Discuss your results.

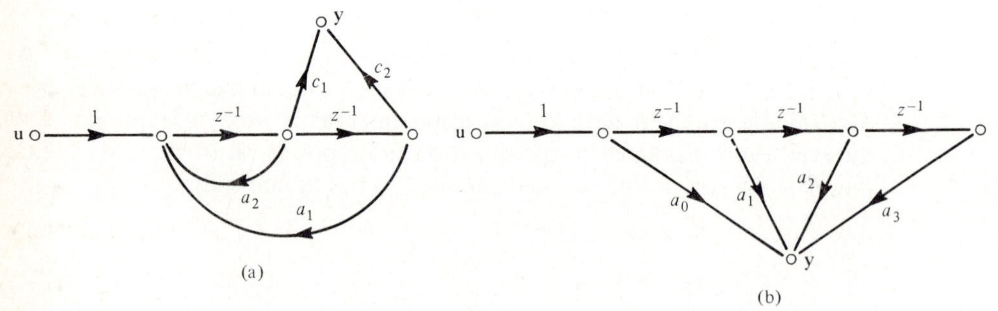

FIGURE 9P.9.

9.10. Consider the following second-order digital filter.

a) Find the output roundoff noise by explicitly finding the unit pulse responses from internal accumulation nodes to the output.
b) Find the output roundoff noise by calculating the $\mathbf{W}$ matrix for this filter. Then use Eq. (9.7.16) to find the output roundoff noise. Which method of calculation do you prefer for this example?

c) Assuming the parameters of the filter can be any real numbers you desire, is this filter l_2 scaled? Explain.

d) If the parameter b_1 were constrained to be unity, is this filter l_2 scaled? Explain.

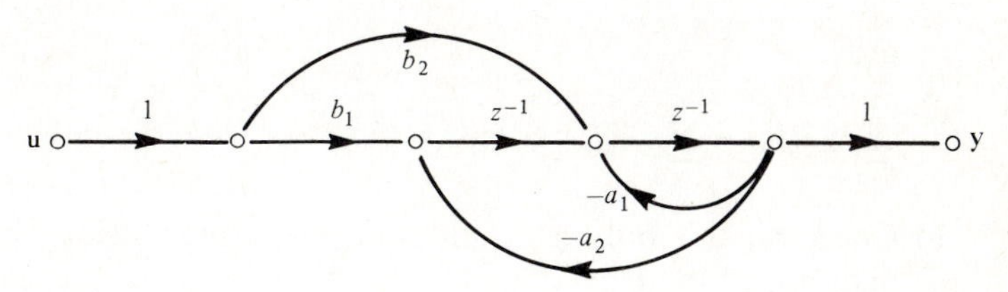

FIGURE 9.P.10.

9.11. Consider the following first-order digital filter.

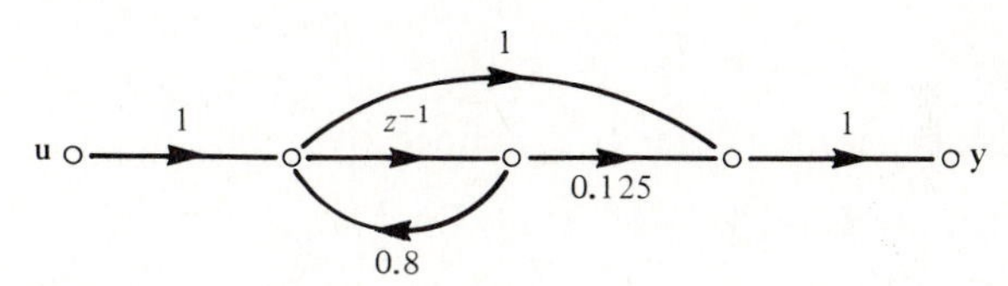

FIGURE 9.P.11.

Assume the unit delay register contains 8 bits and all registers represent numerical values that lie between +1 and −1.

a) Calculate the quantization step size q.

b) Assume **u** is a white Gaussian noise process with zero mean and unity variance. Find the output noise power due to roundoff.

c) For an input **u** with variance $\frac{1}{9}$, what is the probability the variables in the *unit delay register* and the *output register* will overflow?

d) Use l_2 scaling and scale the filter for $\delta = 5$. Redraw the original SFG to include this scaling process. Check your scaling by explicitly calculating the covariance matrix for the scaled filter. What should the diagonal elements of the covariance matrix equal?

9.12. A digital filter with transfer function $H(z) = \dfrac{z^2}{z^2 + z + 0.05}$ has a PSFG as shown in Fig. 9P.12.

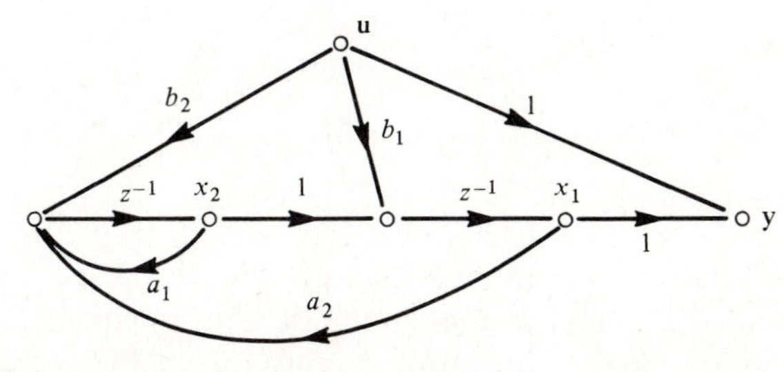

FIGURE 9.P.12.

a) Find the branch values of the PSFG so that the I/O characteristic is the given $H(z)$ and the filter is scaled in the sense that for

$$u(k) = \begin{cases} 1, & k = 0 \\ 0, & k \neq 0 \end{cases}$$

then

$$\sum_{k=0}^{\infty} x_1^2(k) = \sum_{k=0}^{\infty} x_2^2(k) = 1.$$

b) Compute the output roundoff noise power using the branch values determined in a).

9.13. Let the transfer function of a digital filter be $H(z)$ and consider the frequency transformation $F(z) = (z - \alpha)/(1 - \alpha z)$, $|\alpha| < 1$. In terms of state variable $H(z)$ is expressed as

$$H(z) = \mathbf{C}(z\mathbf{I} - \mathbf{A})^{-1}\mathbf{B} + \mathbf{D}.$$

Consider another filter $G(z)$ defined by

$$G(z) = \mathbf{C}'(z\mathbf{I} - \mathbf{A}')^{-1}\mathbf{B}' + \mathbf{D}',$$

where

$$\mathbf{A}' = (\alpha\mathbf{I} + \mathbf{A})(\mathbf{I} + \alpha\mathbf{A})^{-1},$$
$$\mathbf{B}' = (1 - \alpha^2)^{1/2}(\mathbf{I} + \alpha\mathbf{A})^{-1}\mathbf{B},$$
$$\mathbf{C}' = (1 - \alpha^2)^{1/2}\mathbf{C}(\mathbf{I} + \alpha\mathbf{A})^{-1},$$
$$\mathbf{D}' = \mathbf{D} - \alpha\mathbf{C}(\mathbf{I} + \alpha\mathbf{A})^{-1}\mathbf{B}.$$

Show that

a) $G(z) = H(F(z))$, i.e., $G(z)$ is a frequency transformed filter derived from $H(z)$.
b) $\mathbf{K}' = \mathbf{K}$, $\mathbf{W}' = \mathbf{W}$, i.e., under the frequency transformation $F(z)$, $\mathbf{K}$ and $\mathbf{W}$ are invariant.
c) What is the significance of b)?

9.14. The bilinear approximation of a fourth-order lowpass Butterworth filter results in the following poles and zeros for two cascaded second-order sections:

	Poles		Zeros	
Section	Modulus	Angle	Modulus	Angle
1	0.52507	$\pm 10.65^\circ$	1	180° (2 zeros)
2	0.64251	$\pm 27.19^\circ$	1	180° (2 zeros)

$\mathbf{u} \rightarrow \boxed{H_1(z)} \rightarrow \boxed{H_2(z)} \rightarrow \mathbf{y}$, $\quad H(z) = gH_1(z)H_2(z)$

FIGURE 9P.14.

a) Find the minimum roundoff noise filters for these two sections designed in isolation. Normalize the dc gain so that $H(1) = 1.0$.
b) Find the output roundoff noise for the filter sections in isolation.

c) Find the lowest output roundoff noise for the cascaded filters. (Find the best cascade connection.)
d) Check the scaling of the cascaded sections. Rescale if necessary and recompute the output roundoff noise.

9.15. Show that if

$$\mathbf{x}(k) = \mathbf{A}\,\mathbf{x}(k-1),$$
$$y(k) = \mathbf{C}\,\mathbf{x}(k),$$

then

$$\sum_{k=0}^{\infty} y^2(k) = \mathbf{x}^T(0)\,\mathbf{W}\,\mathbf{x}(0),$$

where

$$\mathbf{W} = \mathbf{A}^T\mathbf{W}\mathbf{A} + \mathbf{C}^T\mathbf{C}.$$

9.16. The transfer function of a digital filter $H(z)$ can be expressed as

$$H(z) = D + \mathbf{C}(z\mathbf{I} - \mathbf{A})^{-1}\mathbf{B}.$$

Show that

$$\sum_{k=0}^{\infty} h^2(k) = D^2 + \mathbf{B}^T\mathbf{W}\mathbf{B},$$

where $\mathbf{W}$ is defined in Problem (9.15).

9.17. Let

$$\mathbf{A} = \begin{bmatrix} 0 & 1 \\ -\frac{1}{2} & -1 \end{bmatrix}, \qquad \mathbf{B} = \begin{bmatrix} 0 \\ 1 \end{bmatrix}.$$

On the same graph, plot the following:

a) The boundary of the unit cube: $\{\mathbf{x}: \quad |x_i| \leqslant 1, \quad i = 1, 2\}$.
b) Eight points of the unit-pulse response: $\mathbf{A}^k\mathbf{B}, \quad 0 \leqslant k \leqslant 7$.
c) The image of the unit cube: $\{\mathbf{Ax}: \quad |x_i| \leqslant 1, \quad i = 1, 2\}$.
d) The error ellipsoid: $\{\mathbf{x}: \quad \mathbf{x}^T\mathbf{K}^{-1}\mathbf{x} \leqslant 1\}$.

9.18. The state of a filter evolves according to

$$\mathbf{x}(k+1) = \mathbf{A}\mathbf{x}(k) + \mathbf{B}u(k).$$

Define $\boldsymbol{\phi}(z)$ and $Q(\theta)$ as follows:

$$\boldsymbol{\phi}(z) = (z\mathbf{I} - \mathbf{A})^{-1}\mathbf{B},$$
$$\mathbf{Q}(\theta) = \boldsymbol{\phi}(e^{j\theta})\boldsymbol{\phi}^*(e^{j\theta}).$$

Show that if the input to the filter is given by

$$u(k) = \sum_{l=1}^{m} \alpha_l \cos(k\theta_l + \beta_l),$$

then

$$\mathbf{x}^*(k)\mathbf{K}^{-1}\mathbf{x}(k) \leq 1$$

where

$$\mathbf{K} = m \sum_{l=1}^{m} \left(\frac{\alpha_l}{2}\right)^2 [\mathbf{Q}(\theta_l) + \mathbf{Q}(-\theta_l)]$$

is assumed to be nonsingular.

9.19. Error-feedback is a technique for reducing roundoff noise in IIR filters by increasing the effective word length of the filter. Shown in Fig. 9.P.19 are two error-feedback state space structures

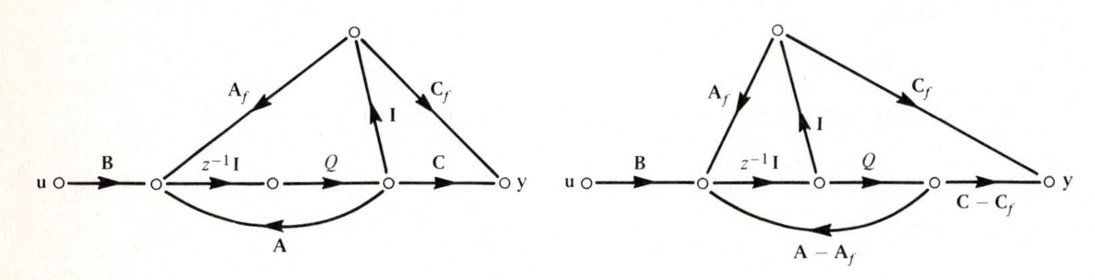

FIGURE 9.P.19. **Two equivalent error-feedback state space filters.**

a) Show they are equivalent structures. The vector $\boldsymbol{\xi}(k)$ contains the quantization errors at each state variable, i.e.,

$$\boldsymbol{\varepsilon}(k) = \mathbf{x}(k) - [\mathbf{x}(k)]_Q.$$

b) Show the noise gain matrix $\hat{\mathbf{W}}$ of the error-feedback structure is

$$\hat{\mathbf{W}} = \hat{\mathbf{C}}^T\hat{\mathbf{C}} + \sum_{k=1}^{\infty} (\mathbf{C}\mathbf{A}^{k-1}\hat{\mathbf{A}})^T(\mathbf{C}\mathbf{A}^{k-1}\hat{\mathbf{A}}),$$

where

$$\hat{\mathbf{C}} = \mathbf{C} - \mathbf{C}_f, \qquad \hat{\mathbf{A}} = \mathbf{A} - \mathbf{A}_f.$$

c) Show that the diagonal elements of $\hat{\mathbf{W}}$ are

$$\hat{W}_{ii} = (\mathbf{a}_i - \boldsymbol{\alpha}_i)^T\mathbf{W}(\mathbf{a}_i - \boldsymbol{\alpha}_i) + (c_i - \xi_i)^2, \qquad i = 1, 2, \ldots, n$$

where $\mathbf{a}_i$ and $\boldsymbol{\alpha}_i$ are the ith columns of $\mathbf{A}$ and $\mathbf{A}_f$, c_i and ξ_i are the ith entries of $\mathbf{C}$ and $\mathbf{C}_f$, respectively.

d) From c), it is clear we can optimize each $\hat{W}_{ii}$, $i = 1, 2, \ldots, n$ independently. What is the best choice for ξ_i, $i = 1, 2, \ldots, n$? The best choice of the α_i, $i = 1, 2, \ldots, n$ results in a quadratic optimization problem. Generally, only integer values for α_i are considered from the set $\{-2, -1, 0, 1, 2\}$. This choice of coefficients for the α_i simplifies the hardware resources needed to implement the feedback.

9.20. In an error-feedback structure if $\mathbf{A} = \mathbf{A}_f$ and $\mathbf{C} = \mathbf{C}_f$, then show that the overall realization is a conventional state space realization in which the multiplications are carried out in two parts. In view of this previous statement, what advantage can we hope to obtain with an error feedback structure?

9.21. Consider a second-order filter with $(\mathbf{K}, \mathbf{W})$ matrices given by

$$\mathbf{K} = \begin{bmatrix} k_0 & k_1 \\ k_1 & k_2 \end{bmatrix}, \qquad \mathbf{W} = \begin{bmatrix} w_0 & w_1 \\ w_1 & w_2 \end{bmatrix}.$$

Define two ellipses by $\mathbf{x}^T\mathbf{K}^{-1}\mathbf{x} = 1$ and $\mathbf{x}^T\mathbf{W}^{-1}\mathbf{x} = 1$, where $\mathbf{x} = [x, y]^T$. Then we can write equations for these two ellipses as

$$k_2x^2 + k_0y^2 - 2k_1xy = \Delta_k,$$
$$w_2x^2 + w_0y^2 - 2w_1xy = \Delta_w.$$

If we plot these two ellipses, they might look like the following:

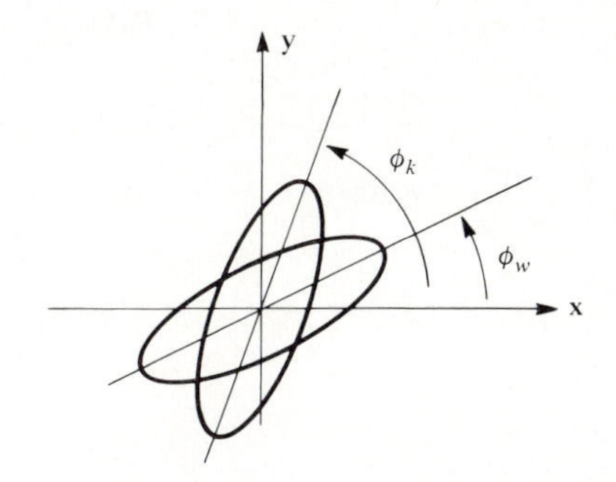

a) Show that

$$\tan(2\phi_k) = \left(\frac{2k_1}{k_0 - k_2}\right), \qquad \tan(2\phi_w) = \left(\frac{2w_1}{w_0 - w_2}\right).$$

b) What orientation of these two ellipses produces the minimum noise structure? What orientation produces the worst case noise structure?

9.22. A normal form structure is sometimes called an antisymmetric structure since $a_{11} = a_{22}$ and $a_{12} = -a_{21}$ in the second-order case (the a_{ij} are the coefficients in the system matrix $\mathbf{A}$). If $(\mathbf{A}_0, \mathbf{B}_0, \mathbf{C}_0)$ is an unscaled normal structure as given in Eq. (9.12.11), then find a transformation that produces an l_2-scaled normal form. Notice that a diagonal transformation cannot be used. (Hint: If $\mathbf{T}$ is normal, then $\mathbf{T}^{-1}\mathbf{A}_0\mathbf{T} = \mathbf{A}_0$.)

9.23. Show that the noise gain of a second-order normal filter can be written in the form

$$\frac{r^2}{2}\left[\frac{1}{1 - |\lambda|^2}\right],$$

where r is given by Eq. (9.12.13) and λ is a complex pole of the filter.

9.24. Show that the noise gain of a minimum noise, second-order filter is

$$\text{Noise gain (MN)} = \text{Noise gain (Normal)} - \frac{r^2}{2}\cos^2(\phi - \xi)\left|\frac{1}{1-\lambda^2}\right|^2,$$

where ϕ and ξ are given in Eqs. (9.12.13) and (9.12.23), respectively. The noise gain of the normal structure is given in Problem 9.23.

9.25. For a second-order filter show that if $\mathbf{A}$ is normal and

$$\mathbf{B}_0 = b\begin{bmatrix}\cos\beta \\ \sin\beta\end{bmatrix}, \qquad \mathbf{C}_0 = c[\cos\gamma, \sin\gamma],$$

then $\mathbf{K}$ and $\mathbf{W}$ can be expressed as

$$\mathbf{K} = \frac{b^2}{2}\begin{bmatrix}\dfrac{1}{1-|\lambda|^2} + \mathrm{Re}(\psi_\beta) & \mathrm{Im}(\psi_\beta) \\ \mathrm{Im}(\psi_\beta) & \dfrac{1}{1-|\lambda|^2} - \mathrm{Re}(\psi_\beta)\end{bmatrix},$$

$$\mathbf{W} = \frac{c^2}{2}\begin{bmatrix}\dfrac{1}{1-|\lambda|^2} + \mathrm{Re}(\psi_\gamma) & -\mathrm{Im}(\psi_\gamma) \\ -\mathrm{Im}(\psi_\gamma) & \dfrac{1}{1-|\lambda|^2} - \mathrm{Re}(\psi_\gamma)\end{bmatrix},$$

where

$$\psi_\beta = e^{j2\beta}/(1-\lambda^2), \qquad \psi_\gamma = e^{-2j\gamma}/(1-\lambda^2)$$

and λ, λ^* are the eigenvalues of $\mathbf{A}$.

9.26. From $\mathbf{K}$ and $\mathbf{W}$ in Problem 9.25 show that the noise gain of a normal structure can be expressed as

$$\text{Noise gain (Normal)} = \frac{b^2c^2}{4}\left\{2\left(\frac{1}{1-|\lambda|^2}\right)^2 + 2\,\mathrm{Re}(\psi_\beta)\,\mathrm{Re}(\psi_\gamma)\right\}.$$

9.27. If the transfer function of an nth-order filter is given by

$$H(z) = \frac{\hat{b}(z)}{\prod_{i=1}^{n}(z-\lambda_i)}, \qquad \lambda_i \text{ all distinct},$$

then show that

$$M_g^n = [\det \mathbf{KW}]^{1/2} = \prod_{i=1}^{n}\left|\frac{\hat{b}(\lambda_i)}{\prod\limits_{\substack{j=1\\ i\neq j}}^{n}(1-\lambda_i\lambda_j)}\right|.$$

This formula allows one to calculate M_g^2, the noise gain of an nth-order filter.

9.28. Read the example of Section 9.15 and design a sixth-order filter of normal, second-order sections in cascade. Design each section in isolation using l_2 scaling. Calculate the overall **K** and **W** of the cascade and determine the noise gain of this structure and compare these results to those given in Section 9.15. [In order to l_2 scale a normal filter $\mathbf{A}_0$, $\mathbf{B}_0$, $\mathbf{C}_0$ as in Eq. (9.12.11) use a normal transformation.]

9.29. The sensitivity of the eigenvalues of the system matrix **A** is sometimes used as a measure of coefficient sensitivity. That is, we find $\partial\lambda_i/\partial a_{jk}$. Show that for the optimal second-order structure these partial derivatives are:

$$\frac{\partial\lambda_i}{\partial a_{11}} = \frac{\partial\lambda_2}{\partial a_{11}} = \frac{\partial\lambda_1}{\partial a_{22}} = \frac{\partial\lambda_2}{\partial a_{22}} = \frac{1}{2},$$

$$\frac{\partial\lambda_1}{\partial a_{12}} = -\frac{\partial\lambda_2}{\partial a_{12}} = -\frac{1}{2}\psi,$$

$$\frac{\partial\lambda_1}{\partial a_{21}} = -\frac{\partial\lambda_2}{\partial a_{21}} = -\frac{1}{2}j\psi.$$

ψ^2 is given by $[P + Q/(P - Q)]$ and P and Q are defined from the transfer function of the filter

$$H(z) = \frac{a}{z - \lambda} + \frac{a^*}{z - \lambda^*}; \quad P = \frac{|a|}{1 - |\lambda|^2}; \quad Q = \mathrm{Im}\left(\frac{a}{1 - \lambda^2}\right).$$

[See Barnes (1984).]

9.30. A global measure of sensitivity is

$$S(\lambda_i, \mathbf{A}) = \left[\sum_{n,m}\sum \left|\frac{\partial\lambda_i}{\partial a_{nm}}\right|^2\right]^{1/2}.$$

a) Show $S(\lambda_i, \mathbf{A}) \geqslant 1$ for any realization.
b) Show

$$S(\lambda_1, \mathbf{A}) = S(\lambda_2, \mathbf{A}) = \frac{1}{2}\left(2 + \psi^2 + \frac{1}{\psi^2}\right)^{1/2}$$

for minimum noise second-order structure.
c) Show

$$S(\lambda, \mathbf{A}) = \frac{P}{\sqrt{P^2 - Q^2}}.$$

(See Problem 9.29 for definitions of ψ, P, and Q.) For the normal form show that $S(\lambda, \mathbf{A})$ obtains its lower bound of unity.

CHAPTER 10

Digital Processing Structures

10.1 Introduction

We have shown in Chapters 5, 8, and 9 that for a given input-output function, there are many ways to break down the task into primitive arithmetic operations. In Chapter 9 we used the freedom in the choice of a structure or algorithm to minimize the adverse effects of finite-length registers. In Chapter 5 we used this freedom in the choice of an algorithm to minimize the number of operations in the calculation of the DFT. There are many structures for a given digital processing problem. What are the fundamental principles in the design of one? How do we design an algorithm in a given application? In this chapter we shall consider these questions as they apply to digital filtering algorithms. However, the underlying principles apply to digital processing algorithms in general.

A structure or algorithm contains the necessary information to characterize the computations that result in a given I/O function. This information includes the definition of internal variables, the order that internal variables are computed, the order in storing and retrieving internal variables, etc. The algorithm can be specified in different ways as we have previously discussed in Chapter 8. These different methods include PSFGs, factored state variable descriptions, and program listings.

The first principle in designing an algorithm is to *identify the constraints imposed by the available computational resources.* We are always restricted to some computational resource, such as a general purpose computer or, perhaps, a small processor. The design of an algorithm is fundamentally influenced by the computational resources available. The problem in algorithm design is to optimize criteria that derive from the available computational resources.

For example, in the computation of the DFT discussed in Chapter 5, we make the implicit assumption that the computation is carried out on a general purpose computer with a sequential or Von Neumann architecture. For this computational resource, the criterion we wish to optimize is the number of operations (and especially multiplications). Why? Because on a sequential machine, the number of operations is directly proportional to the time required to compute the DFT. Suppose in the computation of the DFT the computational resource is an array of microprocessors. The best algorithm in this case is probably not the algorithm that minimizes the number of computations. It is probably more useful to design a highly parallel algorithm and in this way increase total data throughput.

Another example is the computation within a narrow-band, low-pass, digital filter which is needed in a special purpose instrument, say, a spectrum analyzer. In this case, the computational resource is dedicated, special-purpose hardware. Since the filter is incorporated into the instrument, there is no requirement for flexibility in the final algorithm. The criteria imposed in this example may very well include algorithms with good finite register effects. In order to reduce complexity, we may be able to choose filter coefficients that are power-of-2 coefficients. Because we are using dedicated hardware, the criteria we choose to optimize can be much different than that in the case where our computational resource is a general purpose computer. If we restrict our study to large general purpose computers, we are not motivated to dig deeply into algorithmic design. The primary concern for large general purpose (Von Neumann type) computers is merely the number of computations performed per output sample.

What are examples of criteria we wish to optimize in the design of digital filters? We cannot consider all possible situations involving special kinds of computational resources. However, the criteria natural to consider in terms of the examples of this chapter include:

1. Data throughput;
2. Complexity of the algorithm;
3. Finite-register effects;
4. The "geometry" of the algorithm, which includes such things as the length of data paths, the memory organization, modularity and regularity of the design, bus arrangements for ground and power, local connectivity, etc.;
5. Adaptability of the algorithm.

The computational resources available for a given task may include one or more basic computational modules. For example, in Chapter 8 a sum of products of the form shown in Fig. 10.1.1 is considered the basic computation.

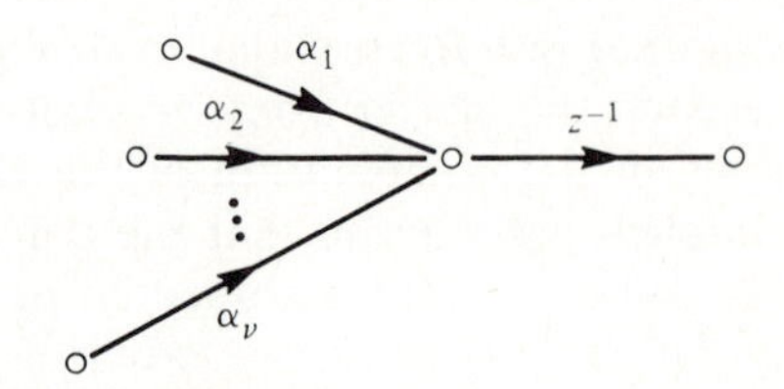

FIGURE 10.1.1 **A sum of products.**

As pointed out in Chapter 8, any digital filter can be formulated in terms of one or more sums of products followed by a simple updating process. The constants in the sum of products modules define the filter characteristic. One could also envision other types of primitives and then construct algorithms based on those modules. For example, some authors have proposed the use of a rotation as the basic computational module. A schematic is shown in Fig. 10.1.2. In this case, the input and output to the computational modules are two element vectors. The computational modules perform simple rotations on the input vector to obtain the output vector. Lattice filters are an example where rotation modules can be used to synthesize a filter. One could, of course, break the rotation module down into simpler primitives. However, depending on the technology used in constructing the computational units, some modules may be "natural" for a given technology.

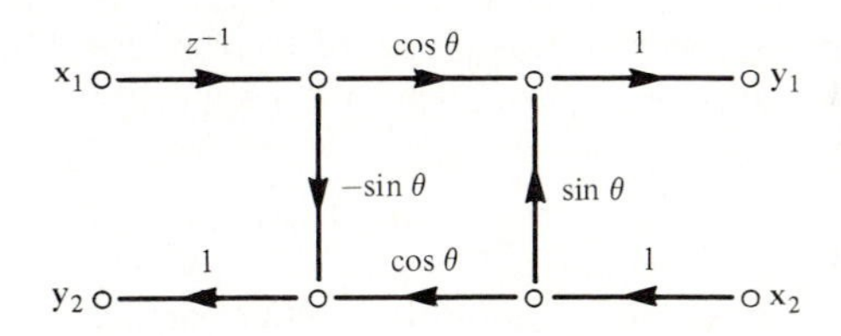

FIGURE 10.1.2 **A rotation module.**

Hardware Implementations for a Sum of Products

The hardware implementation for a module can be an important part in the choice of a basic computational unit. If an implementation is particularly simple in terms of its hardware, then it is a good candidate for a module. An example of this is the use of distributed arithmetic (Burrus 1977; Peled 1974; Zohar 1973) in implementing a sum of products.

Let's consider the implementation of a sum of products in more detail Each product consists of a variable multiplied by a known constant, and thus the basic calculation is of the form

$$y = \sum_{i=1}^{L} \alpha_i x_i = \boldsymbol{\alpha}^T \mathbf{x}, \tag{10.1.1}$$

where $\boldsymbol{\alpha}$ is a column vector of constants and $\mathbf{x}$ is a column vector of variables. Assume that Eq. (10.1.1) is computed digitally with finite length words of B bits. The constraint of a finite word length means there are only a finite number of possible outcomes for y. Assuming that each x_i is represented in unsigned binary, we have

$$x_i = \sum_{j=0}^{B-1} x_i^j 2^{-j}, \qquad x_i^j = 0 \text{ or } 1. \tag{10.1.2}$$

If we assume a signed two's complement representation for x_i, then x_i^0 is the sign bit and the corresponding representation is

$$x_i = -x_i^o + \sum_{j=1}^{B-1} x_i^j 2^{-j}. \tag{10.1.3}$$

In Eq. (10.1.2) x_i^j is the jth bit of x_i. Substituting Eq. (10.1.2) in Eq. (10.1.1), we have

$$y = \sum_{i=1}^{L} \alpha_i \sum_{j=0}^{B-1} x_i^j 2^{-j} \quad = \sum_{j=0}^{B-1} 2^{-j}\left(\sum_{i=1}^{L} \alpha_i x_i^j\right). \tag{10.1.4}$$

The key idea in this discussion is the interpretation of the inside sum over i in Eq. (10.1.4). This sum can be interpreted as a real-valued function ϕ of L binary arguments, i.e.,

$$\phi(b_1, b_2, \cdots, b_L) = \sum_{i=1}^{L} b_i \alpha_i, \qquad b_i = 0, 1. \tag{10.1.5}$$

Since each b_i takes on only one of two values, 0 or 1, there are exactly 2^L possible values for ϕ. For example, suppose $L = 2$. The function ϕ in this case is

$$\phi(b_1, b_2) = b_1\alpha_1 + b_2\alpha_2. \tag{10.1.6}$$

There are only four possible values for ϕ in Eq. (10.1.6). They are $\{\alpha_1, \alpha_1 + \alpha_2, \alpha_2, 0\}$. Assume that we store the 2^L possible values for ϕ in a table (read-only memory). Equation (10.1.4) can be expressed in the form

$$y = \sum_{j=0}^{B-1} 2^{-j}\phi(x_1^j, \cdots, x_L^j). \tag{10.1.7}$$

Equation (10.1.7) suggests the following implementation. The data vector $\mathbf{x}$ is stored in L, B-bit long, registers. This data vector generates addresses for the stored values of ϕ. Thus in Eq. (10.1.6) if the jth bits of the two-data words are both 1, the address (1, 1) is used to address and recall from memory the value $(\alpha_1 + \alpha_2)$. If we interpret Eq. (10.1.7) as a sequential accumulation beginning with the least significant bit, the 2^{-j} multiplier is a right shift of previous accumulated ϕ values. An implementation of Eq. (10.1.7) for $L = 3$ is shown in Fig. 10.1.3. The right shift of previous accumulated ϕ values can be "hardwired" into the structure as indicated schematically in register $R2$ of Fig. 10.1.3.

There are several variations of this basic scheme. If L is large, then the 2^L words required for the ϕ table may be unreasonable. For large L, it is more efficient in terms of memory storage to break the L-fold sum into K smaller sums of M_i terms each. The memory requirement is reduced from 2^L words to $\sum_{i=1}^{K} 2^{M_i}$ words. For example, if $L = 10$, then $2^L = 1024$ words. If we break the 10-term sum into three sums of 3, 3, and 4, the total memory needed is only 32 words. The disadvantage of breaking the L-fold sum into smaller sums, is that one must use more adders to accumulate the partial sums. This means more processing time and some additional hardware.

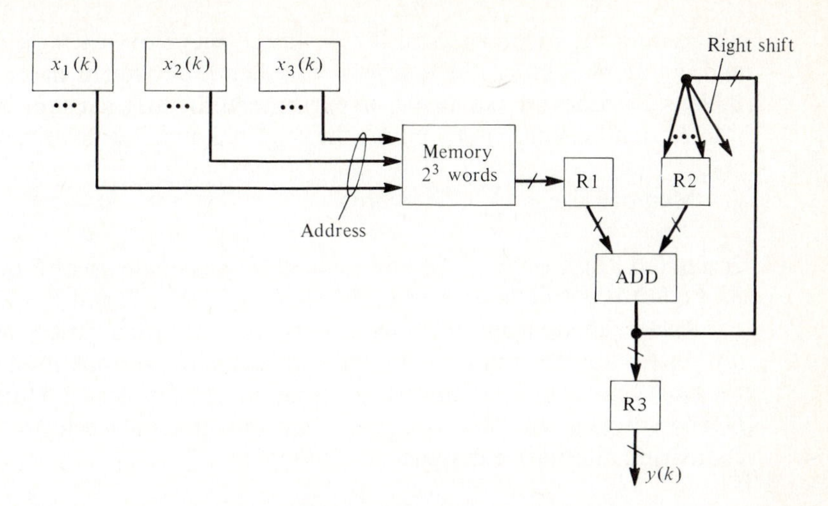

FIGURE 10.1.3 A sum of products implementation using distributed arithmetic.

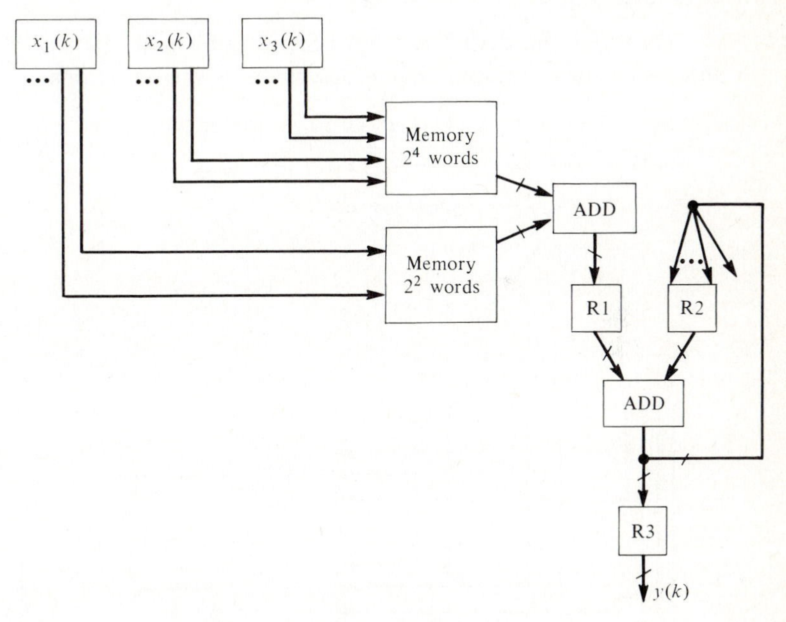

FIGURE 10.1.4 A distributed arithmetic implementation for a sum of products for $L = 3$, $K = 2$, $P = 2$.

Another variation on the basic scheme depicted in Fig. 10.1.3 is to address the ϕ table with more than a single bit from each data variable x_i. This requires a larger table for the ϕ function. If, for example, we use two bits from each data variable x_i to address the ϕ table the memory storage increases from 2^L to 2^{2L} words. The processing time is reduced by a factor of 2. If we assume we address the memory table for ϕ with P bits from each data variable x_i and the number of bits in x_i is $NP = B$, then the required memory for the ϕ table is 2^{PL} words. If we have also partitioned the original sum of L products into K smaller sums of M_i terms, the required memory for the ϕ function is

$$\text{Memory size} = \sum_{i=1}^{K} 2^{PM_i} \text{ words.} \tag{10.1.8}$$

In Eq. (10.1.8) $L = \sum_{i=1}^{K} M_i$ and each address is made up of P bits from each x_i. Figure 10.1.4 depicts an example for $L = 3$, $K = 2$, and $P = 2$.

As we can see from this brief discussion, there are a variety of distributed arithmetic implementations. In order to compare various implementations, we must agree on some measure of hardware complexity. Hardware complexity coupled with the data rate of an implementation form the basis for comparing alternative designs.

EXAMPLE 10.1.1

A distributed arithmetic state variable filter.

Consider the distributed arithmetic implementation of a state variable filter. The state variable filter is described by the equations

$$x_1(k+1) = a_{11}x_1(k) + a_{12}x_2(k) + b_1u(k),$$
$$x_2(k+1) = a_{21}x_1(k) + a_{22}x_2(k) + b_2u(k),$$
$$y(k) = c_1x_1(k) + c_2x_2(k) + du(k).$$

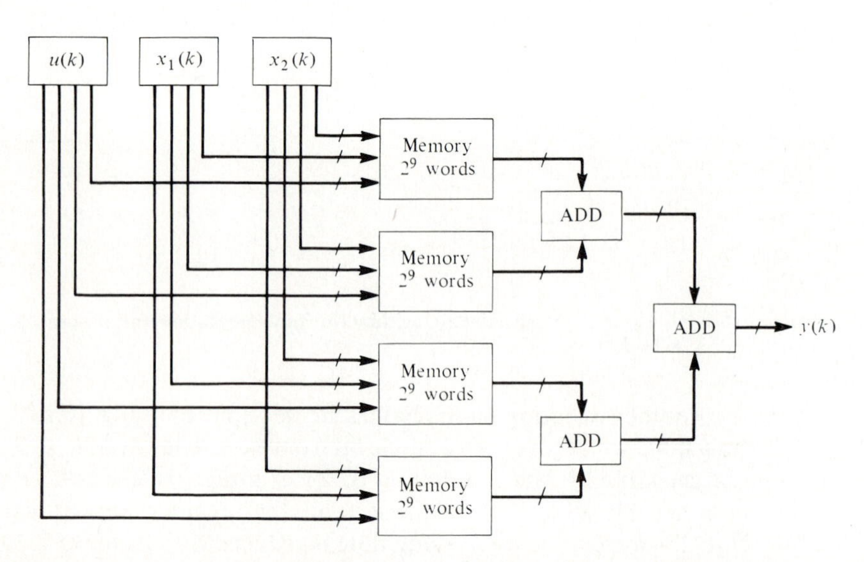

FIGURE 10.1.5 **A distributed arithmetic implementation of one state variable equation.**

Implementation of the filter requires implementation of three third-order inner products. Suppose we have a short word length of $B = 12$ bits. We can implement one of the equations as shown in Fig. 10.1.5. In this scheme, each memory is addressed by three bits from each of the three data words consisting of $x_1(k)$, $x_2(k)$, and $u(k)$. The adders have hardwired-in the current power-of-2 shifts. The memory needed for the ϕ table is $4 \times 2^9 = 2048$ words. The time needed to compute the output is approximately one memory access plus two addition cycles. The total memory needed is $3(2048) = 6144$ words for the three equations. There is very little control, and no monolithic multipliers are required.

Distributed arithmetic is an implementation that relies on the use of memory for simplifying the required hardware. Because our present technology has made possible inexpensive, fast access, and high-density ROM's, this kind of implementation is well-suited for certain applications in digital filtering. We can use this kind of implementation in any particular structure, as the following discussion points out.

10.2 BLOCK PROCESSING DIGITAL FILTERS

The first class of digital filtering structures we shall consider are known as block structures. These are multi-input, multi-output (MIMO) filters, which are equivalent to a single-input, single-output (SISO) filter. These filters offer the following advantages:

1. By increasing parallel computations, the output data rate can be increased;
2. There is a reduction in the average output noise due to roundoff noise as compared to the corresponding SISO filter;
3. Other finite-register effects are also improved;
4. There is a reduction in the number of multipliers when compared to low roundoff state space filters.

Consider an SISO digital filter with state variable equations

$$\begin{bmatrix} \mathbf{x}(k+1) \\ y(k) \end{bmatrix} = \begin{bmatrix} \mathbf{A} & \mathbf{B} \\ \mathbf{C} & D \end{bmatrix} \begin{bmatrix} \mathbf{x}(k) \\ u(k) \end{bmatrix}. \tag{10.2.1}$$

In Eq. (10.2.1) $u(k)$, $y(k)$ are scalars, $\mathbf{x}(k)$ is $n \times 1$, $\mathbf{A}$, $\mathbf{B}$, $\mathbf{C}$, D are $n \times n$, $n \times 1$, $1 \times n$, and 1×1 matrices, respectively. Recall the transfer function for Eq. (10.2.1) is

$$H(z) = D + \mathbf{C}(z\mathbf{I} - \mathbf{A})^{-1}\mathbf{B}. \tag{10.2.2}$$

Suppose instead of processing a single input $u(k)$ to obtain a single output $y(k)$, we process the input sequence in blocks of length L. The state equations for this situation are obtained by merely writing out Eq. (10.2.1) for $k+2$, $k+3, \ldots, k+L$. We obtain

$$\begin{bmatrix} \mathbf{x}(k+L) \\ y(k) \\ y(k+1) \\ \vdots \\ y(k+L-1) \end{bmatrix} = \left[\begin{array}{c|cccc} \mathbf{A}^L & \mathbf{A}^{L-1}\mathbf{B} & \mathbf{A}^{L-2}\mathbf{B} & \cdots & \mathbf{B} \\ \hline \mathbf{C} & D & 0 & \cdots & 0 \\ \mathbf{CA} & \mathbf{CB} & D & \cdots & 0 \\ \vdots & & & \ddots & \\ \mathbf{CA}^{L-1} & \mathbf{CA}^{L-2}\mathbf{B} & \mathbf{CA}^{L-3}\mathbf{B} & \cdots & D \end{array}\right] \begin{bmatrix} \mathbf{x}(k) \\ u(k) \\ u(k+1) \\ \vdots \\ u(k+L-1) \end{bmatrix}$$

$$= \begin{bmatrix} \mathbf{A}' & \mathbf{B}' \\ \mathbf{C}' & \mathbf{D}' \end{bmatrix} \begin{bmatrix} \mathbf{x}(k) \\ u(k) \\ \vdots \\ u(k+L-1) \end{bmatrix}, \tag{10.2.3}$$

where

$$\mathbf{A}' = \mathbf{A}^L, \qquad \mathbf{B}' = [\mathbf{A}^{L-1}\mathbf{B}, \mathbf{A}^{L-2}\mathbf{B}, \ldots, \mathbf{B}],$$

$$\mathbf{C}' = \begin{bmatrix} \mathbf{C} \\ \mathbf{CA} \\ \vdots \\ \mathbf{CA}^{L-1} \end{bmatrix}, \quad \mathbf{D}' = \begin{bmatrix} D & 0 & 0 & \cdots & 0 \\ \mathbf{CB} & D & 0 & \cdots & 0 \\ \mathbf{CAB} & \mathbf{CB} & D & \cdots & 0 \\ \vdots & & & \ddots & \\ \mathbf{CA}^{L-2}\mathbf{B} & \mathbf{CA}^{L-3}\mathbf{B} & & \cdots & D \end{bmatrix}. \tag{10.2.4}$$

An SFG for Eq. (10.2.3) is shown in Fig. 10.2.1. There are several observations we can make concerning the filter in Eq. (10.2.3).

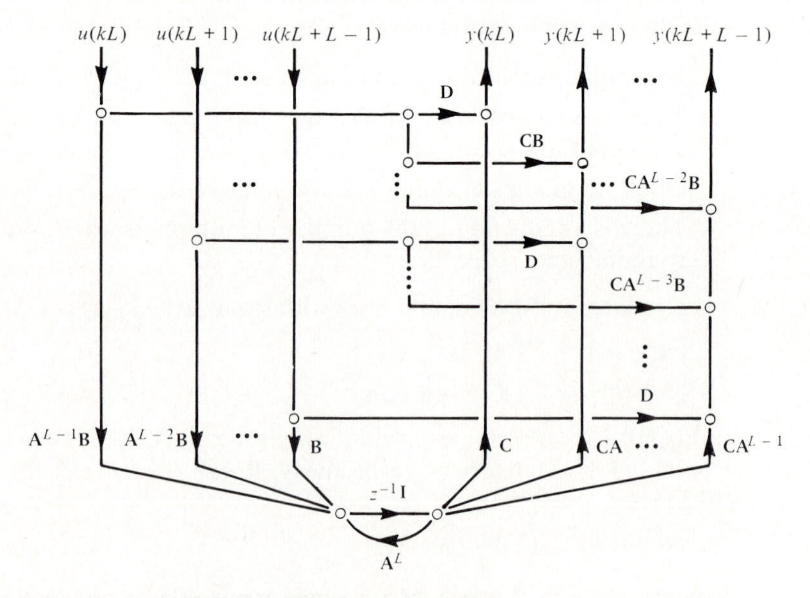

FIGURE 10.2.1 **A SFG for a block processing filter.**

1. If Eq. (10.2.3) can be implemented directly without delay, then the output data rate is increased by a factor of L, the block length.
2. Equation (10.2.3) can be interpreted as a shift-invariant, block-state representation of a periodically time-varying filter of period L, where $\mathbf{A}^L = \mathbf{A}(k + L - 1)\mathbf{A}(k + L - 2), \ldots, \mathbf{A}(k + 1)\mathbf{A}(k)$.
3. Since $\mathbf{A}' = \mathbf{A}^L$, the eigenvalues of $\mathbf{A}'$ are λ_i^L, $i = 1, 2, \ldots, n$, where λ_i are eigenvalues of $\mathbf{A}$. This means that the poles of the block filter are closer to $z = 0$ than the original filter, implying that limit cycle behavior of the block filter can be a great deal better than the original filter.
4. Any nonsingular transformation $\mathbf{T}$ applied to the SISO state-space realization $(\mathbf{T}^{-1}\mathbf{A}\mathbf{T}, \mathbf{T}^{-1}\mathbf{B}, \mathbf{C}\mathbf{T}, D)$ results in a block state-space realization $(\mathbf{T}^{-1}\mathbf{A}'\mathbf{T}, \mathbf{T}^{-1}\mathbf{B}', \mathbf{C}'\mathbf{T}, \mathbf{D}')$.
5. If the SISO state-space realization is a minimal realization of $H(z)$, i.e., it has the minimum number of unit delays, then the resulting block-state realization is also minimal.
6. The number of multiplies for Eq. (10.2.3) is $(n + L)$ inner products of $(n + L)$ variables repeated every L steps. [Some of the $(n + L)$ variables are zero.] This is an average of $(1 + n/L)$ inner products for each step k. This compares favorably with the number of multiplies per output sample for the direct form realization. From Eq. (10.2.3), we find the total number of multipliers per output sample is

$$r = \frac{n(n + L) + L\left(n + \dfrac{L + 1}{2}\right)}{L}. \tag{10.2.5}$$

The optimal length L that minimizes Eq. (10.2.5) is

$$L_{\text{opt}} = \sqrt{2}n. \tag{10.2.6}$$

In this case, Eq. (10.2.5) is

$$r_{\text{opt}} \simeq 3.41n + \tfrac{1}{2}. \tag{10.2.7}$$

In the case of cascaded second-order direct forms (DF), the number of multipliers per output sample (see Section 9.10) is

$$r_{\text{DF}} = 2.5n + 1. \tag{10.2.8}$$

Thus we see that the number of multiplies per output sample for the block filter is close to the minimum. This assumes the SISO (**A**, **B**, **C**, **D**) structure contains nontrivial values in all positions.

Thus far we have established that processing data in blocks has some significant advantages. What are the roundoff noise properties of these filters? We shall assume the roundoff noise occurs only at the outputs of the state variable summing nodes and the output nodes of the filter. We also assume rounding is done after double length accumulation. Our model is

$$\begin{aligned} \mathbf{x}(k+L) &= \mathbf{A}'\mathbf{x}(k) + \mathbf{B}'\mathbf{u}(k) + \mathbf{e}_1(k), \\ y(k) &= \mathbf{C}'\mathbf{x}(k) + \mathbf{D}'\mathbf{u}(k) + \mathbf{e}_2(k). \end{aligned} \tag{10.2.9}$$

We shall assume the covariances of the noise sources $\mathbf{e}_1(k)$ and $\mathbf{e}_2(k)$ are

$$\mathbf{\Gamma}_{e1} = \frac{q^2}{12}\mathbf{I}_N, \qquad \mathbf{\Gamma}_{e2} = \frac{q^2}{12}\mathbf{I}_L. \tag{10.2.10}$$

The correlation matrix of the output is given by

$$\mathbf{R}_y(m) = E(\mathbf{y}(k+mL)\mathbf{y}^T(k)) = \frac{q^2}{12}\left\{\mathbf{I}_L\delta(m) + \sum_{k=0}^{\infty} \mathbf{C}'\mathbf{A}'^{k+mL}(\mathbf{C}'\mathbf{A}'^{k})^T\right\}. \tag{10.2.11}$$

The output noise power for the individual outputs is

$$\sigma_{yj}^2 = [\mathbf{R}_y(0)]_{jj} = \frac{q^2}{12}\left\{1 + \sum_{k=0}^{\infty} \mathbf{C}\mathbf{A}^{kL+j-1}(\mathbf{C}\mathbf{A}^{kL+j-1})^T\right\}, \qquad j=1,2,\ldots,L. \tag{10.2.12}$$

Now

$$\sum_{k=0}^{\infty} \mathbf{C}\mathbf{A}^{kL+j-1}(\mathbf{C}\mathbf{A}^{kL+j-1})^T \leqslant \sum_{k=0}^{\infty} (\mathbf{C}\mathbf{A}^{k})(\mathbf{C}\mathbf{A}^{k})^T. \tag{10.2.13}$$

Inequality (10.2.13) is true because there are fewer terms in the sums on the left-hand side. Define an average noise power (averaged over one block period) as

$$\begin{aligned} \sigma^2 &= \frac{1}{L}\sum_{j=1}^{L} \sigma_{yj}^2 \\ &= \frac{1}{L}\left\{\sum_{j=1}^{L}\left[\sum_{k=0}^{\infty} \mathbf{C}\mathbf{A}^{kL+j-1}(\mathbf{C}\mathbf{A}^{kL+j-1})^T + 1\right]\right\}\frac{q^2}{12} \\ &\leqslant \left\{\frac{1}{L}\left[\sum_{k=0}^{\infty} \mathbf{C}\mathbf{A}^{k}(\mathbf{C}\mathbf{A}^{k})^T + 1\right]\right\}\frac{q^2}{12}. \end{aligned} \tag{10.2.14}$$

Equation (10.2.14) states that the average roundoff error per output sample is reduced by a factor of L as compared to the SISO system. Thus the roundoff noise properties of the block filter are improved in comparison to the original SISO filter.

These properties taken together are attractive advantages for the block processing structure. The major disadvantage of the block processing structure are the two buffers required at the input and output. The input buffer is serial-in and parallel-out, and the output buffer is parallel-in and serial-out. This extra hardware and the associated control is the only additional complexity necessary.

The block filter ($\mathbf{A}'$, $\mathbf{B}'$, $\mathbf{C}'$, $\mathbf{D}'$) properties derive from the underlying choice of ($\mathbf{A}$, $\mathbf{B}$, $\mathbf{C}$, D) for the SISO filter. Thus we have a great deal of freedom

TABLE 10.2.1 The number of multipliers for three classes of block filters.

Filter Structure	Number of Multipliers Per Output	Optimal Values for the Block Length L
Full state-space filter	$\frac{n^2}{L} + 2n + \frac{L+1}{2}$	$n\sqrt{2}$
mth-order cascaded sections	$\left(\frac{m}{L} + 2 + \frac{L+1}{2m}\right)n$	$m\sqrt{2}$
mth-order parallel sections	$\left(\frac{m}{L} + 2 + \frac{L-1}{2m}\right)n + 1$	$m\sqrt{2}$

in how we construct the MIMO block filter. Some examples are the following:

1. A full-scale state-space block structure derived from a full state-space structure (**A**, **B**, **C**, D);
2. A cascade connection which is made up of mth-order subsections in cascade. The mth-order subsections could be direct forms, minimum noise forms, normal forms, etc.;
3. A parallel connection which is made up of mth-order subsections connected in parallel. The mth-order subsections could again be any structure.

The properties of the overall block filter derive from the underlying structure of the SISO filter (**A**, **B**, **C**, D). For each of the structures discussed above the number of multipliers/output are given in Table 10.2.1.

In Table 10.2.1 we assume each of the underlying subsections is a full state structure and that the order of the subsections, m, is an integral divisor of the order of the overall filter, n, i.e., $n = km$, k an integer.

EXAMPLE 10.2.1

A comparison of the number of multiplies per output for three structures.

Consider a comparison in the number of multiplies per output for three structures for an eighth-order filter with second-order sections and optimum block length of $m\sqrt{2}$. Thus we have that $n = 8$, $m = 2$, and $L = 2\sqrt{2} \cong 3$. The number of multiplies per output for the block cascade (BC) structure can be obtained from Table 10.2.1. Thus for block processing, we have

$$r_{\mathrm{BC}} = 8(\tfrac{2}{3} + 2 + \tfrac{4}{4}) \cong 29.$$

The corresponding cascade of second-order direct terms has $2.5n + 1 = 21$ multiplies per output. A cascade of second-order state variable forms (which include minimum noise and normal forms) has $4n + 1 = 33$ multiplies per output. If we were to use parallel connections instead of cascaded sections, the results would be:

1. For the block parallel structure $r_{BP} = 28$,
2. For a parallel connection of second-order direct form structures, $r_{DFP} = 24$,
3. For a parallel connection of second-order state variable structures, $r_{SVP} = 36$.

If we increase the block length to $L = n\sqrt{2} \cong 11$ and use a full state space block processing structure, the number of multipliers per output is

$$r_B = \tfrac{64}{11} + 16 + \tfrac{12}{2} \cong 28.$$

Thus for the block processing structures, the number of multiplies per output does not vary a great deal. Parallel or cascade connections of second-order block processing structures do not markedly affect the number of multiplies per output.

Hardware Considerations

Block processing filter structures possess inherent paralizability. The essential computation is a vector inner product, which can be implemented either using monolithic multipliers or, as discussed in Section 10.1, using distributed arithmetic. A schematic of three possible block processing structures is shown in Fig. 10.2.2. Using distributed arithmetic, the computation of an inner product can be performed in B add-shift steps, where B is the word length of the data variables. With parallel inner-product computing units, we can achieve an L-fold increase in speed. This is obtained by the parallel computation of the entire output block of L-samples in the time of one inner-product computation.

As discussed in Section 10.1, the look-up table memory requirements are 2^R, where R is the inner-product length. In Eq. (10.2.3) we see that there are $(n + L)$ inner products [the number of rows in Eq. (10.2.3)] of maximum length $(n + L)$. Thus the memory storage requirements are somewhat less than $(n + L)2^{n+L}$ words (since there are some zeros in the $(n + L) \times (n + L)$ state matrix). To reduce the storage requirements, the long inner products can be partitioned as discussed previously. The time to obtain P outputs $(P \leq L)$ is given by

$$t \propto \frac{N}{P}(t_0 B), \tag{10.2.15}$$

where N is the number of inner products, P is the number of parallel inner-product units, t_0 is the cycle time for an add and shift, and B is the number of bits in the data variables. The way in which inner products are actually computed will depend on the resources available to the designer. There are an enormous number of possibilities.

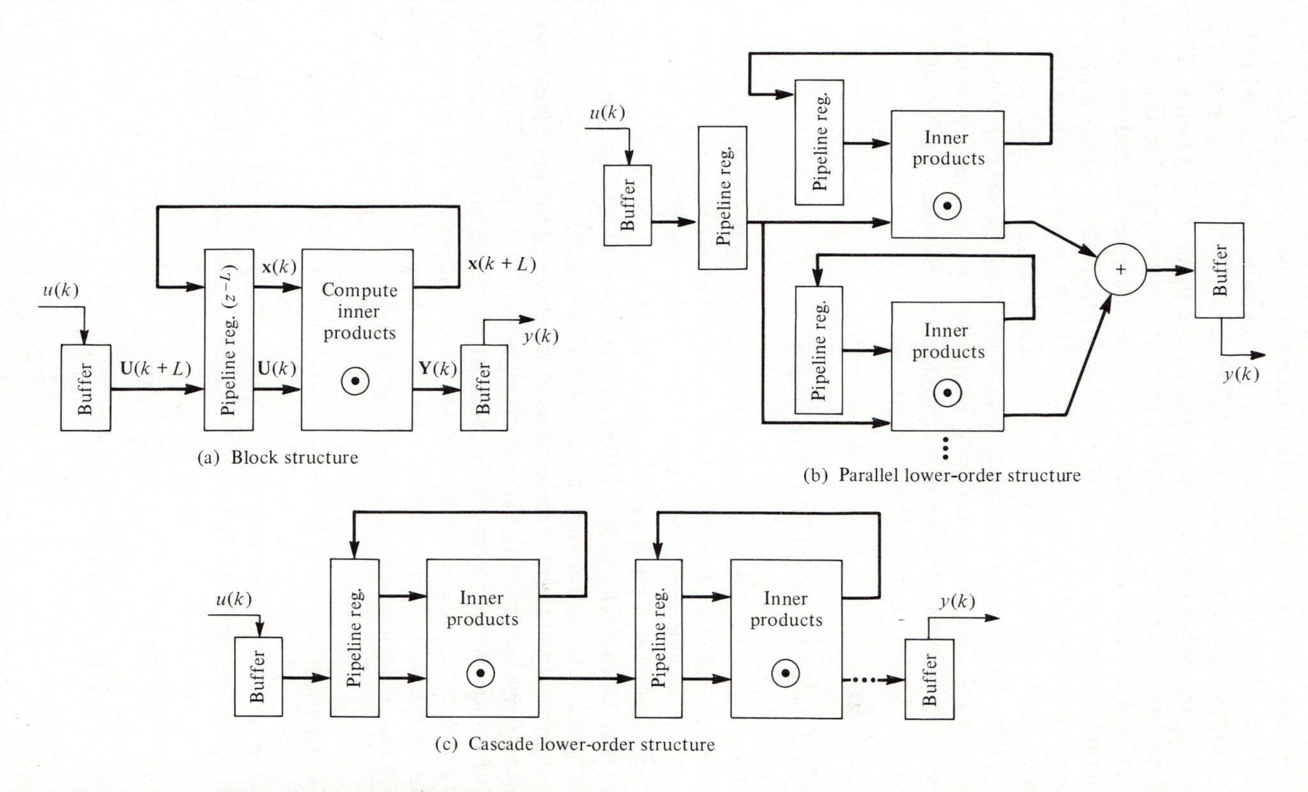

FIGURE 10.2.2 **Three block processing structures.**

Summary

Block processing filters are an attractive alternative for linear digital filters. They can be designed with low roundoff noise, low coefficient sensitivity, and good limit cycle behavior by using the principles discussed in Chapter 9 for the design of the underlying SISO filter. The essential properties of the underlying SISO filter are inherited by the block MIMO filter. The number of multiplies per output compares favorably with direct forms and is approximately 1.3 to 1.5 times that of a direct form (connected as second-order sections in cascade or parallel). The block processing filter formulation results in a parallel algorithm and is well-suited for an implementation that possesses many inner-product computational units working in parallel.

The good finite register effects of the block processing algorithm occur because the eigenvalues of $\mathbf{A}' = \mathbf{A}^L$ are further from the unit circle than those of the original matrix $\mathbf{A}$. This can, in some cases, eliminate limit cycles or, at the very least, reduce their amplitude (Barnes 1980b).

10.3 MULTIRATE DIGITAL FILTERS

Multirate digital signal processing is concerned with problems involving more than one sampling rate in a digital system. The problem is to efficiently compute the output for either a higher or a lower sampling rate. The process of lowering the sampling rate is called *decimation*. Similarly, the process of raising the sampling rate is called *interpolation*. Digital systems that perform sampling rate conversion can be viewed as linear, periodically time-varying systems. Sampling rate conversions in the cases of integer reductions, integer increases, and rational fraction changes can be implemented with systems that are efficient in terms of the number of computations. In order to understand decimation and interpolation, we must consider the spectra of signals that are involved in sampling rate conversions. We shall first consider the process of decimation.

Decimation

Suppose the sampling rate of a data sequence $\mathbf{u}$ is t_0 and we wish to decimate the sampling rate by a factor of M. Thus the new sampling rate is t_0', given by

$$t_0' = Mt_0, \qquad f' = \frac{1}{t_0'} = \frac{1}{Mt_0} = \frac{f}{M}. \tag{10.3.1}$$

The new data sequence is $\mathbf{y}$, where

$$y(k) = u(Mk), \qquad k = 0, \pm 1, \pm 2, \ldots. \tag{10.3.2}$$

This is shown schematically in Fig. 10.3.1. The spectrum of $\mathbf{y}$ in terms of the spectrum of $\mathbf{u}$ is an aliased version of the spectrum of $\mathbf{u}$ since the decimator subsamples and compresses the original sequence $\mathbf{u}$ (see Section 4.6).

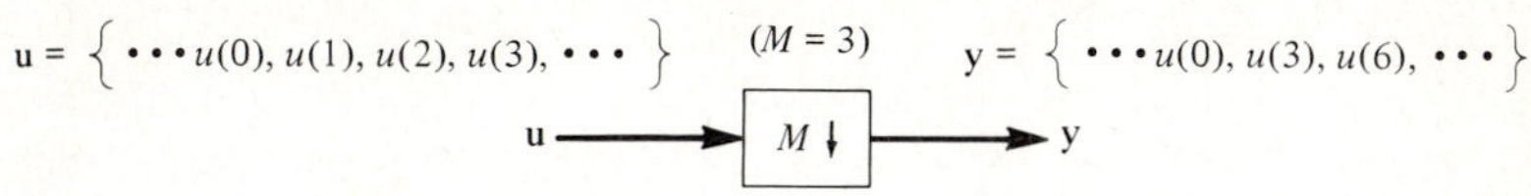

$\mathbf{u} = \{\cdots u(0), u(1), u(2), u(3), \cdots\}$ $(M = 3)$ $\mathbf{y} = \{\cdots u(0), u(3), u(6), \cdots\}$

u → $M\downarrow$ → y

FIGURE 10.3.1 A decimator system.

Thus, using the concepts in Section 4.6, we can express the spectrum of **y** as

$$\mathbf{Y}(e^{j\theta}) = \frac{1}{M} \sum_{l=0}^{M-1} U(e^{-j(2\pi l + \theta)/M}) \tag{10.3.3}$$

A sketch of these spectra is shown in Fig. 10.3.2. One of the applications of a decimator system is narrow-band, low-pass filtering. Suppose we wish to retain π/M of the spectra of **u** to process. Then we can decimate **u** by a factor of M, since the highest frequency we wish to retain is π/M (instead of π). In order to eliminate the aliasing in Eq. (10.3.3), we must precede the subsampling process inherent in decimation by an antialiasing filter with bandwidth π/M. Figure 10.3.3 depicts the various spectra involved in a decimator with an antialiasing filter. The question we wish to answer is how should one efficiently implement such a system. The key element in the decimator is the antialiasing filter since the subsampling process is merely a switch at the frequency f/M. One traditional method of realizing the antialiasing filter is to use a FIR filter, as depicted in Fig. 10.3.4. In the algorithm depicted in Fig. 10.3.4, the filter **h** operates at the high sampling rate f. All the multiplications and additions occur at the frequency f. We can obtain a more efficient algorithm by realizing that the operations of multiplication and subsampling commute. This is shown in Fig. 10.3.5. The proof is obtained by direct evaluation of each output. Using this result, we can interchange the subsampling process and the multiplication by the filter coefficients to obtain an algorithm as shown in Fig. 10.3.6. The advantage of this structure is the multiplies and additions occur at the decimated rate, $f' = f/M$. For every M samples into the filter, one sample

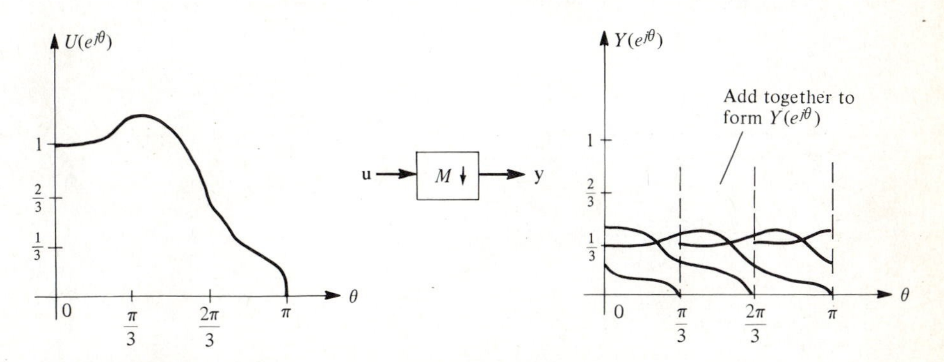

FIGURE 10.3.2 Spectra of the output of a decimator.

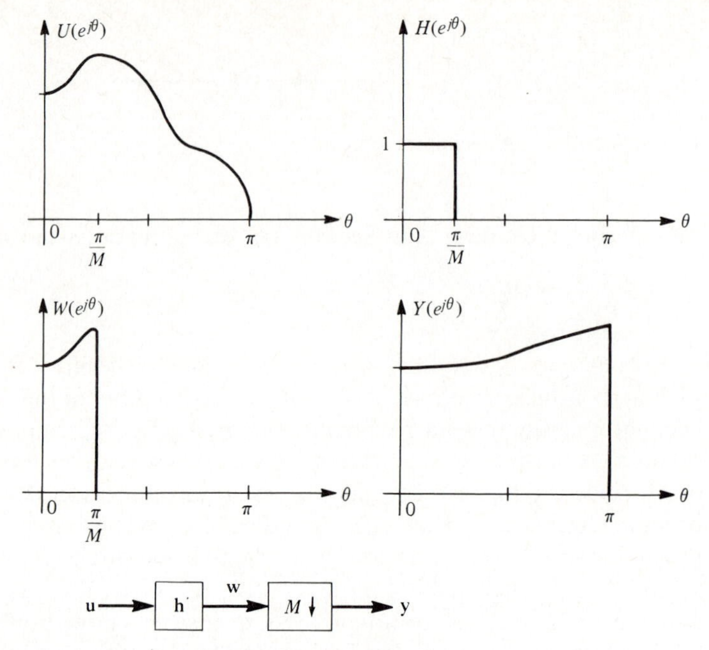

FIGURE 10.3.3 Spectra associated with a decimator.

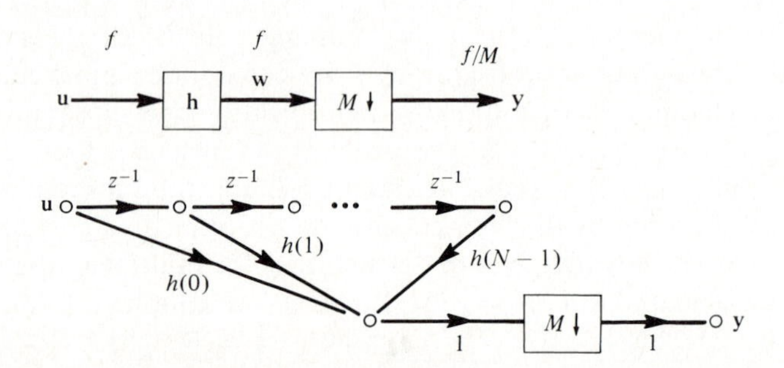

FIGURE 10.3.4 An antialiasing filter and decimator.

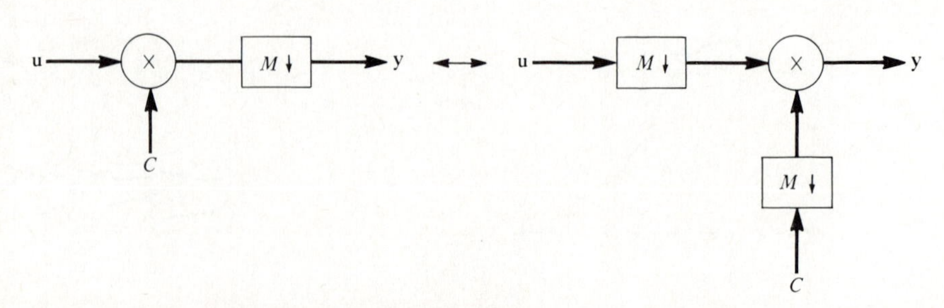

FIGURE 10.3.5 Multiplication and subsampling.

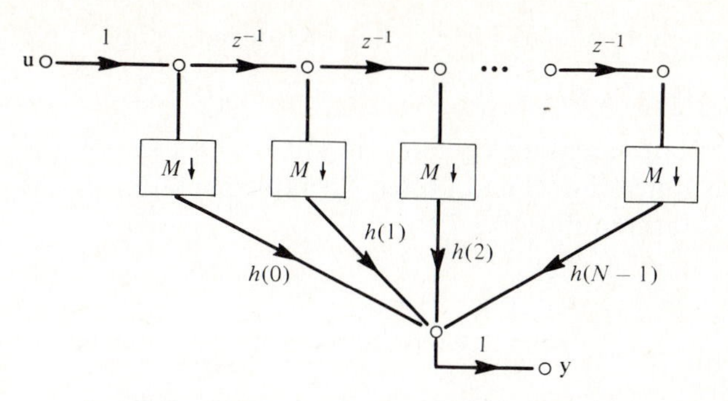

FIGURE 10.3.6 **An efficient FIR antialiasing filter and decimator structure.**

appears at the output, where

$$y(k) = \sum_{l=0}^{N-1} h(l)u(Mk - l). \tag{10.3.4}$$

For a linear phase FIR filter, the computation rate is therefore $n_{\text{FIR}}/2M$, where n_{FIR} is the order of the FIR filter.

A natural question arises as to the use of IIR filters in decimator systems. The most successful implementation uses the block processing algorithms discussed in Section 10.2. These structures offer the designer an excellent alternative to the use of FIR filters in decimator structures. Suppose we let P be the block size. Then the low-pass filter satisfies

$$\begin{aligned} \mathbf{x}(k + P) &= \mathbf{A}'\mathbf{x}(k) + \mathbf{B}'\mathbf{U}(k) \\ \mathbf{Y}(k) &= \mathbf{C}'\mathbf{x}(k) + \mathbf{D}'\mathbf{U}(k) \end{aligned} \tag{10.3.5}$$

where

$$\mathbf{U}(k) = \begin{bmatrix} u(k) \\ u(k+1) \\ \vdots \\ u(k+P-1) \end{bmatrix}, \qquad \mathbf{Y}(k) = \begin{bmatrix} y(k) \\ y(k+1) \\ \vdots \\ y(k+P-1) \end{bmatrix} \tag{10.3.6}$$

$$\mathbf{A}' = \mathbf{A}^P, \quad \mathbf{B}' = [\mathbf{A}^{P-1}\mathbf{B}, \mathbf{A}^{P-2}\mathbf{B}, \ldots, \mathbf{B}], \quad \mathbf{C}' = \begin{bmatrix} \mathbf{C} \\ \mathbf{CA} \\ \vdots \\ \mathbf{CA}^{P-1} \end{bmatrix},$$

$$\mathbf{D}' = \begin{bmatrix} D & 0 & 0 & \cdots & 0 \\ \mathbf{CB} & D & 0 & \cdots & 0 \\ \mathbf{CAB} & \mathbf{CB} & D & \cdots & 0 \\ & \vdots & & \ddots & \\ \mathbf{CA}^{P-2}\mathbf{B} & \cdots & & & D \end{bmatrix} \tag{10.3.7}$$

Assuming the output rate is decimated by a factor $P = M$, we need only compute one sample from each output block $\mathbf{Y}(k)$. Suppose we compute the first sample of each block. Then the output equation reduces to

$$y(k) = \mathbf{C}\mathbf{x}(k) + Du(k), \qquad k = 0, P, 2P, \ldots. \tag{10.3.8}$$

The state updating equation in (10.3.5) and the output equation in (10.3.8) constitute the decimator filter. An implementation of this structure is shown in SFG form in Fig. 10.3.7.

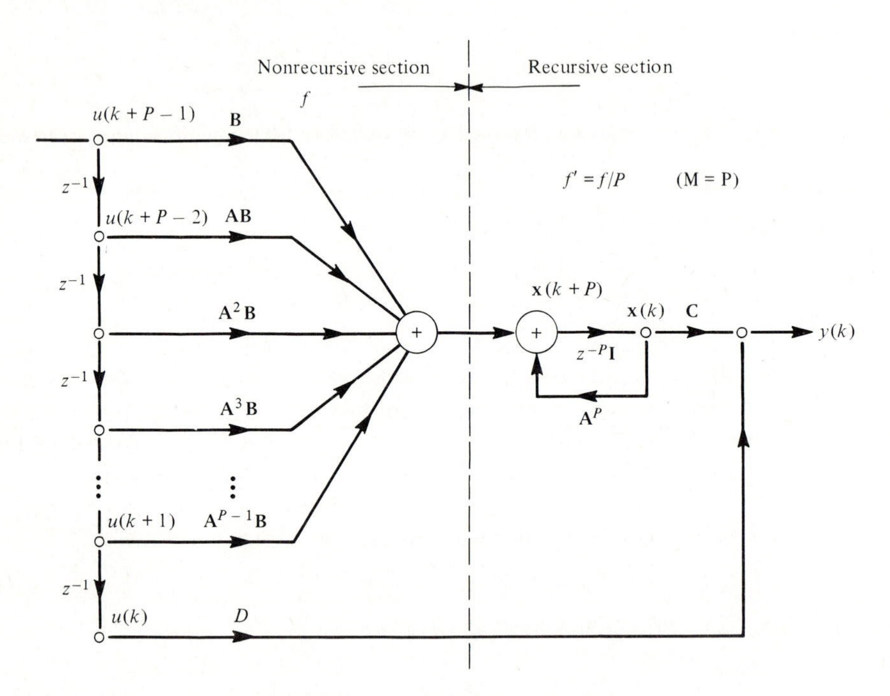

FIGURE 10.3.7 Decimating block processing filter structure.

The structure in Fig. 10.3.7 consists of two sections. The first section is the nonrecursive portion of the filter. The frequency reduction is obtained after summation of $\mathbf{B}'\mathbf{U}(k)$. The most computational intensive portion of the algorithm is the recursive portion of the filter and this computation is performed at the decimated rate, f/M. The state computed every Pth sampling interval is sufficient to generate the output at all future sampling instants. Assuming the block length is equal to the decimation factor, $P = M$, the computation rate of the filter is

$$\mathrm{CR}_{\mathrm{IIR}} = \frac{(n_{\mathrm{IIR}} + 1) + n_{\mathrm{IIR}}(n_{\mathrm{IIR}} + M)}{M} = n_{\mathrm{IIR}} + \frac{(n_{\mathrm{IIR}} + 1)^2}{M}. \tag{10.3.9}$$

Recall for the FIR decimator for this number is $n_{\mathrm{FIR}}/2M$. Thus for the block filter to compete with the FIR decimator, we must have that the ratio

$$\left[\frac{n_{\mathrm{FIR}}}{n_{\mathrm{IIR}}}\right] \gg 2M. \tag{10.3.10}$$

For sharp transitions in the antialiasing filter, this kind of inequality is often satisfied, and so in those applications this type of IIR decimator is useful. As we have discussed in Section 10.2, there are other compelling reasons for using block filters.

EXAMPLE 10.3.1

Decimating block filter computations.

Let's consider the form of the computations performed in a block filter. Assume the block length is $P = 3$ and the decimating factor is $M = 3$. The input is divided into blocks of $P = 3$ samples. For each input block, a state vector is then computed using Eq. (10.3.11). In this particular example,

$$\mathbf{x}(k + P) = \mathbf{A}'\mathbf{x}(k) + \mathbf{B}'\mathbf{U}(k) \tag{10.3.11}$$

the vector $\mathbf{U}(k)$ is 3×1. Since $P = 3$, the states computed are $\mathbf{x}(3)$, $\mathbf{x}(6)$, $\mathbf{x}(9), \ldots,$ etc. The nondecimated output from the block filter occur in blocks of three. We compute these blocks from Eq. (10.3.12).

$$\mathbf{Y}(k) = \mathbf{C}'\mathbf{x}(k) + \mathbf{D}'\mathbf{U}(k). \tag{10.3.12}$$

However, since we require only a decimated output, $M = 3$, we need compute only $y(0)$, $y(3)$, $y(6), \ldots.$ These computations are shown schematically in Fig. 10.3.8.

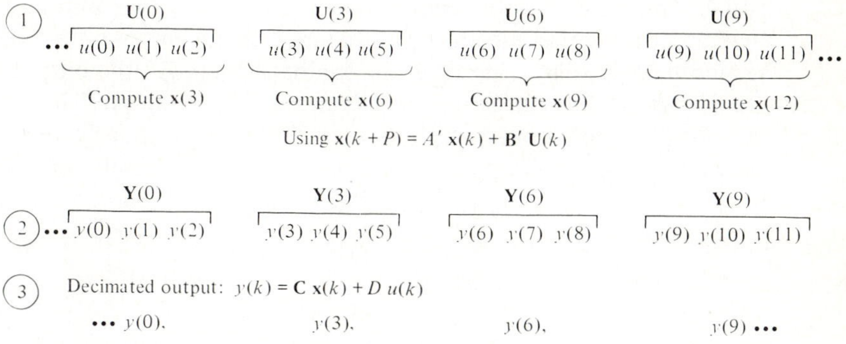

(1) U(0) U(3) U(6) U(9)
... $u(0)$ $u(1)$ $u(2)$ | $u(3)$ $u(4)$ $u(5)$ | $u(6)$ $u(7)$ $u(8)$ | $u(9)$ $u(10)$ $u(11)$ ...
Compute $\mathbf{x}(3)$ Compute $\mathbf{x}(6)$ Compute $\mathbf{x}(9)$ Compute $\mathbf{x}(12)$
Using $\mathbf{x}(k+P) = A'\,\mathbf{x}(k) + \mathbf{B}'\,\mathbf{U}(k)$

(2) Y(0) Y(3) Y(6) Y(9)
... $y(0)$ $y(1)$ $y(2)$ | $y(3)$ $y(4)$ $y(5)$ | $y(6)$ $y(7)$ $y(8)$ | $y(9)$ $y(10)$ $y(11)$

(3) Decimated output: $y(k) = \mathbf{C}\,\mathbf{x}(k) + D\,u(k)$
... $y(0)$, $y(3)$, $y(6)$, $y(9)$ ...

FIGURE 10.3.8 **Computations for a block filter decimator.**

EXAMPLE 10.3.2

A decimation filter obtained by factoring $H(z)$.

Another decimation filter structure can be obtained by factoring the transfer function $H(z)$. Consider

$$H(z) = D + \mathbf{C}(z\mathbf{I} - \mathbf{A})^{-1}\mathbf{B}.$$

Without loss of generality, suppose $D = 0$. Then we have

$$H(z) = \mathbf{C}(z\mathbf{I} - \mathbf{A})^{-1}\mathbf{B} = z^{-1}\mathbf{C}(\mathbf{I} - z^{-1}\mathbf{A})^{-1}\mathbf{B}.$$

Now

$$(\mathbf{I} - z^{-1}\mathbf{A})^{-1} = \sum_{k=0}^{\infty} z^{-k}\mathbf{A}^k.$$

Suppose we factor $(\mathbf{I} - z^{-1}\mathbf{A})^{-1}$ as follows:

$$\begin{aligned}(\mathbf{I} - z^{-1}\mathbf{A})^{-1} &= \mathbf{I} + z^{-1}\mathbf{A} + z^{-2}\mathbf{A}^2 + z^{-3}\mathbf{A}^3 + \cdots \\ &= (I + z^{-1}\mathbf{A} + \cdots + z^{-(M-1)}\mathbf{A}^{M-1}) \\ &\quad \times (\mathbf{I} + z^{-M}\mathbf{A}^M + z^{-2M}\mathbf{A}^{2M} + \cdots) \\ &= \left(\sum_{k=0}^{M-1} z^{-k}\mathbf{A}^k\right)(\mathbf{I} - z^{-M}\mathbf{A}^M)^{-1}.\end{aligned}$$

Let M be the output decimation factor. We can express the transfer function $H(z)$ in the form

$$H(z) = z^{-1}\mathbf{C}(\mathbf{I} - z^{-1}\mathbf{A})^{-1}\mathbf{B} = z^{-1}\mathbf{C}(\mathbf{I} - z^{-M}\mathbf{A}^M)^{-1}\left(\sum_{k=0}^{M-1} z^{-k}\mathbf{A}^k\right)\mathbf{B}.$$

In the preceding transfer function, we identify $z^{-1}\mathbf{C}(\mathbf{I} - z^{-M}\mathbf{A}^M)^{-1}$ as a recursive filter and $(\sum_{k=0}^{M-1} z^{-k}\mathbf{A}^k)\mathbf{B}$ as a nonrecursive filter, as shown in Fig. 10.3.9. The nonrecursive section is a single input, two output filter, and, similarly, the recursive section is a two input, single output filter. The recursive section has M units of delay in the feedback loop. All the output multiplies are performed only once for each decimated output sample. Thus the number of multiplies per input sample is $(2 + 6/M)$, and similarly the number of additions per input sample is $(2 + \frac{3}{4})$. This structure is similar in some respects to the block processing filter in that it consists of a nonrecursive section

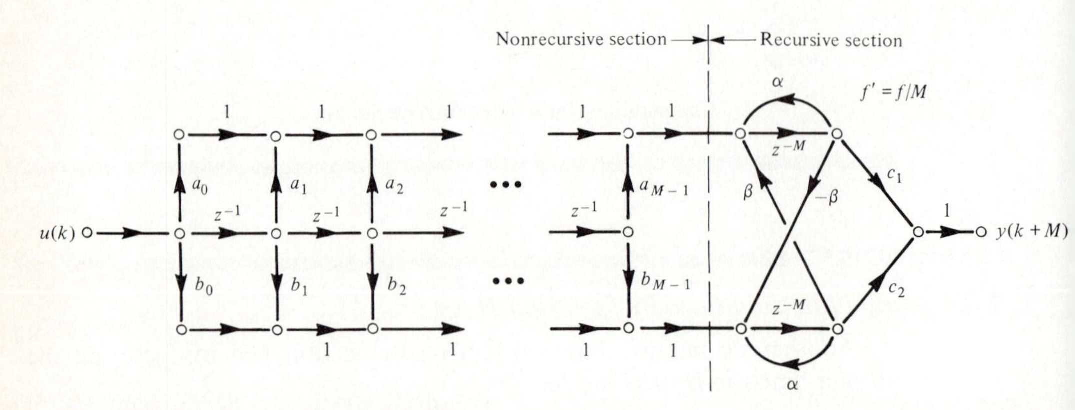

FIGURE 10.3.9 **A decimating filter.**

operating at a rate f and a recursive section operating at a decimated rate of $f' = f/M$. (See Barnes 1980b.)

Interpolator Structures

In the case of interpolation structures the new sample times are related to the original sample time by

$$t_0' = \frac{t_0}{L}, \qquad f' = Lf. \tag{10.3.13}$$

The new sample rate is a factor of L faster than the original. This means we must insert $L - 1$ zeros between the original data samples. Thus the interpolated sequence **w** in terms of the original sequence **u** is given by

$$w(k) = \begin{cases} u(k/L), & k = 0, \pm L, \pm 2L, \ldots \\ 0, & \text{otherwise} \end{cases}. \tag{10.3.14}$$

This is shown schematically in Fig. 10.3.10. The spectrum of **w** in terms of **u** is

$$W(e^{j\theta}) = U(e^{j2\pi f L t_0}) = U(e^{jL\theta}). \tag{10.3.15}$$

Thus the spectrum of **w** is a compressed version of the original spectrum **u** by a factor of L. An example is shown in Fig. 10.3.11.

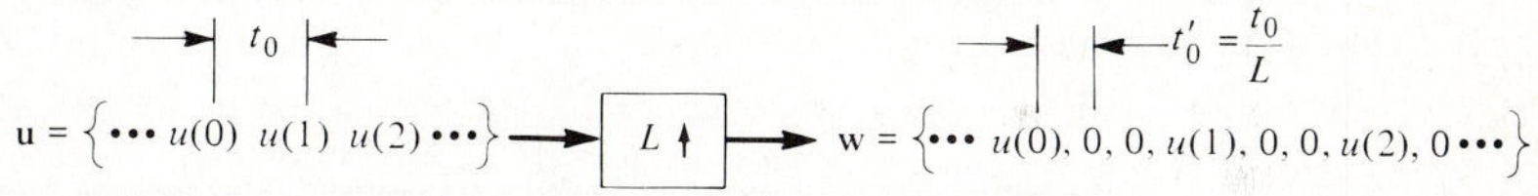

t0
t'0 = t0/L
u = {••• u(0) u(1) u(2) •••}
L ↑
w = {••• u(0), 0, 0, u(1), 0, 0, u(2), 0 •••}

FIGURE 10.3.10 An interpolator.

u(e^jθ)
0
π
2π
θ
w(e^jθ)
L = 3
π/3
2π/3
π
θ

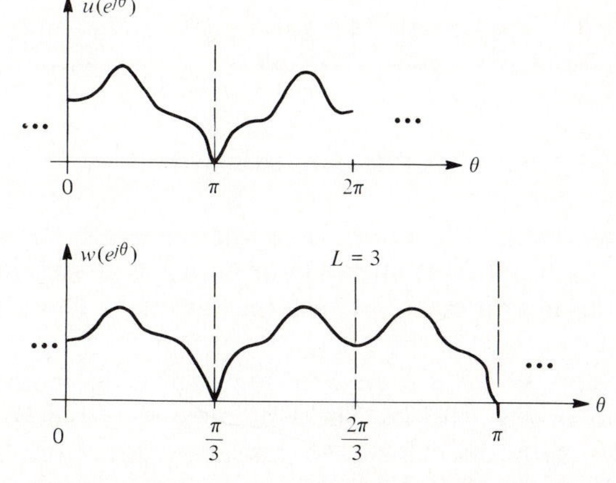

FIGURE 10.3.11 The spectrum of an interpolated output in terms of the original spectrum.

If we wish to filter on the interval π/L without the repeated versions of the original spectrum present above π/L, we need to follow the interpolator with a low-pass filter. The spectral components above π/L carry no additional information. Thus a complete interpolator system consists of the box in Fig. 10.3.10 followed by a low-pass filter, as shown in Fig. 10.3.12.

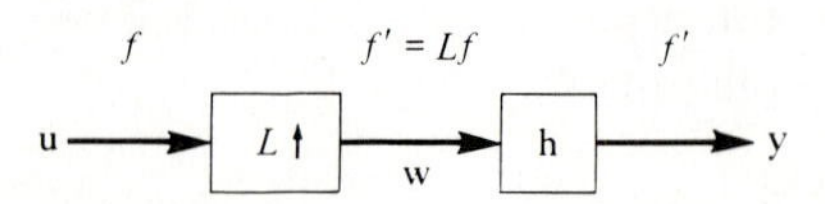

FIGURE 10.3.12 **An interpolator system.**

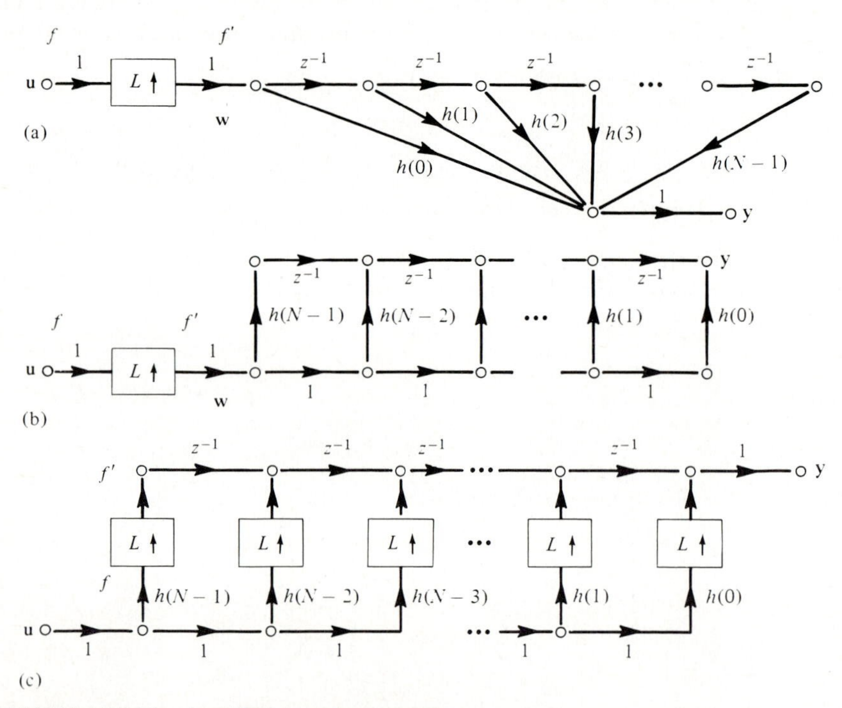

FIGURE 10.3.13 **An FIR interpolator system.**

In Fig. 10.3.12, the interpolator filter processes data at the high rate $f' = Lf$. The key to efficient interpolator design is to design the overall structure so that this filter processes data at the low rate f. This is analogous to the problem we encountered in the decimator system.

Suppose we use an FIR low-pass filter in the interpolator system as shown in Fig. 10.3.13. The FIR filter is a traditional direct form. We can transform the direct form to a transposed direct form as shown in Fig. 10.3.13(b) (see Problem 10.3). Now we commute the sampling rate expander box with the multipliers in the FIR filter in a manner similar to the decimation case to

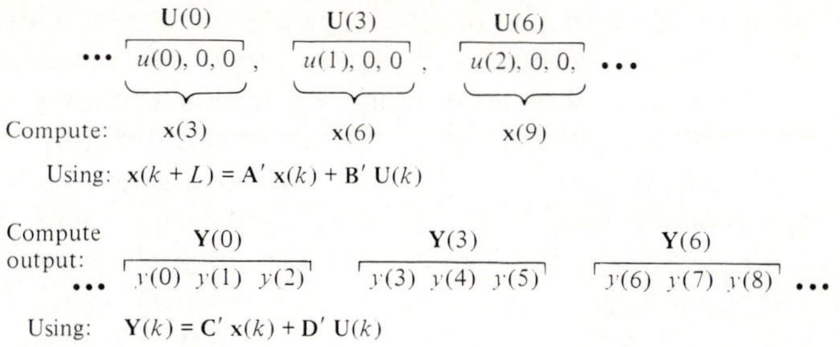

$$\cdots \underbrace{\overbrace{u(0), 0, 0}^{\mathbf{U}(0)}}_{}, \quad \underbrace{\overbrace{u(1), 0, 0}^{\mathbf{U}(3)}}_{}, \quad \underbrace{\overbrace{u(2), 0, 0,}^{\mathbf{U}(6)}}_{} \cdots$$

Compute: $\mathbf{x}(3)$ $\mathbf{x}(6)$ $\mathbf{x}(9)$

Using: $\mathbf{x}(k+L) = \mathbf{A}'\,\mathbf{x}(k) + \mathbf{B}'\,\mathbf{U}(k)$

Compute output:

$$\cdots \overbrace{y(0)\ y(1)\ y(2)}^{\mathbf{Y}(0)} \quad \overbrace{y(3)\ y(4)\ y(5)}^{\mathbf{Y}(3)} \quad \overbrace{y(6)\ y(7)\ y(8)}^{\mathbf{Y}(6)} \cdots$$

Using: $\mathbf{Y}(k) = \mathbf{C}'\,\mathbf{x}(k) + \mathbf{D}'\,\mathbf{U}(k)$

FIGURE 10.3.14 Computations for a block filter interpolator.

obtain the structure shown in Fig. 10.3.13(c). Notice that the multiplications now occur at the lower rate f, and thus we obtain a more efficient interpolator system.

In a manner similar to decimation systems, we can develop interpolator systems using block processing filters. In this case, the input block consists of one nonzero sample and the L output samples are obtained using

$$\mathbf{Y}(k) = \mathbf{C}'\mathbf{x}(k) + \mathbf{D}'\mathbf{U}(k). \tag{10.3.16}$$

The computations are depicted schematically in Fig. 10.3.14 for $L = 3$.

There are two principles that make block processing filters efficient decimators and interpolators: First, the state computation $\mathbf{x}(k + P)$ is performed once for each input block. Second, the matrix multiplies involving $\mathbf{Y}(k)$ or $\mathbf{U}(k)$ are sparse because either $\mathbf{Y}(k)$ or $\mathbf{U}(k)$ contains only one nonzero value. As a result the computation rate of block processing interpolators and decimators is comparable to FIR interpolators and decimators. There is also a great deal of flexibility in the use of block processing filters. The underlying $(\mathbf{A}, \mathbf{B}, \mathbf{C}, D)$ SISO filter can be chosen for excellent finite register effects. We could also choose $(\mathbf{A}, \mathbf{B}, \mathbf{C}, D)$ as an FIR structure or choose sparse realizations of $(\mathbf{A}, \mathbf{B}, \mathbf{C}, D)$ to create a sparse block processing structure. The possibilities are numerous, and the particular structure one should use depends a great deal on the possible applications of the filter.

10.4 ORTHOGONAL FILTER STRUCTURES

Real time applications of digital filtering will inevitably require integrated circuit hardware implementation. As the speed and density of integrated circuits increase, the assumptions that influence hardware design will change. The number of multiplications, for example, is not an adequate measure of complexity.

What are the characteristics of filter structures (or algorithms) desired in large scale integration? It is desirable, for example, to keep data paths local and to avoid a global bus. It is desirable as well to distribute memory and computation over the chip area in a loosely connected array. The individual processors in the array should be as much alike as possible (hopefully identical except for local microcode) so that the design effort is reduced.

The problem then is to design a processing module that can realize an arbitrary filter when connected in a one- or two-dimensional array. The result should have good, if not optimal, finite word length performance. We shall propose such a module design as a candidate solution to this problem. The module involves a 2×2 rotation matrix multiplication, which is the result of forcing orthogonality relative to the state covariance matrix.

A filter structure is *orthogonal* if at any time between computations, all internal variables (except the filter output) are uncorrelated and have unit variance, assuming a white noise input. The unit variance requirement is the l_2 scaling rule (9.6.7), but the pairwise orthogonality is something new. Forcing orthogonality allows the use of rich algebraic properties that provide structures having the following advantages:

1. The scaling rule is automatically met.
2. The roundoff noise gain, while not minimum, is low, and invariant under frequency transformations.
3. Overflow oscillations are impossible.
4. The structures can be manipulated to achieve either high-speed, parallel computation using a sparsely connected array of processor modules or low-speed, single processor realizations.

The disadvantages of orthogonal filter structures are:

1. Since the primitive computational unit is a 2×2 matrix rotation, they require about $8n$ multiplies per output (n is the filter order).
2. Zero-input limit cycles due to internal quantization have been observed to be larger than in other structures.

The development of orthogonal filter structures in the remainder of this section relies heavily on algebraic properties of orthogonal and unitary matrices. We shall proceed as follows:

1. An algebraic characterization of orthogonal structures is given that uses the factored state variable description.
2. The special case of orthogonal, all-pass filters is introduced. These are shown to be intimately connected with orthogonal matrices and have the surprising property that any *connection* of them is yet another orthogonal, all-pass filter ("connection" is used in a precise sense to be defined).
3. A minimal decomposition problem is introduced and solved for orthogonal, all-pass filters.
4. A minimal synthesis of orthogonal filters with arbitrary $H(z)$ is given.
5. The finite word length properties of orthogonal structures is discussed.

Characterization of Orthogonal Filter Structures

A state variable structure

$$\begin{bmatrix} \mathbf{x}(k+1) \\ y(k) \end{bmatrix} = \begin{bmatrix} \mathbf{A} & \mathbf{B} \\ \mathbf{C} & D \end{bmatrix} \begin{bmatrix} \mathbf{x}(k) \\ u(k) \end{bmatrix} \tag{10.4.1}$$

is *orthogonal* if

$$E \begin{bmatrix} \mathbf{x}(k) \\ u(k) \end{bmatrix} [\mathbf{x}^T(k) \quad u(k)] = \begin{bmatrix} \mathbf{K} & 0 \\ 0 & 1 \end{bmatrix} = \mathbf{I}_{(n+1)\times(n+1)}. \tag{10.4.2}$$

This will be true if and only if

$$\mathbf{K} = \mathbf{I} = \mathbf{A}\mathbf{A}^T + \mathbf{B}\mathbf{B}^T. \tag{10.4.3}$$

The assumption behind such a description is that all the variables on the left-hand side of Eq. (10.4.1) are being computed in parallel (or concurrently). If that is not the case, then a more appropriate description is the factored state variable description (8.5.35). This models individual computations that are done separately but in a fixed cyclical order. The equations are

$$\begin{aligned} &\mathbf{q}_0(k) = \begin{bmatrix} \mathbf{x}(k) \\ u(k) \end{bmatrix}, \qquad \mathbf{q}_i(k) = \mathbf{F}_i \mathbf{q}_{i-1}(k), \qquad 1 \leqslant i \leqslant L, \\ &\begin{bmatrix} \mathbf{x}(k+1) \\ y(k) \end{bmatrix} = \mathbf{q}_L(k). \end{aligned} \tag{10.4.4}$$

The state variable description may be recovered by "eliminating" the intermediate computations, with the result that

$$\begin{bmatrix} \mathbf{A} & \mathbf{B} \\ \mathbf{C} & D \end{bmatrix} = \mathbf{F}_L \mathbf{F}_{L-1} \cdots \mathbf{F}_1. \tag{10.4.5}$$

In this context, orthogonality means that

$$\begin{aligned} &E\,\mathbf{q}_i(k)\mathbf{q}_i(k)^T = \mathbf{I}_{(n+1)\times(n+1)}, \qquad 1 \leqslant i \leqslant L-1, \\ &E\,\mathbf{q}_L(k)\mathbf{q}_L(k)^T = \left[\begin{array}{c|c} \mathbf{I}_{n\times n} & \times \\ \hline \times & \times \end{array}\right]. \end{aligned} \tag{10.4.6}$$

Since these covariance matrices satisfy Eq. (9.10.7), the preceding conditions require that

$$\mathbf{F}_i \mathbf{F}_i^T = \begin{cases} \mathbf{I}_{(n+1)\times(n+1)}, & 1 \leqslant i < L, \\ \left[\begin{array}{c|c} \mathbf{I}_{n\times n} & \times \\ \hline \times & \times \end{array}\right], & i = L. \end{cases} \tag{10.4.7}$$

Equation (10.4.7) characterizes orthogonal filters. For $L = 1$, Eqs. (10.4.3) and (10.4.7) are equivalent.

A matrix having real elements and satisfying

$$\mathbf{F}\mathbf{F}^T = \mathbf{F}^T\mathbf{F} = \mathbf{I} \tag{10.4.8}$$

is called an *orthogonal matrix*: its rows (or columns) form an orthonormal basis. The set of all such $N \times N$ matrices forms a multiplicative group, denoted $\mathcal{O}(N)$:

$$\mathbf{FF}^T = \mathbf{I},\ \mathbf{GG}^T = \mathbf{I} \quad \Rightarrow \quad (\mathbf{FG})^T(\mathbf{FG}) = \mathbf{G}^T(\mathbf{F}^T\mathbf{F})\mathbf{G} = \mathbf{G}^T\mathbf{G} = \mathbf{I}. \qquad (10.4.9)$$

The *unitary group*, denoted $\mathcal{U}(N)$, generalizes this to matrices with complex elements:

$$\mathbf{F} \in \mathcal{U}(N) \quad \text{if} \quad \mathbf{FF}^* = \mathbf{F}^*\mathbf{F} = \mathbf{I}_{N \times N}. \qquad (10.4.10)$$

These matrix groups play a key role in the theory of orthogonal filter structures.

Orthogonal All-Pass Filters

An all-pass filter is one whose magnitude response is identically one,

$$|H(e^{j\theta})| = 1 \quad \text{for all real } \theta. \qquad (10.4.11)$$

This can be generalized to filters with m inputs and outputs. The $m \times m$ matrix transfer function $\mathbf{H}(z)$ is *all-pass* if

$$\mathbf{H}(e^{j\theta}) \in \mathcal{U}(m) \quad \text{for all real } \theta. \qquad (10.4.12)$$

All-pass filters may be considered generalized delays or energy storage elements. Although they may seem to have little intrinsic significance, they obey remarkable composition rules that make them useful building blocks. A simple modification can be added to an appropriately chosen orthogonal, all-pass filter to produce an orthogonal filter with arbitrary $H(z)$.

We shall begin with the case $L = 1$.

DEFINITION 1 (orthogonal, all-pass filters):

$$\mathbf{F} = \begin{bmatrix} \mathbf{A} & \mathbf{B} \\ \mathbf{C} & \mathbf{D} \end{bmatrix} \in \mathcal{F}(m, n) \quad \text{if}$$

1. $\mathbf{AA}^T + \mathbf{BB}^T = \mathbf{I}_{(n \times n)}$, (orthogonal) **(10.4.13)**
2. $\mathbf{H}(e^{j\theta}) = \mathbf{D} + \mathbf{C}(e^{j\theta}\mathbf{I} - \mathbf{A})^{-1}\mathbf{B} \quad \in \quad \mathcal{U}(m)$ for all θ (all-pass). **(10.4.14)**

The subset $\mathcal{F}_0(m, n)$ of $\mathcal{F}(m, n)$ contains matrices $\mathbf{F}$ satisfying Eqs. (10.4.13) and (10.4.14) and the additional condition that

$$(\mathbf{A}, \mathbf{B}, \mathbf{C}) \text{ is } \textit{controllable} \text{ and } \textit{observable}. \qquad (10.4.15)$$

These are the *minimal*, orthogonal, all-pass filters.

The minimal filters are quite constrained by Eqs. (10.4.13) and (10.4.14). If $\mathbf{A}$ is strictly stable, then the filter is both observable and controllable. Problems (10.19) and (10.20) provide proof for the following.

PROPOSITION 1. If $\mathbf{F} \in \mathcal{F}(m, n)$, then the following statements are equivalent:

1. $\mathbf{F} \in \mathcal{F}_0(m, n)$.
2. $(\mathbf{A}, \mathbf{B})$ is controllable.

3. $(\mathbf{A}, \mathbf{C})$ is observable.
4. $\det(\lambda\mathbf{I} - \mathbf{A}) = 0 \quad \Rightarrow \quad |\lambda| < 1.$

A key property for our purposes is the surprising correspondence between $\mathscr{F}(m, n)$ (orthogonal, all-pass filters) and $\mathcal{O}(m + n)$ (orthogonal matrices). In Eq. (10.4.7), all the factors except $\mathbf{F}_L$ are orthogonal. For all-pass filters, $\mathbf{F}_L$ is also orthogonal.

PROPOSITION 2.

$$\mathbf{F} \in \mathcal{O}(m + n) \quad \Rightarrow \quad \mathbf{F} \in \mathscr{F}(m, n), \tag{10.4.16}$$

$$\mathbf{F} \in \mathscr{F}_0(m, n) \quad \Rightarrow \quad \mathbf{F} \in \mathcal{O}(m + n). \tag{10.4.17}$$

The proof of this proposition is given in Appendix 10A. The slight difference between the exact converse of Eqs. (10.4.16) and (10.4.17) is the reason for the distinction between $\mathscr{F}$ and $\mathscr{F}_0$. If we work with orthogonal matrices in $\mathcal{O}(m + n)$, we shall also be talking about filters in $\mathscr{F}(m, n)$, minimal or not. It would be simpler to restrict our consideration to minimal filters. Unfortunately, orthogonal matrices are not in one-to-one correspondence with $\mathscr{F}_0$, but rather $\mathscr{F}$.

For any orthogonal matrix $\mathbf{F} \in \mathcal{O}(N)$, there are N possible filters, one each in $\mathscr{F}(m, N - m)$, $1 \leqslant m \leqslant N$. The choice of m depends on how we partition the matrix $\mathbf{F}$, i.e.,

$$\mathbf{F} = \left[\begin{array}{c|c} \mathbf{A} & \mathbf{B} \\ \hline \mathbf{C} & \mathbf{D} \end{array}\right] \begin{array}{l} \updownarrow N - m \\ \updownarrow \; m \end{array} \qquad \begin{array}{cc} \overleftrightarrow{N - m} & \overleftrightarrow{m} \end{array} \tag{10.4.18}$$

The case $m = N$ is a memoryless filter. Figure 10.4.1 is a vector signal flow graph representation of the filter corresponding to $\mathbf{F}$. We take each of the $n = N - m$ outputs and tie them back to n inputs through unit delays to form one of the N possible filters corresponding to $\mathbf{F}$ (and some partition).

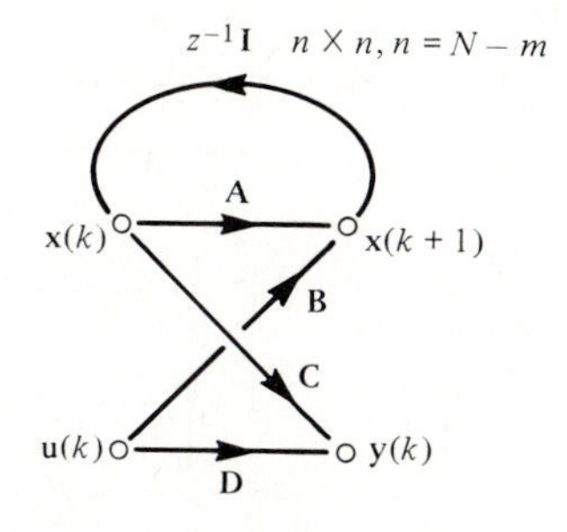

FIGURE 10.4.1 **A vector signal flow graph representing Eq. (10.4.18).**

Examples of Filters in $\mathscr{F}_0$

The simplest nontrivial minimal orthogonal filter is a single unit-delay element. The matrix **F** is

$$\mathbf{F} = \begin{bmatrix} \mathbf{A} & \mathbf{B} \\ \mathbf{C} & \mathbf{D} \end{bmatrix} = \begin{bmatrix} 0 & 1 \\ 1 & 0 \end{bmatrix}. \tag{10.4.19}$$

It is trivial to verify $\mathbf{AA}^T + \mathbf{BB}^T = \mathbf{I}$ and $|H(e^{j\theta})| = |e^{-j\theta}| = 1$.

Next consider a cascade of two unit delays. The matrix **F** is

$$\mathbf{F} = \begin{bmatrix} \mathbf{A} & \mathbf{B} \\ \mathbf{C} & \mathbf{D} \end{bmatrix} = \left[\begin{array}{cc|c} 0 & 1 & 0 \\ 0 & 0 & 1 \\ \hline 1 & 0 & 0 \end{array}\right]: \tag{10.4.20}$$

In this case $\mathbf{AA}^T + \mathbf{BB}^T = \mathbf{I}_{2\times 2}$ and $|H(e^{j\theta})| = |e^{-2j\theta}| = 1$. Notice that Eq. (10.4.20) suggests that a simple cascade of two filters in $\mathscr{F}_0$ is another filter in $\mathscr{F}_0$.

Figure 10.4.2 depicts several simple filters in $\mathscr{F}_0$. In Fig. 10.4.2(c) is

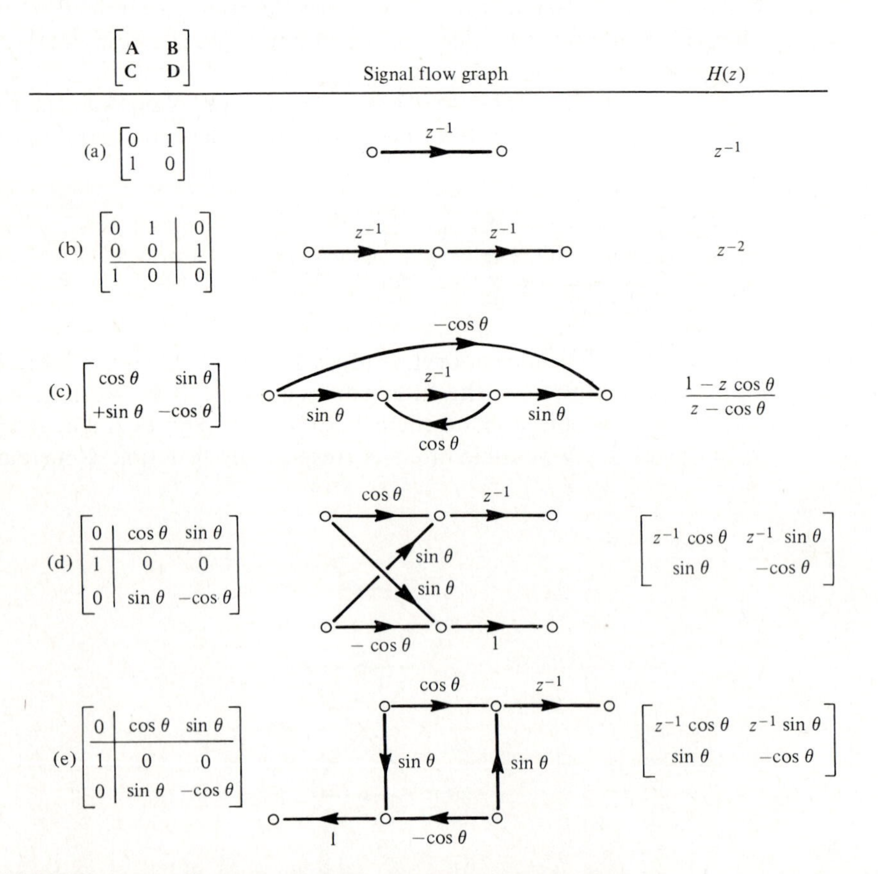

FIGURE 10.4.2 **Examples of simple orthogonal, all-pass filters.**

depicted a first-order, orthogonal, all-pass filter. In (d) and (e) of Fig. 10.4.2, we have identical filters, drawn differently to suggest sections of FIR and IIR lattice filters. The filter section in (e) is the basic module or building block in the normalized lattice filter of Gray and Markel [Gray 1975]. Notice in Fig. 10.4.2(e) that if we feed output 1 back to input 1, thereby eliminating one input and output, we obtain the filter in (c).

Connections of Orthogonal, All-Pass Filters

A set of simple filters in $\mathscr{F}_0$ or $\mathscr{F}$ may be used to create higher-order filters in $\mathscr{F}$ by *connecting any single output in a set of filters to any single input*, thereby eliminating both participants. Some examples are shown in Fig. 10.4.3. Notice that feedback connections are allowed. A remarkable property of filters in $\mathscr{F}$ is summarized in the following statement.

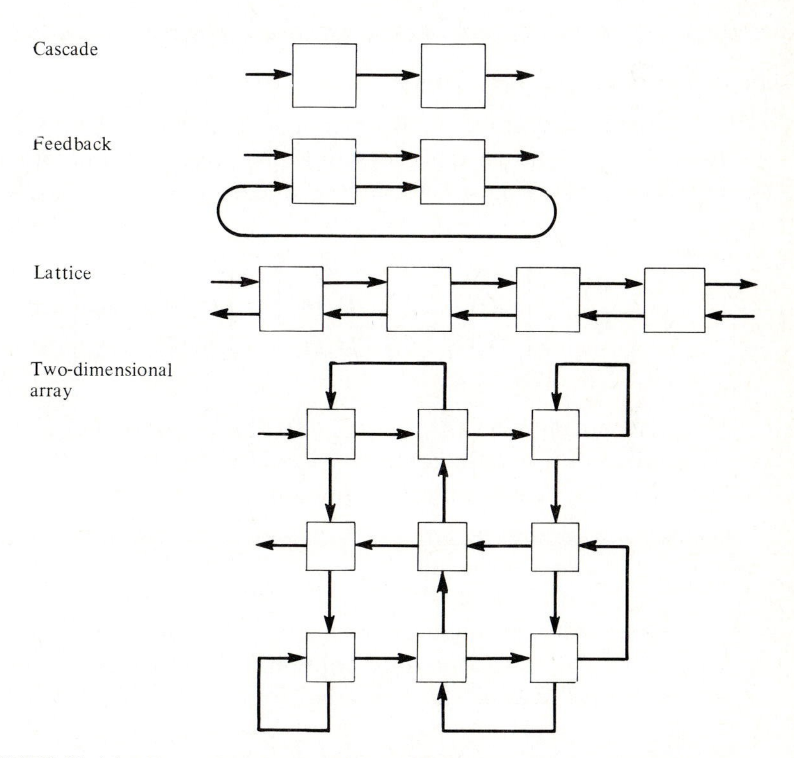

FIGURE 10.4.3 **Examples of connections of filters in $\mathscr{F}$ that are again filters in $\mathscr{F}$.**

PROPOSITION 3. Any connection of filters from $\mathscr{F}$ that does not introduce delay-free loops is again a filter in $\mathscr{F}$. (A connection is defined as connecting any single output in a set of filters to any single input.)

If in connecting orthogonal, all-pass filters a delay-free loop is formed, then it must be eliminated. The resulting filter is again in $\mathscr{F}$. The proof of Proposition 3 is based on the following properties of orthogonal matrices.

1. *$\mathcal{O}(N)$ is a multiplicative group.* **(10.4.21)**

$$2.\ \begin{bmatrix} \mathbf{A} & \mathbf{B} \\ \mathbf{C} & \mathbf{D} \end{bmatrix} \in \mathcal{O}(N) \Rightarrow \begin{bmatrix} \mathbf{A} & 0 & \mathbf{B} \\ 0 & \mathbf{I} & 0 \\ \mathbf{C} & 0 & \mathbf{D} \end{bmatrix} \in \mathcal{O}(N+M). \qquad \textbf{(10.4.22)}$$

3. *If* $\begin{bmatrix} \mathbf{A} & \mathbf{B} \\ \mathbf{C} & \mathbf{D} \end{bmatrix} \in \mathcal{O}(m+n)$ and $\mathbf{I} - \mathbf{D}$ is invertible,

then

$$\mathbf{A} + \mathbf{B}(\mathbf{I} - \mathbf{D})^{-1}\mathbf{C} \quad \in \quad \mathcal{O}(n). \qquad \textbf{(10.4.23)}$$

Cascade connections of filters use properties (10.4.21) and (10.4.22). Feedback connections use property (10.4.23) (see Appendix 10A).

EXAMPLE 10.4.1

Cascade connections of filters from $\mathscr{F}$.

Consider a cascade of two orthogonal, all-pass filters. Suppose the two filters have state variable descriptions given by $(\mathbf{A}_1, \mathbf{B}_1, \mathbf{C}_1, \mathbf{D}_1)$ and $(\mathbf{A}_2, \mathbf{B}_2, \mathbf{C}_2, \mathbf{D}_2)$. Figure 8.4.3 shows that the FSVD for the cascade connection is given by

$$\begin{bmatrix} \mathbf{A}_3 & \mathbf{B}_3 \\ \mathbf{C}_3 & \mathbf{D}_3 \end{bmatrix} = \left[\begin{array}{cc|c} \mathbf{A}_1 & 0 & \mathbf{B}_1 \\ \mathbf{B}_2\mathbf{C}_1 & \mathbf{A}_2 & \mathbf{B}_2\mathbf{D}_1 \\ \hline \mathbf{D}_2\mathbf{C}_1 & \mathbf{C}_2 & \mathbf{D}_2\mathbf{D}_1 \end{array}\right] = \begin{bmatrix} \mathbf{I} & 0 & 0 \\ 0 & \mathbf{A}_2 & \mathbf{B}_2 \\ 0 & \mathbf{C}_2 & \mathbf{D}_2 \end{bmatrix} \begin{bmatrix} \mathbf{A}_1 & 0 & \mathbf{B}_1 \\ 0 & \mathbf{I} & 0 \\ \mathbf{C}_1 & 0 & \mathbf{D}_1 \end{bmatrix}. \qquad \textbf{(10.4.24)}$$

Thus the cascade filter $(\mathbf{A}_3, \mathbf{B}_3, \mathbf{C}_3, \mathbf{D}_3)$ has a matrix representation that is the product of two orthogonal matrices. Thus by Eq. (10.4.21) the cascade is in $\mathscr{F}$. This can be generalized to any order cascade.

A connection of filters (such as a cascade) should properly be described using a factored state variable description. Thus we need to extend Definition 1 to this case. All that is required is that

$$\mathbf{F}_i \quad \in \quad \mathcal{O}(m+n) \quad \text{for } 1 \leqslant i \leqslant L \qquad \textbf{(10.4.25)}$$

to maintain orthogonality at each computation time. The product (10.4.5) will lead to an all-pass filter because of property (10.4.21) and Proposition 2.

EXAMPLE 10.4.2

Block processing filters in $\mathscr{F}$.

Another example of the surprising invariance properties of the set of filters **F** is demonstrated by block processing filters as discussed in Section

10.2. For a block length of 3, the state variable equations are

$$\begin{bmatrix} \mathbf{x}(k+3) \\ y(k) \\ y(k+1) \\ y(k+2) \end{bmatrix} = \begin{bmatrix} \mathbf{A}^3 & \mathbf{A}^2\mathbf{B} & \mathbf{AB} & \mathbf{B} \\ \mathbf{C} & D & 0 & 0 \\ \mathbf{CA} & \mathbf{CB} & D & 0 \\ \mathbf{CA}^2 & \mathbf{CAB} & \mathbf{CB} & D \end{bmatrix} \begin{bmatrix} \mathbf{x}(k) \\ u(k) \\ u(k+1) \\ u(k+2) \end{bmatrix}. \tag{10.4.26}$$

We can factor the state variable matrix in Eq. (10.4.26) as follows:

$$\begin{bmatrix} \mathbf{A} & 0 & 0 & \mathbf{B} \\ 0 & \mathbf{I} & 0 & 0 \\ 0 & 0 & \mathbf{I} & 0 \\ \mathbf{C} & 0 & 0 & D \end{bmatrix} \begin{bmatrix} \mathbf{A} & 0 & \mathbf{B} & 0 \\ 0 & \mathbf{I} & 0 & 0 \\ \mathbf{C} & 0 & D & 0 \\ 0 & 0 & 0 & \mathbf{I} \end{bmatrix} \begin{bmatrix} \mathbf{A} & \mathbf{B} & 0 & 0 \\ \mathbf{C} & D & 0 & 0 \\ 0 & 0 & \mathbf{I} & 0 \\ 0 & 0 & 0 & \mathbf{I} \end{bmatrix} \tag{10.4.27}$$

Now if the original single input, single output filter ($\mathbf{A}$, $\mathbf{B}$, $\mathbf{C}$, D) is orthogonal, all-pass (a member of $\mathscr{F}$), then the factorization in Eq. (10.4.27) [using Eqs. (10.4.21) and (10.4.22)] establishes that the block processing filter is also in $\mathscr{F}$.

EXAMPLE 10.4.3

Feedback connections for filters in $\mathscr{F}$.

Suppose ($\mathbf{A}$, $\mathbf{B}$, $\mathbf{C}$, $\mathbf{D}$) represents a two-input, two-output orthogonal, all-pass filter. What happens to the filter if we introduce feedback by connecting output 2 back to input 2? The state variable equations for the original filter are given by

$$\begin{bmatrix} \mathbf{x}(k+1) \\ y_1(k) \\ y_2(k) \end{bmatrix} = \left[\begin{array}{c|cc} \mathbf{A} & \mathbf{B}_1 & \mathbf{B}_2 \\ \hline \mathbf{C}_1 & \mathbf{D}_{11} & \mathbf{D}_{12} \\ \mathbf{C}_2 & \mathbf{D}_{21} & \mathbf{D}_{22} \end{array}\right] \begin{bmatrix} \mathbf{x}(k) \\ u_1(k) \\ u_2(k) \end{bmatrix}. \tag{10.4.28}$$

We introduce feedback by connecting $y_2(k)$ to $u_2(k)$, i.e., we add the equation

$$y_2(k) = u_2(k). \tag{10.4.29}$$

If we now eliminate these two variables from Eqs. (10.4.28) and (10.4.29), we obtain the state variable description for a filter with feedback as shown in Fig. 10.4.4. Thus we have

$$\begin{aligned} \mathbf{x}(k+1) &= \mathbf{A}\mathbf{x}(k) + \mathbf{B}_1 u_1(k) + \mathbf{B}_2 u_2(k), \\ y_1(k) &= \mathbf{C}_1\mathbf{x}(k) + \mathbf{D}_{11}u_1(k) + \mathbf{D}_{12}u_2(k), \\ y_2(k) &= \mathbf{C}_2\mathbf{x}(k) + \mathbf{D}_{21}u_1(k) + \mathbf{D}_{22}u_2(k), \\ u_2(k) &= y_2(k). \end{aligned} \tag{10.4.30}$$

Substituting $u_2(k)$ for $y_2(k)$, we obtain

$$u_2(k) = (\mathbf{I} - \mathbf{D}_{22})^{-1}[\mathbf{C}_2\mathbf{x}(k) + \mathbf{D}_{21}u_1(k)]. \tag{10.4.31}$$

Assuming ($\mathbf{I} - \mathbf{D}_{22}$) is invertible, we can substitute Eq. (10.4.31) into (10.4.30)

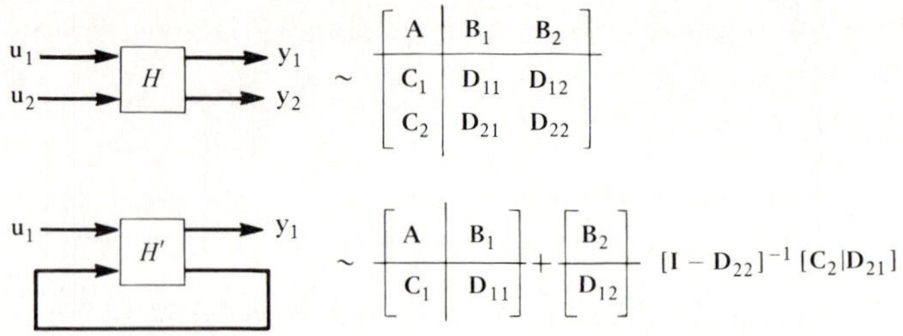

$$u_1, u_2 \rightarrow H \rightarrow y_1, y_2 \sim \left[\begin{array}{c|cc} A & B_1 & B_2 \\ \hline C_1 & D_{11} & D_{12} \\ C_2 & D_{21} & D_{22} \end{array}\right]$$

$$u_1 \rightarrow H' \rightarrow y_1 \sim \left[\begin{array}{c|c} A & B_1 \\ \hline C_1 & D_{11} \end{array}\right] + \left[\begin{array}{c} B_2 \\ \hline D_{12} \end{array}\right] [I - D_{22}]^{-1} [C_2 | D_{21}]$$

FIGURE 10.4.4 **A feedback connection for state variable filter in $\mathscr{F}$.**

to get

$$\begin{bmatrix} \mathbf{x}(k+1) \\ y_1(k) \end{bmatrix} = \left\{ \begin{bmatrix} \mathbf{A} & \mathbf{B}_1 \\ \mathbf{C}_1 & \mathbf{D}_{11} \end{bmatrix} + \begin{bmatrix} \mathbf{B}_2 \\ \mathbf{D}_{12} \end{bmatrix} [\mathbf{I} - \mathbf{D}_{22}]^{-1} [\mathbf{C}_2 \quad \mathbf{D}_{21}] \right\} \begin{bmatrix} \mathbf{x}(k) \\ u_1(k) \end{bmatrix}. \tag{10.4.32}$$

The term in brackets in Eq. (10.4.32) is the state variable representation of $\mathbf{H}'$ in Fig. 10.4.4. To show that this filter is in $\mathscr{F}$, we must show that this matrix is orthogonal. The proof of this result follows directly from Proposition 2 and is proved as a corollary in Appendix 10A. In the feedback connection a delay-free loop is introduced if appropriate elements of $\mathbf{D}_{22}$ are nonzero. However, if the filter is realized from the resulting matrix, i.e., the bracketed term in Eq. (10.4.32), there are no such loops. The computation of the second term on the right-hand side of Eq. (10.4.32) achieves the loop elimination.

As a final example, consider the class of filters generated by frequency transformations, i.e.,

$$G(z) = H(F(z)), \tag{10.4.33}$$

where $F(z)$ is a frequency transformation. If $H(z)$ is in $\mathscr{F}_0$, then $G(z)$ is again in $\mathscr{F}_0$. [See Mullis and Roberts (1976b).]

Decomposition of Orthogonal, All-Pass Filters into Primitives

In the previous section, we showed the set $\mathscr{F}$ of orthogonal, all-pass filters is closed under "proper" connections or transformations. For synthesis, we are more interested in how to decompose high-order filters into simple pieces. The simple pieces we would like to use are more precisely defined in Definition 2.

DEFINITION 2. A filter $\mathbf{F} \in \mathscr{F}$ is *primitive* if $\mathbf{F}$ contains at most four elements which differ from -1, 0, or 1. [Notice that in Fig. 10.4.2 all of the examples are primitive.]

The objective of our synthesis problem is to realize an arbitrary transfer function. Because of the remarkable invariance properties of orthogonal, all-pass filters we would like most of the filter to be orthogonal and all-pass. And, in fact, we shall show by direct computation that any nth-order transfer function can be constructed from an $(n-1)$-st order filter in $\mathscr{F}(2, n-1)$ as shown in Fig. 10.4.5. To construct an arbitrary nth-order transfer function, first build an $(n-1)$-st order two-input, two-output orthogonal, all-pass filter. Then connect one output to one input through a unit delay and form the output $y(k)$ as in Fig. 10.4.5. Assuming for the present that any nth-order transfer function can be realized as in Fig. 10.4.5, we can now concentrate on methods for obtaining orthogonal, all-pass filters with as few primitive elements as possible. We shall use two devices: factorization and orthogonal transformation. The result will be several types of modular filters with good finite register effects due to the orthogonality condition.

Suppose $\mathbf{H}(z)$ satisfies

$$\mathbf{H}(z)\mathbf{H}^T(z^{-1}) \equiv \mathbf{I}. \tag{10.4.34}$$

We wish to construct an orthogonal realization of $\mathbf{H}(z)$, which is the connection of as few primitive elements as possible. We can find a state variable realization $(\mathbf{A}, \mathbf{B}, \mathbf{C}, \mathbf{D})$ for which

$$\mathbf{F} = \begin{bmatrix} \mathbf{A} & \mathbf{B} \\ \mathbf{C} & \mathbf{D} \end{bmatrix} \in \mathcal{O}(m+n). \tag{10.4.35}$$

To find $(\mathbf{A}, \mathbf{B}, \mathbf{C}, \mathbf{D})$ simply take any minimal realization (like a direct form) and then apply a nonsingular transformation so that $\mathbf{K} = \mathbf{I}$. From this realization, additional realizations may be constructed using orthogonal transformations. If $\mathbf{T} \in \mathcal{O}(n)$, then $\mathbf{F}'$ given in Eq. (10.4.36) is again orthogonal. We seek a sparse realization

$$\mathbf{F}' = \begin{bmatrix} \mathbf{A}' & \mathbf{B}' \\ \mathbf{C}' & \mathbf{D}' \end{bmatrix} = \begin{bmatrix} \mathbf{T}^T\mathbf{A}\mathbf{T} & \mathbf{T}^T\mathbf{B} \\ \mathbf{C}\mathbf{T} & \mathbf{D} \end{bmatrix} = \begin{bmatrix} \mathbf{T} & 0 \\ 0 & \mathbf{I} \end{bmatrix}^T \begin{bmatrix} \mathbf{A} & \mathbf{B} \\ \mathbf{C} & \mathbf{D} \end{bmatrix} \begin{bmatrix} \mathbf{T} & 0 \\ 0 & \mathbf{I} \end{bmatrix}. \tag{10.4.36}$$

of $\mathbf{H}(z)$. We may apply any transformation of the form Eq. (10.4.36) and also factor $\mathbf{F}'$ into primitive elements.

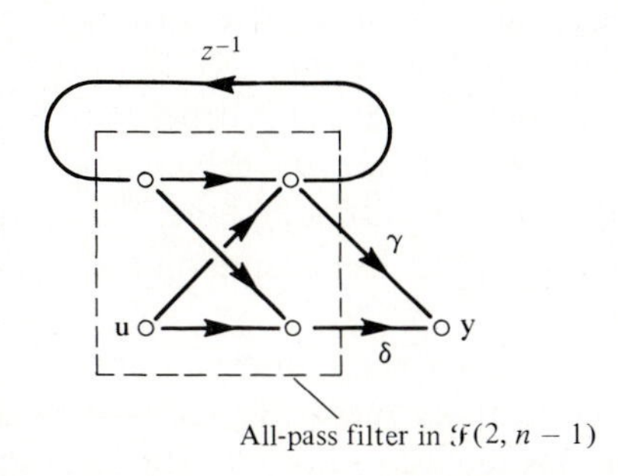

FIGURE 10.4.5 **Construction of an arbitrary transfer function using an appropriate two input/output all-pass filter.**

In order to understand how to transform and factor the matrix **F** in Eq. (10.4.35), we will introduce definitions that we can use in evaluating factorizations of **F**. One such definition is the "depth of **F**."

DEFINITION 3. Let $\mathbf{F} \in \mathcal{O}(N)$. Then

$$depth(\mathbf{F}) \triangleq \min\{d : F_{ij} = 0 \text{ whenever } i - j > d\}.$$

The depth is the number of nonzero diagonals of **F** below the main diagonal. **I** has depth zero.

The next three propositions create the tools we shall use to obtain sparse primitive factors for **F**.

PROPOSITION 4. Let $\mathbf{F} \in \mathcal{O}(n + m)$ be partitioned as in Eq. (10.4.35) with $\mathbf{A}$ $n \times n$. There exists $\mathbf{T} \in \mathcal{O}(n)$ for which $\mathbf{F}'$ has depth m.

Proof: The basic operation used to constructively prove both this and the following two propositions is a rotation applied in a coordinate plane to produce a zero. It is a "normalized Gaussian elimination" procedure.

We can always choose an angle θ (a function of F_{ij} and $F_{i,j+1}$) so that

$$[F_{ij} \quad F_{i,j+1}]\begin{bmatrix} \cos\theta & \sin\theta \\ -\sin\theta & \cos\theta \end{bmatrix} = [0 \quad F'_{i,j+1}] \tag{10.4.37}$$

The angle θ is given by

$$\tan\theta = \frac{F_{ij}}{F_{i,j+1}}. \tag{10.4.38}$$

Consider the case $n = 4$ and $m = 2$ so that $\mathbf{F} \in \mathcal{O}(n + m)$ is 6×6 as shown in Fig. 10.4.6.

$$\mathbf{F}' = \begin{bmatrix} X & X & X & X & X & X \\ X & X & X & X & X & X \\ X & X & X & X & X & X \\ \otimes_6 & X & X & X & X & X \\ \otimes_4 & \otimes_5 & X & X & X & X \\ \otimes_1 & \otimes_2 & \otimes_3 & X & X & X \end{bmatrix} = \mathbf{T}_6^T\mathbf{T}_5^T\mathbf{T}_4^T\mathbf{T}_3^T\mathbf{T}_2^T\mathbf{T}_1^T\mathbf{F}\mathbf{T}_1\mathbf{T}_2\mathbf{T}_3\mathbf{T}_4\mathbf{T}_5\mathbf{T}_6$$

$n = 4,\ m = 2$

FIGURE 10.4.6 **The order in which zeros are created using rotations (which are admissible orthogonal transformations in the sense of Eq. (10.4.36)) is by row.**

The zeros in the southwest corner of $\mathbf{F}'$ in Fig. 10.4.6 are obtained in the order shown by the subscripts. The first zero is obtained by a 2×2 rotation in the (1, 2) coordinate plane. This consists, on the right, of computations on columns 1 and 2 of **F** and, on the left, on the first and second rows of **F** (leaving behind the zero). We continue in this fashion along the bottom row stopping (in this example) with zero number three, which involves a rotation in the (3, 4) coordinate plane. We cannot proceed further along the bottom row because this would require a transformation in coordinates 5 or 6. This is not admissible because Eq. (10.4.36) would be violated. Thus we begin again in row 5 and proceed as before.

The following is to be proved in Problem 10.21.

PROPOSITION 5. If $\mathbf{F} \in \mathcal{O}(N)$ and is upper triangular (depth is zero), then $\mathbf{F}$ is diagonal with diagonal elements of ± 1.

PROPOSITION 6. If $\mathbf{F} \in \mathcal{O}(n + m)$, then there is a transformation of the form in Eq. (10.4.36) for which $\mathbf{F}'$ is the product of q primitive matrices (having at most four noninteger elements), where q is given by

$$q = \sum_{k=1}^{m} (m + n - k) = mn + \frac{m(m-1)}{2}. \tag{10.4.40}$$

Proof: Using Proposition 4, we first bring $\mathbf{F}$ to $\mathbf{F}'$ having depth m. We then continue coordinate rotations on the right only. Referring to Fig. 10.4.7, we create zeros in the order of the subscripts shown. One can check that this order in producing zeros cannot introduce a nonzero number in a previously zeroed entry. The number q in Eq. (10.4.40) is simply the total number of elements that have to be zeroed. The result is that the matrix

$$\mathbf{F}'' = \mathbf{F}'\mathbf{F}_1\mathbf{F}_2 \cdots \mathbf{F}_q \tag{10.4.41}$$

is upper triangular. $\mathbf{F}_k$ is the rotation required to obtain the kth zero. By Proposition 5, $\mathbf{F}''$ is diagonal with diagonal elements ± 1. Thus we can factor the orthogonal matrix $\mathbf{F}'$ as

$$\mathbf{F}' = \mathbf{F}_q^T\mathbf{F}_{q-1}^T \cdots \mathbf{F}_1^T. \tag{10.4.42}$$

In Eq. (10.4.42), we have combined any -1's on the diagonal of $\mathbf{F}''$ into the rotation matrices which constitute the primitive factors of $\mathbf{F}'$. Notice that for $m = 2$, $q = 2n + 1$, which is the number of degrees of freedom in an nth-order filter.

$$F'' = \begin{bmatrix} X & X & X & X & X & X \\ \otimes_9 & X & X & X & X & X \\ \otimes_4 & \otimes_8 & X & X & X & X \\ \bigcirc & \otimes_3 & \otimes_7 & X & X & X \\ \bigcirc & \bigcirc & \otimes_2 & \otimes_6 & X & X \\ \bigcirc & \bigcirc & \bigcirc & \otimes_1 & \otimes_5 & X \end{bmatrix} = F' F_1F_2F_3F_4F_5F_6 F_7 F_8 F_9$$

FIGURE 10.4.7 **The order in which zeros are created (using rotations on the right only) is by diagonal, rather than by row. These rotations are illegal transformations, but lead to a primitive factorization.**

We now have a factorization of any filter in $\mathcal{F}(m, n)$ in terms of q rotations, the primitive elements. For purposes of a modular filter realization, we are interested in alternate factorizations. We shall use Eq. (10.4.42) as a starting point. From our discussion of FSVDs in Chapter 8, Eq. (10.4.42) is a realization with a longest delay-free path of q. We can implement such a realization by time multiplexing a single rotational unit q times for each output. We seek other FSVDs that will trade a small number of time multiplexed modules for a factorization which uses several modules in parallel, thereby reducing the number of factors in Eq. (10.4.42).

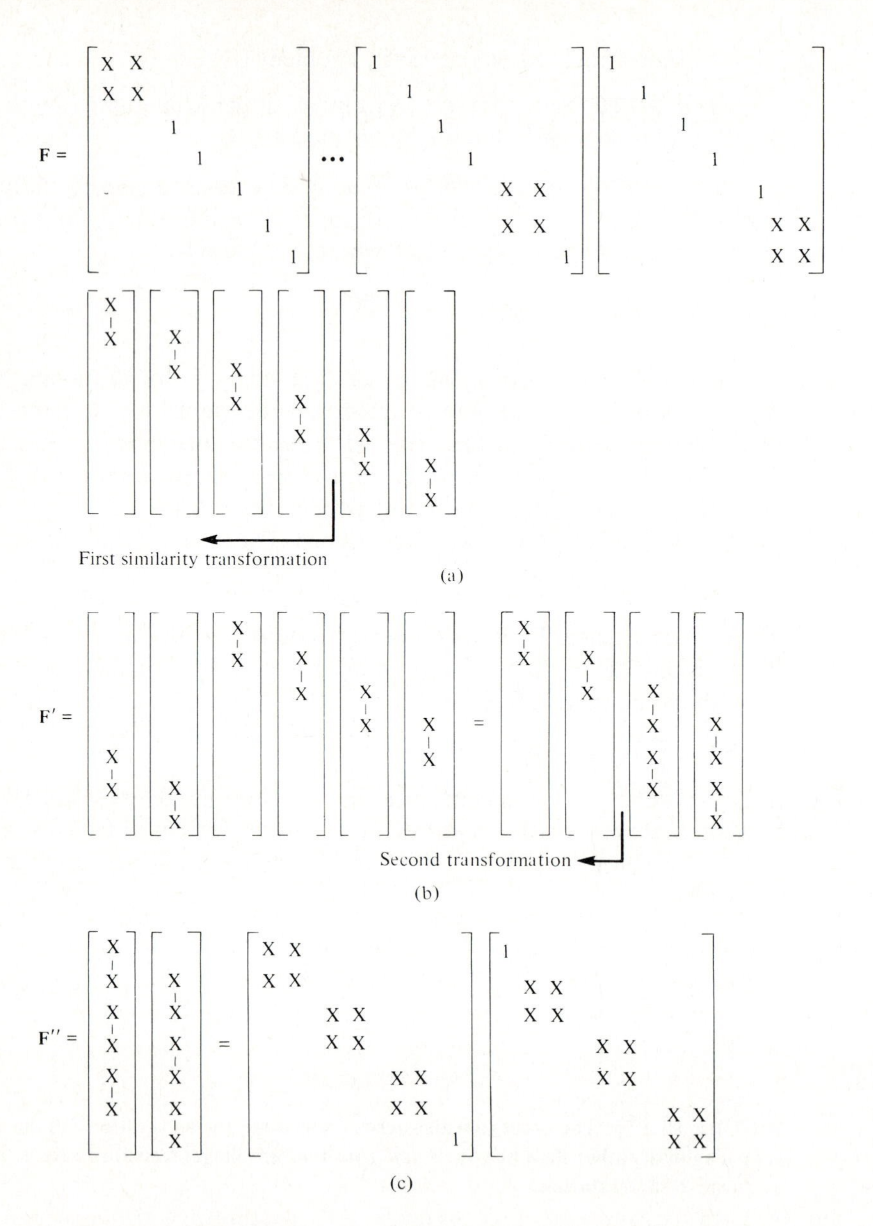

FIGURE 10.4.8 Manipulating the FSVD of Eq. (10.4.42) to obtain *width* = 2, *period* = 2, for the case *m* = 1.

In order to develop alternate factorizations or FSVDs, we have found it convenient to use a schematic shorthand for equations. Consider Fig. 10.4.8. In this example, the number of inputs and outputs is $m = 1$. The factorization in Eq. (10.4.42) has the form shown in part (a) of Fig. 10.4.8. Each of the $n - 1$ matrix factors differs from the identity only in a 2×2 rotation in a coordinate plane. Each of the coordinate planes involve adjacent indices. To represent these factors, we use the notation of the second portion of Fig. 10.4.8(a). Each column represents a matrix which is the identity except in the coordinate plane represented by the x's. Two such factors commute if and only if they have no common x positions, i.e., they "slide" horizontally past each other.

Now consider the similarity transformation indicated by the arrow in Fig. 10.4.8(a). If $\mathbf{T}$ is the matrix formed by multiplying the primitives over the arrow, then $\mathbf{T}$ is in the class required in Eq. (10.4.36) since none of the x positions include the $(n + 1)$-st coordinate. Now form a new matrix $\mathbf{F}'$ as shown in Fig. 10.4.8(b) using

$$\mathbf{F}' = \mathbf{T}^{-1}\mathbf{F}\mathbf{T}. \tag{10.4.43}$$

The operations in Eq. (10.4.43) are equivalent to moving the factors of $\mathbf{T}$ around to the right. (This kind of transformation is equivalent to a rotational permutation of the order of the factors. Such a transformation is admissible so long as the part being moved from left to right contains no x's in the last m coordinate positions.)

On the right-hand side of Fig. 10.4.8(b), we have used commutativity to compact the result. A second transformation as shown results in the final form shown in Fig. 10.4.8(c). This result is a product of two block diagonal matrices.

The series of transformations performed in this example has transformed a highly time-multiplexed structure into a highly parallel structure. The structure corresponding to FSVD at the top of Fig. 10.4.8 has a longest delay-free path or period equal to the number of factors. The structure corresponding to the FSVD at the bottom of Fig. 10.4.8 has a period of 2 and a width of 2. The FSVDs are easily interpreted from the point of view of hardware implementation. For example, the FSVD in Figure 10.4.8(a) implies a single rotation computation is performed sequentially six times for each output. Thus one rotational unit can be used and time-multiplexed to obtain the output. The FSVD in Fig. 10.4.8(c) implies the use of three rotational units, time-multiplexed twice to obtain the output.

We conclude this section with a characterization of the form of the computational units needed and the period required to obtain outputs for orthogonal, all-pass filters.

DEFINITION 4. An FSVD with *period* L and *width* w has the form

$$\mathbf{F} = \mathbf{F}_L\mathbf{F}_{L-1}\cdots\mathbf{F}_1,$$

where each of the L factors $\mathbf{F}_k$ is block diagonal with diagonal blocks of size at most $w \times w$.

PROPOSITION 7. Let $\mathbf{F} \in \mathcal{O}(n + m)$. Then there is a transformation $\mathbf{T} \in \mathcal{O}(n + m)$ which is admissible in the sense of Eq. (10.4.36), for which $\mathbf{F}' = \mathbf{T}^{-1}\mathbf{F}\mathbf{T}$ has an FSVD of period L and width w whenever

$$w + L = 2(m + 1), \qquad 2 \leqslant w \leqslant 2m. \tag{10.4.44}$$

The proof of this proposition requires more words than is worthwhile, but is constructive and can be inferred from examples. In Fig. 10.4.9, an FSVD (in compressed notation) is shown for $m = 1$, $w = 2$, and $L = 4$ satisfying Eq. (10.4.44). In part (a) is an FSVD of Eq. (10.4.42) where commutativity has been used to combine the various rotations as much as possible. A series of transformations as shown in parts (b) and (c) "wraps the factors around a cylinder" leaving a result $m = 2$, $w = 2$, and $L = 4$. In Fig. 10.4.10, a new structure is shown in (a). Pairs of adjacent rotations have been multiplied together to create 3×3 factors as the primitive elements. The series of transformations in (b) and (c) result in a structure with $m = 2$, $w = 2$, $L = 3$. Notice the tradeoff in these two examples between hardware complexity (as measured by w) and the degree of time-multiplexing (as measured by L).

Processor Modules for Width Two Realizations

The basic component of the structures of Figs. 10.4.8 and 10.4.9 is a 2×2 rotation indicated by the joined x's. Consider the structure of Fig. 10.4.11(a), which has period $L = 4$. Each processor module must be assigned one of the rotations for each of the four minor cycles. To simplify interprocessor communication, this assignment should be made as indicated in part (a) of the figure. Notice that the processor must "move" up and down between minor cycles, but that one internal variable is always common.

The processor must contain an arithmetic unit that can do the rotation and a pair of memory registers to save the result. The configuration is shown on the left of part (b) of the figure. During even minor cycle computations, a processor gets one input from its neighbor below while the other input comes from one of its own internal registers. This connection is shown with the dotted lines. A less confusing but equivalent version of the same signal flow graph is shown on the right-hand side. During odd minor cycles, the configuration changes, because one input must come from the neighboring processor above.

The result is a processor module that has an arithmetic unit to do rotations, two storage registers, and a single interprocessor communication path for each of its two neighbors. This path must be bidirectional, since the direction of the data flow is up on even cycles and down on odd cycles. Thus the processor array is one-dimensional with a single bidirectional communication path between processors. The number of processors is roughly one half the filter order, but the "latency" (or time between inputs and outputs) is four minor cycles *no matter how large the filter order*!

Synthesis of Orthogonal Filters—General Case

Let $G(z)$ be the transfer function for an order n filter that is not all-pass. We seek a realization of $G(z)$ that satisfies Eq. (10.4.4), i.e., an orthogonal structure. From any state variable realization, one can obtain a realization for which

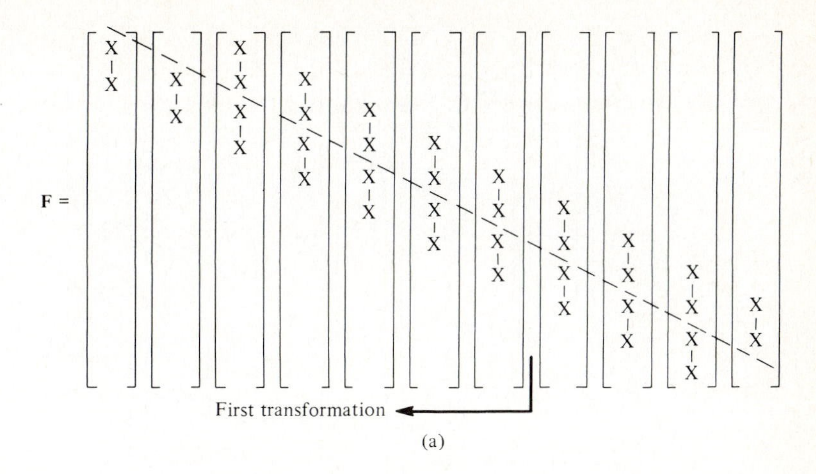

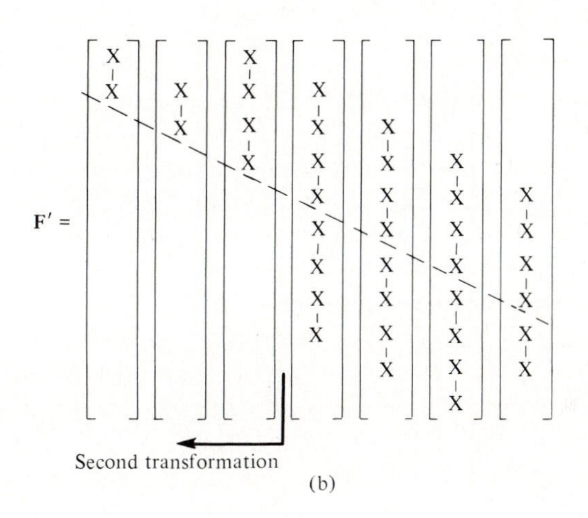

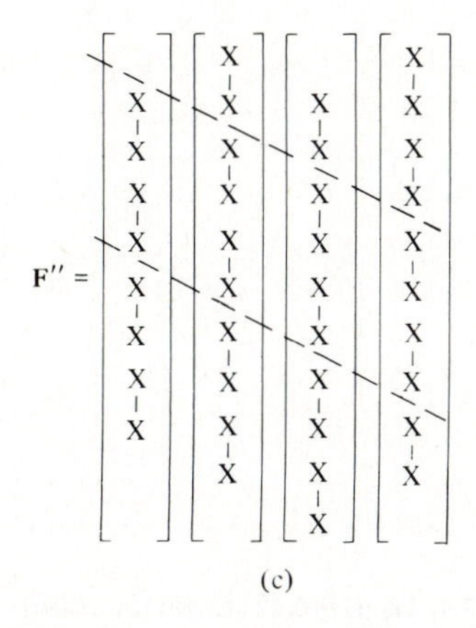

FIGURE 10.4.9 Manipulating a FSVD of Eq. (10.4.42) to obtain *width* = 2, *period* = 4, for the case $m = 2$.

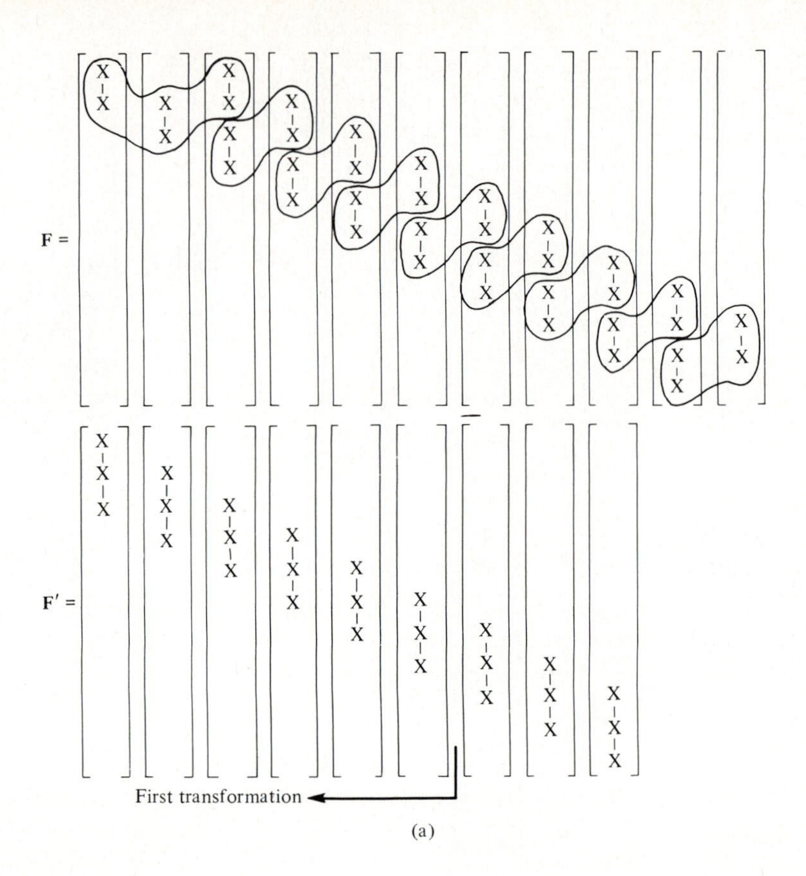

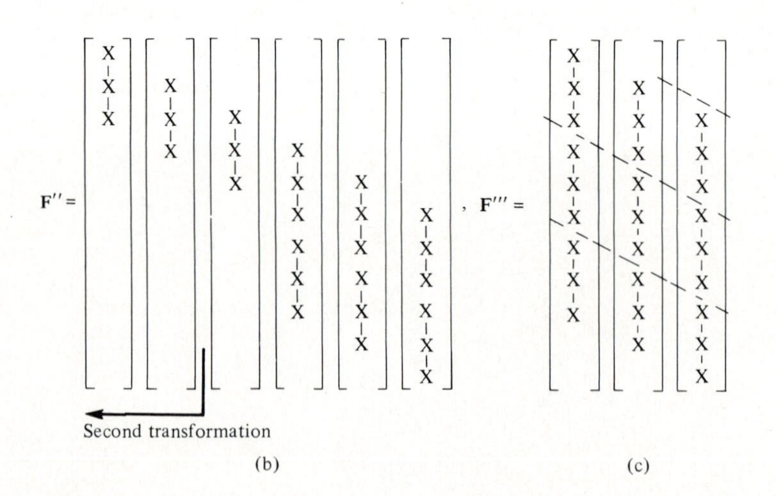

FIGURE 10.4.10 Manipulating a FSVD of Eq. (10.4.42) to obtain *width* = 3, *period* = 3, for the case *m* = 2.

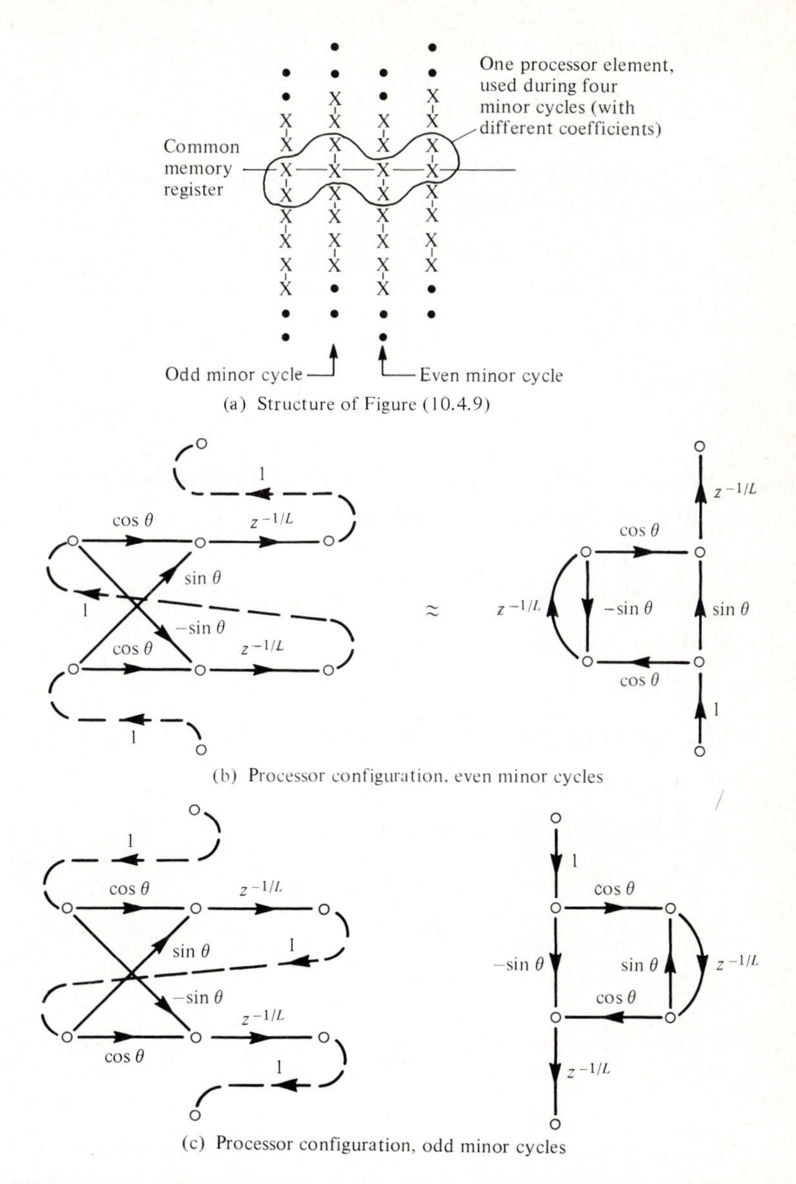

FIGURE 10.4.11 Characterization of processor modules consistent with the FSVDs of Figs. (10.4.8) and (10.4.9).

$\mathbf{K} = \mathbf{I}$. Therefore, suppose that

$$\mathbf{G} = \begin{bmatrix} \mathbf{A} & \mathbf{B} \\ \mathbf{C}_1 & \mathbf{D}_1 \end{bmatrix} \quad \text{realizes} \quad G(z), \quad \mathbf{AA}^T + \mathbf{BB}^T = \mathbf{I}. \tag{10.4.45}$$

The first n rows of $\mathbf{G}$ are orthonormal, but have dimension $n + 1$. Therefore, we can choose one more to complete the set. That is to say that there exists

C, **D** for which

$$\mathbf{F} = \begin{bmatrix} \mathbf{A} & \mathbf{B} \\ \mathbf{C} & \mathbf{D} \end{bmatrix} \in \mathcal{O}(n+1). \tag{10.4.46}$$

The matrices **F** and **G** differ only in the last row. The differences can be measured as follows. Letting

$$[\boldsymbol{\Gamma}, \delta] = [\mathbf{C}_1 \quad \mathbf{D}_1]\mathbf{F}^T, \tag{10.4.47}$$

$$\mathbf{G}_0 = \begin{bmatrix} \mathbf{I} & 0 \\ \boldsymbol{\Gamma} & \delta \end{bmatrix}, \tag{10.4.48}$$

we have

$$\mathbf{G} = \mathbf{G}_0\mathbf{F}. \tag{10.4.49}$$

The extra matrix $\mathbf{G}_0$ converts **F** from an all-pass filter to $G(z)$, but it is not as simple as it might be. Choose

$$\mathbf{T} = \begin{bmatrix} \mathbf{T}_0 & 0 \\ 0 & 1 \end{bmatrix} \in \mathcal{O}(n+1) \tag{10.4.50}$$

so that

$$\boldsymbol{\Gamma}\mathbf{T}_0 = [\;\mathbf{0}\;\; \gamma] \tag{10.4.51}$$

and apply the transformation to Eq. (10.4.49):

$$\mathbf{G}' = \mathbf{T}^{-1}\mathbf{G}\mathbf{T} = (\mathbf{T}^{-1}\mathbf{G}_0\mathbf{T})(\mathbf{T}^{-1}\mathbf{F}\mathbf{T}) = \mathbf{G}_0'\mathbf{F}'. \tag{10.4.52}$$

Then $\mathbf{F}'$ is still orthogonal and

$$\mathbf{G}_0' = \left[\begin{array}{ccc|cc} 1 & & \mathbf{0} & & \\ & \ddots & & & \mathbf{0} \\ \mathbf{0} & & 1 & & \\ \hline & & & 1 & 0 \\ & \mathbf{0} & & \gamma & \delta \end{array}\right] \tag{10.4.53}$$

has only two nontrivial elements. The transformation **T** in Eq. (10.4.50) is admissible in the sense of Eq. (10.4.36) with $m = 1$. We can continue to transform and factor the product in Eq. (10.4.52). However, in order to preserve $\mathbf{G}_0'$, we must restrict transformations to

$$\left[\begin{array}{c|cc} \mathbf{T}_1 & \multicolumn{2}{c}{\mathbf{0}} \\ \hline \mathbf{0} & 1 & 0 \\ & 0 & 1 \end{array}\right] \in \mathcal{O}(n+1). \tag{10.4.54}$$

This is equivalent to operating on $\mathbf{F}'$ as if it belonged to $\mathscr{F}(2, n-1)$. The last $m = 2$ coordinates are to be left alone. This two-input, two-output all-pass filter is the one in Fig. 10.4.5. We can realize

$$\mathbf{F}'' \in \mathscr{F}(2, n-1)$$

with

$$q = 2(n-1) + 1 = 2n - 1$$

rotations, which can be arranged in various ways:

- $L = 2n - 1$ with one, 2×2 processing element,
- $L = 4$ with about $n/2$, 2×2 processors,
- $L = 3$ with about $n/3$, 3×3 processors.

The result gives an orthogonal structure for $H(z)$, with

$$\mathbf{G}'' = (\mathbf{G}_0' \mathbf{F}_L'')\mathbf{F}_{L-1}'' \cdots \mathbf{F}_1''. \tag{10.4.55}$$

(The sparse matrix $\mathbf{G}_0'$ can be combined with $\mathbf{F}_L''$ without increasing the width.)

Overflow Oscillations

All-pass filters can be thought of as energy buffers. They obey the following "energy conservation" law (see Problem 10.17). Let

$$\mathbf{F} = \mathbf{F}_L \cdots \mathbf{F}_1 \in \mathscr{F}_0(m, n). \tag{10.4.56}$$

Then for $k_1 > k_0$,

$$\|\mathbf{x}(k_1)\|^2 = \|\mathbf{x}(k_0)\|^2 + \sum_{k=k_0}^{k_1-1} \|\mathbf{u}(k)\|^2 - \sum_{k=k_0}^{k_1-1} \|\mathbf{y}(k)\|^2. \tag{10.4.57}$$

If we interpret $\|\mathbf{x}\|^2$ as stored energy, then the energy stored at time k_1 is equal to the energy stored at time k_0 plus the energy input minus the energy output. When the input is zero, the energy equality becomes

$$\|\mathbf{x}(k_1)\|^2 = \|\mathbf{x}(k_0)\|^2 - \sum_{k=k_0}^{k_1-1} \|\mathbf{y}(k)\|^2. \tag{10.4.58}$$

This can be used to demonstrate that orthogonal filter structures cannot sustain overflow oscillations.

The ground rules for studying large-scale oscillations were discussed in Sections 9.3 and 9.4. One assumes a nonzero initial state $\mathbf{x}(k)$ and sets the input to zero. The nonlinearity due to register overflow is included in the state variable equations. (Quantization nonlinearities are not included.) The linear equations (10.4.4) become

$$\mathbf{q}_i(k) = f(\mathbf{F}_i \mathbf{q}_{i-1}(k)), \qquad 1 \leqslant i \leqslant L, \tag{10.4.59}$$

where f is the overflow nonlinearity. This function satisfies

$$\|f(\mathbf{q})\|^2 \leqslant \|\mathbf{q}\|^2, \tag{10.4.60}$$

which implies that

$$\|\mathbf{q}_i(k)\|^2 \leqslant \|\mathbf{F}_i \mathbf{q}_{i-1}(k)\|^2 = \|\mathbf{q}_{i-1}(k)\|^2, \tag{10.4.61}$$

since $\mathbf{F}_i$ is norm preserving. Taken together, these inequalities yield

$$\|\mathbf{x}(k+1)\|^2 + y(k)^2 = \|\mathbf{q}_L(k)\|^2 \leqslant \cdots \leqslant \|\mathbf{q}_0(k)\|^2 = \|x(k)\|^2. \tag{10.4.62}$$

[since $u(k) = 0$]. Extrapolation of this inequality in time leads to the conclusion that

$$\|\mathbf{x}(k+n)\|^2 \leqslant \|\mathbf{x}(k)\|^2$$

with equality only if

$$\begin{aligned} &1.\ y(k) = y(k+1) = \cdots = y(k+n-1) = 0, \qquad \text{and} \\ &2.\ \textit{no overflows have occurred.} \end{aligned} \tag{10.4.63}$$

[If an overflow had occurred, then there would have been a decrease in norm in Eq. (10.4.61).] Now if no overflows have occurred the output agrees with the linear filter, which is observable. In this context, Eq. (10.4.63) implies $\mathbf{x}(k) = 0$, which is a contradiction. Therefore, the trajectory must approach the origin, and oscillations cannot be sustained.

General orthogonal filters are realized using Eq. (10.4.49). That structure differs from its associated all-pass filter only in the output equation. The state trajectory must be the same. Consequently, no overflow oscillation is possible.

Noise Gain

Consider an orthogonal state variable structure. Since $\mathbf{K} = \mathbf{I}$, the "noise gain" which appears in the expression (9.9.15) is

$$g_{\text{orth}} = \frac{1}{n}\sum_{i=1}^{n} K_{ii}W_{ii} = \frac{1}{n}\sum_{i=1}^{n} W_{ii} = \frac{1}{n}\operatorname{Tr}(\mathbf{W}) = \frac{1}{n}\operatorname{Tr}(\mathbf{KW}) = \frac{1}{n}\sum_{i=1}^{n}\mu_i^2, \tag{10.4.64}$$

where the μ_i^2 are the second-order modes, or eigenvalues of $\mathbf{KW}$. [See Problem 8.35 for an alternate expression.] For a minimum noise structure,

$$g_{\min} = \frac{1}{n}\sum_{i=1}^{n} K_{ii}W_{ii} = \left[\frac{1}{n}\sum_{i=1}^{n}\mu_i\right]^2. \tag{10.4.65}$$

Thus the difference between g_{orth} and $g_{\min}$ is the difference between a second moment and the square of a first moment, which is the variance. The two are equal only if the μ_i are all equal. Since the second-order modes are invariant under frequency transformation [see Problem 9.13], the noise gain does not depend on the bandwidth for a low-pass filter. To understand how significant this is, one must look at a structure that does not have this property. For an nth-order direct from structure, the noise gain of the filter

$$G(z) = H\left(\frac{z-\alpha}{1-\alpha z}\right) \tag{10.4.66}$$

is

$$g_{\text{direct}}(\alpha) = (1-\alpha^2)^{-2(n-1)}P(\alpha), \tag{10.4.67}$$

where $P(\alpha)$ is a polynomial that is strictly positive for $\alpha^2 \leq 1$. Thus the noise gain of a direct form low-pass filter approaches infinity quite rapidly as the filter bandwidth approaches zero. [Equation (10.4.67) is derived in Mullis and Roberts (1976b).] The flat noise gain property of orthogonal filters was first reported in Gray and Markel (1975).

For the factored state variable structure (10.4.4), one must use the generalization of the roundoff noise variance formula found in Eq. (9.10.11). Using Eqs. (9.10.9) and (9.10.10), and orthogonality, a noise gain upper bound of $L(g_{\text{orth}} + 1)$ may be established. This is also invariant under frequency transformation.

Summary • In this chapter we have presented several examples of algorithms for digital filters. Those algorithms take on very different forms because each algorithm or structure is chosen to optimize criteria that are related to the available computational resources. As we change the resources available to compute a given I/O characteristic, we must also change the algorithms to conform to new constraints. Although we have discussed examples taken from digital filtering, the same principles apply to more general digital processing problems. The criteria we have considered include data throughput, complexity, finite-register effects, and certain "geometric" properties of the algorithm. Generally we cannot optimize all the criteria we would like to simultaneously. Thus the objective in the design of algorithms is to develop a theory of design that allows one to trade off between competing goals. The examples presented here are an attempt at such a theory.

PROBLEMS

10.1. Suppose we wish to calculate $y = \sum_{i=1}^{8} \alpha_i x_i$ using distributed arithmetic. Assume the word length of each x_i is 8 bits. If we use 8 memories and address each memory with 8 address lines, sketch a distributed arithmetic implementation. How many words of memory do the 8 memories contain?

10.2. A second-order digital filter can be described in direct form by the difference equation

$$y(k) = b_0 u(k) + b_1 u(k-1) + b_2 u(k-2) - a_1 y(k-1) - a_2 y(k-2).$$

Design two distributed arithmetic implementations for this recursive specification.

10.3. Show that the two FIR structures in Fig. 10P.3 are equivalent.

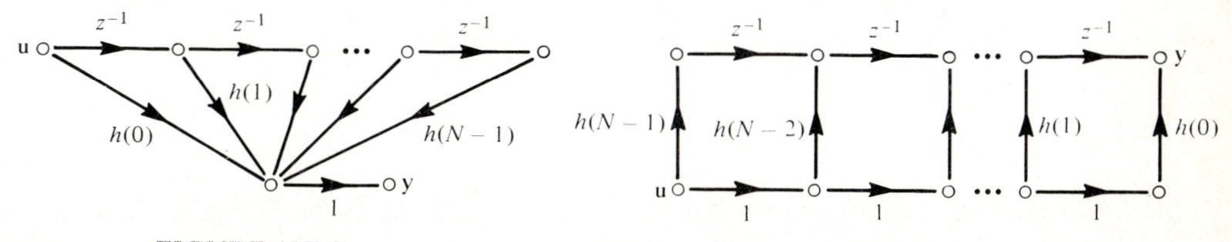

FIGURE 10P.3.

10.4. Consider four alternative distributed arithmetic designs for the calculation of $y = \alpha_1 x_1 + \alpha_2 x_2 + \alpha_3 x_3$. Assume α_1, α_2, α_3 are constants, x_1, x_2, x_3 are variables of b bits. Design four distributed arithmetic implementations and compare the amount of hardware and throughput rate of the designs. The first design is the simple single memory, 1 bit/cycle implementation we used in our introduction of distributed arithmetic. The second design is a two memory, 2 bits/cycle implementation. Memory A is addressed by x_1, say. Memory B is addressed by both x_2 and x_3. The third design consists of b memories, each memory with three address lines—one each from x_1, x_2, and x_3. The memory readout requires a single clock cycle. The fourth design consists of $b/3$ memories (b is assumed to be divisible by 3) with 9 address lines—3 each from x_1, x_2, and x_3. The memory readout again requires only a single clock cycle. Which of these designs is the most efficient? What is your basis of comparison?

10.5. An nth-order SISO state space structure requires $(n + 1)^2$ multiplications and $n(n + 1)$ additions per output sample. Using such a state space structure in a block filter, determine the number of multiplications and additions per output sample for the MIMO state-space filter. Assume a block length of L.

10.6. Assuming an nth-order filter subdivided into mth-order subsections (where n is an integer multiple of m), calculate the number of multiplies for an output sample for mth-order subsections connected in cascade and in parallel. (Assume the underlying SISO filter is a full state-space structure.)

10.7. Expand Fig. 10.2.4 to indicate what each of the elements in the block diagram accomplishes. Explain the operation of the inner product box. How could distributed arithmetic be used in performing these calculations? Sketch a more detailed block diagram of this implementation employing distributed arithmetic.

10.8. Time-multiplexing and parallelism are tradeoffs often made in the design of hardware. As throughput requirements increase in a given application, one must consider parallelism of additional hardware to increase throughput. Consider, for example, the implementation of a direct form FIR filter as shown in Fig. 10P.8. If we have available a single multiplier-accumulator element, describe the sequence of operations we need to perform to calculate the output sequence. Suppose we have three multiplier-accumulator elements. Describe a sequence of operations to obtain the output samples.

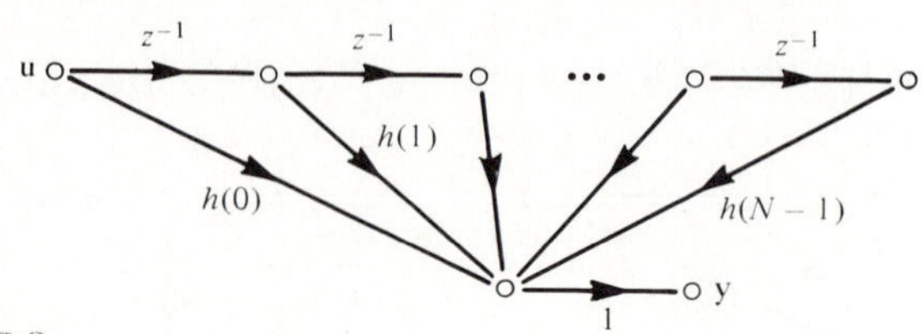

FIGURE 10P.8.

10.9. In the text sampling rate conversions are assumed to be integers. One can generalize that discussion to include conversion by the ratio

$$\frac{t_0'}{t_0} = \frac{M}{L}\,;\, f' = \frac{L}{M}\, f.$$

This conversion can be obtained by cascading two processors. The first increases the sampling rate by L, followed by a decimator with factor M. Interpolation must precede decimation. Why?

In general, the structure is as shown in Fig. 10P.9. Consider the combination $\mathbf{h} = \mathbf{h}_1 * \mathbf{h}_2$. Ideally, what is the frequency response $H(e^{j\theta})$? Assuming $\mathbf{h}$ satisfies the ideal characteristic, what is $Y(e^{j\theta})$?

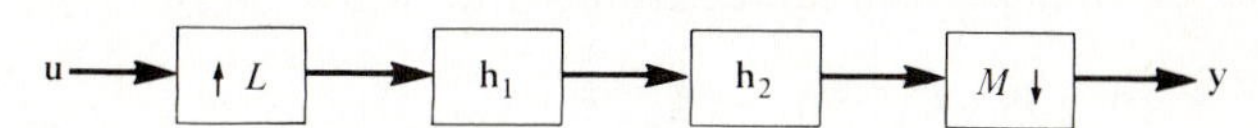

FIGURE 10P.9.

10.10. For large values of decimation M or interpolation L or where $L/M \simeq 1$ but $L \gg 1$, $M \gg 1$, multistage implementations of sampling rate converters are more efficient. We can show, in general, that the number of taps required in an FIR filter is roughly (*constant*) $f_s/\Delta f$, where f_s is sampling frequency, Δf the transition width, and the constant depends on the ripple specifications. Explain why a multistage decimator or interpolator is more efficient.

10.11. Consider a second-order digital filter with state space description

$$\mathbf{x}(k+1) = \mathbf{A}x(k) + \mathbf{B}u(k), \qquad y(k) = \mathbf{C}\mathbf{x}(k) + Du(k).$$

Show that one can rewrite these equations in the form

$$\mathbf{x}(k+1) = \mathbf{A}^L\mathbf{x}(k+1-L) + \sum_{m=0}^{L-1} \mathbf{A}^m\mathbf{B}u(k-m), \qquad y(k) = \mathbf{C}\mathbf{x}(k) + Du(k).$$

This structure can be realized as a cascade of a nonrecursive portion and a recursive part. Sketch an SFG of such a realization. If $\mathbf{A}$ is a normal matrix, then the matrix norm of $\mathbf{A}$ is $\max_i \{|\lambda_i|\}$ where the λ_i are the eigenvalues of $\mathbf{A}$. Furthermore, the matrix norm of $\mathbf{A}^L$ is $\max_i |\lambda_i|^L$. Show that this structure is free of *all* limit cycles if $(\frac{1}{2})^{1/L} > \max_i \{|\lambda_i|\}$. See Fam (1979).

10.12. What angles of pole locations create the poorest zero input limit cycles for a second-order normal form filter? Assume the pole radii are just inside $|z| = 1$. What modification to this structure can be used to reduce these limit cycles?

10.13. Almost any rational transfer function can be realized by an SV realization of the form

$$\begin{bmatrix} \mathbf{A} & \mathbf{B} \\ \mathbf{C} & \mathbf{D} \end{bmatrix} = \left[\begin{array}{cccccc|c} \alpha_0 & \beta_1 & 0 & 0 & & \cdots & b_1 \\ \gamma_1 & \alpha_1 & \beta_2 & 0 & & \cdots & 0 \\ 0 & \gamma_2 & \alpha_2 & \beta_3 & & \cdots & 0 \\ 0 & 0 & \gamma_3 & \alpha_3 & & \cdots & 0 \\ & & & \vdots & & & \\ 0 & 0 & 0 & & & \cdots & 0 \\ \hline c_1 & 0 & 0 & \cdots & & 0 & d \end{array}\right]$$

Show how the parameters can be computed from the transfer function. Give an example of an $H(z)$ that cannot be realized in this way.

10.14. Suppose $H(z)$ is a minimal, stable transfer function and $(\mathbf{A}, \mathbf{B}, \mathbf{C}, \mathbf{D})$ is a direct form realization. Since $H(z)$ is minimal, $(\mathbf{A}, \mathbf{B})$ is controllable and $(\mathbf{A}, \mathbf{C})$ is observable. Find a nonsingular transformation $\mathbf{T}$ for which $(\mathbf{T}^{-1}\mathbf{AT}, \mathbf{T}^{-1}\mathbf{B}, \mathbf{CT}, \mathbf{D})$ is a tridiagonal realization, as in Problem 10.13.

10.15. Suppose $(\mathbf{A}_1, \mathbf{B}_1, \mathbf{C}_1, \mathbf{D})$ and $(\mathbf{A}_2, \mathbf{B}_2, \mathbf{C}_2, \mathbf{D})$ are two tridiagonal realizations of the same transfer function $H(z)$. Since both have the same external description, there is a nonsingular transformation such that

$$\mathbf{T}: (\mathbf{A}_1, \mathbf{B}_1, \mathbf{C}_1, \mathbf{D}) \rightarrow (\mathbf{A}_2, \mathbf{B}_2, \mathbf{C}_2, \mathbf{D}).$$

Show that $\mathbf{T}$ is diagonal. If both $(\mathbf{A}_i, \mathbf{B}_i, \mathbf{C}_i, \mathbf{D})$, $i = 1, 2$ are scaled, show $T_{ii}^2 = 1$.

10.16. The tridiagonal realization contains $3n + 1$ parameters to realize an external description containing $2n + 1$ parameters $[H(z)]$. Can we use the extra n degrees of freedom to improve the roundoff noise performance of the tridiagonal realization? Explain. In order to ascertain the roundoff noise properties of the tridiagonal realization, simulate a sixth-order low-pass Butterworth filter of unity dc gain and a 5% bandwidth. (The optimum noise gain for such a filter is 0.712.)

10.17. Prove the conservation law (10.4.57) for filters in $\mathscr{F}_0(m, n)$.

10.18. Prove that if

$$\begin{pmatrix} \mathbf{A} & \mathbf{B} \\ \mathbf{C} & \mathbf{D} \end{pmatrix} \in \mathscr{F}_0(m, n),$$

then

$$\|\mathbf{x}(k)\|^2 \leqslant \sum_{i=-\infty}^{k-1} \|\mathbf{u}(i)\|^2$$

and the equality is attainable. In view of Eq. (10.4.57), what can be said about the past outputs, when equality is met?

10.19. Prove:

$$\textit{If} \begin{bmatrix} \mathbf{A} & \mathbf{B} \\ \mathbf{C} & \mathbf{D} \end{bmatrix} \in \mathscr{F}_0(m, n),$$

then the eigenvalues of $\mathbf{A}$ satisfy $|\lambda| < 1$.

10.20. Prove:

$$\textit{If} \begin{bmatrix} \mathbf{A} & \mathbf{B} \\ \mathbf{C} & \mathbf{D} \end{bmatrix} \in \mathscr{F}(m, n),$$

and the eigenvalues of $\mathbf{A}$ satisfy $|\lambda| < 1$, then $(\mathbf{A}, \mathbf{B})$ is controllable and $(\mathbf{A}, \mathbf{C})$ is observable.

10.21. Prove Proposition 5 in Section 10.4: If $\mathbf{F} \in \mathcal{O}(N)$ and $\mathbf{F}$ is upper triangular, then $\mathbf{F}$ is diagonal, with diagonal elements of ± 1.

10.22. Proposition 5 in Section 10.4 is not true for infinite matrices. Let

$$H(z) = \sum_{k=0}^{\infty} h(k) z^{-k}$$

be all-pass, and let

$$F_{ij} = h(j - i), \qquad -\infty < i, j < \infty.$$

Then $\mathbf{F}$ is upper triangular. Show that

$$\mathbf{F}^T \mathbf{F} = \mathbf{I},$$

which, for infinite matrices means that $\mathbf{F}$ preserves inner products:

$$(\mathbf{Fx}, \mathbf{Fy}) = (\mathbf{x}, \mathbf{y})$$

for all l_2 sequences $\mathbf{x}$, $\mathbf{y}$.

10.23. Let

$$\begin{bmatrix} \mathbf{A} & \mathbf{B} \\ \mathbf{C} & \mathbf{D} \end{bmatrix} \quad \in \quad \mathscr{F}_0(1, n)$$

and let

$$H(z) = \mathbf{D} + \mathbf{C}(z\mathbf{I} - \mathbf{A})^{-1}\mathbf{B} = \sum_{k=0}^{\infty} h(k) z^{-k}$$

and

$$E_{ij} = h(i + j - 1) \quad \text{for } 1 \leqslant i, j \leqslant \infty.$$

Show that the semi-infinite matrix $\mathbf{E}^2$ is a rank n orthogonal projection by proving the following identities:

$$(\mathbf{E}^2)^2 = \mathbf{E}^2 = (\mathbf{E}^2)^T, \qquad \mathrm{Tr}(\mathbf{E}^2) = n.$$

10.24. Let

$$\begin{bmatrix} \mathbf{A} & \mathbf{B} \\ \mathbf{C} & \mathbf{D} \end{bmatrix} \quad \in \quad \mathscr{F}_0(m, n)$$

and let

$$\mathbf{H}(k) \overset{z}{\longleftrightarrow} \mathbf{H}(z) = \mathbf{D} + \mathbf{C}(z\mathbf{I} - \mathbf{A})^{-1}\mathbf{B}.$$

Show that:

a) $\displaystyle\sum_{k=0}^{\infty} \mathrm{Tr}[\mathbf{H}(k)\mathbf{H}(k)^T] = m.$

b) $\displaystyle\sum_{k=0}^{\infty} k\, \mathrm{Tr}[\mathbf{H}(k)\mathbf{H}(k)^T] = n.$

For SISO filters, one can define the "average delay" as

$$\frac{\left(\sum_{k=0}^{\infty} kh(k)^2\right)}{\left(\sum_{k=0}^{\infty} h(k)^2\right)}.$$

For all-pass filters then, this is exactly the filter order.

10.25.

a) Let $H(z)$ be a filter of order n whose second-order modes are all equal to 1, and has dc gain of $|H(1)| = 1$. Prove that $H(z)$ is all-pass.
b) Show that orthogonal all-pass filters are minimum noise structures.

10.26.

Let $\begin{bmatrix} \mathbf{A} & \mathbf{B} \\ \mathbf{C} & \mathbf{D} \end{bmatrix} \in \mathscr{F}_0(m, n), \qquad \mathbf{H}(z) = \mathbf{D} + \mathbf{C}(z\mathbf{I} - \mathbf{A})^{-1}\mathbf{B}.$

a) Show that

$\mathbf{H}(z)\mathbf{H}^T(z^{-1}) \equiv \mathbf{I}.$

b) Show that if $|z| > 1$, then

$[\mathbf{H}(z)]^*\mathbf{H}(z) < \mathbf{I}.$

10.27. Consider the filter in Fig. 10P.27. Show that the SISO filter $H_2(z)$ whose output is $y_2(k)$ is all-pole. Show that the SISO filter $H_3(z)$ whose output is $y_3(k)$ is all-pass.

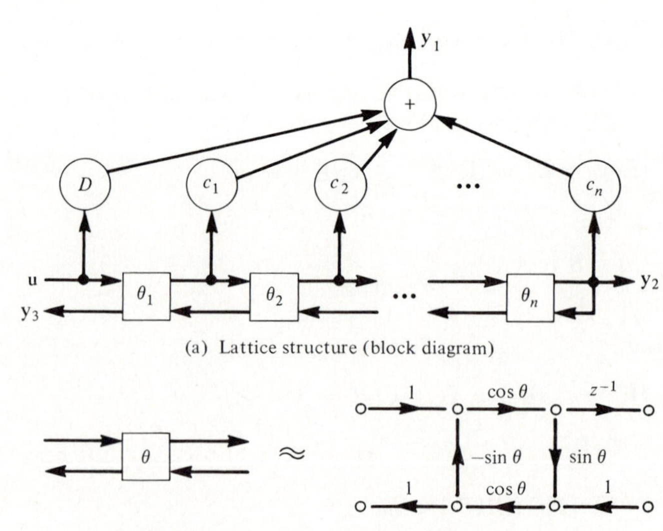

(a) Lattice structure (block diagram)

(b) Lattice section (block diagram and signal flow graph)

10.28. Consider the SISO filter in Fig. 10P.27 whose output is $y_1(k)$. This is the normalized lattice of Gray and Markel (1975). Using the techniques of Section 8.4, show that a factored state variable description for this structure has the form

$$\mathbf{G}_0\mathbf{F}_n\mathbf{F}_{n-1}\cdots\mathbf{F}_1,$$

where each $\mathbf{F}_k$ is primitive (one rotation), and $\mathbf{G}_0$ has the form (10.4.48).

10.29. Replace the lattice sections of Fig. 10P.27 with

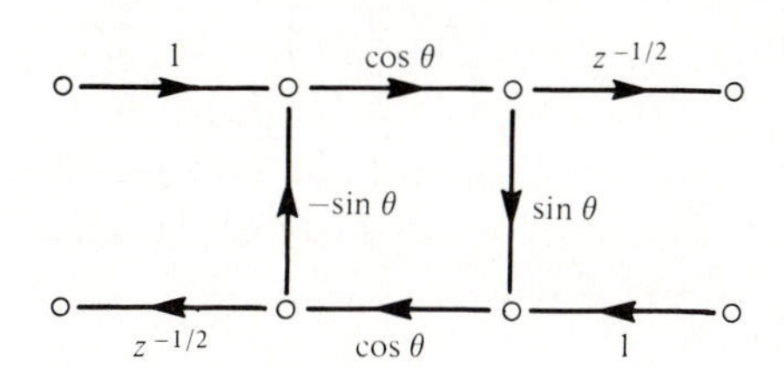

FIGURE 10P.29.

and add an extra half delay to the output y_3.

a) Show that the transfer function $H_3(z)$ is the same after the modification as it was before.

b) Develop a factored state variable description for the modified all-pass filter (ignore outputs y_1 and y_2).

10.30.

a) Show that the two-input, two-output primitives of Fig. 10.4.2(d) cannot be connected in a two-dimensional array [as in Fig. 10.4.3] without delay-free loops.

b) Suppose the blocks in the two-dimensional array are of the form

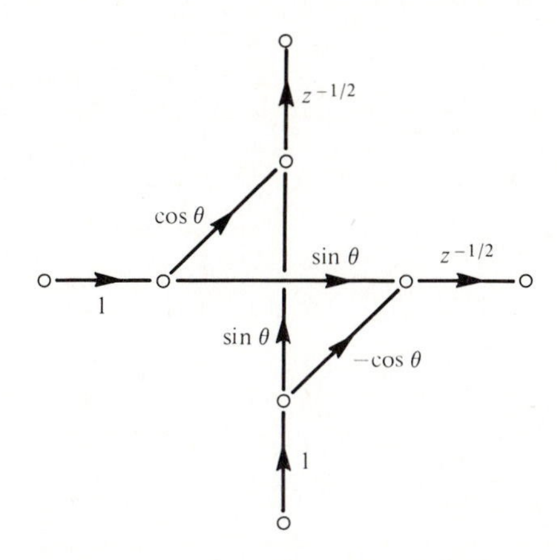

(having the proper orientation). Also, insert two more half delays, one in each of the self loops in the upper right and lower left corners of the array.

Show that every path from input to output, and every loop contain a positive even integer number of half delays.

c) Construct a width-two, period-two factored state variable description equivalent to the signal flow graph constructed in Part b).

10.31. In our discussion of orthogonal filters, we have shown that rotation of a two-dimensional vector is a basic computation. One method of performing a rotation with finite-length registers is to use the CORDIC algorithm. The idea is to realize a rotation by a related application of the following calculation:

$$\begin{bmatrix} x_{i+1} \\ y_{i+1} \end{bmatrix} = \begin{bmatrix} 1 & -\delta_i 2^{-i} \\ \delta_i 2^{-i} & 1 \end{bmatrix} \begin{bmatrix} x_i \\ y_i \end{bmatrix}, \qquad \delta_i = \pm 1.$$

For each iteration of this algorithm, an input vector $\begin{bmatrix} x_i \\ y_i \end{bmatrix}$ is rotated and elongated by how much? How can the δ_i be chosen to approximate a rotation of θ for word lengths of B bits?

Orthogonal Matrices and All-Pass Filters

This appendix contains the proof of Eqs. (10.4.16) and (10.4.17) of Proposition 2, and the feedback property (10.4.23) for orthogonal, all-pass filters.

LEMMA:

$$If \begin{cases} \mathbf{F} = \begin{bmatrix} \mathbf{A} & \mathbf{B} \\ \mathbf{C} & \mathbf{D} \end{bmatrix} \quad \in \quad \mathcal{O}(m+n), \\ \mathbf{U} \in \mathcal{U}(n), \\ (\mathbf{U} - \mathbf{A}) \text{ is invertible,} \end{cases} \tag{10A.1}$$

then

$$\mathbf{G} = \mathbf{D} + \mathbf{C}[\mathbf{U} - \mathbf{A}]^{-1}\mathbf{B} \quad \in \quad \mathcal{U}(m) \tag{10A.2}$$

Proof:

$$\mathbf{I} + \mathbf{A}(\mathbf{U} - \mathbf{A})^{-1} = (\mathbf{U} - \mathbf{A} + \mathbf{A})(\mathbf{U} - \mathbf{A})^{-1} = \mathbf{U}(\mathbf{U} - \mathbf{A})^{-1}. \tag{10A.3}$$

$$\text{Define } \mathbf{\Psi} = \begin{bmatrix} (\mathbf{U} - \mathbf{A})^{-1}\mathbf{B} \\ \mathbf{I} \end{bmatrix}, \qquad (m+n) \times m. \tag{10A.4}$$

Then, using Eq. (10A.3),

$$\mathbf{F\Psi} = \begin{bmatrix} [\mathbf{I} + \mathbf{A}(\mathbf{U} - \mathbf{A})^{-1}]\mathbf{B} \\ \mathbf{G} \end{bmatrix} = \begin{bmatrix} \mathbf{U}(\mathbf{U} - \mathbf{A})^{-1}\mathbf{B} \\ \mathbf{G} \end{bmatrix} = \begin{bmatrix} \mathbf{U} & 0 \\ 0 & \mathbf{G} \end{bmatrix} \mathbf{\Psi}. \tag{10A.5}$$

Since $\mathbf{F}^*\mathbf{F} = \mathbf{I}$, we have

$$\begin{aligned} (\mathbf{F\Psi})^*(\mathbf{F\Psi}) &= \mathbf{\Psi}^*\mathbf{F}^*\mathbf{F\Psi} \\ &= \mathbf{\Psi}^*\mathbf{\Psi} = \mathbf{I} + \mathbf{B}^*[(\mathbf{U} - \mathbf{A})^{-1}]^*(\mathbf{U} - \mathbf{A})^{-1}\mathbf{B}. \end{aligned} \tag{10A.6}$$

Alternately, using Eq. (10A.5),

$$\begin{aligned} (\mathbf{F\Psi})^*(\mathbf{F\Psi}) &= \mathbf{\Psi}^* \begin{bmatrix} \mathbf{U}^* & 0 \\ 0 & \mathbf{G}^* \end{bmatrix} \begin{bmatrix} \mathbf{U} & 0 \\ 0 & \mathbf{G} \end{bmatrix} \mathbf{\Psi} = \mathbf{\Psi}^* \begin{bmatrix} \mathbf{I} & 0 \\ 0 & \mathbf{G}^*\mathbf{G} \end{bmatrix} \mathbf{\Psi} \\ &= \mathbf{G}^*\mathbf{G} + \mathbf{B}^*[(\mathbf{U} - \mathbf{A})^{-1}]^*(\mathbf{U} - \mathbf{A})^{-1}\mathbf{B}. \end{aligned} \tag{10A.7}$$

Equating the right-hand sides of Eqs. (10A.6) and (10A.7) yields

$$\mathbf{G}^*\mathbf{G} = \mathbf{I}, \quad or \quad \mathbf{G} \in \mathscr{U}(m). \tag{10A.8}$$

With

$$\mathbf{U} = e^{j\theta}\mathbf{I} \quad \in \quad \mathscr{U}(n), \tag{10A.9}$$

the lemma establishes Eq. (10.4.16). The dual version of the lemma, which is proved in the same way is

$$If \begin{cases} \mathbf{F} = \begin{bmatrix} \mathbf{A} & \mathbf{B} \\ \mathbf{C} & \mathbf{D} \end{bmatrix} \quad \in \quad \mathscr{O}(m+n), \\ \mathbf{V} \in \mathscr{U}(m), \\ (\mathbf{V} - \mathbf{D}) \text{ is invertible,} \end{cases} \tag{10A.10}$$

then

$$\mathbf{A} + \mathbf{B}(\mathbf{V} - \mathbf{D})^{-1}\mathbf{C} \quad \in \quad \mathscr{U}(n). \tag{10A.11}$$

With $\mathbf{V} = \mathbf{I}$, this establishes Eq. (10.4.23), which in turn establishes the orthogonality of the matrix involved in the feedback example [Eq. (10.4.32)].

ORTHOGONAL ALL-PASS FILTERS HAVE ORTHOGONAL STATE VARIABLE DESCRIPTIONS

***Proof of* (10.4.17):**

We have

$$\mathbf{F} = \begin{bmatrix} \mathbf{A} & \mathbf{B} \\ \mathbf{C} & \mathbf{D} \end{bmatrix} \quad \in \quad \mathscr{F}_0(m, n). \tag{10A.12}$$

Thus we are given orthogonality condition

$$\mathbf{AA}^T + \mathbf{BB}^T = \mathbf{I}, \tag{10A.13}$$

which is one of the four subequations of

$$\begin{bmatrix} \mathbf{A} & \mathbf{B} \\ \mathbf{C} & \mathbf{D} \end{bmatrix} \begin{bmatrix} \mathbf{A}^T & \mathbf{C}^T \\ \mathbf{B}^T & \mathbf{D}^T \end{bmatrix} = \begin{bmatrix} \mathbf{I} & 0 \\ 0 & \mathbf{I} \end{bmatrix}, \tag{10A.14}$$

which is what we want to establish. To complete the set, it is sufficient to prove only that

$$\mathbf{CA}^T + \mathbf{DB}^T = \mathbf{0}, \tag{10A.15}$$

$$\mathbf{CC}^T + \mathbf{DD}^T = \mathbf{I}. \tag{10A.16}$$

Now

$$\mathbf{H}(z) = \mathbf{D} + \mathbf{C}(z\mathbf{I} - \mathbf{A})^{-1}\mathbf{B} \tag{10A.17}$$

is all-pass (by hypothesis). Therefore

$$\mathbf{H}(e^{j\theta})\mathbf{H}(e^{j\theta})^* = \mathbf{I} \tag{10A.18}$$

$$\Rightarrow \quad \mathbf{I} = \frac{1}{2\pi}\int_{-\pi}^{\pi} \mathbf{H}(e^{j\theta})\mathbf{H}(e^{j\theta})^* d\theta = \sum_{k=0}^{\infty} \mathbf{H}(k)\mathbf{H}(k)^T, \tag{10A.19}$$

where

$$\mathbf{H}(k) = \begin{cases} \mathbf{D}, & k = 0 \\ \mathbf{C}\mathbf{A}^{k-1}\mathbf{B}, & k > 0 \end{cases} \quad \overset{z}{\leftrightarrow} \quad \mathbf{H}(z). \tag{10A.20}$$

Therefore Eq. (10A.19) becomes

$$\mathbf{I} = \mathbf{D}\mathbf{D}^T + \mathbf{C}\left[\sum_{k=1}^{\infty} \mathbf{A}^{k-1}\mathbf{B}(\mathbf{A}^{k-1}\mathbf{B})^T\right]\mathbf{C}^T.$$

Now the sum in the brackets is $\mathbf{K} = \mathbf{I}$ (since the filter is controllable and orthogonal). Thus we have proven Eq. (10A.16).

Now for $i > 0$, Eq. (10A.18) implies that

$$\begin{aligned} 0 &= \frac{1}{2\pi}\int_{-\pi}^{\pi} e^{-j\theta i}\mathbf{H}(e^{j\theta})\mathbf{H}(e^{j\theta})^* d\theta = \sum_{k=0}^{\infty} \mathbf{H}(k+i)\mathbf{H}(k)^T \\ &= \mathbf{C}\mathbf{A}^{i-1}\mathbf{B}\mathbf{D}^T + \sum_{k=1}^{\infty} \mathbf{C}\mathbf{A}^{k+i-1}\mathbf{B}\mathbf{B}^T(\mathbf{A}^T)^{k-1}\mathbf{C}^T \\ &= \mathbf{C}\mathbf{A}^{i-1}\mathbf{B}\mathbf{D}^T + \mathbf{C}\mathbf{A}^{i}\mathbf{K}\mathbf{C}^T = \mathbf{C}\mathbf{A}^{i-1}[\mathbf{B}\mathbf{D}^T + \mathbf{A}\mathbf{C}^T]. \end{aligned} \tag{10A.21}$$

Since the filter is observable, this can vanish for all $i > 0$ only if the term in the brackets is zero. This establishes Eq. (10A.15) and completes the proof.

CHAPTER 11

Spectral Estimation

11.1 INTRODUCTION TO SPECTRAL ESTIMATION

"What is the frequency content of this signal?" This question is one of those most frequently addressed in signal processing applications. It is often asked when only a small part of the signal is available, which means that it cannot be answered precisely. *Spectral estimation* is the name given to the collection of methods for supplying imperfect answers to this question.

Before we can attempt an answer, we must define what is meant by "frequency content." For any given band Ω of frequencies, we can isolate that part of the signal that lies in the band Ω by passing it through an ideal bandpass filter whose passband is Ω.

$$x \circ \xrightarrow{H(z)} \circ\, y \tag{11.1.1}$$

$$H(e^{j\theta}) = \begin{cases} 1, & \theta \in \Omega \\ 0, & \text{otherwise} \end{cases}. \tag{11.1.2}$$

The output signal $\mathbf{y}$ in this idealized experiment is what is meant by the phrase "that part of $\mathbf{x}$ which lies in the band Ω. If $\mathbf{y}$ has finite energy, then that energy is "the energy that $\mathbf{x}$ contains in the band Ω." If $\mathbf{x}$ and $\mathbf{y}$ are wide sense stationary with finite power, then the average power of $\mathbf{y}$ is "the power that $\mathbf{x}$ contains in the band Ω."

We will have a complete answer to the frequency content question if we can compute these quantities for any given Ω. This is the purpose of a *spectral density function*, which is a function of the frequency variable θ with the following property:

> *The energy (or power) of the signal in the band Ω is the integral of the density function over Ω.*

We may use this property to construct the density function. There are two classes of signals of interest, finite energy signals, and wide sense stationary random signals. The first kind will have an *energy spectral density*. The second kind will have a *power spectral density*.

Energy Spectral Density (ESD)

Assume that the signal

$$x(k) \quad \overset{DTFT}{\longleftrightarrow} \quad X(e^{j\theta}) \tag{11.1.3}$$

has finite energy

$$\|\mathbf{x}\|^2 = \sum_{k=-\infty}^{\infty} |x(k)|^2 = \frac{1}{2\pi}\int_{-\pi}^{\pi} |X(e^{j\theta})|^2 \, d\theta. \tag{11.1.4}$$

To extract that portion of the energy that lies in a band Ω, we pass $\mathbf{x}$ through the ideal bandpass filter $H(z)$ of Eq. (11.1.2) to get an output $\mathbf{y}$. Then the total energy of $\mathbf{y}$ is that part of the energy of $\mathbf{x}$ which lies in Ω and will be

$$\|\mathbf{y}\|^2 = \frac{1}{2\pi}\int_{-\pi}^{\pi} |Y(e^{j\theta})|^2 \, d\theta. \tag{11.1.5}$$

But

$$\mathbf{y} = \mathbf{h} * \mathbf{x} \quad \overset{DTFT}{\longleftrightarrow} \quad Y(e^{j\theta}) = H(e^{j\theta})X(e^{j\theta}), \tag{11.1.6}$$

and using Eq. (11.1.2), the

$$\text{“energy of } \mathbf{x} \text{ in } \Omega\text{”} = \|\mathbf{y}\|^2 = \frac{1}{2\pi}\int_{\Omega} |X(e^{j\theta})|^2 \, d\theta. \tag{11.1.7}$$

This is the density function property we required, and therefore the function $|X|^2$ is the energy spectral density for the signal $\mathbf{x}$.

Power Spectral Density (PSD)

A nontrivial wide sense stationary signal will not have finite energy, but it can have finite average power. If a zero-mean WSS signal $\mathbf{x}$ is ergodic, then time averages will approach expected values. In particular,

$$\text{average power} = \lim_{L\to\infty} \frac{1}{L}\sum_{k=0}^{L-1} |x(k)|^2 = E\,|x(k)|^2. \tag{11.1.8}$$

Thus the average power is the variance of $x(k)$, or the zeroth element of the autocorrelation sequence

$$r_{xx}(k) = Ex(k+l)x^*(l) \quad \overset{DTFT}{\longleftrightarrow} \quad S_{xx}(\theta). \tag{11.1.9}$$

To find the average power that lies in the band Ω, we pass the signal $\mathbf{x}$ through the bandpass filter $H(z)$ of Eq. (11.1.2). The Wiener-Khintchine theorem (see Section 7.3) yields

$$r_{yy}(k) \quad \overset{DTFT}{\longleftrightarrow} \quad S_{yy}(\theta) = S_{xx}(\theta)|H(e^{j\theta})|^2. \tag{11.1.10}$$

Therefore,

$$\text{“average power in } \mathbf{x} \text{ in the band } \Omega\text{”} = r_{yy}(0) = \frac{1}{2\pi}\int_{-\pi}^{\pi} S_{yy}(\theta)\, d\theta \tag{11.1.11}$$

$$= \frac{1}{2\pi}\int_{\Omega} S_{xx}(\theta)\, d\theta. \tag{11.1.12}$$

[We have used the Wiener-Khintchine relation (11.1.10) and the definition of $H(z)$ in Eq. (11.1.2) to proceed from Eq. (11.1.11) to Eq. (11.1.12).] Therefore, $S_{xx}(\theta)$, which is the DTFT of the autocorrelation sequence for $\mathbf{x}$, is the power spectral density function. That is to say, one can get the average power in the band Ω by integrating this function over Ω.

We are primarily interested in WSS signals and power spectra. There are some parametric families of spectra in which we will be especially interested. Any (measurable) function, however, can be a PSD so long as it enjoys the following essential properties:

$$\textbf{Symmetry}: \quad S(-\theta) = S(\theta) \tag{11.1.13}$$

$$\textbf{Positivity}: \quad S(\theta) \geqslant 0. \tag{11.1.14}$$

(We are assuming real-valued signals.) With these properties in mind, let us consider some specific examples. These examples are interesting because they are associated with a generating mechanism or "model," and we will take care to relate the parameters of the model with the parameters of the spectra.

EXAMPLE 11.1.1

White noise spectra.

The "most random" signal with a given variance σ^2 is a white noise signal. This is defined by the DTFT pair

$$r(k) = \sigma^2\delta(k) \quad \xleftrightarrow{DTFT} \quad S(\theta) = \sigma^2. \tag{11.1.15}$$

Such a signal cannot be predicted using any linear filter with a prediction error variance any less than σ^2. The name "white noise" refers to the fact that $S(\theta)$ has power uniformly distributed among all frequencies, a famous experiment of Sir Isaac Newton having demonstrated that white light is made up of all colors. (Incidentally, Newton was the inventor of the word "spectrum" as it is used in frequency analysis. The word is derived from the Latin word for "apparition.") White noise is a good model for many physical processes. The generation of artificial white noise via "pseudo-random number generators" is a classical problem. See Knuth (1969, Chapter 3).

EXAMPLE 11.1.2
ARMA spectra.

The acronym ARMA stands for *autoregressive moving average* (terms from classical statistics). For us, an ARMA spectrum is one that describes the output of a finite order linear filter driven by white noise. Let this filter be

$$H(z) = \frac{\hat{b}(z)}{\hat{a}(z)}, \tag{11.1.16}$$

where

$$\hat{a}(z) = 1 + a_1 z^{-1} + \cdots + a_n z^{-n}, \tag{11.1.17}$$

$$\hat{b}(z) = b_0 + b_1 z^{-1} + \cdots + b_m z^{-m}. \tag{11.1.18}$$

If the input $\mathbf{u}$ is a white noise signal, then

$$S_{uu}(\theta) \equiv 1. \tag{11.1.19}$$

Therefore, the output signal

$$\mathbf{y} = \mathbf{h} * \mathbf{u} \tag{11.1.20}$$

will have spectrum

$$S_{yy}(\theta) = S_{uu}(\theta)\,|H(e^{j\theta})|^2 = \left|\frac{\hat{b}(e^{j\theta})}{\hat{a}(e^{j\theta})}\right|^2. \tag{11.1.21}$$

(by the Wiener-Khintchine theorem).

There are two special cases of interest. If

$$\hat{b}(z) \equiv b_0, \tag{11.1.22}$$

then the spectrum is called *autoregressive*. If

$$\hat{a}(z) \equiv 1, \tag{11.1.23}$$

then the spectrum is called *moving average*. The filter $H(z)$ will be an "all-pole" filter in the AR case and an FIR filter in the MA case.

There is a pair of useful equations which relate the coefficients in $\hat{a}(z)$ and $\hat{b}(z)$ to the autocorrelation sequence for $\mathbf{y}$ and the unit pulse response sequence. To produce these equations, one starts with the standard difference equation

$$\sum_{l=0}^{n} a_l y(k-l) = \sum_{l=0}^{m} b_l u(k-l). \tag{11.1.24}$$

Let us define

$$h(k) \triangleq Ey(k+l)u(l) \quad \text{(cross-correlation)}, \tag{11.1.25}$$

$$r(k) \triangleq Ey(k+l)y(l) \quad \text{(autocorrelation)}. \tag{11.1.26}$$

We have

$$r(k) \quad \xleftrightarrow{DTFT} \quad S_{yy}(\theta) = \frac{|\hat{b}(e^{j\theta})|^2}{|\hat{a}(e^{j\theta})|^2}. \tag{11.1.27}$$

If we multiply Eq. (11.1.24) on each side by $u(k - i)$ and take expected values, we have

$$\sum_{l=0}^{n} a_l h(i - l) = \sum_{l=0}^{m} b_l \delta(i - l) = b_i. \tag{11.1.28}$$

The z-transform of this equation is

$$\hat{a}(z)H(z) = \hat{b}(z) \Rightarrow H(z) = \frac{\hat{b}(z)}{\hat{a}(z)}. \tag{11.1.29}$$

[We have shown here that the unit pulse response sequence and the cross-correlation sequence in (11.1.25) are one in the same, for white noise inputs.]

If we multiply Eq. (11.1.24) by $y(k - i)$ and take expected values, we have

$$\sum_{l=0}^{n} a_l r(i - l) = \sum_{l=0}^{m} b_l h(l - i). \tag{11.1.30}$$

The z-transform of this equation is

$$\hat{a}(z)R(z) = \hat{b}(z)H(z^{-1}) \Rightarrow R(z) = \frac{\hat{b}(z)}{\hat{a}(z)} H(z^{-1}) = H(z)H(z^{-1}). \tag{11.1.31}$$

Equivalently,

$$S_{yy}(\theta) = R(e^{j\theta}) = |H(e^{j\theta})|^2. \tag{11.1.32}$$

The two time domain equations (11.1.28) and (11.1.30) are fundamental to AR, MA, and ARMA spectra. They are sometimes called extended Yule-Walker equations. From a modeling point of view, producing ARMA signals is not difficult (assuming the existence of a good pseudo-random noise generator to be used for the production of the white noise input).

EXAMPLE 11.1.3

Line spectra.

A "line" spectrum is of the form

$$S(\theta) = \sum_{l=1}^{m} \frac{\pi A_l^2}{2} [\delta(\theta - \theta_l) + \delta(\theta + \theta_l)]. \tag{11.1.33}$$

In other words, it is a combination of Dirac delta functions. As such, all signal power is concentrated on a finite set of points (the numbers θ_l and $-\theta_l$). It is clearly an extreme case, but in the opposite way that white noise is an extreme case. White noise is completely disorganized, and "random" in the sense that it cannot be predicted [the least-squares prediction given any past history is simply $\hat{y}(k+1) = 0$]. On the other hand, a signal that has a line spectrum is highly organized and is not really "random" at all, for we shall see in Section 11.3 that it is completely predictable (with zero error) given a finite past history.

The autocorrelation sequence that is the inverse DTFT of the spectrum in Eq. (11.1.33) is

$$r(k) = \sum_{l=1}^{m} \frac{A_l^2}{2} \cos(k\theta_l); \quad \textbf{(11.1.34)}$$

a combination of sinusoids. How can one generate a signal with such a spectrum? Since the signal can have power only at discrete frequencies, it must consist of sinusoids at those frequencies. Consider the signal

$$y(k) = \sum_{i=1}^{m} A_i \cos(k\theta_i + \phi_i). \quad \textbf{(11.1.35)}$$

The lag products $y(k+l)y(l)$ for this signal involve sum and difference frequencies. When these are averaged over time, only the zero frequency terms remain, to wit,

$$\lim_{L\to\infty} \frac{1}{L} \sum_{l=0}^{L-1} y(k+l)y(l) = \sum_{i=1}^{m} \frac{A_i^2}{2} \cos(k\theta_i). \quad \textbf{(11.1.36)}$$

In this sense, at least, the signal has the required autocorrelation sequence. Notice that we have used *time averaging* here, whereas we used *expected values* for computing the autocorrelation sequence of the output of a filter driven by white noise.

Generating sinusoidal signals involves a direct synthesis (an "oscillator"). It would be inexact and extremely inefficient to try to produce a line spectrum by filtering white noise.

The most popular models for signal spectra involve one of these three examples (white noise, ARMA, lines) or a linear combination of them. In particular, the case of a line spectrum plus white noise has been a popular model for many applications.

Approaches to Spectral Estimation

Let $\mathbf{y}$ be a wide sense stationary signal with power spectral density $S(\theta)$. Suppose that a portion of this signal has been recorded, with L samples.

$$\mathbf{y}[0, L-1] = \{y(0), y(1), \ldots, y(L-1)\}. \quad \textbf{(11.1.37)}$$

This part of the signal, which is known, is called the "data." The problem is to estimate the function $S(\theta)$ given the data. Thus an *estimator* is a mapping that takes the data into a power spectrum:

$$\hat{S}(\theta) = \mathscr{S}(\theta; \mathbf{y}[0, L-1]). \tag{11.1.38}$$

Here, $\mathscr{S}$ is called the estimator and $\hat{S}$ is called the *estimate*. Since the data consist of random variables, the estimate is random. We must judge the estimator with this in mind. There are a number of ways to compare estimators that we shall introduce later at convenient points. For the present, let us consider some approaches to the estimation problem.

Estimation Using the ESD

Knowing only the data $\mathbf{y}[0, L-1]$, we can construct a finite energy signal by extending the data with zeros. Let

$$w(k) = \begin{cases} 1, & 0 \leqslant k \leqslant L-1 \\ 0, & \text{otherwise} \end{cases} \tag{11.1.39}$$

and

$$x(k) = w(k)y(k), \qquad -\infty < k < \infty. \tag{11.1.40}$$

Then $\mathbf{y}$ is an infinite energy power signal, $\mathbf{w}$ is a time limited window (the uniform window), and $\mathbf{x}$ is a time limited signal. Since $\mathbf{x}$ must have finite energy, it has an energy spectral density. We may use this ESD to estimate the PSD of $\mathbf{y}$, with proper scaling. The estimate

$$\hat{S}(\theta) = \frac{1}{L}|X(e^{j\theta})|^2 \tag{11.1.41}$$

is the result, and is called a *periodogram* after Schuster (1898). We shall study the periodogram and some variations of it in Section 11.2.

Indirect Estimation from Autocorrelation Values

It is often the case that the low-order elements of the autocorrelation sequence are independently estimated. One such estimate is

$$\hat{r}(k) = \frac{1}{L}\sum_{l=0}^{L-1-k} y(k+l)y(l), \qquad 0 \leqslant k < L. \tag{11.1.42}$$

Suppose then that $\hat{r}(0)$ through $\hat{r}(n)$ have been obtained ($n < L$). One can often find a unique lowest-order spectrum in a parametric class of spectra that agrees with these values. That spectrum becomes the estimate.

For example, if the spectrum is assumed to be MA, then we may take

$$\hat{S}(\theta) = \sum_{k=-n}^{n} \hat{r}(k)e^{-jk\theta} \tag{11.1.43}$$

(the *Blackman-Tukey estimate*). This is the lowest-order MA spectrum that agrees with $\hat{r}(0)$ through $\hat{r}(n)$.

If the spectrum is assumed to be AR, then we may take

$$\hat{S}(\theta) = \frac{\alpha}{|\sum_{k=0}^{n} a_k e^{-jk\theta}|^2}, \qquad a_0 = 1 \tag{11.1.44}$$

where $(\alpha, a_1, \ldots, a_n)$ are obtained from $(\hat{r}(0), \ldots, \hat{r}(n))$ by Levinson's algorithm (Section 11.3). This AR spectrum will indeed reproduce the given values of the autocorrelation sequence. We will call this the *Levinson estimate.*

There is a unique spectrum consisting of white noise of the proper variance, plus at most n lines on the interval $-\pi$ to π. This is called the *Pisarenko estimate* (Section 11.3).

Every spectral estimate produced by these means must match the values of $\hat{r}(k)$ for $0 \leqslant k \leqslant n$. We are not always able to meet these conditions, however. In fact the Blackman-Tukey estimate (11.1.43) need not be positive everywhere. If it goes negative, then it is unacceptable. But there are infinitely many positive spectra that meet the conditions. Two such spectra may disagree in $\hat{r}(k)$ for $k > n$. Having recognized this, we may pose the general problem of *autocorrelation extension:*

Given $\hat{r}(0), \ldots, \hat{r}(n)$, extend the sequence so that

$$\hat{S}(\theta) = \hat{r}(0) + 2 \sum_{k=1}^{\infty} \hat{r}(k) \cos(k\theta) \geqslant 0, \quad \text{for all } \theta. \tag{11.1.45}$$

Every solution to this problem is a viable spectral estimate, but the problem becomes more interesting when the extension is required to do more. For example, the *maximum entropy extension problem* asks:

Given $\hat{\mathbf{r}}[0, n]$, choose $\hat{\mathbf{r}}[n+1, \infty]$ to maximize $\int_{-\pi}^{\pi} \ln \hat{S}(\theta)\, d\theta$,

subject to the inequality (11.1.45).

This problem was formulated and solved in Burg (1975). The solution is the Levinson estimate (see Section 11.3).

Sinusoidal Data Fitting

If the spectrum is assumed to be a line spectrum, then the data itself is parametric, as well as the spectrum. The signal $\mathbf{y}$ will have the form of Eq. (11.1.35), which is characterized by $3M$ parameters: frequencies, phases, and amplitudes. The problem is then to estimate these parameters given the data. The classical way to do this is to set up a Gaussian least-squares problem:

Given $\mathbf{y}[0, L-1]$ and m, minimize

$$V(A_1, A_2, \ldots, A_m, \theta_1, \ldots, \theta_m, \phi_1, \ldots, \phi_m) = \sum_{k=0}^{L-1} \varepsilon(k)^2$$

where

$$\varepsilon(k) = y(k) - \sum_{i=1}^{m} A_i \cos(k\theta_i + \phi_i). \tag{11.1.46}$$

The function V is a sum of squares and is therefore nonnegative. If it vanishes, then all the errors $\varepsilon(k)$ must be zero, and we have fit the data exactly. One does not expect this to happen, of course, but for the minimizing values the errors will be small. We then rewrite Eq. (11.1.46) to model the data as sinusoids plus white noise (the error sequence will approximate white noise).

The problem with this approach is its high computational cost. The function V is quadratic in the amplitudes but badly nonlinear in the frequencies and phases. Therefore, it is difficult to minimize and for this reason the approach is not often followed. One can modify the least-squares problem and end up with a strictly quadratic function (which is relatively easy to minimize). This modification will be introduced shortly. Although the computation cost then becomes acceptable, the additive noise is falsely modeled. The Gauss least-squares problem remains the honest approach.

Whitening the Data (Analysis/Synthesis)

Suppose the signal $\mathbf{y}$ is passed through a filter which is designed to "whiten" it. In other words, if the filter output is $\mathbf{e}$, then the goal is to achieve

$$S_{ee}(\theta) \equiv \alpha. \tag{11.1.47}$$

Since we do not have all the signal $\mathbf{y}$, but only the data $\mathbf{y}[0, L-1]$, it is reasonable to limit the filter impulse response to have length $< L$. (Without this limitation we could not compute any of the values of $\mathbf{e}$. Thus we are attempting to whiten the data with an FIR filter:

$$\mathrm{y} \circ \xrightarrow{\hat{a}(z)} \circ\ \mathrm{e}, \qquad \hat{a}(z) = 1 + a_1 z^{-1} + \cdots + a_n z^{-n}. \tag{11.1.48}$$

If this is possible, then the Wiener-Khintchine relation will provide the power spectrum of $\mathbf{y}$ and it will necessarily be AR:

$$S_{yy}(\theta) = \frac{S_{ee}(\theta)}{|\hat{a}(e^{j\theta})|^2} = \frac{\alpha}{|\sum_{k=0}^{n} a_k e^{-jk\theta}|^2}. \tag{11.1.49}$$

If S_{yy} is not an AR spectrum, then this approach will produce an AR estimate anyway, which we would hope will approximate the true spectrum.

In order to obtain a spectral estimator, a way must be found to obtain the coefficients of the whitening filter (11.1.48) from the data $\mathbf{y}[0, L-1]$. The usual method for doing this is apparently due to Yule (1927), who was interested in data that was roughly periodic. Assume for the moment that the data is sinusoidal and is given by Eq. (11.1.35). Then provided the order n of the whitening filter is at least twice the number of frequencies in the data, it is possible to completely annihilate the signal $\mathbf{y}$. With $y(k)$ given by Eq. (11.1.35), the filter $\hat{a}(z)$ would be

$$\begin{aligned}\hat{a}(z) &= \prod_{k=1}^{m} (1 - z^{-1}e^{j\theta_k})(1 - z^{-1}e^{-j\theta_k}) \\ &= \prod_{k=1}^{m} (1 - 2z^{-1}\cos(\theta_k) + z^{-2}).\end{aligned} \tag{11.1.50}$$

This filter has zeros on the unit circle at the frequencies of the signal. Conversely, if the filter can annihilate the signal, then it must have these zeros. This suggests that the problem can be modified. Suppose that we design $\hat{a}(z)$ to minimize the output "error" power. If the data is sinusoidal, then we can achieve zero output power, and the frequencies can be obtained by factoring $\hat{a}(z)$. But even if the output error does not vanish, it will be small, and we can view it as a way of "coding" the data if we can recover the signal $\mathbf{y}$ from the error signal $\mathbf{e}$. The difference equation that relates these,

$$\sum_{l=0}^{n} a_l y(k-l) = e(k), \qquad a_0 = 1 \tag{11.1.51}$$

can therefore be viewed in two different ways. It describes the whitening filter in Eq. (11.1.48) when $e(k)$ is the output. This has been called the *analysis* point of view. But it could also describe the situation

$$\mathrm{e} \circ \xrightarrow{\;1/\hat{a}(z)\;} \circ\, \mathrm{y} \tag{11.1.52}$$

where $y(k)$ is the output. This is the *synthesis* point of view. It amounts to a model for the generation of the signal $\mathbf{y}$. If the signal $\mathbf{e}$ is white and has the power spectrum given by Eq. (11.1.47), then $\mathbf{y}$ will be AR and will have the power spectrum given by Eq. (11.1.49).

The key assumption that provides for a spectral estimator is that if we choose $\hat{a}(z)$ to minimize the variance of $e(k)$, then the signal $\mathbf{e}$ is approximately white. Let us consider this assumption after setting up the minimization problem. In Chapter 7, one form of this problem, called the *statistical whitening filter* problem, was introduced.

Given: $\mathbf{r}_{yy}[0, n]$,

Minimize: $V(a_1, \ldots, a_n) = Ee(k)^2.$ **(11.1.53)**

But since we know only the data $\mathbf{y}[0, L-1]$, and not the autocorrelation sequence, we must make do with the following problem:

Given: $\mathbf{y}[0, L-1]$,

Minimize: $V(a_1, \ldots, a_n) = \sum_{k=n}^{L-1} e(k)^2.$ **(11.1.54)**

In both problems, the objective function V is quadratic, and the normal equations are linear [in the coefficients of $\hat{a}(z)$]. Thus the minimization is straightforward.

In what sense is the error signal white? The normal equations associated with Eq. (11.1.53) are equivalent to

$$E[e(k)y(k-l)] = 0, \qquad 1 \leq l \leq n, \tag{11.1.55}$$

which are linear in the coefficients of $\hat{a}(z)$. This says that the present error is uncorrelated with the recent past history of $\mathbf{y}$. It would be more satisfying if it were uncorrelated with its own recent past, i.e.,

$$E[e(k)e(k-l)] = 0, \qquad 1 \leqslant l \leqslant n. \tag{11.1.56}$$

But these equations are quadratic in the coefficients of $\hat{a}(z)$, and are difficult to solve. Indeed, they may have no solution. However, if $e(k)$ is uncorrelated with recent values of $\mathbf{y}$, which in turn depend on past values of $\mathbf{e}$, then $e(k)$ must be approximately uncorrelated with its own past history.

In any event, we have a spectral estimator. To go from the data to the estimate, one computes the parameters by minimizing the function in Eq. (11.1.54) and then forms the spectrum (11.1.49). We shall see in Section 11.3 that this estimator (and some variations that also produce AR spectra) enjoy a number of desirable properties.

Issues

What are the strengths and weaknesses of a given approach? Here are a few generally accepted issues dealing with the quality of an estimator.

POSITIVITY. Does the estimator have the property that the estimate is always nonnegative? For example, the periodogram (11.1.41) is always nonnegative, but the Blackman-Tukey estimate (11.1.43) need not be.

BIAS. An estimate is "biased" if its expected value differs from the true value, and the "bias" of the estimate is the difference between these two. For an estimate of a power spectrum, the bias will be a function of frequency:

$$\text{bias}(\theta) = E\,\hat{S}(\theta) - S(\theta). \tag{11.1.57}$$

No estimator that uses a finite data set can be strictly unbiased, or

$$\text{bias}(\theta) \equiv 0.$$

(See Problem 11.4.) But some estimators generate more bias than others.

VARIANCE. Like bias, the variance of an estimator is a function of frequency:

$$\sigma^2(\theta) = E\,[\hat{S}(\theta) - E\,\hat{S}(\theta)]^2. \tag{11.1.58}$$

If we combine Eqs. (11.1.57) and (11.1.58), we can get an expression for the mean squared error of an estimate:

$$E\,[S(\theta) - \hat{S}(\theta)]^2 = [\text{bias}(\theta)]^2 + \sigma^2(\theta). \tag{11.1.59}$$

CONSISTENCY. Intuitively, an estimator is consistent if it will faithfully reproduce the true spectrum when it is given an infinite amount of data. Let L be the length of the data $\mathbf{y}[0, L-1]$, and assume that we have a sequence of estimates $\hat{S}_L(\theta)$ generated as the data is accumulated. The estimator is said to be "asymptotically unbiased" if

$$\lim_{L \to \infty} \text{bias}_L(\theta) \equiv 0 \tag{11.1.60}$$

and "consistent" if

$$\lim_{L\to\infty} [\text{bias}_L(\theta)]^2 + \sigma_L^2(\theta) \equiv 0. \tag{11.1.61}$$

One can also use weaker notions of bias and consistency, which are appropriate for AR and other parametric estimates. Let

$$\hat{r}(k) \triangleq \frac{1}{2\pi}\int_{-\pi}^{\pi} \hat{S}(\theta)e^{-jk\theta}\, d\theta. \tag{11.1.62}$$

The estimator is "unbiased to lag n" if

$$E\,\hat{r}(k) = r(k), \qquad 0 \leqslant k \leqslant n \tag{11.1.63}$$

and "consistent to lag n" if

$$\lim_{L\to\infty} E\,[r(k) - \hat{r}_L(k)]^2 = 0, \qquad 0 \leqslant k \leqslant n. \tag{11.1.64}$$

STABILITY. The issue of stability arises only in the analysis/synthesis approach to spectral estimation. If $\hat{a}(z)$ is the whitening filter of Eq. (11.1.48), then the model for the generation of the data given in Eq. (11.1.52) involves $1/\hat{a}(z)$. If this synthesis model is unstable, then the power spectral estimate of Eq. (11.1.49) is inappropriate. Therefore one must demand that the inverse of the whitening filter be stable. Surprisingly, this is often the case for a number of popular AR methods discussed in Section 11.3.

COMPUTATION. This issue deals not so much with the quality of an estimate, but the expense incurred in getting it. If two estimators are roughly equivalent in quality, then the one that is the easiest to compute is to be preferred.

11.2 THE PERIODOGRAM AND MA SPECTRA

The periodogram was introduced in Section 11.1. It is an estimate for the power spectral density function of a WSS signal **y**. But it is constructed from an energy spectral density of a finite length windowed version of **y**. Its form is

$$\hat{S}(\theta) = \frac{1}{L}\left|\sum_{k=0}^{L-1} y(k)w(k)e^{-jk\theta}\right|^2 \tag{11.2.1}$$

$$= \sum_{k=-\infty}^{\infty} \hat{r}(k)e^{-jk\theta} \tag{11.2.2}$$

where

$$\hat{r}(k) = \begin{cases} \dfrac{1}{L}\displaystyle\sum_{j=0}^{L-k-1} y(k+j)y(j)w(k+j)w(j), & |k| < L \\ 0, \quad |k| \geqslant L. \end{cases} \tag{11.2.3}$$

From these two ways of writing $\hat{S}$, we may view it either as an estimator that uses the data $\mathbf{y}[0, L-1]$ directly [Eq. (11.2.1)], or indirectly [Eq. (11.2.2)] from estimates of the autocorrelation sequence (11.2.3). Since the transform pair

$$\hat{r}(k) \xleftrightarrow{DTFT} \hat{S}(\theta) \tag{11.2.4}$$

satisfies

$$\hat{r}(k) = 0 \quad \text{for } |k| \geqslant L, \tag{11.2.5}$$

the periodogram yields MA estimates.

Bias

The bias of the periodogram is the amount that $E\hat{S}(\theta)$ differs from $S(\theta)$. In the time domain, applying the expectation operator to Eq. (11.2.3) yields

$$E\,\hat{r}(k) = \begin{cases} r(k)\left[\dfrac{1}{L}\displaystyle\sum_{j=0}^{L-k-1} w(j)w(k+j)\right], & |k| < L \\ 0, & |k| \geqslant L \end{cases}$$

$$= r(k)\left[\frac{1}{L}\sum_{j=-\infty}^{\infty} w(j)w(k+j)\right], \tag{11.2.6}$$

since $w(j) = 0$ outside the window $0 \leqslant j \leqslant L - 1$. Letting

$$w(k) \xleftrightarrow{DTFT} W(e^{j\theta}), \tag{11.2.7}$$

then

$$\frac{1}{L}\sum_{j=-\infty}^{\infty} w(j)w(k+j) \xleftrightarrow{DTFT} \frac{1}{L}|W(e^{j\theta})|^2. \tag{11.2.8}$$

(The right-hand side is the energy spectral density of the window function.) The product in time of Eq. (11.2.6) becomes convolution in frequency; to wit,

$$E\,\hat{S}(\theta) = \frac{1}{L}(S * |W|^2)(\theta)$$

$$= \frac{1}{2\pi}\int_{-\pi}^{\pi} S(\theta - \phi)\frac{1}{L}|W(e^{j\phi})|^2\, d\phi. \tag{11.2.9}$$

This relation characterizes the bias of the periodogram. The expected value of $\hat{S}(\theta)$ is close to the true spectrum $S(\theta)$ when

$$\frac{1}{L}|W(e^{j\theta})|^2 \approx 2\pi\delta(\theta),$$

because of the convolution in Eq. (11.2.9).

The most often used window is the rectangular window

$$w(k) = \begin{cases} 1, & 0 \leqslant l \leqslant L-1 \\ 0, & \text{otherwise} \end{cases} \xleftrightarrow{DTFT} W(e^{j\theta}) = e^{-j[(L-1)/2]\theta}\left[\frac{\sin\left(\dfrac{L\theta}{2}\right)}{\sin\left(\dfrac{\theta}{2}\right)}\right]. \tag{11.2.10}$$

The normalized magnitude squared function is then

$$\frac{1}{L}\sum_{j=-\infty}^{\infty} w(j)w(k+j) = \begin{cases} 1-\dfrac{|k|}{L}, & |k| < L \\ 0, & |k| \geq L \end{cases}$$

$$\updownarrow \text{DTFT}$$

$$\frac{1}{L}|W(e^{j\theta})|^2 = \frac{1}{L}\left[\frac{\sin\left(\dfrac{L\theta}{2}\right)}{\sin\left(\dfrac{\theta}{2}\right)}\right]^2. \quad \textbf{(11.2.11)}$$

This function is shown in Fig. 11.2.1. It is a reasonable approximation for a δ function; its height is proportional to the length of the data, and its width is inversely proportional. Thus the approximation improves as L is increased. In the time domain, the lag estimates corresponding to the rectangular window are

$$\hat{r}(k) = \frac{1}{L}\sum_{j=0}^{L-k-1} y(k+j)y(j), \qquad 0 \leq k < L, \quad \textbf{(11.2.12)}$$

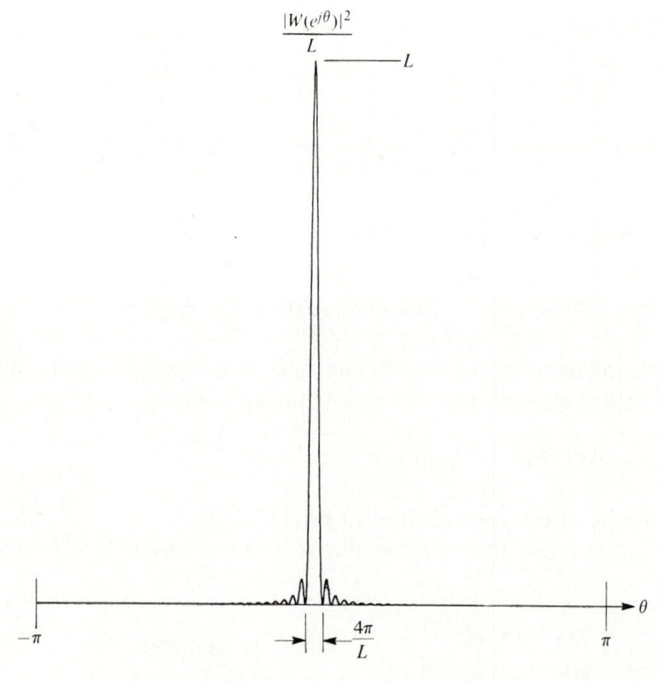

FIGURE 11.2.1 A plot of $|W(e^{j\theta})|^2/L$ for the rectangular window ($L = 64$).

having expected value

$$E\,\hat{r}(k) = \begin{cases} \left(1 - \dfrac{|k|}{L}\right) r(k), & |k| < L \\ 0, & |k| \geq L. \end{cases} \tag{11.2.13}$$

These estimates are biased by the triangular time window of Eq. (11.2.11).

Consider the consequences of the bias relation (11.2.9) for two extreme examples of spectra. If the spectrum is white, then there is no bias.

$$S(\theta) \equiv \sigma^2 \;\Rightarrow\; E\,\hat{S}(\theta) = S(\theta). \tag{11.2.14}$$

This is because the smoothing associated with the convolution in Eq. (11.2.9) will leave a constant spectrum unchanged. However, if the spectrum is a line spectrum, then each line will be replaced with a shifted version of the magnitude squared function shown in Fig. 11.2.1. For example,

$$\begin{aligned} S(\theta) &= 2\pi[\delta(\theta - \theta_0) + \delta(\theta + \theta_0)] \\ \Rightarrow\; E\,\hat{S}(\theta) &= \frac{1}{L}\left|W(e^{j(\theta-\theta_0)})\right|^2 + \frac{1}{L}\left|W(e^{j(\theta+\theta_0)})\right|^2. \end{aligned} \tag{11.2.15}$$

Each line is smoothed and spread out to occupy an interval of width $4\pi/L$. This places a limit on resolution. We cannot hope to resolve two lines closer than about $2\pi/L$ radians apart. Furthermore, some of the signal power has drifted out to occupy the sidelobes of the function in Fig. 11.2.1. These sidelobes may create the impression of spectral peaks, which do not exist in the original spectrum. This phenomenon has been called *leakage*.

EXAMPLE 11.2.1.
Periodograms for MA plus lines.

There are three spectral density functions of interest in our discussion of bias: the true spectrum S, the periodogram $\hat{S}$ (which depends on the data and is therefore random), and $E(\hat{S})$ (which is not random). In order to compare these for a specific example, we need both a known spectrum and a way to generate data. A spectrum consisting of MA and lines allows relatively easy generation of data, and for this reason, we have chosen the following scheme:

White noise, variance 1 u → $H(z)$ → y
Sinusoids v → 1 → y
(11.2.16)

$$\begin{cases} H(z) = (1 - z^{-1})(1 + z^{-1})^3, \\ v(k) = 2\cos\left(\dfrac{\pi k}{2}\right) + 2\cos\left(\dfrac{11\pi k}{20}\right). \end{cases} \tag{11.2.17}$$

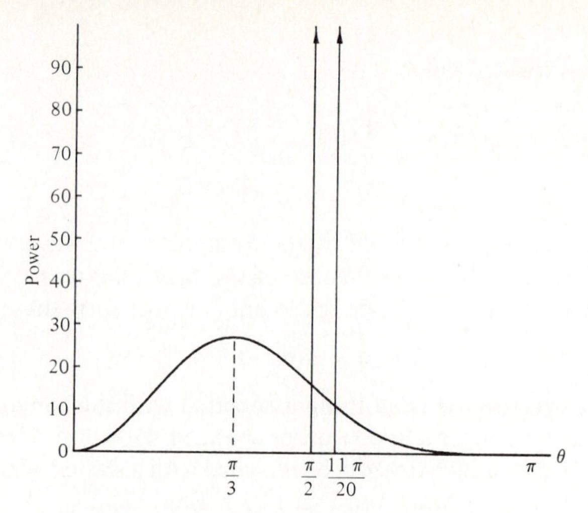

FIGURE 11.2.2 The power spectrum of Example (11.2.1).

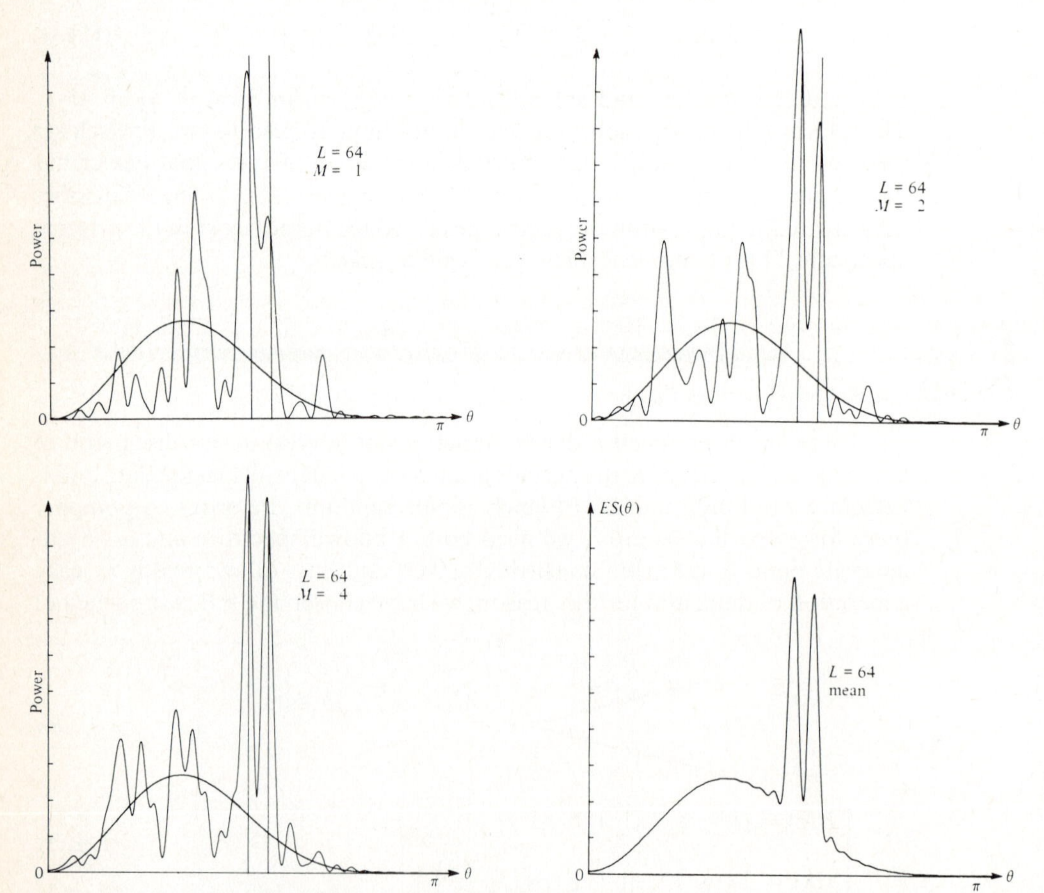

FIGURE 11.2.3 Averages of M periodograms for consecutive data sets of length L, and the mean periodogram (rectangular window).

The resulting power spectrum is given by Eq. (11.2.18) and is shown in Fig. (11.2.2).

$$S(\theta) = 16 \sin^2(\theta)\,(1 + \cos\theta)^2 + 2\pi\left[\delta\left(\theta - \frac{\pi}{2}\right) + \delta\left(\theta + \frac{\pi}{2}\right) + \delta\left(\theta - \frac{11\pi}{20}\right) + \delta\left(\theta + \frac{11\pi}{20}\right)\right]. \tag{11.2.18}$$

The MA portion of the spectrum has power 10, and the sinusoidal portion has power 4. Thus each sinusoid has one seventh of the total power.

Figure 11.2.3 exhibits periodograms or averages of periodograms for a 256-point data set generated via Eq. (11.2.16). The rectangular window in

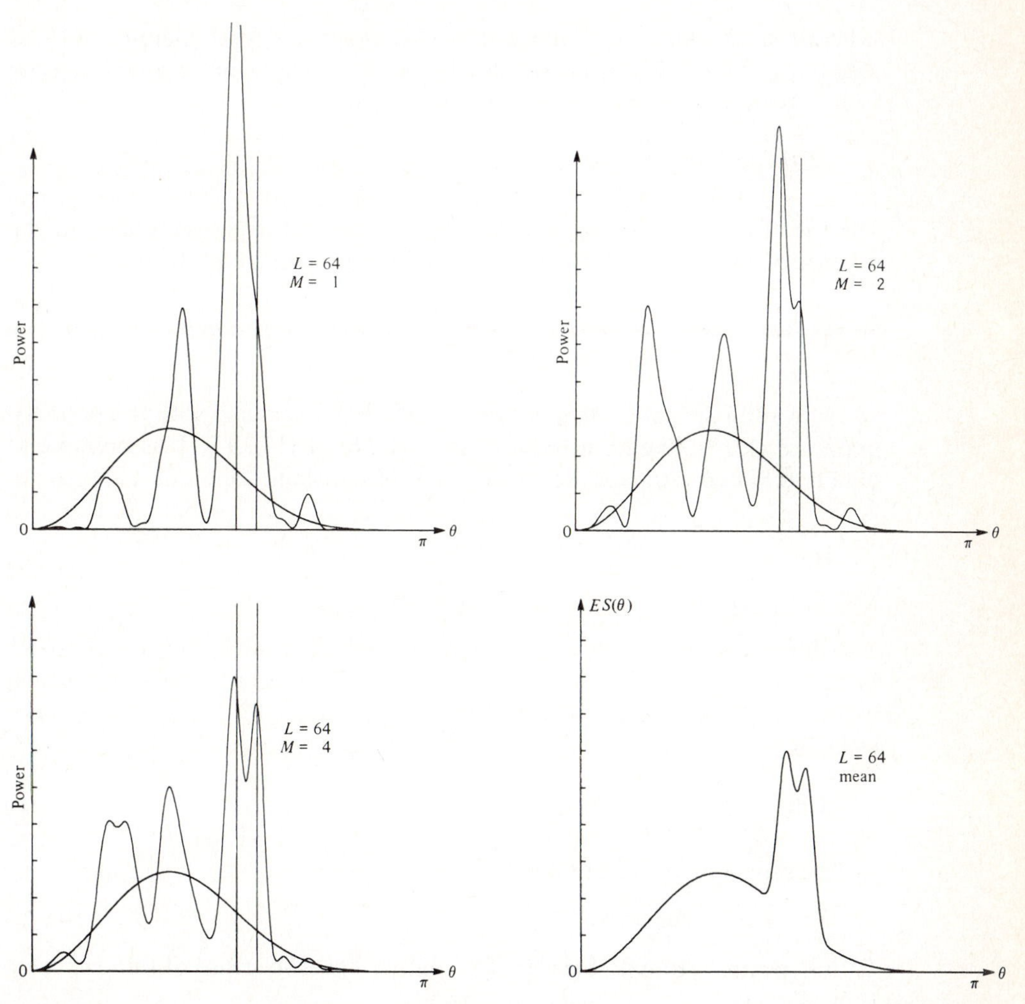

FIGURE 11.2.4 Averages of M periodograms for consecutive data sets of length L, and the mean periodogram (Hanning window).

Eq. (11.2.10) was used to weight 64-point portions of the data. In the upper left-hand corner, only one periodogram (based on the first 64 points) is shown. Then averages of two and then four periodograms are shown. The one with $M = 4$ uses all 256 data values. The true spectrum is also shown on the same plots. The averaging gradually reduces the anomalies (variance). As more and more periodograms are averaged, the result approaches the mean $E(\hat{S})$, which is shown in the lower right-hand corner.

Figure 11.2.4 differs from Fig. 11.2.3 only in the choice of window. The normalized Hanning window

$$w(k) = \begin{cases} \beta\left[1 + \cos\left(\dfrac{2k - L + 1}{L + 1}\pi\right)\right], & 0 \leqslant k \leqslant L - 1 \\ 0, & \text{otherwise} \end{cases}$$

(where β is chosen so that the window sequence has total energy $= L$) was used in Fig. 11.2.4. This window has a broader main lobe but smaller sidelobes and this can be seen in the mean spectrum

$$E\,\hat{S}(\theta) = \frac{1}{L}(S * |W|^2)(\theta).$$

The use of the window has considerably smoothed the periodograms, at the expense of a loss in resolution (evidenced in the treatment of the lines).

The autocorrelation sequence estimates $\hat{r}(k)$ associated with the periodogram are clearly biased, in view of Eq. (11.2.6) or (11.2.13). It is possible to obtain unbiased estimates, of course, at least for small values of k. These are

$$\hat{r}(k) = \begin{cases} \dfrac{1}{L - k} \sum_{j=0}^{L-k-1} y(k + j)y(j), & 0 \leqslant k \leqslant N \\ \hat{r}(-k), & -N \leqslant k < 0. \end{cases} \tag{11.2.19}$$

The difference between this estimate and the biased estimate (11.1.2) is only in the normalization. Now we must take $N < L$, however, because the sum in Eq. (11.2.19) is empty if $N \geqslant L$. Suppose that we use these to form a spectral estimate

$$\hat{S}(\theta) = \sum_{k=-N}^{N} \hat{r}(k)e^{-jk\theta}. \tag{11.2.20}$$

To be consistent, we must take

$$\hat{r}(k) = 0, \qquad |k| > N. \tag{11.2.21}$$

Is this spectral estimate biased? The answer is yes, because of the truncated sum. We have

$$E\,\hat{r}(k) = r(k)q(k), \tag{11.2.22}$$

where

$$q(k) = \begin{cases} 1 & |k| \leqslant N \\ 0, & \text{otherwise} \end{cases} \quad \overset{DTFT}{\longleftrightarrow} \quad Q(e^{j\theta}) = \frac{\sin((N + \frac{1}{2})\theta)}{\sin(\frac{1}{2}\theta)}. \tag{11.2.23}$$

Applying the DTFT to the time domain bias characterization (11.2.22) yields

$$\begin{aligned} E\,\hat{S}(\theta) &= (S * Q)(\theta) \\ &= \frac{1}{2\pi} \int_{-\pi}^{\pi} S(\phi) Q(e^{j(\theta-\phi)})\, d\phi. \end{aligned} \tag{11.2.24}$$

This equation is similar to Eq. (11.2.9), but with a different smoothing function. Resolution is improved (for $N \approx L$) since the width of the main lobe of Q is roughly $2\pi/L$ rather than the $4\pi/L$ of the periodogram. However the function Q is alternately positive and then negative, and it can happen that the estimate $\hat{S}$ in Eq. (11.2.20), together with its expected value in Eq. (11.2.24), can be negative for some values of θ. This is the price that one must pay.

The lag estimates in Eq. (11.2.19) are unbiased. However, the estimate for the entire sequence includes Eq. (11.2.21), which is biased. This leads to the time domain bias characterization in Eq. (11.2.22). No spectral estimator that has only a finite amount of data $\mathbf{y}[0, L-1]$ can be entirely unbiased. However, there is a nontrivial class of spectra for which the estimate in Eq. (11.2.20) is unbiased, namely,

$$E\,\hat{S}(\theta) \equiv S(\theta) \quad \Leftrightarrow \quad S(\theta) \text{ is MA of order} \leqslant N.$$

For the periodogram, on the other hand,

$$E\,\hat{S}(\theta) \equiv S(\theta) \quad \Leftrightarrow \quad S(\theta) \equiv \sigma^2 \quad \text{(white noise)}.$$

Variance

The mean and variance functions for the periodogram (11.2.1) are given by

$$S_W(\theta) \triangleq E\,\hat{S}(\theta) = \frac{1}{2\pi L} \int_{-\pi}^{\pi} S(\phi)\, |W(e^{j(\theta-\phi)})|^2\, d\phi \tag{11.2.25}$$

and

$$\sigma_W^2(\theta) = E\,\hat{S}(\theta)^2 - S_W^2(\theta). \tag{11.2.26}$$

The troublesome part of the variance function is the term

$$E\,\hat{S}(\theta)^2 = \frac{1}{L^2} \sum_k \sum_l \sum_m \sum_n E[y(k)y(l)y(m)y(n)] w(k)w(l)w(m)w(n) e^{-j(k+m-l-n)\theta}. \tag{11.2.27}$$

Whereas the mean S_W involves only second moments (i.e., the autocorrelation sequence), the variance involves fourth-order moments. Until now, no moments of order higher than two have mattered. But computing the variance of the periodogram requires higher-order moments and therefore the variance depends on the distribution of the signal $\mathbf{y}$ as well as its spectrum. Thus an in-depth study of variance would have to take the distribution functions into account.

Let us consider what is perhaps the most important case: wide sense stationary Gaussian signals. Since the distributions of Gaussian random variables are parameterized by the mean (which we take to be zero) and variance, then the moments of all orders will depend only on the second-order moments. Thus the power spectrum (or autocorrelation sequence) characterizes all probabilistic properties of a Gaussian signal. In particular, the troublesome fourth-order moments in Eq. (11.2.27) become

$$E[y(k)y(l)y(m)y(n)] \\ = r(k-m)r(l-n) + r(l-m)r(k-n) + r(k-l)r(m-n), \tag{11.2.28}$$

where $\mathbf{r}$ is the autocorrelation sequence for $\mathbf{y}$.

Substitution of Eq. (11.2.28) into Eq. (11.2.27) results in the following expression for the variance (after some patient bookkeeping).

$$\sigma_W^2(\theta) = S_W^2(\theta) + \left| \frac{1}{2\pi L} \int_{-\pi}^{\pi} S(\phi) W(e^{j(\theta-\phi)}) W(e^{j(\theta+\phi)})\, d\phi \right|^2. \tag{11.2.29}$$

This result is dismaying, since the variance satisfies

$$\sigma_W^2(\theta) \geqslant S_W^2(\theta) \quad \text{for all } L, \tag{11.2.30}$$

i.e., it does not approach zero as L goes to infinity.

EXAMPLE 11.2.2

Behavior of the periodogram for white Gaussian noise.

Let $\mathbf{y}$ be wide sense stationary, Gaussian, and white:

$$r(k) = \sigma^2 \delta(k) \xleftrightarrow{DTFT} S(\theta) \equiv \sigma^2. \tag{11.2.31}$$

Consider a periodogram formed from a portion of the signal, using the rectangular window (11.2.10). The mean spectral estimate will be

$$E\,\hat{S}(\theta) = S_W(\theta) = \sigma^2 = S(\theta). \tag{11.2.32}$$

The covariance will be [using Eqs. (11.2.31) and (11.2.28)]

$$E[\hat{S}(\theta_1) - \sigma^2][\hat{S}(\theta_2) - \sigma^2] \\ = \frac{\sigma^4}{L^2}[|W(e^{j(\theta_1+\theta_2)})|^2 + |W(e^{j(\theta_1-\theta_2)})|^2]. \tag{11.2.33}$$

In particular, the variance will be

$$\sigma_W^2(\theta) = \sigma^4 + \frac{\sigma^4}{L}\left[\frac{1}{L}|W(e^{j2\theta})|^2\right]. \tag{11.2.34}$$

Now the bracketed term in Eq. (11.2.34) is the same function as the one displayed in Fig. 11.2.1, except for the change of scale resulting from the 2 in the exponent. With this in mind, we see that the covariance (11.2.33) is essentially zero unless $\theta_1 = \theta_2$. Thus $\hat{S}(\theta)$ has positive mean and variance, but is essentially uncorrelated (in θ). A plot of it for large L would resemble white noise. What we would have liked, of course, was the constant value of σ^2. Therefore the periodogram is *not* a consistent estimate for S. On the other hand, the autocorrelation estimates

$$\hat{r}(k) \xleftrightarrow{DTFT} \hat{S}(\theta) \tag{11.2.35}$$

are easily shown to have mean and variances

$$E\,\hat{r}(k) = \sigma^2\delta(k) = r(k)$$

$$E\,\hat{r}(k)^2 - [E\,\hat{r}(k)]^2 = \begin{cases} \dfrac{2\sigma^4}{L}, & k = 0 \\ \dfrac{L-|k|}{L^2}\sigma^4, & 0 < |k| < L \\ 0, & |k| \geq L. \end{cases} \tag{11.2.36}$$

Since these variances approach zero as L approaches infinity, each of the estimates $\hat{r}(k)$ **is** consistent. It follows therefore that the part of $\hat{S}(\theta)$ that produces the variance term σ^4 must be very rapidly varying for large L [since it does not corrupt the Fourier coefficients $\hat{r}(k)$].

Bias Versus Variance

The issue of consistency is rather subtle for the periodogram. Let $\hat{S}_L(\theta)$ be a sequence of periodograms based on data sets (or equivalently windows) of increasing length. Let

$$\hat{r}_L(k) \xleftrightarrow{DTFT} \hat{S}_L(\theta). \tag{11.2.37}$$

To be consistent, we require that the estimate be asymptotically unbiased, i.e.

$$S_{W,L}(\theta) = E\,\hat{S}_L(\theta) \to S(\theta) \quad \text{as } L \to \infty, \qquad (11$$

and that the variance approach zero,

$$\sigma^2_{W,L}(\theta) \to 0 \quad \text{as } L \to \infty.$$

The inequality (11.2.30) makes it impossible to achieve bot

The problem is that the spectrum is a function and dimensional. We are attempting to measure an infi with finite dimensional data, and we must be prep compromise usually involves a tradeoff between

There are two general approaches to the r and averaging. "Smoothing" can mean c sequence **w** to reduce variance, or it c window sequence to the autocorrela but it will also further smooth the mean spectral estimate and therefore increase bias. But the bias/variance tradeoff is much more evident when periodograms are averaged.

Welch's Method

The following method of averaging was proposed by Welch (1967). Consider a data set $\mathbf{y}[0, N-1]$ taken from a wide sense stationary signal $\mathbf{y}$. Let

$$x_m(k) = y(k + mD) \tag{11.2.40}$$

be a shifted version of the data, and consider the overlapping short records

$$\{\mathbf{x}_m[0, L-1], \qquad 0 \leqslant m \leqslant M-1\}$$

where

$$(M-1)D + L = N. \tag{11.2.41}$$

$\mathbf{y}[0, N-1]$

$y(0)$ $y(1)$ ••• $y(D)$ ••• $y(L-1)$ ••• $y(D+L-1)$ ••• $y(N-L)$ ••• $y(N-1)$

$\mathbf{x}_0[0, L-1]$

$\mathbf{x}_1[0, L-1]$

$\mathbf{x}_{M-1}[0, L-1]$

ion of the data $\mathbf{y}[0, N-1]$ into M records

ing the M periodograms obtained by

$$\ldots k\theta \Big|^2 \Big\}. \tag{11.2.42}$$

y the window W which has length L,
sfies Eq. (11.2.25). But the variance
grams that are averaged. If successive
e variance would be inversely pro-
are to some extent correlated and so
n suggests the overlap be such that

Roughly then, resolution [the width of $W(e^{j\theta})$] is inversely proportional to L. This is a measure of bias. Variance is inversely proportional to M. With the overlap specified, the total data length N is proportional to the product $L \times M$. Thus for a given data length, one must divide N in order to achieve some balance between bias and variance. An improvement in one must be purchased with a degradation of the other.

Use of the Fast Fourier Transform

In practice one does not compute the entire periodogram (i.e., for all values of θ) but merely samples of it. A set of equally spaced samples of $\hat{S}(\theta)$ can most efficiently be obtained by using a fast Fourier transform algorithm for the DFT. Let $N \geqslant L$ be a power of two, and let

$$\theta_0 = \frac{2\pi}{N}. \tag{11.2.43}$$

Then

$$\begin{aligned} \hat{S}(n\theta_0) &= \frac{1}{L}\left|\sum_{k=0}^{L-1} y(k)w(k)e^{-jkn\theta_0}\right|^2 \\ &= \frac{1}{L}\left|\sum_{k=0}^{N-1} x(k)W_N^{-kn}\right|^2 \\ &= \frac{|X(n)|^2}{L}, \end{aligned} \tag{11.2.44}$$

where

$$W_N = e^{j2\pi/N}, \tag{11.2.45}$$

$$x(k) = \begin{cases} y(k)w(k), & 0 \leqslant k \leqslant L-1 \\ 0, & L \leqslant k \leqslant N-1 \end{cases}, \tag{11.2.46}$$

$$\mathbf{x}[0, N-1] \xleftrightarrow{N\text{-point } DFT} X[0, N-1]. \tag{11.2.47}$$

Thus one can get samples of the periodogram by weighting the data, extending the data to length N by zero-padding, performing a length N FFT, and then normalizing the magnitude squared transform.

It should be understood that increasing the size of the DFT in no way improves the "resolution" (bias) of the spectral estimate. It serves only to obtain more samples of the periodogram. This point is illustrated in Fig. 11.2.6. The periodogram is the continuous curve, while the use of the FFT (with $N = L = 64$) provides only the samples shown by the vertical lines. Increasing N (and zero-padding the data) would decrease the sample interval. This would allow for more accurate representation of the true heights of peaks and valleys.

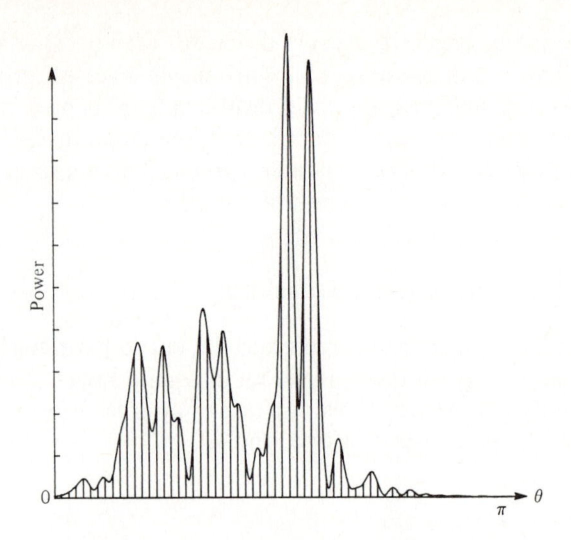

FIGURE 11.2.6 **Samples of the periodogram obtained by using the Fast Fourier Transform. The sample density is increased by zero-padding the data.**

11.3 AUTOREGRESSIVE MODELS

Autoregressive (or AR) models for spectral estimates can arise in several ways. First, one can simply specify an AR form and demand that the parameters be chosen to conform with $\hat{\mathbf{r}}[0, n]$ (estimates of the autocorrelation sequence). Second, one can require the spectrum that has maximum entropy and agrees with $\hat{\mathbf{r}}[0, n]$; the result will be AR. Third, if one adopts the analysis/synthesis approach to spectral estimation with an FIR whitening filter, the resulting estimate must be AR. There are some intriguing theoretical issues that connect these ideas and lead to algorithms which have found application in other areas (including speech coding and filter structures).

The model for an AR signal involves an all-pole filter driven by white noise:

$$u \circ \xrightarrow{1/\hat{a}(z)} \circ\, y \tag{11.3.1}$$

$$\hat{a}(z) \triangleq 1 + a_1 z^{-1} + \cdots + a_n z^{-n}. \tag{11.3.2}$$

The input and output spectra are

$$S_{uu}(\theta) = \alpha, \tag{11.3.3}$$

$$S_{yy}(\theta) = \frac{\alpha}{|\hat{a}(e^{j\theta})|^2}. \tag{11.3.4}$$

The standard difference equation that relates the input and output sequences is

$$\mathbf{a} * \mathbf{y} = \mathbf{u}, \quad \text{or} \quad \sum_{l=0}^{n} a_l y(k-l) = u(k). \tag{11.3.5}$$

This equation may be used to produce the AR version of Eq. (11.1.30), which is

$$\sum_{l=0}^{n} a_l r(k-l) = \alpha\,\delta(k), \qquad \text{for } k \geqslant 0. \tag{11.3.6}$$

Of special interest in the following is the matrix version of this equation, over the range $0 \leqslant k \leqslant n$.

$$[1 \quad a_1 \quad \cdots \quad a_n] \begin{bmatrix} r(0) & r(1) & \cdots & r(n) \\ r(1) & \ddots & \ddots & \vdots \\ \vdots & \ddots & \ddots & r(1) \\ r(n) & \cdots & r(1) & r(0) \end{bmatrix} = [\alpha \quad \bigcirc \quad]$$

or

$$\mathbf{aR} = [\alpha \quad \bigcirc \quad]. \tag{11.3.7}$$

The matrix **R** in Eq. (11.3.7) was introduced in Section 7.3. Since the entries of **R** depend only on the difference of the indices,

$$R(i, j) = E\, y(k-i)y(k-j) = r(i-j), \tag{11.3.8}$$

it is Toeplitz. It must also be symmetric and positive semidefinite. There are two ways to see this [which correspond to the different ways of representing inner products appearing in Eq. (7.3.15)]. Let

$$\psi_n(z) \triangleq \begin{bmatrix} 1 \\ z^{-1} \\ z^{-2} \\ \vdots \\ z^{-n} \end{bmatrix}. \tag{11.3.9}$$

Then we can rewrite Eq. (11.3.2) as

$$\hat{a}(z) = \mathbf{a}\psi_n(z). \tag{11.3.10}$$

The two representations for **R** are

$$\mathbf{R} = \frac{1}{2\pi}\int_{-\pi}^{\pi} \psi_n(e^{j\theta})[\psi_n(e^{j\theta})]^* S(\theta)\, d\theta \tag{11.3.11}$$

$$= E\begin{bmatrix} y(k) \\ y(k-1) \\ \vdots \\ y(k-n) \end{bmatrix} [y(k)\ y(k-1) \cdots y(k-n)]. \tag{11.3.12}$$

These provide alternate expressions for the row vector **aR**, which appears in Eq. (11.3.7), and the scalar $\mathbf{aRa}^T$:

$$\mathbf{aR} = \frac{1}{2\pi}\int_{-\pi}^{\pi} S(\theta)\hat{a}(e^{j\theta})[\psi_n(e^{j\theta})]^* \, d\theta$$
$$= E\left\{\sum_{l=0}^{n} a_l y(k-l)[y(k)\ y(k-1)\cdots y(k-n)]\right\}, \qquad (11.3.13)$$

$$\mathbf{aRa}^T = \frac{1}{2\pi}\int_{-\pi}^{\pi} S(\theta)|\hat{a}(e^{j\theta})|^2 \, d\theta = E\left[\sum_{l=0}^{n} a_l y(k-l)\right]^2. \qquad (11.3.14)$$

Since

$$\mathbf{aRa}^T \geqslant 0,$$

no matter how **a** is chosen, the matrix **R** must be positive semidefinite.

Whitening and Prediction Problems

The analysis/synthesis approach to spectral estimation was introduced in Section 11.1. This involves passing the signal **y** through an FIR filter $\hat{a}(z)$ which is chosen to produce an output signal **e**, which is approximately uncorrelated. Of course, what is actually the case is that the filter parameters are chosen to minimize output power subject to the constraint $a_0 = 1$. One can view this in two ways, which are depicted in the signal flow graph representations (11.3.15) and (11.3.16). [See Makhoul (1975) for a tutorial on linear prediction.]

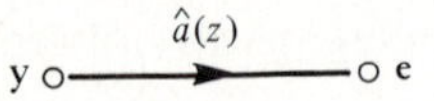

$\hat{a}(z)$

y o——→——o e (whitening filter) **(11.3.15)**

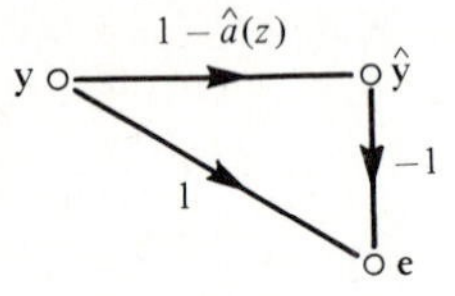

$1 - \hat{a}(z)$

y o——→——o $\hat{y}$ (prediction filter)

1 ↘ ↓ −1 **(11.3.16)**

o e (prediction error)

Since only a finite portion of the signal **y** is available, the minimization of error power must be approximated in some way, and there turn out to be quite a number of ways that this can be done.

Let

$$w(k) = \begin{cases} 1, & 0 \leqslant k \leqslant L-1 \\ 0, & \text{otherwise} \end{cases} \qquad (11.3.17)$$

be the window that corresponds to the available data $\mathbf{y}[0, L-1]$. The error signal that one can compute then is given by

$$e(k) = \sum_{l=0}^{n} a_l y(k-l) w(k-l). \qquad (11.3.18)$$

This is the "forward prediction error" for a linear estimate $\hat{y}(k)$ based on the past values $\mathbf{y}[k-n, k-1]$. The time reversal of the signal **y** has the same spectrum as does $\mathbf{y}$. Therefore, we could also consider a "backward prediction error" signal

$$\tilde{e}(k) = \sum_{l=0}^{n} a_l y(k + l - n) w(k + l - n), \tag{11.3.19}$$

which is the error in a linear prediction of $y(k - n)$ given future values $\mathbf{y}[k - n + 1, k]$. For an infinite amount of data, the variances of both $\mathbf{e}$ and $\tilde{\mathbf{e}}$ would be minimized with the same FIR filter $\hat{a}(z)$. For a finite data set, there may be a small discrepancy. This leads to the notion of minimizing the sum of the backward and forward prediction error powers.

Because of the data window, both error signals are zero outside the range

$$0 \leqslant k \leqslant L + n - 1,$$

and both involve less than the full n summands if $k < n$ or $k > L - 1$. Because one cannot minimize the actual error power without knowing the autocorrelation sequence for $\mathbf{y}$, one must resort to minimizing a sum of squares of the error signal. Let

$$V(\mathbf{a}) = \sum_{k=k_0}^{k_1} e(k)^2, \tag{11.3.20}$$

$$\tilde{V}(\mathbf{a}) = \sum_{k=k_0}^{k_1} \tilde{e}(k)^2. \tag{11.3.21}$$

There are four cases common in the literature:

$k_0 = 0, \quad k_1 = L + n - 1$ autocorrelation method
$k_0 = 0, \quad k_1 = L - 1$ prewindowed
$k_0 = n, \quad k_1 = L + n - 1$ postwindowed
$k_0 = n, \quad k_1 = L - 1$ autocovariance method.

The functions V and $\tilde{V}$ are quadratic in the vector $\mathbf{a}$. To see this, it helps to write equations in matrix form. For each of the four choices of k_0 and k_1, let us form a data matrix $\mathbf{Y}$ which is $(n + 1) \times (k_1 - k_0 + 1)$ as follows:

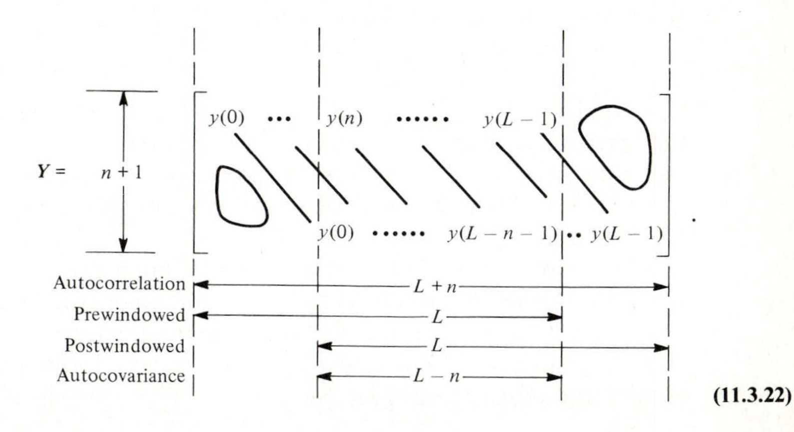

(11.3.22)

We have

$$[e(k_0)\cdots e(k_1)] = \mathbf{aY}, \tag{11.3.23}$$

where $\mathbf{Y}$ is chosen appropriately. Therefore

$$V(\mathbf{a}) = \|\mathbf{aY}\|^2 = \mathbf{aYY}^T\mathbf{a}^T \tag{11.3.24}$$

$$\tilde{V}(\mathbf{a}) = \|\mathbf{aJY}\|^2 = \mathbf{aJYY}^T\mathbf{J}^T\mathbf{a}^T, \tag{11.3.25}$$

where

$$\mathbf{J} = \mathbf{J}^T = \begin{bmatrix} 0 & & 1 \\ & ⋰ & \\ 1 & & 0 \end{bmatrix} \quad \text{is } (n+1) \times (n+1). \tag{11.3.26}$$

The matrix $\mathbf{J}$ turns the row vector $\mathbf{a}$ end for end, thereby reindexing the sum in Eq. (11.3.18) to that of Eq. (11.3.19).

The equation that $\mathbf{a}$ must satisfy to minimize a quadratic function is called the *normal equation.* There are two cases of interest.

Normal Equation for Forward Errors:

$$V(\mathbf{a}) \text{ is minimized} \quad \Leftrightarrow \quad \mathbf{aYY}^T = [\alpha \; \mathbf{0}]. \tag{11.3.27}$$

Normal Equation for Forward and Backward Errors:

$$V(\mathbf{a}) + \tilde{V}(\mathbf{a}) \text{ is minimized}$$

$$\Leftrightarrow \quad \mathbf{a}[\mathbf{YY}^T + \mathbf{JYY}^T\mathbf{J}] = [\alpha \; \mathbf{0}] \tag{11.3.28}$$

These normal equations are similar to those encountered in Chapter 7. The unknowns are the $n + 1$ parameters $\alpha, a_1, \ldots, a_n$. The significance of the parameter α is

$$\alpha = \min_{\mathbf{a}} V(\mathbf{a}), \quad \text{or} \quad \min_{\mathbf{a}} [V(\mathbf{a}) + \tilde{V}(\mathbf{a})]. \tag{11.3.29}$$

It is revealing to compare these data-dependent normal equations with those of the asymptotic case, when the autocorrelation sequence for $\mathbf{y}$ is known.

Normal Equation for Asymptotic Case:

$$E\, e(k)^2 = \mathbf{aRa}^T \text{ is minimized} \quad \Leftrightarrow \quad \mathbf{aR} = [\alpha \; \mathbf{0}]. \tag{11.3.30}$$

This equation is of course identical to Eq. (11.3.7), but one should not forget that the sources for these equations are different. Equation (11.3.7) relates the autocorrelation sequence to the model parameters for an AR spectrum, while Eq. (11.3.30) is the result of a quadratic minimization problem involving the whitening of a given WSS signal $\mathbf{y}$.

One should regard the data dependent normal equations (11.3.27) and (11.3.28) as approximations to the asymptotic case (11.3.30), which provides what we shall later show is the best AR approximation to the true spectrum. But since there are still four choices for quadratic forms involving the data, let us discuss their relative merits.

Autocorrelation Method. The elements of the matrix $\mathbf{YY}^T$ are inner products of two rows of $\mathbf{Y}$. Knowing this, it is not hard to see that

$$(\mathbf{YY}^T)_{ij} = L\hat{r}(i-j) \tag{11.3.31}$$

for the autocorrelation case, where $\hat{r}(k)$ is the biased estimate of $r(k)$ in Eq. (11.2.12). Therefore $\mathbf{YY}^T$ is Toeplitz. Now if $\mathbf{YY}^T$ is Toeplitz, then it commutes with the matrix $\mathbf{J}$ in Eq. (11.3.26), and since $\mathbf{J}^2 = \mathbf{I}$, we have

$$\mathbf{YY}^T = \mathbf{JYY}^T\mathbf{J} \quad \Rightarrow \quad \tilde{V}(\mathbf{a}) = V(\mathbf{a}) \tag{11.3.32}$$

for this case. Therefore it makes no sense to consider backward errors in the autocorrelation case, since the result will be the same. Once the $\hat{r}(k)$ have been computed, the most efficient way to solve the normal equations is Levinson's algorithm, which will be presented later on in this section. This is possible only because $\mathbf{YY}^T$ is Toeplitz.

One drawback of this case lies in the bias of the estimates of the autocorrelation sequence. Because the expected value of $\hat{r}(k)$ is not the true value $r(k)$, applying the expectation operator to the normal equation (11.3.27) does not yield a scalar multiple of Eq. (11.3.30). The difference decreases as L increases for a given model order n.

To get a spectral estimate using this method, form $\hat{\mathbf{r}}[0, n]$ from the data $\mathbf{y}[0, L-1]$ using Eq. (11.2.12). Solve the normal equations (11.3.27) using Levinson's algorithm. Then

$$\hat{S}(\theta) = \frac{1}{L}\frac{\alpha}{|\hat{a}(e^{j\theta})|^2}. \tag{11.3.33}$$

Prewindowed or Postwindowed. In this case, neither $\mathbf{YY}^T$ nor $E(\mathbf{YY}^T)$ is Toeplitz. Since $V(\mathbf{a})$ and $\tilde{V}(\mathbf{a})$ are not now the same functions, the forward/backward problem would yield a different estimate than the forward problem.

Autocovariance Method. In this case,

$$\mathbf{YY}^T \neq \mathbf{JYY}^T\mathbf{J} \tag{11.3.34}$$

and neither of these matrices are Toeplitz. However,

$$E(\mathbf{YY}^T) = E(\mathbf{JYY}^T\mathbf{J}) = (L-n)\mathbf{R} \tag{11.3.35}$$

is not only Toeplitz, but a scalar multiple of what is really wanted. The bias inherent in Eq. (11.3.31) is gone. Because of Eq. (11.3.34), it makes sense to use the forward/backward problem in this case. This can be seen as an attempt to bring the quadratic form a little nearer to Toeplitz, since

$$\mathbf{Q} = \mathbf{YY}^T + \mathbf{JYY}^T\mathbf{J} \tag{11.3.36}$$

would satisfy

$$\mathbf{QJ} = \mathbf{JQ} \tag{11.3.37}$$

even though it is still not Toeplitz. For this reason, the Levinson algorithm cannot be used. There are, however, some more complicated fast algorithms for solving the normal equations (11.3.28). See Friedlander (1982) and Marple (1980).

To form a spectral estimate from the solution of Eq. (11.3.28), set

$$\hat{S}(\theta) = \frac{1}{2(L - n)} \frac{\alpha}{|\hat{a}(e^{j\theta})|^2}. \tag{11.3.38}$$

EXAMPLE 11.3.1

Asymptotic AR models for MA plus lines.

Consider the spectrum of Eq. (11.2.18) consisting of a fourth-order MA background plus two sinusoidal components (four lines). Using the known autocorrelation sequence for this spectrum, one can produce the asymptotic

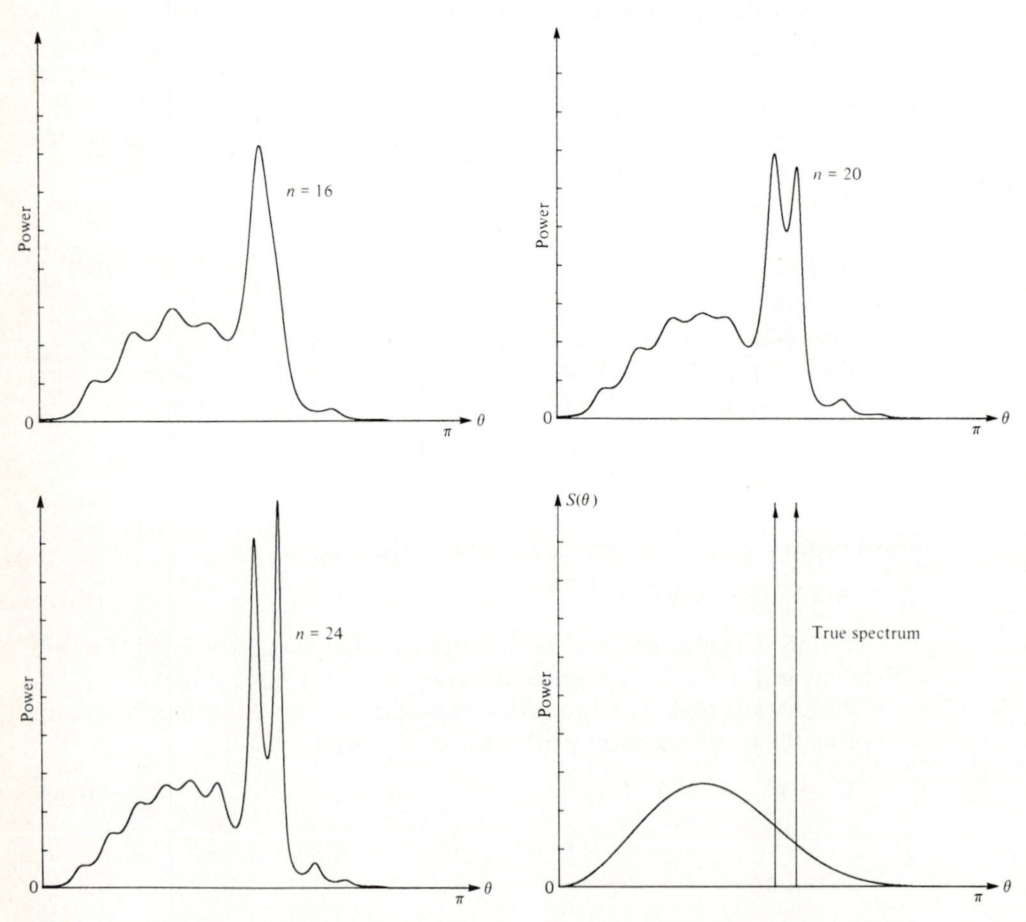

FIGURE 11.3.1 The spectrum of Example 11.2.1 and three-order n AR models, obtained by using true values of $r[0, n]$.

AR model for any order n, using the normal equations (11.3.30). These AR spectra should be considered the best one can hope for, since the exact values of $r(k)$ cannot be obtained from finite data records. In Fig. 11.3.1 are shown four spectra: the true spectrum and the AR models with $n = 16, 20, 24$. For the case $n = 16$, the two lines have not been resolved in the sense that there are two separate peaks. (More information could be obtained by factoring the polynomial $\hat{a}(z)$ to get the exact pole locations.)

It is apparent from this example that AR spectra tend to be "peaky." An AR spectrum can produce a value at one point that is greater than the value at a nearby point only by placing a pole near the circle. Each peak requires

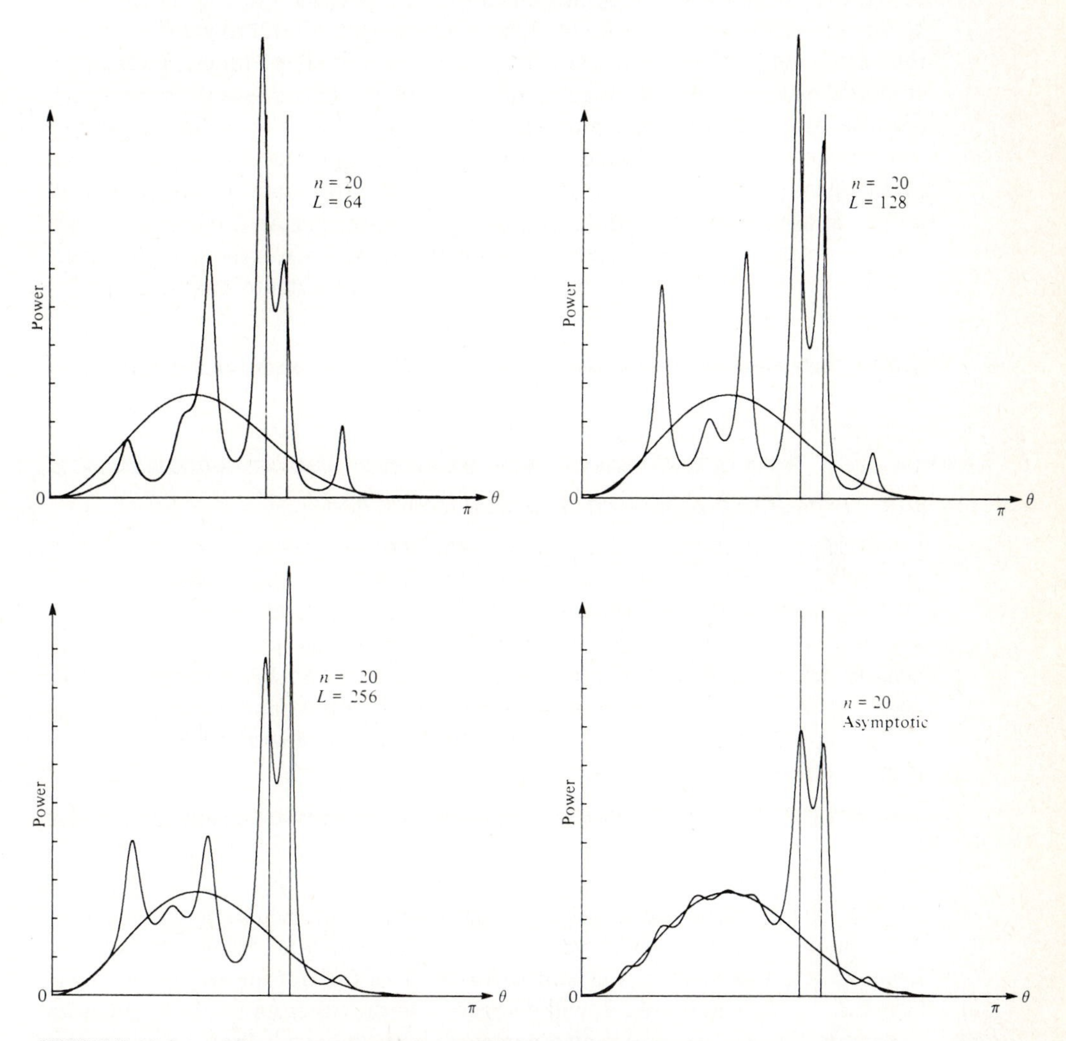

FIGURE 11.3.2 **Three autocorrelation method (or Levinson) AR estimates of order $n = 20$ based on data $y[0, L-1]$. The asymptotic model is approached as $L \to \infty$. The true spectrum is superimposed on each plot.**

at least one pole. Small values of the spectrum can only be obtained by making the constant numerator small, and if α is small, then the poles must be closer to the circle to bring the function back up.

EXAMPLE 11.3.2

AR estimates using autocorrelation method.

Using the same data set that was used for the periodograms of Example 11.2.1, the estimator (11.3.33) produced the results shown in Fig. 11.3.2. This estimate is equivalent to using the biased estimator (11.1.42) to get $\hat{\mathbf{r}}[0, n]$ and then applying the Levinson algorithm to obtain the AR parameters. One can compare each one of the three estimates with its periodogram counterpart because they use the same portion of the total data. The order of each is $n = 20$. The spectral anomalies correspond roughly to anomalies in the periodograms, suggesting that they are inherent in the data. As the length of data increases, these false peaks and valleys smooth out. As L increases, these estimates approach the asymptotic spectrum that corresponds to the true values of $\mathbf{r}[0, 20]$. For point of reference, the true spectrum has been superimposed on each plot.

EXAMPLE 11.3.3

AR estimates using forward/backward covariance method.

Using the estimate (11.3.38) on the same data set yields the results shown in Fig. 11.3.3. The covariance choice of $\mathbf{Y}$ in Eq. (11.3.22) avoids the bias inherent in the autocorrelation method. This plus the combination of forward and backward error energy minimization apparently produces somewhat peakier AR estimates. (Some of the peaks have in fact been clipped in the figure.) The peaks still tend to settle as the length of the data increases.

Computing the points to be plotted in AR (or in general ARMA) models is made efficient by the use of the algorithm in Fig. 2.5.6.

The choice of AR model order n should depend on the length of the data L and some measure of how close the model "fits" the data. About all that one has for such a measure is the sample variance of the output error, which we may call $\hat{\alpha}_n$. As n increases, $\hat{\alpha}_n$ will decrease. On the other hand, the number of representative error samples also decreases with the result that the estimated variance becomes meaningless. High-order AR models for short data records tend to introduce many spurious features. This phenomenon is the AR version

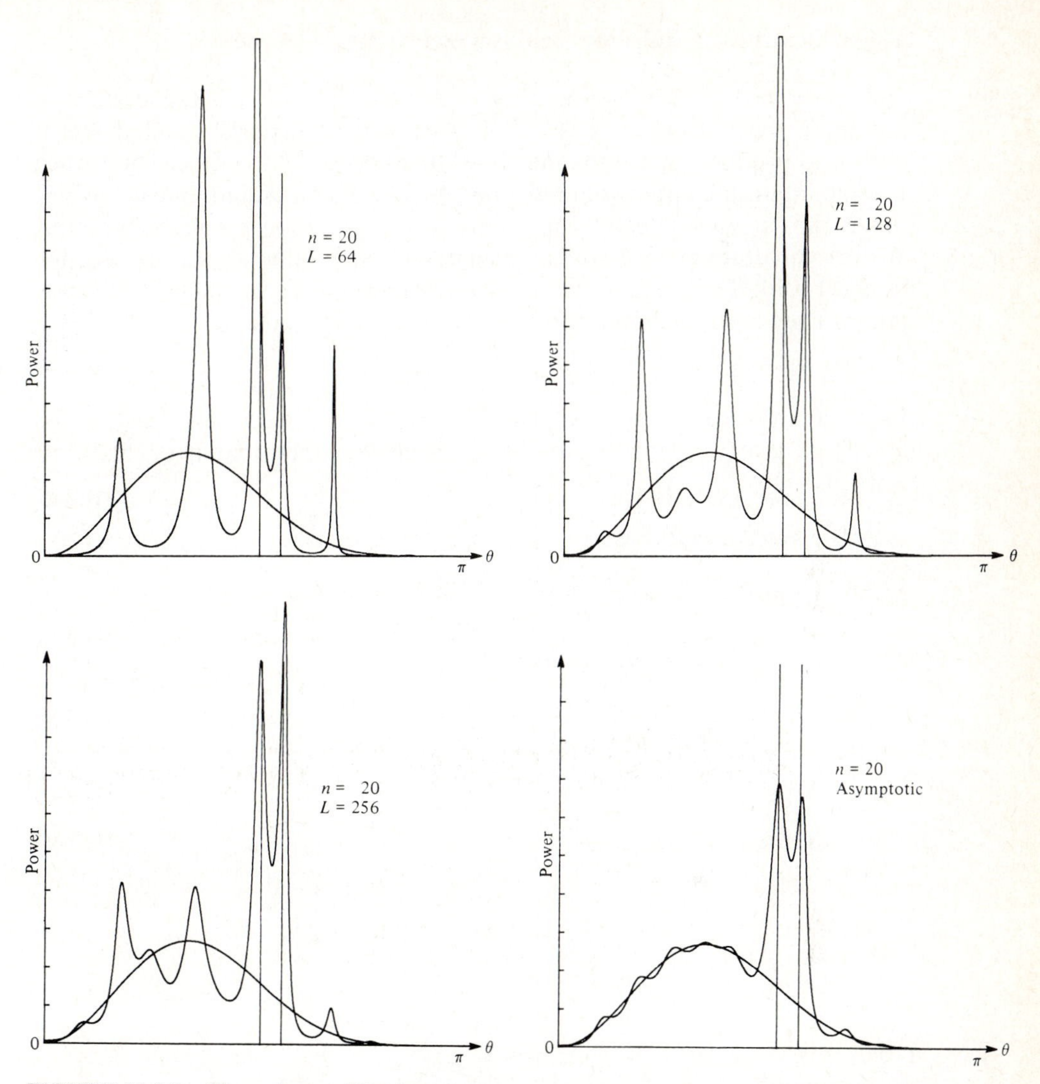

FIGURE 11.3.3 Three forward/backward autocovariance method AR estimates of order $n = 20$ based on data $y[0, L-1]$. The asymptotic model is approached as $L \to \infty$. The true spectrum is also shown.

of the "bias versus variance" dilemma. Some proposed rules for order selection involve the minimization (over n) of the following criteria:

$$\hat{\alpha}_n \left[\frac{L + n + 1}{L - n - 1}\right] \quad \text{(Akaike 1970)},$$

$$\ln(\hat{\alpha}_n) + \frac{2n}{L} \quad \text{(Akaike 1974)},$$

$$\left[\frac{1}{L}\sum_{j=1}^{n}\frac{1}{\hat{\alpha}_j}\right] - \frac{1}{\hat{\alpha}_n} \quad \text{(Parzen 1974)}.$$

Theoretical Issues Relating Analysis and Synthesis Cases

One can interpret the equation $\mathbf{a} * \mathbf{y} = \mathbf{u}$ in two ways. If $\mathbf{y}$ is the input, then the appropriate diagram is (11.3.15). This is the "analysis" case, if $\hat{a}(z)$ is chosen to produce an approximately white output. On the other hand when $\mathbf{u}$ is the input, the appropriate diagram is (11.3.1). Assuming that $\mathbf{u}$ is white, this is the "synthesis" case. The synthesis interpretation produces the Yule-Walker equations (11.3.7) while the analysis case produces the normal equations (11.3.30). These equations are identical in form but also differ in interpretation. They both relate the two sets of $n + 1$ parameters:

$$\mathbf{r}[0, n] \quad \text{versus} \quad (\alpha, \mathbf{a}[1, n]). \tag{11.3.39}$$

In the analysis case, $\mathbf{r}[0, n]$ is assumed known. In the synthesis case, α and $\mathbf{a}[1, n]$ are assumed known. Thus the equations should be written to reflect this:

$$\mathbf{aR} = \alpha \psi^T \quad \text{(analysis)}, \tag{11.3.40}$$

$$\mathbf{rQ} = \alpha \psi^T \quad \text{(synthesis)}. \tag{11.3.41}$$

The matrices and row vectors in these equations are

$$\mathbf{a} = [1 \quad a_1 \quad a_2 \quad \cdots \quad a_n], \tag{11.3.42}$$

$$\mathbf{R} = \begin{bmatrix} r(0) & & r(n) \\ & \diagdown\diagdown\diagdown & \\ r(n) & & r(0) \end{bmatrix}, \tag{11.3.43}$$

$$\psi^T = [1 \quad 0 \quad \cdots \quad 0], \tag{11.3.44}$$

$$\mathbf{r} = [r(0) \quad r(1) \cdots r(n)], \tag{11.3.45}$$

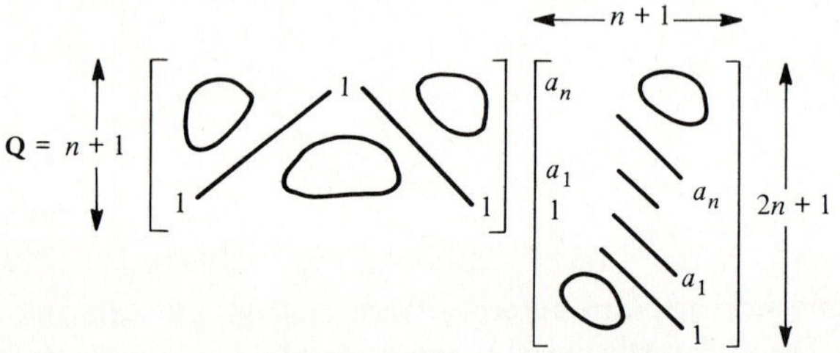

(11.3.46)

The matrix $\mathbf{R}$ is Toeplitz and positive semidefinite. The matrix $\mathbf{Q}$ is the *Jury matrix* (Jury 1964, p. 297).

The two equations (11.3.40) and (11.3.41) are algebraically equivalent. But the parameter sets (11.3.39) cannot be arbitrarily chosen. In the analysis case, the matrix $\mathbf{R}$ must be positive definite:

$$\mathbf{R} > \mathbf{0}; \quad \text{i.e.,} \quad \mathbf{c} \neq \mathbf{0} \;\Rightarrow\; \mathbf{cRc}^T > 0. \tag{11.3.47}$$

In the synthesis case, there is a stability requirement:

$$\alpha > 0 \quad \text{and} \quad \hat{a}(\lambda) = 0 \;\Rightarrow\; |\lambda| < 1. \tag{11.3.48}$$

This suggests some key questions.

Uniqueness/Consistency: If $\mathbf{R}$ satisfies conditions (11.3.47), must there exist a unique solution to Eq. (11.3.40)? If condition (11.3.48) is satisfied, must there exist a unique solution to Eq. (11.3.41)?

Stability/Positivity: If $\mathbf{R}$ is positive, must the solution to Eq. (11.3.40) be stable? If the stability requirement (11.3.48) is satisfied, must the solution to Eq. (11.3.41) produce a positive matrix $\mathbf{R}$?

The following theorems answer all four of these questions in the affirmative. In this discussion and throughout the remainder of this section, we shall need the following notation:

$$\mathbf{R}(m) \triangleq \begin{bmatrix} r(0) & r(1) & & r(m) \\ r(1) & \ddots & \ddots & \\ & \ddots & \ddots & r(1) \\ r(m) & & r(1) & r(0) \end{bmatrix}, \quad \text{Toeplitz, } (m+1) \times (m+1) \tag{11.3.49}$$

THEOREM 11.3.1. If $\mathbf{R}$ is positive definite, then a unique solution α, $\mathbf{a}[1, n]$ exists to Eq. (11.3.40).

Proof. A necessary and sufficient condition for the positivity of $\mathbf{R}$ is Sylvester's condition

$$\det \mathbf{R}(m) > 0, \qquad 0 \leqslant m \leqslant n \tag{11.3.50}$$

(see Appendix 7A). The solution is

$$\alpha = \frac{\det \mathbf{R}(n)}{\det \mathbf{R}(n-1)}, \tag{11.3.51}$$

$$[a_1 \quad \cdots \quad a_n] = -[r(1) \quad \cdots \quad r(n)]\mathbf{R}^{-1}(n-1). \tag{11.3.52}$$

Note that $\mathbf{R}(n-1)$ is invertible because of condition (11.3.50). The formula (11.3.51) is called Gram's formula (Cheney 1966).

THEOREM 11.3.2. If the stability condition (11.3.48) holds, then a unique solution $\mathbf{r}[0, n]$ exists to Eq. (11.3.41).

Proof. A unique solution exists if and only if $\det(\mathbf{Q}) \neq 0$. This can be expressed in terms of the roots of the polynomial $\hat{a}(z)$ as follows. Let

$$\hat{a}(z) = \prod_{k=1}^{n} (1 - z^{-1}\lambda_k). \tag{11.3.53}$$

Then

$$\det \mathbf{Q}(\mathbf{a}) = \prod_{k=1}^{n} \prod_{l=k}^{n} (1 - \lambda_k \lambda_l). \tag{11.3.54}$$

(See Problem 11.9.) The determinant can vanish only if one root is the reciprocal of itself or another root. But this is impossible if all roots have modulus less than one.

THEOREM 11.3.3. If $\mathbf{R} > \mathbf{0}$, then the unique solution to (11.3.40) satisfies the stability condition (11.3.48).

Proof. That $\alpha > 0$ follows from the inequality (11.3.50) and Eq. (11.3.51). To show that the roots of $\hat{a}(z)$ satisfy $|\lambda| < 1$, we will use the Liapunov stability theorem discussed in Section 8.5. To meet the conditions of the theorem, we must find matrices $(\mathbf{A}, \mathbf{B}, \mathbf{K})$, for which

$$\det(z\mathbf{I} - \mathbf{A}) = z^n \hat{a}(z) = \prod_{k=1}^{n} (z - \lambda_k), \tag{11.3.55}$$

$$(\mathbf{A}, \mathbf{B}) \text{ is controllable,} \tag{11.3.56}$$

$$\mathbf{K} \text{ is } n \times n \text{ and positive definite,} \tag{11.3.57}$$

$$\mathbf{K} = \mathbf{AKA}^T + \mathbf{BB}^T. \tag{11.3.58}$$

Let

$$\psi = \psi_n(\infty) = \begin{bmatrix} 1 \\ \bigcirc \end{bmatrix}, \quad (n+1) \times 1 \tag{11.3.59}$$

$$\mathbf{N} = \left[\begin{array}{ccc|c} & \bigcirc & & 0 \\ \hline 1 & & \bigcirc & \\ & \ddots & & \bigcirc \\ \bigcirc & & 1 & \end{array}\right], \quad (n+1) \times (n+1) \tag{11.3.60}$$

$$\mathbf{A} = \left[\begin{array}{cccc} -a_1 & & & -a_n \\ \hline 1 & & \bigcirc & \multicolumn{1}{|c}{} \\ & \ddots & & \multicolumn{1}{|c}{\bigcirc} \\ \bigcirc & & 1 & \multicolumn{1}{|c}{} \end{array}\right], \quad (n \times n) \tag{11.3.61}$$

$$\mathbf{B} = \sqrt{\alpha}\begin{bmatrix} 1 \\ \bigcirc \end{bmatrix}, \quad (n \times 1). \tag{11.3.62}$$

Then Eq. (11.3.55) follows using properties of companion matrices. Controllability follows from the fact that the controllability matrix has the form

$$[\mathbf{B}, \mathbf{AB}, \ldots, \mathbf{A}^{n-1}\mathbf{B}] = \sqrt{\alpha}\begin{bmatrix} 1 & \diagdown & \diagdown \\ & \ddots & \diagdown \\ \bigcirc & & 1 \end{bmatrix}. \tag{11.3.63}$$

The interesting part is Eq. (11.3.58). It is easy to show that

$$\mathbf{N}(\mathbf{I} - \psi\mathbf{a}) = \left[\begin{array}{c|c} 0 & \bigcirc \\ \hline \bigcirc & \mathbf{A} \end{array}\right] \tag{11.3.64}$$

and that

$$\sqrt{\alpha}\mathbf{N}\psi = \left[\begin{array}{c} 0 \\ \hline \mathbf{B} \end{array}\right]. \tag{11.3.65}$$

At this point we can bring in the normal equations (11.3.40), written in the form

$$\mathbf{aR} = \alpha\psi^T. \tag{11.3.66}$$

Then

$$\mathbf{N}(\mathbf{I} - \psi\mathbf{a})\mathbf{R}(\mathbf{I} - \psi\mathbf{a})^T\mathbf{N}^T = \mathbf{NRN}^T - \alpha\mathbf{N}\psi\psi^T\mathbf{N}^T \tag{11.3.67}$$

This equation contains (11.3.58) in the lower right $n \times n$ corner. To see this, we use the fact that since $\mathbf{R} = \mathbf{R}(n)$ is Toeplitz,

$$\mathbf{R}(n) = \left[\begin{array}{c|c} \times & \times\!\!\!\times \\ \hline \mathsf{X} & \mathbf{R}(n-1) \end{array}\right] = \left[\begin{array}{c|c} \mathbf{R}(n-1) & \mathsf{X} \\ \hline \times\!\!\!\times & \times \end{array}\right]. \tag{11.3.68}$$

Now use Eqs. (11.3.68), (11.3.65) and (11.3.64) to rewrite Eq. (11.3.67).

$$\left[\begin{array}{c|c} 0 & \bigcirc \\ \hline \bigcirc & \mathbf{A} \end{array}\right]\left[\begin{array}{c|c} \times & \times\!\!\!\times \\ \hline \mathsf{X} & \mathbf{R}(n-1) \end{array}\right]\left[\begin{array}{c|c} 0 & \bigcirc \\ \hline \bigcirc & \mathbf{A} \end{array}\right]$$

$$= \left[\begin{array}{c|c} 0 & \bigcirc \\ \hline \bigcirc & \mathbf{R}(n-1) \end{array}\right] - \left[\begin{array}{c} 0 \\ \hline \mathbf{B} \end{array}\right][0 \mid \mathbf{B}^T]. \tag{11.3.69}$$

Thus

$$\mathbf{AR}(n-1)\mathbf{A}^T = \mathbf{R}(n-1) - \mathbf{BB}^T. \tag{11.3.70}$$

This is Eq. (11.3.58), with $\mathbf{K} = \mathbf{R}(n-1)$.

THEOREM 11.3.4. If the stability conditions (11.3.48) hold, then the solution $\mathbf{r}[0, n]$ to Eq. (11.3.41) will produce a positive definite $\mathbf{R}$.

Proof. The stable synthesis filter $1/\hat{a}(z)$ driven by white noise of variance α will produce an output $\mathbf{y}$ with spectrum (11.3.4). Then Eq. (11.3.11) provides a representation for $\mathbf{R}$. Let

$$\mathbf{c} = [c_0 \quad c_1 \quad \cdots \quad c_n] \neq \mathbf{0} \tag{11.3.71}$$

$$\hat{c}(z) = \mathbf{c}\psi_n(z) = \sum_{k=0}^{n} c_k z^{-k}. \tag{11.3.72}$$

Combine Eqs. (11.3.4), (11.3.11) and (11.3.72) to get

$$\mathbf{cRc}^T = \frac{1}{2\pi}\int_{-\pi}^{\pi} \alpha \left|\frac{\hat{c}(e^{j\theta})}{\hat{a}(e^{j\theta})}\right|^2 d\theta. \tag{11.3.73}$$

The integrand must be positive except for at most n values of θ [the possible zeros of $\hat{c}(z)$]. Therefore, $\mathbf{cRc}^T > 0$.

Lattices and the Levinson Algorithm

As a consequence of the preceding theorems, we know that given any positive definite Toeplitz matrix $\mathbf{R}$, there is a unique stable AR model that will reproduce $\mathbf{r}[0, n]$. To find the parameters of the model, one must solve the asymptotic normal equations (11.3.40). Now there is an elegant fast algorithm for solving these equations due to N. Levinson. This appeared (albeit not in its present form) in an expository paper on the Wiener Prediction problem (Levinson 1947). Levinson called this algorithm a "mathematically trivial procedure." Perhaps. But the algorithm exposes a lattice structure that has been found to be extremely useful in a wide variety of signal processing applications, including spectral estimation, voice coding, and filter structures. The algorithm was rediscovered and improved by Durbin (1960).

Levinson's algorithm iterates on the order m of the problem. In other words, it computes the solution to the problem of size $(m + 1)$ from the solution to the problem of size m. Let $\mathbf{R}(m)$ be the $(m + 1) \times (m + 1)$ Toeplitz matrix whose first row is $\mathbf{r}[0, m]$ [see Eq. (11.3.49)], and let

$$\mathbf{a}(m) = [1 \quad a_{1m} \quad a_{2m} \quad \cdots \quad a_{mm}]. \tag{11.3.74}$$

Then the normal equation of size m is

$$\mathbf{a}(m)\mathbf{R}(m) = [\alpha_m \quad \mathbf{0} \quad]. \tag{11.3.75}$$

Two properties of symmetric Toeplitz matrices are exploited by the algorithm. First,

$$\text{Property 1:} \quad \mathbf{J}(m)\mathbf{R}(m) = \mathbf{R}(m)\mathbf{J}(m), \tag{11.3.76}$$

where $\mathbf{J}(m)$ is the $(m + 1) \times (m + 1)$ matrix which turns vectors upside down [see Eq. (11.3.26)]. Second, the matrices $\mathbf{R}(m)$ are nested inside the largest, which is $\mathbf{R}(n)$, via

$$\text{Property 2:}\quad \mathbf{R}(m+1) = \left[\begin{array}{c|c} \mathbf{R}(m) & \begin{matrix} r(m+1) \\ \vdots \\ r(1) \end{matrix} \\ \hline r(m+1)\cdots r(1) & r(0) \end{array}\right]. \tag{11.3.77}$$

An obvious guess at the solution of the problem of size $m + 1$ is

$$[\mathbf{a}(m), 0]$$

which is obtained by simply appending a zero. However, this misses the target slightly:

$$\begin{aligned} [\mathbf{a}(m), 0]\mathbf{R}(m+1) &= [\mathbf{a}(m)\mathbf{R}(m), \gamma_m] \quad \text{(Property 2)} \\ &= [\alpha_m \bigcirc \gamma_m] \quad \text{from Eq. (11.3.75)} \end{aligned} \tag{11.3.78}$$

where

$$\gamma_m = \sum_{k=0}^{m} a_{km} r(m+1-k). \tag{11.3.79}$$

Now define the reversal of $\mathbf{a}(m)$ to be

$$\tilde{\mathbf{a}}(m) \triangleq \mathbf{a}(m)\,\mathbf{J}(m) = [a_{mm} \quad \cdots \quad a_{1m} \quad 1]. \tag{11.3.80}$$

Multiply Eq. (11.3.78) on the right by $\mathbf{J}(m + 1)$ and use Property 1 to get

$$[\mathbf{a}(m), \quad 0]\mathbf{R}(m+1)\,\mathbf{J}(m+1) = [\alpha_m \bigcirc \gamma_m]\,\mathbf{J}(m+1)$$

or

$$[\mathbf{a}(m), \quad 0]\,\mathbf{J}(m+1)\,\mathbf{R}(m+1) = [\gamma_m \bigcirc \alpha_m]$$

or

$$[0, \quad \tilde{\mathbf{a}}(m)]\,\mathbf{R}(m+1) = [\gamma_m \bigcirc \alpha_m] \tag{11.3.81}$$

We need only take the appropriate linear combination of Eqs. (11.3.78) and (11.3.81) to get a solution to the problem of size $(m + 1)$. Define the "reflection coefficient"

$$c_{m+1} \triangleq \frac{-\gamma_m}{\alpha_m}, \tag{11.3.82}$$

which is chosen so that

$$\mathbf{a}(m+1) \triangleq [\mathbf{a}(m), \quad 0] + c_{m+1}[0, \quad \tilde{\mathbf{a}}(m)] \tag{11.3.83}$$

satisfies

$$\begin{aligned} \mathbf{a}(m+1)\mathbf{R}(m+1) &= [\alpha_m \bigcirc \gamma_m] - \frac{\gamma_m}{\alpha_m}[\gamma_m \bigcirc \alpha_m] \\ &= [\alpha_{m+1} \bigcirc], \end{aligned} \tag{11.3.84}$$

where

$$\alpha_{m+1} = \alpha_m - \frac{\gamma_m^2}{\alpha_m} = \alpha_m[1 - c_{m+1}^2]. \tag{11.3.85}$$

We summarize:

LEVINSON-DURBIN ALGORITHM

Given: n, $\mathbf{r}[0, n]$

To compute: $\{\alpha_m, a_{k,m}, c_m: 1 \leq m \leq n, 1 \leq k \leq m\}$ **satisfying** (11.3.75)

Initialization: $\alpha_0 = r(0)$

$a_{00} = 1$

Body: For $m = 0$ to $n - 1$, do

$$\gamma_m = \sum_{k=0}^{m} r(m+1-k)a_{km}$$

$$c_{m+1} = -\gamma_m/\alpha_m$$

$$\alpha_{m+1} = \alpha_m(1 - c_{m+1}^2)$$

$$a_{m+1,m} = 0$$

For $k = 0$ to $m + 1$, do

$$a_{k,m+1} = a_{km} + c_{m+1}a_{m+1-k,m}$$

(end loop on k)

(end loop on m)

Levinson's algorithm is a fast algorithm for the Cholesky (or triangular) decomposition of $\mathbf{R}(n)$. Define

$$\mathbf{T}(n) = \begin{bmatrix} 1 & & \times \\ & \ddots & \\ 0 & & 1 \end{bmatrix}, \qquad T_{ij}(n) = a_{j-i,n-i} \quad \text{for } 0 \leq i, j \leq n. \tag{11.3.86}$$

Using the fact that $\mathbf{R}(m)$ forms the upper left $(m + 1) \times (m + 1)$ corner of $\mathbf{R}(n)$, the set of Eqs. (11.3.75) for $m \leq n$ gives

$$\mathbf{T}(n)\mathbf{R}(n) = \begin{bmatrix} \alpha_n & & 0 \\ & \ddots & \\ \times & & \alpha_0 \end{bmatrix}, \tag{11.3.87}$$

where the elements below the diagonal are of no interest to us. Now consider

$$\mathbf{T}(n)\mathbf{R}(n)\mathbf{T}(n)^T = \mathbf{T}(n)(\mathbf{T}(n)\mathbf{R}(n))^T$$

$$= \begin{bmatrix} 1 & & \times \\ & \ddots & \\ 0 & & 1 \end{bmatrix} \begin{bmatrix} \alpha_n & & \times \\ & \ddots & \\ 0 & & \alpha_0 \end{bmatrix} = \begin{bmatrix} \alpha_n & & \times \\ & \ddots & \\ 0 & & \alpha_0 \end{bmatrix}. \tag{11.3.88}$$

The left-hand side of this equation is obviously symmetric, however, and therefore the right-hand side must be symmetric. Since the elements below the diagonal are zero, so must the elements above the diagonal. Thus with

$$\mathbf{D}(n) = \operatorname{diag}\{\alpha_n, \alpha_{n-1}, \ldots, \alpha_0\}, \tag{11.3.89}$$

$$\begin{cases} \mathbf{D}(n) = \mathbf{T}(n)\mathbf{R}(n)\mathbf{T}(n)^T \\ \mathbf{R}(n) = \mathbf{T}(n)^{-1}\mathbf{D}(n)[\mathbf{T}(n)^{-1}]^T, \\ \mathbf{R}(n)^{-1} = \mathbf{T}(n)^T\mathbf{D}(n)^{-1}\mathbf{T}(n). \end{cases} \tag{11.3.90}$$

The matrix $\mathbf{T}(n)$ has determinant one. Therefore, for each m

$$\begin{aligned} \det(\mathbf{R}(m)) &= \det[\mathbf{T}(m)^{-1}\mathbf{D}(m)\mathbf{T}(m)^{-T}] \\ &= \det \mathbf{D}(m) \\ &= \prod_{k=0}^{m} \alpha_k. \end{aligned} \tag{11.3.91}$$

Writing this another way, we obtain the classical result known as Gram's theorem (Cheney 1966):

$$\alpha_n = \min\{\mathbf{a}\mathbf{R}(n)\mathbf{a}^T: \quad a_0 = 1\} = \frac{\det \mathbf{R}(n)}{\det \mathbf{R}(n-1)}. \tag{11.3.92}$$

Now the positivity of $\mathbf{R}(n)$ guarantees that each α_k must be positive. (The converse is also true; this is in fact Sylvester's condition.) This forces two important properties, both of which follow from Eq. (11.3.85), assuming that all $\mathbf{R}(n)$ are strictly positive:

$$0 < \alpha_{m+1} \leqslant \alpha_m \quad \text{for all } m \geqslant 0, \tag{11.3.93}$$

$$c_m^2 < 1 \quad \text{for all } m > 0. \tag{11.3.94}$$

The reflection coefficients must have magnitude less than one. The fact that the sequence α_m is nonincreasing is not surprising. The error variance for a whitening filter of order $(m + 1)$ cannot be greater than the error variance for a filter of order m.

The Lattice Form of the Levinson Algorithm

Levinson's algorithm produces the parameters in the triangular decomposition of $\mathbf{R}(n)$ of Eq. (11.3.90) using about n^2 multiplies. Normally, such a decomposition would require the order of n^3 multiplies. The reduction is possible because $\mathbf{R}$ is Toeplitz, and the sequence of solutions $(\alpha_m, \mathbf{a}(m))$ is updated using the underlying structure of Toeplitz matrices. This update mechanism is made clearer when we consider the whitening filters

$$\hat{a}(m, z) = \mathbf{a}(m)\psi_m(z) = \sum_{k=0}^{m} a_{km} z^{-k} \tag{11.3.95}$$

$$\begin{aligned}\tilde{a}(m, z) = \tilde{\mathbf{a}}(m)\psi_m(z) &= \sum_{k=0}^{m} a_{m-k,m} z^{-k} \\ &= z^{-m}\hat{a}(m, z^{-1}).\end{aligned} \tag{11.3.96}$$

The Levinson recursion (11.3.83) becomes

$$\hat{a}(m+1, z) = \hat{a}(m, z) + z^{-1} c_{m+1} \tilde{a}(m, z), \tag{11.3.97}$$

and this forces a recursion for the backward prediction filters:

$$\begin{aligned}\tilde{a}(m+1, z) &= z^{-(m+1)}\hat{a}(m+1, z^{-1}) \\ &= z^{-(m+1)}[\hat{a}(m, z^{-1}) + z c_{m+1} z^m \hat{a}(m, z)] \\ &= z^{-1}\tilde{a}(m, z) + c_{m+1}\hat{a}(m, z).\end{aligned} \tag{11.3.98}$$

In matrix form, these two updates become

$$\begin{bmatrix} \hat{a}(m+1, z) \\ \tilde{a}(m+1, z) \end{bmatrix} = \begin{bmatrix} 1 & z^{-1} c_{m+1} \\ c_{m+1} & z^{-1} \end{bmatrix} \begin{bmatrix} \hat{a}(m, z) \\ \tilde{a}(m, z) \end{bmatrix}. \tag{11.3.99}$$

This has a signal flow graph description shown in Fig. 11.3.4(a). When these "lattice sections" are joined together, the result is the forward lattice of Fig. 11.3.4(b). The node signals in this graph are the forward and backward prediction errors of all orders less than or equal to n:

$$e_m(k) = \sum_{l=0}^{m} a_{lm} y(k-l), \qquad a_{0m} = 1 \tag{11.3.100}$$

$$\tilde{e}_m(k) = \sum_{l=0}^{m} a_{m-l,m} y(k-l), \qquad a_{0m} = 1. \tag{11.3.101}$$

The transfer function from $\mathbf{y}$ to $\mathbf{e}_m$ in this graph is $\hat{a}(m, z)$ and the transfer function from $\mathbf{y}$ to $\tilde{\mathbf{e}}_m$ is $\tilde{a}(m, z)$. In Fig. 11.3.4(c), the graph has been completely reduced, leaving only the input $\mathbf{y}$ and the two outputs $\mathbf{e}_n$ and $\tilde{\mathbf{e}}_n$.

The lattice structure comes about because the matrix $\mathbf{R}$ is Toeplitz, which in turn is a consequence of the stationarity of the signal $\mathbf{y}$. The prediction errors $\mathbf{e}_m$ and $\tilde{\mathbf{e}}_m$ are sufficient statistics for computing the errors $\mathbf{e}_{m+1}$ and $\tilde{\mathbf{e}}_{m+1}$. One would have expected a more complicated dependence. In fact, we

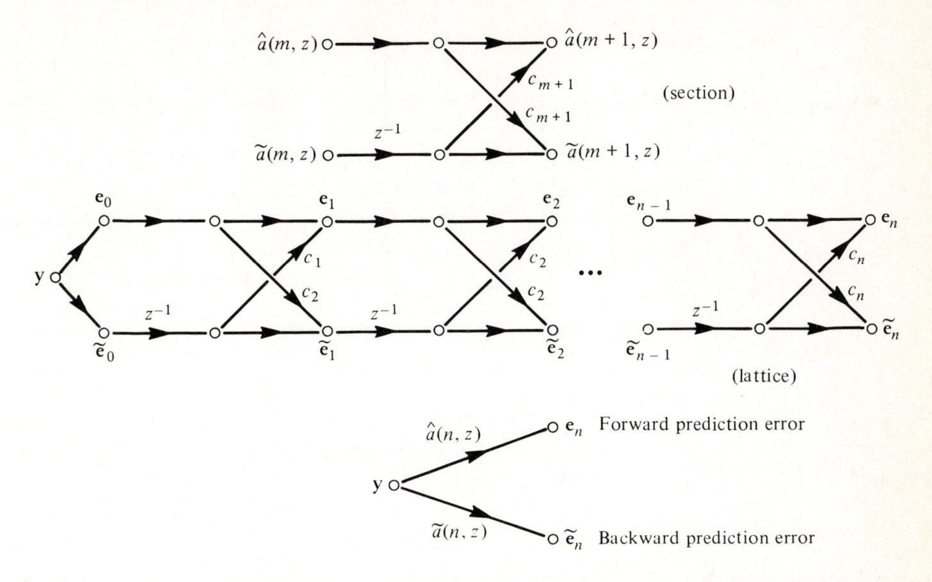

FIGURE 11.3.4 The forward or "analysis" lattice derived from the Levinson algorithm.

have a situation where a kind of local minimization is equivalent to global minimization. One can choose the reflection coefficient c_{m+1} to minimize the variance of $e_m(k) + c_{m+1}\tilde{e}_m(k)$. This is local minimization. But if we put all the local minimization problems together, as in the lattice, then we have achieved a solution to the larger problem of minimizing the variance of $e_n(k)$ based on $\mathbf{y}[k-n, k]$. This fact is exploited by the Burg algorithm, which we shall discuss shortly.

One can get a lattice structure for the synthesis filter $1/\hat{a}(n, z)$ by operating on the lattice for the analysis filter $\hat{a}(n, z)$. The result is shown in Fig. 11.3.5. All the signals in the synthesis lattice have exactly the same algebraic relationships as do their counterparts in the analysis lattice. Only the interpretation has been changed. In the analysis lattice $\mathbf{y}$ is an input, whereas it is an output in the synthesis lattice. Problem 11.26 develops the synthesis lattice.

Inverting the Levinson Algorithm

There are three sets of $n + 1$ parameters that are equivalent in the sense that any one of them can be used to compute the other two. These are:

$$\begin{cases} \mathbf{r}[0, n] & \text{with} \quad \mathbf{R}(n) > 0 \\ \alpha_n, \mathbf{a}[1, n] & \text{with} \quad \alpha_n > 0,\ \hat{a}(n, \lambda) = 0 \Rightarrow |\lambda| < 1 \\ r(0), \mathbf{c}[1, n] & \text{with} \quad r(0) > 0,\ c_k^2 < 1 \text{ for } 1 \leqslant k \leqslant n. \end{cases} \tag{11.3.102}$$

Each set has its own provisions. With the autocorrelation parameters, it is the positivity of the matrix $\mathbf{R}(n)$ and so on. The provisions for one set will force the accompanying provisions for the other two. Schematically, the Levinson

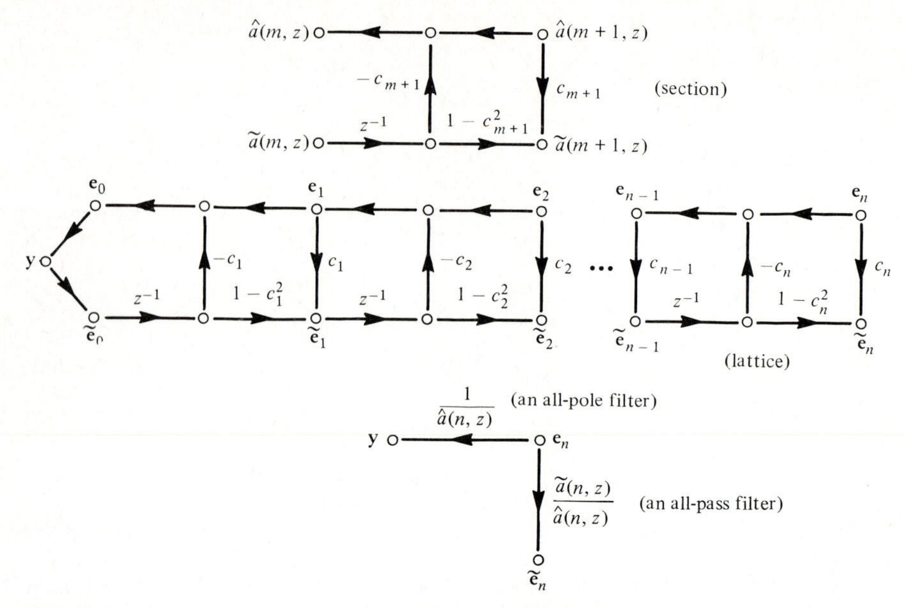

FIGURE 11.3.5 The backward or "synthesis" lattice that results from inverting the forward lattice.

algorithm performs the following task:

$$
\mathbf{r}[0, n] \begin{cases} \nearrow \alpha_n, \mathbf{a}[1, n] \\ \searrow r(0), \mathbf{c}[1, n] \end{cases} \tag{11.3.103}
$$

To start with another set of parameters, we must "invert" the algorithm.

If we multiply both sides of the matrix Levinson recursion (11.3.99) by the inverse of the matrix involved, we will get

$$
\begin{bmatrix} \hat{a}(m, z) \\ \tilde{a}(m, z) \end{bmatrix} = \frac{1}{1 - c_{m+1}^2} \begin{bmatrix} 1 & -c_{m+1} \\ -c_{m+1}z & z \end{bmatrix} \begin{bmatrix} \hat{a}(m+1, z) \\ \tilde{a}(m+1, z) \end{bmatrix}. \tag{11.3.104}
$$

The reflection coefficient cannot be arbitrary in this equation. Either one of the following requirements

$$
\begin{aligned} a_{0m} &= 1 \\ a_{m+1,m} &= 0 \end{aligned} \tag{11.3.105}
$$

will force

$$
c_{m+1} = a_{m+1,m+1}. \tag{11.3.106}
$$

We may combine Eq. (11.3.106), the top half of Eq. (11.3.104), and Eq. (11.3.85) to produce the following algorithm.

JURY (REFLECTION COEFFICIENT) ALGORITHM

Given: α_n, $\mathbf{a}[1, n]$, n

To compute: $r(0)$, $\mathbf{c}[1, n]$, $\{a_{k,m}, \alpha_m : 0 \leqslant m \leqslant n, 0 \leqslant k \leqslant m\}$ **satisfying** (11.3.104)

Initialization: $a_0 = 1$

$a_{kn} = a_k \quad \text{for } 0 \leqslant k \leqslant n$

Body: For $m = n$ Down to 1, Do

$c_m = a_{mm}$

$\alpha_{m-1} = \alpha_m/(1 - c_m^2)$

For $k = 0$ to m, Do

$$a_{k,m-1} = \frac{a_{km} - c_m a_{m-k,m}}{1 - c_m^2}$$

(end loop on k)

(end loop on m)

$r(0) = \alpha_0$

We have titled this algorithm the *Jury algorithm* because it is essentially the same as the Jury tabular stability test. The polynomial $\hat{a}(z)$, whose coefficients are inputs in this algorithm will satisfy the stability condition (11.3.48) if and only if $c_m^2 < 1$ for $1 \leqslant m \leqslant n$. The Jury algorithm will decide this without having to factor the polynomial. The algorithm actually computes all the polynomials $\hat{a}(m, z)$ and the α_m as well as the reflection coefficients. These may be used in turn to compute the autocorrelation sequence.

COMPUTATION OF THE AUTOCORRELATION SEQUENCE

Given: $n, \{\alpha_m, a_{km}: \quad 0 \leqslant k \leqslant m, 0 \leqslant m \leqslant n\}$

To compute: $r[0, n]$

Initialization: $r(0) = \alpha_0$

$r(1) = -\alpha_0 a_{1,1}$

Body: For $m = 1$ to $n - 1$, Do

$$r(m + 1) = -\alpha_m a_{m+1,m+1} - \sum_{k=1}^{m} a_{km} r(m + 1 - k)$$

(end loop on m)

This algorithm is based on Eqs. (11.3.79) and (11.3.82). Taken together, the last two algorithms solve the synthesis equation (11.3.41). Furthermore, it is a fast algorithm for solving the matrix covariance equation (11.3.70). This in turn can be modified with a suitable coordinate transformation to solve the general equation

$$\mathbf{K} = \mathbf{AKA}^T + \mathbf{BB}^T \tag{11.3.107}$$

for the matrix $\mathbf{K}$. [The transformation involved is found in Problem 8.20. See Problem 11.29.] This computation is fundamental to scaling, roundoff noise, and other problems. Seen yet another way, the algorithms provide a fast means of computing the formidable integral

$$r(k) = \frac{1}{2\pi} \int_{-\pi}^{\pi} \frac{\alpha_n e^{-jk\theta}}{|\sum_{l=0}^{n} a_l e^{-jl\theta}|^2} \, d\theta. \tag{11.3.108}$$

The Burg Spectral Estimator

The forward and backward error signals in the analysis' lattice of Fig. 11.3.4 satisfy the equations

$$\begin{cases} e_m(k) = e_{m-1}(k) + c_m \tilde{e}_{m-1}(k-1) \\ \tilde{e}_m(k) = c_m e_{m-1}(k) + \tilde{e}_{m-1}(k-1) \end{cases} \tag{11.3.109}$$

which follow from the Levinson recursion (11.3.99). With $\mathbf{y}$ as input, the reflection coefficients are chosen to minimize the variances of all these error signals. (This is global minimization.) Therefore the reflection coefficient c_m will minimize the variance of either $\mathbf{e}_m$ or $\tilde{\mathbf{e}}_m$, using Eq. (11.3.109). (This is local minimization.) The Burg estimator (1967) assumes the lattice structure inherent in Eq. (11.3.109) and chooses the reflection coefficients to minimize error energy when only a finite data record $\mathbf{y}[0, L-1]$ is available. To avoid a kind of windowing that would involve assuming values for $\mathbf{y}$ outside the data interval, the error signals are only used in the range

$$m \leqslant k \leqslant L - 1. \tag{11.3.110}$$

The Burg rule is to choose c_m so that

$$\begin{aligned} V_m &= \sum_{k=m}^{L-1} [e_m(k)^2 + \tilde{e}_m(k)^2] \\ &= \min_{c_m} \sum_{k=m}^{L-1} [(e_{m-1}(k) + c_m \tilde{e}_{m-1}(k-1))^2 + (c_m e_{m-1}(k) + \tilde{e}_{m-1}(k-1))^2]. \end{aligned} \tag{11.3.111}$$

In other words, the sum of the energies of the forward and backward errors (over the interval $[m, L-1]$) is minimized. There are other rules that might be used; see Makhoul (1975). The function to be minimized here is quadratic and has the form

$$V_m = \min_{c_m} [(1 + c_m{}^2) d_m + 4 c_m f_m], \tag{11.3.112}$$

where

$$d_m = \sum_{k=m}^{L-1} [e_{m-1}(k)^2 + \tilde{e}_{m-1}(k-1)^2], \tag{11.3.113}$$

$$f_m = \sum_{k=m}^{L-1} e_{m-1}(k)\, \tilde{e}_{m-1}(k-1). \tag{11.3.114}$$

The minimum is attained at

$$c_m = -2\frac{f_m}{d_m} \tag{11.3.115}$$

and has the value

$$V_m = d_m(1 - c_m^2) \tag{11.3.116}$$

[which is reminiscent of Eq. (11.3.85)]. A comparison of the definitions of V_m and d_m in Eqs. (11.3.111) and (11.3.113) shows that

$$\begin{aligned} d_{m+1} &= V_m - e_m(m)^2 - \tilde{e}_m(L-1)^2 \\ &= d_m(1 - c_m^2) - e_m(m)^2 - \tilde{e}_m(L-1)^2. \end{aligned} \tag{11.3.117}$$

This observation saves the expense of computing the sums of squares in Eq. (11.3.113). The Burg algorithm combines Eqs. (11.3.109), (11.3.114), (11.3.115), and (11.3.117) with appropriate initialization.

BURG ALGORITHM

Given: L, $\mathbf{y}[0, L-1]$

To compute: α, $\mathbf{c}[1, n]$ parameterizing the Burg power spectrum estimate

Initialization: $d_0 = 2 \sum_{k=0}^{L-1} y(k)^2$

$c_0 = 0$

For $k = 0$ to $L-1$, Do

$e_0(k) = y(k)$

$\tilde{e}_0(k) = y(k)$

(end loop on k)

Body: For $m = 1$ to n, Do

$d_m = (1 - c_{m-1}^2)d_{m-1} - e_{m-1}^2(m-1) - \tilde{e}_{m-1}^2(L-1)$

$f_m = \sum_{k=m}^{L-1} e_{m-1}(k)\, \tilde{e}_{m-1}(k-1).$

$c_m = -2f_m/d_m$

For $k = m$ to $L - 1$, Do

$$e_m(k) = e_{m-1}(k) + c_m \tilde{e}_{m-1}(k-1)$$

$$\tilde{e}_m(k) = c_m e_{m-1}(k) + \tilde{e}_{m-1}(k-1)$$

(end loop on k)

(end loop on m)

$$\alpha = d_n/[2(L-n)]$$

The Burg algorithm provides reflection coefficients. The polynomial coefficients may be obtained from these by using the Levinson recursion (11.3.99) or simply editing the Levinson algorithm to remove the lines that compute the reflection coefficients. The Burg algorithm does not solve any normal equations of the form (11.3.27) or (11.3.28). However as L is increased, the parameters provided by either the Burg algorithm or the solution to the normal equation will approach the parameters for the asymptotic normal equations (which use the true autocorrelation sequence).

EXAMPLE 11.3.4

AR estimates using the Burg algorithm.

Using the data set from Examples 11.2.1, 11.3.2, and 11.3.3, the Burg algorithm produces the results shown in Fig. 11.3.6. The results are remarkably similar to the forward/backward covariance estimates.

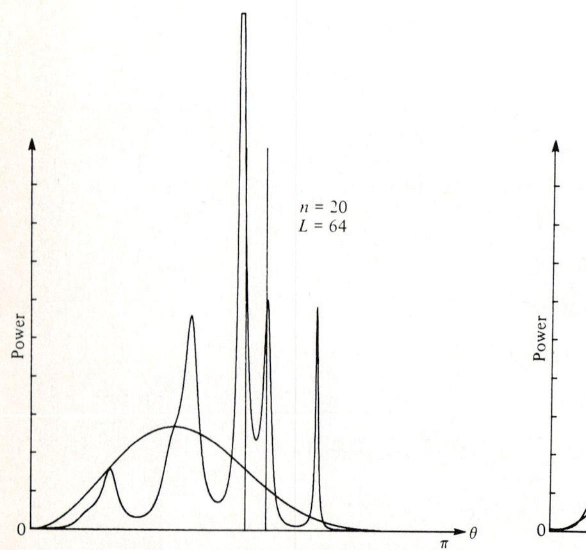

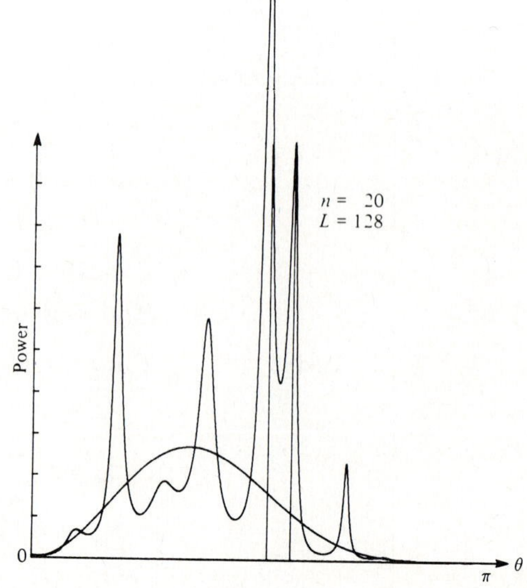

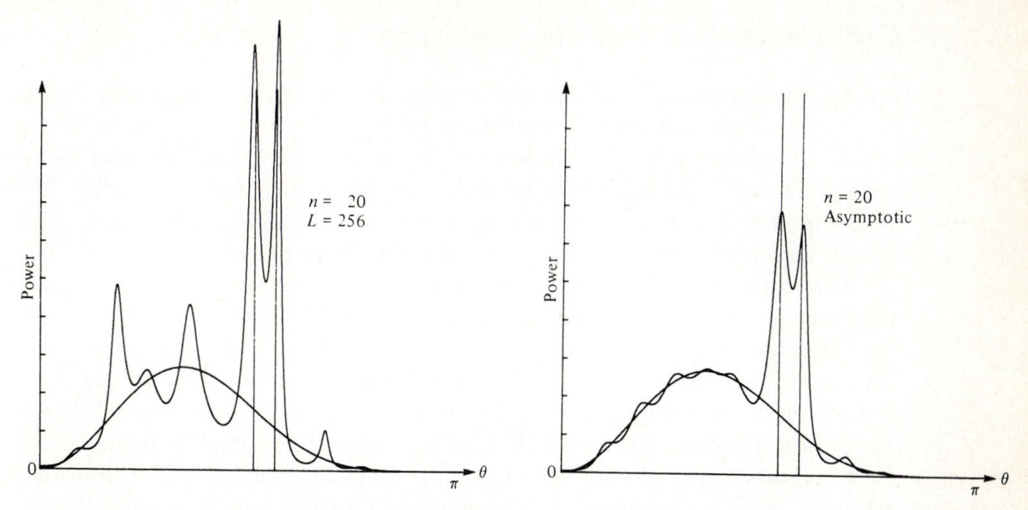

FIGURE 11.3.6 Three Burg AR estimates of order $n = 20$ based on data $y[0, L - 1]$. The asymptotic AR model and the true spectrum are also shown.

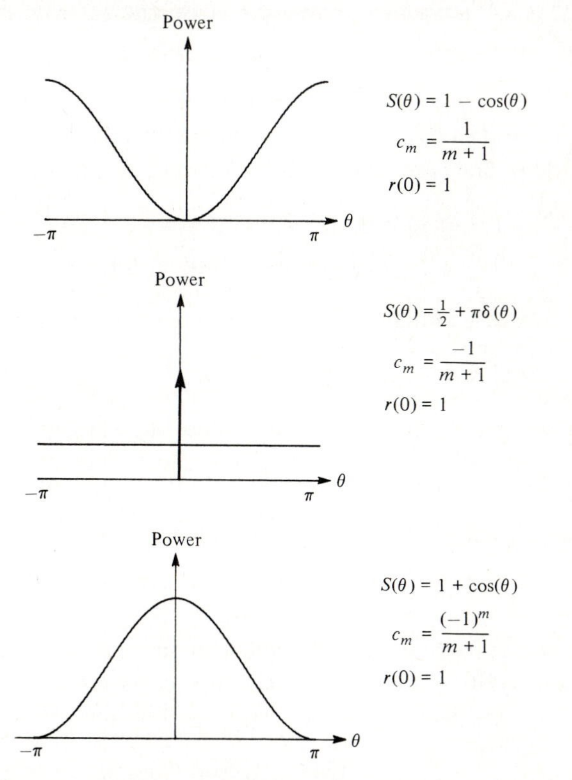

FIGURE 11.3.7 Three power spectra and their associated reflection coefficient sequences.

Coding Spectra with Reflection Coefficients

An AR model of order n is fully parameterized by the $(n + 1)$ parameters $r(0)$, $\mathbf{c}[1, n]$. Since the reflection coefficients satisfy $c_k^2 < 1$, they are automatically scaled. One application of this has been speech coding (Markel and Gray 1980). For our purposes, one can produce a signal having the required spectrum by using the synthesis lattice of Fig. 11.3.5 directly. (This is essentially the speech reconstruction, except that the input is not really white.)

The coding is quite inscrutable, however. One simply cannot look at the sequence of reflection coefficients and guess what $S(\theta)$ looks like. To support this statement, consider the three spectra in Fig. 11.3.7. Each of these spectra have reflection coefficients that differ only in sign. To get the second from the first, all signs are changed. To get the third, every other sign is changed. [The computation of the reflection coefficients in this example draws on Table 7.2.1.] These examples also show that one cannot infer the presence of lines in the spectrum by considering only the magnitudes of the reflection coefficients, since the second spectrum has a line, while the others do not.

EXAMPLE 11.3.5

Reflection coefficients for a sinusoid in white noise.

An important category of spectra in applications is associated with signals consisting of sinusoids in a white noise background. Consider the simplest case of one sinusoid. Then the power spectrum is

$$\begin{aligned} r(k) &= \sigma^2\delta(k) + \cos(k\theta_0) \xleftrightarrow{DTFT} \\ S(\theta) &= \sigma^2 + \pi[\delta(\theta - \theta_0) + \delta(\theta + \theta_0)]. \end{aligned} \tag{11.3.118}$$

In the case of high SNR ($\sigma^2 \ll 1$), the whitening filter should be approximately

$$\hat{a}(z) \approx 1 - 2\rho\cos(\theta_0)z^{-1} + \rho^2 z^{-2}, \tag{11.3.119}$$

which would completely annihilate the sinusoid if $\rho = 1$. (The zeros of this filter are at $\rho e^{i\theta_0}$ and $\rho e^{-j\theta_0}$.) The reflection coefficients for this polynomial are

$$\begin{aligned} c_1 &= \frac{-2\rho\cos(\theta_0)}{1 + \rho^2} \approx -\cos(\theta_0) \\ c_2 &= \rho^2 \approx 1 \end{aligned} \tag{11.3.120}$$

But it is more interesting when the noise power is not negligible. Because the matrix $\mathbf{R}(n) - \sigma^2\mathbf{I}$ has rank two, the reflection coefficients for the spectrum (11.3.118) may be derived (with considerable patience). The result is

$$c_n = -2\sin(\theta_0)\frac{(2\sigma^2 + n)\cos(n\theta_0)\sin(\theta_0) - \sin(n\theta_0)\cos(\theta_0)}{(2\sigma^2 + n)^2\sin^2(\theta_0) - \sin^2(n\theta_0)}. \tag{11.3.121}$$

In particular,

$$c_1 = \frac{-1}{1+\sigma^2}\cos(\theta_0)$$

$$c_2 = 1 - \sigma^2\left[\frac{1 + 2\cos^2(\theta_0) + \sigma^2}{\sin^2(\theta_0) + \sigma^4 + 2\sigma^2}\right], \tag{11.3.122}$$

which should be compared with Eq. (11.3.120). It is interesting that for large n, the reflection coefficients are not negligible; indeed,

$$c_n \approx \frac{-2\cos(n\theta_0)}{n} \tag{11.3.123}$$

even if $\sigma^2 \ll 1$.

The Singular Case and Line Spectra

We have assumed before that the matrices $\mathbf{R}(n)$ were all positive definite. What happens in the borderline case that $\mathbf{R}(n)$ is positive semidefinite but singular? The answer is that $S(\theta)$ is a line spectrum. This fact is the basis for estimators that assume a line spectral component (such as the Pisarenko model). The precise connection is stated in the following theorem.

THEOREM 11.3.5.

$$\begin{cases} \mathbf{R}(n-1) > \mathbf{0} \\ \det \mathbf{R}(n) = 0 \end{cases} \Leftrightarrow \begin{cases} S(\theta) = \sum_{k=1}^{n} \frac{\pi A_k^2}{2}\,\delta(\theta - \theta_k), \\ A_k \neq 0, \qquad 1 \leqslant k \leqslant n. \end{cases} \tag{11.3.124}$$

Before the theorem is proved, consider some equivalent sets of conditions. If $S(\theta)$ is a line spectrum consisting of n lines (which is twice the number of sinusoids, plus possible single lines at $\theta = 0$ or $\theta = \pi$), then

$$\det \mathbf{R}(k) \begin{cases} >0 & \text{for } k < n \\ =0 & \text{for } k \geqslant n \end{cases} \tag{11.3.125}$$

$$\alpha_k \begin{cases} >0 & \text{for } k < n \\ =0 & \text{for } k \geqslant n \end{cases} \tag{11.3.126}$$

$$c_k^2 \begin{cases} <1, & \text{for } k < n \\ =1, & \text{for } k = n \\ \text{arbitrary}, & \text{for } k > n. \end{cases} \tag{11.3.127}$$

This phenomenon was observed in Example 11.3.5, with $n = 2$, as σ^2 approached zero.

Proof. We shall need only the representation (11.3.11) for $\mathbf{R}(n)$ in terms of $S(\theta)$ and its consequences. First, if a nontrivial solution exists to the homogeneous equation

$$\mathbf{aR}(n) = [\bigcirc], \qquad \mathbf{a} = [a_0, \ldots, a_n] \neq [\bigcirc]. \tag{11.3.128}$$

Then [using Eq. (11.3.14)],

$$0 = \mathbf{aR}(n)\mathbf{a}^T = \frac{1}{2\pi}\int_{-\pi}^{\pi} S(\theta)\,|\hat{a}(e^{j\theta})|^2\,d\theta \tag{11.3.129}$$

implies that $S(\theta)$ can be nonzero only on zeros of $\hat{a}(z)$ which are on the unit circle. Thus $S(\theta)$ is a line spectrum with no more than n lines.

Second, if $S(\theta)$ is a line spectrum with exactly n lines, then there exists a nontrivial solution to the homogeneous equation (11.3.128), namely,

$$\hat{a}(z) = \mathbf{a}\psi_n(z) = \prod_{k=1}^{n} (1 - z^{-1}e^{j\theta_k}) \tag{11.3.130}$$

[as can be seen using Eq. (11.3.14) again].

Now suppose that the left set of conditions (11.3.124) are true. Then $S(\theta)$ must be a line spectrum with at most n lines. But it cannot have fewer than n or a nontrivial solution to

$$\mathbf{bR}(n-1) = [\bigcirc], \qquad \mathbf{b} \neq [\bigcirc] \tag{11.3.131}$$

would exist, which would contradict the condition that $\mathbf{R}(n-1) > \mathbf{0}$.

On the other hand, if $S(\theta)$ is of the form on the right-hand side of Eq. (11.3.124), then $\mathbf{a}$ constructed via Eq. (11.3.130) will satisfy the homogeneous equation (11.3.128), which implies that det $\mathbf{R}(n) = 0$. But if $\mathbf{R}(n-1)$ were not positive definite (but positive semidefinite), it would have a zero eigenvalue, thereby guaranteeing a solution to Eq. (11.3.131), which would in turn imply that $S(\theta)$ is a line spectrum with at most $n-1$ lines. That would be a contradiction. Hence $\mathbf{R}(n-1) > \mathbf{0}$.

Consider the computational problem inherent in Eq. (11.3.124). If $\mathbf{r}[0, n]$ satisfies these conditions, how does one obtain the parameters of the line spectrum? A casual inspection indicates the presence of $2n$ parameters: n angles and n amplitudes. Since $S(\theta)$ must be real and even, however, each angle θ_k not at 0 or π must have a twin $\theta_m = -\theta_k$ with equal amplitude $A_m = A_k$. These considerations reduce the effective number of parameters to n. The procedure for obtaining these parameters is as follows.

STEP 1. Find a solution to the homogeneous equation (11.3.128). Levinson's algorithm may be used, even though $\alpha_n = 0$ should result. Then factor $\hat{a}(z)$ as in Eq. (11.3.130) to obtain the angles.

STEP 2. The inverse DTFT of the line spectrum in Eq. (11.3.124) has the form

$$r(k) = \sum_{l=1}^{n} e^{jk\theta_l}x_l \tag{11.3.132}$$

with

$$x_l = \frac{A_l^2}{4}. \tag{11.3.133}$$

The system of equations (11.3.132), with $0 \leqslant k \leqslant n - 1$, is Vandermonde; and since the roots of $\hat{a}(z)$ are distinct, the system is nonsingular. The solution must be positive and real because of Eq. (11.3.133).

This procedure requires significantly more effort than anything we have done so far because of the need to factor a polynomial. This step is also quite sensitive to numerical inaccuracy. If the conditions of the theorem are met, then the roots of $\hat{a}(z)$ must *all* be on the circle. But numerical errors may cause deviations from this. One notes however that if $\hat{a}(z)$ satisfies Eq. (11.3.130), then its coefficients must be symmetric, i.e., $a_{n-k} = \pm a_k$. See Problem 11.8.

The Pisarenko Model

Given $\mathbf{r}[0, n]$, and assuming that $\mathbf{R}(n) > \mathbf{0}$, a unique stable AR model of order n exists that matches $\mathbf{r}[0, n]$. The coefficients of the model can be obtained by the Levinson algorithm. But there are many other possible models. Before we indicate why AR models should be preferred, let us consider one alternative: lines plus white noise.

PROBLEM. Given $\mathbf{r}[0, n]$ with $\mathbf{R}(n) > \mathbf{0}$, obtain the parameters of the spectrum

$$S(\theta) = \sigma^2 + \sum_{k=1}^{n} \frac{\pi A_k^2}{2} \delta(\theta - \theta_k) \tag{11.3.134}$$

to match $\mathbf{r}[0, n]$.

It is not yet clear whether a solution exists or, if it does exist, whether it is the only possible solution. The existence and uniqueness of a solution to this problem were established in Pisarenko (1973). Once the procedure is revealed, it can be used to generate a variety of estimators that produce a "white noise plus lines" spectrum. One need only obtain estimates $\hat{\mathbf{r}}[0, n]$ and proceed.

The Pisarenko model in Eq. (11.3.134) may be considered a limiting case of another family of models, namely, white noise + AR of order n. Now the white noise component of a spectrum affects only $r(0)$, and therefore only the diagonal of $\mathbf{R}(n)$. From the trivial identity

$$\mathbf{R}(n) = \underset{\text{(white noise)}}{\sigma^2 \mathbf{I}} + \underset{\text{(AR of order } n)}{[\mathbf{R}(n) - \sigma^2 \mathbf{I}]}, \tag{11.3.135}$$

we see that it is possible to get a spectral model that is the sum of a white noise component with power σ^2 plus an order n AR component, provided only that $\mathbf{R}(n) - \sigma^2 \mathbf{I}$ is positive definite. This condition limits the range of acceptable values of σ^2. Let the eigenvalues of $\mathbf{R}(n)$ be

$$\lambda_0 \geqslant \lambda_1 \geqslant \cdots \geqslant \lambda_n > 0. \tag{11.3.136}$$

Then the eigenvalues of $\mathbf{R}(n) - \sigma^2\mathbf{I}$ will be

$$\lambda_0 - \sigma^2 \geqslant \lambda_1 - \sigma^2 \geqslant \cdots \geqslant \lambda_n - \sigma^2.$$

These must all be positive, for $\mathbf{R}(n) - \sigma^2\mathbf{I}$ to be positive definite. Therefore the range of permissible values of σ^2 is

$$0 \leqslant \sigma^2 < \lambda_n = \text{minimum eigenvalue of } \mathbf{R}(n). \tag{11.3.137}$$

EXAMPLE 11.3.6

White noise plus AR family of models.

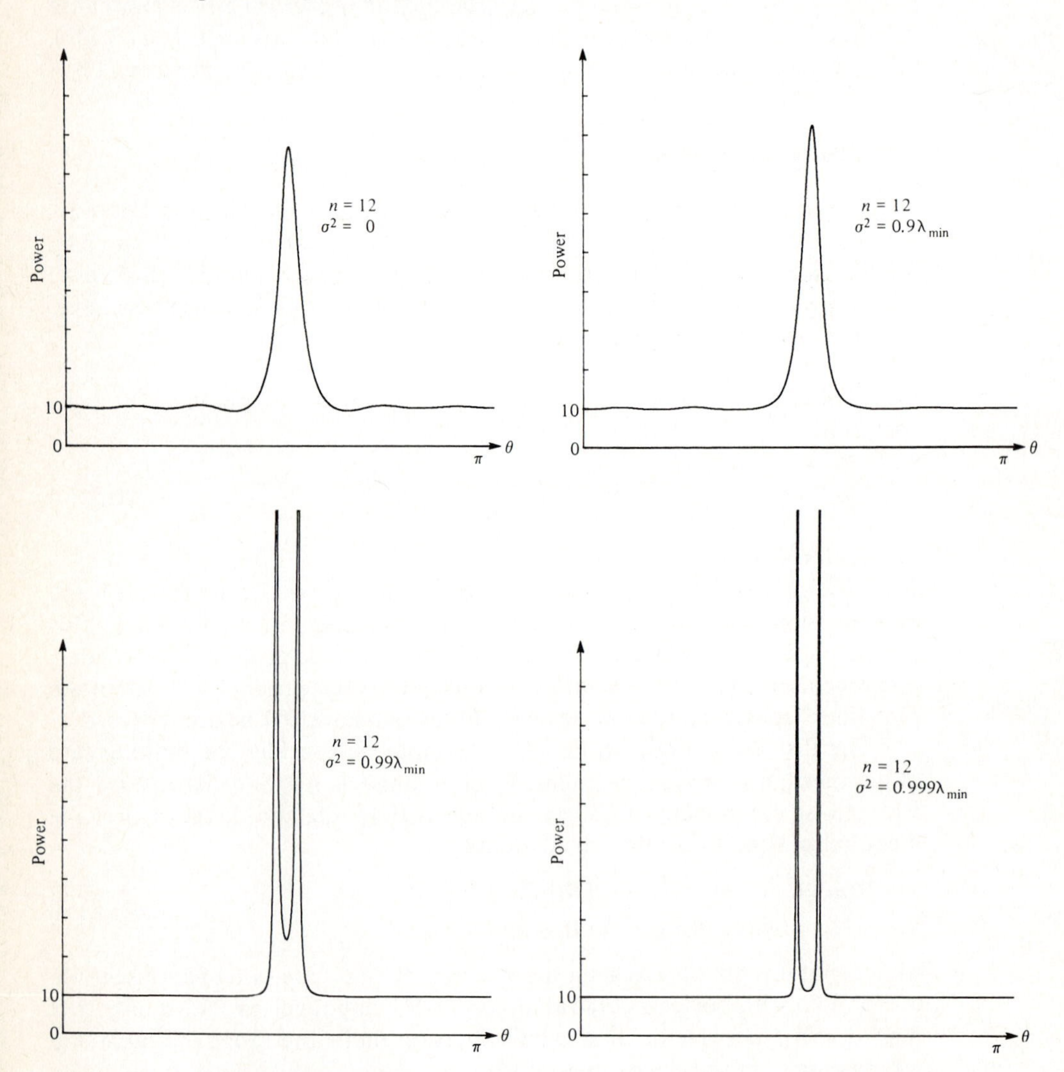

FIGURE 11.3.8 White noise (variance σ^2) plus order 12 AR models for a spectrum that contains white noise and two sinusoids (four lines).

As σ^2 increases, the AR part of the model resulting from the decomposition (11.3.135) gets more and more peaky. Consider two spectra. The first is truly white noise plus lines. The second is the spectrum of Example 11.2.1 which has lines, but a colored MA background. For the first case, let

$$S(\theta) = 10.0 + 2\pi\left[\delta\left(\theta + \frac{\pi}{2}\right) + \delta\left(\theta - \frac{\pi}{2}\right) + \delta\left(\theta + \frac{11\pi}{20}\right) + \delta\left(\theta - \frac{11\pi}{20}\right)\right].$$

Figure 11.3.8 exhibits four white noise plus AR models, with different values of σ^2. For this spectrum $\lambda_{\min} = 10$ for any $n \geqslant 4$. The AR models in the figure all have order 12. Note that as σ^2 increases, the AR part of the model gets

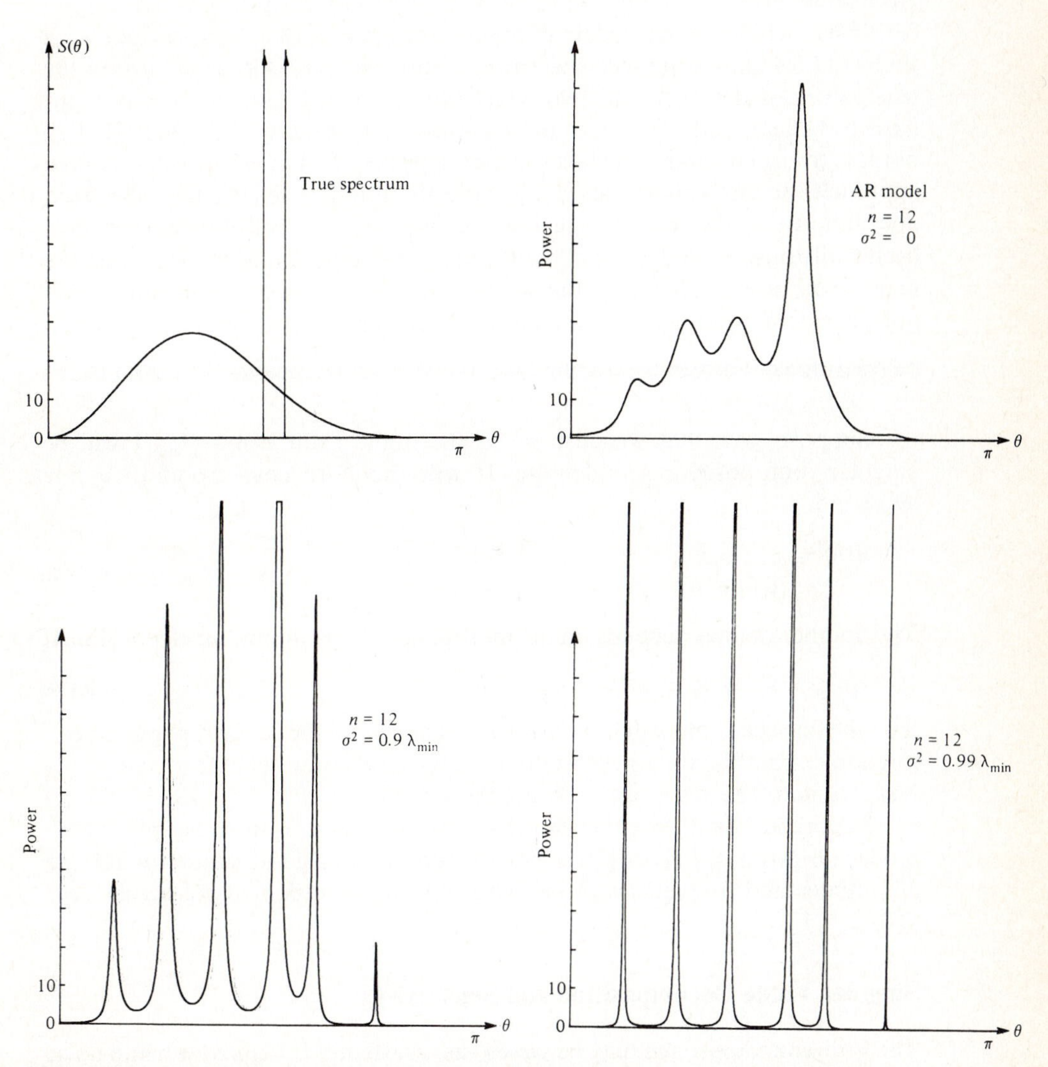

FIGURE 11.3.9 **White noise plus AR models for the MA plus lines example spectrum.**

peakier, but not significantly so until σ^2 gets very close to $\lambda_{\min}$. The peaks that result correspond to the lines in the spectrum, and there are only four of them (the other eight poles apparently approach the origin). The parameters of the AR model were obtained by running the Levinson algorithm on

$$\{r(0) - \sigma^2, r(1), \ldots, r(n)\}$$

with $\sigma^2 < 10$. The white noise level is then added to the AR part to produce the combined model.

Consider now the spectrum of Example 11.2.1 which has the same lines, but an MA background. Both of these spectra have the same background power. But since the MA background has zeros, one expects the minimum eigenvalue of $\mathbf{R}(n)$ to approach zero as n increases. For $n = 12$, $\lambda_{\min} = 0.0109498$, which is four orders of magnitude smaller than the previous $\lambda_{\min}$. Figure 11.3.9 shows the true spectrum, a pure order 12 AR model (using the true values of $\mathbf{r}[0, 12]$), and two white noise plus AR models. It is perhaps hard to believe that all four of these spectra have exactly the same $\mathbf{r}[0, 12]$, but it is true. (The diagram clips the higher peaks.) Note that all 12 poles now approach the circle as σ^2 nears $\lambda_{\min}$, and the line part of the true spectrum does not attract the poles as much as one would like. (Both frequencies have been badly missed; and if one chose the highest two peaks, they would miss the mark even more.) Therefore the white noise plus lines decomposition does not provide a good estimate for lines in a colored background.

Suppose now that we push σ^2 to the limit. Then $\mathbf{R}(n) - \lambda_{\min}\mathbf{I}$ will be singular, but positive semidefinite. It will therefore correspond to a line spectrum.

$$\mathbf{R}(n) = \underset{\text{(white noise)}}{\lambda_n \mathbf{I}} + \underset{\text{(singular} \Rightarrow \text{line spectrum)}}{[\mathbf{R}(n) - \lambda_n \mathbf{I}]} \tag{11.3.138}$$

The number of lines depends on the multiplicity of the minimum eigenvalue. If

$$\lambda_{m-1} > \lambda_m = \lambda_{m+1} = \cdots = \lambda_n > 0, \tag{11.3.139}$$

then the line spectrum will have m lines. One may use the procedure previously given to extract the parameters of the line spectral component from $r(0) - \lambda_n$, $r(1), \ldots, r(m)$. The result is the Pisarenko model.

In practice, the Pisarenko estimate, which forces a "white noise plus lines" model on estimates $\hat{\mathbf{r}}[0, n]$, has not proved to be a good estimator for the amplitudes and frequencies of known sinusoidal components of actual data $\mathbf{y}[0, L - 1]$.

Singular Value Decomposition and Sensitivity

The Pisarenko approach may be viewed as an attempt to remove a white noise component from the autocorrelation matrix $\mathbf{R}$. The context is "signal plus noise" with sinusoidal signals. If one makes this assumption about the data, then it is naturally tempting to try to remove the noise from the signal before estimating the parameters of the line spectrum. This brings up an important question: How sensitive is the estimate to the underlying noise? If the sensitivity is low, then there is no need to pursue this approach. But that is not the case, as was shown in an analysis of the whitening filter approach (Kumaresan and Tufts 1981). The new tool used here is the singular value decomposition (SVD) of a matrix.

In order to use this framework, let us recast the problem:

$$\text{Minimize } \|\mathbf{aY}\|^2, \quad \text{subject to } \mathbf{a} = [1, \mathbf{b}],$$

where

$$\mathbf{Y} = \begin{bmatrix} \mathbf{g} \\ \mathbf{F} \end{bmatrix}.$$

The solution is to take

$$\mathbf{b} = \mathbf{b}^0 \triangleq -\mathbf{g}\mathbf{F}^T(\mathbf{F}\mathbf{F}^T)^{-1}, \tag{11.3.140}$$

as may be seen by completing the square:

$$\begin{aligned} \|\mathbf{aY}\|^2 &= \|\mathbf{g} + \mathbf{bF}\|^2 \\ &= \|\mathbf{g}\|^2 - \|\mathbf{b}^0\mathbf{F}\|^2 + \|(\mathbf{b} - \mathbf{b}^0)\mathbf{F}\|^2. \end{aligned} \tag{11.3.141}$$

The matrix $\mathbf{Y}$ contains the data $\mathbf{y}[0, L-1]$, and is therefore random. This data is combined, using Eq. (11.3.140), to get the filter parameters in $\mathbf{b}^0$, but in a way that obscures the sensitivity issue. Assuming that the data has a sinusoidal component and a white noise component, then

$$\mathbf{F} = \mathbf{F}_{\text{signal}} + \mathbf{F}_{\text{noise}}. \tag{11.3.142}$$

How sensitive is $\mathbf{b}^0$ to the small changes in $\mathbf{F}$ due to the presence of $\mathbf{F}_{\text{noise}}$?

The SVD of a matrix is a very useful representation to use for perturbation analysis. The singular value decomposition of $\mathbf{F}$ is the factorization

$$\mathbf{F} = \mathbf{U\Sigma V}^T = \sum_{k=1}^{n} \sigma_k \mathbf{U}^{(k)}[\mathbf{V}^{(k)}]^T, \tag{11.3.143}$$

where

$$\begin{cases} \mathbf{U} \text{ is } n \times n \text{ and orthogonal,} & \mathbf{U}^T\mathbf{U} = \mathbf{I} \quad n \times n \\ \mathbf{V} \text{ is } m \times m \text{ and orthogonal,} & \mathbf{V}\mathbf{V}^T = \mathbf{I} \quad m \times m \\ \mathbf{\Sigma} = \left[\begin{array}{ccc|c} \sigma_1 & & 0 & \\ & \ddots & & 0 \\ 0 & & \sigma_n & \end{array}\right] & \text{is } n \times m, \\ \sigma_1 \geqslant \sigma_2 \geqslant \cdots \geqslant \sigma_n \geqslant 0. \end{cases} \tag{11.3.144}$$

[See Klemma and Laub (1980).] Since

$$\mathbf{F}\mathbf{F}^T = \mathbf{U}\boldsymbol{\Sigma}\boldsymbol{\Sigma}^T\mathbf{U}^T$$

$$= \mathbf{U}\begin{bmatrix} \sigma_1^2 & & 0 \\ & \ddots & \\ 0 & & \sigma_n^2 \end{bmatrix}\mathbf{U}^T, \tag{11.3.145}$$

the singular values are seen to be the positive square roots of the eigenvalues of $\mathbf{F}\mathbf{F}^T$, and therefore the $\boldsymbol{\Sigma}$ portion of the SVD is unique. The columns of $\mathbf{U}$ are eigenvectors of $\mathbf{F}\mathbf{F}^T$;

$$(\mathbf{F}\mathbf{F}^T)\mathbf{U}^{(k)} = \sigma_k^2\mathbf{U}^{(k)} \tag{11.3.146}$$

and here uniqueness depends on the multiplicity of σ_k. Similarly, the columns of $\mathbf{V}$ are eigenvectors of $\mathbf{F}^T\mathbf{F}$.

Let us now use the SVD of $\mathbf{F}$ to represent $\mathbf{b}^0$. Using the properties (11.3.144), Eq. (11.3.140) reduces to

$$\mathbf{b}^0 = -\mathbf{g}\mathbf{V}\begin{bmatrix} 1/\sigma_1 & & 0 \\ & \ddots & \\ 0 & & 1/\sigma_n \\ \hline & 0 & \end{bmatrix}\mathbf{U}^T = \sum_{k=1}^{n} \frac{-1}{\sigma_k}\mathbf{g}\mathbf{V}^{(k)}[\mathbf{U}^{(k)}]^T. \tag{11.3.147}$$

The appearance of the reciprocal of the singular value σ_k in this representation is disconcerting, because we expect the smallest singular values to come from the noise component alone. If this is true, then the vector $\mathbf{b}^0$ is quite noise sensitive.

What do we expect the singular values to be for the case of a single sinusoid in white noise? Take the "autocovariance case" of Eq. (11.3.22) to define $\mathbf{Y}$. Then $\mathbf{F}_{\text{signal}}$ will be $n \times (L - n)$ and have rank two. This can be seen by factoring the matrix into rank two factors. Let

$$\begin{aligned}(\mathbf{F}_{\text{signal}})_{kl} &= A\cos((n - k + l)\theta_0 + \phi_0) \\ &= \beta e^{-j(k-l)\theta_0} + \beta^* e^{j(k-l)\theta_0} \\ \text{for} \quad 1 \leqslant k \leqslant n, &\quad 0 \leqslant l \leqslant L - n - 1,\end{aligned} \tag{11.3.148}$$

where

$$\beta = \frac{A}{2}e^{j(n\theta_0 + \phi_0)}. \tag{11.3.149}$$

This produces the factorization

$$\mathbf{F}_{\text{signal}} = \begin{bmatrix} e^{-j\theta_0} & e^{j\theta_0} \\ e^{-j2\theta_0} & e^{j2\theta_0} \\ \vdots & \vdots \\ e^{-jn\theta_0} & e^{jn\theta_0} \end{bmatrix}\begin{bmatrix} \beta & 0 \\ 0 & \beta^* \end{bmatrix}\begin{bmatrix} 1 & e^{j\theta_0} & \cdots & e^{j(L-n-1)\theta_0} \\ 1 & e^{-j\theta_0} & \cdots & e^{-j(L-n-1)\theta_0} \end{bmatrix}$$

$$= \mathbf{PBQ}. \tag{11.3.150}$$

The columns of **P** are eigenvectors of $\mathbf{F}_{\text{signal}}\mathbf{F}^*_{\text{signal}}$.

$$[\mathbf{F}_{\text{signal}}\mathbf{F}^*_{\text{signal}}]\mathbf{P} = \mathbf{P}[\mathbf{BQQ^*B^*P^*P}]. \tag{11.3.151}$$

The nonzero eigenvalues are the eigenvalues of the 2×2 matrix

$$\Lambda = \mathbf{B}(\mathbf{QQ}^*)\mathbf{B}^*(\mathbf{P}^*\mathbf{P}). \tag{11.3.152}$$

These may be determined by

$$\begin{aligned}\lambda_1\lambda_2 &= \det(\Lambda)\\ &= \left(\frac{A}{2}\right)^4\left[(L-n)^2 - \left[\frac{\sin((L-n)\theta_0)}{\sin(\theta_0)}\right]^2\right]\left[n^2 - \left[\frac{\sin(n\theta_0)}{\sin(\theta_0)}\right]^2\right]\end{aligned} \tag{11.3.153}$$

and

$$\begin{aligned}\lambda_1 + \lambda_2 &= \mathrm{Tr}(\Lambda)\\ &= \left(\frac{A}{2}\right)^2\left[2n(L-n) + \frac{2\cos((L-2)\theta_0 + 2\phi_0)\sin((L-n)\theta_0)\sin(n\theta_0)}{[\sin(\theta_0)]^2}\right].\end{aligned} \tag{11.3.154}$$

Now let us take stock. In the noise-free case, there should be only two nonzero singular values, satisfying Eq. (11.3.153) and (11.3.154). With added white noise, we can estimate the average contribution of the noise to $\mathbf{FF}^T$ by considering

$$E(\mathbf{F}_{\text{noise}}\mathbf{F}^T_{\text{noise}}) = (L-n)\sigma^2\mathbf{I}. \tag{11.3.155}$$

Thus the eigenvalues of

$$E(\mathbf{FF}^T) = \mathbf{F}_{\text{signal}}\mathbf{F}^T_{\text{signal}} + (L-n)\sigma^2\mathbf{I} \tag{11.3.156}$$

are

$$\lambda_1 + (L-n)\sigma^2,\ \lambda_2 + (L-n)\sigma^2,\ (L-n)\sigma^2, \ldots, (L-n)\sigma^2. \tag{11.3.157}$$

Thus there are two dominant singular values and $(n-2)$ smaller singular values of the same size. In the actual random matrix $\mathbf{FF}^T$ there should still be two dominant singular values, but the others will fluctuate about the mean $(L-n)\sigma^2$. The smaller of these can produce large perturbations in $\mathbf{b}^0$, as can be seen from Eq. (11.3.147). It is these perturbations that cause the artificial peaks so noticeable in AR spectral estimates from small data sets.

The remedy in the sinusoid plus white noise case is clear. The matrix $\mathbf{F}_{\text{signal}}$ should have rank two. A property of the SVD of **F** is that the best rank two approximation for **F** (in the Euclidean operator norm) is obtained by simply truncating the sum in Eq. (11.3.143) to two terms. This is equivalent to zeroing out all but the two largest singular values. This, in turn, leads to truncating the sum in Eq. (11.3.147) to two terms, thereby eliminating the wild perturbations due to the small singular values. Thus the estimation of the signal parameters involves the revised vector

$$\hat{\mathbf{b}}^0 = -\sum_{k=1}^{2}\frac{1}{\sigma_k}\mathbf{g}\mathbf{V}^{(k)}[\mathbf{U}^{(k)}]^T. \tag{11.3.158}$$

(In the case of m sinusoids, the sum involves $2m$ terms.) The results, as may be seen in Kumaresan and Tufts (1981) are remarkable. Whereas the n poles associated with $\mathbf{b}^0$ are random with large variance, the n poles associated with $\hat{\mathbf{b}}^0$ are well organized. Two of these are very near $e^{j\theta_0}$ and $e^{-j\theta_0}$ while the other $n - 2$ are observed to be approximately equally spaced around a circle well within the unit circle. These provide a suitable AR approximation to the white noise background. Thus the AR estimate obtained from $\hat{\mathbf{b}}^0$ has one prominent peak above a relatively level background.

If θ_0 is not close to 0 or π, then Eqs. (11.3.153) and (11.3.154) may be approximated by

$$\lambda_1 \lambda_2 = \left(\frac{A}{2}\right)^4 (L - n)^2 n^2, \tag{11.3.159}$$

$$\lambda_1 + \lambda_2 = \left(\frac{A}{2}\right)^2 2n(L - n). \tag{11.3.160}$$

The solution to these equations is

$$\lambda_1 = \lambda_2 = \left(\frac{A}{2}\right)^2 n(L - n). \tag{11.3.161}$$

The ratio of the dominant singular values to the small ones would then be

$$\frac{\left(\frac{A}{2}\right)^2 n(L - n) + \sigma^2(L - n)}{\sigma^2(L - n)} = 1 + n \frac{\left(\frac{A}{2}\right)^2}{\sigma^2}, \tag{11.3.162}$$

which is

$$1 + \frac{1}{2}(\text{model order}) \frac{(\text{power in sinusoid})}{(\text{power in noise})}.$$

This suggests that the model order n should be chosen large even though $n = 2$ would seem to suffice for locating a single spectral line. However, with n too large, $L - n$ becomes small and this leads to some undesirable end effects. The choice

$$n \approx \frac{L}{2}$$

would lead to a roughly square matrix $\mathbf{Y}$. Kumaresan and Tufts recommend

$$n \approx \frac{L}{3}$$

and doing the SVD for a forward/backward version of $\mathbf{Y}$.

The Maximum Entropy Property

An often used justification for the use of AR spectral models is that they are solutions to the maximum entropy extension problem (see Section 11.1). This result is due to J. Burg (1967, 1975). Given $\mathbf{r}[0, n]$, we seek the spectrum that is consistent with these values and that maximizes the functional

$$\mathscr{H}_0(S) \triangleq \frac{1}{2\pi} \int_{-\pi}^{\pi} \ln S(\theta)\, d\theta. \tag{11.3.163}$$

Approaching this problem using the calculus of variations leads to an AR functional form for $S(\theta)$, at which point the Levinson algorithm may be invoked to produce the model parameters. But there is another approach that provides more intuition.

If the data vector $\mathbf{y}[0, L-1]$ is Gaussian, with mean zero and covariance $\mathbf{R}(L-1)$, then it has a probability density function

$$f(\mathbf{y}) = (2\pi)^{-L/2}[\det \mathbf{R}(L-1)]^{-1/2} \exp[-\tfrac{1}{2}\mathbf{y}^T\mathbf{R}^{-1}(L-1)\mathbf{y}]. \tag{11.3.164}$$

This density function has "average entropy"

$$-\frac{1}{L}\int f(\mathbf{y}) \ln f(\mathbf{y})\mathbf{dy} = \frac{1}{2}\ln(2\pi e) + \frac{1}{2L}\ln\det \mathbf{R}(L-1). \tag{11.3.165}$$

The dependence on the power spectrum is in the second term, and using Eq. (11.3.91) this is

$$\frac{1}{2L}\ln\det \mathbf{R}(L-1) = \frac{1}{2L}\ln \prod_{k=0}^{L-1} \alpha_k = \frac{1}{2}\left[\frac{1}{L}\sum_{k=0}^{L-1} \ln \alpha_k\right]. \tag{11.3.166}$$

Thus the average entropy involves the arithmetic mean of the logarithms of the prediction error variances. Consider a second functional

$$\mathscr{H}_1(S) \triangleq \lim_{L\to\infty} \frac{1}{L}\sum_{k=0}^{L-1} \ln \alpha_k. \tag{11.3.167}$$

Having introduced two "entropy" functionals, we will consider two questions. What is the relation between them, and how does one maximize $\mathscr{H}_1(S)$?

The maximization turns out to be very simple. Since the error variance sequence is nonincreasing, $\alpha_k \leqslant \alpha_n$ for all $k > n$. Therefore

$$\mathscr{H}_1(S) \leqslant \ln \alpha_n \tag{11.3.168}$$

with equality if and only if $\alpha_k = \alpha_n$ for all $k > n$. But this is not hard to do. From the relation $\alpha_{k+1} = \alpha_k(1 - c_{k+1}^2)$, the upper bound (11.3.168) (which is forced on us by the known quantities $\mathbf{r}[0, n]$) is attained by setting

$$c_k = 0 \quad \text{for } k > n. \tag{11.3.169}$$

In other words, maximizing $\mathscr{H}_1$ is equivalent to extending the reflection coefficient sequence with zeros. From the lattice equation (11.3.97), we see that this forces

$$\hat{a}(k, z) \equiv \hat{a}(n, z) \quad \text{for } k > n. \tag{11.3.170}$$

Thus the sequence of AR models "sticks" at the nth one, and the limit is an nth-order AR model. Note also that the maximum entropy spectrum (in the sense of $\mathscr{H}_1$) will also have the greatest asymptotic prediction error variance.

Can the functionals $\mathscr{H}_0$ and $\mathscr{H}_1$ be reconciled? To show that they are the same, we shall make use of the following theorem (Grenander and Szëgo 1958).

THEOREM 11.3.6. Let $S(\theta)$ be a power spectrum,

$$r(k) \overset{DTFT}{\longleftrightarrow} S(\theta),$$

and let $\mathbf{R}(n)$ have eigenvalues

$$\lambda_0(n) \geqslant \lambda_1(n) \geqslant \cdots \geqslant \lambda_n(n) > 0.$$

If f is a continuous real function, then

$$\lim_{n\to\infty} \frac{1}{n+1} \sum_{k=0}^{n} f(\lambda_k(n)) = \frac{1}{2\pi} \int_{-\pi}^{\pi} f(S(\theta))\, d\theta. \tag{11.3.171}$$

This theorem is quite useful. A loose interpretation is that the eigenvalues of $\mathbf{R}(n)$ are approximated by samples of the power spectrum. For example, if $S(\theta)$ is white, then $\mathbf{R}(n) = \sigma^2\mathbf{I}$ has eigenvalues all equal to σ^2. If $S(\theta)$ is a line spectrum with two lines, then indeed the rank of $\mathbf{R}(n)$ is 2. Thus it has only two nonzero eigenvalues; the rest are zero. And in this case most of $S(\theta)$ is indeed zero.

To show that $\mathscr{H}_0 = \mathscr{H}_1$, we invoke the theorem with $f(\lambda) = \ln \lambda$, and use the fact that

$$\det \mathbf{R}(n) = \prod_{k=0}^{n} \lambda_k(n) = \prod_{k=0}^{n} \alpha_k. \tag{11.3.172}$$

Take the logarithm of this equation and substitute into Eq. (11.3.171). The left-hand side is $\mathscr{H}_1$, and the right-hand side is $\mathscr{H}_0$.

Now α_n is the error variance for an optimal one step linear prediction based on the last n values of a signal. Thus the error variance for the infinite order, or Wiener, predictor is

$$\alpha_\infty = \lim_{n\to\infty} \alpha_n = e^{\mathscr{H}_1} = e^{\mathscr{H}_0} = \exp\left[\frac{1}{2\pi} \int_{-\pi}^{\pi} \ln S(\theta)\, d\theta\right]. \tag{11.3.173}$$

(This uses the monotonicity of the sequence $\boldsymbol{\alpha}$ in the relation of α_∞ to $\mathscr{H}_1$.) This result is due to Szëgo (Grenander and Szëgo 1958). Thus the maximum entropy spectrum will also have the greatest asymptotic prediction error variance. Using this model is therefore consistent with the philosophy that if one plans for the worst, then he will not be disappointed.

A Classification of Spectra

The behavior of the prediction error variance sequence can be used to categorize power spectra, as in Fig. 11.3.10. The list of categories may be taken to be in decreasing order of suitability for AR modeling. It is, conversely, increasing in predicatability. Thus white noise (category 0) is not in the least

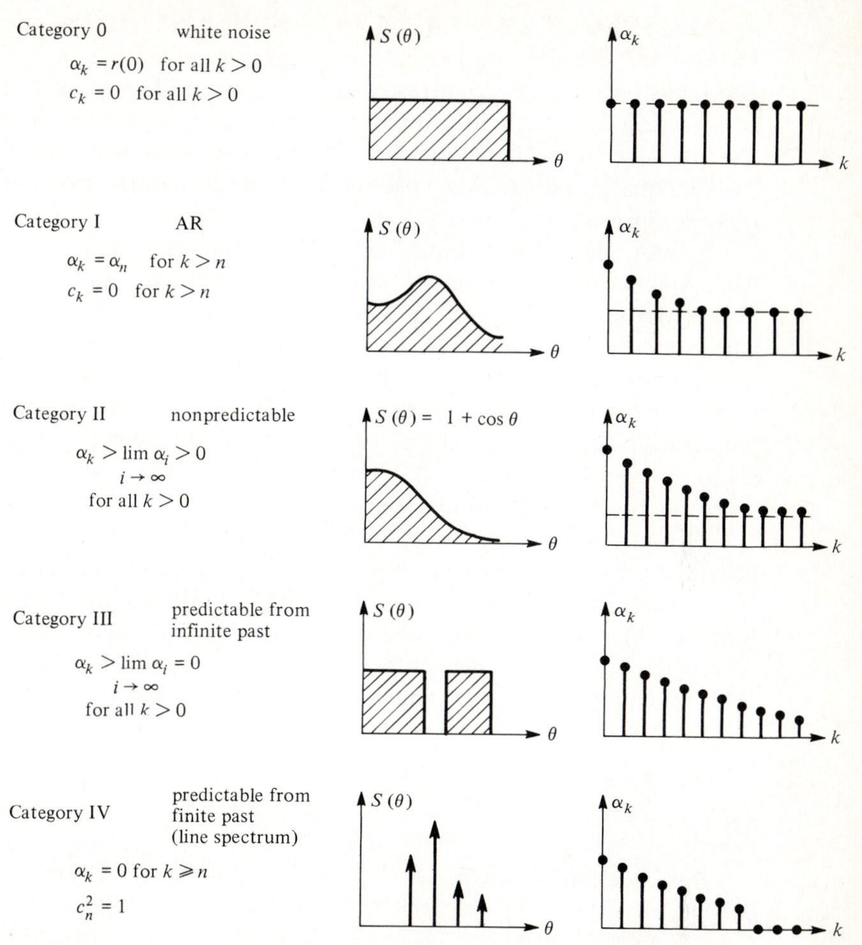

FIGURE 11.3.10 A categorization of spectra base on the behavior of the error variance sequence $\{\alpha_k\}$. The example spectra are not necessarily representative of the entire category.

predictable, but can be modeled by a zero-order AR model. Category I contains the finite-order AR spectra. In this category, the Wiener prediction filter needs only a finite past history of the signal, the rest being superfluous. In Category II, the Wiener filter needs the entire past history, and the sequence of α_k's approaches a positive limit, but does not reach it in finite length. In Category III, the sequence of error variances approaches zero. Thus the signal is perfectly predictable from an infinite past history. Notice that, in the example, the notch can be made as narrow as one cares to make it. The spectrum is very close to white noise. And yet, because of Eq. (11.3.173), it is completely predictable while white noise is completely unpredictable! Finally, if the sequence α_k hits zero with finite length, then the signal is perfectly predictable from a past history of that length and no more. This is the singular case, and the spectrum must be a line spectrum.

11.4 ARMA SPECTRA AND AN ANALYSIS/SYNTHESIS POINT OF VIEW

There has been much recent research activity in ARMA spectral estimation. No unifying approach has apparently been presented. (See the Suggested Readings for Spectral Estimation.) In this section is a study of the Yule-Walker equations for ARMA spectra based on a generalization of the analysis/synthesis approach.

ARMA spectra were introduced in Example 11.1.2. For an order (m, n) ARMA spectrum there are $n + m + 1$ parameters: $\mathbf{a}[1, n]$ and $\mathbf{b}[0, m]$. For the ARMA estimation problem, one must devise a means of computing these parameters from the data $\mathbf{y}[0, L - 1]$. There are many ways to do this. For example, one can ask for an order (m, n) ARMA spectrum that agrees with known or estimated values of $\mathbf{r}[0, m + n]$. If a solution exists, then it may be found using the development of Problem 11.44. There may, however, be no solution, even when $\mathbf{R}(m + n)$ is positive definite. Thus, the class of ARMA spectra is not complete in this sense.

The analysis/synthesis approach (which is so consistent for AR models) generalizes to ARMA models, but not in the way one would like. It requires information that cannot easily be obtained from the data. It is more suitable to the problem of modeling a linear system given both input and output data. Nevertheless, the approach can be used to study the ARMA class. The "Yule-Walker" equations (11.1.28) and (11.1.30) are fundamental to ARMA models.

Let $\mathbf{u}$ be a unit variance white noise signal and let

$$\mathbf{y} = \mathbf{h} * \mathbf{u}. \tag{11.4.1}$$

Then

$$E\, y(i + k)u(i) = h(k), \tag{11.4.2}$$

$$E\, y(i + k)y(i) = r(k) = \sum_{i=0}^{\infty} h(k + i)h(i). \tag{11.4.3}$$

If $H(z)$ is rational, having the form

$$H(z) = \frac{\hat{b}(z)}{\hat{a}(z)} = \frac{b_0 + \cdots + b_m z^{-m}}{1 + a_1 z^{-1} + \cdots + a_n z^{-n}}, \tag{11.4.4}$$

then the Yule-Walker equations (11.1.28) and (11.1.30) relate the parameters $\mathbf{a}[1, n]$, $\mathbf{b}[0, m]$, $\mathbf{r}[0, n]$, $\mathbf{h}[0, m]$ as follows.

$$\mathbf{a}\, \mathbf{H}(m, n) = \mathbf{b}, \tag{11.4.5}$$

$$\mathbf{a}\, \mathbf{R}(n) = \mathbf{b}\, \mathbf{H}^T(m, n), \tag{11.4.6}$$

where

$$\mathbf{a} = [1, a_1, \ldots, a_n], \tag{11.4.7}$$

$$\mathbf{b} = [b_0, b_1, \ldots, b_m], \tag{11.4.8}$$

$$\mathbf{R}(n) \triangleq E\begin{bmatrix} y(k) \\ \vdots \\ y(k-n) \end{bmatrix}[y(k)\cdots y(k-n)] = \begin{bmatrix} r(0) & & r(n) \\ & \ddots & \\ r(n) & & r(0) \end{bmatrix}, \tag{11.4.9}$$

$$\mathbf{H}(m, n) \triangleq E\begin{bmatrix} y(k) \\ \vdots \\ y(k-n) \end{bmatrix}[u(k)\cdots u(k-m)]$$

$$= \begin{bmatrix} h(0) & \ddots & h(m) \\ 0 & \ddots & \\ & & h(m-n) \end{bmatrix} \quad (\text{Toeplitz}, n+1 \times m+1). \tag{11.4.10}$$

Analysis

In analogy to the whitening problem of Section 11.3, there is a whitening problem that produces equations like the Yule-Walker equations (11.4.5) and (11.4.6). The problem is that both signals **u** and **y** are necessary. Construct a whitener for **y** as follows:

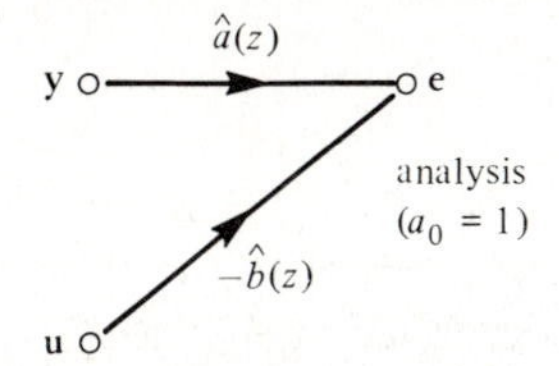

$$e(k) = \sum_{i=0}^{n} a_i y(k-i) - \sum_{i=0}^{m} b_i u(k-i). \tag{11.4.11}$$

We need not assume that the input/output relation (11.4.1) is valid, but we do require that **u** is unit variance white noise and that the cross-correlation of **u** and **y** and the autocorrelation of **y** satisfy Eqs. (11.4.9) and (11.4.10). The problem is to whiten **y** by minimizing the quadratic function

$$V(\mathbf{a}, \mathbf{b}) = E\, e(k)^2$$

$$= [\mathbf{a} \quad \mathbf{b}]\begin{bmatrix} \mathbf{R}(n) & -\mathbf{H}(m, n) \\ -\mathbf{H}^T(m, n) & \mathbf{I} \end{bmatrix}\begin{bmatrix} \mathbf{a}^T \\ \mathbf{b}^T \end{bmatrix} \tag{11.4.12}$$

subject to the constraint $a_0 = 1$. The normal equations are

$$[\mathbf{a} \quad \mathbf{b}]\begin{bmatrix} \mathbf{R}(n) & -\mathbf{H}(m, n) \\ -\mathbf{H}^T(m, n) & \mathbf{I} \end{bmatrix} = [\alpha \quad 0], \tag{11.4.13}$$

where

$$\alpha = E\, e(k)^2 \tag{11.4.14}$$

is the minimum possible error variance. These equations relate the two sets of parameters

$$\{\alpha, \mathbf{a}[1, n], \mathbf{b}[0, m]\} \quad \text{and} \quad \{\mathbf{r}[0, n], \mathbf{h}[0, m]\},$$

and are almost identical to the Yule-Walker equations, if we put Eq. (11.4.13) into the form

$$\mathbf{a}\,\mathbf{H}(m, n) = \mathbf{b}, \tag{11.4.15}$$

$$\mathbf{a}\,\mathbf{R}(\mathbf{n}) = \mathbf{b}\,\mathbf{H}^T(m, n) + [\alpha \bigcirc]. \tag{11.4.16}$$

They become identical when the error variance α vanishes. Thus α can be regarded as a measure of distance from ARMA.

Synthesis

The synthesis version of Eq. (11.4.11) has the form

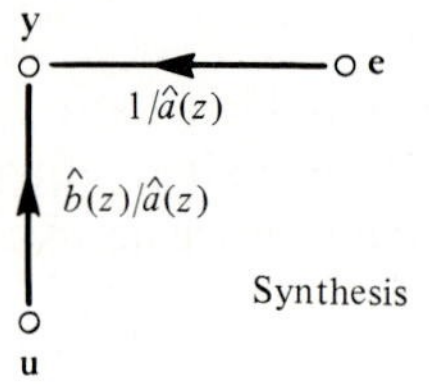

y

e

$1/\hat{a}(z)$

$\hat{b}(z)/\hat{a}(z)$

Synthesis

u

$$\sum_{i=0}^{n} a_i y(k-i) = \sum_{i=0}^{m} b_i u(k-i) + e(k). \tag{11.4.17}$$

Equations (11.4.11) and (11.4.17) are identical except in the selection of outputs. If we assume that **u** and **e** are uncorrelated white noise signals with variances

$$\begin{aligned} E\,u(k)^2 &= 1, \\ E\,e(k)^2 &= \alpha, \end{aligned} \tag{11.4.18}$$

then

$$\hat{S}_{yy}(\theta) = \left|\frac{\hat{b}(e^{j\theta})}{\hat{a}(e^{j\theta})}\right|^2 + \frac{\alpha}{|\hat{a}(e^{j\theta})|^2}. \tag{11.4.19}$$

With $\alpha = 0$, we have the ARMA spectrum consistent with the Yule-Walker equations (11.4.5) and (11.4.6).

There are a number of theoretical issues that this version of analysis/synthesis suggests. One is an extension problem.

Extension Problem

Given $\mathbf{h}[0, m]$, $\mathbf{r}[0, n]$, find extensions of these sequences for which

$$r(0) + 2\sum_{k=1}^{\infty} r(k)\cos(k\theta) \equiv \left|\sum_{k=0}^{\infty} h(k)e^{-jk\theta}\right|^2. \tag{11.4.20}$$

The existence of solutions to this problem is characterized in the following theorem.

THEOREM 11.4.1: Let

$$\mathbf{K} = \mathbf{R}(n) - \mathbf{H}(m, n)\mathbf{H}^T(m, n). \tag{11.4.21}$$

(This matrix depends only on $\mathbf{h}[0, m]$, $\mathbf{r}[0, n]$.) Then

1. If $\mathbf{x}^T\mathbf{K}\mathbf{x} < 0$ for some $\mathbf{x}$, then there is no solution to the extension problem.
2. If $\mathbf{K} \geqslant \mathbf{0}$ but $\det \mathbf{K} = 0$, then there may or may not be a solution. If there is a solution, then it is unique and corresponds to an ARMA (n', m') model with $n' \leqslant n$, $m' \leqslant m$. (This is the $\alpha = 0$ case.)
3. If $\mathbf{K} > \mathbf{0}$, there are infinitely many solutions.

[The proofs of this theorem, and the next one appear in Mullis and Roberts (1976c), and Inouye (1983).]

Another issue is whether the analysis and synthesis descriptions are consistent.

THEOREM 11.4.2. Assume that the matrix $\mathbf{K}$ in Eq. (11.4.21) is positive definite. Then there exists a unique solution $(\alpha, \mathbf{a}, \mathbf{b})$ to the normal equations (11.4.16) and (11.4.15), and

1. The synthesis model is stable, i.e., $\hat{a}(\lambda) = 0 \Rightarrow |\lambda| < 1$.
2. The power spectrum of Eq. (11.4.19) matches $\mathbf{r}[0, n]$.
3. The transfer function $\hat{b}(z)/\hat{a}(z)$ matches $\mathbf{h}[0, m]$.

Application

The analysis problem (11.4.11) is not directly appropriate to power spectrum estimation, since it requires either input data or cross-correlation measurements. These are impossible if output data alone is available. In the system identification context, where an I/O model relating $\mathbf{u}$ and $\mathbf{y}$ is sought, the problem is appropriate. The parameters $\mathbf{h}[0, m]$ may be indirectly estimated in a number of ways. One can, for example, extract the front end of the unit-pulse response of a higher-order AR model (Scharf and Luby 1979). In general, one can produce the unit-pulse response by a spectral factorization of any $\hat{S}(\theta)$.

These estimates can then be combined in a number of ways. As a corollary of Theorem 11.4.2, one can show that if σ^2 and μ^2 are chosen so that

$$\mathbf{K} = \mathbf{R}(n) - \sigma^2\mathbf{I} - \mu^2\mathbf{H}(m, n)\mathbf{H}^T(m, n) > \mathbf{0} \tag{11.4.22}$$

and α, $\mathbf{a}[1, n]$, $\mathbf{b}[0, m]$ are obtained from

$$\begin{cases} \mathbf{a}\,\mathbf{K} = [\alpha \;\bigcirc\;], & a_0 = 1 \\ \mathbf{b} = \mathbf{a}\,\mathbf{H}(m, n) \end{cases} \tag{11.4.23}$$

then $\hat{a}(z)$ will be stable, and the estimate

$$\hat{S}(\theta) = \sigma^2 + \mu^2 \left|\frac{\hat{b}(e^{j\theta})}{\hat{a}(e^{j\theta})}\right|^2 + \frac{\alpha}{|\hat{a}(e^{j\theta})|^2} \tag{11.4.24}$$

will match $\mathbf{r}[0, n]$. This estimate has separate white noise, ARMA, and AR components. The combination, of course, is ARMA. With $\mu^2 = \sigma^2 = 0$, an AR estimate is obtained. The Pisarenko estimate is approached as σ^2 increases to the minimum eigenvalue of $\mathbf{R}(n)$, with $\mu^2 = 0$.

Summary • The frequency content of a signal is an important property of the signal. This chapter provides a brief look into this important problem. We have considered several methods for estimating the spectral content of a signal. The literature contains a vast array of methods. The choice of a particular method depends on the application and the data length.

PROBLEMS

11.1. Let $\mathbf{y}$ be white Gaussian noise with variance σ^2. Let

$$\hat{r}(0) = \frac{1}{L} \sum_{k=0}^{L-1} y(k)^2.$$

How big should L be so that the variance of $\hat{r}(0)$ satisfies

$$E\,\hat{r}(0)^2 - [E\,\hat{r}(0)]^2 < 10^{-2}\sigma^4?$$

11.2. a) Show that if $S(\theta) \geqslant 0$, then $r(1)^2 \leqslant r(0)^2$.

b) Under what conditions will the order one Blackman-Tukey estimate

$$\hat{S}(\theta) = \hat{r}(0) + 2\hat{r}(1)\cos\theta$$

be nonnegative?

c) Give an example of a spectrum $S(\theta)$ for which $E\,\hat{S}(\theta)$ is negative for some values of θ, where $\hat{S}(\theta)$ is the Blackman-Tukey estimate.

11.3. Let W be the window function in Eq. (11.2.10). Characterize the set of all spectral density functions for which

$$\frac{1}{2\pi L} \int_{-\pi}^{\pi} S(\theta - \phi)|W(e^{j\phi})|^2\, d\phi \equiv S(\theta).$$

11.4. Let $\hat{S}(\theta) = \mathscr{S}(\theta;\ \mathbf{y}[0, L-1])$ be some estimator. Show that the estimate must be biased, i.e.,

$$E\,\hat{S}(\theta) - S(\theta),$$

cannot be identically zero for all power spectra if L is finite.

11.5. Many commercial "Fourier analyzers" employ exponential averaging of periodograms: $\hat{S}_0(\theta) = 0$,

$$\hat{S}_n(\theta) = a\hat{S}_{n-1}(\theta) + \frac{1-a}{L}\left|\sum_{k=0}^{L-1} w(k)y(k)e^{-jk\theta}\right|^2.$$

Assuming that successive periodograms are uncorrelated, compute the mean and variance of $\hat{S}_n(\theta)$ for Gaussian signals $(0 < a < 1)$.

11.6. Let the signal $\mathbf{y}$ consist of m real sinusoids as in Eq. (11.1.35). Let $n = 2m$ be the order of a linear prediction filter for this signal using Eqs. (11.3.27) or (11.3.28), where $\mathbf{Y}$ is the $(n + 1) \times (L - n)$ covariance case matrix of Eq. (11.3.22).

a) (Forward only.) If $L = 4m$, the zeros of $\hat{a}(z)$ obtained using Eq. (11.3.27) must be at the frequencies of the sinusoids.

b) (Forward backward.) If $L = 3m$, then the normal equations (11.3.28) will give the desired results.

(Problem contributed by Allan Steinhardt.)

11.7. Find the general form of a 3×3 matrix $\mathbf{Q}$ satisfying $\mathbf{QJ} = \mathbf{JQ}$. Must this matrix be Toeplitz?

11.8. Show that if

$$\hat{a}(z) = \sum_{k=0}^{n} a_k z^{-k} = \prod_{k=1}^{n} (1 - z^{-1}\lambda_k)$$

has real coefficients and $|\lambda_k| = 1$ for all k, then $\mathbf{a} = \pm\mathbf{aJ}$, i.e., $a_k = \pm a_{n-k}$. Show that the converse is not necessarily true. Show that the equation

$$\mathbf{aQ} = \mathbf{b}$$

will have a solution satisfying $\mathbf{aJ} = \mathbf{a}$ if $\mathbf{JQ} = \mathbf{QJ}$ and $\mathbf{bJ} = \mathbf{b}$.

11.9. Derive Eq. (11.3.54). (This is difficult.)

11.10. Suppose $\mathbf{a}$, $\mathbf{R}$, and α are related by Eq. (11.3.30) and that

$$\hat{a}(z) = \mathbf{a}\psi_n(z) = \prod_{k=1}^{n} (1 - z^{-1}\lambda_k).$$

Show that

$$\det \mathbf{R}(n) = \frac{\alpha^{n+1}}{\prod_{k=1}^{n} \prod_{j=1}^{n} (1 - \lambda_k \lambda_j^*)}.$$

11.11. Let

$$r(k) = \begin{cases} n - |k|, & |k| \leqslant n \\ 0, & \text{otherwise} \end{cases}.$$

Find the Cholesky decomposition of $\mathbf{R}(n)$ of the form (11.3.90). (The result can be used to test a coding of the Levinson algorithm.)

11.12. Produce a Levinson-type algorithm to solve the equation

$$\mathbf{aR} = \mathbf{b}$$

for $\mathbf{a}$ when $\mathbf{b}$, $\mathbf{R}$ are known and $\mathbf{R}$ is Toeplitz and symmetric.

11.13. Repeat problem 12 except $\mathbf{R}$ is Toeplitz but not necessarily symmetric.

11.14. Let $S_1(\theta)$ and $S_2(\theta)$ be two spectra with error variance sequences $\{\alpha_{1,n}\}$ and $\{\alpha_{2,n}\}$. Let $S(\theta) = S_1(\theta) + S_2(\theta)$ have sequence $\{\alpha_n\}$. True or false: $\alpha_n \leqslant \alpha_{1,n} + \alpha_{2,n}$ for all n.

11.15. Suppose $S_1(\theta) \leqslant S_2(\theta)$ for all θ. True or false: $\alpha_{1,n} \leqslant \alpha_{2,n}$ for all n.

11.16. Suppose $S_2(\theta) = S_1(\theta) + \delta(\theta + \theta_0) + \delta(\theta - \theta_0)$. True or false: $\alpha_{2,n+2} \leqslant \alpha_{1,n}$ for all n.

11.17. Show that the polynomials $\hat{a}(m, z)$ of Eq. (11.3.95), which are generated from an autocorrelation sequence

$$r(k) \xleftrightarrow{DTFT} S(\theta)$$

are orthogonal in the sense that

$$\frac{1}{2\pi}\int_{-\pi}^{\pi} \hat{a}(m, e^{j\theta})\, \hat{a}(n, e^{-j\theta})\, S(\theta)\, d\theta = \alpha_n \delta_{mn}.$$

[Use Eqs. (11.3.13) and (11.3.30).] Then compute, for $0 \leqslant m \leqslant n \leqslant k$

$$\frac{1}{2\pi}\int_{-\pi}^{\pi} \frac{\hat{a}(m, e^{j\theta})\hat{a}(n, e^{-j\theta})}{|\hat{a}(k, e^{j\theta})|^2}\, d\theta.$$

11.18. Let $\hat{S}_n(\theta)$ be the nth-order AR model for $S(\theta)$ whose coefficients satisfy Eq. (11.3.30). Show that

$$\frac{1}{2\pi}\int_{-\pi}^{\pi} \frac{S(\theta)}{\hat{S}_n(\theta)}\, d\theta = 1.$$

11.19. Find explicit formulas for the reflection coefficients $\mathbf{c}[1, n]$ in terms of $\mathbf{a}[1, n]$ for the cases $n = 2,3$ (assuming $a_0 = 1$).

11.20. Establish these reflection coefficient identities:

a) $$1 - c_{n+1}^2 = \frac{\det \mathbf{R}(n+1) \det \mathbf{R}(n-1)}{[\det \mathbf{R}(n)]^2}$$

b) $$c_n = \frac{(-1)^n}{\det \mathbf{R}(n-1)} \det \begin{bmatrix} r(1) & & r(n) \\ & \ddots & \\ r(n-2) & & r(1) \end{bmatrix}$$

c) for $n > 1$,

$$c_n = -\frac{1}{\alpha_n}\left\{ r(n) - [r(1) \cdots r(n-1)]\mathbf{R}^{-1}(n-2) \begin{bmatrix} r(n-1) \\ \vdots \\ r(1) \end{bmatrix} \right\}.$$

11.21. Let

$$\begin{cases} \mathbf{Q}(c, \theta) \triangleq \begin{bmatrix} e^{j\theta/2} & ce^{-j\theta/2} \\ ce^{j\theta/2} & e^{-j\theta/2} \end{bmatrix}, \\ \mathbf{P}_0(\theta) = \mathbf{I}, \\ \mathbf{P}_m(\theta) = \mathbf{Q}(c_m, \theta)\mathbf{P}_{m-1}(\theta). \end{cases}$$

Show that the AR spectrum parameterized by $r(0)$, $\mathbf{c}[1, n]$ is

$$S_n(\theta) = 2r(0) \det[\mathbf{P}_n(\theta)] \Big/ \left\| \mathbf{P}_n(\theta) \begin{bmatrix} 1 \\ 1 \end{bmatrix} \right\|^2.$$

11.22. Suppose $S(\theta)$ is a spectrum whose reflection coefficient sequence is periodic with period m and nonzero. Using the notation of Problem 11.21, let

$$\mathbf{P}(\theta) = \mathbf{P}_m(\theta) = \mathbf{Q}(c_m, \theta) \cdots \mathbf{Q}(c_1, \theta).$$

Show that the trace of $\mathbf{P}(\theta)$ is real and that

$$S(\theta) = 0 \quad \Leftrightarrow \quad [\tfrac{1}{2}\operatorname{Tr}\mathbf{P}(\theta)]^2 > \prod_{k=1}^{m} (1 - c_k^2).$$

11.23. Suppose $c_n = c$ for all n. Characterize the set $\{\theta : S(\theta) = 0\}$.

11.24. Capon estimate (1969):
Given $\mathbf{r}[0, n]$, to estimate $S(\theta_0)$ (at the single frequency θ_0) obtain the FIR filter

$$\hat{a}(z) = \mathbf{a}\psi_n(z)$$

to minimize

$$V(\mathbf{a}, \theta_0) = \frac{1}{2\pi} \int_{-\pi}^{\pi} S(\theta)\, |\hat{a}(e^{j\theta})|^2 \, d\theta$$

subject to

$$\hat{a}(e^{j\theta_0}) = 1, \qquad (a_0 \text{ need not be 1, and } a_k \text{ need not be real}).$$

Intuitively, $\hat{a}(z)$ will approximate a narrow bandpass filter with center frequency θ_0. The Capon spectral estimate is then

$$\hat{S}_c(\theta_0) = (n + 1)V(\mathbf{a}, \theta_0).$$

a) Show that

$$\hat{S}_c(\theta) = (n + 1)[\psi_n^*(e^{j\theta})\mathbf{R}^{-1}(n)\psi_n(e^{j\theta})]^{-1}.$$

b) Show that the one-step prediction error variance satisfying the normal equations (11.3.30) is

$$\alpha_n = [\psi_n^*(\infty)\mathbf{R}^{-1}(n)\psi_n(\infty)]^{-1}$$

c) and that the nth-order AR spectrum resulting from Eq. (11.3.30) is

$$\hat{S}_n(\theta) = \frac{\psi_n^*(\infty)\mathbf{R}^{-1}(n)\psi_n(\infty)}{|\psi_n^*(e^{j\theta})\mathbf{R}^{-1}(n)\psi_n(\infty)|^2}.$$

d) Show that

$$\hat{S}_c(\theta) = \left[\frac{1}{n + 1} \sum_{k=0}^{n} \hat{S}_k(\theta)^{-1}\right]^{-1}$$

[Hint: use Eq. (11.3.90).]

e) What is $\hat{S}_c(\theta)$ when $S(\theta) = \sigma^2$ (white noise)?

11.25. Consider a whitening filter of the form

$$e(k) = y(k) + a_1 y(k - 1) + a_3 y(k - 3),$$

where the parameters are chosen to minimize $E\, e^2(k)$. To do the minimization, one needs $\mathbf{r}[0, 3]$. Must the estimate

$$\hat{S}(\theta) = \frac{\alpha}{|1 + a_1 e^{-j\theta} + a_3 e^{-j3\theta}|^2}$$

reproduce $\mathbf{r}[0, 3]$? (Give a proof or a counterexample.)

11.26 a) Derive the backward lattice section of Fig. 11.3.5 from the forward lattice section of Fig. 11.3.4.

b) Justify the claim that the connection of backward lattice sections does indeed produce a partial inversion of the whole forward lattice. In other words, prove that the algebraic relationships between the signals in the lattices of Figs. 11.3.4 and 11.3.5 are identical.

11.27. Compute the covariance of the vector

$$[e_0(k), \tilde{e}_0(k), e_1(k), \tilde{e}_1(k), \ldots, e_n(k), \tilde{e}_n(k)]$$

of variables appearing in Fig. 11.3.4 assuming that the reflection coefficients are chosen to whiten $\mathbf{y}$ (i.e., are computed using the Levinson algorithm).

11.28. Prove the following two results:

a) If $S(\theta)$ is MA with $r(k) = 0$ for $k > n$, and $\mathbf{R}(n)$ is circulant, then the eigenvalues of $\mathbf{R}(n)$ are

$$\lambda_k = \frac{r(0) + S\left(\dfrac{2\pi k}{n+1}\right)}{2}, \qquad 0 \leqslant k \leqslant n.$$

b) If $S(\theta)$ is AR of order n and $\mathbf{R}(n)$ is circulant, then the eigenvalues are

$$\lambda_k = \left[\alpha S\left(\frac{2\pi k}{n+1}\right)\right]^{1/2}, \qquad 0 \leqslant k \leqslant n.$$

11.29. Develop an algorithm for the efficient computation of the solution $\mathbf{K}$ to

$$\mathbf{K} = \mathbf{A}\mathbf{K}\mathbf{A}^T + \mathbf{B}\mathbf{B}^T$$

using Problem 8.20 and the Jury and Autocorrelation sequence algorithms. You may assume the existence of a subroutine that computes the coefficients of the characteristic polynomial of $\mathbf{A}$. The key is Eq. (11.3.70).

11.30. (Bias of frequency estimates using AR models). Construct the polynomial

$$\hat{a}(z) = 1 + c_1(1 + c_2)z^{-1} + c_2 z^{-2} = (1 - z^{-1}re^{j\phi})(1 - z^{-1}re^{-j\phi})$$

using Eq. (11.3.122) with $\theta_0 = \pi/4$. Plot the function $\phi(\sigma^2)$ for $0 \leqslant \sigma^2 \leqslant 2.0$. [One would like $\phi(\sigma^2) \equiv \theta_0$.] Repeat for $\theta_0 = \pi/2$, $\theta_0 = \pi/5$.

11.31. Suppose $\mathbf{R}(n)$ satisfies Eq. 11.3.11 for some power spectrum $S(\theta)$. Let

$$\mathbf{c} = [1 \quad c_1 \quad \cdots \quad c_m]$$
$$\mathbf{b} = [1 \quad b_1 \quad \cdots \quad b_{n-m}]$$
$$\mathbf{a} = [1 \quad a_1 \quad \cdots \quad a_n] = \mathbf{b} * \mathbf{c}, \qquad \text{i.e.,} \qquad a_k = \sum_{i=0}^{k} c_i b_{k-i}$$

Show that $\mathbf{c}\,\mathbf{R}(m) = \mathbf{0}$ implies $\mathbf{a}\,\mathbf{R}(n) = \mathbf{0}$. What consequences does this have for the "singular case"?

11.32. Let

$$\mathbf{v}(\theta) \triangleq e^{(jn\theta/2)}\psi_n(e^{j\theta}),$$

$$W(\theta) = \frac{\sin\left(\dfrac{n+1}{2}\theta\right)}{\sin(\frac{1}{2}\theta)}.$$

Show that

$$\mathbf{R}(n)\mathbf{v}(\theta) = \frac{1}{2\pi}\int_{-\pi}^{\pi} W(\theta-\phi)S(\phi)\mathbf{v}(\phi)\,d\phi.$$

For large n, and continuous S, one can approximate the integral as

$$\mathbf{R}(n)\mathbf{v}(\theta) \approx S(\theta)\mathbf{v}(\theta),$$

which is an eigenvalue-eigenvector relation. This is consistent with Theorem 11.3.6.

11.33. Let the eigenvalues of $\mathbf{R}(n)$ be given by Eq. (11.3.136). Prove that

$$\min_{\theta} S(\theta) \leqslant \lambda_n,$$

$$\max_{\theta} S(\theta) \geqslant \lambda_0.$$

Under what conditions can the equality be met?

11.34. Let

$$S(\theta) = \begin{cases} 1, & |\theta| < \dfrac{\pi}{2} \\ 0, & \text{otherwise.} \end{cases}$$

Estimate the eigenvalues of $\mathbf{R}$ (256).

11.35. Do lines change the entropy of a spectrum? Let

$S_\varepsilon(\theta) = S(\theta) + P_\varepsilon(\theta)$, where

$$P_\varepsilon(\theta) = \begin{cases} \dfrac{1}{\varepsilon}, & |\theta - \theta_0| < \pi\varepsilon \\ \dfrac{1}{\varepsilon}, & |\theta + \theta_0| < \pi\varepsilon \\ 0, & \text{otherwise.} \end{cases}$$

Show that

$$\lim_{\varepsilon \to 0} \mathscr{H}_0(S_\varepsilon) = \mathscr{H}_0(S).$$

11.36. Let

$$H(z) = \sum_{k=0}^{\infty} h(k)z^{-k}$$

be stable and stably invertible (or minimum phase). Show that with

$$S(\theta) = |H(e^{j\theta})|^2,$$

the entropy is

$$\mathscr{H}_0(S) = \ln(h(0)^2).$$

What does this correspond to for an AR model? What is the entropy for the spectrum of Example 11.2.1? If the spectrum is ARMA, how does the entropy depend on the poles and zeros?

11.37. Given

$$r(0) = 1,$$

$$r(n) = \cos(\pi/k),\ k \text{ a positive integer},$$

find a lowest-order MA spectra that agrees with $r(0)$, $r(n)$.
(Problem contributed by Allan Steinhardt.)

11.38. Suppose the matrix $\mathbf{A}$ in Eq. (11.3.70) has distinct eigenvalues. Let

$$\mathbf{A}\mathbf{v}(k) = \lambda_k \mathbf{v}(k), \qquad 1 \leqslant k \leqslant n$$

where $\|\mathbf{v}(k)\| = 1$.

a) What is $\mathbf{v}(k)$?
b) Express $\mathbf{R}(n-1)$ in terms of the eigenvalues and eigenvectors of $\mathbf{A}$.

11.39. (Generalization of the Pisarenko model to colored background noise.) Given $\mathbf{r}[0, n]$, with $\mathbf{R}(n) > \mathbf{0}$, show that there exists a unique spectrum of the form

$$\hat{S}(\theta) = \sigma^2 S_0(\theta) + \sum_{k=1}^{n} \frac{\pi A_k^2}{2} \delta(\theta - \theta_k),$$

which matches $\mathbf{r}[0, n]$, where $S_0(\theta)$ is a known spectrum. How does one compute the model parameters?

11.40. Suppose the signal $\mathbf{y}$ is known to have spectrum

$$S(\theta) = \sigma^2 S_0(\theta) + \frac{\pi A^2}{2} [\delta(\theta - \theta_0) + \delta(\theta + \theta_0)],$$

where $S_0(\theta)$ is known but σ^2, A, θ_0 are unknown. How could one modify the SVD approach to accommodate the colored background?

11.41. Given $\mathbf{R}(n) > \mathbf{0}$, must there exist an order n FIR filter $\hat{a}(z)$ as in Eq. (11.3.15) for which

$$E\, e(k + l)e(l) = 0 \quad \text{for } 1 \leqslant k \leqslant n?$$

Give a proof or a counterexample.

11.42. Given $\mathbf{R}(n) > \mathbf{0}$, there always exists an order n AR model that matches $\mathbf{r}[0, n]$. Must there exist an order n MA model? Must there exist an ARMA model with m zeros and $n - m$ poles?

11.43. (Generalized analysis/synthesis)
Let $\mathbf{y}$ be a WSS signal. Statistics on this signal relative to the n-dimensional state variable system

$$\mathbf{x}(k+1) = \mathbf{A}x(k) + \mathbf{B}y(k), \qquad \text{where } \mathbf{A} \text{ is stable and } \mathbf{A}, \mathbf{B} \text{ is controllable,}$$

are measured:

$$\begin{bmatrix} r(0) & \mathbf{L}^T \\ \mathbf{L} & \mathbf{K} \end{bmatrix} = E\begin{bmatrix} y(k) \\ \mathbf{x}(k) \end{bmatrix}[y(k) \quad \mathbf{x}(k)^T].$$

a) Choose $\mathbf{C}$ to minimize

$$V(\mathbf{C}') = E(y(k) + \mathbf{C}'\mathbf{x}(k))^2$$

and compute

$$\alpha = V(\mathbf{C}).$$

The analysis or whitening filter is then

$$H(z) = 1 + \mathbf{C}(z\mathbf{I} - \mathbf{A})^{-1}\mathbf{B}.$$

The synthesis filter is (see Problem 8.28)

$$\frac{1}{H(z)} = 1 - \mathbf{C}[z\mathbf{I} - \mathbf{A} + \mathbf{BC}]^{-1}\mathbf{B},$$

and the spectral estimate is

$$\hat{S}(\theta) = \frac{\alpha}{|H(e^{j\theta})|^2}.$$

b) Prove that the synthesis filter is stable by showing that

$$\mathbf{K} = (\mathbf{A} - \mathbf{BC})\mathbf{K}(\mathbf{A} - \mathbf{BC})^T + \alpha\mathbf{BB}^T.$$

c) Show that $\hat{S}(\theta)$ reproduces the statistics $r(0)$, $\mathbf{L}$, $\mathbf{K}$.
d) How can $\mathbf{A}$, $\mathbf{B}$ be chosen to reduce to the case studied in Section 11.3?

11.44. Let

$$\mathbf{R}(k, n) \triangleq \begin{bmatrix} r(k) & & r(k+n) \\ & \ddots & \\ r(k-n) & & r(k) \end{bmatrix}, \qquad \text{Toeplitz, } (n+1) \times (n+1).$$

Suppose that an ARMA spectrum of the form (11.1.21) exists that agrees with $\mathbf{r}[0, m+n]$. Show that

a) $\mathbf{aR}(m, n) = [(b_0 b_m), \bigcirc\ ]$

b) $\hat{b}(z)\hat{b}(z^{-1}) = \mathbf{a}\left[\sum_{k=-m}^{m} z^{-k}\mathbf{R}(k, n)\right]\mathbf{a}^T$

Must the right-hand side of b) be nonnegative for every $\mathbf{a}$ when $z = e^{j\theta}$?

11.45. Which of the following AR estimation methods will always produce a stable $\hat{a}(z)$, assuming $\mathbf{YY}^T > \mathbf{0}$?

a) Autocorrelation method.
b) Autocovariance method.
c) Burg method.

11.46. Let

$$r(k) = \sigma^2\delta(k) + 1.$$

Find c_n and $\hat{a}(n, z)$ for all $n > 0$, and compute the value of the order n AR spectrum at $\theta = 0$.

11.47. Let

$$r(k) = \begin{cases} \sigma^2 + 1, & k = 0 \\ \dfrac{\sin(k\theta_0)}{k\theta_0}, & k \neq 0 \end{cases}, \text{ where } \theta_0 = \frac{\pi}{6}.$$

a) Compute $\lim_{n\to\infty} \alpha_n$ as a function of σ^2.
b) Plot the order 12 AR spectrum for $\sigma^2 = 10^{-3}$ and the true spectrum together.

11.48 How can one simulate a WSS signal, and start it up so that it "looks" stationary? Let $\mathbf{r}$ be a given autocorrelation sequence. Let $\mathbf{u}$ be a white noise signal with variance one. Find time-varying coefficients $a_{k,m}$, $b_{0,m}$ so that with

$$y(m) = 0 \qquad \text{for } m < 0,$$

$$y(m) = b_{0,m}u(m) - \sum_{k=1}^{m} a_{k,m}\, y(m-k), \quad \text{for } m \geqslant 0,$$

we will have

$$E\, y(k+j)y(j) = r(k)$$

for all $k \geqslant 0$, $\quad j \geqslant 0$. See (Scharf 1981.)

Research Problem

Given $\lambda_0 \geqslant \lambda_1 \geqslant \lambda_2 \cdots \geqslant \lambda_n > 0$. Must there exist a Toeplitz matrix $\mathbf{R}(n)$ with these numbers as its eigenvalues?

Research Problem

Suppose $\mathbf{R}(n) > 0$. Does there exist a spectrum $\hat{S}(\theta)$ that is the sum of a line spectrum with at most m lines and an AR spectrum of order $n - m$? If so, is the decomposition unique?

References

Akaike, H., "On a semi-automatic power spectrum estimation procedure," *Proc. 3rd Hawaii Int. Conf. Syst. Sci.*, Part 2, pp. 974–977, 1970

Akaike, H., "A new look at the statistical model identification," *IEEE Trans. Autom. Control*, Vol. AC-19, pp. 716–723, Dec. 1974.

Bachman, George, *Elements of Abstract Harmonic Analysis*, Academic Press, New York, 1964.

Barnes, C. W., "On the design of optimal state-space realizations of second-order digital filters," *IEEE Trans. Circuits Syst.*, Vol. CAS-31, pp. 602–608, July 1984.

Barnes, C. W. and A. T. Fam, "Minimum norm recursive digital filters that are free of overflow limit cycles," *IEEE Trans. Circuits Syst.*, Vol. CAS-24, pp. 569–574, Oct. 1977.

Barnes, C. W. and S. Shinnaka, "Finite word effects in block state realizations of fixed point digital filters," *IEEE Trans. Circuits Syst.*, Vol. CAS-27, pp. 345–349, May 1980a.

Barnes, C. W. and S. H. Leung, "Use of transversal-recursive structures for efficient realization of low-noise digital filters with decimated output," *IEEE Trans. Acoust., Speech, Signal Process.*, Vol. ASSP-28, pp. 645–651, 1980b.

Bennett, W. R., "Spectra of quantized signals," *Bell Syst. Tech. J.*, Vol. 27, pp. 446–472, July 1948.

Blackman, R. B., *Linear Data-Smoothing and Prediction in Theory and Practice*, Addison-Wesley, Reading, MA, 1965.

Blahut, R. E., *Fast Algorithms for Digital Signal Processing*, Addison-Wesley. Reading, MA, 1985.

Bomar, B. W., "New second-order state-space structures for realizing low roundoff noise digital filters," *IEEE Trans. Acoust., Speech, Signal Process.*, Vol. ASSP-33, pp. 106–110, Feb. 1985.

Bracewell, R., *The Fourier Transform and Its Applications*, McGraw-Hill, New York, 1978.

Brophy, F. and A. C. Salazar, "Considerations of the Pade approximate technique in the synthesis of recursive digital filters," *IEEE Trans. Audio Electroacoust.*, Vol. AU-21, pp. 500–505, Dec. 1973.

Burg, J. P., "Maximum entropy spectral analysis," in *Proc. 37th Meeting Soc. of Exploration Geophysicists*, 1967.

Burg, J. P., "Maximum entropy spectral analysis," *Ph.D. dissertation*, Dept. Geophysics, Stanford University 1975.

Burrus, C. S., "Digital filter structures described by distributed arithmetic," *IEEE Trans. Circuits Syst.*, Vol. CAS-24, pp. 674–680, Dec. 1977a.

Burrus, C. S., "Index mappings for multidimensional formulation of the DFT and convolution," *IEEE Trans. Acoustics, Speech, Signal Process.*, Vol. ASSP-25, pp. 239–242, June 1977b.

Capon, J., "High-resolution frequency—Wavenumber spectrum analysis," *Proc. IEEE*, Vol. 57, pp. 1408–1418, Aug. 1969.

Cheney, E. W., *Introduction to Approximation Theory*, McGraw-Hill, New York, 1966.

Childers, D. G. (ed.), *Modern Spectrum Analysis*, IEEE Press, New York, 1978.

Churchill, R. V., *Introduction to Complex Variables and Applications*, McGraw-Hill, New York, 1948.

Conte, S. D. and C. de Boor, *Elementary Numerical Analysis*, McGraw-Hill, New York, 1972.

Cooley, J. W. and J. N. Tukey, "An algorithm for the machine calculation of complex Fourier series," *Math. Comput.*, Vol. 19, No. 2, pp. 297–301, April 1965.

Digital Signal Processing Committee of IEEE ASSPS, *Programs for Digital Signal Processing*, IEEE, New York, 1979.

Durbin, J., "The fitting of time series models," *Rev. Instrum. Int. Stat.*, Vol. 28, 1960.

Ebert, P. M., J. E. Mazo, and M. G. Taylor, "Overflow oscillations in digital filters," *Bell Syst. Tech. J.*, Vol. 48, pp. 3021–3030, Nov. 1969.

Fam, A. T. and C. W. Barnes, "Non-minimal realizations of fixed point digital filters that are free of all finite wordlength limit cycles," *IEEE Trans. Acoust., Speech, Signal Process.*, ASSP-27, pp. 149–153, April 1979.

Farden, D. C. and L. L. Scharf, "Statistical design of nonrecursive digital filters," *IEEE Trans. Acoust., Speech, Signal Process.*, Vol. ASSP-22, pp. 188–196, June 1974.

Fourier, J., *Theorie analytique de la chaleur*, 1822, Dover reprint of English translation, 1955.

Franchitti, J. C., "All-pass filter interpolation and frequency transformation problems," *M.S. thesis*, Dept. of Electrical and Computer Engineering, Univ. of Colorado, Boulder, CO, 1985.

Friedlander, B., "Lattice methods for spectral estimation," *Proc. IEEE*, Vol. 70, pp. 990–1017, Sept. 1982.

Frost, O. L., "Power-spectrum estimation," *Proc. 1976 NATO Advanced Study Institute on Signal Processing with Emphasis on Underwater Acoustics*, La Spezia, Italy, 1976.

Gable, R. A., "On asymmetric FIR interpolators with minimum Lp error," *Proc. Int. Conf. Acoust., Speech, Signal Process.*, pp. 256–259, 1980.

Good, I. J., "The relationship between two fast Fourier transforms," *IEEE Trans. Comput.*, Vol. C-20, pp. 310–317, 1971.

Gray, A. H. and J. D. Markel, "Digital lattice and ladder filter synthesis," *IEEE Trans. Acoust., Speech, Signal Process.*, Vol. AU-21, pp. 491–500, Dec. 1973.

Gray, A. H. and J. D. Markel, "A normalized digital filter structure," *IEEE Trans. Acoust., Speech, Signal Process.*, Vol. ASSP-23, pp. 258–277, June 1975.

Gray, A. H. and J. D. Markel, "A computer program for designing digital elliptic filters," *IEEE Trans. Acoust., Speech, Signal Process.*, Vol. ASSP-24, pp. 529–538, Dec. 1976.

Gray, R. M., "On the asymptotic eigenvalue distribution of Toeplitz matrices," *IEEE Trans. Inf. Theory*, Vol. IT-18, pp. 725–730, Nov. 1972.

Grenander, O. and G. Szego, *Toeplitz Forms and Their Applications*, University of California, Berkeley, 1958.

Guillemin, E. A., *Synthesis of Passive Networks*, Wiley, New York, 1957.

Haykin, S., *Communication Systems*, Wiley, New York, 1983.

Heideman, M. T., D. Johnson, and C. S. Burrus, "Gauss and the history of the Fast Fourier Transform," *IEEE ASSP Magazine*, Vol. 1, No. 4, pp. 14–21, Oct. 1984.

Herrmann, O. and H. W. Schuessler, "Design of nonrecursive digital filters with minimum phase," *Electron. Lett.*, Vol. 6, pp. 329–330, 1970a.

Herrmann, O., "On the design of nonrecursive digital filters with linear phase," *Electron. Lett.*, Vol. 6, pp. 328–329, 1970b.

Heyliger, G., "Design of numerical filters," *4th Ann. Conf. Proc. of the Allerton Conf. on Circuits and Systems*, pp. 175–185, Oct. 1966.

Heyliger, G., "Simple design parameters for Chebyshev arrays and filters," *IEEE Trans. Audio Electroacoust.*, Vol. AU-18, pp. 502–503, Dec. 1970.

Hofstetter, E., A. V. Oppenheim, and J. Siegel, "A new technique for the design of nonrecursive digital filters," *Digital Signal Processing*, L. R. Rabiner and C. M. Rader (eds.), IEEE, New York, 1972.

Inouye, Y., "Some notes on the second-order interpolation problem of digital filters," *IEEE Trans. on Acoust., Speech, and Signal Process.*, Vol. ASSP-31, No. 1, Feb. 1983, pp. 209–212.

Jackson, L. B., "Roundoff noise bounds derived from coefficient sensitivities for digital filters," *IEEE Trans. Circuits Syst.*, Vol. CAS-23, pp. 481–485, Aug. 1976.

Jackson, L. B., "An analysis of limit cycles due to multiplication rounding in recursive digital (sub)filters," *Proc. of 7th Ann. Allerton Conf. on Circuit and System Theory*, pp. 69–78, 1978.

Jackson L. B., "Limit cycles in state-space structures for digital filters," *IEEE Trans. Circuits Syst.*, Vol. CAS-26, pp. 67–68, Jan. 1979.

Jackson, L. B., A. G. Lindgren, and Y. Kim, "Optimal synthesis of second-order state structures for digital filters," *IEEE Trans. Circuits Syst.*, Vol. CAS-26, pp. 149–155, March 1977.

Jaynes, E. T., "On the rationale of maximum-entropy methods," *Proc. IEEE*, Vol. 70, pp. 939–953, Sept. 1982.

Jury, E. I., *Theory and Application of the z-Transform Method*, Wiley, New York, 1964.

Kaiser, J. F., "Digital Filters," Chap. 7 in *Systems Analysis by Digital Computer*, F. F. Kuo & J. F. Kasier (eds.), Wiley, New York, 1966.

Kay, S. M. and S. L. Marple, "Spectrum analysis—A modern perspective," *Proc. IEEE*, Vol. 69, pp. 1380–1419, Nov. 1981.

Klemma, V. C. and A. J. Laub, "The singular value decomposition: Its computation and some applications," *IEEE Trans. Autom. Control*, Vol. AC-25, pp. 164–176, April 1980.

Kline, M., *Mathematical Thought from Ancient to Modern Times*, Oxford University, New York, 1972.

Knuth, D. E., *The Art of Computer Programming*, Vol. 2: *Seminumerical Algorithms*, Addison-Wesley, Reading, MA, 1969.

Lam, H. Y-F., *Analog and Digital Filters*, Prentice-Hall, Englewood Cliffs, NJ, 1979.

Larson, J. H., and B. O. Shubert, *Probabilistic Models in Engineering Sciences*, Vol. II, Wiley, New York, 1979.

Levinson, N., "The Wiener rms error criterion in filter design and prediction," *J. Math. Phys.*, Vol. 25, 1947.

Long, L. J. and T. N. Trick, "An absolute bound on limit cycles due to roundoff errors in digital filters," *IEEE Trans. Audio Electroacoust.*, Vol. AU-21, pp. 27–30, Feb. 1973.

McClellan, J. H. and T. W. Parks, "A unified approach to the design of optimum FIR linear phase digital filters," *IEEE Trans. Circuit Theory*, Vol. CT-20, pp. 697–701, Nov. 1973a.

McClellan, J. H., T. W. Parks, and L. R. Rabiner, "A computer program for designing optimum FIR linear phase digital filters," *IEEE Trans. Audio Electroacoust.*, Vol. AU-21, pp. 506–526, Dec. 1973b.

Makhoul, J., "Linear prediction: A tutorial review," *Proc. IEEE*, Vol. 62, pp. 561–580, April 1975.

Markel, J. D. and A. H. Gray, Jr., *Linear Prediction of Speech*, Springer-Verlag, New York, 1976.

Marple, S. L., "A new autoregressive spectrum analysis algorithm," *IEEE Trans. Acoust., Speech, Signal Process.*, Vol. ASSP-28, pp. 441–454, Aug. 1980.

Mills, W. L., C. T. Mullis, and R. A. Roberts, "Low Roundoff Noise and Normal Realizations of Fixed Point IIR Digital Filters," *IEEE Trans. on Acoust. Speech, Signal Process.*, Vol. ASSP-29, pp. 893–903, Aug. 1981.

Mullis, C. T. and R. A. Roberts, "Synthesis of minimum roundoff noise fixed point digital filters," *IEEE Trans. Circuits Syst.*, Vol. CAS-23, pp. 551–562, Sept. 1976a.

Mullis, C. T. and R. A. Roberts, "Roundoff noise in digital filters: Frequency transformations and invariants," *IEEE Trans. Acoust., Speech, Signal Process.*, Vol. ASSP-24, pp. 538–550, Dec. 1976b.

Mullis, C. T. and R. A. Roberts, "The use of second-order information in the approximation of discrete-time linear systems," *IEEE Trans. on Acoustics, Speech, and Signal Processing*, Vol. ASSP-24, pp. 226–238, June 1976c.

Noble, B., *Applied Linear Algebra*, Prentice-Hall, Englewood Cliffs, NJ, 1969.

Ore, O., *Number Theory and Its History*, McGraw-Hill, New York, 1948.

Orfanidis, S. J., *Optimum Signal Processing, An Introduction*, Macmillan, New York, 1985.

Papoulis, A., *The Fourier Integral and Its Applications*, McGraw-Hill, New York, 1962.

Parker, S. R. and S. F. Hess, "Limit-cycle oscillations in digital filters," *IEEE Trans. Circuit Theory*, Vol. CT-18, pp. 687–696, Nov. 1971.

Parks, T. W. and J. H. McClellan, "Chebyshev approximation for nonrecursive digital filters with linear phase," *IEEE Trans. Circuit Theory*, Vol. CT-19, pp. 189–194, March 1972.

Parzen, E., "Some recent advances in time series modelling," *IEEE Trans. Autom. Control*, Vol. AC-19, pp. 723–730, Dec. 1974.

Peled, A. and B. Liu, "A new hardware realization of digital filters," *IEEE Trans. Acoust., Speech, Signal Process.*, Vol. ASSP-22, pp. 456–462, Dec. 1974.

Pisarenko, V. F., "The retrieval of harmonics from a covariance function," *Geophys. J. R. Astron. Soc.*, Vol. 33, pp. 347–366, 1973.

Proc. IEEE, Special Issue on Spectral Estimation, Vol. 70, 1982.

Rabiner, L. R., "Linear program design of finite impulse response digital filters," *IEEE Trans. Audio Electroacoust.*, Vol. AU-20, pp. 280–288, Oct. 1972.

Rabiner, L. R. and B. Gold, *Theory and Application of Digital Signal Processing*, Prentice-Hall, Englewood Cliffs, NJ, 1975.

Rader, C. M. and B. Gold, "Effects of parameter quantization on the poles of a digital filter," *Proc. IEEE*, Vol. 55, pp. 688–689, May 1967.

Robinson, E. A., *Statistical Communication and Detection*, Hafner, New York, 1976.

Robinson, E. A., "An historical perspective of spectrum estimation," *Proc. IEEE*, Vol. 70, pp. 885–907, Sept. 1982.

Sandberg, I. W. and J. F. Kaiser, "A bound on limit cycles in fixed-point implementations of digital filters," *IEEE Trans. Audio Electroacoust.*, Vol. AU-10, pp. 110–112, June 1972.

Scharf, L. and Luby, J., "Statistical Design of Autoregressive—Moving Average Digital Filters," *IEEE Transactions on Acoust., Speech and Signal Process.*, Vol. ASSP-27, pp. 240–247, June 1979.

Scharf, L., Gueguen, C., and Dugre, J., "Parametric Spectrum Modelling: A Signal Processing Perspective," *1st IEEE Workshop on Spectrum Analysis*, McMaster University, Hamilton, Ontario, Aug. 1981.

Schuster, A., "On the investigation of hidden periodicities with application to a supposed 26-day period of meteorological phenomena," *Terr. Magn.*, Vol. 3, pp. 13–41, 1898.

Singleton, R. C., "A method for computing the fast Fourier transform with auxiliary memory and limited high speed storage," *IEEE Trans. Audio Electroacoust.*, Vol. AU-15, pp. 91–97, June 1967.

Steiglitz, K., "Computer-aided design of recursive digital filters," *IEEE Trans. Audio Electroacoust.*, Vol. AU-18, pp. 123–129, June 1970.

Truxal, J. G., *Automatic Feedback Control Synthesis*, McGraw-Hill, New York, 1955.

Tufts, D. W. and R. Kumaresan, "Frequency estimation of multiple sinusoids: making linear prediction preform like maximum likelihood," *Proc. IEEE*, Vol. 70, pp. 975–989, Sept. 1982.

Van Valkenburg, M. E., *Analog Filter Design*, Holt, Rinehart and Winston, New York, 1982.

Welch, P. D., "The use of fast Fourier transform for the estimation of power spectra: A method based on time averaging over short, modified periodograms," *IEEE Trans. Audio Electroacoust.*, Vol. AU-15, pp. 70–73, June 1967.

Widrow, B., "A study of rough amplitude quantization by means of Nyquist sampling theory," *IRE Trans. Circuit Theory*, Vol. CT-3, pp. 266–276, Dec. 1956.

Widrow, B. and S. D. Stearns, *Adaptive Signal Processing*, Prentice-Hall, Englewood Cliffs, NJ, 1985.

Winograd, S., "On computing the discrete Fourier transform," *Proc. Natl. Acad. Sci. USA*, Vol. 73, pp. 1005–1006, 1976.

Wylie, C. R., Jr., *Advanced Engineering Mathematics*, McGraw-Hill, New York, 1966.

Yule, G. E., "On a method of investigating periodicities in disturbed series, with special reference to Wolfer's sunspot numbers," *Philos. Trans. R. Soc. London*, Ser. A, Vol. 226, pp. 267–298, July 1927.

Zohar, S., "New hardware realizations of nonrecursive digital filters," *IEEE Trans. Comput.*, Vol. C-22, pp. 328–347, April 1973.

Suggested Readings

The following is a brief list of additional readings for the subjects listed. It is by no means exhaustive and represents only a small portion of the available work on the subjects listed.

Discrete-Time Signals/Systems

Gabel, R. A. and R. A. Roberts, *Signals and Linear Systems*, Wiley, New York, 1986.

Mayhan, R. J., *Discrete-Time and Continuous-Time Linear Systems*, Addison-Wesley, Reading, MA, 1984.

Oppenheim, A. V. and A. S. Willsky, *Signals and Systems*, Prentice-Hall, Englewood Cliffs, NJ, 1983.

Fourier Analysis

See all of the preceding references for an introduction to Fourier analysis of discrete-time signals and systems.

Bracewell, R., *The Fourier Transform and Its Applications*, McGraw-Hill, New York, 1965.

Papoulis, A., *The Fourier Integral and Its Applications*, McGraw-Hill, New York, 1962.

DFT/Convolution Fast Algorithms

Blahut, R. E., *Fast Algorithms for Digital Signal Processing*, Addison-Wesley, Reading, MA, 1985.

Burrus, C. S. and T. W. Parks, *DFT/FFT and Convolution Algorithms*, Wiley, New York, 1985.

Elliott, D. F. and R. Rao, *Fast Algorithms, Analyses, and Applications*, Academic, NY 1983.

McClellan, J. H. and C. M. Radar, *Number Theory in Digital Signal Processing*, Prentice-Hall, Englewood Cliffs, NJ, 1979.

Nussbaumer, H. J., *Fast Fourier Transform and Convolution Algorithms*, 2nd ed., Springer-Verlag, Berlin, 1982.

Digital Filters

Lam, H. Y-F., *Analog and Digital Filters*, Prentice-Hall, Englewood Cliffs, NJ, 1979.

Oppenheim, A. V. and R. W. Schafer, *Digital Signal Processing*, Prentice-Hall, Englewood Cliffs, NJ, 1975.

Rabiner, L. R. and B. Gold, *Theory and Application of Digital Signal Processing*, Prentice-Hall, Englewood Cliffs, NJ, 1975.

Tretter, S. A., *Introduction to Discrete-Time Signal Processing*, Wiley, New York, 1976.

Van Valkenburg, M. E., *Analog Filter Design*, Holt, Rinehart and Winston, New York, 1982.

Least-Squares Filters

Kailath, T., "An innovations approach to least squares estimation. Part I: Linear filtering in additive white noise," *IEEE Trans. Autom. Control*, Vol. AC-13, pp. 646–655, Dec. 1968.

Kailath, T., "A view of three decades of linear filtering theory," *IEEE Trans. Inf. Theory*, Vol. IT-20, pp. 146–181, March 1974.

Orfanidis, S. J., *Optimum Signal Processing, An Introduction*, Macmillan, New York, 1985.

Robinson, E. A., *Statistical Communication and Detection*, Hafner, New York, 1976.

Finite Register Effects

Barnes, C. W., "On the design of optimal state-space realizations of second-order digital filters," *IEEE Trans. Circuits Syst.*, Vol. CAS-31, pp. 602–608, July 1984.

Barnes, C. W. and A. T. Fam, "Minimum norm recursive digital filters that are free of overflow limit cycles," *IEEE Trans. Circuits Syst.*, Vol. CAS-27, pp. 569–574, Oct. 1980.

Jackson, L. B., "Roundoff noise bounds derived from coefficient sensitivities for digital filters," *IEEE Trans. Circuits Syst.*, Vol. CAS-23, pp. 481–485, Aug. 1976.

Jackson, L. B., A. G. Lindgren, and Y. Kim, "Optimal synthesis of second-order state structures for digital filters," *IEEE Trans. Circuits Syst.*, Vol. CAS-26, pp. 149–155, March 1979.

Mills, W. L., C. T. Mullis, and R. A. Roberts, "Digital filter realizations without overflow oscillations," *IEEE Trans. Acoust., Speech, Signal Process.*, Vol. ASSP-26, pp. 334–338, Aug. 1978.

Mills, W. L., C. T. Mullis, and R. A. Roberts, "Low roundoff noise and normal realizations of fixed point IIR digital filters," *IEEE Trans. Acoust. Speech, Signal Process.*, Vol. ASSP-29, pp. 893–903, Aug. 1981.

Mullis, C. T. and R. A. Roberts, "Synthesis of minimum roundoff noise fixed point digital filters," *IEEE Trans. Circuits Syst.*, Vol. CAS-23, pp. 551–562, Sept. 1976.

Mullis, C. T. and R. A. Roberts, "Roundoff noise in digital filters: Frequency transformations and invariants," *IEEE Trans. Acoust., Speech, Signal Process.*, Vol. ASSP-24, pp. 538–550, Dec. 1976.

Digital Filter Structures

Barnes, C. W. and S. Shinnaka, "Block-shift invariance and block implementation of discrete-time filters," *IEEE Trans. Circuits Syst.*, Vol. CAS-27, pp. 667–672, Aug. 1980.

Burrus, C. S., "Block implementation of digital filters," *IEEE Trans. Circuit Theory*, Vol. CT-18, pp. 697–701, Nov. 1971.

Burrus, C. S., "Digital filter structures described by distributed arithmetic," *IEEE Trans. Circuits Syst.*, Vol. CAS-24, pp. 674–680, Dec. 1977.

Crochiere, R. E. and L. R. Rabiner, *Multirate Digital Signal Processing*, Prentice-Hall, Englewood Cliffs, NJ, 1983.

Henrot, D., "On modularity and computational parallelism in digital filter implementations," Ph.D. thesis, Department of Electrical and Computer Engineering, University of Colorado, Boulder, 1983.

Mullis, C. T. and R. A. Roberts, "Digital processing algorithms for VLSI," *VLSI Signal Processing*, IEEE, New York, pp. 158–161, 1984.

Zeman, J. and A. G. Lindgren, "Fast digital filters with low roundoff noise," *Proc. of the 1979 Asilomar Conf. on Circuits, Systems, and Computers*, Pacific Grove, CA, pp. 257–262, Nov. 1979.

Spectral Estimation

Blackman, R. B. and J. W. Tukey, *The Measurement of Power Spectra*, Dover reprint, New York, 1959.

Childers, D. G. (ed.), *Modern Spectrum Analysis*, IEEE Press Selected Reprint Series, New York, 1978.

Frost, O. L., "Power-spectrum estimation," *Proc. 1976 NATO Advanced Study Inst. on Signal Processing with Emphasis on Underwater Acoustics*, La Spezia, Italy, Aug. 30–Sept. 11, 1976.

Kay, S. M. and S. L. Marple, Jr., "Spectrum analysis—A modern perspective," *Proc. IEEE*, Vol. 69, pp. 1380–1419, Nov. 1981.

Robinson, E. A., "A historical perspective of spectrum estimation," *Proc. IEEE*, Vol. 70, pp. 885–907, Sept. 1982.

Fast Algorithms For Toeplitz and Related Equations

Astrom, K. J., E. I. Jury, and R. G. Agniel, "A numerical method for the evaluation of complex integrals," *IEEE Trans. Autom. Control*, Vol. AC-15, pp. 468–471, Aug. 1970.

Durbin, J., "The fitting of time series models," *Rev. Int. Stat. Inst.*, Vol. 28, pp. 233–244, 1960.

Friedlander, B., "Lattice methods for spectral estimation," *Proc. IEEE*, Vol. 70, pp. 990–1017, Sept. 1982.

Friedlander, B., M. Morf, T. Kailath, and L. Ljung, "New inversion formulas for matrices classified in terms of their distances from Toeplitz matrices," *Linear Algebra and Its Applications*, Vol. 27, pp. 31–60, 1979.

Lee, D. T. L., M. Morf, and B. Friedlander, "Recursive least squares ladder estimation algorithms," *IEEE Trans. Acoust., Speech, Signal Process.*, Vol. ASSP-29, pp. 627–641, June 1981.

LeRoux, J. and C. Gueguen, "A fixed point computation of the partial correlation coefficients," *IEEE Trans. Acoust., Speech, Signal Process.*, Vol. 25, pp. 257–259, June 1977.

Levinson, N., "The Wiener RMS error criterion in filter design and prediction," *J. Math. Phys.*, Vol. 25, pp. 261–278, Jan. 1947.

Makhoul, J., "Stable and efficient lattice methods for linear prediction," *IEEE Trans. Acoust., Speech, Signal Process.*, Vol. ASSP-25, pp. 423–428, Oct. 1977.

Marple, S. L., "A new autoregressive spectrum analysis algorithm," *IEEE Trans. Acoust., Speech, Signal Process.*, Vol. ASSP-28, pp. 441–454, Aug. 1980.

Mullis, C. T. and R. A. Roberts, "The use of second-order information in the approximation of discrete-time linear systems," *IEEE Trans. Acoust., Speech, Signal Process.*, Vol. ASSP-24, June 1976.

Rissanen, J., "Algorithms for triangular decomposition of block Hankel and Toeplitz matrices with application to factoring positive matrices," *Mathematics of Computation*, Vol. 23, pp. 147–154, Jan. 1973.

Wiggens, R. A. and E. A. Robinson, "Recursive solution to the multivariate filtering problem," *J. Geophys. Res.*, Vol. 70, April 1965.

Index

Properties of the discrete Fourier transform

Linearity: $\mathbf{V}[\alpha\mathbf{f} + \beta\mathbf{g}] = \alpha\mathbf{V}\,\mathbf{f} + \beta\mathbf{V}\,\mathbf{g}$

$$\alpha f(k) + \beta g(k) \quad \xleftrightarrow{DFT:N} \quad \alpha F(n) + \beta G(n)$$

Orthogonality properties: $\mathbf{V}^*\mathbf{V} = N\,\mathbf{I}$

$$\sum_{k=0}^{N-1} f(k)[g(k)]^* = \frac{1}{N}\sum_{n=0}^{N-1} F(n)[G(n)]^*$$

$$\sum_{k=0}^{N-1} |f(k)|^2 = \frac{1}{N}\sum_{n=0}^{N-1} |F(n)|^2$$

Time shift property: $\mathbf{V}\mathbf{R}^{-1} = \mathbf{D}^{-1}\mathbf{V}$

$$f(k - m \bmod N) \quad \xleftrightarrow{DFT:N} \quad W_N^{-mn}F(n)$$

$$\sum_{m=0}^{N-1} g(m)f(k - m \bmod N) \quad \xleftrightarrow{DFT:N} \quad F(n)G(n)$$

Frequency shift property: $\mathbf{V}\mathbf{D} = \mathbf{R}^{-1}\mathbf{V}$

$$f(k)W_N^{km} \quad \xleftrightarrow{DFT:N} \quad F(n - m \bmod N)$$

$$f(k)g(k) \quad \xleftrightarrow{DFT:N} \quad \frac{1}{N}\sum_{m=0}^{N-1} G(m)F(n - m \bmod N)$$

Symmetry properties

$$\mathrm{Im}[f(k)] = 0 \quad \Leftrightarrow \quad F(-n \bmod N) = [F(n)]^*$$

$$f(-k \bmod N) = [f(k)]^* \quad \Leftrightarrow \quad \mathrm{Im}[F(n)] = 0$$

$$\mathrm{Re}[f(k)] = 0 \quad \Leftrightarrow \quad F(-n \bmod N) = -[F(n)]^*$$

$$f(-k \bmod N) = -[f(k)]^* \quad \Leftrightarrow \quad \mathrm{Re}[F(n)] = 0$$

$$H(z) = d + \frac{q_1 z^{-1} + q_2 z^{-2}}{1 + p_1 z^{-1} + p_2 z^{-2}}$$

$\big\downarrow \mathbf{T}_N$

Unscaled normal form: $(\mathbf{A}_0, \mathbf{B}_0, \mathbf{C}_0, D)$; $\quad a_{11} = a_{22}$

$\big\downarrow \mathbf{R}\left(\frac{\phi}{2}\right) = \mathbf{T}_R$, rotation

Unscaled normal form: $(\mathbf{A}_1, \mathbf{B}_1, \mathbf{C}_1, D)$; $a_{11} = a_{22}$; $\quad b_1c_1 = b_2c_2$

$\big\downarrow \mathbf{T}_S$, diagonal

l_2 scaled minimum noise form: $(\mathbf{A}_2, \mathbf{B}_2, \mathbf{C}_2, D)$

A design for minimum noise, second-order filters